To Ursula

—JAB

To Mellony

—JGS

PRENTICE HALL INTERNATIONAL SERIES
IN INDUSTRIAL AND SYSTEMS ENGINEERING

W. J. Farbrycky and J. H. Mize, Editors

Contents

Preface

This book describes and develops stochastic models of a variety of different types of manufacturing systems such as flow lines, job shops, transfer lines, flexible manufacturing systems, flexible assembly systems, cellular systems. These models evaluate the performance, address issues in the design, control and operation of these systems, and provide an understanding of how different components of a manufacturing system can be coordinated.

We believe that this is the first comprehensive treatment of stochastic models of manufacturing systems. We have provided extensive bibliographies of the relevant literature, including the pioneering papers on models of the various types of systems. The book contains much new material. There are new treatments of classical issues such as workload allocation and new models such as of assembly systems with strict job sequence requirements. We have also developed a new framework to describe the interaction between information and material flow in manufacturing. The formal representation of this framework provides the basis for the analysis of different ways of coordinating material flow in manufacturing systems.

The book is primarily intended for master's and doctorate's students in industrial engineering, operations research, and operations management. We also hope that both industrial and academic researchers interested in understanding manufacturing systems will find it a valuable reference on issues, models, and insights. A basic knowledge of stochastic modeling, such as is obtained from a senior-level undergraduate course or a first course at the master's level, is a desirable prerequisite. Part of the material in the book has been used, at various stages of its evolution, as class notes in graduate courses taught at the University of Waterloo and the University of California, Berkeley.

We would like to thank Rhonda Righter, Li Zhuang, and Xao-Gao Liu for their thorough reading of the manuscript and their helpful suggestions. We also want to express our appreciation to Li Zhuang and Xao-Gao Liu for providing most of the computational results. We are also indebted to a significant number of people, in particular Mike Magazine and Ken McKay at the University of Waterloo, and Sridhar Seshadri at the University of California, Berkeley, for their interest at various stages of this project. Leo Hanifin of the University of Detroit; Drazen Kostelski, Cindy Morey, and Jane Ragotte of General Motors of Canada; Barrie Beatty of Allen-Bradley Canada; and Steve Weiner of Ford Motor Company have each in their own way taught us much about the issues and problems with recent developments in manufacturing systems. Special thanks go to David D. Yao of Columbia University, Abraham Seidman of William E. Simon Graduate School–University of Rochester, and Richard F. Serfozo of Georgia Institute of Technology—all reviewers of Prentice Hall—for their valuable comments and insights.

John A. Buzacott's research was supported by the Natural Sciences and Engineering Research Council of Canada via operating grants and the strategic grants on "Modelling and Implementation of Just-in-Time Cells" and "Robust Planning and Scheduling," and by the Manufacturing Research Corporation of Ontario. During this project J. George Shanthikumar was supported in part by National Science Foundation grant ECS-8811234, by the Sloan Foundation grant for the "Consortium on Competitiveness and Cooperation," and by the Sloan Foundation Grant for the study on "Competitive Semiconductor Manufacturing."

John A. Buzacott

J. George Shanthikumar

1

Discrete Part Manufacturing Systems

1.1 INTRODUCTION

Manufacturing systems consist of machines and work stations where operations such as machining, forming, assembly, inspection, and testing are carried out on parts, items, subassemblies, and assemblies to create products that can be delivered to customers. Further essential components of any manufacturing system are the material handling and storage devices. They enable items to move from one work station to another, ensure that appropriate parts are available for assembly, and enable work to be held or stored until it can enter machines or work stations for processing. The tasks at the machines or work stations can be performed by human operators using written or verbal instructions as to how to perform the task, or they can be automated, that is, performed under automatic control and guidance with no human intervention. Our focus is on *discrete part* manufacturing systems, where each item processed is distinct. Such systems are normal in the mechanical, electrical, and electronics industries making products such as cars, refrigerators, electric generators, or computers. Systems that process fluids, like those found in the chemical and metallurgical industries will not be considered, although sometimes in these industries fluids are processed in distinct batches or packets, and, if such a packet is taken as the unit of manufacture, then the system can be considered to process discrete parts. For simplicity we will call an item, part, subassembly, or assembly processed by a machine or work station a *job*.

1

1.2 CLASSIFICATION

Volume and diversity of jobs.   Although each job is distinct, different jobs can be in all respects identical and in particular have the same processing requirements. If the manufacturing system only processes one type of job, then some aspects of its design and operation, in particular the material handling, will be simplified because all jobs can then be handled the same way. If the system processes many different types of jobs, however, instructions for each type will be required, and control will tend to be more complex. As well as the diversity of the types of jobs manufactured by the system, the total volume of jobs produced will also have a key impact on the design and operation of a manufacturing system. We will refer to the diversity of job types manufactured by the system as its *scope* and the total volume of jobs as the *scale* of the system.

1.3 DIFFERENT TYPES OF SYSTEMS

1.3.1 Flow Lines and Job Shops

Although a manufacturing system with the capability for a large scope and a large scale would promise the potential to cope with changes in the requirements of customers, the design of the product, or the manufacturing processes, such a system would tend to be difficult to control and would be uneconomical to operate. Until recently manufacturing systems had evolved toward emphasizing either scope or scale but not both. The two traditional forms of organizing manufacturing systems are the *job shop* and the *flow line*. The *job shop* consists of a variety of different types of machines, some of which can perform operations on different types of jobs, although this may require some set-up or change-over time between job types. Material handling is such that different types of jobs can visit machines in different sequences. The job shop has scope capability, but the problems associated with controlling the movement of jobs and the change-over of machines result in limited ability to produce high volumes efficiently and economically. By contrast the *flow line* requires all jobs to visit machines and work centers in the same sequence, thus simplifying material handling but drastically limiting the scope of the manufacturing system. The simple material handling combined with the limited scope makes it easier to control work flow and instruct machines and workers on their tasks, thus enabling high volumes to be produced economically.

1.3.2 Automatic Transfer Lines

Many flow lines produce only a single type of product, and with increasing volume it becomes attractive to automate individual machines and to replace human operators by automatic devices and machines. The *automatic transfer line* goes one step farther. Not only are the machines and the material handling from one machine to the next automated but all machines are linked so that they begin their tasks simultaneously and so that the material movement is synchronized. In this way the number of jobs in process can be kept small, and extremely high productivity is potentially possible. Hence the automatic transfer line is characterized by very high scale but very low scope because only one product can be produced at a time, and change-over can be long and expensive.

1.3.3 Flexible Transfer Lines

Initially, when automatic transfer lines came into widespread use in the 1950s, the instructions on how to perform the tasks at a machine were embodied in physical information storage devices such as cams, jigs, and other fixtures. This has been called "hard automation" and obviously made it difficult to change instructions. As electronic devices for storage and processing of information were developed, combined with numerical control (that is, transducers that convert digital information into physical motion of tools), changing the instructions on performing tasks became simpler. This resulted in the development of automated *flexible transfer lines* and *flexible flow lines*, distinguished by whether synchronized movement of jobs was retained or not. Because in flexible flow lines job movement is not synchronized, it is essential to provide some storage space between machines; otherwise job movement will always be determined by the slowest machine. Flexible transfer lines and flexible flow lines have greater scope than traditional automatic transfer lines, in that they can manufacture variants of a product. Unless the variation between jobs is very small, however, there may still be a requirement to change tools between jobs, and hence their scale potential tends to be less than automatic transfer lines.

1.3.4 Flexible Machining Systems

Transfer lines are essentially automated flow lines, and thus all jobs have to visit all the machines in the same sequence. If numerically controlled machines are combined with a material handling system that enables different jobs to visit machines in different sequences then the resulting system is known as a *flexible manufacturing system* (FMS). Such systems were first implemented in the 1970s for machining tasks so FMS is sometimes understood to stand for a flexible machining system. An FMS is essentially an automated job shop. It has reasonable scope, although initially the scope was limited by the ability to store or deliver the required tools to individual machines. Also it has reasonable scale potential, although the fixed costs associated with computer-based control of job movement in the system and computer-based numerical control of individual machines mean that FMSs are not economical at low total production rates.

1.3.5 Flexible Assembly Systems

The nature of assembly tasks is such that it is difficult to automate them economically unless the tasks have unusual characteristics such as size, or weight, or temperature, chemical or radiation hazards. Beginning in the clothing and shoe industries and then spreading to the electronic industry in the early 1980s, it was recognized that *flexible assembly systems* with automated job movement to assembly, inspection, and test stations, and often also linked to automated job identification systems, resulted in significant improvement in work flow and control in assembly systems producing a variety of different jobs, even though many individual tasks are still performed by human operators. Some flexible assembly systems enable jobs to move between any pair of work stations, whereas others, such as those introduced in the mid-1980s by the automobile industry to replace the traditional assembly line, generally have a series structure, but with paralleling of work stations and some feedback loops so that jobs can be readily reprocessed if they do not meet required quality standards.

1.3.6 Multiple Cell Systems

Usually, the introduction of automation of machines, work stations, and material handling to discrete part manufacturing systems proceeds in stages. Each segment of the overall system that is automated creates an *island of automation*. Hence, to begin with, the overall system would consist of several separate islands of automation. In some cases these islands may eventually be linked up into one closely coupled integrated system (similar to the continuous process systems in the chemical and metallurgical industries). The need to achieve a manufacturing system with adequate scope, however, makes closely coupled systems less desirable in discrete part manufacturing because they may not have sufficient flexibility. An alternative path of development is to consider the overall system as a *cellular manufacturing system* made up of *manufacturing cells* where each cell has specific capabilities in terms of its function, scale and scope. The coupling between cells need not necessarily be tight, and the number, capability, and configuration of the cells will change as the production requirements evolve over time. Thus it is important to be able to understand how the capabilities of the individual cells and the method of cell coordination determine the system performance.

1.3.7 Material Handling Systems

Apart from the machines and work stations a key component of the manufacturing system is the *material handling system*. It is important to understand how different types of material handling systems behave. Apart from linking machines, the material handling system also acts as a temporary store for jobs in process, and links dedicated storage facilities or warehouses into the manufacturing system.

1.4 MANAGEMENT DECISIONS

Management decisions for manufacturing are concerned with answering the questions: What to make? How to make? Where to make? When to make? How much to make? Traditionally most manufacturing organizations were set up so that "What to make?" was decided by *design engineering* (and marketing). Design engineering created the engineering drawings, specifications, and sometimes prototypes of the product. Next, on receiving the information describing the product, "How to make?" was decided by *manufacturing engineering*. They determined the manufacturing processes, machines, tools, jigs, and fixtures, and wrote the process specifications. They fixed the sequence of operations and identified the inspection and test requirements. They recommended any new plant or facilities that might be necessary to produce the product at the required volume and quality levels. Once the manufacturing process and methods were established it was then up to *production management* to decide "Where to make?", "When to make?", and "How much to make?". Production management ensured that the required work force, both in number and skills, was available. They were responsible for having all parts and raw materials ordered and delivered, and then they had to see that the required production occurred on the available facilities, and products were delivered to customers as promised and scheduled. Recently, it has come to be recognized that in large organizations this strict separation of decision making and rigid sequencing of decisions does not contribute to overall manufacturing effectiveness. So the attempt is now being made to consider all the questions at the same time, and, in particular, closely integrate design and manufacturing engineering. Automated manufacturing systems also have the potential to defer decisions relating to the details on how to make the product until the product is in production (e.g., the choice between alternative operation sequences can be made to depend on machine or worker utilization). So eventually there will also be a closer integration of manufacturing engineering decisions and production management decisions.

Management decisions for manufacturing differ substantially in their time horizon, that is, the amount of time into the future over which the decision will have impact. The acquisition of a new plant may have a time horizon of 10 to 20 years, whereas the choice of job to process next on a machine may have a time horizon of only a few hours. Some manufacturing engineering decisions tend to have long time horizons and involve the commitment of considerable resources, particularly those relating to the acquisition of new equipment. These decisions then affect other decisions because the performance characteristics of the equipment and the manufacturing system which incorporates the equipment set significant constraints on the production management decisions. The time horizon for production management decisions is usually short, at most 12 to 18 months for ordering long lead time raw materials, or hiring and training skilled workers. Most production management decisions relate to planning capacity and work loads, and have a 3- to 6-month time horizon or less.

1.4.1 Strategic, Tactical, and Operational Decisions

Based on the time horizon and the extent of the system impacted by the decision it is sometimes useful to distinguish between strategic, tactical, and operational decisions. *Strategic* decisions are those with a long time horizon (more than a year or two) and involve major commitments of the organization's resources. In manufacturing, strategic decisions relate to the size and location of plants and major facilities, the diversity of products, the choice of technology, and the degree of automation. *Tactical* decisions have an intermediate time horizon of between 3 and 18 months or a lesser scope. Determining work force size or planned production rates are examples of tactical production management decisions, whereas the capacity of in-process inventory banks or the layout of the plant are examples of tactical manufacturing engineering decisions. *Operational* decisions are typically those made every day or every week, such as decisions on the release of jobs to manufacturing, or the allocation of jobs and workers to machines.

The increasing level of automation in manufacturing has as its goal not only the automation of the physical tasks but also the incorporation of many operational decisions into the software of an information, decision, and control system. Thus it is essential to understand how the equipment and the physical characteristics of the machines and work stations interact with the software of information and control. As for tactical decisions, although they are not routine, they nevertheless require effective support so that the implications of different courses of action can be explored. Preferably, because these decisions will have to be made reasonably frequently, this support should be in the form of computer software that can be used by the manufacturing engineer or the production manager. Strategic decisions usually involve the commitment of large resources. They too require support to explore alternatives, but, even more, there is the need to develop an awareness in those responsible for the decisions as to the basic tradeoffs involved and the key parameters that will determine the performance of the chosen manufacturing system. All this emphasizes the need to have a formal and structured approach to developing an understanding of what determines the performance of a manufacturing system, and then incorporating this understanding in software that can be used to support decisions about the system.

1.4.2 Disturbances, Variety, and Hierarchy

Any attempt at understanding how manufacturing systems behave and what can be done to improve their performance has to address several issues. In particular, it is necessary to consider the impact of Murphy's Law: "If anything can go wrong, it will." That is, manufacturing systems rarely perform exactly as expected and predicted. Something is always going wrong: customers may change their orders, overall demand may increase, equipment may break down, workers may not show up for work, parts may be found to be defective, and so on. Responding to these disturbances occupies a significant amount of the time of manufacturing managers.

When something goes wrong in a manufacturing system, the manufacturing manager has to identify what can be done to alleviate the problem and restore the system to the desired condition. If the manager has few options then it is less likely that an effective response can be made. Options may be limited for a variety of reasons. The problem may be new and outside the manager's previous experience, so the manager may not be aware of available options. Alternatively, the manager may not have access to the required information, and may not have the responsibility to deal with the situation. Also, the physical design and layout may prohibit certain actions. For example, there may be no available space to store jobs in front of a failed machine while it is being repaired. Because there is a range or variety of possible disturbances to respond effectively there has to be a variety of actions available. This is known as Ashby's

Principle of Requisite Variety: "Only variety can destroy variety." For everything that can go wrong with a system there should be an effective response.

Provision of effective responses requires considering the frequency and magnitude of all the disturbances of concern. If the disturbance only affects a small part of the system, then it is usually better to deal with the disturbance close to where it occurred. If the disturbance occurs very frequently, and it is not possible to eliminate its causes, then simple routine responses are preferable so that the managers are not overloaded and the response can be delegated to lower level staff. This suggests that effective responses should be organized into a hierarchy, with the lowest level dealing with disturbances that are either limited in scope or where, because of their frequency, a programmed response is appropriate. At the middle level of the hierarchy, there is greater freedom for creative response but still within constraints. The information about the problem may be filtered by the lower levels and may reflect the inability of the lower level to deal with it within their permitted range of responses. At high levels in the hierarchy, the problem is typically either one affecting the whole system or one that has arisen outside the system. It is clear that the design of the hierarchy requires deciding what range of responses are to be available at each level and the way in which information is to be filtered as it is passed up the hierarchy.

In reality discrete part manufacturing will involve several different manufacturing units, even if there is not a cellular structure. Each unit may be provided with its own hierarchy to respond to disturbances. The different hierarchies should be coupled and information shared for effective control so that if a disturbance in one unit affects another then the other system can develop an appropriate anticipatory response. An alternative approach to minimizing the effects of disturbances in one unit on other units is to provide buffers between them.

Even so, to solve a problem with a manufacturing system, it is necessary to define the boundary of the system. Within the boundary it can be assumed that the management has some control over what is going on; outside the boundary they have no control and have to accept whatever impacts on their system.

1.4.3 Goals, Performance Measures, and Targets

To choose between alternative courses of actions and guide managerial decision making, the manager has to have appropriate criteria. At the strategic level, these criteria are usually based on rather broad organizational goals. For example, likely goals are the survival and growth of the organization, and the creation of wealth and other benefits for the owners, managers, workers, customers, the community, or even the nation as a whole. For tactical and operational decisions, however, these general goals have to be converted into more concrete performance criteria that are susceptible to measurement and tracking over time. Further, top management may set specific targets for these performance measures that managers and workers have to aim at achieving.

The usual performance measures in the manufacturing context relate to *production volume*, *quality of output*, *cost* and *customer service*. The production volume may be measured by the dollar value or total quantity, but it would usually also be expressed in terms of the mix of products produced. Traditionally, at least for low- and middle-level management, production volume is the dominant performance criterion. Increasing consumer concerns have led to an increased emphasis on the quality of the product measured by the extent to which the product meets specifications. Also, customer service, measured by the ability to make promised delivery to customers, has also become an important performance criterion. For upper management, the dominant concern in evaluating manufacturing activities has been cost, although quality and service targets must be met. Although some components of cost, such as raw material consumption, are to a large extent independent of the design and operation of the manufacturing system, those components that are linked to production volume, quality, or the product range will be strongly linked to the system design and operation. For example, the number, capability, and location of test and inspection points is related to the outgoing quality, whereas the size and location of in-process inventories is related to the production volume and customer service. To make decisions about manufacturing systems and develop appropriate responses to disturbances it is important to have a means of understanding the nature of these relationships and how the performance goals are impacted by features of the system design and operation.

1.5 MODELS

Traditionally, manufacturing system designers relied on experience and rules of thumb to identify the effect of design parameters and control rules on the performance. The complexity of automated systems, their high initial cost, and the often lengthy time required to bring them up to their designed production targets have resulted in designers using formal models of the system to assess performance.

1.5.1 Nature of Models

A model is based on a representation of the operation of the system. To the user of a model, the model is either a mathematical formula or a computer program that when supplied with the numerical values of various parameters will make a numerical prediction of the system performance measures. By changing the values of a parameter the user can gain insight into its influence on performance. The model may also enable the optimal values of the parameters to be found, where optimal means that they optimize the performance of the system as described by the model. Models are intended to support management decisions about the system, so there is rarely one single model that will support all decisions. Rather, different decisions require different models because different aspects of the design and operation of the manufacturing system will be critical to the model's success in reliably and accurately predicting performance.

1.5.2 Basic Approach to Modeling

The process of modeling involves the following steps:

   1. *Identify the issues to be addressed.* First, ascertain the needs of the user. The major concerns of management have to be elucidated, and the decisions that the model is required to support have to be identified. As mentioned earlier the decisions can be strategic, tactical, or operational, and the modeler has to ensure that management can express its goals, performance measures, and targets. The modeling activity can be initiated by the need to make a clear-cut decision such as the acquisition of a manufacturing system for a new product line, or it can develop because management seeks new insights into festering problems of low productivity and poor customer service in an existing plant. A wide variety of issues are usually raised at this stage. Many of them are impossible to define, measure, and model, and some are irrelevant. Nevertheless, the modeler should become aware of these apparently tangential issues as often in the course of the modeling process new insights are gained and, of course, the apparently tangential can be found to be central.

2. *Learn about the system.* Identify the components of the system, such as the machines, material handling and storage, and the data collection and control systems. Determine the characteristics of the jobs and the target volumes, quality, and cost. This step requires close contact with the system designers and managers. If at all possible the modeler should observe the manufacturing system in action or at least talk to knowledgeable and experienced people who have designed and operated similar systems. It is by no means uncommon for the modeler to recognize that management perceptions of the nature of the problem are not supported by observation and that different issues and different approaches should be considered. It is useful to review any existing models to identify their capabilities, their shortcomings, and the level of familiarity of managers with the use of modeling approaches to help in manufacturing system design or operation.

3. *Choose a modeling approach.* Various types of modeling approaches can be used, ranging from formal mathematical models through computer simulations to the development of a "toy" system in which toy parts physically move from one toy machine to another. The choice of modeling approach is determined by the time and cost budget for model development and the anticipated way in which the model will be used. Often, a variety of different approaches will be used. The modeler may end up delivering a simulation model as the end product of the modeling activity, but in the process of developing and validating the simulation, extensive use may be made of analytical models. Further, there are a variety of software packages available that simplify the task of developing models of specific manufacturing situations so the modeler has to evaluate their appropriateness for the issues with which he or she is concerned.

4. *Develop and test the model.* This step requires obtaining data on the parameters of the model. Often the lack of desirable data forces the model to be substantially simplified. If the system being modeled does not yet exist then the available data is usually limited, and it will be difficult to get even estimates of the average values of parameters. It is nearly impossible to get information about the distributions of random variables such as the time to failure of a machine, so "reasonable" assumptions have to be made. Even if the system exists, it is often the case that despite voluminous performance reports the information about the system is not collected in a way that is appropriate for the model. Collecting data from the system itself is often expensive and unreliable because of possible negative reactions by operators to the observer. Thus a necessary assessment that the modeler has to make is the level of detail and accuracy required in the data.

5. *Verify and validate the model.* The model has to be checked to see that it is a reasonably correct representation of reality. This involves two aspects. Verification is the process of ensuring that the model results are correct for the assumptions made in developing the model (i.e., that the model is internally consistent). Verification involves mathematical and logical techniques. If the model contains approximations then the model has to be compared with another model to assess the adequacy of the approximations. Validation is the process of ensuring that the assumptions of the model and the results of the model are an accurate representation of the real manufacturing system. The modeler has to make sure that the assumptions of the model are clearly stated and that they seem reasonable to people who understand the situation being modeled.

6. *Develop a model interface for the user.* If the model is to be of value in making decisions it has to be provided with some interface so that it can be used by manufacturing managers. This requires the modeler to either incorporate the model into a decision support system, or to present the model and its implications in a way that manufacturing managers can understand. The user has to be convinced that the model is adequate for the decision that the model is required to support.

7. *Experiment with the model.* This requires exploring the impact of changes in model parameters and developing understanding of the factors influencing performance of the system so that the manager can be confident in the decisions made using the model. With simulation models in particular the modeler should consider using formal experimental designs to ensure both economy and reliability in obtaining information from the model.

8. *Present the results.* Using the model the manager should come up with a recommended course of action. This may have to be presented to higher-level management, and the role of the model in aiding the decision may have to be explained. Alternatively, the model and the results of its use will be presented in a report or paper that, in addition to describing the model itself, should explain what the model can do, what it cannot do, and how accurate its predictions are.

1.5.3 Context of Modeling

There are two contexts in which manufacturing system modeling occurs. Modeling in industry occurs to support decisions that usually are about specific systems. This means that there tends to be pressure to include in the model all the detail that various managers think is relevant to the design and operation of the specific system, even though some of this detail may be irrelevant to the particular decision that the model is to support. Further, as pointed out by Feltner and Weiner [6], the model development is often viewed as supporting only one specific decision, whereas in practice, once developed, the model is likely to be used and reused many times whenever a similar decision comes up. As a result industry modeling may suffer from a limited model design that does not adequately allow for changes in parameters and operating rules.

The other context in which manufacturing modeling occurs is within academia. The modeling effort is then motivated by a desire to understand how manufacturing systems behave, and in particular how some of the new concepts for manufacturing improve system performance. Further, because manufacturing system modeling attempts to describe complex systems that are subject to a wide range of possible disturbances, the process of model development and model solution has considerable intellectual challenge.

These differing contexts for the modeling activity mean that industry practitioners and academics are inclined to focus on the shortcomings of each other's approaches to implementing the modeling process. To academics, industry is perceived as not validating its models carefully enough. Academics claim that as long as they look right, industry takes the models as valid. They point out that there is a lack of exposure of industry modeling activity to third-party evaluation and appraisal. Further, much industry modeling activity makes assumptions that are poorly documented and rarely communicated to the model users. Conversely, to industry, academic models are usually limited in their realism. Much detail about the real system is omitted from the model, sometimes just to achieve tractability or even "elegance," and it often appears as if the modeler is unaware of what has been omitted. Rarely do academics validate their models in specific industrial settings. This can lead to conclusions being drawn from the model that are overgeneral and that may not apply to the real system that the model is intended to represent. Further, academics often do not focus on the crucial issues. Often performance measures are inappropriate, and operating rules are unrealistic and impractical.

1.5.4 Types of Models

For discrete part manufacturing there are three types of models in common use.

- *Physical* models represent the real system by another physical system, in which parts move from one machine to another and the machines perform processing operations on the parts. The major difference from the real system is that the model uses a different dimensional scale, so a large system will occupy a table top. Physical models can use toy size components, but may be provided with a control system that employs the same logic as the real system. Physical models are excellent as a means of educating production management and workers about the control of the system, but they do not lend themselves to assessing the long run behavior of the system as it is difficult to represent the statistical properties of events such as machine failures.

- *Simulation* models represent the events that could occur as a system operates by a sequence of steps in a computer program. This means that the logical relationships that exist between events can be described in detail. The probabilistic nature of many events, such as machine failure, can be represented by sampling from a distribution representing the pattern of occurrence of the event. Thus to represent the typical behavior of the system it is necessary to run the simulation model for a sufficiently long time, so that all events can occur a reasonable number of times. Simulation models can be provided with an interactive graphic display to demonstrate the movement of jobs and material handling devices. This can be of great value in communicating the assumptions of the model to manufacturing engineers and others.

- *Analytical* models describe the system using mathematical or symbolic relationships. These are then used to derive a formula or to define an algorithm or computational procedure by which the performance measures of the system can be calculated. Analytical models can also be used to demonstrate properties of various operating rules and control strategies. Sometimes it is not possible, within a reasonable amount of computer time or space, to obtain the performance measure from the relationships describing the system. Further assumptions that modify these relationships then have to be made. The resulting model is then approximate rather than exact. Testing the approximation may then require a simulation model, so approximate models are only useful if they are easy to use and provide insight into what determines the system behavior.

1.5.5 Why Model?

Models can be developed for a variety of reasons, in particular:

- *Understanding*. The model is used to explain why and how. The model is intended to help develop insight. Sometimes the model just indicates the direction of influence of some variable on performance (e.g., as the variable increases in value, performance may improve). Alternatively, the model can be quite complex, even though its major function is to explain why the system behaves in certain ways.

- *Learning*. As well as providing insight a model may be intended to teach manufacturing managers or workers about the factors that determine performance. The model may omit many features of the real system and focus on those aspects that are considered crucial for those people responsible for effective operation of the system.

- *Improvement*. The model is used to improve system design and operation. Changes in parameters and rules can be explored, and factors critical for achieving performance targets can be identified. To make sure that conclusions drawn from the model will apply to the real system, such models pay particular emphasis to the adequacy with which they describe the system behavior.

- *Optimization models*. Given a model that predicts performance as a function of various parameters, an optimization model determines the optimal combination of these parameters. This usually means that the optimization problem is formulated as a mathematical programming problem, generally with a mixture of integer and continuous variables.

- *Decision making*. The model is to be used to aid decisions about either the design or operation of the system. The model has to be able to discriminate the effects of different courses of action and project their impact over time. Sometimes the model may be built into an integrated decision support system so it can be called up for use as required.

1.5.6 Requirements of Models

Models are based on certain assumptions about the system and its components, and how the system is going to be operated. Then there are the assumptions about the nature of the disturbances that will impact the system and the range of responses to these disturbances. The hierarchy of control and the flow of information also have to be represented in the model. In developing a model there are several considerations.

- *Complexity versus simplicity.* Modeling involves compromises in deciding how much detail to represent. More detail enables the model to more precisely represent reality. It also means, however, that the model will be more difficult to verify and validate, it will be more difficult for users to understand, and it will take longer to develop. Conversely, a simple model may not represent the system adequately, and thus it may give inaccurate predictions, and omit key decisions or useful responses to disturbances.

- *Flexibility.* A model may be used to support decision making as the system evolves over time. From initial concept through planning, detailed design, installation, and operation, there is a need for models to support decisions. Although no one single model will support all decisions it is desirable for a model to be useful at several different stages in the evolution of the system. This means that the model should permit changes in the system modeled. Some of these changes may relate to the structure of the system such as the number of machines, the way in which they are connected by the material handling system, or the way in which the control hierarchy is set up. Other changes may relate to the values of parameters such as the frequency of machine failures or the demand rate. A model or a modeling approach has to be evaluated with respect to the ease of making both these sorts of changes.

- *Data requirements.* Although there is often a great deal of data available in manufacturing, it is rare that the data is in the form required by the model. At the planning stage there is often doubt as to the applicability of data collected from different systems in different environments, whereas when the system is operational the data may well only apply over a limited range of operating conditions. Thus a model should use the least amount of data required to make adequate predictions, and an important component of the validation of a model is assessing the sensitivity of the model to errors in the data.

- *Transparency.* Because the model has to be accepted by its users it is desirable that the assumptions and procedures used in the model be reasonably transparent to others beside the model developer. The developer should be able to convince the user that the model is a reasonably accurate representation of reality. Along with this goes the requirement for adequate documentation of the model and how to use it, written in language appropriate to the user's education and experience.

- *Efficiency.* Models can consume significant resources, both in their development and in their use. Modeling approaches differ in their requirements on the knowledge, skill, and elapsed time required for development. Because most models will be implemented on a computer, such issues as the storage requirements and running time can also be important. Of course, developments in computing keep changing what is considered acceptable.

- *User interface.* If a model is going to be of real value it should be usable by manufacturing engineers and managers, as well as by the model developer. A user interface is essential to guide the user in the correct use of the model, ensuring that it is clear what data he or she should provide, and avoiding any ambiguity in interpreting the results.

1.5.7 Analytical versus Simulation Models

To compare analytical and simulation models with respect to how well they meet the preceding requirements it is necessary to describe the two modeling approaches. Simulation models can be developed in several different ways. One is to try and include all the required detail of the system from the beginning. This tends to result in long development and debugging time, and usually makes the resultant code unreadable to all but the developer. Another approach is to develop models of the system components, and, after each component model has been debugged, tested and validated, link the components two at a time, test, and validate, then three at a time, test, and validate, and so on. Sometimes there are already existing simulation models of various subsystems. These may be too detailed for the overall system model but often it is possible to devise ways of summarizing the performance of the subsystem model and replacing it by a simple subsystem component model that can be used as part of the overall model. This approach is called the micro-macro approach and is useful in developing models of very large systems.

Simulation models tend to be difficult for others beside the developer to verify because they are usually written in a simulation language such as GPSS, SLAM, SIMAN, SIMULA, or SMALLTALK, and only experts in the simulation language can read the code. The models can be validated by looking at extreme cases in which the system's performance can be easily predicted (e.g., when it is assumed that machines never fail). Sometimes they can be validated by comparison with analytical models, but this is often not easy, even if the analytical model has the same assumptions as the simulation model. Analytical models typically determine the expected value of some performance criterion, whereas simulation models determine a time-limited sample of performance. Thus some statistical test is required to determine whether the sample is consistent with the analytical results.

Analytical models either give a formula or define a computational procedure. If they give a computational procedure then it is necessary to consider the computational complexity of the procedure. It is by no means uncommon for the computational requirements to increase exponentially with the number of machines or the number of different types of jobs. This often limits the size of the system and the range of parameter values for which the procedure can be used. Often the procedure can only be used on "toy" systems. It is then difficult to extrapolate results from the toy system to systems of realistic size. Although approximate analytical models are often useful, such models can only be validated by means of a simulation.

1.5.8 Comparison of Merits

- *Complexity*. Simulation models can describe considerable complexity, and, apart from limitations in the time required to run the model, they can be used for whatever level of complexity is desired. Analytical models are usually quite limited in the complexity of the system that they can describe. Typically the computational complexity increases exponentially with the size of the system. Also some control rules result in very complex models. Approximate analytical models can handle larger systems, but because of the difficulty in developing and testing approximations, the range of manufacturing systems for which proven approximate models exist is small.

- *Flexibility*. A well-designed simulation model should enable a variety of related system structures to be analyzed. It is usually essential to define the sort of changes to be modeled before developing the simulation, however, so that the user can make the changes rather than the model developer. Changes in parameter values may also have to be considered in advance because some simulation languages do not enable such changes to be made readily by the user. By contrast, it is usually very easy to change the parameters of an analytical model. Changes in the structure of the system usually require a totally new model, however, and it is not uncommon for apparently minor changes in structure to change a model from one where efficient solution is possible to one where solution is very difficult.

- *Data requirements*. Most analytical models have minimal data requirements because they tend to be fairly simple descriptions of the system. Simulation models can require large amounts of data and so they may not be easy to implement. Thus it is important to consider data availability at the design stage.

- *Transparency*. Simulation models are usually not at all transparent to their user. It may be possible to describe the logic of the model to the user, and this should be done whenever possible. Only an experienced programmer can "read" the simulation code, however. Analytical models are usually transparent to other modelers who have the appropriate mathematical skills, but these skills are rarely found in manufacturing managers, so the model may be far from transparent to them.

- *Efficiency*. One of the advantages of simulation modeling in an industrial context is that the time required to develop a model can usually be assessed reasonably accurately if the simulation group has had adequate experience. Analytical models are much more unpredictable in the time and effort required to develop an adequate model, and there is a greater variability between individual modelers. Once the analytical model is developed it usually does not require much time to use it to get results, particularly if the model is communicated in the form of a computer algorithm. Simulation models can require substantial time, both computer time and elapsed time, if a variety of changes in parameter values are to be explored.

- *User interface.* Simulation models require a user interface, preferably visual and interactive, so that they can be used by managers and engineers. Such systems are available and are of increasing value. Analytical models need to be incorporated within a decision support system for them to be used effectively. The decision support system should incorporate a model management system to guide the user in the choice of the appropriate model and the selection of the required data.

1.5.9 Need for Both Analytical and Simulation Capability

Effective modeling of discrete part manufacturing systems requires both analytical and simulation modeling capability. It is often necessary to use approximations in analytical models so simulation models are required to test these approximations. Even if no approximations are required, model development is often aided if a simulation study is carried out to identify key parameters and significant assumptions. Because of the limited complexity of analytical models, the modeling effort can then focus on the essentials that the model should address. Further, the simulation model can suggest hypotheses about the nature of the optimal control rules that can then be explored using analytical modeling techniques.

Even when the size and complexity of the system being modeled mean that simulation modeling is the only appropriate modeling technique, analytical models can still perform an essential role in validating parts or special cases of the simulation model. For example, it sometimes occurs that in developing a simulation model of a discrete part manufacturing system assumptions are made about the sequence of part flows in and out of inventory banks that change the effective capacity of the banks by plus or minus one. Such an error can only be detected if there is an analytical model to compare with the simulation results. Analytical models may suggest the general shape of relationships between performance and parameters, and this can be useful in validating the simulation. Thus it is essential for groups involved in modeling discrete part manufacturing systems to have both analytical and simulation expertise.

1.6 MANAGERS AND MODELERS

Because the model, or at least its results, are going to be used by managers responsible for the effective acquisition, installation, and use of the manufacturing system, it is essential that managers believe in the model. They have to be convinced that the model describes adequately how the system behaves and that conclusions drawn from the model are valid. It is also desirable that they believe in the internal consistency of the model, that is, that the calculation of its performance predictions are correct. This often requires, however, that the manager have detailed understanding of the techniques of model formulation and solution. This is still rare in manufacturing.

Any model contains limitations. It may have approximations that limit the range of its validity, some assumptions may be restrictive and of doubtful validity, and the measures of performance may be inappropriate for some decisions. The model developer or the model itself has to convey these limitations to the model user or manager who will use the model results. If the model is developed for a specific situation then the model user should participate in the model development. When a general model is used the general principles of the model must be conveyed to its users.

There is often the potential for conflict between the modeler and the manager because the manager wants the model to describe all the reality that he sees while the modeler usually wants to include only those aspects that he considers critical. Further, the modeler will be further constrained by the need to solve the model and hence may also omit some details to get results. The manager has to understand the limitations of the modeling approach, and the modeler has to be able to convince the manager that what has been omitted does not significantly affect the conclusion drawn using the model.

A model has to be appropriate to the situation that it models and the decisions that are supported by its use. Thus a key part of the skill of the modeler lies in the ability to observe and describe reality. In models used to support tactical and operational decisions the hardware and the software of the system is typically well documented, so the reality is fairly well defined. When models are developed to support strategic decisions, however, there is often no way reality can be observed in an objective way. Such decisions can involve manager's hunches and perceptions about relationships. Often he or she is not able to express these relationships in a quantitative way. The modeler's skill then has to be used to elucidate these relationships so that they can be captured in the model.

1.7 SCOPE OF THE BOOK

This book develops and describes analytical models of discrete part manufacturing systems. In particular, it focuses on models to predict and optimize performance, as well as to gain insight into the factors influencing performance of manufacturing systems. Such systems include machines, work centers, material handling, storage, information collection, and control.

The models developed here are not concerned with describing individual processes, and the physical and chemical principles that determine how and why a process operates. We will need to know how long the equipment or people that use the process require to perform the task and how effectively the task is performed from a quality viewpoint. We will also need to know the nature of the disturbances that influence time and quality, and in particular their frequency, duration, and pattern of occurrence. We will need to understand what information is needed to perform the task and also how information about the occurrence of disturbances is acquired and processed.

The need to describe disturbances and variability means that the emphasis of the book is on stochastic models. Because of the complexity of the systems we model, we will frequently have to use approximate models, so we will present the results of simulation studies to validate them. Many of the models address issues relating to the control of the movement of jobs through the system, and we will try and demonstrate control policies that can be proved to optimize performance.

The book is organized around the different types of manufacturing systems: flow lines, job shops, automatic transfer lines, flexible machining systems, flexible assembly systems, and cellular systems. Further, because of the importance of the material handling system that links processes, we will incorporate it into our models of the manufacturing systems whenever needed. The overall control of work flow in a manufacturing system is crucial to its effective integration into the operations of the firm so we have several chapters that address this and related issues.

BIBLIOGRAPHY

[1] R. U. AYRES. Technology forecast for CIM. *Manufacturing Review*, 2:43–52, 1989.

[2] G. BOOTHROYD. *Fundamentals of Metal Machining and Machine Tools*. McGraw-Hill, New York, 1975.

[3] J. A. BUZACOTT. Flexible models of flexible manufacturing systems. In F. Archetti, M. Lucertini, and P. Serafini, editors, *Operations Research Models in Flexible Manufacturing Systems*, pages 115–122, Springer, New York, 1989.

[4] J. A. BUZACOTT. Modelling manufacturing systems. *Robotics and Computer-Integrated Manufacturing*, 2:25–32, 1985.

[5] J. A. BUZACOTT. Validation of manufacturing system models. In J. A. White and I. W. Pence, Jr., editors, *Progress in Material Handling and Logistics*, volume 1, pages 237–245, Springer, New York, 1989.

[6] C. E. FELTNER and S. A. WEINER. Models, myths and mysteries in manufacturing. *Industrial Engineering*, 17:70–79, July 1985.

[7] S. B. GERSHWIN, R. R. HILDEBRANT, R. SURI, and S. K. MITTER. A control theorist's perspective on recent trends in manufacturing systems. *IEEE Control Systems Magazine*, 6:3–15, April 1986.

[8] M. P. GROOVER. *Automation, Production Systems, and Computer-Aided Manufacturing*. Prentice-Hall, Englewood Cliffs, NJ, 1980.

[9] K. HITOMI. *Manufacturing Systems Engineering*. Taylor and Francis, London, 1979.

[10] A. M. LAW and S. W. HAIDER. Selecting simulation software for manufacturing applications: practical guidelines and software survey. *Industrial Engineering*, 21:33–46, May 1989.

[11] A. M. LAW and M. G. McCOMAS. Pitfalls to avoid in the simulation of manufacturing systems. *Industrial Engineering*, 21:28–31, May 1989.

[12] R. SURI. Manufacturing system modeling: its role and current issues. In J. A. White and I. W. Pence, Jr., editors, *Progress in Material Handling and Logistics*, volume 1, pages 209–221, Springer, New York, 1989.

[13] R. SURI. A new perspective on manufacturing systems analysis. In W. D. Compton, editor, *Design and Analysis of Integrated Manufacturing Systems*, pages 118–133, National Academy Press, Washington, DC, 1988.

2

Evolution of Manufacturing System Models: An Example

2.1 SOLVING PROBLEMS USING MODELS

In this chapter we will illustrate the basic steps involved in the development and evolution of models through a simple example. In particular, we consider the *machine interference* problem.

Example 2.1 **(Machine Interference)**

A machine shop contains several automatic machines. These machines only require attention when a tool must be changed, a machine breaks down, or the machine has run out of material. Management is concerned about staffing levels (i.e., how many machines should be assigned to one operator). They recognize that with too few machines assigned to one operator, the operator will often be idle, whereas with too many machines there will be interference (i.e., several machines will be stopped waiting for the operator to come and fix them).

Now suppose that at this point management approaches the analyst to see how the analyst can help them solve the problem. The analyst would then typically go through the eight steps described in Chapter 1, Section 1.5.2. In this process the analyst would determine that there are two aspects of performance of concern: the performance of the operators and the performance of the machines. The measures of system performance that are appropriate for this example are

1. *Operator utilization ρ.* Total time the operator is engaged in servicing machines divided by the total time for which the operator is paid.

2. *Net production rate TH.* Actual production achieved from the system (or from each individual machine) divided by the time during which the system is observed.

3. *Machine efficiency or availability η.* Ratio of the net production rate to the gross production rate, where the gross production rate is the rate that would be achieved if no machine ever required attention from the operator. (Note that in calculating the gross production rate it would be normal to allow for the effect of external interruptions such as power failures, material shortages, lack of work, and strikes.)

In this chapter we will primarily focus on the modeling approach involved in step 4 of the eight-step procedure discussed in Chapter 1.

2.2 MODELING APPROACH

In the remainder of this chapter we are concerned with describing a variety of models for the machine interference example presented earlier. We begin with simple models that consider just a few aspects of the situation. Of course, determining what those aspects should be demands skill on the part of the analyst. As increased understanding is gained the situation represented by the model can be made more complex and the effect of more factors included.

2.2.1 Long-Run Analysis

One approach that is valuable to use when setting out to develop models of a new situation is to look at the typical pattern of behavior of the system over some long period. For example, the change in the number of machines not working as time goes on could be expected to follow a pattern such as in Figure 2.1.

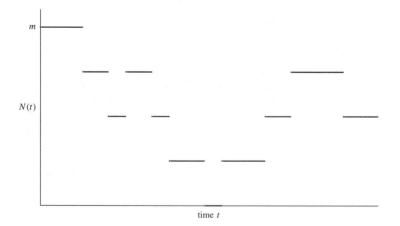

Figure 2.1   Number of Machines Failed versus Time

Let m be the total number of machines and let P_n be the proportion of time n machines are down (determined as the total time exactly n machines are down divided by the total time t over which the system is observed).

If the machines are identical and each has a gross production rate of h units per hour then the total production during $(0, t]$ is equal to $\sum_{n=0}^{m}(m-n)P_n th$. Therefore if TH is the net production rate and G is the gross production rate, one has

$$TH = \sum_{n=0}^{m}(m-n)P_n h, \tag{2.1}$$

and

$$G = mh \tag{2.2}$$

Then from equations 2.1 and 2.2 one sees that the system efficiency, η, is given by

$$\eta = \frac{TH}{G} = \frac{TH}{mh} = 1 - \sum_{n=0}^{m}\frac{nP_n}{m} \tag{2.3}$$

The operator utilization, ρ, is given by

$$\rho = \sum_{n=1}^{m}P_n = 1 - P_0 \tag{2.4}$$

Suppose that the mean time to repair a machine is $1/\mu$. Then $\overline{D}(t)$, the average number of repairs completed during $(0, t]$, is given by

$$\overline{D}(t) = \mu\rho t \tag{2.5}$$

Now if the mean time to failure of a machine is $1/\lambda$, then $\overline{A}(t)$, the average number of machine failures during $(0, t]$, is given by

$$\overline{A}(t) = \lambda\sum_{n=0}^{m}(m-n)P_n t$$

$$= \lambda\eta m t \tag{2.6}$$

where the second inequality follows from equation 2.3. Because in a long time period it would be expected that the number of machine failures is approximately equal to the number of machine repairs, that is, $\overline{A}(t) \cong \overline{D}(t)$, it follows from equations 2.5 and 2.6 that

$$\eta = \rho\frac{\mu}{m\lambda} \quad \text{and} \quad TH = \rho\frac{\mu h}{\lambda} \tag{2.7}$$

where the second equality follows from equation 2.3. The quantity $1/(m\lambda)$ can be interpreted as the mean time between stoppages when all machines are working.

Next, using equations 2.3, 2.4, and 2.7, one can determine the average number of machines, $\overline{L}$, waiting for service as

$$\overline{L} = \sum_{n=1}^{m} (n-1) P_n$$

$$= \sum_{n=1}^{m} n P_n - \rho$$

$$= m(1 - \eta) - \rho$$

$$= m - \rho \left(1 + \frac{\mu}{\lambda}\right). \tag{2.8}$$

Also, the total time that machines will be down during $(0, t]$ is equal to $\sum_{n=1}^{m} n P_n t = m(1 - \eta)t$ (see equation 2.3). Thus, the average duration of a down time of a machine is, from equation 2.6,

$$\overline{T} = \frac{m(1 - \eta)t}{A(t)}$$

$$= \frac{(1 - \eta)}{\eta \lambda} \tag{2.9}$$

The average number of machines down is, from equation 2.3,

$$\overline{N} = \sum_{n=1}^{m} n P_n = m(1 - \eta) \tag{2.10}$$

Comparing equations 2.9 and 2.10 one sees that

$$\overline{N} = \lambda^* \overline{T} \tag{2.11}$$

where $\lambda^* = A(t)/t = \lambda \eta m$ (see equation 2.6) is the average number of machine failures per unit time, and from equation 2.8 $\overline{L} = \overline{N} - \lambda^*/\mu$. Hence,

$$\overline{L} = \lambda^* \overline{W} \tag{2.12}$$

where $\overline{W} = \overline{T} - 1/\mu$ is the average waiting time of a failed machine until its repair begins.

This type of analysis, based on just counts and long-run averages, usually yields several relationships between the different parameters. In Chapter 3 we will develop these relationships formally for general single stage systems. Particularly, we will see that the relationship (called Little's formula) given in equations 2.11 and 2.12 holds for very general systems. Also, it can give bounds on performance measures. For example, because the operator utilization cannot exceed 1, it follows from equation 2.7 that

$$TH = \rho \frac{\mu}{\lambda} h \leq \frac{\mu h}{\lambda} \tag{2.13}$$

irrespective of the number of machines assigned to an operator.

As another example, consider the average duration of a machine downtime, $\overline{T}$. Clearly $\overline{T} \geq 1/\mu$. Then from equations 2.3 and 2.9 one obtains

$$\eta \leq \frac{\mu}{\mu + \lambda} \quad \text{and} \quad TH \leq \frac{m\mu}{\mu + \lambda} h \qquad (2.14)$$

Now combining equations 2.13 and 2.14 one has

$$TH \leq \begin{cases} \dfrac{m\mu h}{\mu + \lambda}, & m \leq 1 + \dfrac{\mu}{\lambda} \\[2ex] \dfrac{\mu h}{\lambda}, & m \geq 1 + \dfrac{\mu}{\lambda} \end{cases} \qquad (2.15)$$

Sometimes this type of analysis even gives the desired measures of performance; examples of this will be given in Chapter 6 on transfer lines. Even though it might sometimes appear questionable to deal just in terms of long run averages, usually more formal analysis methods justify the results obtained using this approach. This will be illustrated in Chapter 3.

In the case of the machine interference problem it can be seen that all the performance measures depend on ρ, the operator utilization. We cannot determine ρ itself from consideration of the long term averages, however.

2.2.2 Deterministic Model

It is often valuable to consider the special case in which all times are deterministic. For example, if m machines are assigned to an operator and if $m \leq 1 + \mu/\lambda$ then it will be seen that there would never be more than one machine down at any time. To see this assume that machine j fails at time $\frac{1}{m}\left(\frac{1}{\lambda} + \frac{1}{\mu}\right)(j-1)$, $j = 1, 2, \ldots, m$. Then because $m \leq 1 + \mu/\lambda$ is equivalent to $\frac{1}{m}\left(\frac{1}{\lambda} + \frac{1}{\mu}\right) \geq \frac{1}{\mu}$ the first m machine failures will be repaired without any interference. That is, the first repair completion of machine j will occur at time $\frac{1}{m}\left(\frac{1}{\lambda} + \frac{1}{\mu}\right)(j-1) + \frac{1}{\mu}$. Thus the second failure of machine j will occur at time $\frac{1}{m}\left(\frac{1}{\lambda} + \frac{1}{\mu}\right)(j-1) + \left(\frac{1}{\lambda} + \frac{1}{\mu}\right)$, $j = 1, 2, \ldots, m$. This repeats the failure pattern that occurred during $\left[0, \frac{1}{\lambda} + \frac{1}{\mu}\right)$. Then by induction it can be seen that the kth failure (and repair completion) of machine j will occur at times $(k-1)\left(\frac{1}{\lambda} + \frac{1}{\mu}\right) + \frac{1}{m}\left(\frac{1}{\lambda} + \frac{1}{\mu}\right)(j-1)$, (and $\left((k-1)\left(\frac{1}{\lambda} + \frac{1}{\mu}\right) + \frac{1}{m}\left(\frac{1}{\lambda} + \frac{1}{\mu}\right)(j-1) + \frac{1}{\mu}\right)$) for $j = 1, 2, \ldots, m$; $k = 1, 2, \ldots$. When $m \geq 1 + \mu/\lambda$, suppose the first repair completion time of machine j is $(j-1)\frac{1}{\mu}$, $j = 1, 2, \ldots, m$. Then the second machine j failure will occur at time $(j-1)\frac{1}{\mu} + \frac{1}{\lambda}$. Because $\frac{1}{\lambda} \leq (m-1)\frac{1}{\mu}$, this time is less than or equal to $m\frac{1}{\mu} + (j-2)\frac{1}{\mu}$, the time at which (as an induction hypothesis we assume that) the second repair completion of machine $j-1$ occurs. Then the second repair completion of machine j will occur at time $m\frac{1}{\mu} + (j-2)\frac{1}{\mu} + \frac{1}{\mu} = m\frac{1}{\mu} + (j-1)\frac{1}{\mu}$. Continuing this analysis one sees that the kth repair completion of machine j will occur at time $(k-1)m\frac{1}{\mu}+(j-1)\frac{1}{\mu}$, $j = 1, 2, \ldots, m$; $k = 1, 2, \ldots$. In this case the operator is kept busy all the time. Hence we have

$$
\rho = \begin{cases} \dfrac{m\lambda}{\mu + \lambda}, & m \leq 1 + \dfrac{\mu}{\lambda} \\[2ex] 1, & \text{otherwise} \end{cases}
$$

$$
\eta = \begin{cases} \dfrac{\mu}{\mu + \lambda}, & m \leq 1 + \dfrac{\mu}{\lambda} \\[2ex] \dfrac{\mu}{m\lambda}, & \text{otherwise} \end{cases}
$$

$$
TH = \begin{cases} \dfrac{m\mu h}{\mu + \lambda}, & m \leq 1 + \dfrac{\mu}{\lambda} \\[2ex] \dfrac{\mu h}{\lambda}, & \text{otherwise} \end{cases} \tag{2.16}
$$

These results for deterministic times are an upper bound on the performance measures of systems with random failures and repair times (see equation 2.15). As we have seen, with deterministic times it is possible to achieve a staggering of machine down times to minimize machine interference and hence to maximize the throughput.

2.2.3 Binomial Approximation

The deterministic model gives results that would only be achieved if it were possible to coordinate the operation of the different machines. The next modeling approach that might be tried is to assume that the machine failures and repairs occur independently. That is, we ignore the interference of machines when they are down. With one machine it is known that, irrespective of the distribution of run time and down time, the machine efficiency $\eta(1)$ (the proportion of time the machine is working) is given by

$$
\eta(1) = \frac{\mu}{\mu + \lambda} \tag{2.17}
$$

Note that if $m = 1$, $\eta(1)$ is the system efficiency as well. Whenever needed, the parameter values of the system under consideration will be explicitly shown as argument(s) of the performance measures. For example $\eta(m)$ is then the system efficiency with m machines.

Now if it is assumed that the machines operate independently, and the probability a machine is operating at any randomly chosen instant of time is given by p, then the number of machines operating at any point in time would be determined by the binomial distribution.

That is, the probability n machines are down, p_n, is given by

$$
p_n = \binom{m}{n} (1 - p)^n p^{m-n}. \tag{2.18}
$$

Hence from equations 2.4, 2.7 and 2.18 one has

$$p_0 = p^m$$

$$\rho = 1 - p^m \tag{2.19}$$

and $$\eta = (1 - p^m)\frac{\mu}{m\lambda} \tag{2.20}$$

The simplest approach is to set $p = \eta(1)$ given in equation 2.17.

Example 2.2

If $\mu = 1$, $\lambda = 1/2$, $m = 3$, then $p = 2/3$ and $\rho = 0.703$, while $\eta = 0.469$.

The shortcoming with this approach is that machine interference will result in the down time being extended over the actual repair time. Because $\overline{T}$ is the mean down time of a machine it would be more reasonable to set $p = \frac{1/\lambda}{\overline{T}+1/\lambda}$.

From equation 2.9 it can be seen that

$$\frac{1/\lambda}{\overline{T} + 1/\lambda} = \eta$$

So setting $p = \frac{1/\lambda}{\overline{T}+1/\lambda}$ is equivalent to setting $p = \eta$. Then it follows that because $\rho = 1 - \eta^m$ and $\eta = \rho\mu/(m\lambda)$, (see equation 2.20) the system efficiency must satisfy the equation

$$\eta = \frac{\rho\mu}{m\lambda} = \frac{\mu}{m\lambda}(1 - \eta^m)$$

Example 2.3

Continuing example 2.2 η must satisfy the equation

$$\eta = \frac{2}{3}(1 - \eta^3)$$

which has the solution $\eta = 0.5536$ and hence $\rho = 0.8304$.

Although the first approximation is usually quite poor unless there is little machine interference, the second approximation is sometimes quite good.

2.2.4 Detailed Long-Run Analysis

Another approach to the problem is to do a more detailed long-run analysis. Returning to Figure 2.1 it can be seen that over some long period $(0, t]$ the mean number of times the number of machines down changes from n to $n + 1$, $C_n(t)$, must equal the mean number of times the number of machines down changes from $n + 1$ to n, $R_{n+1}(t)$.

Now in time $(0, t]$, the total time exactly n machines are down will be $P_n t$, and the total production during this time will be $(m - n)P_n th$. Thus if the mean time to stoppage of a machine is $1/\lambda$ it appears reasonable to say that

$$C_n(t) = (m - n)P_n t\lambda. \tag{2.21}$$

Similarly, the total time $n+1$ machines are down is $P_{n+1}t$ so with mean repair time $1/\mu$ it appears reasonable to say that

$$R_{n+1}(t) = P_{n+1}t\mu \tag{2.22}$$

Hence, setting $C_n(t) = R_{n+1}(t)$ one obtains from equations 2.21 and 2.22,

$$P_{n+1} = (m - n)\frac{\lambda}{\mu}P_n, \quad n = 0, 1, \ldots, m - 1 \tag{2.23}$$

Thus solving equation 2.23 one obtains

$$P_n = \frac{m!}{(m - n)!}\left(\frac{\lambda}{\mu}\right)^n P_0, \quad n = 1, \ldots, m - 1 \tag{2.24}$$

and

$$P_0 = \frac{1}{\sum_{n=0}^{m}\frac{m!}{(m-n)!}\left(\frac{\lambda}{\mu}\right)^n}$$

Example 2.4

Continuing with our examples 2.2 and 2.3, this approach gives

$$P_0 = \frac{1}{1 + \frac{3}{2} + \frac{6}{4}} = 0.2105$$

Hence from equations 2.4 and 2.7 one sees that

$$\rho = 0.7895$$

$$\eta = 0.5263$$

Obviously, the correctness of these results depends on whether the statements "it is reasonable to say that . . ." earlier are correct. In general these need not be true. Under certain assumptions on the repair and failure times, however, we will see that the preceding results are valid. In particular, we will see in Section 2.2.6 that the previous results are valid when the times to failure and repair times are exponentially distributed.

Further understanding of machine interference requires that we use a more formal mathematical approach and clarify the validity of assumptions. Let $\tau^{(j)} = \{\tau_n^{(j)}, n = 1, 2, \ldots\}$ be the sequence of machine up times (i.e., the time to failures) of machine j, $j = 1, \ldots, m$ and $\mathbf{S} = \{S_n, n = 1, 2, \ldots\}$ be the sequence of machine repair times. Further analysis requires that we make specific assumptions concerning the distribution of the time to failure and time to repair a machine.

2.2.5 Sample Path Analysis

First we will see how a formal sample path analysis may be carried out. Such a sample path analysis will usually shed light on the dependence of the system performance on the parameters of the system such as the number of machines, mean time to failure

$E[\tau_n^{(j)}] = 1/\lambda$, mean repair time $E[S_n] = 1/\mu$, and so on. For ease of illustration, we will consider the case in which the failure times are all deterministic and equal to $1/\lambda$. In addition, we will assume that initially all the machines are down at time zero. Let $D_n^{(1)}$ be the time at which the nth machine failure occurs after time zero, and $D_n^{(2)}$ be the time at which the nth machine repair is completed. We set $D_0^{(j)} = 0$, $j = 1, 2$. Let $D^{(j)}(t) = \sup\{n : D_n^{(j)} \leq t, n = 0, 1, \ldots\}$, $j = 1, 2$. $D^{(1)}(t)$ is the number of machines failed during $(0, t]$ and $D^{(2)}(t)$ is the number of repair completions during $(0, t]$. Then the number of failed machines at time t is

$$N(t) = m + D^{(1)}(t) - D^{(2)}(t), \quad t \geq 0$$

where we have assumed $N(0) = m$. Because the expected time spent on repairing the first n machine failures is n/μ and the expected time of the nth machine repair completion is $E[D_n^{(2)}]$ it is easily seen that

$$\rho = \lim_{n \to \infty} \frac{n}{\mu E[D_n^{(2)}]} \tag{2.25}$$

Then from equation 2.7 it is clear that

$$\eta = \lim_{n \to \infty} \frac{n}{m\lambda E[D_n^{(2)}]} \tag{2.26}$$

and

$$TH = \lim_{n \to \infty} \frac{nh}{\lambda E[D_n^{(2)}]} \tag{2.27}$$

In addition the average number of machines working is,

$$m\eta = \lim_{n \to \infty} \frac{n}{\lambda E[D_n^{(2)}]} \tag{2.28}$$

Because the machine up times are deterministic and are the same for all machines, the order of repairs and failures of the different machines will be the same. Specifically

$$D_n^{(1)} = D_n^{(2)} + 1/\lambda, \quad n = 1, 2, \ldots. \tag{2.29}$$

Since there are m machines already at the repair facility at time zero, the nth machine failure will be the $(n + m)$th repair completion. Hence

$$D_n^{(2)} = D_{n-1}^{(2)} + S_n, \quad n = 1, \ldots, m \tag{2.30}$$

and

$$D_{n+m}^{(2)} = \max\{D_{n+m-1}^{(2)}, D_n^{(1)}\} + S_{n+m}, \quad n = 1, 2, \ldots. \tag{2.31}$$

From the preceding set of equations 2.29 to 2.31 it is easily seen that the repair completion and machine failure times are increasing and convex functions of the service times. That is, we can write

$$D_n^{(j)} = \phi_n^{(j)}(S_k, k = 1, \ldots, n), \quad j = 1, 2 \tag{2.32}$$

where $\phi_n^{(j)}(\mathbf{x})$ is an increasing and convex function of $\mathbf{x}$, $j = 1, 2$. Then by Jensen's inequality

$$E[\phi_n^{(j)}(S_k, \ k = 1, \ldots, n)] \geq \phi_n^{(j)}(E[S_k], \ k = 1, \ldots, n), \quad j = 1, 2 \qquad (2.33)$$

From equation 2.33 it can be seen that the expected machine repair completion times are smaller under deterministic repair times, and thus the operator utilization is higher (see equation 2.25). In addition, the efficiency of the system, the number of working machines, and throughput are higher (see equations 2.26 to 2.28). Recall that when all the machine repair times are the same, the operator utilization for the deterministic model is explicitly known (see Section 2.2.2).

Throughout the remaining part of this chapter we will assume that the machine up times, $\{\tau_n^{(j)}\}$, $j = 1, \ldots, m$, and the machine repair times, $\{S_n\}$, are all mutually independent sequences of independent and identically distributed (iid) random variables. Different modeling techniques and solution approaches can be used depending on the specific distributional forms that can be assumed for $\tau_n^{(j)}$, $j = 1, \ldots, m$ and S_n.

2.2.6 Markov Models

Suppose that the machine up times $\{\tau_n^{(j)}\}$ are exponentially distributed with mean $1/\lambda$ and suppose the times to repair $\{S_n\}$ have an exponential distribution with mean $1/\mu$. That is

$$F_\tau(x) = P\{\text{time to failure} \leq x\} = 1 - e^{-\lambda x}, \quad x \geq 0$$

and

$$F_S(x) = P\{\text{time to repair} \leq x\} = 1 - e^{-\mu x}, \quad x \geq 0$$

Throughout this book, for a given sequence $\{X_n, n = 1, 2, \ldots\}$ of random variables X will be a generic random variable representing the limiting distribution of this sequence, F_X will be its distribution function, $\overline{F}_X = 1 - F_X$ its survival function and f_X its density function. That is, $F_X(x) = P\{X \leq x\} = \lim_{n \to \infty} P\{X_n \leq x\}$. If this sequence is a collection of iid random variables, then X will be the usual generic random variable.

Let $N(t)$ be the number of machines down at time t. Because the machine failure times are exponentially distributed with the same mean and the repair times are also exponentially distributed, it is easily seen that $\{N(t), t \geq 0\}$ is a continuous-time Markov process on the state space $\{0, \ldots, m\}$. Indeed it is a birth-death process. Define

$$p(n, t) = P\{N(t) = n\}, \quad n = 0, \ldots, m; \ t \geq 0$$

and

$$p_n = \lim_{t \to \infty} p(n, t), \quad n = 0, \ldots, m$$

That is, $p(n, t)$ is the probability that n machines are down at time t, and p_n is the probability n machines are down under stationary conditions. Then by considering the system at times t and $t + \Delta t$, where Δt is small and asking how the number of machines down at time t can change during Δt, we get

$$p(n, t + \Delta t) = p(n, t)(1 - (m - n)\lambda \Delta t - \mu \Delta t)$$
$$+ p(n - 1, t)(m - n + 1)\lambda \Delta t + p(n + 1, t)\mu \Delta t$$
$$+ o(\Delta t), \quad n = 1, \ldots, m - 1$$
$$p(0, t + \Delta t) = p(0, t)(1 - m\lambda \Delta t) + p(1, t)\mu \Delta t + o(\Delta t)$$
$$p(m, t + \Delta t) = p(m, t)(1 - \mu \Delta t) + p(m - 1, t)\lambda \Delta t + o(\Delta t) \qquad (2.34)$$

where $o(\Delta t)$ are the terms that have the property $\lim_{\Delta t \to 0} \frac{o(\Delta t)}{\Delta t} = 0$. Rearranging the terms in equation 2.34 one can obtain

$$p(n, t + \Delta t) - p(n, t) = p(n, t)(-(m - n)\lambda \Delta t - \mu \Delta t)$$
$$+ p(n - 1, t)(m - n + 1)\lambda \Delta t$$
$$+ p(n + 1, t)\mu \Delta t + o(\Delta t), \quad n = 1, \ldots, m - 1$$
$$p(0, t + \Delta t) - p(0, t) = p(0, t)(-m\lambda \Delta t) + p(1, t)\mu \Delta t + o(\Delta t)$$
$$p(m, t + \Delta t) - p(m, t) = p(m, t)(1 - \mu \Delta t) + p(m - 1, t)\lambda \Delta t + o(\Delta t) \qquad (2.35)$$

Now dividing both sides of the set of equations 2.35 by Δt and taking the limit as $\Delta t \to 0$ one gets (the so-called Chapman-Kolmogorov forward differential equations)

$$\frac{dp(n, t)}{dt} = -((m - n)\lambda + \mu)p(n, t) + (m - n + 1)\lambda p(n - 1, t)$$
$$+ \mu p(n + 1, t), \quad n = 1, \ldots, m - 1$$
$$\frac{dp(0, t)}{dt} = -(m\lambda + \mu)p(0, t) + \mu p(1, t)$$
$$\frac{dp(m, t)}{dt} = -\mu p(m, t) + \lambda p(m - 1, t) \qquad (2.36)$$

Now, although we could solve the preceding set of equations, often we would only be interested in the equilibrium or steady-state solutions where the derivatives on the left-hand side are all zero. Thus we obtain

$$((m - n)\lambda + \mu)p_n = (m - n + 1)\lambda p_{n-1} + \mu p_{n+1}, \quad n = 1, \ldots, m - 1$$
$$m\lambda p_0 = \mu p_1$$
$$\mu p_m = \lambda p_{m-1} \qquad (2.37)$$

Observe that the left-hand side of equation 2.37 is the rate of probability flow out of state n and the right-hand side is the rate of flow into state n from other states ($n = 0, \ldots, m$). Therefore these equations can be written down *directly* by equating the rates of probability

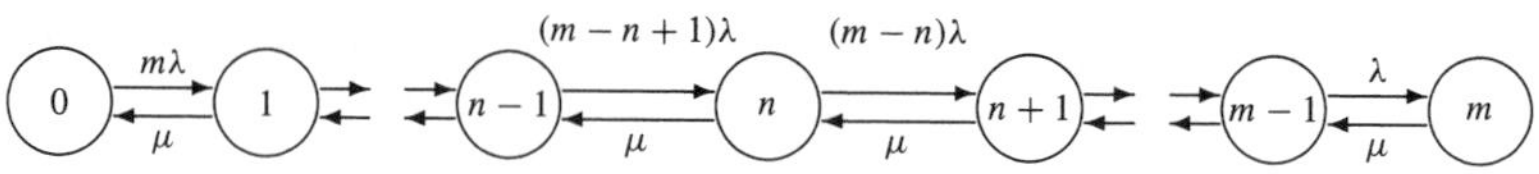

Figure 2.2   Transition Rate Diagram

inflow and outflow from each state of the Markov process $\{N(t), t \geq 0\}$. This step can be greatly facilitated by the aid of the *transition rate diagram* of this process shown in Figure 2.2.

When the time dependent probabilities, $p(\cdot, t)$, are not needed, it is customary to use the transition rate diagram to obtain the steady state balance equations of a Markov process rather than to set it up as we did here.

Now if we add the equations for p_0 and p_r ($r = 1, \ldots, n$) in (2.37) we obtain

$$(m - n)\lambda p_n = \mu p_{n+1}, \quad n = 0, \ldots, m - 1 \tag{2.38}$$

Observe that the left hand side of the above equation is the rate of probability flow out of the compound state $\{0, \ldots, n\}$ and the right hand side is the rate of probability flow into the compound state $\{0, \ldots, n\}$ from outside. This is the *level crossing balance equation*. These too can be directly written down from the transition rate diagram by appropriately choosing the compound states.

Solving (2.38) with the normalization condition

$$\sum_{n=0}^{m} p_n = 1,$$

we get

$$p_n = \frac{m!}{(m - n)!} \hat{\rho}^n p_0, \quad n = 1, 2, \ldots, m,$$

$$p_0 = \frac{1}{\sum_{n=0}^{m} \frac{m!}{(m-n)!} \hat{\rho}^n}, \tag{2.39}$$

where $\hat{\rho} = \lambda/\mu$. Let

$$G(\hat{\rho}, k) = \sum_{n=0}^{k} \frac{k!}{(k - n)!} \hat{\rho}^n, \quad k = 0, 1, \ldots \tag{2.40}$$

Then it can be shown that

$$G(\hat{\rho}, k) = 1 + k\hat{\rho} G(\hat{\rho}, k - 1), \quad k = 0, 1, \ldots \tag{2.41}$$

with $G(\hat{\rho}, -1) = 0$. From equations 2.4, 2.7, and 2.39 to 2.41 one sees that

$$\eta = \frac{\rho}{m\hat{\rho}} = \frac{1}{m\hat{\rho}} \left(1 - \frac{1}{G(\hat{\rho}, m)} \right)$$

$$= \frac{G(\hat{\rho}, m - 1)}{G(\hat{\rho}, m)} \tag{2.42}$$

Thus equation 2.41 provides a basis for an efficient computational algorithm to calculate the efficiency, η, and other performance measures of this system.

Under the assumption of exponential times to failures and repair, the preceding model and analysis easily extend to the multiple-operator case. This we shall discuss in Section 2.3.2.

We will now see how the preceding results can be applied in selecting the number of machines allocated to a single operator.

Example 2.5

Suppose we wish to select the number of machines to allocate to a single operator. Let w be the lease or purchase cost per machine per unit time. If the machines are purchased, we assume that the purchase cost is converted to a per unit time basis. Also let r be the profit per item produced. Then, if we allocate m machines to a single operator, the profit per unit time is $rTH - mw$ and from equation 2.7, given by

$$R(m) = m\eta(m)hr - mw, \quad m = 0, 1, \ldots \quad (2.43)$$

where $\eta(m)$ is the efficiency of the system with m machines allocated to one operator. It can be verified that $R(m)$ is a concave function of m. Therefore the number of machines that maximizes the profit $R(m)$ per unit time is (using equation 2.42),

$$m^* = \min\{m : R(m+1) - R(m) \leq 0, m = 0, 1, \ldots\}$$

$$= \min\{m : \frac{(m+1)G(\hat{\rho}, m)}{G(\hat{\rho}, m+1)}$$

$$- \frac{mG(\hat{\rho}, m-1)}{G(\hat{\rho}, m)} \leq \frac{w}{hr}, \quad m = 0, 1, \ldots\} \quad (2.44)$$

The relationship for $G(\hat{\rho}, m)$ and $G(\hat{\rho}, m-1)$ given in equation 2.41 can be recursively used starting with $m = 1$ to obtain m^*.

2.3 GENERALIZATIONS

At this point there are two directions that development of further models can take. One is to keep the same basic framework and use general distributions for the operating time and repair time of the machines, and the other is to incorporate other features of the real world into the models.

In the first type of generalization one moves from looking at what can be recognized as a $M/M/1 : m$ queue in the language of queueing theory to queues where the exponential distributions (i.e. the M's) are replaced by other distributions, such as Erlang E_k, deterministic D, or general G distributions. (Probability distributions commonly used in the modeling of manufacturing systems are summarized in Appendix A.) The basic structure of the model is unaltered.

The other direction of development is to attempt to make the model more realistic by including more features of the real situation. In particular, the emphasis might be on the activities of the operator. In real situations operators often have other duties besides repair of failures so it is important to consider the effect of these on the system.

We look first at the approaches of generalizing the distributions.

2.3.1 Erlang Distributions

Erlang time to failure.   We consider first the case in which the operating time (i.e., the time to failure) is Erlang-k and the repair time is exponential. That is, the $E_k/M/1 : m$ queue. Suppose the mean operating time is $1/\lambda$. Then an Erlang up time is equivalent to the following. Divide the up time into k phases and assume that the times in each phase have identical exponential distributions with mean $1/k\lambda$. After completion of phase i a machine moves to phase $i+1$, $i = 1, \ldots, k-1$. At the completion of phase k, the machine fails, and the operator is alerted of its failure.

Let $M_i(t)$ be the number of machines operating in phase i at time t, $i = 1, \ldots, k$. Then $N(t) = m - \sum_{i=1}^{k} M_i(t)$ is the number of machines down at time t. Based on the exponentiality assumption of the time spent at each phase and of the repair times, it is clear that $\{(M_1(t), \ldots, M_k(t)), t \geq 0\}$ is a continuous time Markov process on the state space $\{(l_1, \ldots, l_m) : 0 \leq l_i \leq m, i = 1, \ldots, k$ and $0 \leq \sum_{i=1}^{k} l_i \leq m\}$. Define

$$p(l_1, \ldots, l_k) = \lim_{t \to \infty} P\{M_i(t) = l_i, \ i = 1, \ldots, k\}$$

That is, $p(l_1, l_2, \ldots, l_k)$ is the probability that l_i machines are operating in phase i ($i = 1, 2, \ldots, k$) and $m - l$ are down where $l = \sum_{i=1}^{k} l_i$. The steady-state balance equations for this Markov process are

$$(lk\lambda + \mu)p(l_1, \ldots, l_k) = \mu p(l_1 - 1, l_2, \ldots, l_k)$$

$$+ \sum_{i=2}^{k}(l_{i-1} + 1)k\lambda p(l_1, \ldots, l_{i-1} + 1, l_i - 1, \ldots, l_k)$$

$$+ (l_k + 1)k\lambda p(l_1, l_2, \ldots, l_k + 1), \quad l = 0, 1, \ldots, m - 1$$

$$mk\lambda p(l_1, \ldots, l_k) = \mu p(l_1 - 1, l_2, \ldots, l_k)$$

$$+ \sum_{i=2}^{k}(l_{i-1} + 1)k\lambda p(l_1, \ldots, l_{i-1} + 1, l_i - 1, \ldots, l_k), \ l = m$$

$$\mu p(0, \ldots, 0) = k\lambda p(0, \ldots, 0, 1) \tag{2.45}$$

It can be shown that the solution to equation 2.45 is

$$p(l_1, \ldots, l_k) = \frac{(\mu/k\lambda)^l}{l_1! .. l_k!} p(0, \ldots, 0) \tag{2.46}$$

Using equation 2.46 and

$$p(l) = \sum_{\ell_1} \sum_{\ell_1} \cdots \sum_{\ell_k} p(l_1, \ldots, l_k)$$

where the sum is over all positive l_i such that $\sum_{i=1}^{k} l_i = l$, it can be shown that

$$p(l) = \frac{(\lambda/\mu)^l}{l!} p(0, \ldots, 0) \tag{2.47}$$

That is, the solution for the probability distribution of the number of machines working is independent of k. It is the same solution as the $M/M/1 : m$ queue. Indeed it can be shown that the same solution is obtained for general distributions of the operating times of machines. That is, all the preceding results for the $M/M/1 : m$ system apply to the $G/M/1 : m$ system. To determine the performance of this system we only need to know the mean time to failure of a machine and not the other moments of the time to failure. It is important to note that the insensitivity of the performance to the second and higher moments of the time to failure need not be true in other models.

Erlang repair time.   Now suppose the machine up times have an exponential distribution, and the repair times have an Erlang distribution with k phases. That is, we consider the $M/E_k/1 : m$ queue. Let $N(t)$ be the number of machines down and $X(t)$ be the phase of the current repair if any of the machines are down, and let $X(t) = 0$ otherwise, at time t. Then $\{(N(t), X(t)), t \geq 0\}$ is a continuous time Markov process on the state space $\{(0,0), (n,r), r = 1, \ldots, k; n = 1, \ldots, m\}$. Define

$$p(n,r) = \lim_{t \to \infty} P\{N(t) = n, X(t) = r\}$$

Then, $p(n,r)$ is the probability n machines are down and the repair is in phase r, $r = 1, 2, \ldots, k$. The steady-state balance equations for this Markov process are

$$((m-n)\lambda + k\mu)p(n,r) = (m-n+1)\lambda p(n-1,r)$$
$$+ k\mu p(n, r-1), \quad n = 1, \ldots, m, \ r = 2, \ldots, k$$
$$((m-n)\lambda + k\mu)p(n,1) = (m-n+1)\lambda p(n-1,1)$$
$$+ k\mu p(n+1,k), \quad n = 2, \ldots, m-1$$
$$k\mu p(m,1) = \lambda p(m-1,1)$$
$$m\lambda p(0,0) = k\mu p(1,k)$$
$$p(0,r) = 0, \quad r = 1, \ldots, k$$
$$((m-1)\lambda + k\mu)p(1,1) = k\mu p(2,k) + m\lambda p(0,0) \tag{2.48}$$

These equations do not have a simple closed form solution, so it is necessary to develop a computational scheme to find the performance measures. Because there are $mk + 1$ equations, direct solution of the equations would be the simplest approach. If m or k are very large, however, it would be possible to develop a computational procedure that takes account of the sparse coefficient matrix and its special structure.

2.3.2 General Repair Distributions

Suppose that there is just one operator. If the machines have exponential (operating) times to failure then it is possible to determine the probability $a_{lr}(s)$ that r out of l machines fail during a time interval $(0, s]$ given that no repair of a failed machine is possible during $(0, s]$.

$$a_{lr}(s) = \binom{l}{r} e^{-(l-r)\lambda s}(1 - e^{-\lambda s})^r \qquad (2.49)$$

Now suppose that the operator begins repair on a machine at an instant when l machines are working, that is, $m - l$ machines are down. Let $\overline{B}_{m-l}$ be the expected time from this instant until the instant when the operator has completed all repairs on failed machines (i.e., until the operator next has no machines waiting repair). During this time he will have to repair the $m - l$ failed machines plus any further failures that occur during the time he is working on repair of failed machines. Now if the first repair takes time s, then conditioning on the number of machine failures during $(0, s]$ one sees that

$$\overline{B}_{m-l}(s) = s + \sum_{r=0}^{l} a_{lr}(s)\overline{B}_{m-l-1+r} \qquad (2.50)$$

Unconditioning this repair time s with respect to the repair time distribution F_S from equation 2.50 one sees that

$$\overline{B}_{m-l} = \frac{1}{\mu} + \sum_{r=0}^{l} a_{lr}\overline{B}_{m-l-1+r} \qquad (2.51)$$

where from equation 2.49,

$$a_{lr} = \int_0^\infty \binom{l}{r} e^{-(l-r)\lambda s}(1 - e^{-\lambda s})^r dF_S(s) \qquad (2.52)$$

Rewriting equation 2.51 in detail we have

$$\overline{B}_N = \frac{1}{\mu} + \overline{B}_{N-1}$$

$$\overline{B}_{N-1} = \frac{1}{\mu} + a_{10}\overline{B}_{N-2} + a_{11}\overline{B}_{N-1}$$

$$\overline{B}_{N-2} = \frac{1}{\mu} + a_{20}\overline{B}_{N-3} + a_{21}\overline{B}_{N-2} + a_{22}\overline{B}_{N-1}$$

$$\overline{B}_1 = \frac{1}{\mu} + a_{N-1,0}0 + a_{N-1,1}\overline{B}_1 + \cdots \qquad (2.53)$$

Divide the operation of the system into periods when the operator is idle and periods when the operator is busy. The average length of a busy period is thus $\overline{B}_1$, and the average length of an idle period is $1/m\lambda$. Therefore

$$\rho = \frac{\overline{B}_1}{\overline{B}_1 + 1/m\lambda} \tag{2.54}$$

(Note that this solution approach assumes that the order of repair does not matter.) Thus solving equation 2.53 for $\overline{B}_1$, the performance measures of the system can be obtained from equations 2.7 and 2.54.

Example 2.6

Continuing examples 2.2 to 2.4 with constant repair times, if the duration of repair is constant and equal to $1/\mu$, then from equation 2.52 one has

$$a_{10} = e^{-\hat{\rho}}$$

$$a_{11} = 1 - e^{-\hat{\rho}}$$

$$a_{21} = \binom{2}{1} e^{-\hat{\rho}} (1 - e^{-\hat{\rho}})$$

$$a_{22} = (1 - e^{-\hat{\rho}})^2, \quad \text{and so on}$$

Recall that we defined $\hat{\rho} = \lambda/\mu$. If $m = 3$ and equations 2.53 for the $\overline{B}_i$'s are solved it is found that

$$\overline{B}_1 = (e^{3\hat{\rho}} - e^{2\hat{\rho}} + e^{\hat{\rho}})/\mu$$

Hence if $\mu = 1$, $\lambda = 1/2$, then $\overline{B}_1 = 3.412$, whence $\rho = 0.8365$ and $\eta = 0.5577$. It can be seen that there is not a particularly significant improvement in performance through changing the repair distribution from exponential to deterministic. Recall that in the case of exponentially distributed repair times, $\rho = 0.7895$ and $\eta = 0.5263$.

It is possible to obtain a general solution for equations 2.53 for the $\overline{B}_i$'s. For deterministic repair times it is

$$\overline{B}_1 = \left(1 + \sum_{r=1}^{m-1} \binom{m-1}{r} \prod_{i=1}^{r} (e^{i\hat{\rho}} - 1) \right) / \mu \tag{2.55}$$

For general repair times write

$$\tilde{F}_S(\lambda i) = \int_0^\infty e^{-\lambda i t} dF_S(t) dt$$

and

$$f_i = \frac{1 - \tilde{F}_S(\lambda i)}{\tilde{F}_S(\lambda i)}$$

Then it can be shown that

$$\overline{B}_1 = \left(1 + \sum_{j=1}^{m-1} \binom{m-1}{j} \prod_{i=1}^{j} f_i \right) / \mu \tag{2.56}$$

Let C_S^2 be the squared coefficient of variation of the repair times. That is, if $E[S^2]$ is the second moment of the repair times, then $C_S^2 = \frac{E[S^2]}{E[S]^2} - 1$. Hence

$$\tilde{F}_S'(0) = -1/\mu$$
$$\tilde{F}_S''(0) = E[S^2] = (1 + C_S^2)/\mu^2 \tag{2.57}$$

Now, because

$$\tilde{F}_S(s) = \tilde{F}_S(0) + s\tilde{F}_S'(0) + \frac{s^2}{2!}\tilde{F}_S''(0) + \frac{s^3}{3!}\tilde{F}_S'''(0) + \cdots \tag{2.58}$$

we have from equations 2.57 and 2.58

$$\tilde{F}_S(\lambda i) = 1 - i\hat{\rho} + \frac{\hat{\rho}^2 i^2}{2!}(1 + C_S^2) + \cdots \tag{2.59}$$

If $\hat{\rho} < 1$ the terms in $\hat{\rho}^3$ and so on will be small and can be neglected.

Thus the performance will be principally determined by the two parameters $\hat{\rho}$ and C_S^2. Therefore if we use a distribution with the same first two moments there will be little difference in performance. For example, we could use the Erlang distribution for $C_S^2 \le 1/2$ or 1. In fact $\hat{\rho}$ as a function of C_S^2 is almost a straight line over the range 0 to 1.

2.3.3 Multiple Operators

Suppose there are c operators available to attend to failed machines, and the machine failure and repair times are exponentially distributed with means $1/\lambda$ and $1/\mu$, respectively. If any of the c operators can repair a failed machine, that is, repair is pooled, and all the operators are equally competent so they have the same distribution of service time, then the steady-state or equilibrium equations become

$$m\lambda p_0 = \mu p_1$$
$$((m-r)\lambda + r\mu)p_r = (m-r+1)\lambda p_{r-1} + (r+1)\mu p_{r+1}, \quad r = 1, \ldots, c-1$$
$$((m-n)\lambda + c\mu)p_n = (m-n+1)\lambda p_{n-1} + c\mu p_{n+1}, \quad n = c, \ldots, m$$
$$c\mu p_m = \lambda p_{m-1} \tag{2.60}$$

with solution

$$p_n = \frac{m!}{(m-n)!n!}\hat{\rho}^n p_0$$
$$= \binom{m}{n}\hat{\rho}^n p_0, \quad n = 0, 1, \ldots, c$$
$$p_n = \frac{m!}{(m-n)!\, c!}\frac{c^c}{c!}\left(\frac{\hat{\rho}}{c}\right)^n p_0, \quad n = c, \ldots, m \tag{2.61}$$

p_0 can be found from the requirement that the sum of all probabilities equals 1.

2.4 MORE FEATURES OF ACTUAL SITUATION

2.4.1 Dissimilar Machines

With machines that are nonidentical the model becomes more complex. It is now necessary to specify precisely which machines are failed and which machines are being repaired. Suppose that the machines are labeled with index j, $j = 1, 2, \ldots, m$. Assume that the machines have exponential distributions of time to failure with the failure rate of machine j being λ_j.

Next, assume that the repair time of a machine has an exponential distribution. In the case in which the repair rate of the machines is identical and equal to μ it is possible to develop a model with a simple solution. Suppose that the machines are repaired in the order of failure. Then the state of the system will be specified by $i_1, i_2, \ldots, i_n$ where i_k denotes the machine with position k in the queue of machines waiting repair. If there is one operator the machine under repair will be machine i_1.

Then it can be shown that

$$p(i_1, i_2, \ldots, i_n) = \frac{1}{\mu^n} \left(\prod_{k=1}^{n} \lambda_{i_k} \right) p(0), \tag{2.62}$$

where $p(0)$ is the probability that no machine is failed.

Note that all permutations of $i_1, i_2, \ldots, i_n$ have the same probability in equation 2.62, thus the probability that machines $i_1, \ldots, i_n$ are failed is given by

$$P(i_1, i_2, \ldots, i_n) = \sum_{\text{all permutations}} p(i_1, i_2, \ldots, i_n)$$

$$= \frac{r!}{\mu^n} \prod_{k=1}^{n} \lambda_{i_k} p(0). \tag{2.63}$$

$p(0)$ can be found from the requirement that the sum of all probabilities equals 1.

The same solution for the $P(i_1, i_2, \ldots, i_n)$ is obtained if service is in random order, that is, on completion of repair the operator selects the next machine to repair at random from those waiting repair. Other sequences of service, however, that take account of the features of specific machines (e.g., priority rules that take account of the failure rates of machines) will give different results.

A simple solution also exists for the case in which there are c operators who repair in order of failure. The state is then $(i_1, i_2, \ldots, i_c, i_{c+1}, \ldots, i_n)$ where index k for $k = c+1, \ldots, n$ denotes the sequence of failure of machines, whereas for $k = 1, 2, \ldots, c$ the index just denotes which machines are under repair. It can be shown that

$$p(i_1, i_2, \ldots, i_c, i_{c+1}, \ldots, i_n) = \frac{1}{\mu^n c^{n-c}} \left(\prod_{k=1}^{n} \lambda_{i_k} \right) p(0), \quad n = c+1, \ldots, m$$

$$p(i_1, i_2, \ldots, i_n) = \frac{1}{\mu^n} \left(\prod_{k=1}^{n} \lambda_{i_k} \right) p(0), \quad n = 1, \ldots, c \tag{2.64}$$

2.4.2 Ancillary Work

There are various ways in which ancillary work could be assumed to occur.

1. The operator leaves his job whenever he feels like it, irrespective of what he is doing.
2. The operator only leaves when all repairs are complete, but he may come back at any time (not related to the state of the machines).
3. The operator only does ancillary work when he has no repairs to perform and stops it as soon as a machine fails.

Obviously in case 3 all the above results as appropriate apply unchanged. The maximum fraction of time that the operator can devote to ancillary work will be $1 - \rho$. For the other cases we will outline some quick fixes of the previous results that can be used as approximations. Alternatively one may develop formal models as illustrated before and analyze them.

In case 1 it is necessary to develop a new model in which the set of states of the system is augmented by states corresponding to the operator being absent. Two possible extremes can be used to bound the solution, however. If the times the operator is working and absent have means W and I respectively, then if W and I are very large, that is, concentrated work, then

$$\text{Operator efficiency} = \frac{W}{W + I} \rho(\hat{\rho})$$

where $\rho(\hat{\rho})$ is the operator efficiency of the system without operator absenteeism and $\hat{\rho} = \lambda/\mu$. If W and I, however, are very short, that is, spread work,

$$\text{Operator efficiency} = \frac{W}{W + I} \rho(\rho')$$

where $\rho(\rho')$ is the operator efficiency of the system without operator absenteeism, but with an increased mean repair time of $1/\mu' = 1/[\mu(1 - A)]$; $\rho' = \lambda/\mu'$, with $A = I/(I + W)$ being the proportion of time the operator spends on ancillary work.

Interruptions can also be related to the cumulative production rather than time, for example, if the interruption represents time to go and collect more material.

2.4.3 Patrolling

Another direction of altering the model to make it more realistic is to assume that the operator patrols the machines in a given sequence. Machine failures may not be signaled to him, so he has to check each machine to see that it is working correctly. The time he spends at a machine depends on whether he finds it working or failed. With patrolling it is possible for a machine to be down for some time, waiting for the operator to reach it even though he is not occupied in repair.

In another variant of the model where failures are signaled, the repair time is augmented to allow for the time it takes for the operator to walk from one machine to the next. He will then repair the nearest machine next. In situations in which the machines are not all at the same location this model would be worthwhile.

2.4.4 Spare Machines

A different direction for extending the model is to assume that not all machines are required all the time. That is, the number of working machines must not exceed some given number. There will be two queues in the system: machines waiting for repair and machines waiting to be put into service.

Thus, one can view the system as being equivalent to two phases: the repair phase with its queue and the working phase with its queue. The customers are the machines that circulate in the system, going from the repair phase to the working phase and back to the repair phase.

Now to model the service process it will be assumed that when there are n_i customers in phase i the service rate is $\mu_i(n_i)$. This is a means by which we can represent a single server ($\mu_i(n_i) = \mu_i$, $n_i \geq 1$; $\mu_i(0) = 0$ for all i) or c_i servers ($\mu_i(n_i) = n_i\mu_i$ for $n_i \leq c_i$, $\mu_i(n_i) = c_i\mu_i$ for $n_i \geq c_i$). Then it can be shown that the state probabilities are given by

$$p(n_1, n_2) = \frac{K(m)}{\prod_{i=1}^{n_1} \mu_1(i) \prod_{j=1}^{n_2} \mu_2(j)} \text{ for } n_1 + n_2 = m, 0 \leq n_1 \leq m \qquad (2.65)$$

where the constant $K(m)$ must be determined from the requirement that the sum of all probabilities equals 1. This product form solution is a special case of results that will be discussed in Chapter 8.

Obviously once the customers go from one server to another it becomes reasonable to extend the model to a network of servers with customers flowing from one server to another in accordance with some routing probability and perhaps with transit time between the servers. Such a model is no longer concerned with describing features of the machine interference problem, however, and its application will be to other sorts of situations remote from the original.

2.5 CONCLUSIONS

It can be seen how the modeling process has proceeded by successive refinement until in one direction the model has become something much broader and more general, and it is no longer closely related to the original problem. This problem mirrors much of the experience with other modeling in manufacturing. First keeping the model structure the same but looking at more general distributions sometimes indicates that the results are not too dependent on distribution. To show this often requires complex numerical procedures to be developed. Changing the model structure and trying to investigate more aspects of reality sometimes yields much complexity, but often this complexity can be reduced through the use of approximation techniques that try and extend other results by suitably modifying parameters. Sometimes through generalizing the model it happens that much more general and more powerful results are obtained, and the model turns out to fit a wide range of situations.

2.6 BIBLIOGRAPHICAL NOTE

Machine interference was the first manufacturing system issue for which stochastic models were developed. Initial attempts using the binomial approximation can be found in Wright et al. [17] and Jones [9]. The first correct solution of the $M/M/1 : m$ machine interference problem is due to Palm [14] [13], whereas the $M/D/1 : m$ solution is due to Ashcroft [1]. Much interesting early work in developing models to capture the actual issues in machine minding was done by Benson and Cox [3] [4] with the most accessible report of this work in Cox and Smith [7]. Barlow has a comprehensive review of the models themselves in [8]. Early generalizations of the machine interference problem can be found in Koenigsberg [10] [11] and Benson and Gregory [5], whereas an application of the model to a somewhat different manufacturing context (the soaking pit/rolling mill system of a steel mill) is in Buzacott and Callahan [6]. An extensive listing of references on the machine interference problem can be found in Stecke and Aronson [16].

PROBLEMS

2.1 Consider a collection of m machines attended by a single operator. Suppose that the machine failure times are exponentially distributed with mean $1/\lambda$ and the repair times are exponentially distributed with mean $1/\mu$. The operator, once idle, will wait until k failed machines become available for repair before initiating a repair. Once the process is initiated, the operator will continue repairing until all failed machines are repaired. Once there are no more failed machines to be repaired, the operator will become idle again and the preceding cycle will be repeated. Let $N(t)$ be the number of failed machines in the system at time t.

   (a) Is $\{N(t), t \geq 0\}$ a Markov process?

   (b) If the answer to part a is yes, then write down the balance equation for the steady-state probability distribution p_n, $n = 0, 1, \ldots, m$, of $\{N(t), t \geq 0\}$. Plot the average number of failed machines as a function of k.

   (c) Suppose a cost of w is incurred per unit time per failed machine and suppose s is the start-up cost of the repair process for each start-up. Find the optimal start-up level k (i.e., the number of failed machines at the start-up) that will minimize the total cost of failed machines and start-ups.

2.2 Consider problem 1 with the following start-up policy. As soon as the operator becomes idle, the operator will leave the system for a random amount of time that is exponentially distributed with mean $1/\gamma$. At the time of return, the operator will check to see if there is at least one failed machine. If there is none the operator will take off for a random amount of time, exponentially distributed with mean $1/\gamma$. Otherwise (i.e., if there is at least one failed machine at the time of the operator's return), the operator will initiate the repair process and continue repairing until there are no more failed machines. Let $N(t)$ be the number of failed machines at time t. With this start-up policy answer parts a to c of problem 1.

2.3 Consider a manufacturing facility consisting of m machines. Suppose only one machine is used for manufacturing at any time. All other machines are kept as spares. Suppose the machines fail only when they are used. The time to machine failures are exponentially distributed with mean $1/\lambda$. Suppose we have a single operator, and the times to repair have a general distribution F_S.
   (a) If at time 0, there is exactly one failed machine and its repair was just initiated, find the average time taken for the operator to become free (i.e., idle).
   (b) Use the results from part a to find the operator utilization and system efficiency.
   (c) Suppose we have a choice between two operators. Operator 1 takes exactly $1/\mu$ units of time to repair a failed machine, whereas operator 2 takes a random amount of time with mean $1/\mu$ to repair a machine. To maximize the system efficiency, which operator should be used? Use a sample path analysis to justify your answer.

2.4 Suppose we have assigned c operators to m machines. The times to machine failure and repair times are all exponentially distributed with means $1/\lambda$ and $1/\mu$, respectively. Let $N(t)$ be the number of failed machines at time t and p_n, $n = 0, 1, \ldots, m$, be its stationary distribution. Find p_n, $n = 0, 1, \ldots, m$, and the average number of failed machines $E[N(c)]$ as a function of the number of operators, c.
   (a) Is $E[N(c)]$ a decreasing and convex function of c?
   (b) If each operator costs s per unit time and a failed machine costs w per unit time, find the optimal number of operators that will minimize the total cost of operators and failed machines.

2.5 Suppose we have a set of m_i machines assigned to operator i, $i = 1, 2$. The time to machine failures and the repair times are probabilistically identical in both systems. Show that if pooling the machines and operators will not affect the machine failure rates and repair rates, then pooling will produce higher operator utilization, system efficiency, and a higher throughput than if the systems are operated separately.

2.6 A set of m machines is attended by a single operator. The times to machine failure are exponentially distributed with mean $1/\lambda$. With probability α the failure of a machine is of type 1 and with probability $1 - \alpha$ the failure is of type 2. The time to repair a type i failure is exponentially distributed with mean $1/\mu_i$, $i = 1, 2$. Without loss of generality assume that $\mu_1 > \mu_2$. Find the operator utilization and system efficiency if
   (a) The operator will always serve a machine with a type 1 failure, if available, before serving a machine with a type 2 failure, and if
   (b) The operator will always serve a machine with a type 2 failure, if available, before serving a machine with a type 1 failure.
   (c) Compare the system performances in parts a and b.

2.7 A manufacturing system consists of m machines, and the system operates if and only if all the m machines are working. Suppose the time to failure of machine j is exponentially distributed with mean $1/\lambda_j$, $j = 1, \ldots, m$. As soon as one machine fails, the operation of the other machines will be forced to stop. A machine not operating is assumed not to fail. If the time needed to repair a failed machine j is random with mean $1/\mu_j$ and squared coefficient of variation $C_{S_j}^2$, $j = 1, \ldots, m$, find the system efficiency η, and the operator utilization ρ.

2.8 Show that the throughput of m machines attended by a single operator with exponentially distributed times to failure and repair times, is given by

$$TH = \frac{\mu h}{\lambda} \left(\frac{\sum_{n=1}^{m} \frac{m!}{(m-n)!} \left(\frac{\lambda}{\mu}\right)^n}{\sum_{n=0}^{m} \frac{m!}{(m-n)!} \left(\frac{\lambda}{\mu}\right)^n} \right)$$

Here $1/h$ is the production cycle time of a machine, $1/\lambda$ is the mean time to failure and $1/\mu$ is the mean repair time. Suppose it is possible to decrease the failure rate λ by decreasing the production rate h. However, the ratio λ/h remains a constant, say equal to 1, for all possible choices of the pair of parameters (λ, h). Find the optimal cycle time $1/h$ that will maximize the throughput, TH.

BIBLIOGRAPHY

[1] H. ASHCROFT. The productivity of several machines under the care of one operator. *J. ROY. Stat. Soc.*, B12:145–151, 1950.

[2] R. E. BARLOW. Repairman problems. In *Studies in Applied Probability and Management Science*, pages 18–33, Stanford University Press, Stanford, 1962.

[3] F. BENSON. Further notes on the productivity of machines requiring attention at random intervals. *J. Roy. Stat. Soc.*, B14:200–210, 1952.

[4] F. BENSON and D. R. COX. The productivity of machines requiring attention at random intervals. *J. Roy. Stat. Soc.*, B13:65–82, 1951.

[5] F. BENSON and G. GREGORY. Closed queuing systems: a generalization of the machine interference model. *J. Roy. Stat. Soc.*, B23:385–393, 1961.

[6] J. A. BUZACOTT and J. R. CALLAHAN. The capacity of the soaking pit-rolling mill complex in steel production. *INFOR*, 9:87–95, 1971.

[7] D. R. COX and W. L. SMITH. *Queues*. Methuen, London, 1961.

[8] W. FELLER. *An Introduction to Probability Theory and Its Applications*, volume 1. John Wiley and Sons, New York, 1950.

[9] D. JONES. A simple way to figure machine downtime. *Factory Management and Maintenance*, 104:118–121, October 1946.

[10] E. KOENIGSBERG. Cyclic queues. *Operational Research Quarterly*, 9:22–35, 1958.

[11] E. KOENIGSBERG. Finite queues and cyclic queues. *Operations Research*, 8:246–253, 1960.

[12] P. NAOR. On machine interference. *J. Roy. Stat. Soc.*, B18:280–287, 1956.

[13] C. PALM. Assignment of workers in servicing automatic machines. *J. Industrial Eng.*, 9:28–42, 1958. Originally appeared in Swedish as [14].

[14] C. PALM. The distribution of repairmen in servicing automatic machines (in Swedish). *Industritidningen Norden*, 75:75–80, 90–94, 119–123, 1947.

[15] G. H. REYNOLDS. An $M/M/m/n$ queue or the shortest distance priority machine interference problem. *Operations Research*, 23:325–341, 1975.

[16] K. E. STECKE and J. E. ARONSON. Review of operator/machine interference models. *Int. J. Prod. Res.*, 23:129–151, 1985.

[17] W. R. WRIGHT, W. G. DUVALL, and H. A. FREEMAN. Machine interference: two solutions of a problem raised by multiple machine units. *Mechanical Engineering*, 58:510–514, 1936.

3

Single-Stage
"Produce-to-Order" Systems

3.1 INTRODUCTION

The simplest manufacturing system is a single machine, worker, or test facility. The machine (or worker or tester) has to perform a task or set of tasks on jobs. Throughout this book we will assume that the jobs are discrete. Each job will require a processing time or service time that need not be the same for all jobs and that may vary between successive repetitions of identical jobs. The actual time required may be known in advance of processing, but sometimes the time cannot be predicted with any accuracy in advance of processing because it will depend on problems encountered while processing the job. For example, in machining gears for helicopters, the dimensional stability is significantly affected by weather conditions and microscopic material variations. Successive repetitions of identical jobs can require machining times that may vary by as much as a factor of 10. Human operators show a characteristic variability in the time required to perform repetitive tasks, and although the amount of variability depends on the level of skill and training of the operator, it can never be eliminated. The impact of this variability will be discussed further in Chapter 5.

In this chapter we will consider a slight generalization of the single machine, which we call a single-stage system. A single-stage system consists of a group of c ($c \in \{1, 2, \ldots\}$) identical machines. A job can be processed by any one of the machines, but only one machine is required to complete the required tasks. Further, the service time of a job does not depend on the machine on which the job is processed.

To process a job at a machine it will be necessary to have available both the physical raw materials, parts, or product, and the written or verbal instructions on the specific tasks to be performed on the job. Further, these instructions not only indicate what is to be done on the job, they also authorize the machine to perform the task. Therefore, when we talk about a job, we mean the combination of the physical entity and the job instructions or job order, and unless both are present the job is not available for processing.

Thus, in front of the group of parallel machines there will be two sorts of storage, one where the physical raw materials, parts, or product are stored, and the other where the information on the job orders and instructions are filed. Traditionally, the job order was often attached to the raw material, but in systems where the information processing is automated the two sorts of storage are quite separate and distinct. In most plants the physical storage is very evident because of the space that it occupies, whereas the information storage is hidden, at worst in a pile of paper on the foreman's desk. Very often there is also output storage as well as input storage. For output storage there will again be the physical store for the product and the file for the completed job orders. These stores and files may have limited space to hold jobs.

No job can be processed at a machine unless there is both the necessary physical material and the job order. In addition, however, there also have to be procedures or rules, or what we will call *protocols* that specify which job of those available will be processed next by a machine when the machine becomes free. These protocols are often quite simple, such as if there is a job available the machine that has been idle longest should process it, whereas if no machine is available the machine that becomes free first should process the job that has been waiting for processing the longest (i.e., the first-come first-served [FCFS] service protocol). Many other service protocols are used in manufacturing, however, and it is by no means uncommon for no clearly defined protocol to be used in that the decision is left to the whim of the machine operator. Nevertheless, we will make some general assumptions about the nature of the protocols that would be used in a manufacturing environment.

1. *No preemption.* Once the processing of a job is initiated on a machine it remains on that machine until processing is complete. Of course, in reality there are the occasional rush jobs that have such high priority that the job in process is taken off the machine. This we will ignore. The no preemption rule, however, does not exclude the situation in which it is necessary to set the job aside because problems are encountered in its processing, and it is necessary to seek further instructions. When these additional instructions arrive the job is treated as a new arrival.

2. *No holding machines idle.* It will also be assumed that if there are available machines and available jobs no machine will be held idle. Again, in reality a machine may be kept idle to process a possible rush job. If this just means that one or two machines are always set aside for rush jobs then our models can be used, but if the rule for determining whether to hold a machine idle depends on the number of jobs or the anticipated arrival time of the rush job then different models will be required.

3. *No interruptions to service.* Once the machine begins processing a job it will continue processing the job until the processing is complete. This assumption is much less restrictive than it appears, however. If the interruption is determined by the nature of the job or the amount of time spent serving this job then the interruption can be included as a component of the service time. It is only when the interruption depends on what is happening at the other machines, or is created by the arrival or departure of jobs from the input or output store that this assumption becomes relevant.

Our discussion of single-stage job shops will be divided into two parts.

1. In this chapter we consider "produce-to-order" systems in which jobs, both the physical material and the job order, arrive from outside the shop, and the process that generates the arrivals is independent of whatever is happening in the shop, and its associated input and output stores. In this situation it is not necessary to make any distinction between the physical job and the job order when modeling the system.

2. In the next chapter we consider "produce-to-stock" systems in which the raw material arrives from outside the system, and there is always sufficient material available; however, the arrival of the job orders at the input store (and hence the authorization to proceed with the manufacture) is influenced by what is happening in the shop, and its associated output or input stores.

3.1.1 Issues

In designing and operating a single-stage produce-to-order system there are several management concerns. The first concern is capacity. Given the processing facilities what is the maximum rate of order receipt that can be tolerated such that all orders can eventually be satisfied? Next, is there any way of operating the shop so that this capacity can be increased without acquiring new facilities or investing in faster machines? Suppose the order receipts are less than capacity, how are performance measures such as the level of work in process or the delay in filling the orders influenced by the characteristics of the service process at machines or the pattern of order receipts? What service protocols improve performance? Are there good service protocols that do not require more information about the service requirements of individual jobs to be collected and used? How does the particular performance measure, such as average work-in-process inventory or average delay in filling customer orders, determine the choice of the appropriate protocol? Suppose certain customers' orders are given higher priority for processing. How does this influence performance? Is there a priority assignment that works to the advantage of all customers?

These are the issues that will be addressed in this chapter. In particular we want to understand how the stochastic nature of service and arrivals influences the system. To develop this understanding we will develop a wide variety of different models of varying degrees of complexity and mathematical sophistication. So, apart from addressing the managerial issues, the goal of the chapter is to introduce the reader to several different modeling techniques. Further, it will be found that many of the models and their results will be used in later chapters as essential components in modeling more complex manufacturing systems. For those readers who are already well versed in queueing theory there will be benefit in seeing how this theory can be used to provide insight into the managerial issues described earlier. For those readers with a more limited knowledge of queueing theory the chapter will also serve to introduce them to several different modeling approaches that will prove to be of considerable value.

Present

3.2 CAPACITY AND GENERAL PERFORMANCE RELATIONSHIPS

Jobs arrive at a single-stage job shop with c ($c \in \{1, 2, \ldots\}$) identical multipurpose machines according to an arbitrary arrival process $\{A_n, n = 1, 2, \ldots\}$. A_n is the arrival time of the nth job: $0 < A_1 \leq A_2 \leq \cdots$. Jobs can perhaps arrive more than one at a time, but we assume that the number arriving in any finite time interval is uniformly finite. The nth job to arrive will be identified by the label n, and it will keep this label until it departs from the system. The service time needed to process job n is $S_n, n = 1, 2, \ldots$.

We will first look at some basic identities among the performance measures of this job shop with few or no additional assumptions.

3.2.1 Capacity

Let D_n be the time instant at which the nth job departure occurs. Then the number of jobs departed during $(0, t]$ is

assumes
one server　　$? \rightarrow$

least upper bound

$$D(t) = \sup\{n : D_n \leq t, n = 0, 1, \ldots\} \tag{3.1}$$

where we set $D_0 = 0$. (Note that $D(t)$ includes all those jobs that may depart at time t). If $N(t)$ is the number of jobs in the system at time t, then

$$N(t) = N(0) + A(t) - D(t), \quad t \geq 0 \tag{3.2}$$

where

$$A(t) = \sup\{n : A_n \leq t, n = 0, 1, \ldots\} \tag{3.3}$$

with $A_0 = 0$ is the number of jobs arrived during $(0, t]$. Suppose

$$\frac{A(t)}{t} \rightarrow \lambda \text{ uniformly as } t \rightarrow \infty \tag{3.4}$$

where λ is the job arrival rate and

$$\beta = \lim_{t \to \infty} \frac{D(t)}{t} \tag{3.5}$$

where β is the job departure rate. The value of λ and β need not be the same. If the number of jobs in the system is uniformly finite for all time $t \geq 0$, however, then they are the same. That is, the following flow balance will hold:

$$\text{arrival rate} = \text{departure rate} \tag{3.6}$$

Indeed, what we need is that $\lim_{t\to\infty} N(t)/t = 0$, which is the case when $N(t)$ is uniformly finite. To see this divide equation 3.2 by t and take the limit as $t \to \infty$. A sufficient condition for the uniform finiteness of the number of jobs in the system at all time $t \geq 0$ is therefore crucial. It will be seen that this condition is

$$\lambda < c\mu$$

where $1/\mu$ is the average service time. We assume that

$$\frac{1}{k}\sum_{n=1}^{k} S_n \to \frac{1}{\mu} \quad \text{uniformly as } k \to \infty \tag{3.7}$$

Under this condition we will also see that at least one server will repeatedly become idle. In particular if $I(t)$ is the total idle time of all c servers during $(0, t]$, we will see that

$$I(t) = \int_0^t \left(c - \sum_{i=1}^{c} R_i(\tau)\right) d\tau \to \infty \quad \text{as } t \to \infty \tag{3.8}$$

where for $i = 1, \ldots, c$ and $t \geq 0$,

$$R_i(t) = \begin{cases} 1 & \text{if server } i \text{ is busy at time } t \\ 0 & \text{if server } i \text{ is idle at time } t \end{cases}$$

Then $\int_0^t R_i(\tau)\,d\tau$ is the total service time provided by server i during $(0, t]$ and the total work load in the system at time t is

$$V(t) = V(0) + \sum_{n=1}^{A(t)} S_n - \int_0^t \sum_{i=1}^{c} R_i(\tau)d\tau, \qquad t \geq 0 \tag{3.9}$$

Now suppose that none of the servers ever become idle. That is $R_i(t) = 1, t \geq 0, i = 1, \ldots, c$. Then

$$V(t) = V(0) + \sum_{n=1}^{A(t)} S_n - ct \geq 0, \quad t \geq 0 \tag{3.10}$$

Dividing equation 3.10 by t and taking the limit as $t \to \infty$ one sees that

$$\lim_{t\to\infty} \frac{1}{t}\sum_{n=1}^{A(t)} S_n = \lim_{t\to\infty} \frac{A(t)}{t}\frac{1}{A(t)}\sum_{n=1}^{A(t)} S_n = \frac{\lambda}{\mu} \geq c \tag{3.11}$$

Observe that equation 3.11 disagrees with the condition $\lambda < c\mu$. Therefore the assumption that all c servers are busy all the time must be wrong. Hence there exists a finite time t such that $\sum_{i=1}^{c} R_i(t) < c$, that is, at least one server must then be idle. Let I_1 be the first time when at least one server becomes idle, that is, $I_1 = \inf\{t : \sum_{i=1}^{c} R_i(t) < c, t \geq 0\}$. Then $I_1 < \infty$. Now let B_1 be the time epoch after I_1 when all servers become busy. If no such B_1 exists we have $N(t) \leq c - 1$ for all $t \geq I_1$. This therefore establishes that

$N(t)$ is uniformly finite for all t. If $B_1 < \infty$ then as before we can obtain a time epoch $I_2 > B_1$ such that at least one server becomes idle at time I_2 and $I_2 - I_1 < \infty$. Also let B_2 be the time epoch after I_2 when all c servers become busy. Now either $B_2 = \infty$ or, there exists I_3 and B_3 such that $B_2 < I_3 \leq B_3$ with $I_3 - I_2 < \infty$ and so on. If there exists a finite K such that $B_{K-1} < \infty$ and $B_K = +\infty$ then $N(t) \leq c - 1$, $t \geq B_{K-1}$ and therefore the number in the system is uniformly finite. If no such K exists, then

$$N(t) \leq c - 1 + A(t) - A(I_n) \quad I_n \leq t < I_{n+1} \tag{3.12}$$

Because $I_{n+1} - I_n < \infty$ and the convergence in equations 3.4 and 3.7 is uniform, one sees that the sequence $\{(I_{n+1} - I_n), n = 0, 1, 2, \ldots\}$ is uniformly finite. Because the number of customers arrived during finite time intervals is uniformly finite, from equation 3.12 it can be seen that, for all t, $N(t)$ is uniformly finite. Thus we have shown that if $\lambda < c\mu$ the number in the system will be uniformly finite.

On the contrary if $\lambda > c\mu$, then we will see that $N(t) \to \infty$ as $t \to \infty$. Suppose the nth job departs at time $D_{K(n)}$. That is, the $K(n)$th departure is job n. The total work provided by the c servers during $(0, t]$ is greater than or equal to $\sum_{n=1}^{A(t)} S_n I_{\{K(n) \leq D(t)\}}$, where $I_{\{\cdot\}}$ is an indicator function that takes the value one if "·" is true and zero otherwise. Because the maximum work that could be delivered during $(0, t]$ is ct,

$$\sum_{n=1}^{A(t)} S_n I_{\{K(n) \leq D(t)\}} \leq ct \tag{3.13}$$

Note that

$$\sum_{n=1}^{A(t)} S_n = \sum_{n=1}^{A(t)} S_n I_{\{K(n) \leq D(t)\}} + \sum_{n=1}^{A(t)} S_n I_{\{K(n) > D(t)\}} \tag{3.14}$$

and that $\sum_{n=1}^{A(t)} I_{\{K(n) > D(t)\}}$ is less than or equal to the number of jobs in the system at time t. Because the work load brought in by each job is uniformly finite, if $\lim_{t \to \infty} N(t)$ is finite, so is $\lim_{t \to \infty} \sum_{n=1}^{A(t)} S_n I_{\{K(n) > D(t)\}}$. Dividing equation 3.14 by t and taking the limit as $t \to \infty$, one obtains

$$\lim_{t \to \infty} \frac{1}{t} \sum_{n=1}^{A(t)} S_n I_{\{K(n) \leq D(t)\}} = \lim_{t \to \infty} \frac{1}{t} \sum_{n=1}^{A(t)} S_n = \frac{\lambda}{\mu} \tag{3.15}$$

From equations 3.13 and 3.15 it is clear that $\lambda \leq c\mu$. This violates our condition $\lambda > c\mu$, however. Therefore our assumption that $\lim_{t \to \infty} N(t)$ is uniformly finite cannot be true and therefore $N(t) \to \infty$ as $t \to \infty$. The preceding analysis does not necessarily imply that $\beta \leq c\mu$. It is indeed possible for the departure rate to be greater than $c\mu$ when $\lambda > c\mu$. This will only be achieved by eventually letting the number in the system grow without limit (i.e., with $N(t) \to \infty$ as $t \to \infty$). For this reason we call $c\mu$ the *capacity* of the system. That is, we define the capacity such that any arrival rate below the capacity can be handled so that the number in the system is always finite. Any arrival rate larger than the capacity will produce infinitely many customers in the system. It is also the value beyond which any higher departure rate can only be achieved by allowing the queue length to grow without limit. Note that capacity is *not* the maximum rate at which jobs can depart the system.

Example 3.1

The following simple example illustrates the difference between the capacity of the system and the job departure rate. Suppose jobs arrive at the system every $1/\lambda$ units of time and require a service of length 2 with probability 1/2 and service of length 4 with probability 1/2. There is only a single server present in the system who serves customers according to the shortest processing time service protocol. Suppose the number of jobs requiring a processing time 2 (say type 1 jobs) present in the system is uniformly finite. Hence the departure rate of type 1 jobs from the system should be equal to its arrival rate $\lambda/2$. The fraction of server's time utilized by type 1 jobs is then $(2)(\lambda/2) = \lambda$. Obviously we need this to be less than 1. Hence the fraction of server's time utilized by jobs requiring a processing time of 4 (say type 2 jobs) is at most $1 - \lambda$. Because the arrival rate of type 2 jobs is $\lambda/2$, the departure rate of type 2 jobs is equal to $\min\{(1 - \lambda)/4, \lambda/2\}$. The total departure rate is then $\lambda/2 + \min\{(1 - \lambda)/4, \lambda/2\}$, $0 < \lambda < 1$. For $\lambda \geq 1$, the number of type 1 jobs present in the system grows beyond limit and the departure rate is equal to 1/2. Figure 3.1 illustrates the relationship between the departure rate and the arrival rate for this example. Observe that because the average service time is 3 (i.e., $\mu = 1/3$), the capacity of this system is 1/3.

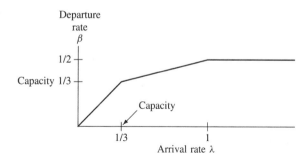

Figure 3.1   Job Departure Rate versus Job Arrival Rate

Recall that the capacity is determined based on the following general assumptions:

- The average arrival rate λ is the uniform limit of $A(t)/t$.
- The average service time $1/\mu$ is the uniform limit of $(1/k) \sum_{n=1}^{k} S_n$, and it is fixed and unaffected by the arrival rate or the service protocol.
- The c servers are always available to render service whenever jobs are present in the system.

Though we assume the existence of λ, we do not, however, require the arrival process to be independent of the service process. Therefore the arrival process could be regulated by the dynamics (such as the number of jobs) in the system. Hence we see that no input control can increase the capacity of the system unless it violates the second assumption or, equivalently, selects jobs for input to the system on the basis of their service requirements. Observe that the second assumption does not permit us to have service times that depend on the dynamics of the system. For example, if the service time depends on the waiting time of a job then the average service time of a job will be affected by the arrival process and therefore in such a case the capacity cannot be prespecified without the knowledge of the input process and the service protocol.

3.2.2 Little's Formula

We next look at an identity, called Little's formula, that relates the average time spent in the system to the average number of jobs in the system. Let

$$T_n = D_{K(n)} - A_n \tag{3.16}$$

be the time spent in the system by the nth customer, $n = 1, 2, \ldots$. We will assume that $T_n, n = 1, 2, \ldots$ is uniformly finite. Suppose the customer average time spent in the system, $\overline{T} = \lim_{k \to \infty} (1/k) \sum_{n=1}^{k} T_n$ exists. That is,

$$\overline{T} = \lim_{k \to \infty} \frac{1}{k} \sum_{n=1}^{k} (D_{K(n)} - A_n) \tag{3.17}$$

Let us also assume that the time average number of jobs in the system given by

$$\overline{N} = \lim_{t \to \infty} \frac{1}{t} \int_0^t N(\tau) d\tau$$

exists. We will next see that under these assumptions, we have *Little's formula*.

$$\overline{N} = \lambda \overline{T} \tag{3.18}$$

Let $Z_n(t)$ be an indicator variable that takes the value 1 if job n is in the system at time t and zero otherwise. Then $Z_n(t) = 1$, $A_n \le t < D_{K(n)}$ and $N(\tau) = \sum_{n=1}^{A(t)} Z_n(t)$, $0 \le \tau \le t$. Hence

$$\int_0^t N(\tau) d\tau = \int_0^t \left(\sum_{n=1}^{A(t)} Z_n(\tau) \right) d\tau = \sum_{n=1}^{A(t)} \{ t \wedge D_{K(n)} - A_n \} \tag{3.19}$$

where $a \wedge b$ is the minimum of a and b. $K^{-1}(n)$ is the label of the nth departure. Then $A_{K^{-1}(n)}$ is the arrival time of the nth departure, and therefore

$$\sum_{n=1}^{D(t)} (D_n - A_{K^{-1}(n)}) \le \sum_{n=1}^{A(t)} \{ t \wedge D_{K(n)} - A_n \} \le \sum_{n=1}^{A(t)} \{ D_{K(n)} - A_n \} \tag{3.20}$$

Because $\lim_{t \to \infty} (1/t) \sum_{n=1}^{A(t)} \{ D_{K(n)} - A_n \} = \lim_{t \to \infty} (A(t)/t) \cdot (1/A(t)) \sum_{n=1}^{A(t)} \{ D_{K(n)} - A_n \} = \lambda \overline{T}$, from equations 3.17 and 3.19 and the second inequality in equation 3.20, it is clear that $\overline{N} \le \lambda\overline{T}$. Let t' be the time by which all the jobs arrived during $(0, t]$ will have left the system, that is, $t' = \sup\{D_{K(n)} : n = 1, 2, \ldots, A(t)\}$. Then

$$\sum_{n=1}^{A(t)}\{D_{K(n)} - A_n\} \le \sum_{n=1}^{D(t')}\{D_n - A_{K^{-1}(n)}\} \tag{3.21}$$

By assumption T_n is uniformly finite for all n and therefore $t' - t$ is uniformly finite and in particular $t' \to \infty$ as $t \to \infty$. Therefore dividing equation 3.21 by t and taking the limit as $t \to \infty$ we see that

$$\lambda\overline{T} \le \lim_{t\to\infty}\frac{1}{t}\sum_{n=1}^{D(t')}\{D_n - A_{K^{-1}(n)}\} = \lim_{t\to\infty}\frac{1}{t}\sum_{n=1}^{D(t)}\{D_n - A_{K^{-1}(n)}\} \tag{3.22}$$

Now from equation 3.19, and the first inequality in equations 3.20 and 3.22 it follows that $\lambda\overline{T} \le \overline{N}$. This combined with the previous observation $\overline{N} \le \lambda\overline{T}$ implies that $\overline{N} = \lambda\overline{T}$. Note that other than the uniform finiteness of the time spent by jobs in the system and the existence of the appropriate limits $\overline{N}$ and $\overline{T}$ no additional assumptions are needed for this identity. Particularly observe that absolutely no assumption regarding the availability of the server is made. Therefore even if the servers are interrupted (e.g., because of machine failures or the servicing of jobs from other job streams) the preceding identity with respect to the job stream we have specified is still applicable.

In reality, however, we may never observe the limits. It is therefore of interest to see whether a similar identity exists for finite time t. Indeed as we have seen from equations 3.19 and 3.20 we have

$$\frac{1}{t}\sum_{n=1}^{D(t)}(D_n - A_{K^{-1}(n)}) \le \frac{1}{t}\int_0^t N(\tau)d\tau = \frac{1}{t}\sum_{n=1}^{A(t)}\{t \wedge D_{K(n)} - A_n\} \le \frac{1}{t}\sum_{n=1}^{A(t)}\{D_{K(n)} - A_n\} \tag{3.23}$$

The preceding inequality can be used to check the consistency of data collected from a real system or from a simulation of a hypothetical system. To apply $\overline{N} = \lambda\overline{T}$ we should verify that the T_n's are uniformly finite. We will see that when $\lambda < c\mu$, we have this condition satisfied. From equation 3.9 one sees that

$$V(t) - V(A_n) \le \sum_{k=n+1}^{A(t)} S_k - \int_{A_n}^t \sum_{i=1}^c R_i(\tau)d\tau, \quad t \ge A_n \tag{3.24}$$

We have already seen that when $\lambda < c\mu$, $V(t)$ is uniformly finite for all t. Then dividing equation 3.24 by t and taking the limit as $t \to \infty$ we see that

$$\lim_{t\to\infty}\frac{1}{t}\int_{A_n}^t \sum_{i=1}^c R_i(\tau)d\tau \le \frac{\lambda}{\mu} < c \tag{3.25}$$

Therefore

$$\int_{A_n}^t \left(c - \sum_{i=1}^c R_i(\tau)\right)d\tau \to \infty \quad \text{as } t \to \infty \tag{3.26}$$

Because a server may not be idle while a customer is waiting for service we know that job n will be receiving service whenever a server is idling during $(A_n, D_{K(n)})$. Hence

$$\int_{A_n}^{D_{K(n)}} \left(1 - \frac{1}{c} \sum_{i=1}^{c} R_i(\tau) \right) d\tau \leq S_n \tag{3.27}$$

From equation 3.27 and the uniform finiteness of S_n, it is evident that there exists a $t' < \infty$ such that $t' - A_n < \infty$ and $\int_{A_n}^{t'} (1 - \frac{1}{c} \sum_{i=1}^{c} R_i(\tau)) d\tau \geq S_n$. Hence $D_{K(n)} - A_n \leq t' - A_n < \infty$. This combined with the uniform convergence in equations 3.4 and 3.7 implies that T_n is uniformly finite for all n. Note that from equation 3.25 it follows that the average number of busy servers is λ/μ.

3.2.3 Performance Observed at Arrival, Departure, and Arbitrary Times

The number of jobs in the system varies over time and as a result the number seen by an arrival need not be the same as the number of jobs in the system at an arbitrary time. The average number of jobs seen by an arrival under general conditions is, however, equal to the average number of jobs left behind by departures. For this define

$$N_n^a = N(A_n-) + \sum_{k=1}^{n-1} I_{\{A_k = A_n\}} \tag{3.28}$$

as the number of jobs seen by the nth arrival. Note that $\sum_{k=1}^{n-1} I_{\{A_k = A_n\}}$ is the number of jobs arrived at the same time as the nth job, but tagged with an index smaller than n. Similarly

$$N_n^d = N(D_n-) - \sum_{k=1}^{n-1} I_{\{D_k = D_n\}} \tag{3.29}$$

be the number of jobs left behind by the nth departure. Note that $\sum_{k=1}^{n-1} I_{\{D_k = D_n\}}$ is the number of jobs departed before job n, but at the same time as job n. Let A_n^l and D_n^l be the nth time epochs where an arrival sees l jobs in the system and a departure leaves behind l jobs in the system, respectively. If no such time exists we set it equal to $+\infty$. Also we set $A_1^l = 0$ if $l < N(0)$ (i.e., all $N(0)$ arrive together at time zero). Then the number of arrivals during $(0, t]$ that see l jobs in the system is

$$\sum_{n=1}^{A(t)} I_{\{N_n^a = l\}} = \sup\{n : A_n^l \leq t, n = 1, 2, \ldots\} \tag{3.30}$$

and the number of departures during $(0, t]$ that will leave behind l jobs in the system is

$$\sum_{n=1}^{D(t)} I_{\{N_n^d = l\}} = \sup\{n : D_n^l \leq t, n = 1, 2, \ldots\} \tag{3.31}$$

Because the process $\{N(t), t \geq 0\}$ increases by one at each job arrival and decreases by one at each job departure, we have $A_1^l \leq D_1^l \leq A_2^l \leq D_2^l \leq \ldots$. Hence $D_n^l \leq t$ implies $A_n^l \leq t$ and $D_n^l > t$ implies $A_{n+1}^l > t$. Therefore

$$0 \leq \sum_{n=1}^{A(t)} I_{\{N_n^a = l\}} - \sum_{n=1}^{D(t)} I_{\{N_n^d = l\}} \leq 1 \tag{3.32}$$

Now dividing equation 3.32 by t and taking the limit as $t \to \infty$

$$\lambda r^a(l) = \beta r^d(l), \quad l = 0, 1, \ldots \tag{3.33}$$

where

$$r^a(l) = \lim_{k \to \infty} \frac{1}{k} \sum_{n=1}^{k} I_{\{N_n^a = l\}} \tag{3.34}$$

is the fraction of jobs that on their arrival see l jobs in the system and

$$r^d(l) = \lim_{k \to \infty} \frac{1}{k} \sum_{n=1}^{k} I_{\{N_n^d = l\}} \tag{3.35}$$

is the fraction of departing jobs that leave behind l jobs in the system.

It is worth noting that the left-hand side of equation 3.33 is the rate of upcrossing of the process $\{N(t), t \geq 0\}$ from a level less than or equal to l to a level greater than l and the right-hand side of equation 3.33 is the rate of downcrossings from a level greater than l to a level less than or equal to l. Hence equation 3.33 can be viewed as a balance equation for the rates of upcrossings and downcrossings (the *level crossing balance equation*). If $\beta = \lambda$ then from equation 3.31 we have

$$r^a(l) = r^d(l), \quad l = 0, 1, \ldots . \tag{3.36}$$

Recall that for a single-stage system we have shown that when $\lambda < c\mu$, the flow balance holds (i.e., $\lambda = \beta$), and therefore equation 3.36 holds as well.

Next, we will look at the average number of jobs seen by arrivals and departures. Let

$$\overline{N}^a = \lim_{k \to \infty} \frac{1}{k} \sum_{n=1}^{k} N_n^a \tag{3.37}$$

be the average number of jobs seen by an arrival and

$$\overline{N}^d = \lim_{k \to \infty} \frac{1}{k} \sum_{n=1}^{k} N_n^d \tag{3.38}$$

be the average numbers of jobs left behind by a departing job. Because $N_n^a = \sum_{l=1}^{\infty} l I_{\{N_n^a = l\}}$, substituting this in equation 3.36 and taking the limit into the summation we get

$$\overline{N}^a = \sum_{l=1}^{\infty} \lim_{k \to \infty} \frac{1}{k} \sum_{n=1}^{k} l I_{\{N_n^a = l\}} = \sum_{l=1}^{\infty} l r^a(l) \tag{3.39}$$

Similarly from equation 3.38 we get

$$\overline{N}^d = \sum_{l=1}^{\infty} l r^d(l) \tag{3.40}$$

Therefore when $r^a(l) = r^d(l), l = 0, 1, \ldots$, we have

$$\overline{N}^a = \overline{N}^d \tag{3.41}$$

It is crucial to observe that in general the time average number of jobs, $\overline{N}$, in the system need not be the same as $\overline{N}^a$ and $\overline{N}^d$. If the jobs arrive according to a Poisson process, however, and the service protocols do not use information about the future arrival times of the jobs, it is well known that $\overline{N} = \overline{N}^a$. This is generally called the Poisson arrivals see time averages (PASTA) property (see Wolff [48]).

To obtain more detailed performance measures we need to obtain additional information on the arrival and service processes, and the specification of the service protocols. Because a precise representation of the actual arrival and service processes is usually impossible, we test the appropriateness of using standard stochastic processes for such representations. In this respect we would test whether

1. The arrival process is a renewal process.
2. The service times of jobs form a sequence of iid random variables.
3. The arrival process and the service times are mutually independent.

In subsequent modeling of the single-stage system we will assume that these three conditions are satisfied.

3.3 SINGLE-STAGE SINGLE-CLASS SYSTEMS

3.3.1 *M/M/*1 Model

We will focus on a model of a single-stage job shop with a single machine. In this shop jobs are selected for service without using information about their processing requirements. For example, FCFS or last-come first-served (LCFS) are two service protocols that satisfy this assumption. It will be assumed that the arrivals form a Poisson process, and the service times are exponentially distributed. Hence we model the dynamics of the system by an $M/M/1$ queueing system.

Birth-death model.   The number of jobs in the system $\{N(t), t \geq 0\}$ forms a Markov (birth-death) process on the state space $\mathcal{S} = \{0, 1, 2, \ldots\}$. The transition rate diagram of $\{N(t), t \geq 0\}$ is shown in Figure 3.2.

The probability flow balance equations (rate out of state = rate into the same state) for states 0 and n ($n = 1, 2, \ldots$) are, respectively,

$$\lambda p(0) = \mu p(1) \tag{3.42}$$

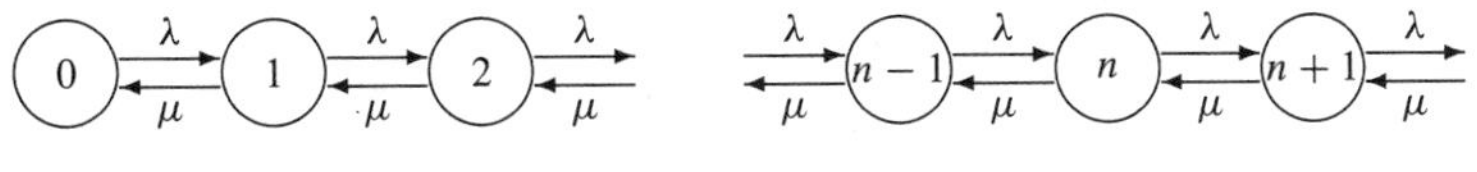

Figure 3.2   Transition Rate Diagram

and

$$(\lambda + \mu)p(n) = \lambda p(n-1) + \mu p(n+1), \quad n = 1, 2, \ldots \tag{3.43}$$

where

$$p(n) = \lim_{t \to \infty} P\{N(t) = n\}, \quad n = 0, 1, \ldots \tag{3.44}$$

is the steady-state probability that there are n jobs in the system (provided that this limit exists). From equation 3.43 one sees that

$$\lambda p(n-1) - \mu p(n) = \lambda p(n) - \mu p(n+1), \quad n = 1, 2, \ldots \tag{3.45}$$

and from equations 3.42 and 3.45 with $n = 1$ it can be observed that

$$\lambda p(n) = \mu p(n+1), \quad n = 0, 1, \ldots \tag{3.46}$$

If we relate the rates of probability inflow and outflow from the compound states $\{0, 1, \ldots, n\}$ we would obtain equation 3.46 directly. In the analysis of some Markov processes it may be advantageous to use such a balance equation rather than the balance equations for individual states. From equation 3.46 it directly follows that

$$p(n+1) = \frac{\lambda}{\mu} p(n), \quad n = 0, 1, \ldots$$

$$= \left(\frac{\lambda}{\mu}\right)^{n+1} p(0), \quad n = 0, 1, \ldots \tag{3.47}$$

Now using the total probability law $\sum_{n=0}^{\infty} p(n) = 1$, we get   *steady state prob.*

$$p(n) = (1 - \rho)\rho^n, \quad n = 0, 1, \ldots, \tag{3.48}$$

provided $\rho = \lambda/\mu < 1$ (i.e., $\lambda < \mu$). This condition agrees with the general condition $\lambda < c\mu$ we derived earlier for the uniform finiteness of $\{N(t), t \geq 0\}$. The expected number of jobs in the system under steady state is then

$$E[N] = \sum_{n=0}^{\infty} np(n) = \sum_{n=0}^{\infty} n(1 - \rho)\rho^n \tag{3.49}$$

$$= \frac{\rho}{1 - \rho}, \quad \rho < 1$$

Because the Markov process $\{N(t), t \geq 0\}$ has a steady-state distribution when $\rho < 1$, the time-average measures of this process, independent of the initial state, are the same as the corresponding stationary measures. Specifically

$$\overline{N} = \lim_{t\to\infty} \frac{1}{t} \int_0^t N(\tau)d\tau = E[N]$$

and

$$r(n) = \lim_{t\to\infty} \frac{1}{t} \int_0^t I_{\{N(\tau)=n\}}d\tau = p(n), \quad n = 0, 1, \ldots$$

Because we have a Poisson arrival process, the time-average performance measures are the same as the arrival epoch averages. That is, $r^a(n) = r(n) = r^d(n)$, $n = 0, 1, \ldots$, and $\overline{N} = \overline{N}^a = \overline{N}^d$. Therefore using the equivalence of the pointwise limits to the time averages we get,

$$p^a(n) = p(n) = p^d(n), n = 0, 1, \ldots. \tag{3.50}$$

and

$$E[N] = E[N^a] = E[N^d] \tag{3.51}$$

where $N_k^a = N(A_k-)$ and $N_k^d = N(D_k)$ because jobs arrive and depart one at a time, $p^a(n) = \lim_{k\to\infty} P\{N_k^a = n\}$ and $p_n^d = \lim_{k\to\infty} P\{N_k^d = n\}$, $n = 0, 1, \ldots$.

SOJOURN TIMES **Flow time.** Next, we will obtain the distribution of the job flow times through the job shop when operated under FCFS service protocol (and jobs leave the shop as soon as they finish service). Suppose a tagged job sees n jobs on its arrival. One of these jobs will be partially processed and the other $n - 1$ will be waiting in the queue. So the flow time of the tagged job is the sum of the processing times of the $n - 1$ waiting jobs, its processing time, and the remaining service time of the job in service. Because the service times are exponentially distributed the remaining service time will also be exponentially distributed with mean $1/\mu$. Hence the distribution of the flow time of the tagged job conditioned on it seeing n jobs on its arrival is Erlang-$(n + 1)$ with mean $(n+1)/\mu$. If we therefore take an arbitrary job that arrives to the system under stationary conditions and tag it, the distribution of the number of jobs it sees is given by $(1 - \rho)\rho^n$, $n = 0, 1, \ldots$. Hence the distribution of the flow time of the tagged job is

$$F_T(x) = \sum_{n=0}^{\infty}(1-\rho)\rho^n \sum_{l=n+1}^{\infty} \frac{e^{-\mu x}(\mu x)^l}{l!}$$

$$= \sum_{l=1}^{\infty}\sum_{n=0}^{l-1}(1-\rho)\rho^n \frac{e^{-\mu x}(\mu x)^l}{l!}$$

$$= \sum_{l=1}^{\infty}(1-\rho^l)\frac{e^{-\mu x}(\mu x)^l}{l!}$$

$$= 1 - e^{-\mu(1-\rho)x}, \quad x \geq 0 \tag{3.52}$$

It is interesting to note that the flow time is exponentially distributed with mean

$$E[T] = \frac{1}{\mu(1-\rho)}, \quad \rho < 1 \tag{3.53}$$

The flow time is magnified by a factor $1/(1 - \rho)$, which is increasing and convex in ρ. As $\rho \to 1$, $E[T] \to \infty$, signifying that even a slight increase in the arrival rate or a decrease in the service capacity in a heavily loaded (i.e., ρ close to 1) job shop may have severe adverse effects on its performance. Because there can be benefits in either increasing the arrival rate (owing to increased revenues) or decreasing the service capacity (owing to a decrease either in capital investment or the operating cost) there should be a tradeoff between the arrival rate and the service capacity. The following cost model illustrates how this tradeoff could be determined using the performance measures obtained so far.

Example 3.2

Suppose m alternative machines with different processing rates are available. From these m machines one has to be selected. The average time needed to process a job on machine j is $1/\mu_j$, and the cost of obtaining machine j is c_j per unit time. The reward received per serviced job is v, and the cost of spending time in the system is w per job per unit time. We will assume that $v > w/\mu_j$. Otherwise machine j will never be used. We will first look at the optimal arrival rate that maximizes the profit per unit time for a given machine when machine j is selected. Then using the optimal profit we will choose the machine that maximizes the overall profit. Let $R_j(\lambda)$ be the profit per unit time obtained when the arrival rate is λ and the machine j is used for service. Then

$$R_j(\lambda) = \lambda v - E[N]w - c_j = \lambda v - \frac{\lambda w}{\mu_j - \lambda} - c_j, \quad 0 < \lambda < \mu_j \tag{3.54}$$

The optimization problem is then

$$\max\{R_j(\lambda) : 0 < \lambda < \mu_j; j = 1, 2, \dots, m\}$$

First for a fixed j consider $\max\{R_j(\lambda) : 0 < \lambda < \mu_j\}$. It is easily verified that $R_j(\lambda)$ is concave in λ. Hence the optimal λ^* that maximizes $R_j(\lambda)$ satisfies

$$\frac{d}{d\lambda} R_j(\lambda)\Big|_{\lambda=\lambda^*} = v - \frac{\mu_j w}{(\mu_j - \lambda)^2} = 0 \tag{3.55}$$

Therefore

$$\lambda^* = \mu_j - \sqrt{\mu_j \frac{w}{v}} > 0$$

is the optimal input rate (recall that we assume $v > w/\mu_j$). The corresponding profit is

$$R_j(\lambda_j^*) = (\sqrt{\mu_j v} - \sqrt{w})^2 - c_j \tag{3.56}$$

Machine j^* that maximizes $(\sqrt{\mu_j v} - \sqrt{w})^2 - c_j$ is the one that should be used with the desired arrival rate of λ^* for this machine.

Departure process. We will now look at the stationary distribution of the interdeparture times. If a departing job leaves behind an empty system the next departure will take place after a time equal in distribution to the sum of two independent exponential random variables, one with mean $1/\lambda$ (corresponding to the interarrival time) and the other with mean $1/\mu$ (corresponding to the service time). Conversely, if a customer leaves behind at least one customer, the next departure will occur at a time equal in distribution to an exponential distribution with mean $1/\mu$ (corresponding to the service time). Under steady state conditions the probabilities of these two events are $1 - \rho$ and ρ, respectively (see equation 3.48). The steady-state distribution of the interdeparture times is then given by

$$F_D(x) = (1 - \rho) \int_0^x (1 - e^{-\mu(x-y)}) \lambda e^{-\lambda y} dy + \rho(1 - e^{-\mu x})$$

$$= 1 - e^{-\lambda x}, \quad x \geq 0 \tag{3.57}$$

That is, the stationary interdeparture times have an exponential distribution same as that of an interarrival time. Indeed it can be established that the stationary departure process from the $M/M/1$ queueing system is a Poisson process with rate λ (e.g., see Walrand [45]).

3.3.2 M/G/1 Model

In many real job shop systems it has been observed that the Poisson process is a good representation of the arrival process. Exponential distributions may not always be good representations of the service times. In this section we will see how useful results on the performance measures of the job shop can be obtained even when only a nonstandard distribution fits well the real processing times. To facilitate this we will develop and analyze a model of the job shop without any specific assumptions about the distribution of service time. We represent the service time distribution by F_S. The systems under study are assumed to be operated according to a service protocol (such as FCFS or LCFS), that is service time independent. Because the remaining service time of a job currently in service (if any) at time t will in general depend on the attained service time, the information on the number of jobs in the system at time t is insufficient to determine the future evolution of $N(t)$. Hence $\{N(t), t \geq 0\}$ is not a Markov process.

Embedded Markov chain model. One way to overcome the preceding problem is to model the system dynamics by observing the number of jobs in the system just after each job departure. At these time epochs the attained service of the job, if any, is zero, and hence the future is predictable based on just the number in the system. Consequently $\{N_n^d, n = 1, 2, \ldots\}$ forms a Markov chain on $\mathcal{N}_+ = \{0, 1, 2, \ldots\}$.

Let X_n be the number of jobs arrived during the service time $S_{K^{-1}(n)}$ of the nth job to depart. Because the arrival process is Poisson, service times are iid, and the service protocol is independent of the service times $\{X_n, n = 1, 2, \ldots\}$, forming a sequence of iid random variables. Conditioning on the service times it is seen that

$$P\{X_n = k\} = \int_0^\infty P\{X_n = k | S_{K^{-1}(n)} = x\} dF_S(x)$$

$$= \int_0^\infty \frac{e^{-\lambda x} (\lambda x)^k}{k!} dF_S(x), \quad k = 0, 1, \ldots \tag{3.58}$$

Let $\tilde{F}_S(s)$ be the Laplace-Stieltjes transform (LST) of the service time distribution F_S, that is, $\tilde{F}_S(s) = \int_0^\infty e^{-sx} dF_S(x)$. Then from equation 3.58 it can be seen that

$$P\{X_n = k\} = (-1)^k \lambda^k \tilde{F}_S^{(k)}(\lambda)/k!, \quad k = 0, 1, \ldots$$

where $\tilde{F}_S^{(k)}(\lambda)$ is the kth derivative of $\tilde{F}_S(s)$ at $s = \lambda$. Note that $P\{X_n = 0\} = \tilde{F}_S(\lambda)$. Therefore, if $\tilde{F}_S$ is explicitly known these probabilities may be directly computed using the preceding relationship.

If the $(n-1)$th departing job leaves behind an empty system (i.e., $N_{n-1}^d = 0$), then the nth departing job will leave behind only those jobs that arrived during its service (i.e., $N_n^d = X_n$). Conversely, if $N_{n-1}^d \geq 1$, then the number of jobs left behind by the nth departing job will include the $N_{n-1}^d - 1$ of those who were present at the service initiation and the X_n jobs that arrived during service (i.e., $N_n^d = N_{n-1}^d - 1 + X_n$). Therefore

$$N_n^d = [N_{n-1}^d - 1]^+ + X_n, \quad n = 1, 2, 3, \ldots \qquad (3.59)$$

Then the transition probability matrix $\mathbf{P} = (p_{ij})$ of the Markov chain $\{N_n^d, n = 1, 2, \ldots\}$ is given by

$$p_{ij} = P\{N_n^d = j | N_{n-1}^d = i\}$$

$$= \begin{cases} P\{X = j - i + 1\}, & i \geq 1, j \geq i - 1 \\ P\{X = j\}, & i = 0, j \geq 0 \\ 0, & \text{otherwise} \end{cases} \qquad (3.60)$$

where $P\{X = k\} = P\{X_n = k\}, k = 0, 1, 2, \ldots$. If the stationary distribution $\mathbf{p}^d = (p^d(0), p^d(1), \ldots)$ of $\{N_n^d, n = 0, 1, 2, \ldots\}$ exists then $\mathbf{p}^d = \mathbf{p}^d \mathbf{P}$. The corresponding set of linear equations are

$$p^d(j) = p^d(0) P\{X = j\} + \sum_{i=1}^{j+1} p^d(i) P\{X = j - i + 1\}, \quad j = 0, 1, 2, \ldots \qquad (3.61)$$

Multiplying both sides of equation 3.61 by z^j and summing over $j = 0, 1, \ldots$ one obtains

$$\tilde{p}^d(z) = p^d(0)\tilde{Q}(z) + \frac{1}{z}\tilde{p}^d(z)\tilde{Q}(z) - \frac{1}{z}p^d(0)\tilde{Q}(z) \qquad (3.62)$$

where

$$\tilde{p}^d(z) = \sum_{j=0}^\infty z^j p^d(j) \qquad (3.63)$$

$$\tilde{Q}(z) = \sum_{j=0}^\infty z^j P\{X = j\} \qquad (3.64)$$

are the moment generating functions (MGF) of N^d and X, respectively. Solving equation 3.62 for $\tilde{p}^d(z)$ one obtains

$$\tilde{p}^d(z) = \frac{(z-1)\tilde{Q}(z)p^d(0)}{z - \tilde{Q}(z)} \qquad (3.65)$$

Now to obtain the only unknown $p^d(0)$ in equation 3.65 we use the law of total probability (i.e., $\sum_{j=0}^{\infty} p^d(j) = \lim_{z \to 1} \tilde{p}^d(z) = 1$). Taking this limit in equation 3.65 by applying L'Hôpital's rule, we get

$$p^d(0) = 1 - \lim_{z \to 1} \frac{d}{dz} \tilde{Q}(z) = 1 - E[X] \tag{3.66}$$

From equations 3.58 and 3.64 one sees that

$$\tilde{Q}(z) = \sum_{j=0}^{\infty} z^j \int_0^{\infty} e^{-\lambda x} \frac{(\lambda x)^j}{j!} dF_S(x)$$

$$= \int_0^{\infty} \left(\sum_{j=0}^{\infty} e^{-\lambda x} \frac{(\lambda z x)^j}{j!} \right) dF_S(x)$$

$$= \int_0^{\infty} e^{-(\lambda - \lambda z)x} dF_s(x)$$

$$= \tilde{F}_S(\lambda - \lambda z) \tag{3.67}$$

where $\tilde{F}_S(\lambda - \lambda z)$ is the value of the LST of F_S evaluated at $s = \lambda - \lambda z$. Then

$$E[X] = \lim_{z \to 1} \frac{d}{dz} \tilde{F}_S(\lambda - \lambda z) = \lambda \lim_{s \to 0} \left\{ -\frac{d}{ds} \tilde{F}_S(s) \right\} = \lambda E[S]$$

Therefore

$$p^d(0) = 1 - \rho, \quad \rho < 1 \tag{3.68}$$

where $\rho = \lambda E[S]$ is the server utilization. Substituting equations 3.67 and 3.68 in 3.65 we obtain

$$\tilde{p}^d(z) = \frac{(1 - \rho)(z - 1)\tilde{F}_S(\lambda - \lambda z)}{z - \tilde{F}_S(\lambda - \lambda z)}, \quad \rho < 1 \tag{3.69}$$

The first and second moments of the number of jobs in the system at a departure epoch are then

$$E[N^d] = \lim_{z \to 1} \frac{d}{dz} \tilde{p}^d(z)$$

$$= \frac{\lambda^2 E[S^2]}{2(1 - \rho)} + \rho \tag{3.70}$$

$$E[(N^d)^2] = \lim_{z \to 1} \left\{ \frac{d^2}{dz^2} + \frac{d}{dz} \right\} \tilde{p}^d(z)$$

$$= \frac{\lambda^3 E[S^3]}{3(1 - \rho)} + \frac{\lambda^4 E[S^2]^2}{2(1 - \rho)^2} \tag{3.71}$$

$$+ \frac{\lambda^3 E[S]E[S^2]}{6(1 - \rho)} + \frac{\lambda^2 E[S^2](3 - 2\rho)}{2(1 - \rho)} + \rho$$

When $\rho < 1$, the Markov process $\{N_n^d, n = 1, 2, \ldots\}$ has a stationary distribution, and therefore the time and customer average performance measures are the same as the corresponding stationary performance measures. Hence $\tilde{p}^a(z) = \tilde{p}^d(z)$. In addition because we have Poisson arrival processes $\tilde{p}(z) = \tilde{p}^a(z)$ where $\tilde{p}(z)$ and $\tilde{p}^a(z)$ are the MGFs of N and N^a, respectively.

Flow time.   Now assuming that the service protocol is FCFS we will obtain the stationary flow time distribution. Under FCFS we observe that the number of jobs left behind in the system by the nth departure are those jobs arrived during the flow time T_n of the nth departing job. Therefore because the MGF of the number of Poisson arrivals with rate λ during a random time interval of length Y is $\tilde{F}_Y(\lambda - \lambda z)$ (e.g., see equation 3.67), we have

$$\tilde{p}^d(z) = \tilde{F}_T(\lambda - \lambda z) \tag{3.72}$$

Substituting $z = (\lambda - s)/\lambda$ in equation 3.72 from equation 3.69 we find that the LST of the flow time is given by

$$\tilde{F}_T(s) = \frac{(1 - \rho)s\tilde{F}_S(s)}{s - \lambda(1 - \tilde{F}_S(s))} \tag{3.73}$$

The first two moments of the flow time are

$$E[T] = \lim_{s \to 0}\{-\frac{d}{ds}\tilde{F}_T(s)\}$$

$$= \frac{\lambda E[S^2]}{2(1 - \rho)} + E[S] \tag{3.74}$$

$$E[T^2] = \lim_{s \to 0}\{\frac{d^2}{ds^2}\tilde{F}_T(s)\}$$

$$= \frac{\lambda E[S^3]}{3(1 - \rho)} + \frac{\lambda^2 E[S^2]^2}{2(1 - \rho)^2}$$

$$+ \frac{\lambda E[S^2]E[S]}{1 - \rho} + E[S^2] \tag{3.75}$$

By taking the limit of the nth derivative of equation 3.73 it follows that

$$E[T^n] = E[S^n] + \frac{\lambda}{(n + 1)(1 - \rho)}\sum_{k=2}^{n+1}\binom{n + 1}{k}E[S^k]E[T^{n+1-k}], \quad n = 1, 2, \ldots$$

It can be readily verified that Little's formula $E[N] = \lambda E[T]$ holds in this case. Alternatively, Little's formula could have been used to obtain $E[T]$ from $E[N]$ or vice versa.

Number in the queue and waiting time.   In certain applications it will be of interest to know the number in the queue and their waiting time. For example, the cost incurred per unit of time waiting may differ from the cost incurred per unit time in service.

Because the number in the queue $L = [N - 1]^+$ and the flow time $T_n = W_n + S_n$, $n = 1, 2, \ldots$, it follows that the MGF $\tilde{p}^q(z)$ of L is related to $\tilde{p}^d(z)$ by $\tilde{p}^d(z) = z(\tilde{p}^q(z) - p^d(0)) + p^d(0)$, and the LST $\tilde{F}_W(s)$ of W is related to $\tilde{F}_T(s)$ by $\tilde{F}_T(s) = \tilde{F}_W(s)\tilde{F}_S(s)$. Therefore from equations 3.69 and 3.73 we obtain

$$\tilde{p}^q(z) = \frac{(1 - \rho)(z - 1)}{z - \tilde{F}_S(\lambda - \lambda z)} \tag{3.76}$$

and

$$\tilde{F}_W(s) = \frac{(1 - \rho)s}{s - \lambda(1 - \tilde{F}_S(s))} \tag{3.77}$$

The first moment of the number of jobs in the queue and the waiting time are

$$E[L] = \frac{\lambda^2 E[S^2]}{2(1 - \rho)} \tag{3.78}$$

and

$$E[W] = \frac{\lambda E[S^2]}{2(1 - \rho)} \tag{3.79}$$

respectively. It can easily be verified that $E[L] = \lambda E[W]$, thus showing that Little's formula is applicable even to a subsystem that in this case is the "queue."

Mean value analysis (MVA).   When only the mean performance measures are of interest, it is sufficient to apply a MVA technique. In such an analysis one works with the mean value rather than the distribution of the system states. Taking the expectations on both sides of equation 3.59 one gets

$$E[N_n^d] = E[N_{n-1}^d] - (1 - P\{N_{n-1}^d = 0\}) + E[X_n] \tag{3.80}$$

Taking the limit as $n \to \infty$ and assuming the existence of steady state we get

$$p^d(0) = 1 - E[X] = 1 - \rho \tag{3.81}$$

Now squaring both sides of equation 3.59 and taking expectations one finds that

$$
\begin{aligned}
E[(N_n^d)^2] &= E[(N_{n-1}^d)^2] - 2E[N_{n-1}^d I_{\{N_{n-1}^d \geq 1\}}] + E[I_{\{N_{n-1}^d \geq 1\}}^2] \\
&\quad + 2E[N_{n-1}^d X_{n-1} - I_{\{N_{n-1}^d \geq 1\}} X_n] + E[X_n^2] \\
&= E[(N_{n-1}^d)^2] - 2E[N_{n-1}^d] + E[I_{\{N_{n-1}^d \geq 1\}}] \\
&\quad + 2E[N_{n-1}^d - I_{\{N_{n-1}^d \geq 1\}}]E[X_n] + E[X_n^2]
\end{aligned} \tag{3.82}
$$

Taking the limit of equation 3.82 as $n \to \infty$ and solving for $E[N^d]$ we obtain

$$E[N^d] = \frac{E[X^2] - E[I_{\{N^d \geq 1\}}]}{2(1 - E[X])} + E[I_{\{N^d \geq 1\}}]$$

$$= \frac{\lambda^2 E[S^2]}{2(1 - \rho)} + \rho \tag{3.83}$$

The last equality follows from equation 3.81 and $E[X^2] = \lambda^2 E[S^2] + \lambda E[S]$.

3.3.3 *PH/PH/1* Model

When the arrival process is not Poisson it becomes difficult to provide results for the general case. Therefore if Poisson process is not a good fit for the arrival process one should test whether a special class of arrival process called the *PH*-renewal process will be a good fit. This class of renewal process is very large and is very likely to fit the real data provided it is renewal. The *PH*-renewal process is characterized by the interrenewal times that have phase-type (*PH*) distribution. Therefore obtaining a *PH*-renewal process representation for the arrival process requires finding a *PH*-distribution to fit the interarrival times. In this section we focus on job shop systems where *PH* distributions are found to fit both the interarrival and service times.

Phase-type distributions for service times may naturally arise if, for example, the service to the job involves performing several tasks where each task takes an exponentially distributed amount of time. To make it clear, suppose there are m different types $\{1, \ldots, m\}$ of tasks that need to be performed on a job to complete its service. Suppose the first task that needs to be performed on the job is of type j with probability α_j, $j = 1, \ldots, m$. The time it takes to complete task j has an exponential distribution with mean $1/\mu_j$, $j = 1, \ldots, m$. When task j is completed a job may require the service of task i with probability p_{ji}, $i = 1, \ldots, m$ or may have completed its service and thus leave the system with probability $1 - \sum_{i=1}^{m} p_{ji}$, $j = 1, \ldots, m$. Suppose we initiated a service to a job at time 0 and let $X(t)$ be the type of task it is being serviced with at time t, if it is still in service; otherwise set $X(t) = m + 1$. Then $\{X(t), t \geq 0\}$ is a continuous time Markov process with an infinitesimal generator matrix $\mathbf{Q}$, where

$$\mathbf{Q} = \begin{bmatrix} \mathbf{T} & -\mathbf{Te} \\ \mathbf{0} & 0 \end{bmatrix} \tag{3.84}$$

$\mathbf{T}$ is an $m \times m$ matrix with $T_{ij} = \mu_i p_{ij}$, $j \neq i = 1, \ldots, m$; $T_{ii} = -\mu_i + \mu_i p_{ii}$, $i = 1, \ldots, m$ and $\mathbf{e} = (1, \ldots, 1)'$ is a column m-vector of ones. The initial probability distribution of $X(0)$ is $(\boldsymbol{\alpha}, 0)$ (i.e., $P\{X(0) = i\} = \alpha_i$, $i = 1, \ldots, m$). State $m + 1$ is an absorbing state for $\{X(t), t \geq 0\}$, and the job service time S is equal to the time it takes for $X(t)$ to reach the absorbing state $m+1$; that is, $S = \inf\{t : X(t) = m+1, t \geq 0\}$. Because from the theory of Markov processes one has $P\{X(t) = j\} = [\boldsymbol{\alpha}e^{\mathbf{T}t}]_j$, $j = 1, \ldots, m$; $t \geq 0$, it is clear that

$$P\{S > x\} = \sum_{j=1}^{m} P\{X(x) = j\} = \boldsymbol{\alpha}e^{\mathbf{T}x}\mathbf{e}, \quad x \geq 0 \tag{3.85}$$

Therefore the distribution function of S is

$$F_S(x) = 1 - \boldsymbol{\alpha} e^{\mathbf{T}x} \mathbf{e}, \quad x \geq 0$$

and its density function is

$$f_S(x) = \boldsymbol{\alpha} e^{\mathbf{T}x} (-\mathbf{T}\mathbf{e}), \quad x \geq 0$$

Note that the phase-type distribution is fully specified by $(\boldsymbol{\alpha}, \mathbf{T})$, and hence this is called the representation of F_S, f_S, or S. It can now be verified that

$$E[S] = -\boldsymbol{\alpha} \mathbf{T}^{-1} \mathbf{e}$$

Even if the existence of these tasks is not explicitly observable, if a phase-type distribution is a good fit for the service (or interarrival) times, one may imagine the existence of virtual tasks and use it for modeling the service (or interarrival) times by a Markov process. Procedures for obtaining the representation $(\boldsymbol{\alpha}, \mathbf{T})$ for a phase-type distribution from a given set of data or distribution function are referenced in Appendix A.

Suppose the distributions selected for the arrival process and the service times are represented by $(\boldsymbol{\alpha}, \mathbf{A})$ and $(\boldsymbol{\beta}, \mathbf{B})$ respectively. Then the arrival rate λ is equal to $(-\boldsymbol{\alpha}\mathbf{A}^{-1}\mathbf{e})^{-1}$, and the service rate μ is equal to $(-\boldsymbol{\beta}\mathbf{B}^{-1}\mathbf{e})^{-1}$. Therefore the finiteness of the stationary number of jobs in the system is guaranteed by the condition

$$-\boldsymbol{\alpha}\mathbf{A}^{-1}\mathbf{e} > -\boldsymbol{\beta}\mathbf{B}^{-1}\mathbf{e}.$$

Quasi birth-death (Markov) process model. As before in the $M/G/1$ model, the number of jobs in the system does not form a Markov process. Because the interarrival times and service times are of phase type we can imagine the existence of two hypothetical Markov processes evolving concurrently with the number of jobs in the system. Let $\{X(t), t \geq 0\}$ and $\{Y(t), t \geq 0\}$ be the Markov processes on $\{1, 2, \ldots, m_a + 1\}$ and $\{1, 2, \ldots, m_b + 1\}$, respectively, that represent the arrival and service processes. Here m_a and m_b are the sizes of the two vectors $\boldsymbol{\alpha}$ and $\boldsymbol{\beta}$, respectively. When $X(t) = i$, the interarrival time process at time t is in state i, and the time to the next arrival is the first passage time from state i to state $m_a + 1$. Once the process reaches state $m_a + 1$ an arrival occurs and instantaneously the state is reset according to the probability vector $(\boldsymbol{\alpha}, 0)$ so that the interarrival time process for subsequent arrivals can continue. As a consequence state $m_a + 1$ will not be occupied by the process $\{X(t), t \geq 0\}$. When $Y(t) = j$ the service process is in state j and the remaining service time of the job in service at time t is the first passage time from state j to state $m_b + 1$. At the time when this process reaches state $m_b + 1$, service completion occurs. If there are any jobs in the system at this time a job will be taken for processing, and the state of the process is instantaneously reset according to the probability vector $(\boldsymbol{\beta}, 0)$. Conversely, if there are no jobs available in the system the process remains in state $m_b + 1$ until an arrival occurs. At the time of this arrival the state of the process is reset according to the probability vector $(\boldsymbol{\beta}, 0)$. Therefore the combined process $\{(N(t), X(t), Y(t)), t \geq 0\}$ is a Markov process on the state space $\mathcal{S} = \{(0, i, m_b + 1), i = 1, \ldots, m_a; (n, i, j), i = 1, \ldots, m_a; j =$

$1, \ldots, m_b; n = 1, 2, \ldots\}$. Partition $\mathcal{S}$ into the compound states $\{\mathbf{0}, \mathbf{1}, \mathbf{2}, \ldots\}$ associated with the number of jobs in the system, where $\mathbf{0} = ((0, i, m_b + 1), i = 1, \ldots, m_a)$ and $\mathbf{n} = ((n, i, j), i = 1, \ldots, m_a, j = 1, \ldots, m_b), n = 1, 2, \ldots$. With respect to this partitioning the transition rate matrix $\mathbf{R}$ of this Markov process is of the following tridiagonal form (and hence called quasi birth-death) because at most one job may arrive or depart at any given time epoch.

$$\mathbf{R} = \begin{bmatrix} \mathbf{R}_1' & \mathbf{R}_0' & \mathbf{0} & \mathbf{0} & \mathbf{0} & \cdots \\ \mathbf{R}_2' & \mathbf{R}_1 & \mathbf{R}_0 & \mathbf{0} & \mathbf{0} & \cdots \\ \mathbf{0} & \mathbf{R}_2 & \mathbf{R}_1 & \mathbf{R}_0 & \mathbf{0} & \cdots \\ \mathbf{0} & \mathbf{0} & \mathbf{R}_2 & \mathbf{R}_1 & \mathbf{R}_0 & \cdots \\ \cdots & \cdots & \cdots & \cdots & \cdots & \cdots \end{bmatrix} \tag{3.86}$$

where

$$\mathbf{R}_0' = \begin{bmatrix} (-\mathbf{Ae})\alpha\beta_1 & (-\mathbf{Ae})\alpha\beta_2 & \cdots & (-\mathbf{Ae})\alpha\beta_{m_b} \end{bmatrix} \tag{3.87}$$

is an $m_a \times (m_a m_b)$ matrix and

$$\mathbf{R}_0 = \begin{bmatrix} -\mathbf{Ae}\alpha & 0 & \cdots & 0 \\ 0 & -\mathbf{Ae}\alpha & \cdots & 0 \\ . & . & \cdots & 0 \\ . & . & \cdots & 0 \\ . & . & \cdots & 0 \\ . & . & \cdots & -\mathbf{Ae}\alpha \end{bmatrix} \tag{3.88}$$

is a block diagonal $(m_a m_b) \times (m_a m_b)$ matrix.

$$\mathbf{R}_1' = \mathbf{A} - (\mathbf{A})_D \tag{3.89}$$

and

$$\mathbf{R}_1 = \begin{bmatrix} \mathbf{A} - (\mathbf{A})_D & b_{12}\mathbf{I} & \cdots & b_{1m_b}\mathbf{I} \\ b_{21}\mathbf{I} & \mathbf{A} - (\mathbf{A})_D & \cdots & b_{2m_b}\mathbf{I} \\ \cdots & \cdots & \cdots & \cdots \\ \cdots & \cdots & \cdots & \cdots \\ \cdots & \cdots & \cdots & \cdots \\ b_{m_b1}\mathbf{I} & b_{m_b2}\mathbf{I} & \cdots & \mathbf{A} - (\mathbf{A})_D \end{bmatrix} \tag{3.90}$$

where $(\mathbf{A})_D$ is the diagonal matrix with the diagonal entries $\mathbf{A}$. That is, if $\mathbf{A} = (a_{ij})_{i,j=1,\ldots,m_a}$, then

$$(\mathbf{A})_D = \begin{pmatrix} a_{11} & 0 & \cdots & 0 \\ 0 & a_{22} & \cdots & 0 \\ \cdots & \cdots & \cdots & \cdots \\ 0 & 0 & \cdots & a_{m_a m_a} \end{pmatrix}$$

Note that we remove the diagonal entries from $\mathbf{A}$ because it is an infinitesimal generator and not a rate matrix.

$$\mathbf{R}_2' = \begin{bmatrix} (-\mathbf{Be})_1\mathbf{I} \\ (-\mathbf{Be})_2\mathbf{I} \\ \cdots \\ (-\mathbf{Be})_{m_b}\mathbf{I} \end{bmatrix} \tag{3.91}$$

is a matrix of size $(m_a m_b) \times m_a$ and

$$
\mathbf{R}_2 =
\begin{bmatrix}
(-\mathbf{Be})_1 \beta_1 \mathbf{I} & (-\mathbf{Be})_1 \beta_2 \mathbf{I} & \cdots & (-\mathbf{Be})_1 \beta_{m_b} \mathbf{I} \\
(-\mathbf{Be})_2 \beta_1 \mathbf{I} & (-\mathbf{Be})_2 \beta_2 \mathbf{I} & \cdots & (-\mathbf{Be})_2 \beta_{m_b} \mathbf{I} \\
\cdots & \cdots & \cdots & \cdots \\
\cdots & \cdots & \cdots & \cdots \\
(-\mathbf{Be})_{m_b} \beta_1 \mathbf{I} & (-\mathbf{Be})_{m_b} \beta_2 \mathbf{I} & \cdots & (-\mathbf{Be})_{m_b} \beta_{m_b} \mathbf{I}
\end{bmatrix}
\tag{3.92}
$$

is a $(m_a m_b) \times (m_a m_b)$ matrix. Here I is an identity matrix of size $m_a \times m_a$.

Let $\mathbf{p} = \{p(s), s \in \mathcal{S}\}$ be the steady-state probability vector of the Markov process $\{(N(t), X(t), Y(t)), t \geq 0\}$, assuming its existence. Then $\mathbf{pQ} = \mathbf{0}$ where $\mathbf{Q}$ is the infinitesimal generator corresponding to $\mathbf{R}$ (i.e., $\mathbf{Q} = \mathbf{R} - (\mathbf{Re})_D$). Therefore, with the same partitioning of the matrix $\mathbf{Q}$ as that of $\mathbf{R}$ (see equation 3.86), we have

$$
\mathbf{p}(0)\mathbf{Q}'_1 + \mathbf{p}(1)\mathbf{Q}'_2 = \mathbf{0}
$$

$$
\mathbf{p}(0)\mathbf{Q}'_0 + \mathbf{p}(1)\mathbf{Q}_1 + \mathbf{p}(2)\mathbf{Q}_2 = \mathbf{0}
$$

$$
\mathbf{p}(n-1)\mathbf{Q}_0 + \mathbf{p}(n)\mathbf{Q}_1 + \mathbf{p}(n+1)\mathbf{Q}_2 = \mathbf{0}, \quad n = 2, 3, \ldots
\tag{3.93}
$$

where $\mathbf{p}(0) = (p(0, i, m_b+1), i = 1, \ldots, m_a)$ and $\mathbf{p}(n) = (p(n, i, j) \, i = 1, \ldots, m_a; j = 1, \ldots, m_b)$, $n = 1, 2, \ldots$ are probability vectors for the compound state corresponding to an empty system and n jobs in the system, respectively. Here $\mathbf{Q}'_0 = \mathbf{R}'_0$, $\mathbf{Q}_0 = \mathbf{R}_0$, $\mathbf{Q}'_1 = \mathbf{R}'_1 + (\mathbf{A})_D + (\mathbf{B})_D$, $\mathbf{Q}_1 = \mathbf{R}_1 + (\mathbf{A})_D + (\mathbf{B})_D$, $\mathbf{Q}'_2 = \mathbf{R}'_2$ and $\mathbf{Q}_2 = \mathbf{R}_2$.

Consider the set of equations obtained from equation 3.93 for $n = k, k+1, \ldots$. In the matrix form it is

$$
(\mathbf{p}(k), \mathbf{p}(k+1), \ldots)\mathbf{U} = (-\mathbf{p}(k-1)\mathbf{Q}_0, \mathbf{0}, \mathbf{0}, \ldots), \quad k = 2, 3, \ldots
\tag{3.94}
$$

where

$$
\mathbf{U} =
\begin{bmatrix}
\mathbf{Q}_1 & \mathbf{Q}_0 & \mathbf{0} & \mathbf{0} & \cdots \\
\mathbf{Q}_2 & \mathbf{Q}_1 & \mathbf{Q}_0 & \mathbf{0} & \cdots \\
\mathbf{0} & \mathbf{Q}_2 & \mathbf{Q}_1 & \mathbf{Q}_0 & \cdots \\
\cdots & \cdots & \cdots & \cdots & \cdots
\end{bmatrix}.
$$

Let $\mathbf{V}$ be the inverse of $\mathbf{U}$ such that $\mathbf{UV} = \mathbf{I}$. Then if we set $\mathbf{G} = -\mathbf{Q}_0 \mathbf{V}_{11}$, where $\mathbf{V}_{11}$ is the upper left block of size $(m_a m_b) \times (m_a m_b)$ of the matrix $\mathbf{V}$,

$$
\mathbf{p}(k) = \mathbf{p}(k-1)\mathbf{G}
\tag{3.95}
$$

Observe that equation 3.95 is true for any $k = 2, 3, \ldots$. Therefore

$$
\mathbf{p}(n) = \mathbf{p}(1)\mathbf{G}^{n-1}, \quad n = 1, 2, \ldots.
\tag{3.96}
$$

Substituting equation 3.96 in equation 3.93 suggests the following matrix quadratic equation for $\mathbf{G}$:

$$
\mathbf{Q}_0 + \mathbf{GQ}_1 + \mathbf{G}^2 \mathbf{Q}_2 = \mathbf{0}
\tag{3.97}
$$

Without any special structure for the $\mathbf{Q}$'s it is not possible to solve equation 3.97 explicitly for $\mathbf{G}$. Therefore we will use the following iterative relationship to obtain the solution. Let $\mathbf{G}^{(0)} = \mathbf{0}$, and define

$$\mathbf{G}^{(r)} = [\mathbf{Q}_0 + (\mathbf{G}^{(r-1)})^2 \mathbf{Q}_2](-\mathbf{Q}_1)^{-1}, \quad r = 1, 2, \dots . \tag{3.98}$$

We will next see that $\mathbf{G}^{(r)} \to \mathbf{G}$ as $r \to \infty$. Because $\mathbf{G}$ is non-negative (see Neuts [31] for a probabilistic interpretation of $\mathbf{G}$), $\mathbf{G}^{(0)} \le \mathbf{G}$. It will be established below that $\mathbf{G}^{(0)} \le \mathbf{G}^{(1)} \le \mathbf{G}^{(2)} \le \cdots \le \mathbf{G}$. To prove this by induction we assume that $\mathbf{G}^{(r-1)} \le \mathbf{G}^{(r)} \le \mathbf{G}$. Then from equation 3.98 we get

$$\mathbf{G}^{(r+1)} - \mathbf{G}^{(r)} = [(\mathbf{G}^{(r)})^2 - (\mathbf{G}^{(r-1)})^2]\mathbf{Q}_2(-\mathbf{Q}_1)^{-1} \ge 0 \tag{3.99}$$

because $\mathbf{Q}_2 \ge \mathbf{0}$, $(-\mathbf{Q}_1)^{-1} \ge \mathbf{0}$, and by the induction assumption $(\mathbf{G}^{(r)})^2 - (\mathbf{G}^{(r-1)})^2 \ge \mathbf{0}$. In addition

$$\mathbf{G} - \mathbf{G}^{(r+1)} = [\mathbf{G}^2 - (\mathbf{G}^{(r)})^2]\mathbf{Q}_2(-\mathbf{Q}_1)^{-1} \ge \mathbf{0} \tag{3.100}$$

Therefore $\mathbf{G}^{(r)} \le \mathbf{G}^{(r+1)} \le \mathbf{G}$ for all $r = 0, 1, \dots$. Observe that the limit of $\mathbf{G}^{(r)}$ as $r \to \infty$ satisfies the matrix quadratic equation 3.97. Because $\mathbf{G}$ is the minimal non-negative solution of equation 3.97 (see Neuts [31]), we see that $\mathbf{G}^{(r)} \to \mathbf{G}$ as $r \to \infty$. Substituting equation 3.96 into equation 3.93 along with the normalization equations we get

$$\mathbf{p}(0)\mathbf{Q}_1' + \mathbf{p}(1)\mathbf{Q}_2' = \mathbf{0}$$

$$\mathbf{p}(0)\mathbf{Q}_0' + \mathbf{p}(1)[\mathbf{Q}_1 + \mathbf{G}\mathbf{Q}_2] = \mathbf{0}$$

$$\mathbf{p}(0)\mathbf{e} + \mathbf{p}(1)[\mathbf{I} - \mathbf{G}]^{-1}\mathbf{e} = 1 \tag{3.101}$$

We can solve equation 3.101 to obtain $\mathbf{p}(0)$ and $\mathbf{p}(1)$ and use equation 3.95 to obtain the remaining probability vectors $\mathbf{p}(n)$, $n = 2, 3, \dots$. Detailed analysis of the $PH/PH/1$ queue is given in Neuts [31].

The mean number of jobs in the system is then given by

$$E[N] = \mathbf{p}(1) \sum_{n=1}^{\infty} n\mathbf{G}^{n-1}\mathbf{e} = \mathbf{p}(1)[\mathbf{I} - \mathbf{G}]^{-2}\mathbf{e} \tag{3.102}$$

Therefore from Little's formula the average waiting time of a customer is

$$E[W] = (-\alpha\mathbf{A}^{-1}\mathbf{e})(\mathbf{p}(1)[\mathbf{I} - \mathbf{G}]^{-2}\mathbf{e}) \tag{3.103}$$

3.3.4 *GI/G/1* Model

It may happen that none of the distributions described previously fit the interarrival and service times. Rather than present models for all possible combinations of interarrival and service time distributions, in this section we will present approximations and bounds for the performance measures of a single-stage job shop without making any specific assumptions about the interarrival and service time distributions. Consequently, for a general job shop we will use a $GI/G/1$ queueing model that does not make any specific assumption on the interarrival and service time distributions.

Suppose jobs are serviced on a FCFS basis. The time at which the service to job n can be initiated is the maximum of the $(n-1)$th departure time and the nth arrival time, that is, $\max\{D_{n-1}, A_n\}$. Because the service time of job n is S_n, we have

$$D_n = \max\{D_{n-1}, A_n\} + S_n, \qquad n = 1, 2, \ldots. \qquad (3.104)$$

Therefore the flow time $T_n = D_n - A_n$ of job n is given by

$$
\begin{aligned}
T_n &= [D_{n-1} - A_n]^+ + S_n \\
    &= [D_{n-1} - A_{n-1} - (A_n - A_{n-1})]^+ + S_n \\
    &= [T_{n-1} - \tau_n]^+ + S_n \qquad (3.105)
\end{aligned}
$$

where $\tau_n = A_n - A_{n-1}$ is the interarrival time between the $(n-1)$th and the nth job arrival times (see Figure 3.3). Note that if $D_{n-1} < A_n$ or equivalently $T_{n-1} < \tau_n$, the server will be idling during (D_{n-1}, A_n) for a duration $I_n = (A_n - D_{n-1}) = (\tau_n - T_{n-1})$. Otherwise the server will be processing job n immediately after the departure of job $n-1$. In this case we set $I_n = 0$. That is,

$$I_n = [\tau_n - T_{n-1}]^+, \qquad n = 1, 2, \ldots \qquad (3.106)$$

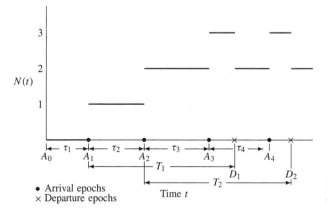

• Arrival epochs
× Departure epochs

Figure 3.3   Sample Path of Number of Jobs in System.

Subtracting equation 3.106 from 3.105 we get

$$T_n - I_n = T_{n-1} - \tau_n + S_n, \qquad n = 1, 2, \ldots \qquad (3.107)$$

We will assume $E[T_n] \to E[T] < \infty$ as $n \to \infty$. Then taking the expectations of both sides of equation 3.107 and taking the limit as $n \to \infty$ we get

$$E[I] = E[\tau] - E[S] \qquad (3.108)$$

For this to be proper we require $E[\tau] > E[S]$. We have earlier seen that for $\{T_n, n = 1, 2, \ldots\}$ to be uniformly finite, however, we need $\lambda < \mu$, or equivalently, we need $E[\tau] > E[S]$. From here on we will assume that this condition is satisfied. Now rewriting equation 3.107 as

$$T_n - S_n - I_n = T_{n-1} - \tau_n \tag{3.109}$$

and squaring both sides of equation 3.109 we get

$$T_n^2 - 2T_n S_n + S_n^2 - 2(T_n - S_n)I_n + I_n^2 = T_{n-1}^2 + \tau_n^2 - 2T_{n-1}\tau_n \tag{3.110}$$

Because $T_n - S_n$ and S_n are independent we have $E[(T_n - S_n)S_n] = E[T_n S_n - S_n^2] = (E[T_n] - E[S_n])E[S_n]$. Therefore $E[T_n S_n] = E[T_n]E[S_n] + \text{var}[S_n]$. Observe that $T_n - S_n > 0 \Leftrightarrow I_n = 0$ and $T_n - S_n = 0 \Leftrightarrow I_n > 0$. Therefore we have $(T_n - S_n)I_n = 0$. Taking expectations of both sides of equation 3.110 and using these observations we obtain

$$E[T_n^2] - 2E[T_n]E[S_n] - 2\text{var}[S] + E[S^2] + E[I_n^2] = E[T_{n-1}^2] + E[\tau_n^2] - 2E[T_{n-1}]E[\tau_n] \tag{3.111}$$

Suppose $E[T_n^2] \to E[T^2] < \infty$ as $n \to \infty$. Then taking the limit of equation 3.111 as $n \to \infty$ results in

$$
\begin{aligned}
E[T] &= \frac{E[\tau^2] + 2\text{var}(S) - E[S^2] - E[I^2]}{2(E[\tau] - E[S])} \\
&= \frac{E[\tau^2] - 2E[\tau]E[S] + E[S^2] - E[I^2]}{2(E[\tau] - E[S])} + E[S] \tag{3.112}
\end{aligned}
$$

The mean flow time therefore can be obtained as soon as $E[I^2]$ is determined. In the general context it is not possible to obtain $E[I^2]$. We will therefore obtain bounds and approximations for $E[I^2]$.

First, we will illustrate that for Poisson arrival process equation 3.112 will lead to the exact expression for the mean flow time. When the interarrival times are exponentially distributed $\{I_n \,|\, I_n > 0\}$, which is equal to $\{\tau_n - T_{n-1} \,|\, \tau_n - T_{n-1} > 0\}$, has an exponential distribution with mean $E[\tau]$. Therefore

$$
\begin{aligned}
E[I_n^k] &= E[I_n^k \,|\, I_n > 0]P\{I_n > 0\} + E[I_n^k \,|\, I_n = 0]P\{I_n = 0\} \\
&= (k!)(E[\tau])^k P\{I_n > 0\}, \quad k = 1, 2, \ldots \tag{3.113}
\end{aligned}
$$

Hence we have (using equation 3.113 for $k = 1$ and equation 3.108)

$$P\{I_n > 0\} = \frac{E[\tau] - E[S]}{E[\tau]} \tag{3.114}$$

and (using equation 3.113 for $k = 2$ and equation 3.114)

$$
\begin{aligned}
E[I^2] &= 2E[\tau]^2 \frac{(E[\tau] - E[S])}{E[\tau]} \\
&= 2E[\tau]^2 - 2E[\tau]E[S] \tag{3.115}
\end{aligned}
$$

Substituting equation 3.115 into equation 3.112 we obtain

$$E[T] = \frac{E[\tau^2] - 2E[\tau]E[S] + E[S^2] - (2E[\tau]^2 - 2E[\tau]E[S])}{2(E[\tau] - E[S])} + E[S]$$

$$= \frac{\lambda E[S^2]}{2(1 - \rho)} + E[S] \tag{3.116}$$

where $\lambda = 1/E[\tau]$ and $\rho = E[S]/E[\tau]$. Observe that equation 3.116 agrees with our previous results for the $M/G/1$ queue.

Next we obtain a bound on $E[I^2]$ without any assumption on the arrival process. Because for any random variable X, $E[X^2] \geq E[X]^2$ we have $E[I^2] \geq E[I]^2 = (E[\tau] - E[S])^2$. Therefore from equation 3.112 we get

$$E[T] \leq \frac{E[\tau^2] - 2E[\tau]E[S] + E[S^2] - (E[\tau] - E[S])^2}{2(E[\tau] - E[S])} + E[S]$$

$$= \frac{\text{var}(\tau) + \text{var}(S)}{2(E[\tau] - E[S])} + E[S] \tag{3.117}$$

Let $C_a^2 = \text{var}(\tau)/E[\tau]^2$ and $C_S^2 = \text{var}(S)/E[S]^2$ be the squared coefficient of variation of the interarrival and service times, respectively. Then equation 3.117 can be rewritten in the form

$$E[T] \leq \frac{C_a^2 + \rho^2 C_S^2}{2\lambda(1 - \rho)} + E[S] \tag{3.118}$$

Using the observation that $I_n = [\tau_n - T_{n-1}]^+$ we will now obtain a better lower bound on $E[I^2]$ than $E[I]^2$. Specifically we will obtain the lower bound $(C_a^2 + 1)E[I]^2$, which is larger than $E[I]^2$. To obtain this, first let R be a non–negative random variable with survival function $\overline{F}_R$, and consider

$$\frac{E[\{(R - x)^+\}^2]}{\{E[(R - x)^+]\}^2} = \frac{\int_x^\infty 2(y - x)\overline{F}_R(y)dy}{\left(\int_x^\infty \overline{F}_R(y)dy\right)^2}, \quad x \geq 0 \tag{3.119}$$

Taking the derivative of equation 3.119 with respect to x, one finds that

$$\frac{d}{dx}\left\{\frac{E[\{(R - x)^+\}^2]}{\{E[(R - x)^+]\}^2}\right\} =$$

$$\frac{2(\overline{F}_R(x))^2}{\left(\int_x^\infty \overline{F}_R(y)dy\right)^3}\left\{\int_x^\infty 2(y - x)\frac{\overline{F}_R(y)}{\overline{F}_R(x)}dy - \left(\int_x^\infty \frac{\overline{F}_R(y)}{\overline{F}_R(x)}dy\right)^2\right\} \geq 0, x \geq 0 \tag{3.120}$$

where the last inequality follows from the observation that the term within the double brackets is the variance of $\{R - x | R > x\}$. From equation 3.120 one sees that $E[\{(R - x)^+\}^2]/E[(R - x)^+]^2$ is increasing in x. Hence

$$E[\{(R - x)^+\}^2] \geq \frac{E[R^2]}{E[R]^2}E[(R - x)^+]^2, \quad x \geq 0 \tag{3.121}$$

Getting back to the bound for $E[I^2]$, conditioning on T_{n-1} one has from equation 3.121,

$$E[I_n^2] = E[E[\{(\tau_n - T_{n-1})^+ | T_{n-1}\}^2]]$$

$$\geq E\left[\frac{E[\tau^2]}{E[\tau]^2} E[(\tau_n - T_{n-1})^+ | T_{n-1}]^2\right]$$

$$\geq \frac{E[\tau^2]}{E[\tau]^2} E[I_n]^2 \qquad (3.122)$$

where the last inequality follows from $E[X^2] \geq E[X]^2$ when X is defined by $X = E[(\tau_n - T_{n-1})^+ | T_{n-1}]$. Now taking the limit of equation 3.122 as $n \to \infty$ one sees that $E[I^2] \geq (1 + C_a^2)E[I]^2$ and feeding this into equation 3.112 we obtain

$$E[T] \leq \frac{\rho(2 - \rho)C_a^2 + \rho^2 C_S^2}{2\lambda(1 - \rho)} + E[S] \qquad (3.123)$$

Observe that this bound is always better than the bound given in equation 3.118 because for all $\rho < 1$, we have $\rho(2 - \rho) < 1$.

Departure process.   Next we will look at the interdeparture times of jobs from the job shop. Recall that the departure time D_n of job n is given by $D_n = [D_{n-1} \vee A_n] + S_n$. Therefore the interdeparture time Δ_n between jobs $n - 1$ and n is given by

$$\Delta_n = D_n - D_{n-1} = [A_n - D_{n-1}]^+ + S_n = [\tau_n - T_{n-1}]^+ + S_n = I_n + S_n \qquad (3.124)$$

where the last equality follows from equation 3.106. Therefore taking expectations of both sides of equation 3.124 and the limit as $n \to \infty$ we get

$$E[\Delta] = E[\tau] \qquad (3.125)$$

This is not surprising because under the condition $E[\tau] > E[S]$ we have already seen that the departure rate $\beta(= 1/E[\Delta])$ will be the same as the arrival rate $\lambda(= 1/E[\tau])$.

Because I_n and S_n are independent, from equation 3.124 one has

$$\text{var}(\Delta) = \text{var}(I) + \text{var}(S) \qquad (3.126)$$

and consequently the squared coefficient of variation C_d^2 of the departure process is

$$C_d^2 = \frac{E[I^2]}{E[\tau]^2} - (1 - \rho)^2 + \rho^2 C_S^2 \qquad (3.127)$$

We can now relate C_d^2 to $E[T]$ through equation 3.112.

$$C_d^2 = C_a^2 + 2\rho^2 C_S^2 + 2\rho(1 - \rho) - E[T]2\lambda(1 - \rho) \qquad (3.128)$$

When the arrival process is Poisson $E[I^2] = 2E[\tau]^2 - 2E[\tau]E[S]$. Therefore for $M/G/1$ models one obtains from equation 3.127,

$$C_d^2 = 1 - \rho^2 + \rho^2 C_S^2 \qquad (3.129)$$

To obtain bounds or approximations for C_d^2 we may use equation 3.127, and the bounds or approximations for $E[I^2]$. Alternatively, we may use equation 3.128, and the bounds or approximations for $E[T]$. First consider the bound $E[I^2] \geq E[I]^2 = E[\tau]^2(1 - \rho)^2$. From equation 3.127 we obtain

$$C_d^2 \geq \rho^2 C_S^2 \tag{3.130}$$

Using the bound $E[I^2] \geq (E[\tau^2]/E[\tau]^2)E[I]^2 = E[\tau]^2(C_a^2 + 1)(1 - \rho)^2$ (see equation 3.122) one gets

$$C_d^2 \geq (1 - \rho)^2 C_a^2 + \rho^2 C_S^2. \tag{3.131}$$

It is often possible to identify certain characteristics of the interarrival or service time distributions. For example, it is often justified to assume that the remaining mean service time of a job decreases as the amount of attained service increases. This property "$E[S - x \mid S > x]$ decreasing in x" is called the decreasing mean residual life (DMRL) property. Nonparametric properties of this type for different distributions are summarized in Appendix C. Using such insights on the interarrival and service times one can improve the bounds on $E[T]$ developed so far. To develop these bounds we will use the following property (given by equation 3.134) of a non-negative random variable R with decreasing mean residual life. Consider

$$\frac{E[\{(R - x)^+\}^2]}{E[(R - x)^+]} = \frac{\int_x^\infty 2(y - x)\overline{F}_R(y)dy}{\int_x^\infty \overline{F}_R(y)dy}, \quad x \geq 0 \tag{3.132}$$

Taking the derivative of equation 3.132 with respect to x one obtains

$$\frac{d}{dx}\left\{\frac{E[\{(R - x)^+\}^2]}{E[(R - x)^+]}\right\} = \frac{\overline{F}_R(x)\int_x^\infty 2(y - x)\overline{F}_R(y)dy - 2\left(\int_x^\infty \overline{F}_R(y)dy\right)^2}{\left(\int_x^\infty \overline{F}_R(y)dy\right)^2} \tag{3.133}$$

Now consider the first term in the numerator of equation 3.133:

$$\overline{F}_R(x)\int_x^\infty 2(y - x)\overline{F}_R(y)dy = \overline{F}_R(x)\int_x^\infty \left(\frac{1}{\overline{F}_R(z)}\int_z^\infty \overline{F}_R(y)dy\right)\overline{F}_R(z)dz$$

$$\leq \left(\int_x^\infty \overline{F}_R(y)dy\right)\left(\int_x^\infty \overline{F}_R(z)dz\right)$$

where the last inequality follows because "R is DMRL" is equivalent to "$(1/\overline{F}_R(z)) \times \int_z^\infty \overline{F}_R(y)dy$, is decreasing in z." Then from equation 3.133 one has

$$\frac{d}{dx}\left\{\frac{E[\{(R - x)^+\}^2]}{E[(R - x)^+]}\right\} \leq 0$$

and therefore

$$\frac{E[\{(R - x)^+\}^2]}{E[(R - x)^+]} \leq \frac{E[R^2]}{E[R]}, \quad x \geq 0 \tag{3.134}$$

Similarly if R has the increasing mean residual life (IMRL) property

$$\frac{E[\{(R-x)^+\}^2]}{E[(R-x)^+]} \geq \frac{E[R^2]}{E[R]}, \qquad x \geq 0 \tag{3.135}$$

Now suppose τ has the DMRL property. Then from equations 3.106 and 3.134 we get

$$
\begin{aligned}
E[I_n^2] &= E[E[\{(\tau_n - T_{n-1})^+\}^2 | T_{n-1}]] \\
&\leq E\left[\frac{E[\tau^2]}{E[\tau]} E[(\tau_n - T_{n-1})^+ | T_{n-1}]\right] \\
&= \frac{E[\tau^2]}{E[\tau]} E[I_n] \tag{3.136}
\end{aligned}
$$

Substituting the limit of equation 3.136 as $n \to \infty$ into equation 3.112 we get

$$E[T] \geq \frac{\rho(C_a^2 - 1 + \rho) + \rho^2 C_S^2}{2\lambda(1-\rho)} + E[S] \tag{3.137}$$

Observe that the difference between the upper bound from equation 3.123 and the lower bound from equation 3.137 is $(1 + C_a^2)E[S]/2$. Because a random variable with decreasing mean residual life has a squared coefficient of variation less than one, $(1 + C_a^2)E[S]/2 \leq E[S]$. Therefore if we approximate the mean flow time by the average of these two bounds the absolute value of the error in the approximation will be less than or equal to $E[S]/2$ for any $GI/G/1$ queue with DMRL interarrival time.

The upper bound on C_d^2 corresponding to equation 3.137 is

$$C_d^2 \leq C_a^2(1-\rho) + \rho^2 C_S^2 + \rho(1-\rho) \tag{3.138}$$

Here the difference between the upper and lower bounds is $\rho(1-\rho)(1 + C_a^2)$, which is less than or equal to 1/2. When the server utilization is either very small or large, the difference between the upper and lower bounds will be very small.

When the interarrival time has the IMRL property analogously one obtains

$$E[T] \leq \frac{\rho(C_a^2 - 1 + \rho) + \rho^2 C_S^2}{2\lambda(1-\rho)} + E[S] \tag{3.139}$$

and

$$C_d^2 \geq C_a^2(1-\rho) + \rho^2 C_S^2 + \rho(1-\rho) \tag{3.140}$$

For this case the above upper bound for $E[T]$ is smaller than the one given by equation 3.123 and the above lower bound for C_d^2 is larger than that given by equation 3.131. It can be shown (using the results for $M^X/G/1$ queue given in Section 4.3) that the mean time spent in a $GE/G/1$ queueing system with generalized exponential interarrival time is given by the right-hand side of equation 3.139. Generalized exponential distribution is a mixture of an exponential distribution with a degenerate distribution concentrated on zero (see Appendix A), that is,

$$F_\tau(x) = \frac{2}{1 + C_a^2} e^{-(2/(1+C_a^2))\lambda x}, \qquad x \geq 0$$

so that $E[\tau] = 1/\lambda$ and $C_\tau^2 = C_a^2$. Because generalized exponential has the IMRL property we see that the bounds given in equations 3.139 and 3.140 are tight.

Little's formula can be now used to obtain bounds for the average number of jobs in the $GI/G/1$ queueing system with FCFS service protocols. For example, from equations 3.123 and 3.137 it can be seen that if the interarrival time has a DMRL distribution, then

$$\frac{\rho C_a^2 - \rho(1 - \rho) + \rho^2 C_S^2}{2(1 - \rho)} + \rho \leq E[N] \leq \frac{\rho(2 - \rho)C_a^2 + \rho^2 C_S^2}{2(1 - \rho)} + \rho \qquad (3.141)$$

It is useful to note that the difference between the upper and lower bounds is $\rho(C_a^2 + 1)/2$, which is in this case less than or equal to ρ. Hence either one of these bounds or a convex combination of these two can be used as an approximation for $E[N]$. Because the service times are iid any service protocol that does not use any information on the service times will not affect the probabilistic evolution of the number in the system. Hence these bounds are also applicable to the average number in the system under those service protocols that do not use the information on service times (e.g., LCFS, RANDOM). Then by Little's results, it is also clear that the average flow time $E[T]$ is unaffected by such protocols and hence the bounds developed for the FCFS service protocols are applicable to these cases as well.

Approximations.   We will next use the previous bounds and some heuristic reasoning to develop approximations for the mean flow time and the squared coefficient of variation of the interdeparture times for a $GI/G/1$ model. Observe that the upper bounds given in equations 3.118 and 3.123 are not tight even for the $M/G/1$ model. Therefore if we wish to use these upper bounds to approximate the mean flow time we should at least multiply them by a factor such that they will be exact for the $M/G/1$ model. The factor $\rho^2(1 + C_S^2)/(1 + \rho^2 C_S^2)$ will do this for equation 3.118, and the factor $\rho(1 + C_S^2)/(2 - \rho + \rho C_S^2)$ will do this for equation 3.123. The two approximations for $E[T]$ are then $\hat{T}_1$ and $\hat{T}_2$ where

$$\hat{T}_1 = \left\{ \frac{\rho^2(1 + C_S^2)}{1 + \rho^2 C_S^2} \right\} \left\{ \frac{(C_a^2 + \rho^2 C_S^2)}{2\lambda(1 - \rho)} \right\} + E[S] \qquad (3.142)$$

$$\hat{T}_2 = \left\{ \frac{\rho(1 + C_S^2)}{2 - \rho + \rho C_S^2} \right\} \left\{ \frac{\rho(2 - \rho)C_a^2 + \rho^2 C_S^2}{2\lambda(1 - \rho)} \right\} + E[S] \qquad (3.143)$$

Extensive empirical testing has shown that these approximations work very well when $C_a^2 \leq 2$. When the squared coefficient of variation of the interarrival time is very large the previous approximations can be very poor. This is to be expected because the range of the exact mean flow time for different distributions of interarrival times with a fixed mean and large coefficient of variation can be very large. Indeed, even with the restriction of IMRL interarrival distributions (which have $C_a^2 \geq 1$), the range of the mean time spent in a $GI/G/1$ queueing system is $([\rho^2(1 + C_S^2)]/[2\lambda(1 - \rho)] + E[S]$, $[\rho(C_a^2 - 1 + \rho) + \rho^2 C_S^2]/[2\lambda(1 - \rho)] + E[S])$. The lower limit is the mean flow time in an $M/G/1$ queueing system (see equation 3.116), and the upper limit is the mean flow time in a $GE/G/1$ queueing system (see equation 3.139). Therefore it can be seen that with the first two moments of the interarrival and service specified, the mean flow time can vary by as much as $[\rho(C_a^2 - 1)]/[2\lambda(1 - \rho)]$ with a maximum percentage error of $[C_a^2 - 1]/[\rho(C_S^2 - 1)]$ in the mean waiting time. For example when $\lambda = .8$, $C_a^2 = 16$, $\mu = 1$, $C_S^2 = 1$, (i.e., $\rho = .8$), the error (in absolute value) in any approximation that uses only the first two moments will be *at least* 18.75 in comparison to a maximum (minimum) flow time of 42.5 (5). Therefore it is clear that attempting to approximate the performance of a queueing system with an arrival process that is highly variable (i.e., large value of C_a^2) with only the first two moments is futile.

We have seen before that the upper and lower bounds for DMRL interarrival times are tight. Here $C_a^2 \leq 1$. Therefore for $C_a^2 \leq 1$ we propose the approximation that averages the upper and lower bounds by factors $1 - C_a^2$ and C_a^2, respectively. The corresponding approximation is

$$\hat{T}_3 = \frac{\rho C_a^2(1 - (1 - \rho)C_a^2) + \rho^2 C_S^2}{2\lambda(1 - \rho)} + E[S] \tag{3.144}$$

Observe that the approximation in equation 3.144 is exact for both $M/G/1$ and $D/D/1$ models. It is a good approximation for $GI/G/1$ models with $C_a^2 \leq 1$. For systems with $C_a^2 \geq 1$, the preceding approximation should not be used because it leads to poor performance.

Corresponding approximations for $E[N]$ and C_d^2 can be obtained using Little's formula and equation 3.128. For easy reference we list these approximations in Table 3.1. Numerical results of these approximations with the exact results for the $E_k/E_l/1$ queueing

TABLE 3.1 APPROXIMATIONS FOR GI/G/1 QUEUEING SYSTEMS

Approx.	Mean number in system $E[N]$	Squared coefficient of variation of interdeparture times C_d^2
1	$\left\{\dfrac{\rho^2(1+C_S^2)}{1+\rho^2 C_S^2}\right\}\left\{\dfrac{C_a^2+\rho^2 C_S^2}{2(1-\rho)}\right\} + \rho$	$(1-\rho^2)\left\{\dfrac{C_a^2+\rho^2 C_S^2}{1+\rho^2 C_S^2}\right\} + \rho^2 C_S^2$
2	$\left\{\dfrac{\rho(1+C_S^2)}{2-\rho+\rho C_S^2}\right\}\left\{\dfrac{\rho(2-\rho)C_a^2+\rho^2 C_S^2}{2(1-\rho)}\right\} + \rho$	$1-\rho^2 + \rho^2 C_S^2 + (C_a^2 - 1)\left\{\dfrac{(1-\rho^2)(2-\rho)+\rho C_S^2(1-\rho)^2}{2-\rho+\rho C_S^2}\right\}$
3	$\dfrac{\rho^2(C_a^2+C_S^2)}{2(1-\rho)} + \dfrac{(1-C_a^2)C_a^2\rho}{2} + \rho$	$(1-\rho)(1+\rho C_a^2)C_a^2 + \rho^2 C_S^2$

system for various values of C_a^2, C_S^2 and ρ are given in Tables 3.2 and 3.3. From these results we see that approximation 3 overestimates the exact results, whereas for the most part approximations 1 and 2 underestimate the exact results. All three approximations are, however, very good and as the server utilization gets very close to 1 (say for $\rho = .99$), they lead almost to the same value.

TABLE 3.2 APPROXIMATE AND
EXACT MEAN NUMBER OF JOBS IN
A *GI/G/*1 QUEUEING SYSTEM FOR
SERVER UTILIZATION $\rho = .9$

		C_a^2		
C_S^2	Method	.25	.33	.5
.25	App. 1	2.805	3.156	3.858
	App. 2	2.810	3.161	3.861
	App. 3	3.010	3.363	4.050
	Exact	2.768	3.119	3.825
.33	App. 1	3.111	3.465	4.174
	App. 2	3.118	3.471	4.179
	App. 3	3.347	3.700	4.388
	Exact	3.095	3.448	4.157
.5	App. 1	3.732	4.092	4.813
	App. 2	3.742	4.101	4.819
	App. 3	4.022	4.375	5.063
	Exact	3.753	4.109	4.823
1	App. 1	5.644	6.017	6.762
	App. 2	5.659	6.030	6.773
	App. 3	6.047	6.400	7.088
	Exact	5.745	6.106	6.830

We will now consider approximating the steady-state probability distribution ($p(n)$, $n = 0, 1, 2, \ldots$) of the number of jobs in the system. Because the time-average probability that the server is idle is $1 - \rho$, we have $p(0) = 1 - \rho$. Now assuming that the remaining probabilities have a geometric form (i.e., $p(n) = a\sigma^{n-1}, n = 1, 2, \ldots$) and using the requirement that the total probability should be one, we have $a = \rho(1 - \sigma)$. This approximated distribution has a mean of $\rho/(1 - \sigma)$. Suppose we choose to approximate the mean number of jobs in the system by $\hat{N}$ (where $\hat{N}$ could be one of the three approximations specified in Table 3.1). Then requiring that this approximation be the same as the mean of the approximate distribution of the number of jobs in the system we have

$$p_n \approx \begin{cases} 1 - \rho, & n = 0 \\ \rho(1 - \sigma)\sigma^{n-1}, & n = 1, 2, \ldots \end{cases} \tag{3.145}$$

where $\sigma = (\hat{N} - \rho)/\hat{N}$.

TABLE 3.3 APPROXIMATE AND EXACT
MEAN NUMBER OF JOBS IN A *GI/G*/1
QUEUEING SYSTEM FOR SERVER
UTILIZATION $\rho = .95$

C_S^2	Method	C_a^2		
		.25	.33	.5
.25	App. 1	5.328	6.095	7.629
	App. 2	5.331	6.098	7.631
	App. 3	5.551	6.320	7.838
	Exact	5.292	6.059	7.597
.33	App. 1	6.045	6.816	8.358
	App. 2	6.049	6.820	8.361
	App. 3	6.304	7.072	8.590
	Exact	6.033	6.802	8.344
.5	App. 1	7.491	8.269	9.823
	App. 2	7.497	8.274	9.827
	App. 3	7.808	8.576	10.094
	Exact	7.520	8.292	9.839
1	App. 1	11.884	12.675	14.256
	App. 2	11.893	12.683	14.262
	App. 3	12.320	13.089	14.606
	Exact	11.997	12.775	14.331

3.4 MULTIPLE SERVERS

In this section we will study single-stage dynamic job shops that have more than one server. Let c be the number of servers in the system. We will assume that all these c servers are functionally identical. As before, depending on the fit between the interarrival, service times and standard probability distributions we will consider different models of the system.

3.4.1 *M/M/c* model

Here we consider the model for a job shop with interarrival and service times that can be represented by exponential distributions. For the $M/M/c$ model the stationary distribution of the number of jobs in the system can be obtained in the same manner as we did for the $M/M/1$ model. Writing the level crossing rate balance equations for the compound states $\{0, 1, \ldots, n\}$ of the process $\{N(t), t \geq 0\}$ we get

$$\lambda p(n) = r(n)\mu p(n+1), \qquad n = 0, 1, \ldots \tag{3.146}$$

where $r(n)\mu = \min\{n, c\}\mu$ is the rate at which service completions occur when there are n jobs in the system. Solving equation 3.146 with the normalizing condition $\sum_{n=0}^{\infty} p(n) = 1$, one obtains

$$p(n) = \left(\frac{\lambda}{\mu}\right)^n \prod_{i=1}^{n} \frac{1}{r(i)} \bigg/ \left\{\sum_{k=0}^{\infty} \left(\frac{\lambda}{\mu}\right)^k \prod_{i=1}^{k} \frac{1}{r(i)}\right\}, \qquad n = 0, 1, \ldots \qquad (3.147)$$

Equivalently if we define $f(n) = f(n-1)\lambda/(r(n)\mu)$, $n = 1, 2\ldots$, with $f(0) = 1$, we find that

$$p(n) = f(n)p(0) \qquad (3.148)$$

Therefore from equation 3.148 one finds that

$$p(0) = 1 \bigg/ \sum_{n=0}^{\infty} f(n) \qquad (3.149)$$

The condition needed for the existence of the stationary distribution is $\sum_{n=0}^{\infty} f(n) < \infty$, which in this case is equivalent to $\lambda < c\mu$. It can be routinely checked to see that the result in equation 3.147 also holds for an $M/M(n)/1$ model whose service rate is $r(n)\mu$ whenever there are n jobs in the system. Simplifying equation 3.147 for the $M/M/c$ model using the substitution $r(n) = \min\{n, c\}$ we obtain

$$p(n) = \begin{cases} \frac{(c\rho)^n}{n!} p(0) & n = 0, 1, \ldots, c-1 \\ \frac{c^c \rho^n}{c!} p(0) & n = c, c+1, \ldots \end{cases} \qquad (3.150)$$

where $p(0) = 1/\{\sum_{k=0}^{c-1}(c\rho)^k/k! + (c\rho)^c/[(1-\rho)c!]\}$ and $\rho = \lambda/c\mu$. The average number of jobs in the system is then

$$E[N] = \sum_{n=0}^{\infty} np(n)$$

$$= \sum_{n=1}^{c-1} \frac{(c\rho)^n}{(n-1)!} p(0) + \frac{c^c}{c!} \sum_{n=c}^{\infty} n\rho^n p(0)$$

$$= c\rho + \left\{\frac{(c\rho)^c}{c!} \left(\frac{\rho}{(1-\rho)^2}\right)\right\} p(0) \qquad (3.151)$$

From equation 3.150 it can be verified that $P\{N = n | N \geq c\} = (1-\rho)\rho^{n-c}$, $n = c, c+1, \ldots$. This is to be expected because the total service rate whenever all servers are busy is a constant $c\mu$. Therefore whenever all servers are busy the number of jobs in the $M/M/c$ system behaves like the number of jobs in a $M/M/1$ system with service rate $c\mu$. The average number of jobs in the system given $N \geq c$ is then $c + \rho/(1-\rho)$. The average number of jobs in the queue given $N \geq c$ is $\rho/(1-\rho)$. Because $P\{N \geq c\} = (c^c \rho^c p(0))/[c!(1-\rho)]$, the average number of jobs in the queue is

$$E[L] = \frac{(c\rho)^c}{c!} \left(\frac{\rho}{(1-\rho)^2}\right) p(0) \qquad (3.152)$$

Because the average number of busy servers is $c\rho$ the number of jobs in the system should be $E[L] + c\rho$, which agrees with equation 3.151. The arrival process is Poisson, therefore by the PASTA property the number of customers seen by an arrival has the same distribution as N.

Waiting time.   We will next derive the distribution function of the waiting time W of an arbitrary job arriving at the system under steady state. If this tagged job on its arrival sees n jobs in the system, then it has to wait for the service completion of $k = (n + 1 - c)^+$ jobs. Because jobs will be serviced at a total rate of $c\mu$ when there are at least c jobs in the system, the time taken for the tagged job to begin service will have an Erlang-k distribution with mean $k/c\mu$. That is, if N^a is the number of jobs seen by the tagged job

$$P\{W > x | N^a = n\} = \begin{cases} 0, & n = 0, 1, \ldots, c - 1 \\ \sum_{l=0}^{n-c} \frac{e^{-c\mu x}(c\mu x)^l}{l!}, & n = c, c + 1, \ldots. \end{cases} \quad (3.153)$$

Because N^a has the same distribution as N one has from equations 3.150 and 3.153,

$$P\{W > x\} = \sum_{n=c}^{\infty} P\{W > x | N^a = n\}(1 - \rho)\rho^{n-c} P\{N \geq c\}$$

$$= e^{-(c\mu - \lambda)x} P\{N \geq c\}$$

$$= e^{-(c\mu - \lambda)x} \left\{ \frac{c^c \rho^c}{c!(1 - \rho)} \right\} p(0), \quad x > 0 \quad (3.154)$$

Note that $P\{W = 0\} = 1 - \lim_{x \to 0} P\{W > x\} = 1 - (c^c \rho^c / [c!(1 - \rho)]) p(0) = P\{N < c\}$ and

$$E[W] = \left(\frac{1}{c\mu - \lambda} \right) \left(\frac{c^c \rho^c}{c!(1 - \rho)} \right) p(0) \quad (3.155)$$

The average flow time is then

$$E[T] = \left(\frac{1}{c\mu - \lambda} \right) \left(\frac{c^c \rho^c}{c!(1 - \rho)} \right) p(0) + \frac{1}{\mu} \quad (3.156)$$

It can be verified that the stationary departure process from the $M/M/c$ queueing system is a Poisson process with rate λ.

3.4.2 *GI/G/c* Model

As pointed out earlier, the arrival processes to job shops can often be represented by Poisson processes. Therefore a $M/G/c$ model for a job shop would be valuable because we do not need to make any specific assumptions about the service time distributions. Unfortunately unlike the $M/G/1$ model, the solution to the $M/G/c$ model cannot be explicitly obtained. Therefore we will develop approximations for the performance measures of the $GI/G/c$ model. Let $E[W]_{GI/G/c}$, $c = 1, 2, \ldots$, be the average waiting time of an arbitrary job in a $GI/G/c$ queue with fixed interarrival time distributions and with service time distributions at each server the same as that of cS. If S has a distribution with mean $1/\mu$ and variance C_S^2/μ^2 then cS has mean c/μ and variance $c^2 C_S^2/\mu^2$, that is, the squared coefficient of variation is unchanged through this scaling. Also the server utilization ρ is unchanged. Then the ratio $E[W]_{GI/G/c}/E[W]_{GI/G/1}$ is the ratio by which the mean waiting time changes by going from one server to c servers. We assume that this ratio is fairly insensitive to the interarrival and service time distributions. Then we may approximate $E[W]_{GI/G/c}$ as follows

$$E[W]_{GI/G/c} \approx \frac{E[W]_{M/M/c}}{E[W]_{M/M/1}} E[W]_{GI/G/1}, \quad c = 2, 3, \ldots \tag{3.157}$$

Then

$$E[T]_{GI/G/c} \approx \frac{E[W]_{M/M/c}}{E[W]_{M/M/1}} E[W]_{GI/G/1} + E[S] \tag{3.158}$$

Hence from Little's formula one obtains

$$E[L]_{GI/G/c} \approx \frac{E[L]_{M/M/c}}{E[L]_{M/M/1}} E[L]_{GI/G/1} \tag{3.159}$$

and

$$E[N]_{GI/G/c} \approx \frac{E[L]_{M/M/c}}{E[L]_{M/M/1}} E[L]_{GI/G/1} + c\rho \tag{3.160}$$

Now any of the approximations developed earlier for $E[W]_{GI/G/1}$ or $E[L]_{GI/G/1}$ can be used in equations 3.157 to 3.160 to obtain corresponding approximations for the $GI/G/c$ queueing system. For example if we use $\hat{T}_3$ for $C_a^2 \leq 1$, we get the following approximation for $E[T]_{GI/G/c}$:

$$\frac{C_a^2(1 - (1 - \rho)C_a^2)/\rho + C_S^2}{2} E[W]_{M/M/c} + E[S] \tag{3.161}$$

where $E[W]_{M/M/c}$ is given by equation 3.155.

3.5 PRIORITY QUEUES

So far we have modeled the arrival process by a single class of job streams. In a real job shop, however, the number of different types of jobs that need processing can be very large. In most cases, it is sufficient to model the arrival process with multiple job types by an aggregate single job class. Details of an aggregation procedure will be given in Chapter 7. If the issue we are trying to focus on involves a service process that distinguishes certain job types, however, then we must model the arrival process by a multiple job class stream. The job classes we use in our model may be created either based on the processing requirement or based on a priority assigned by the manager.

In this section we will develop and analyze such models. First we will look at a basic conservation identity for multiple-class single-stage job shops.

3.5.1 Conservation Identity

Suppose k classes of jobs arrive at a single-stage job shop with c servers. The nth class l job arrives at time $A_n^{(l)}$. That is $\{A_n^{(l)}, n = 1, 2, \ldots\}$ is the sequence of arrival epochs for class l jobs $(l = 1, \ldots, k)$. The processing time of the nth class l job is $S_n^{(l)}$, $n = 1, 2, \ldots; l = 1, \ldots, k$. The arrival processes and the service times for each class of jobs are mutually independent. The service times for each class form an iid sequence.

Let $\lambda^{(l)} = \lim_{t \to \infty} A^{(l)}(t)/t$ be the average arrival rate of class l jobs, where $A^{(l)}(t) = \sup\{n : A_n^{(l)}(t), n = 1, 2, \ldots\}$, $A_0^{(l)} = 0$ is the number of type l jobs arrived up to time t.

Let $V_S^j(t)$ be the remaining service requirement of the job in service at server j at time t and $V_W(t)$ be the total service requirement of all jobs waiting in the queue at time t. Then the total work load in the system at time t is $V(t) = \sum_{j=1}^c V_S^j(t) + V_W(t)$. Observe that when jobs are processed according to a nonpreemptive service protocol

$$E[V_W(t)] = \sum_{l=1}^k E[L^{(l)}(t)]E[S^{(l)}] \tag{3.162}$$

where $L^{(l)}(t)$ is the number of type l jobs in the queue at time t. A sample path of the process $V_S^j(t)$ is shown in Figure 3.4. Each time server j initiates service on a job, say the nth job in class l, the process $\{V_S^j(t), t \geq 0\}$ jumps by an amount $S_n^{(l)}$. This process will then decrease at a rate of one as long as the server is busy and will remain at zero when the server is idle. The area under this process contributed by the jump of magnitude $S_n^{(l)}$ is $(1/2)(S_n^{(l)})^2$ except for the last jump. If a type l job is in service at time t, the contribution is $(1/2)(S_n^{(l)})^2 - (1/2)(V_S^{(l)}(t))^2$ (see Figure 3.4). Therefore if $\hat{D}(t)$ is the number of jobs service initiated during $(0, t]$, then

$$\sum_{j=1}^c \int_0^t V_S^j(x)dx = \sum_{n=1}^{\hat{D}(t)} \frac{1}{2}(\hat{S}_n)^2 - \sum_{j=1}^c \frac{1}{2}(V_S^j(t))^2, \quad t \geq 0 \tag{3.163}$$

where $\hat{S}_n$ is the service time of the nth serviced job. Now dividing both sides of equation 3.163 by t and taking the limit as $t \to \infty$ we get

$$\sum_{j=1}^c E[V_S^j] = \sum_{l=1}^k \frac{1}{2}\beta^{(l)}E[(S^{(l)})^2] \tag{3.164}$$

where $\beta^{(l)}$ is the average departure rate of class l jobs. When $\sum_{l=1}^k \lambda^{(l)}E[S^{(l)}] < c$, the arrival and departure rates will be the same. That is $\beta^{(l)} = \lambda^{(l)}, l = 1, \ldots, k$. Then from equations 3.162 and 3.164 one obtains

$$E[V] = \sum_{l=1}^k E[L^{(l)}]E[S^{(l)}] + \sum_{l=1}^k \frac{1}{2}\lambda^{(l)}E[(S^{(l)})^2] \tag{3.165}$$

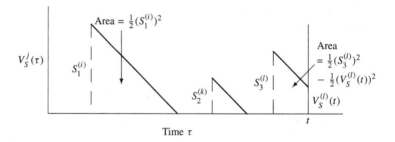

Figure 3.4 Workload of Jobs at Server j

provided $\sum_{l=1}^{k} \lambda^{(l)} E[S^{(l)}] < c$. When there is only a single server available $V(t)$ will be the same no matter what the service protocol is, as long as it is work conserving. Hence for a single server job shop $\sum_{l=1}^{k} E[L^{(l)}] E[S^{(l)}]$ is constant, independent of the work conserving service protocol in use. Because $E[T^{(l)}] = (1/\lambda^{(l)}) E[L^{(l)}] + E[S^{(l)}]$ by Little's formula, one also sees that

$$\sum_{l=1}^{k} \rho^{(l)} E[T^{(l)}] = \text{constant} \qquad (3.166)$$

Here $\rho^{(l)} = \lambda^{(l)} E[S^{(l)}]$.

This *conservation identity* for multiple class $GI/G/1$ models shows that any service protocol that improves the performance (i.e., average flow time) of a particular job class may do so only at the expense of worsening the performance of some other job class.

We will next illustrate how the previous conservation identity can be used to assign priorities in a two-class single-stage job shop. Let w_l be the cost per unit time of being in the system for a class l job, $l = 1, 2$. Let $T^{(l:j)}$ be the mean flow time of a class l job when class j jobs are assigned the highest priority, $l = 1, 2$; $j = 1, 2$. The corresponding mean cost per unit time is $TC_j = \sum_{l=1}^{2} \lambda_l w_l E[T^{(l:j)}] = \sum_{l=1}^{2} w_l \mu_l \rho^{(l)} E[T^{(l:j)}]$, $j = 1, 2$. Then from equation 3.166 it follows that

$$TC_1 - TC_2 = (w_1 \mu_1 - w_2 \mu_2) \rho^{(1)} (E[T^{(1:1)}] - E[T^{(1:2)}])$$

Therefore if $w_1 \mu_1 > w_2 \mu_2$ assigning the higher priority to class 1 jobs will reduce the mean cost per unit time. Hence the optimal priority assignment that minimizes the mean cost rate is such that higher priority is given to the class of jobs that has the higher $w\mu$ value.

3.5.2 Multiple Class *M/G/*1 Model with Nonpreemptive Priority (NPP)

We will focus our attention on job shops where the arrival processes $\{A_n^{(l)}, n = 1, 2, \ldots\}$ $l = 1, \ldots, k$ are mutually independent Poisson processes with rates $\lambda^{(l)}$, $l = 1, \ldots, k$.

The service protocol in this job shop is such that service to a class l ($l = 1, \ldots, k$) job is initiated only if there are no class r jobs, $r = 1, 2, \ldots, l - 1$ in the system. Within each class jobs are serviced according to the FCFS protocol. Tag a class l ($l = 1, 2, \ldots, k$) job in the system at time zero. The time at which this job enters service will depend on the initial work load in the system. This initial delay (say B_0) will consist of the remaining service time of any job in service, the service time of all higher priority jobs (i.e., class $1, 2, \ldots, l - 1$) and the service time of all class l jobs that arrived before our tagged job. We will call the time taken to initiate the service of the tagged job the l-delayed busy period with initial delay B_0. Let $B^{(l)}$ be this delayed busy period. Observe that the length of this delayed busy period will be unaffected by the service protocol used among the initial delay and the higher-priority jobs as long as higher priority is given to these jobs over our tagged job. Suppose the initial delay is $x + y$ and let $\tilde{F}_{B^{(l)}(x+y)}(s)$ be the LST of the l-delayed busy period, say $B^{(l)}(x + y)$, and let $b^{(l)}(x + y)$ be its mean. Now, consider serving the first x units of work followed by the higher-priority jobs until the system becomes empty of any higher-priority jobs. The distribution of this period is then the same as that of an l-delayed busy period with an initial delay x. Now service the remaining y units of the initial delay, followed by the higher-priority jobs. Because the arrival process is Poisson the time needed for this second portion of service will have the same distribution as that of a l-delayed busy period with an initial delay y. Poisson arrival process guarantees that these two delayed busy periods are independent. Therefore

$$b^{(l)}(x + y) = b^{(l)}(x) + b^{(l)}(y), \quad x, y \geq 0 \tag{3.167}$$

and

$$\tilde{F}_{B^{(l)}(x+y)}(s) = \tilde{F}_{B^{(l)}(x)}(s)\tilde{F}_{B^{(l)}(y)}(s), \quad x, y \geq 0 \tag{3.168}$$

The only solution to equation 3.167 is $b^{(l)}(x) = rx$ for some constant r. The expected work load of higher-priority jobs in the system at the end of service of the initial delay of x units is $\lambda_h E[S_h]x$ owing to Poisson arrivals of higher-priority jobs, say with rate λ_h, and owing to each higher-priority job bringing an average work load of, say, $E[S_h]$. The expected duration of the l-delayed busy period with the initial delay having a mean $\lambda_h E[S_h]x$ is then $r\lambda_h E[S_h]x$. Therefore

$$rx = x + r\lambda_h E[S_h]x \tag{3.169}$$

Solving equation 3.169 for r one finds that

$$b^{(l)}(x) = x/(1 - \rho_h) \tag{3.170}$$

where $\rho_h = \lambda_h E[S_h]$.

Now consider equation 3.168. Because the probability of a higher-priority job arrival in Δ time units is $\lambda_h \Delta + o(\Delta)$,

$$\tilde{F}_{B^{(l)}(\Delta)}(s) = \lambda_h e^{-s\Delta} \tilde{F}_{B^{(l)}}(s) + (1 - \lambda_h \Delta)e^{-s\Delta} + o(\Delta) \tag{3.171}$$

where $B^{(l)}$ is the delayed busy period with an initial delay equal in distribution to the service time of a high-priority job. We will call this just the l-busy period. Now combining equations 3.168 and 3.171 for $y = \Delta$ one gets

$$\tilde{F}_{B^{(l)}(x+\Delta)}(s) = \lambda_h \Delta e^{-s\Delta} \tilde{F}_{B^{(l)}}(s) \tilde{F}_{B^{(l)}(x)}(s) + (1 - \lambda_h \Delta) e^{-s\Delta} \tilde{F}_{B^{(l)}(x)}(s) + o(\Delta), \quad x \geq 0 \tag{3.172}$$

and hence, rearranging the terms in equation 3.172, we get

$$\tilde{F}_{B^{(l)}(x+\Delta)}(s) - \tilde{F}_{B^{(l)}(x)}(s) = -\{s + \lambda_h - \lambda_h \tilde{F}_{B^{(l)}}(s)\} \Delta \tilde{F}_{B^{(l)}(x)}(s) + o(\Delta), \quad x \geq 0 \tag{3.173}$$

Dividing equation 3.173 by Δ and taking the limit as $\Delta \to 0$ we get

$$\frac{d}{dx} \tilde{F}_{B^{(l)}(x)}(s) = -\{s + \lambda_h - \lambda_h \tilde{F}_{B^{(l)}}(s)\} \tilde{F}_{B^{(l)}(x)}(s), \quad x \geq 0 \tag{3.174}$$

Solving equation 3.174 with the boundary condition $\tilde{F}_{B^{(l)}(0)}(s) = 1$ one obtains

$$\tilde{F}_{B^{(l)}(x)}(s) = e^{-\{s + \lambda_h - \lambda_h \tilde{F}_{B^{(l)}}(s)\}x}, \quad x \geq 0 \tag{3.175}$$

Because $\tilde{F}_{B^{(l)}}(s) = \int_0^\infty \tilde{F}_{B^{(l)}(x)}(s) dF_{S_h}(x)$, where $F_{S_h}(x)$ is the distribution function of the service times of higher-priority jobs, from equation 3.175 one finds that $\tilde{F}_{B^{(l)}}(s)$ is the solution to

$$\tilde{F}_{B^{(l)}}(s) = \tilde{F}_{S_h}(s + \lambda_h - \lambda_h \tilde{F}_{B^{(l)}}(s)) \tag{3.176}$$

One may easily verify that the mean length $E[B^{(l)}(x)]$ of the l-delay busy cycle computed from its transform in equation 3.176 will agree with equation 3.170. The higher moments of $B^{(l)}(x)$ can be computed from equation 3.176.

Mean value analysis.   We will now use a mean value analysis to compute the mean flow time of each class of job. First consider the highest priority (i.e., class 1) jobs. Suppose the time average number of class 1 jobs in the queue is $E[L^{(l)}]$. Because class 1 jobs arrive according to a Poisson process, the number of class 1 jobs it sees will have an average $E[L^{(1)}]$. In addition the average remaining service of the job, if any, in service at its arrival is $\lambda E[S^2]/2$, where $\lambda E[S^2] = \sum_{l=1}^k \lambda^{(l)} E[(S^{(l)})^2]$, (see equation 3.164). The mean waiting time of this class 1 job is then the mean time of the initial delay, that is,

$$E[W^{(1)}] = E[L^{(1)}]E[S^{(1)}] + \frac{\lambda E[S^2]}{2} \tag{3.177}$$

Applying Little's formula to class 1 jobs (i.e., $\lambda^{(1)} E[W^{(1)}] = E[L^{(1)}]$) and solving equation 3.177 we obtain

$$E[L^{(1)}] = \frac{\lambda^{(1)} \lambda E[S^2]}{2(1 - \rho^{(1)})} = \frac{\lambda^{(1)} W_0}{1 - \rho_1} \tag{3.178}$$

where we set $W_0 = \lambda E[S^2]/2$, $\lambda_r = \sum_{l=1}^r \lambda^{(l)}$ and $\rho_r = \sum_{j=1}^r \rho^{(l)}$, $r = 1, \ldots, k$.

Next consider a class 2 job. The initial delay encountered by a class 2 job has the mean $E[L^{(2)}]E[S^{(2)}] + E[L^{(1)}]E[S^{(1)}] + W_0$. The corresponding 2-delayed busy period has a mean $1/[1 - \rho_1]\{E[L^{(2)}]E[S^{(2)}] + E[L^{(1)}]E[S^{(1)}] + W_0\}$. Because $E[L^{(2)}] = \lambda^{(2)}E[W^{(2)}]$, as before we find that

$$E[L^{(2)}] = \frac{\lambda^{(2)}W_0}{(1 - \rho_1)(1 - \rho_2)} \tag{3.179}$$

Continuing this way we get

$$E[L^{(l)}] = \frac{\lambda^{(l)}W_0}{(1 - \rho_l)(1 - \rho_{l-1})}, \quad l = 1, 2, \ldots, k \tag{3.180}$$

where $\rho_0 = 0$. The mean flow time and the number of class l jobs in the system are

$$E[T^{(l)}] = \frac{\lambda E[S^2]}{2(1 - \rho_l)(1 - \rho_{l-1})} + E[S^{(l)}], \quad l = 1, \ldots, k \tag{3.181}$$

$$E[N^{(l)}] = \frac{\lambda^{(l)}\lambda E[S^2]}{2(1 - \rho_l)(1 - \rho_{l-1})} + \rho^{(l)}, \quad l = 1, \ldots, k \tag{3.182}$$

Optimal priority assignment.   When the cost of being in the system for each class of job is specified, we can use the preceding results to obtain the optimal assignment of priorities such that the total expected cost of being in the system for all job classes is minimized. We illustrate this next. Let w_l be the cost per unit time of being in the system for a type l job ($l = 1, 2, \ldots, k$). When we assign the priority index $\pi(l)$ for type l jobs, they will be called class $\pi(l)$ jobs and class l will have higher priority over class $l + 1, \ldots, k$ jobs. Therefore we seek to find the optimal priority index $\pi(l)$, $l = 1, \ldots, k$ for a given set of k different types of jobs. Without loss of generality assume that $w_l/E[S_l] \geq w_{l+1}/E[S_{l+1}], l = 1, \ldots, k-1$. We will then show that the priority index $\pi(l) = l, l = 1, 2, \ldots, k$ will minimize the total costs. For a priority assignment π define $\lambda_\pi^{(l)} = \lambda^{\pi^{-1}(l)}$, $E[S_\pi^{(l)}] = E[S^{\pi^{-1}(l)}]$ and $\rho_l^\pi = \sum_{r=1}^{l} \lambda^{\pi^{-1}(r)}E[S^{\pi^{-1}(r)}], l = 1, 2, \ldots, k$. Then the objective is to minimize the total cost:

$$TC(\pi) = \sum_{l=1}^{k} \frac{w_\pi^{(l)}\lambda_\pi^{(l)}W_0}{(1 - \rho_{l-1}^\pi)(1 - \rho_l^\pi)} + \sum_{l=1}^{k} w_\pi^{(l)}\lambda_\pi^{(l)}E[S_\pi^{(l)}] \tag{3.183}$$

where $w_\pi^{(l)} = w_{\pi^{-1}(l)}$ and $\rho_l^\pi = \sum_{j=1}^{l} \lambda_\pi^{(j)}E[S_\pi^{(j)}], l = 1, \ldots, k$. Observe that the second term in equation 3.183 is independent of the priority assignment π. Suppose for some $l, \pi^{-1}(l) > \pi^{-1}(l + 1)$, that is, a job with a higher-type index ($\pi^{-1}(l)$) is assigned a lower-class index (i.e., higher priority) over a job with a lower-type index $\pi^{-1}(l + 1)$. Consider an alternative priority assignment π' such that $\pi'(r) = \pi(r)$, $r \neq l, l+1$, $\pi'(l) = \pi(l + 1)$ and $\pi'(l + 1) = \pi(l)$. The difference in the total cost can then be shown to be given by

$$
\begin{aligned}
& T C\ (\pi) - T C(\pi') \\
&= \left\{ \frac{w_\pi^{(l)} \lambda_\pi^{(l)}}{\left(1 - \rho_{l-1}^\pi\right)\left(1 - \rho_l^\pi\right)} + \frac{w_\pi^{(l+1)} \lambda_\pi^{(l+1)}}{\left(1 - \rho_l^\pi\right)\left(1 - \rho_{l+1}^\pi\right)} \right. \\
& \left. - \frac{w_\pi^{(l+1)} \lambda_\pi^{(l+1)}}{\left(1 - \rho_{l-1}^\pi\right)\left(1 - \rho_{l-1}^\pi - \rho_\pi^{(l+1)}\right)} - \frac{w_\pi^{(l)} \lambda_\pi^{(l)}}{\left(1 - \rho_{l-1}^\pi - \rho_\pi^{(l+1)}\right)\left(1 - \rho_{l+1}^\pi\right)} \right\} W_0 \\
&= -W_0 \left\{ \frac{w_\pi^{(l)}}{E\left[S_\pi^{(l)}\right]} - \frac{w_\pi^{(l+1)}}{E\left[S_\pi^{(l+1)}\right]} \right\} \rho_\pi^{(l)} \rho_\pi^{(l+1)} (1 - \rho_{l-1}^\pi + 1 - \rho_{l+1}^\pi) \\
&\geq 0
\end{aligned}
\tag{3.184}
$$

because $w_{\pi^{-1}(l)}/E[S^{\pi^{-1}(l)}] \leq w_{\pi^{-1}(l+1)}/E[S^{\pi^{-1}(l+1)}]$ (recall $\pi^{-1}(l) > \pi^{-1}(l+1)$). We see that the $w\mu$ rule that assigns decreasing priorities (i.e., increasing class indices) in decreasing order of $w_l \mu_l$ ($= w_l/E[S^{(l)}]$) minimizes the total cost of time spent in the system by all types of jobs. Observe that when the cost rate w_l is the same for all job types, the $w\mu$ rule is the same as assigning priorities in decreasing order of μ_l's. Equivalently, priorities are assigned according to the shortest expected processing time (SEPT). Thus the SEPT priority assignment minimizes the total average number of jobs in the system and therefore also the total average flow time of all types of jobs. This result is in agreement with the result that the shortest processing time (SPT) service protocol minimizes the average flow time in a static job shop with deterministic processing times. When the processing times are deterministic, $S^{(l)}$ is a fixed constant for each type l job ($l = 1, \ldots, k$), and the $w\mu$ rule implies that the SPT service protocol minimizes the average flow time of jobs when jobs arrive according to a Poisson process. This optimality of the SPT service protocol need not be true when job arrivals do not conform to a Poisson process.

The $w\mu$ rule is easily justified intuitively when the service times are exponentially distributed. Among all jobs available for processing at time zero, suppose we choose a type l job. The probability that service to this job will be completed during $(0, \Delta]$ is $\mu_l \Delta + o(\Delta)$. Therefore the reduction in the total expected cost of being in the system for all jobs during $(\Delta, 2\Delta]$ is $w_l \mu_l \Delta + o(\Delta)$. A greedy approach to minimizing the total expected cost is to choose the job with the highest $w_l \mu_l$.

Level crossing analysis.   To obtain the distribution function of the flow time of class l jobs, we will first look at a sample path of the virtual waiting time of a class l job. The virtual waiting time of a class l job at time t is the waiting time of a class l job if *it were to arrive* at time t. We start the system empty at time zero. If our virtual class l job were to arrive before the first job arrival epoch, our virtual job would enter into service immediately. Therefore the virtual waiting time is zero until the first job arrives. If the virtual job were to arrive just after the first job arrived, then its waiting time will have the same distribution as an l-delayed busy period with initial delay equal to the service time of the first job. The LST of this waiting time is $\tilde{F}_J(s) = \tilde{F}_S(s + \lambda_h - \lambda_h \tilde{F}_{B^{(l)}}(s))$ where $\tilde{F}_{B^{(l)}}(s) = \tilde{F}_{S_h}(s + \lambda_h - \lambda_h \tilde{F}_{B^{(l)}}(s))$ (see equation 3.175). The virtual waiting time will therefore jump by this amount at the arrival epoch of the first job. Subsequently the virtual waiting time will drop at a rate of 1 until it becomes 0 or a class l job arrives. If our virtual job arrives just after a class l job arrives, the virtual waiting time will jump by an amount whose magnitude will have the same distribution as a l-delayed busy period with an initial delay equal to the service time of the class l job. The LST of this jump is $\tilde{F}_{J_l}(s) = \tilde{F}_{S^{(l)}}(s + \lambda_h - \lambda_h \tilde{F}_{B^{(l)}}(s))$ (follows from equation 3.175). Such jumps will occur every time a class l job enters the system. In between such jumps the virtual waiting time will drop at a rate of 1 until it hits 0. Whenever this happens, either the system is empty or there are jobs with larger class indices than l. If there are any such jobs available then the virtual waiting time will jump by an amount equal in distribution to the l-delayed busy period with initial delay equal to the processing time of that job. The LST of an arbitrary jump corresponding to these jumps is $\tilde{F}_{J_s}(s) = \tilde{F}_{S_s}(s + \lambda_h - \lambda_h \tilde{F}_{B^{(l)}}(s))$ where S_s is the service time of an arbitrary job from classes with indices larger than l.

The mean time for the system to become empty from the time of the first arrival is the mean duration of a $(k + 1)$-delayed busy period with an initial delay equal to the service time of an arbitrary job. This mean is equal to $E[S]/(1 - \rho)$. The mean time to the first arrival is $1/\lambda$. Therefore the mean total duration of a busy cycle is $1/\lambda + E[S]/(1 - \rho) = 1/\{\lambda(1 - \rho)\}$. The rate of upcrossings over level x caused by the first jump in each busy cycle is $\lambda(1 - \rho)\overline{F}_J(x)$. These upcrossings can be caused by either higher-priority, lower-priority, or class l jobs. Because the probability that the first arrival is a high-priority job is λ_h/λ and the corresponding magnitude of the jump has a LST $\tilde{F}_{J_h}(s) = \tilde{F}_{S_h}(s + \lambda_h - \lambda_h \tilde{F}_{B^{(l)}}(s))$, the rate of upcrossings at the beginning of the busy period owing to higher-priority jobs is $\lambda_h(1 - \rho)\overline{F}_{J_h}(x)$. The rate of upcrossings over level x caused by the arrivals of class l jobs is $\lambda^{(l)} \int_0^x \overline{F}_{J_l}(x - y) dF_V(y)$ where F_V is the stationary distribution of the virtual waiting time process. The rate of upcrossings over level x caused by the lower-priority jobs is $\lambda_s \overline{F}_{J_s}(x)$. Equating the rate of upcrossings and downcrossings, which is $f_V(x)$ one gets

$$f_V(x) = \lambda_h(1 - \rho)\overline{F}_{J_h}(x) + \lambda^{(l)} \int_0^x \overline{F}_{J_l}(x - y) dF_V(y) + \lambda_s \overline{F}_{J_s}(x), \quad x \geq 0 \qquad (3.185)$$

Now taking the LST of both sides of equation 3.185 and solving for $\tilde{F}_V(s)$ we get

$$\tilde{F}_V(s) = \frac{\lambda_h(1 - \rho)(1 - \tilde{F}_{J_h}(s)) + \lambda_s(1 - \tilde{F}_{J_s}(s)) + sF_V(0)}{s - \lambda^{(l)}(1 - \tilde{F}_{J_l}(s))} \qquad (3.186)$$

Using $\lim_{s \to 0} \tilde{F}_V(s) = 1$, one finds that $F_V(0) = 1 - \rho$. Then

$$\tilde{F}_V(s) = \frac{(1 - \rho)(s + \lambda_h - \lambda_h \tilde{F}_{J_h}(s)) + \lambda_s(1 - \tilde{F}_{J_s}(s))}{s - \lambda^{(l)} + \lambda^{(l)} \tilde{F}_{J_l}(s)} \qquad (3.187)$$

By PASTA the waiting time of a class l job will then have a LST given by equation 3.187. The higher moments of the waiting time can be derived from equation 3.187.

In particular the first and second moments are

$$E\left[W^{(l)}\right] = \frac{\lambda E[S^2]}{2(1-\rho_l)(1-\rho_{l-1})}$$

$$E\left[\left(W^{(l)}\right)^2\right] = \frac{\lambda E[S^3]}{(1-\rho_l)^2(1-\rho_{l-1})} + \frac{\lambda E[S^2]}{2(1-\rho_{l-1})^2(1-\rho_l)}$$

$$+ \left\{ \frac{\lambda_{l-1} E[S^2(l-1)]}{1-\rho_{l-1}} + \frac{\lambda_{l-1} E[S^2(l-1)]}{1-\rho_l} + \frac{\lambda^{(l)} E[(S^{(l)})^2]}{1-\rho_l} \right\} \quad (3.188)$$

where $E[S^2(l)] = 1/\lambda_l \sum_{r=1}^{l} \lambda^{(r)} E[(S^{(r)})^2]$, $l = 1, \ldots, k$.

3.5.3 Multiple Class *GI/G/c* Models with NPP Service Protocols

A single-stage job shop with c servers and several classes of job arrivals is considered in this section. The jobs from each class arrive according to a renewal process, and the service times are iid for each class. It is not possible to obtain explicit performance measures for such a system. Therefore we will provide approximations based on those available for the single class $GI/G/c$ model. Let $E[W^{(l)}]_{GI/G/c:NPP}$ be the mean waiting time of a class l job in the $GI/G/c$ model with the NPP discussed in Section 3.5.2. Then we use the following approximations for the mean waiting time:

$$E[W^{(l)}]_{GI/G/c:NPP} \approx \frac{E[W^{(l)}]_{M/G/1:NPP}}{E[W]_{M/G/1:FCFS}} E[W]_{GI/G/c:FCFS}$$

$$= \left\{ \frac{1-\rho}{(1-\rho_l)(1-\rho_{l-1})} \right\} E[W]_{GI/G/c:FCFS} \quad (3.189)$$

Now approximating this further using the approximation for $E[W]_{GI/G/c:FCFS}$ (see equation 3.157) we get

$$E[W^{(l)}]_{GI/G/c:NPP} \approx \left\{ \frac{1-\rho}{(1-\rho_l)(1-\rho_{l-1})} \right\} \left\{ \frac{E[W]_{M/M/c}}{E[W]_{M/M/1}} \right\} E[W]_{GI/G/1:FCFS} \quad (3.190)$$

Similarly one obtains

$$E[L^{(l)}]_{GI/G/c:NPP} \approx \left\{ \frac{\lambda^{(l)}(1-\rho)}{\lambda(1-\rho_l)(1-\rho_{l-1})} \right\} \left\{ \frac{E[L]_{M/M/c}}{E[L]_{M/M/1}} \right\} E[L]_{GI/G/1:FCFS} \quad (3.191)$$

The approximation for the performance of an arbitrary job is then,

$$E[W]_{GI/G/c:NPP} \approx \left[\sum_{l=1}^{k} \left\{ \frac{\lambda^{(l)}(1-\rho)}{\lambda(1-\rho_l)(1-\rho_{l-1})} \right\} \right] \left\{ \frac{E[W]_{M/M/c}}{E[W]_{M/M/1}} \right\} E[W]_{GI/G/1:FCFS}$$

$$(3.192)$$

and

$$E[L]_{GI/G/c:\text{NPP}} \approx \left[\sum_{l=1}^{k} \left\{ \frac{\lambda^{(l)}(1-\rho)}{\lambda(1-\rho_l)(1-\rho_{l-1})} \right\} \right] \left\{ \frac{E[L]_{M/M/c}}{E[L]_{M/M/1}} \right\} E[L]_{GI/G/1:\text{FCFS}}$$

(3.193)

Thus we have related an approximation for the multiple-server job shop with NPP service protocol to that of a single-server job shop with FCFS service protocol. We will use the same relationship for other service protocols as well.

3.5.4 Job Shops with Service Time–Dependent Protocols

We will consider three service protocols: SPT, truncated shortest processing time (SPTT $-\alpha$), and two-class NPP (2C-NP-α). Under SPT service protocols, jobs are assigned class indices equal to their processing times. Therefore a job with the lowest service time gets the highest priority. Under SPTT-α service protocols, jobs requiring processing times greater than or equal to α are given a class index α. Any other job (i.e., a job requiring a processing time less than α) is given a class index equal to its processing time. Under 2C-NP-α service protocol jobs requiring service times less (greater) than or equal to α are given a priority index 1(2). The jobs are then served according to the priority protocol discussed earlier. Let $E[W^{(p)}]_{GI/G/c:\text{SPT}}$, $E[W^{(p)}]_{GI/G/c:\text{SPTT}-\alpha}$, and $E[W^{(p)}]_{GI/G/c:\text{2C-NP}-\alpha}$ be the mean waiting time of a job with processing time requirement p in a $GI/G/c$ model with SPT, SPTT-α, and 2C-NP-α service protocols, respectively. Suppose the mean service times are absolutely continuous. Then from equation 3.188 and limiting arguments, one gets

$$E[W^{(p)}]_{M/G/1:\text{SPT}} = \frac{\lambda E[S^2]}{2(1-\rho_p)^2}, \qquad p \geq 0$$

(3.194)

where $\rho_p = \lambda \int_0^p dF_S(x)$.

$$E[(W^{(p)})^2]_{M/G/1:\text{SPT}} = \frac{\lambda E[S^3]}{3(1-\rho_p)^3} + \frac{\lambda^2 E[S^2]E[S_p^2]}{(1-\rho_p)^4}$$

(3.195)

where $E[S_p^2] = \int_0^p x^2 dF_S(x)$.

$$E[W^{(p)}]_{M/G/1:\text{SPTT}-\alpha} = \begin{cases} \frac{\lambda E[S^2]}{2(1-\rho_p)^2}, & 0 \leq p < \alpha \\ \frac{\lambda E[S^2]}{2(1-\rho_\alpha)(1-\rho)}, & p \geq \alpha \end{cases}$$

(3.196)

$$E\left[(W^{(p)})^2\right]_{M/G/1:\text{SPTT}-\alpha} = \begin{cases} \frac{\lambda E[S^3]}{3(1-\rho_p)^3} + \frac{\lambda^2 E[S^2]E[S_p^2]}{2(1-\rho_p)^4}, & 0 \leq p \leq \alpha \\ \frac{\lambda E[S^3]}{3(1-\rho_\alpha)^2(1-\rho)} + \frac{\lambda^2 E[S^2]^2}{2(1-\rho_\alpha)^2(1-\rho)^2} + \frac{\lambda^2 E[S^2]E[S_\alpha^2]}{2(1-\rho_\alpha)^3(1-\rho)}, & p > \alpha \end{cases}$$

(3.197)

$$E\left[W^{(p)}\right]_{M/G/1:2\text{C–NP}-\alpha} = \begin{cases} \frac{\lambda E[S^2]}{2(1-\rho_\alpha)}, & 0 \le p \le \alpha \\ \frac{\lambda E[S^2]}{2(1-\rho_\alpha)(1-\rho)}, & \alpha < p \end{cases} \tag{3.198}$$

$$E\left[\left(W^{(p)}\right)^2\right]_{M/G/1:2\text{C–NP}-\alpha} = \begin{cases} \frac{\lambda E[S^3]}{3(1-\rho_\alpha)} + \frac{\lambda^2 E[S^2]E[S^2(\alpha)]}{2(1-\rho_\alpha)^2}, & 0 \le p \le \alpha \\ \frac{\lambda E[S^3]}{(1-\rho_\alpha)^2(1-\rho)} + \frac{\lambda^2 E[S^2]E[S^2(\alpha)]}{(1-\rho_\alpha)^3(1-\rho)} + \frac{2\lambda^2 E[S^2]^2}{(1-\rho_\alpha)^2(1-\rho)^2}, & \alpha < p \end{cases}$$
$$\tag{3.199}$$

The corresponding mean waiting times of an arbitrary job are

$$E[W]_{M/G/1:\text{SPT}} = \int_0^\infty \frac{\lambda E[S^2]}{2(1-\rho_p)^2} dF_S(p) \tag{3.200}$$

$$E[W]_{M/G/1:\text{SPTT}-\alpha} = \int_0^\alpha \frac{\lambda E[S^2]}{2(1-\rho_p)^2} dF_S(p) + \frac{\lambda E[S^2]}{2(1-\rho_\alpha)(1-\rho)}\overline{F}_S(\alpha) \tag{3.201}$$

and

$$E[W]_{M/G/1:2\text{C–NP}-\alpha} = \frac{\lambda E[S^2]}{2(1-\rho_\alpha)} F_S(\alpha) + \frac{\lambda E[S^2]}{2(1-\rho_\alpha)(1-\rho)}\overline{F}_S(\alpha) \tag{3.202}$$

Using the approximation developed for $E[W]_{GI/G/c:\text{NPP}}$ we now get

$$E[W]_{GI/G/c:\text{SPT}} = (1-\rho)\left\{\int_0^\infty \frac{dF_S(p)}{(1-\rho_p)^2}\right\}\left\{\frac{E[W]_{M/M/c}}{E[W]_{M/M/1}}\right\} E[W]_{GI/G/1:\text{FCFS}} \tag{3.203}$$

$$E[W]_{GI/G/c:\text{SPT}-\alpha} = \left\{(1-\rho)\left[\int_0^\infty \frac{dF_S(p)}{(1-\rho_p)^2}\right] + \frac{\overline{F}_S(\alpha)}{1-\rho_\alpha}\right\}$$
$$\left\{\frac{E[W]_{M/M/c}}{E[W]_{M/M/1}}\right\} E[W]_{GI/G/1:\text{FCFS}}$$
$$\tag{3.204}$$

and

$$E[W]_{GI/G/c:2\text{C–NP}-\alpha} = \left\{\frac{(1-\rho)F_S(\alpha)}{1-\rho_\alpha} + \frac{\overline{F}_S(\alpha)}{1-\rho_\alpha}\right\}\left\{\frac{E[W]_{M/M/c}}{E[W]_{M/M/1}}\right\} E[W]_{GI/G/1:\text{FCFS}}$$
$$\tag{3.205}$$

For quadratic cost function of flow times, the expected costs per job were computed using the preceding results for a large number of $M/E_k/1$ queueing systems. For the SPTT-α and 2C-NP-α service protocols the optimalvalue of α that minimizes the waiting cost is obtained and used to compute the minimum waiting cost. The results are summarized in Table 3.4.

TABLE 3.4 EXPECTED COST $E[T] + 0.05E[T^2]$ OF TIME SPENT IN $M/E_k/1$ QUEUEING SYSTEM UNDER DIFFERENT SERVICE PROTOCOLS, FOR VARIOUS VALUES OF k AND SERVER UTILIZATION ρ

ρ	k	Service protocol			
		SPT	SPTT	2C-NP	FCFS
0.6	1	2.494	2.444	2.557	3.125
	2	2.156	2.156	2.217	2.514
	5	2.005	2.004	2.030	2.173
	10	1.970	1.966	1.978	2.064
0.9	1	16.142	12.023	12.102	20.000
	2	14.238	10.115	10.095	13.506
	5	14.051	9.064	9.023	10.168
	10	14.778	8.723	8.696	9.148
0.99	1	8111.293	728.325	726.007	1100.000
	2	7395.730	532.246	531.075	639.007
	5	7885.360	404.893	404.572	422.008
	10	8874.023	356.681	356.602	359.608

Based on the numerical results the following observations can be made.

1. SPTT-α and 2C-NP-α always perform better than FCFS.
2. In general 2C-NP-α is a much better scheduling discipline under a wide range of conditions.

3.6 IMPLICATIONS

Some of the implications of the models developed for single-stage produce-to-order manufacturing systems are

1. The capacity of the system is entirely determined by the number of servers and the *average* service time. It is unaffected by the variability in service times and is invariant under any service protocol as long as the service protocol is work conserving.
2. Different performance measures of the system are tightly coupled. (1) The mean number of jobs in the system is always proportional to the mean flow time with the proportionality constant equal to the job arrival rate. This allows us to infer

the mean flow time simply by collecting information on the average number of jobs in the system and the job arrival or departure rates. (2) The number of jobs seen by an arriving job has the same distribution as the number of jobs left behind by departing jobs. In general, however, the number of jobs present in the system at an arbitrary time point need not be the same as the number of jobs seen by an arrival or the number of jobs left behind by a departing job. When the job arrival process is Poisson these three random variables will have the same distribution.

3. Though the variability in service and interarrival times does not affect the capacity of the system, their performance can be significantly affected. When the interarrival times have a regular distribution (in particular one with decreasing mean residual times), however, the higher moments of the service and interarrival times do not significantly affect the mean performance of these systems. Hence the average performance can be evaluated by simply collecting information on the first two moments of the service and interarrival times. If the interarrival times have a highly variable distribution (e.g., hyperexponential), even the *average* performance of the system can be significantly affected by the third and higher moments of the interarrival times. In such a case performance predicted using only the first two moments of the interarrival times should be used very cautiously.

4. In a single job class system any nonpreemptive service protocol that does not use the information on the service times cannot improve the performance of the system over FCFS service protocol. In particular the probabilistic evolution of the number of jobs in the system is unaffected by any such service protocols.

5. In a multiple job class system, any improvement in the mean flow time of one job class owing to a service protocol that does not use information on the service times will always degrade the mean flow time performance of jobs from at least one other job class.

3.7 BIBLIOGRAPHICAL NOTE

Some parts of the material presented in this chapter, such as the analysis of the $M/M/1$, $M/G/1$, and $M/M/c$ queueing systems, can be picked up from any elementary textbook on queueing theory (e.g., see Cooper [7], Gross and Harris [14], Kleinrock [22]). Neuts [31] should be consulted for a thorough analysis of the $PH/PH/1$-type queueing system. Some parts of this chapter are heavily influenced by the book by Wolff [49]. In our presentation, however, we have paid less attention to the technical details, such as in the proof of the finiteness of the number of jobs in the system. Interested readers should consult Loynes [26].

The applicability under general settings of Little's formula [25], was established by Stidham [42]. A recent review of this formula and its extensions are given in Whitt [47]. Level crossing analysis of discrete state processes, useful in relating the arrival epoch, departure epochs, and other time epochs is described in Shanthikumar and Chandra [39]. The analysis used in Section 3.2.3 is a simplified version of it. The level crossing analysis of continuous state processes was independently introduced in Brill and Posner [3], and Cohen [6]. The extension of this approach to multiple class queueing systems (with server vacations) is due to Shanthikumar [32][34]. The analysis of Section 3.5 is based on this extension.

The upper bound in equation 3.118 was derived by Kingman [49], and the improved upper bound in equation 3.123 is due to Daley [9]. The lower bound in equation 3.137 was first derived by Marshall [28] for interarrival times with increasing failure rates, and its validity in the case of DMRL interarrival times was established by Daley (e.g., see Stoyan [84]). The tightness of the $M/G/1$ lower bound and the $GE/G/1$ upper bound for the mean time spent in a $GI/G/1$ queueing system with IMRL interarrival times is established in Whitt [46].

Approximation 3.143 is due to Marchal [27], and see Shore [41] and Whitt [46] for other approximations for $GI/G/c$ queues. The analysis and approximations for $GI/G/1$ queues with SPT and other service time–dependent service protocols can be found in Shanthikumar [26] [35], and Shanthikumar and Buzacott [28].

PROBLEMS

3.1 Jobs arrive at a single-stage single-server job shop according to a Poisson process with rate λ. The processing times are iid with a two-phase hyperexponential distribution

$$F_S(x) = 1 - p_1 e^{-\mu_1 x} - p_2 e^{-\mu_2 x}, \quad x \geq 0$$

$(p_1 + p_2 = 1)$. What is
(a) The capacity of this system?
(b) The maximum job departure rate, say $\beta^*(\lambda)$ as a function of λ? Show that

$$\beta^*(\lambda) < \max\{\mu_1, \mu_2\}$$

and

$$\beta^*(\lambda) \to \max\{\mu_1, \mu_2\} \quad \text{as} \quad \lambda \to \infty$$

3.2 Several types of jobs arrive at a single-stage c-server job shop. The management has decided to only accept orders for a preselected set of r types of jobs at a fixed ratio of $p_1 : p_2 : \cdots : p_r$, $\sum_{j=1}^n p_j = 1$. Suppose the mean processing time of type j job is $1/\mu_j$, $j = 1, \ldots, r$.
(a) What is the capacity of this system?
(b) Can this capacity be increased by changing the ratio of the job-type mix?
(c) If the answer to part b is yes, find the maximum capacity of this job shop when you are allowed to change the ratio of the job-type mix.

3.3 Consider the system described in problem 2. Suppose the revenue for processing a type j job is v_j. The management wants to find the best ratio of job-type mix.
(a) Suppose the input rates of all types of jobs are unlimited. What is the optimal ratio of job-type mix that will maximize the revenue?

(b) Suppose the maximum (demand) rate of type j jobs is λ_j^u, $j = 1, \ldots, m$. Find the optimal input rates of each type of job that will maximize the revenue.

3.4 Honest Stereo Repair shop proudly advertises that the average service time to repair a stereo system (i.e., starting from the time you brought the system for repair and ending with the time it is repaired and returned to you) is two days. Unfortunately no data on this time is available at their stores. Logs indicate that they have on the average 10 customers arriving per day and about 80 stereo systems either being repaired or awaiting repair on any given day.

(a) Do you believe Honest Stereo Repair shop's ad?

(b) If the answer to part a is yes, explain how you arrived at the answer.

(c) If the answer to part b is no, what do you think would be the average service (flow) time?

3.5 A vice-president (VP) of a company who owns a job shop is interested in obtaining the time-average performance of the job shop. For this, the VP is considering making several surprise visits to the job shop. Because of limited time availability, the VP cannot visit the facility every day, but only (at a rate of) once a week (of five days) for the next ten weeks. The VP's able assistant suggests that the VP should visit for ten consecutive days and claims that this strategy will give a good estimate of the time average.

Conversely, the head of the OR analysis group at the company recommends the VP should toss a (loaded) coin everyday. If the coin shows heads (with probability 0.2), then the VP should visit the job shop. Otherwise the VP should not visit the job shop. Thus on the average the VP will visit once a week. Which of the two inspection strategies will you recommend? Provide justifications for your answer.

3.6 Consider the Example considered in Section 3.3.1. Suppose the cost of waiting, w per unit time is levied only while the job is waiting but not levied while the job is in service. Choose the machine that will maximize the overall profit.

3.7 Use equations 3.69 and 3.73 derived for the $M/G/1$ queueing system to recover the results equations 3.48 and 3.52 derived for the $M/M/1$ queueing system.

3.8 Consider a single-stage single-server job shop where jobs arrive according to Poisson process with rate λ. The processing times are iid with Erlang-k distribution and mean $1/\mu$. Find the transition rate matrix of this system using the $PH/PH/1$ queue representation (see Section 3.3.3). Verify that the mean number of jobs derived using this representation agrees with the results of the $M/G/1$ model.

3.9 Suppose jobs arrive at a single-stage single-server job shop according to a renewal process with a two-phase hyperexponential interarrival time distribution (H_2).

$$F_\tau(x) = 1 - pe^{-\lambda_1 x} - (1-p)e^{-\lambda_2 x}, \quad x \geq 0 \ (\lambda_1 > \lambda_2)$$

with mean $1/\lambda$ and squared coefficient of variation C_a^2. The service times are exponentially distributed. Use the representation of $PH/PH/1$ queueing system for this $H_2/M/1$ queueing system.

(a) Find the transition rate matrix $\mathbf{R}$ (equation 3.86).

(b) Solve equation 3.97 for $\mathbf{G}$ explicitly and hence obtain $\mathbf{p}(n)$, $n = 0, 1, \ldots$, explicitly from equations 3.101 and 3.96.

(c) Using the results in part b show that when λ and C_a^2 are fixed, the mean number in the system is increasing as a function of λ_1.

    i. In particular, the maximum mean waiting time is attained when $\lambda_1 \to \infty$, that is, for the $GE/M/1$ queueing system with generalized exponential distribution (see Appendix A).

    ii. In particular, the minimum mean waiting time is attained when $\lambda_1 \to \lambda$, that is, for the $M/M/1$ queueing system.

3.10 Show that the mean number of jobs in a single-stage job shop with c machines, Poisson job arrival process and exponentially distributed processing time is decreasing and convex.

3.11 Suppose we have a choice of either purchasing c machines that work at a nominal rate of one unit/unit time or one machine that works at an inflated rate of c units/unit time. That is, a job requiring S_n units of work will be processed in S_n/c units of time. Let $E[L_c]$ ($E[L_1]$) be the mean number of jobs in queue in the system if we choose the c servers (single server).

    (a) Show that

$$E[L_c] \geq E[L_1] - \left(\frac{c-1}{c}\right)\left(\frac{1+C_S^2}{2}\right)\rho$$

    where C_S^2 is the squared coefficient of variation of the service time, and ρ is the server utilization.

    (b) If the work load of each job has a new better than used in expectation distribution, show that

$$E[L_c] \leq E[L_1] + \left(\frac{c-1}{c}\right)\left(c - \frac{1+C_S^2}{2}\rho\right)$$

3.12 Consider the optimal priority assignment in Section 3.5. Suppose we wish to minimize the total waiting cost subject to the constraint that the mean flow time of a type l job should not exceed $\hat{w}^{(l)}$, $l = 1, \ldots, k$. Find the optimal priority assignment.

3.13 Carry out the level crossing analysis of Section 3.5 for job shops with Poisson job arrival process and SPT service protocol. Compare your results with those in Section 3.5.4.

3.14 Explain why the optimized 2C-NP-α service protocol will *always* outperform the FCFS service protocol and why the optimized SPTT-α service protocol will *always* outperform the SPT service protocol.

3.15 It is usually expected that the performance of a manufacturing system can be improved by reducing the variability. Observe that in Table 3.4, the expected cost in the $M/G/1$ (SPT) queueing system, under high utilization (of .99), increases as the service time becomes more regular (i.e., as k increases from 2 to 10). Explain this phenomenon.

BIBLIOGRAPHY

[1] S. L. ALBIN. Approximating a point process by a renewal process. II: Superposition arrival processes to queues. *Operations Research*, 32:1133–1162, 1984.

[2] T. ALTIOK. On the phase-type approximations of general distributions. *IIE Transactions*, 17:110–116, 1985.

[3] P. H. BRILL and M. J. M. POSNER. Level crossings in point processes applied to queues: single-server case. *Operations Research*, 25:662–674, 1977.

[4] P. J. BURKE. The output of a queueing system. *Operations Research*, 4:699–704, 1956.

[5] A. COBHAM. Priority assignment in waiting lines. *Operations Research*, 2:70–76, 1954.

[6] J. W. COHEN. On up- and down-crossings. *J. Appl. Prob.*, 14:405–410, 1977.

[7] R. B. COOPER. *Introduction to Queueing Theory*, 2nd edition. Edward Arnold, London, 1981.

[8] D. R. COX and W. L. SMITH. On the superposition of renewal processes. *Biometrika*, 41:91–99, 1954.

[9] D. J. DALEY. Inequalities for moments of tails of random variables, with queueing applications. *Z. Wahrsch. Verw. Gebiete*, 41:139–143, 1977.

[10] P. D. FINCH. The output process of the queueing system $M/G/1$. *J. Roy. Statist. Soc. [B]*, 21:375–380, 1959.

[11] F. G. FOSTER. On the stochastic matrices associated with certain queueing processes. *Ann. Math. Statist.*, 24:355–360, 1953.

[12] D. P. GAVER. Imbedded Markov chain analysis of a waiting–line process in continuous time. *Ann. Math. Stat.*, 30:698–720, 1959.

[13] D. P. GAVER JR. A waiting line with interrupted service, including priorities. *J. Roy. Statist. Soc. [B]*, 24:73–90, 1962.

[14] D. GROSS and C. M. HARRIS. *Fundamentals of Queueing Theory*. John Wiley and Sons, New York, 1974.

[15] U. HERZOG, L. WOO, and K. M. CHANDY. Solution of queueing problems by a recursive technique. *IBM J. Res. Dev.*, 19:295–300, 1975.

[16] P. HOKSTAD. Approximations for the $M/G/m$ queue. *Operations Research*, 26:510–523, 1978.

[17] N. K. JAISWAL. *Priority Queues*. Academic Press, New York, 1968.

[18] D. G. KENDALL. Some problems in the theory of queues. *J. Roy. Statist. Soc. [B]*, 13:151–173, 1951.

[19] D. G. KENDALL. Stochastic processes occurring in the theory of queues and their analysis by the method of the imbedded Markov chain. *Ann. Math. Statist.*, 24:338–354, 1953.

[20] H. KESTEN and J. T. RUNNENBERG. Priority in waiting line problems. *Prok. Akad. Wet. Amst. [A]*, 60:161–200, 1957.

[21] J. F. C. KINGMAN. Some inequalities for the $GI/G/1$ queue. *Biometrika*, 49:315–324, 1962.

[22] L. KLEINROCK. *Queueing Systems*, volume 1: *Theory*. John Wiley and Sons, New York, 1975.

[23] A. M. LEE and P. A. LONGTON. Queueing processes associated with airline passenger check-in. *Operat. Res. Quart.*, 10:56–71, 1957.

[24] D. V. LINDLEY. The theory of queues with a single server. *Proc. Camb. Phil. Soc.*, 48:277–289, 1952.

[25] J. D. C. LITTLE. A proof of the queueing formula: $L = \lambda W$. *Operations Research*, 9:383–387, 1961.

[26] R. M. LOYNES. The stability of a queue with non–independent interarrival and service times. *Proc. Camb. Phil. Soc.*, 58:497–520, 1962.

[27] W. C. MARCHAL. An approximate formula for waiting time in single-server queues. *AIIE Trans.*, 8:473–474, 1976.

[28] K. T. MARSHALL. Some inequalities in queueing. *Operations Research*, 16:651–665, 1968.

[29] K. T. MARSHALL. Some relationships between the distributions of waiting time, idle time and interoutput time in the $GI/G/1$ queue. *SIAM J. Appl. Math.*, 16:324–327, 1968.

[30] R. G. MILLER, JR. Priority queues. *Ann. Math. Stat.*, 31:86–103, 1960.

[31] M. F. NEUTS. *Matrix-Geometric Solutions in Stochastic Models—an Algorithmic Approach.* John Hopkins University Press, Baltimore, 1981.

[32] J. G. SHANTHIKUMAR. Analysis of priority queue with server control. *Opsearch*, 27:183–192, 1984.

[33] J. G. SHANTHIKUMAR. *Approximate queueing models of dynamic job shops.* PhD thesis, Department of Industrial Engineering, University of Toronto, 1979.

[34] J. G. SHANTHIKUMAR. Level crossing analysis of priority queues and a conservation identity for vacation models. *Naval Research Logistics*, 36:797–806, 1989.

[35] J. G. SHANTHIKUMAR. On reducing time spent in $M/G/1$ systems. *Eur. J. Operat. Res.*, 9:286–294, 1982.

[36] J. G. SHANTHIKUMAR. Some analyses on the control of queues using level crossing of regenerative processes. *J. Appl. Prob.*, 17:814–821, 1980.

[37] J. G. SHANTHIKUMAR and J. A. BUZACOTT. On the approximations for the single server queue. *Int. J. Prod. Res.*, 18:761–773, 1980.

[38] J. G. SHANTHIKUMAR and J. A. BUZACOTT. Open queueing networks of dynamic job shops. *Int. J. Prod. Res.*, 14:255–266, 1981.

[39] J. G. SHANTHIKUMAR and M. J. CHANDRA. Application of level crossing analysis to discrete state processes in queueing systems. *Naval Res. Logistics Q.*, 29:593–608, 1982.

[40] H. SHORE. Simple approximations for the $GI/G/c$ queue-I: the steady-state probabilities. *J. Opl. Res. Soc.*, 39:279–284, 1988.

[41] H. SHORE. Simple approximations for the $GI/G/c$ queue: II. The moments, the inverse distribution function and the loss function of the number in the system and of the queue delay. *J. Opl. Res. Soc.*, 39:381–391, 1988.

[42] S. STIDHAM. A last word on $L = \lambda W$. *Operations Research*, 22:417–422, 1974.

[43] D. STOYAN. *Comparison Methods for Queues and Other Stochastic Models.* John Wiley and Sons, New York, 1983. English edition, edited and revised by D. J. Daley.

[44] H. C. TIJMS, M. H. VAN HOORN, and A. FEDERGRUEN. Approximations for the steady state probabilities in the $M/G/c$ queue. *Adv. Appl. Prob.*, 13:186–206, 1981.

[45] J. WALRAND. *An Introduction to Queueing Networks.* Prentice Hall, Englewood Cliffs, NJ, 1988.

[46] W. WHITT. The Marshall and Stoyan bounds for $IMRL/G/1$ queues are tight. *Oper. Res. Letters*, 1:209–213, 1982.

[47] W. WHITT. A review of $L = \lambda W$ and extensions. *Queueing Systems*, 9:235–268, 1991.

[48] R. W. WOLFF. Poisson arrivals see time averages. *Operations Research*, 30:223–231, 1982.

[49] R. W. WOLFF. *Stochastic Modeling and the Theory of Queues.* Prentice Hall, Englewood Cliffs, NJ, 1989.

[50] D. D. YAO. Refining the diffusion approximations for the $M/G/m$ queue. *Operations Research*, 33:1266–1277, 1985.

[51] U. YECHIALI and P. NAOR. Queueing problems with heterogeneous arrivals and service. *Operations Research*, 19:722–734, 1971.

4

Single-Stage "Produce-to-Stock" Systems

4.1 INTRODUCTION

4.1.1 Motivation

The principal motivation for operating a manufacturing system on a produce-to-stock basis is usually the desire to improve the level of service to customers and gain the competitive advantages that result from improved service. Produce-to-stock operation should reduce the delay in filling customer orders, and also availability of items in stock may lead to increased sales. Produce-to-stock manufacture may also result in savings in manufacturing costs, in particular in costs associated with setting up or starting up the machines or processes. It also may be perceived as resulting in a more uniform or predictable use of facilities leading to either lower labor costs or higher productivity.

Produce-to-stock operation, however, usually implies that there is an inventory of items available to fill customer orders as they arrive. This inventory will occupy space and will have to be maintained and kept secure from theft or pilferage. Usually of greater significance is the fact that the inventory requires an investment of money that could be used elsewhere in the firm. Thus, in deciding whether to produce to stock, the firm has to weigh the advantages of improved service, and perhaps lower manufacturing costs against the costs and other problems associated with keeping inventories.

4.1.2 System Characteristics

Produce-to-stock operation requires that careful consideration be given to the characteristics of customer demand and of the manufacturing process. We will review several aspects that have to be considered.

Production variety.   If it is intended that several different products be made using common facilities then it is important to determine the interrelationships between the demand and manufacture of the products. That is, it is necessary to know whether different products are to be produced simultaneously or one at a time, and whether they will be demanded together (e.g., in sets) or independently. When, for example, two products A and B are produced simultaneously then for every unit of product A k units of product B are produced. This is often the case in chemical production, but it can also happen in other types of manufacture, such as where a press stamps out simultaneously both a stator and a rotor lamination for an electric motor from a sheet of steel. More common is the situation in which only one of the products can be produced at a time on a machine, and there is a change-over cost or time between products. Demands for the products would be related if a customer usually orders one each of all products produced on the facility. At the other extreme demands for the different products may be independent.

Pattern of demand.   Two aspects of the demand pattern are significant: (1) the pattern of arrival of customers and (2) the pattern of demand for items by a given customer. Customer arrivals may be nonstationary (e.g., they may exhibit trends with time and seasonal variations) as well as vary with such external factors as the level of economic activity. Next, arrival patterns may be influenced by the firm through pricing and promotion policies. We shall, however, restrict our analysis to demand patterns that are stationary (although this does not rule out considering demand patterns whose parameters are influenced by inventory level). Each customer may require just one item of one product or the number of items demanded by a customer may be variable, depending on extraneous factors or sometimes on price or the level of inventory. Further, a customer may require more than one product at a time.

Predictability of demand.   If the arrival time and the quantity of items required by each customer are known sufficiently far in advance then it might be possible to meet all demand as it occurs yet keep no inventory. Thus the predictability of demand is an important factor. Because we will be restricting our discussion to stationary demand patterns, by predictability we do not mean just knowledge of the mean level of demand or other summary descriptors of the demand process, but rather the ability to predict partially or fully the actual realization of demand.

Manufacturing capability.   All manufacturing processes are to some extent unreliable or uncertain. That is, machines can fail, tools can break, or operators can be absent. Furthermore, even when the facility is producing, some or all of the items produced may fail to meet specification and have to be scrapped or reworked. This yield loss can be quite significant in some industries such as very large scale integrated (VLSI) chip fabrication. Sometimes the uncertainty of production is complicated by the difficulty and cost of testing products so that defects are not identified until after delivery to the customer.

Change-over and set-up.   In multiple product facilities change-over between products can require a significant time. Also there may be costs associated with starting up or shutting down the facility, or there may be a need to shut down the facility for cleaning, maintenance, or to replace worn tools after some given total quantity of items has been produced. This may lead to production in batches, where the batch size can be fixed and constant, or it can vary between successive runs of a given product.

4.2 ISSUES IN SYSTEM DESIGN AND OPERATION

4.2.1 Design

Before manufacture of an item of a product can begin, we require that both the physical raw material, parts, and necessary tools *and* the production authorization are present. In this chapter we will assume that sufficient raw material, parts, and tools are always available so whether production can occur is solely determined by the presence of the production authorization. By *system design* we mean the specification of (1) the rules that determine how production authorizations are generated and are made available at the manufacturing facility, and (2) the protocols that determine when set-up or change-over of machines occurs. In this chapter we restrict our discussion to single-stage manufacturing facilities, that is, a group of c identical machines, any of which is capable (perhaps after appropriate set-up) of making any one of the products produced by the facility. Completed items of each product are kept in an output store. As customers arrive their demands are met by delivering to them items from the output store. If all demands cannot be met immediately then two alternatives will be considered: *lost sales*, where the customer accepts what is available, if any, and leaves immediately, *back-logged demand*, where the customer waits until all his required demand is eventually met.

Production authorization (PA) cards.   We will initially assume that when each item is produced by the manufacturing facility a tag is associated with the item and thus for every item in the output store there is a tag. When an item of a given product is delivered to a customer the tag is removed, and this then becomes the production authorization or PA card for that product. There are a variety of rules that could be used for determining when the PA cards would be transmitted to the manufacturing facility. We will consider three different rules for a given product.

1. Immediate transmittal to the facility as soon as a PA card is generated.

2. Transmittal of batches of fixed size q as soon as at least q cards have been accumulated.

3. Transmittal of a batch of all available PA cards once at least q have been accumulated.

Note that if customer demands are of unit size there is essentially no difference between rules 2 and 3.

Usually, the presence of one PA card at the manufacturing facility authorizes the production of one item. If there is a yield loss in manufacture, however, then the PA card could authorize manufacture of more than one item; alternatively, when an item is found to be defective it will immediately create a PA card. The preceding rules for transmittal of PA cards can then be used.

Generalized PA systems.   Rather than having the number of tags (and hence the number of PA cards) the same as the maximum inventory permitted in the output store, a more general approach is to divorce the tags from the physical inventory so that the maximum physical inventory is not the same as the maximum number of tags at the output store. When a customer arrives each demand for an item results in (1) the issuance of one physical item from inventory to the customer and (2) the conversion of one tag into one PA card. If the maximum number of tags is less than the maximum physical inventory then there can be a "tag backlog," that is, demands waiting for tags to convert to PA cards, even though the physical demand has been met. If the maximum number of tags is greater than the maximum physical inventory then there can be an "inventory backlog," that is, customers waiting for demands to be met even though the backlogged demands have resulted in PA cards being released. These generalized systems would be advantageous if machine processing rates depend on workload (e.g., rates decline when too many PA cards are waiting at the machine), and they may also be useful in some multiproduct systems in influencing the occurrence of machine change-overs.

Another major advantage of divorcing the number of tags from the maximum inventory is that it is then possible to convert tags to PA cards not just when demands arrive but when information on demand arrivals in the future becomes available. For example, a customer arrival may be signalled at a time of τ units before the actual arrival. If tags are converted to PA cards as soon as the information signal is received then it might be possible to keep a lower final inventory and still meet all demand. Indeed, if production takes less than τ units of time and there is no interference between the production of different items, all demand can be met immediately as it occurs, and there would be no inventory uncommitted to a specific customer.

Set-up and change-over.   If the system only makes a single product then there are a limited number of rules that can be used to decide when to set up or start up the facility. If set-up or start-up requires negligible time or cost then the simplest rule is for the machine to produce whenever PA cards are available. Otherwise the following options are possible:

- Start up once there are at least q PA cards present, continue producing as long as there are any PA cards, then shut down once there are no PA cards and the facility becomes idle.

- If PA cards arrive in batches, set up for each batch.

If the system produces several different products then there are various protocols that could be used for deciding when to change over between products.

FCFS. Process PA cards in order of arrival at the facility. Thus, there would be a change-over every time a product j PA card follows a product i PA card and $i \neq j$ in a facility consisting of a single machine. If PA cards arrive in batches (all PA cards in a batch are for the same product), however, then this will assure that each run of a product is at least equal to the size of the batch.

Alternating Priority (AP). Once a machine is set up for a given product then it continues producing that product as long as there are any PA cards for that product available. When there are no PA cards for the current product then the machine changes over to the product of the first PA card in order of arrival at the facility.

NPP. Suppose there is a predefined priority listing for the different products produced by the facility. Then under the priority protocol the facility will produce the product of the highest priority for which PA cards are present until either no PA cards for that product are left, when it will change over to the product of the next highest priority, or until a PA card for a higher-priority product arrives, when the facility will change over to that product as soon as the item currently in process is complete. Rather than a predetermined priority listing, priorities could also be set dynamically. For example, if a PA card is generated because an item is found to be defective then that product could be given high priority.

4.2.2 Operation

Once the rules for PA card generation and transmittal and the protocols for determining set-up and change-over have been decided, it is necessary to decide on the parameters of the rules. These usually can be changed relatively easily, in accordance with changes in demand or the requirements of a particular manufacturing location. The major parameters are the maximum inventory level permitted and the batch sizes for transmittal of PA cards. Their optimal values can generally be found using cost models analogous to classical inventory theory, where the cost components are inventory holding costs, set-up and change-over costs, and shortage or run-out costs. The cost models require the determination of such performance indicators as (1) the average finished product inventory, (2) the average backlogged demand, (3) the average lead time to fill an order, (4) the probability a demand has to be backlogged, and (5) the frequency of set-up or change-over. We will develop models that, given the rules and protocols, will enable these performance indicators to be determined.

4.3 SINGLE-PRODUCT-TYPE PRODUCE-TO-STOCK SYSTEMS

Consider a single-stage manufacturing system that produces items of a single product type to stock. Completed items are kept in a store from which customer demands are met. Customers arrive according to a process $\{A_n, n = 1, 2, \ldots\}$; A_n is the arrival time of the nth customer. The number of items (i.e., the quantity) required by the nth customer is X_n, $n = 1, 2, \ldots$. If a customer's demand cannot be met from available stock, the customer may leave and satisfy his demand elsewhere (the lost sales case). Alternatively, he may choose to wait until his demand can be satisfied, thus allowing the backlogging of his order. The manufacturing process of items involves the transformation of the raw material by processing it on a single machine. Items are processed one at a time and the processing time of the nth item is S_n, $n = 1, 2, \ldots$.

Even though the system is intended to produce to stock, the machines need not be working on items as soon as raw material becomes available. The way in which raw material will be taken for processing will depend on the type of PA mechanism. The type of model needed and the form of the solution approach will depend on the PA mechanism implemented. We will next formulate and analyze models for different types of PA mechanisms and compare their merits.

4.3.1 Target-Level PA Mechanism

Consider the PA mechanism that stops production as soon as the number of items in the store reaches a target level, say Z. Production authorization, in this case, is transmitted to the manufacturing facility only when the number of completed items is fewer than this target. A systematic implementation of this production authorization mechanism is as follows:

> Each item in the store will have a tag. Each time an item is taken from the store by a demand, the tag attached to it is removed and transmitted to the manufacturing facility as a PA card. Consequently, if the inventory in the output store is at the target level Z (i.e., the store is full), there will be no PA cards available for the facility, thus forcing all machines in the facility to be idle. When the output store is empty, and if backlogging of demands is permitted, the customer demands will be kept as a queue of backlogs. Observe that the backlogged customer demand will not initiate the release of a PA card to the manufacturing facility. If backlogging is not permitted the customer demands that cannot be met from the items in the output store are lost.

Single machine with unit demand and backlogging.   Initially, we will assume that there is only a single machine to process the items and that customer demands are backlogged. We assume that there are Z tags available in the system. When there are r finished items in the output store, r of these tags are attached, one for each of the finished items. The remaining $Z - r$ tags will be available at the machine acting as PA cards. Suppose the output store is initially full at time zero, and that there is unlimited raw material available in front of the machine at all times. In addition we look at the case where $X_n = 1$, $n = 1, 2, \ldots$ (i.e., each customer requires only one item). At time A_1 the first customer arrives and takes an item from the store and leaves the system. The tag on that item is immediately removed and sent to the machine as a PA card. Hence at time $D_1 = A_1 + S_1$ the first newly processed item will be received at the output store. Now suppose the nth item is received at the output store at time D_n, $n = 1, 2, \ldots$. If $A_{n+1} > D_n$, the store is full at time D_n, and therefore the machine will begin processing the $(n + 1)$th item only at time A_{n+1}. In this case $D_{n+1} = A_{n+1} + S_{n+1}$. Conversely, if $A_{n+1} \leq D_n$, there is at least one PA card available at the machine at time D_n. Hence $D_{n+1} = D_n + S_{n+1}$. Combining these two cases we get

$$D_{n+1} = \{D_n \vee A_{n+1}\} + S_{n+1}, \quad n = 1, 2, \ldots \tag{4.1}$$

$\{D_n, n = 1, 2, \ldots\}$ given by equation 4.1 is the same as the departure process of a single server queueing system with arrival times $\{A_n, n = 1, 2, \ldots\}$ and service times $\{S_n, n = 1, 2, \ldots\}$ and FCFS service protocol provided the queueing system is empty at time zero. Let $I(t)$ be the inventory, that is, the number of finished items, in the output store, $R(t)$ be the number of items delivered to customers, $B(t)$ be the number of customers backlogged at time t, and $C(t)$ be the number of PA cards available at the machine at time t. Then

$$I(t) = Z - C(t) = Z + D(t) - R(t) \tag{4.2}$$

$$R(t) = \min\{Z + D(t), A(t)\} \tag{4.3}$$

$$B(t) = A(t) - R(t) \tag{4.4}$$

$$C(t) = \min\{A(t) - D(t), Z\} \tag{4.5}$$

where $A(t)$ is the number of customers that arrived during $(0, t]$ and $D(t) = \sup\{n : D_n \leq t, n = 0, 1, \ldots\}$ is the number of items produced during $(0, t]$. Let $N(t)$ be the number of jobs in the single server queueing system described earlier. Then

$$N(t) = A(t) - D(t)$$

and from equations 4.2 and 4.4 one gets

$$I(t) - B(t) = Z + D(t) - A(t) = Z - N(t) \tag{4.6}$$

Because $I(t) > 0$ implies $B(t) = 0$ and $B(t) > 0$ implies $I(t) = 0$, one has from equation 4.6,

$$I(t) = \{Z - N(t)\}^+ \tag{4.7}$$

and

$$B(t) = \{N(t) - Z\}^+ \tag{4.8}$$

Then from equation 4.5 one gets

$$C(t) = \min\{N(t), Z\} \tag{4.9}$$

It is now not difficult to see that

$$B(t) + C(t) = N(t), \quad t \geq 0 \tag{4.10}$$

Therefore, it is sufficient for us to study the process $\{N(t), t \geq 0\}$ to obtain information on $\{I(t), t \geq 0\}$, $\{B(t), t \geq 0\}$ and $\{C(t), t \geq 0\}$.

M/M/1 Model.   As an example of an application of this analogy, we will first consider a $M/M/1$ model for this system when the customer arrival process is Poisson with rate λ and the processing times are iid and exponentially distributed with mean $1/\mu$. Using $P\{N = n\} = (1 - \rho)\rho^n$, $n = 0, 1, \ldots$, for $\rho < 1$ (see Section 3.3.1) we get, from equations 4.7 to 4.9

$$P\{B = n\} = \begin{cases} 1 - \rho^{Z+1}, & n = 0 \\ (1 - \rho)\rho^{n+Z}, & n = 1, 2, \ldots \end{cases} \tag{4.11}$$

$$P\{C = n\} = \begin{cases} (1 - \rho)\rho^n, & n = 0, \ldots, Z - 1 \\ \rho^Z, & n = Z \end{cases} \tag{4.12}$$

$$P\{I = n\} = \begin{cases} \rho^Z, & n = 0 \\ (1 - \rho)\rho^{Z-n}, & n = 1, 2, \ldots, Z \end{cases} \tag{4.13}$$

Also,

$$E[B] = \frac{\rho^{Z+1}}{1 - \rho} \tag{4.14}$$

$$E[C] = \frac{\rho}{1 - \rho}(1 - \rho^Z) \tag{4.15}$$

$$E[I] = Z - \frac{\rho}{1 - \rho}(1 - \rho^Z) \tag{4.16}$$

If the inventory carrying cost is k_1 per item per unit time, and the backlogging cost is k_2 per item per unit time, the total cost rate $TC(Z)$ for this target level PA mechanism is

$$TC(Z) = k_1 \left(Z - \frac{\rho}{1 - \rho}(1 - \rho^Z) \right) + k_2 \frac{\rho^{Z+1}}{1 - \rho} \tag{4.17}$$

It is easily verified that $TC(Z)$ is convex in Z and that $\Delta TC(Z) = TC(Z) - TC(Z-1)$ is given by

$$\Delta TC(Z) = k_1 - (k_1 + k_2)\rho^Z \tag{4.18}$$

Hence the optimal target level Z^* is the integer $\lfloor \hat{Z} \rfloor$ or $\lceil \hat{Z} \rceil$ where $\hat{Z} = \ln(k_1/[k_1 + k_2])/\ln \rho$, which minimizes $TC(Z)$.

Because the customers arrive according to a Poisson process the number of customers backlogged and the number of items in the output store seen by a customer on its arrival will have the same distribution as B and I specified earlier in equations 4.11 and 4.13. Therefore

$$P\{\text{a customer is backlogged}\} = P\{I = 0\} = \rho^Z \tag{4.19}$$

In addition, if L is the time to fill a demand (fill time) under stationary conditions

$$
P\{L > x\} = P\left\{ \sum_{n=1}^{B+I_{\{N \geq Z\}}} S_n > x \right\}
$$

$$
= \sum_{r=0}^{\infty} \sum_{k=0}^{r} \frac{e^{-(\mu x)}(\mu x)^k}{k!}(1 - \rho)\rho^{Z+r}
$$

$$
= \rho^Z e^{-(\mu - \lambda)x}, \quad x \geq 0 \tag{4.20}
$$

Therefore

$$
P\{L = 0\} = 1 - \rho^Z \tag{4.21}
$$

and

$$
E[L] = \frac{\rho^Z}{\mu - \lambda} \tag{4.22}
$$

GI/G/1 Model. In general, if the customer arrivals form a renewal process and the processing times are iid, we can model the system by a $GI/G/1$ model and use the result in section 3.3.4.

Let $\hat{N}$ be the approximate average number of jobs in the system computed for this $GI/G/1$ queueing system and $\sigma = (\hat{N} - \rho)/\hat{N}$. Then we approximate the distribution of B, C, and I by

$$
P\{B = n\} \approx \begin{cases} 1 - \rho\sigma^Z, & n = 0 \\ \rho(1 - \sigma)\sigma^{n-1+Z}, & n = 1, 2, \ldots \end{cases}
$$

$$
P\{C = n\} \approx \begin{cases} 1 - \rho, & n = 0 \\ (1 - \rho)\rho\sigma^{n-1}, & n = 1, \ldots, Z - 1 \\ \rho\sigma^{Z-1}, & n = Z \end{cases}
$$

$$
P\{I = n\} \approx \begin{cases} \rho\sigma^{Z-1}, & n = 0 \\ (1 - \rho)\rho\sigma^{Z-n-1}, & n = 1, \ldots, Z - 1 \\ 1 - \rho & n = Z \end{cases}
$$

Single machine with unit demand and lost sales. We consider a single-machine produce-to-stock system in which customer demands not met by items from the output store are lost. The nth customer arrives at time A_n and demands one item (i.e., $X_n = 1$), $n = 1, 2, \ldots$. The processing times $\{S_n, n = 1, 2, \ldots\}$ of the items are iid and mutually independent of $\{A_n, n = 1, 2, \ldots\}$. Let C_n denote the customer whose demand is met by the nth item delivered from the output store. Because the customer demands not met by items from the output store are lost, C_n will usually be greater than n. Suppose D_n is the time at which the nth processed item is transferred from the manufacturing facility to the output store. Then if the $(n + 1)$th delivery to a customer occurs at time A_k, the inventory in the store just before to this customer's arrival must be positive, that is, the number of transfers from the manufacturing facility to the output store cannot be less than $n + 1 - Z$, or $A_k \geq D_{n+1-Z}$. Hence it follows that

$$C_{n+1} = \min\{k : A_k \geq D_{n+1-Z}, k \geq C_n + 1\}$$

and

$$D_{n+1} = \max\{D_n, A_{C_{n+1}}\} + S_{n+1}, \quad n = 0, 1, \ldots$$

where $D_{-n} = 0$, $n = 0, 1, \ldots$, and $C_0 = 0$. If $D(t) = \sup\{n : D_n \leq t, n = 0, 1, \ldots\}$ and $R(t) = \sup\{n : A_{C_n} \leq t, n = 0, 1, \ldots\}$, then $N(t) = R(t) - D(t)$ is the number of jobs in a single server queueing system $GI/G/1/Z$ with arrival process $\{A_n, n = 1, 2, \ldots\}$, service process $\{S_n, n = 1, 2, \ldots\}$ and with a buffer size limit Z on the number of jobs in the system at any time. Jobs finding the buffer full are rejected and lost from the system forever. The performance of the produce-to-stock manufacturing facility can be obtained from the performance of the $GI/G/1/Z$ queueing system in the same way as we did for the case of backlogging. Specifically we have

$$I(t) = Z - N(t), \quad t \geq 0$$

and

$$C(t) = N(t), \quad t \geq 0$$

Define the service level SL as the fraction of customer demands met, that is, $SL = \lim_{t \to \infty} R(t)/A(t)$. To find SL we need the stationary probability that an arrival sees an empty output store. Let $N_n^- = N(A_{n^-})$, $n = 1, 2, \ldots$. Then

$$R(t) = \sum_{n=1}^{A(t)} I_{\{Z - N_n^- \geq 1\}}$$

and

$$SL = 1 - \lim_{n \to \infty} P\{N_n^- = Z\} \qquad (4.23)$$

As an illustration we will first consider the $M/M/1/Z$ model before presenting the results for the $GI/G/1/Z$ case.

M/M/1/Z Model.   Customers arrive according to a Poisson process with rate λ that is independent of the sequence of processing times at the machine. The processing times are iid with exponential distribution that has a mean $1/\mu$. Then $\{N(t), t \geq 0\}$ is a birth-death process on the state space $S = \{0, \ldots, Z\}$. Let $p(n) = \lim_{t \to \infty} P\{N(t) = n\}$, $n \in S$ be its stationary distribution. Then solving the stationary balance equations of $\{N(t), t \geq 0\}$ one finds that

$$p(n) = \frac{(1 - \rho)\rho^n}{1 - \rho^{Z+1}}, \quad n = 0, \ldots, Z \qquad (4.24)$$

where $\rho = \lambda/\mu$. Then the distribution of the inventory and the number of PA cards at the machine are

$$P\{I = n\} = \frac{(1 - \rho)\rho^{Z-n}}{1 - \rho^{Z+1}}, \quad n = 0, \ldots, Z \qquad (4.25)$$

and

$$P\{C = n\} = \frac{(1 - \rho)\rho^n}{1 - \rho^{Z+1}}, \quad n = 0, \ldots, Z \tag{4.26}$$

The corresponding means are

$$E[I] = \frac{(1 - \rho)\left(Z + \rho^{Z+1}\right) - \rho\left(1 - \rho^{Z+1}\right)}{(1 - \rho)\left(1 - \rho^{Z+1}\right)} \tag{4.27}$$

and

$$E[C] = \frac{\rho\left(1 - \rho^{Z+1}\right) - (1 - \rho)(Z + 1)\rho^{Z+1}}{(1 - \rho)\left(1 - \rho^{Z+1}\right)} \tag{4.28}$$

To find the service level observe that because Poisson arrivals see time averages, $P\{N^- = n\} = p(n)$, $n = 0, \ldots, Z$. Therefore

$$SL = 1 - p(Z) = \frac{1 - \rho^Z}{1 - \rho^{Z+1}} \tag{4.29}$$

Suppose the inventory carrying cost is k per item per unit time, and the profit per item (excluding the inventory carrying cost) is v. Then the total profit per unit time is

$$
\begin{aligned}
TP(Z) &= v \cdot \lambda SL - k \cdot E[I] \\
&= v \cdot \lambda \left(\frac{1 - \rho^Z}{1 - \rho^{Z+1}}\right) - k \cdot \left\{\frac{(1 - \rho)\left(Z + \rho^{Z+1}\right) - \rho\left(1 - \rho^{Z+1}\right)}{(1 - \rho)\left(1 - \rho^{Z+1}\right)}\right\}
\end{aligned}
$$

Now the value of Z that maximizes $TP(Z)$ can be obtained.

M/G/1/Z Model.   Customers arrive according to a Poisson process with rate λ and the item processing times $\{S_n, n = 1, 2, \ldots\}$ are iid with distribution function F_S. Let N_n^+ be the number of jobs in an $M/G/1/Z$ queueing system just after the nth service completion epoch. Then $\{N_n^+, n = 1, 2, \ldots\}$ forms a Markov chain on the state space $S = \{0, \ldots, Z - 1\}$ with transition probability matrix $\mathbf{P} = (p_{ij})_{i,j \in S}$.

$$p_{0j} = a_j, \quad j = 0, 1, \ldots, Z - 2$$

$$p_{0,Z-1} = 1 - \sum_{j=0}^{Z-2} p_{0j}$$

$$p_{ij} = 0, \quad j = 0, \ldots, i - 2; i = 2, \ldots, Z - 1$$

$$p_{ij} = a_{j+1-i}, \quad j = i - 1, i, \ldots, Z - 2; i = 1, 2, \ldots, Z - 1$$

$$p_{i,Z-1} = 1 - \sum_{j=0}^{Z-2} p_{ij}$$

where

$$a_k = \int_0^\infty \frac{e^{-\lambda t}(\lambda t)^k}{k!} dF_S(t)$$

is the probability that k jobs arrive during a service time, $k = 0, 1, \ldots$. If $\mathbf{p}^+$ is the stationary probability vector of $\{N_n^+, n = 1, 2, \ldots\}$ the corresponding probability flow balance equations for the states $n = 0, \ldots, Z - 2$ are

$$p^+(n) = p^+(0)a_n + \sum_{j=1}^{n+1} p^+(j)a_{n+1-j}, \quad n = 0, \ldots, Z - 2$$

These $Z - 1$ equations along with the normalizing equation $\sum_{n=0}^{Z-1} p^+(n) = 1$ can be solved to obtain $\mathbf{p}^+$. Observe that when $\rho < 1$ the above $Z - 1$ equations are exactly the same as the balance equations for the $M/G/1$ queue for states $n = 0, \ldots, Z - 2$. Hence if $\mathbf{p}_\infty$ is the stationary probability distribution of the number of jobs in an $M/G/1$ queue, one sees that

$$p^+(n) = \frac{p_\infty(n)}{1 - \overline{P}_\infty(Z)}, \quad n = 0, \ldots, Z - 1$$

where $\overline{P}_\infty(Z) = \sum_{n=Z}^{\infty} p_\infty(n)$.

Now let $\mathbf{p}$ be the time average probability distribution of the number of jobs in the $M/G/1/Z$ queue. Then the rate at which customers are accepted into the system is $\hat{\lambda} = \lambda(1 - p(Z))$. Because arrivals and departures occur one at a time the probability distribution of the number of jobs seen on arrival by an *accepted* job is the same as $\mathbf{p}^+$. Then the rate of upcrossings over level n is

$$\hat{\lambda} p^+(n) = \lambda p(n), \quad n = 0, \ldots, Z - 1$$

So

$$p(n) = p_\infty(n) \frac{1 - p(Z)}{1 - \overline{P}_\infty(Z)}, \quad n = 0, \ldots, Z - 1 \tag{4.30}$$

The server utilization of the $M/G/1/Z$ queueing system is $1 - p(0)$, and the rate at which work is lost is $(\lambda - \hat{\lambda})/\mu = \rho p(Z)$. Adding these two up we should have

$$1 - p(0) + \rho p(Z) = \rho \tag{4.31}$$

Then solving for $p(Z)$ using equation 4.31 and equation 4.30 for $n = 0$ we get

$$p(Z) = \frac{(1 - \rho)\overline{P}_\infty(Z)}{1 - \rho \overline{P}_\infty(Z)}. \tag{4.32}$$

Therefore

$$p(n) = \begin{cases} \frac{p_\infty(n)}{1 - \rho \overline{P}_\infty(Z)}, & n = 0, \ldots, Z - 1 \\ \frac{(1-\rho)\overline{P}_\infty(Z)}{1 - \rho \overline{P}_\infty(Z)}, & n = Z \end{cases} \tag{4.33}$$

Hence we have

$$P\{I = n\} = \begin{cases} \frac{(1-\rho)\overline{P}_\infty(Z)}{1 - \rho \overline{P}_\infty(Z)}, & n = 0 \\ \frac{p_\infty(Z-n)}{1 - \rho \overline{P}_\infty(Z)}, & n = 1, \ldots, Z \end{cases} \tag{4.34}$$

and

$$SL = 1 - p(Z) = \left(\frac{1 - \overline{P}_\infty(Z)}{1 - \rho \overline{P}_\infty(Z)} \right) \tag{4.35}$$

GI/G/1/Z Model. Suppose the customers arrive according to a renewal process, and the item processing times are iid. To obtain the probability distribution of N we will use the approximation for the $GI/G/1$ queue. The relationship between the distribution of N in the $M/M/1/Z$ model and that for the $M/M/1$ queueing system can be explained as follows: The functional form of the distribution of N in the $M/M/1/Z$ model is the same as that of the normalized distribution of N in the $M/M/1$ model. For $\lambda < \mu$, let $\hat{N} = \hat{N}_{M/M/1}(\lambda, \mu)$ be the mean number of jobs in a $M/M/1$ queue with arrival rate λ and service rate μ. Then the value of ρ needed for the distribution of N is obtained by setting $\rho = (\hat{N} - \rho)/\hat{N}$. If $\lambda > \mu$, let $\hat{N}_R = \hat{N}_{M/M/1}(\mu, \lambda)$ be the mean number of jobs in the (reversed) $M/M/1$ queue with arrival rate μ and service rate λ. Here $\rho = \hat{N}_R/(\hat{N}_R - 1/\rho)$. Suppose $\hat{N}(\lambda, C_a^2, \mu, C_S^2)$ is an approximation selected (see Chapter 3) for the average number of jobs in the $GI/G/1$ queueing system when $\rho < 1$ and $\hat{N}_R = \hat{N}(\mu, C_S^2, \lambda, C_a^2)$ is the approximation selected for the average number of jobs in a $GI/G/1$ queueing system when $\rho > 1$. Then adapting the structure discussed earlier for the $M/M/1/Z$ model results, we approximate the probability distribution of N by

$$p(n) \approx \begin{cases} \frac{1-\rho}{1-\rho\sigma^Z}, & n = 0 \\ \frac{\rho(1-\sigma)\sigma^{n-1}}{1-\rho\sigma^Z}, & n = 1, \ldots, Z \end{cases} \tag{4.36}$$

where $\sigma = (\hat{N} - \rho)/\hat{N}$ when $\rho < 1$, $\sigma = \hat{N}_R/(\hat{N}_R - 1/\rho)$ when $\rho > 1$. For $\rho = 1$, we use the limiting result of equation 4.36 as $\rho \to 1$. Then

$$P\{I = n\} \approx \begin{cases} \frac{\rho(1-\sigma)\sigma^{Z-n-1}}{1-\rho\sigma^Z}, & n = 0, \ldots, Z - 1 \\ \frac{1-\rho}{1-\rho\sigma^Z}, & n = Z, \end{cases}$$

and

$$P\{C = n\} \approx \begin{cases} \frac{1-\rho}{1-\rho\sigma^Z}, & n = 0 \\ \frac{\rho(1-\sigma)\sigma^{n-1}}{1-\rho\sigma^Z}, & n = 1, \ldots, Z \end{cases}$$

Observe that the fraction of time the server is busy is $1 - p(0)$, and therefore the customer departure rate is $\mu(1 - p(0))$. If SL is the service level, the rate at which customers are accepted into the $GI/G/1/Z$ system is $\lambda.SL$. Now equating the customer departure rate to the rate at which customers are accepted from equation 4.36 we get

$$SL \approx \frac{1 - \sigma^Z}{1 - \rho\sigma^Z}$$

Note that if the approximation $\hat{N}$ selected for the $GI/G/1$ queue is exact for $M/M/1$, then all the preceding approximations are exact for the $M/M/1/Z$ model.

Single machine with interruptible demand.   So far we have assumed that customers arrive at the system even when the output store is empty. When customer demands are generated by a single source, the generation of customers (demands) may be turned off, as soon as the output buffer becomes empty and turned back on when it is no longer empty. That is, once an arrival results in the output store becoming empty the arrival generation process is switched off and no more arrivals can be generated as long as the store is empty. Not until the store receives a processed item can the arrival generation process be switched on again. In such a case the customer arrival process is described by the sequence $\{\tau_n, n = 1, 2, \ldots\}$ of times needed to generate customers. We will next analyze this *stopped arrival* produce-to-stock system.

As before let D_n be the time at which the nth processed item is transferred from the manufacturing facility to the output store, and let A_n be the arrival time of the nth demand. Then

$$A_{n+1} = \max\{A_n, D_{n-Z+1}\} + \tau_{n+1}$$

and

$$D_{n+1} = \max\{D_n, A_{n+1}\} + S_{n+1}, \quad n = 0, 1, \ldots$$

where $D_{-n} = 0$, $n = 0, 1, \ldots$, and $A_0 = 0$. Let $A(t) = \sup\{k : A_k \le t, k = 0, 1, \ldots\}$ be the number of customers arrived during $(0, t]$, $D(t) = \sup\{k : D_k \le t, k = 0, 1, \ldots\}$ and let $N(t)$ be the number of jobs in a $GI/G/1/Z$: stopped arrival queueing system with interarrival times $\{\tau_n, n = 1, 2, \ldots\}$, service times $\{S_n, n = 1, 2, \ldots\}$ and finite buffer capacity Z. Then

$$I(t) = Z - N(t), \quad t \ge 0$$

For this system we define the service level SL as the ratio of the actual customer demand generated to the maximum number of customer demands, that could have been generated. This is then equal to the fraction of time the input process is actively generating demands, which is equal to fraction of time the inventory level in the output store is greater than zero. That is,

$$SL = 1 - p(Z) \tag{4.37}$$

When the interarrival times are exponentially distributed it can be verified that the arrival process $\{A_n, n = 1, 2, \ldots\}$ has the same distributional properties as $\{A_{C_n}, n = 1, 2, \ldots\}$ in the $M/G/1/Z$ model. Therefore the results derived earlier for the $M/G/1/Z$ lost arrival model are applicable for the $M/G/1/Z$ stopped arrival model.

GI/M/1/Z Stopped Arrival Model.   Now consider the case where the service times are exponentially distributed. In this case $\{N_n^-, n = 1, 2, \ldots\}$ forms a Markov chain on the state space $\mathcal{S} = \{0, \ldots, Z - 1\}$. The transition probability matrix $\mathbf{P} = (p_{ij})_{i,j \in \mathcal{S}}$, where $p_{ij} = P\{N_{n+1}^- = j | N_n^- = i\}$ is given by

$$p_{ij} = 0, \quad j = i + 2, \ldots, Z - 1; i = 0, \ldots, Z - 3$$

$$p_{ij} = \int_{t=0}^{\infty} \frac{e^{-\mu t} (\mu t)^{i+1-j}}{(i+1-j)!} dF_\tau(t), \quad ; j = 1, \ldots, i+1; i = 0, \ldots, Z - 2$$

$$p_{Z-1,j} = \int_{t=0}^{\infty} \frac{e^{-\mu t} (\mu t)^{Z-1-j}}{(Z-1-j)!} dF_\tau(t), \quad j = 1, \ldots, Z - 1$$

and

$$p_{i0} = 1 - \sum_{j=1}^{Z-1} p_{ij}, \quad i = 0, \ldots, Z - 1$$

The steady-state probability vector $\mathbf{p}^-$ of N^- can then be obtained by solving $\mathbf{p}^- \mathbf{P} = \mathbf{0}$ and $\mathbf{p}^- \mathbf{e} = 1$. In the case of Poisson arrival process we have

$$p^-(n) = \frac{(1-\rho)\rho^n}{1 - \rho^Z}, \quad n = 0, \ldots, Z - 1$$

where $\rho = \lambda/\mu$ and $\lambda = 1/E[\tau]$. Note that $p^-(n)$ here is the probability that an *accepted* job in an $M/M/1/Z$ queueing system sees n jobs in the system at its arrival epoch, $n = 0, \ldots, Z - 1$. This is the same as the probability distribution of the number of jobs seen by an *arrival* in an $M/M/1/Z - 1$ queueing system.

This equivalence between the $M/M/1/Z$ stopped arrival and $M/M/1/Z-1$ queueing system extends to the general renewal arrival process case as well. To see this let $\hat{N}_n^-$ be the number of jobs seen by the nth arrival at a $GI/M/1/Z - 1$ queue with iid interarrival times $\{\tau_n, n = 1, 2, \ldots\}$, service rate μ, and buffer capacity $Z - 1$. It can be easily seen that $\{\hat{N}_n^-, n = 1, 2, \ldots\}$ is a Markov chain on $\mathcal{S} = \{0, \ldots, Z - 1\}$, and its transition probability matrix is the same as that for $\{N_n^-, n = 1, 2, \ldots\}$. Therefore *the steady-state probability distribution of the number of jobs seen by an arrival in a $GI/M/1/Z$ stopped arrival queueing system is the same as that of the number of jobs seen by an arrival in a $GI/M/1/Z - 1$ queueing system.*

If an arrival sees n jobs in the system, the average time taken for the next arrival in the stopped arrival system is $1/\lambda$ if $n \leq Z - 2$ and $1/\lambda + 1/\mu$ if $n = Z - 1$. Therefore the average customer arrival rate $\hat{\lambda}$ is given by

$$\hat{\lambda} = 1 / \left\{ \frac{1}{\lambda} + \frac{1}{\mu} p^-(Z - 1) \right\}$$

$$= \lambda \left\{ \frac{1}{1 + \rho p^-(Z - 1)} \right\}$$

Therefore

$$SL = 1 / \{1 + \rho p^-(Z - 1)\}$$

Let $N(t)$ be the number of jobs in the $GI/M/1/Z$ stopped arrival queue at time t. Now equating the rate of upcrossings ($\hat{\lambda} p^-(n - 1)$) and downcrossings ($\mu p(n)$) over level n one finds that

$$p(n) = \begin{cases} \frac{1 - \rho + \rho p^-(Z-1)}{1 + \rho p^-(Z-1)}, & n = 0 \\ \frac{\rho p^-(n-1)}{1 + \rho p^-(Z-1)}, & n = 1, \ldots, Z \end{cases} \qquad (4.38)$$

Hence we have

$$P\{I = n\} = \begin{cases} \frac{\rho p^-(Z-n-1)}{1 + \rho p^-(Z-1)}, & n = 0, \ldots, Z - 1 \\ \frac{1 - \rho + \rho p^-(Z-1)}{1 + \rho p^-(Z-1)}, & n = Z \end{cases} \qquad (4.39)$$

GI/G/1/Z Stopped Arrival Model.   For this model we obtain approximations for
$\mathbf{p}^-$, $\mathbf{p}$ and hence for $P\{I = n\}$ and SL using the properties observed for the $M/G/1/Z$
and $GI/M/1/Z$ stopped arrival models. First we adopt the following relationships
(equations 4.30 and 4.32) from the $M/G/1/Z$ stopped arrival model for the case $\rho < 1$:

$$p(n) = p_\infty(n) \left\{ \frac{1 - p(Z)}{1 - \overline{P}_\infty(Z)} \right\}, \quad n = 0, \ldots, Z - 1$$

and

$$p(Z) = \frac{(1 - \rho)\overline{P}_\infty(Z)}{1 - \rho\overline{P}_\infty(Z)}$$

Now using the approximation $p_\infty(0) = (1-\rho)$ and $p_\infty(n) = \rho(1-\sigma)\sigma^{n-1}, n = 1, 2, \ldots$
where $\sigma = (\hat{N} - \rho)/\hat{N}$ and $\hat{N}(\lambda, C_a^2, \mu, C_S^2)$ is an approximation chosen for the average
number of jobs in a $GI/G/1$ queue, one gets

$$p(n) \approx \begin{cases} \dfrac{1-\rho}{1-\rho^2\sigma^{Z-1}}, & n = 0 \\[2mm] \dfrac{\rho(1-\sigma)\sigma^{n-1}}{1-\rho^2\sigma^{Z-1}}, & n = 1, \ldots, Z - 1 \\[2mm] \dfrac{(1-\rho)\rho\sigma^{Z-1}}{1-\rho^2\sigma^{Z-1}}, & n = Z \end{cases} \tag{4.40}$$

When $\rho > 1$, consider a $(GI/G/1/Z)_R$: stopped arrival queueing system with
interarrival times $\{S_n, n = 1, 2, \ldots\}$ and service times $\{\tau_n, n = 1, 2, \ldots\}$. Let $N_R(t)$ be
the number of jobs in it at time t, and assume that $N_R(0) = Z$. It can then be verified
that

$$Z - N_R(t) = N(t), \quad t \geq 0$$

Therefore if $\mathbf{p}_R$ is the stationary probability distribution of N_R,

$$p(n) = p_R(Z - n), \quad n = 0, \ldots, Z$$

Therefore when $\rho > 1$ we consider the approximation $\hat{N}_R = \hat{N}(\mu, C_S^2, \lambda, C_a^2)$ for the
average number of jobs in the $(GI/G/1)_R$ system and approximate $\mathbf{p}$ by

$$p(n) \approx \begin{cases} \dfrac{(1-\rho_R)\rho_R\sigma_R^{Z-1}}{1-\rho_R^2\sigma_R^{Z-1}}, & n = 0 \\[2mm] \dfrac{\rho_R(1-\sigma_R)\sigma_R^{Z-1-n}}{1-\rho_R^2\sigma_R^{Z-1}}, & n = 1, \ldots, Z - 1 \\[2mm] \dfrac{1-\rho_R}{1-\rho_R^2\sigma_R^{Z-1}}, & n = Z \end{cases} \tag{4.41}$$

where $\rho_R = \mu/\lambda = 1/\rho$ and $\sigma_R = (\hat{N}_R - \rho_R)/\hat{N}_R$. Substituting $\rho = 1/\rho_R$ and $\sigma = 1/\sigma_R$
in equation 4.41 one sees that

$$p(n) \approx \begin{cases} \dfrac{1-\rho}{1-\rho^2\sigma^{Z-1}}, & n = 0 \\[2mm] \dfrac{\rho(1-\sigma)\sigma^{n-1}}{1-\rho^2\sigma^{Z-1}}, & n = 1, \ldots, Z - 1 \\[2mm] \dfrac{(1-\rho)\rho\sigma^{Z-1}}{1-\rho^2\sigma^{Z-1}}, & n = Z \end{cases} \tag{4.42}$$

Observe that approximation 4.42 is the same as that for the case of $\rho < 1$. Except that here σ is defined by $\hat{N}_R/(\hat{N}_R - 1/\rho)$, where $\hat{N}_R$ is the approximate number of jobs in the $(GI/G/1)_R$ system. Taking the limit as $\rho \to 1$ and $\sigma \to 1$ and defining $v = \lim_{\rho \to 1} d\sigma/d\rho$ one gets

$$p(n) \approx \begin{cases} \frac{1}{2+(Z-1)v}, & n = 0 \\ \frac{v}{2+(Z-1)v}, & n = 1, \ldots, Z-1 \\ \frac{1}{2+(Z-1)v}, & n = Z \end{cases} \tag{4.43}$$

For example, if we use the approximation $\hat{N}(\lambda, C_a^2, \mu, C_S^2) = \lambda \hat{T}_1$ given by equation 3.142 we get, after some algebra

$$v = 2/\left(C_a^2 + C_S^2\right) \tag{4.44}$$

Observe that when $C_a^2 + C_S^2 > 2$ we get the boundary state probabilities (i.e., $p(0)$ and $p(Z)$) to be larger than the internal state probabilities. When $C_a^2 + C_S^2 < 2$, the opposite is true.

The corresponding approximations for $P\{I = n\}$ and SL are then for $\rho \neq 1$

$$P\{I = n\} \approx \begin{cases} \frac{(1-\rho)\rho\sigma^{Z-1}}{1-\rho^2\sigma^{Z-1}}, & n = 0 \\ \frac{\rho(1-\sigma)\sigma^{Z-1-n}}{1-\rho^2\sigma^{Z-1}}, & n = 1, \ldots, Z-1 \\ \frac{1-\rho}{1-\rho^2\sigma^{Z-1}}, & n = Z \end{cases} \tag{4.45}$$

and for $\rho = 1$

$$P\{I = n\} \approx \begin{cases} \frac{1}{2+(Z-1)v}, & n = 0 \\ \frac{v}{2+(Z-1)v}, & n = 1, \ldots, Z-1 \\ \frac{1}{2+(Z-1)v}, & n = Z \end{cases} \tag{4.46}$$

Then from equation 4.37 one has

$$SL \approx \frac{1 - \rho\sigma^{Z-1}}{1 - \rho^2\sigma^{Z-1}}, \quad \rho \neq 1 \tag{4.47}$$

$$SL \approx \frac{1 + (Z-1)v}{2 + (Z-1)v}, \quad \rho = 1 \tag{4.48}$$

or using the approximation given by equation 4.44

$$SL \approx \frac{C_a^2 + C_S^2 + 2(Z-1)}{2(C_a^2 + C_S^2 + Z - 1)} \tag{4.49}$$

when $\rho = 1$.

Because the maximum rate at which demands can be met is the service rate μ, an upper bound on the service level for $\lambda \geq \mu$ or $\rho \geq 1$ is $\mu/\lambda = 1/\rho$. Observe that the

TABLE 4.1 ADEQUACY OF SERVICE-LEVEL APPROXIMATION FOR $GI/G/1/Z$ STOPPED ARRIVAL QUEUE FOR $C_a^2 = 0$, $C_S^2 = 1$

ρ	Service level	Z								
		1	2	3	4	5	6	7	8	9
0.5	Exact	.6667	.9366	.9876	.9975	.9995	.9999			
	approx.	.6667	.9465	.9896	.9979	.9996	.9999			
0.8	Exact	.5555	.8135	.9072	.9487	.9700	.9820	.9890	.9932	.9957
	approx.	.5555	.8317	.9154	.9527	.9722	.9832	.9897	.9936	.9960
1.0	Exact	.5000	.7311	.8237	.8696	.8966	.9143	.9268	.9362	.9434
	approx.	.5000	.7500	.8333	.8750	.9000	.9167	.9286	.9375	.9444
1.25	Exact	.4444	.6403	.7174	.7530	.7718	.7825	.7890	.7930	.7955
	approx.	.4444	.6575	.7259	.7574	.7742	.7840	.7899	.7935	.7958
2.0	Exact	.3333	.4519	.4864	.4962	.4989	.4997	.4999	.5000	
	approx.	.3333	.4617	.4897	.4971	.4992	.4998	.4999	.5000	

TABLE 4.2 ADEQUACY OF SERVICE-LEVEL APPROXIMATION $C_a^2 = 1/3$, $C_S^2 = 1/3$

ρ	Service level	Z					
		1	2	3	4	5	6
0.5	Sim.	.6663	.9432	.9928	.9989	.9999	1.0000
	approx.	.6667	.9462	.9895	.9978	.9996	.9999
0.8	Sim.	.5563	.8390	.9348	.9704	.9864	.9938
	approx.	.5555	.8658	.9411	.9710	.9850	.9920
1.0	Sim.	.4992	.7595	.8583	.9002	.9250	.9395
	approx.	.5000	.8000	.8750	.9091	.9286	.9412
1.25	Sim.	.4446	.6729	.7475	.7744	.7882	.7954
	approx.	.4444	.6926	.7529	.7768	.7880	.7937
2.0	Sim.	.3327	.4723	.4976	.5013	.4991	.5002
	approx.	.3333	.4331	.4947	.4989	.4998	.5000

approximation given above for SL has a limiting value as $Z \to \infty$, which is equal to $1/\rho$.

Tables 4.1 to 4.3 compare these approximations for the service level with the exact service level in a $D/M/1/Z$ stopped arrival system and to simulations of an

TABLE 4.3 ADEQUACY OF SERVICE-LEVEL APPROXIMATION $C_a^2 = 2$, $C_S^2 = 2$

ρ	Service level	\multicolumn{9}{c}{Z}								
		1	2	3	4	5	6	7	8	9
0.5	Sim.	.6659	.8114	.8822	.9236	.9492	.9652	.9751	.9804	.9876
	approx.	.6667	.7826	.8538	.8998	.9304	.9513	.9657	.9757	.9828
0.8	Sim.	.5509	.6890	.7663	.8087	.8407	.8688	.8895	.9028	.9217
	approx.	.5555	.6638	.7356	.7865	.8242	.8532	.8760	.8943	.9093
1.0	Sim.	.5002	.6230	.6914	.7388	.7662	.7893	.8081	.8295	.8438
	approx.	.5000	.6000	.6667	.7143	.7500	.7778	.8000	.8182	.8333
1.25	Sim.	.4429	.5558	.6103	.6509	.6733	.7112	.7123	.7247	.7411
	approx.	.4444	.5310	.5885	.6292	.6594	.6826	.7000	.7155	.7274
2.0	Sim.	.3351	.4056	.4402	.4572	.4781	.4787	.4907	.4905	.4932
	approx.	.3333	.3913	.4269	.4499	.4652	.4756	.4828	.4879	.4914

$E_3/E_3/1/Z$ stopped arrival system and a $C_2/C_2/1/Z$ stopped arrival system. Note that the approximation is good but that it is worst when $Z = 2$.

Note that all these approximations are exact for the $M/M/1/Z$ stopped arrival models provided $\hat{N}(\lambda, 1, \mu, 1) = \rho/(1 - \rho)$ for $\rho < 1$.

Single or multiple machines with bulk demand.   The idea of modeling the given single-machine manufacturing system with this PA mechanism by a queueing system can be extended to general single-stage manufacturing systems. For this consider a queueing system consisting of a waiting (dispatch) area and a service facility. Customers arrive at the queueing system according to the arrival process $\{A_n, n = 1, 2, \ldots\}$. The number of tasks brought in by the nth customer is X_n, $n = 1, 2, \ldots$. Each customer on its arrival enters the waiting area. If the number of tasks in the service facility is less than Z, tasks from the waiting area are sent into the service facility one at a time until there are Z tasks in the service facility. The service mechanism inside will duplicate the manufacturing process of the system being considered. For example, if we have c parallel machines and unlimited raw material in a single-stage manufacturing system, the service mechanism needed for the queueing system is simply c parallel servers. The service facility here consists of the c servers and $C(t)$ tasks in it at time t. (Recall $C(t) \le Z$ is the number of PA cards released to the manufacturing facility). The number of tasks waiting in the waiting area is the same as the number of items backlogged $B(t)$ at time t. Then, $N(t) = B(t) + C(t)$ is the number of tasks in the queueing system at time t. As before, because $B(t) > 0$ if and only if $C(t) = Z$,

$$B(t) = \{N(t) - Z\}^+, \quad t \ge 0 \tag{4.50}$$

and

$$C(t) = \{N(t) \wedge Z\}, \quad t \geq 0 \tag{4.51}$$

When the customer arrival process is renewal, service times are iid, the number of items demanded X_n, $n = 1, 2, \ldots$ are iid *and* all these three processes are mutually independent, $N(t)$ is the number of tasks in a bulk arrival $GI^X/G/c$ queueing system.

We will next consider some special cases of this model before we present an approximation for the $GI^X/G/c$ model.

$M^X/G/1$ Model.   Here we consider the situation in which customers arrive according to a Poisson process with rate λ. Let N_n^+ be the number of tasks in the system just after the service completion of the nth task and let Y_{n+1} be the number of tasks arrived during the service of the $(n + 1)$th task. Then

$$N_{n+1}^+ = \begin{cases} N_n^+ - 1 + Y_{n+1}, & \text{if } N_n^+ > 0, \ n = 1, 2, \ldots \\ \hat{X}_n - 1 + Y_{n+1}, & \text{if } N_n^+ = 0 \end{cases} \tag{4.52}$$

where $\hat{X}_n$ is the number of tasks brought to the system by the first customer that entered the system after the nth task departure. $\hat{X}_n$ will have the same distribution as X. We have

$$E\left[z^Y\right] = \tilde{F}_S(\lambda(1 - \tilde{F}_X(z))) = \tilde{F}_Y(z) \tag{4.53}$$

where $\tilde{F}_X(z) = E\left[z^X\right]$ is the moment generating function of X. Taking the MGF of the left- and right-hand sides of equation 4.52 one gets

$$\tilde{F}_{N^+}(z) = \tilde{F}_Y(z) \left(\frac{1}{z} P\{N^+ = 0\}(\tilde{F}_X(z) - 1) + \frac{1}{z}\tilde{F}_{N^+}(z)\right) \tag{4.54}$$

Solving equation 4.54 for $\tilde{F}_{N^+}(z)$ with the condition $\tilde{F}_{N^+}(1) = 1$, we get

$$\tilde{F}_{N^+}(z) = \frac{(\tilde{F}_X(z) - 1)(1 - \rho)\tilde{F}_Y(z)}{E[X](z - \tilde{F}_Y(z))} \tag{4.55}$$

where $\rho = \lambda E[X]E[S]$. The mean number of tasks in the system at a task service completion epoch is

$$E[N^+] = \frac{\lambda^2 E[X]^2 E[S^2] + \lambda E[X^2]E[S] - \lambda E[X]E[S]}{2(1 - \rho)} + \frac{E[X^2] - E[X]}{2E[X]} + \rho \tag{4.56}$$

The probability distribution of the number of tasks seen by an arriving *task* is the same as F_{N^+} (see Chapter 3). Because *tasks* do not arrive according to a Poisson process, however, the distribution F_N of the number of tasks present in the system at an arbitrary time need not be the same as F_{N^+}. Because the *customer* arrival process is Poisson, the distribution F_{N^-} of the number of tasks seen by an arriving *customer* is the same as F_N (i.e., $F_N(n) = F_{N^-}(n)$, $n = 0, 1, \ldots$). We will next use the fact that the customer arrival process is Poisson and apply level crossing analysis to obtain the distribution F_{N^-} of the number of tasks in the system seen by an arriving customer and then obtain $F_N(n) = F_{N^-}(n)$, $n = 0, 1, \ldots$.

At each customer arrival epoch, the number of tasks in the system may jump from a level less than or equal to n to a level higher than n. The probability of such an upcrossing is $\sum_{l=0}^{n} \overline{F}_X(n - l) f_{N^-}(l)$. Downcrossing to any level may occur only at a task-service completion epoch. The probability of a downcrossing to level n at a task-service completion epoch is $f_{N^+}(n)$; hence the *rate* of downcrossing to level n is $f_{N^+}(n)$ multiplied by the rate of service completions $\lambda E[X]$. Figure 4.1 shows the level crossings.

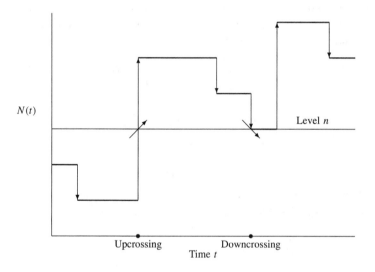

Figure 4.1   Level Crossings in a $M^X/G/1$ Queue

Now equating the rates of upcrossings and downcrossings we obtain

$$\lambda \sum_{l=0}^{n} \overline{F}_X(n - l) f_{N^-}(l) = \lambda E[X] f_{N^+}(n), \quad n = 0, 1, \ldots \tag{4.57}$$

Therefore, taking the MGF of both sides of equation 4.57 one gets

$$\tilde{F}_{N^-}(z) = \frac{E[X](1 - z)\tilde{F}_{N^+}(z)}{1 - \tilde{F}_X(z)} = \frac{(z - 1)(1 - \rho)\tilde{F}_Y(z)}{z - \tilde{F}_Y(z)} \tag{4.58}$$

Then

$$E[N] = E[N^-] = E[N^+] - \frac{E[X^2] - E[X]}{2E[X]}$$

$$= \frac{\lambda^2 E[X]^2 E[S^2] + \lambda E[X^2]E[S] - \lambda E[X]E[S]}{2(1 - \rho)} + \rho \tag{4.59}$$

The expected remaining service time of a task in service, if any, at an arbitrary time point is $E[S^2]/2E[S]$. Then the total work load in the system at an arbitrary time epoch is

$$(E[N]-\rho)E[S]+\rho E[S^2]/2E[S] = \lambda(E[X]E[S^2]+E[X^2]E[S]^2 - E[X]E[S]^2)/2(1-\rho)$$

Observe that this is the average work load in an $M/\hat{G}/1$ system with arrival rate λ, mean service time $E[X]E[S]$ and second moment of service time $E[X]E[S^2]+E[X^2]E[S]^2 - E[X]E[S]$. That is, the mean and second moment of this service time are the mean and second moment of the distribution of $\sum_{n=1}^{X} S_n$. Indeed, an alternative approach to solve the $M^X/G/1$ model is to analyze the $M/\hat{G}/1$ queue with arrival rate λ and service time equal in distribution to $\sum_{n=1}^{X} S_n$. The work load process for this queue and the $M^X/G/1$ queue will be identical.

The $M/\hat{G}/1$ queue can be viewed as a queueing system for jobs where each job is the aggregate of the tasks brought in by each customer. We will consider this representation to obtain approximations for single-stage manufacturing systems with renewal customer arrival process.

$GI^X/G/1$ Model.   We consider the same scenario as earlier except that the customer arrival process is renewal and not necessarily Poisson. The $GI/\hat{G}/1$ queue models the dynamics of the jobs, where each job is the aggregate of the tasks brought in by each customer, and has an arrival process $\{A_n, n = 1, 2, \ldots\}$ and service times $\hat{S}_n = \sum_{j=1}^{X_n} S_{nj}, n = 1, 2, \ldots$. The mean and the squared coefficient of variation of the service times $\hat{S}$ are then $E[X]E[S]$ and $C_X^2 + (1/E[X])C_S^2$. Let $\hat{W}_{GI/G/1}(\lambda, \mu, C_a^2, C_S^2)$ be any one of the approximations for the average waiting time in queue for a $GI/G/1$ queue with mean interarrival time $1/\lambda$, mean service time $1/\mu$, squared coefficient of variation of interarrival time C_a^2, and squared coefficient of variation of service time C_S^2. Then the average waiting time of a job in the $GI/\hat{G}/1$ system is approximated by $w_0 = \hat{W}_{GI/G/1}(\lambda, 1/[E[X]E[S]], C_a^2, C_X^2 + (1/E[X])C_S^2)$. The average number of *jobs waiting in the queue* can be approximated by λw_0. Because each job in queue consists of an average of $E[X]$ tasks, the average number of *tasks* in the system corresponding to those *jobs waiting in the queue* is $\lambda E[X]w_0$. The average number of *tasks* in the system corresponding to the *jobs in service*, if any, is $(E[X^2] + E[X])/2E[X]$. Because the probability that a job is in service is $\rho = \lambda E[X]E[S]$, we can approximate the average number of tasks, $E[N]$, in the system by

$$n_0 = \lambda E[X]w_0 + \frac{\lambda(E[X^2] + E[X])E[S]}{2} \tag{4.60}$$

Observe that when $C_a^2 = 1$ the preceding approximation agrees with the results for Poisson arrival process, provided the approximation $\hat{W}_{GI/G/1}$ selected is exact for $M/G/1$ queues.

Therefore, the distribution of the number of tasks in the $GI^X/G/1$ system is approximated by

$$P\{N = n\} \approx \begin{cases} (1-\rho), & n = 0 \\ \rho(1-\sigma)\sigma^{n-1}, & n = 1, 2, \ldots \end{cases} \tag{4.61}$$

where

$$\sigma = (n_0 - \rho)/n_0 \qquad (4.62)$$

is chosen such that average of the approximated distribution is equal to n_0.

With this approximation we can carry out an analysis similar to the $M/M/1$ analysis that we did earlier in this section and obtain an approximation for the optimal threshold level Z^*.

$GI^X/G/c$ Model. The approximation technique illustrated in this section can be applied to the single-stage multiple machine manufacturing system operated under the target level PA mechanism. For this, we will first approximate the $GI^X/G/c$ model by c $GI/\hat{G}/1$ systems each with arrival process $\{A_n, n = 1, 2, \ldots\}$ and service times $\hat{S}_n = 1/c \sum_{j=1}^{X_n} S_{nj}, n = 1, 2, \ldots$. The approximated work load in the system is then $c\hat{W}_{GI/G/1}(\lambda, c/[E[S]E[X]], C_a^2, C_X^2 + 1/(E[X])C_S^2)$. The average number in the system can be obtained from this approximation (see Section 3.4.2).

Produce-to-stock with yield losses.   We will now study the effects of yield losses on the performance of produce-to-stock manufacturing systems. Let q be the probability that a completed item is defective. Two scenarios for the detection of defective items will be considered: (1) defects are detected at the manufacturing facility and defective items are reprocessed until free of defects, and (2) defects of defective items are discovered only at delivery to the customer and the item is then discarded.

Detection at Manufacturing Facility.   Suppose the reprocessing times are iid with distribution the same as the original processing times. Then the total processing time needed to produce a good item has a distribution equal to the geometric sum of iid service times $\{S_n, n = 1, 2, \ldots\}$. Let $\{\hat{S}_n, n = 1, 2, \ldots\}$ be these processing times. The LST of $\hat{S}$ is given by

$$\tilde{F}_{\hat{S}}(s) = \frac{(1-q)\tilde{F}_S(s)}{1 - q\tilde{F}_S(s)}$$

If $N(t)$ is the number of jobs in a $GI^X/\hat{G}/1$ queueing system with arrival process $\{A_n, n = 1, 2, \ldots\}$, bulk size $\{X_n, n = 1, 2, \ldots\}$, and iid service times $\{\hat{S}_n, n = 1, 2, \ldots\}$, then $I(t) = Z - N(t), t \geq 0$. Note that when S is exponentially distributed with mean μ, $\hat{S}$ has an exponential distribution with mean $\mu/(1 - q)$.

Detection at Output Store.   Suppose the defective items are detected at the output store by customers, and these defective items are then discarded. Let $\hat{X}_n$ be the number of items needed to find X_n good items. Then we need to produce $\hat{X}_n$ items to meet the X_n items demanded by the nth customer, that is, $\hat{X}_n = \sum_{k=1}^{X_n} Y_k$, where Y_k is the number of items that have to be produced to find the kth good item. Because we need to produce a geometrically distributed number of items with mean $1/(1 - q)$ to obtain one good item, the moment generating function of $\hat{X}$ is given by

$$g_{\hat{X}}(z) = g_X\left(\frac{(1-q)z}{1 - qz}\right)$$

Alternatively, if we knew $g_{\hat{X}}(z')$ then

$$g_X(z') = g_{\hat{X}} \left(\frac{z'}{1 - q + qz'} \right)$$

Let $N(t)$ be the number of jobs in a $GI^{\hat{X}}/G/1$ queueing system with arrival process $\{A_n, n = 1, 2, \ldots\}$, batch size $\{\hat{X}_n, n = 1, 2, \ldots\}$ and service times $\{S_n, n = 1, 2, \ldots\}$. Then if $N(t) \leq Z$ (i.e., there are no backlogged demands), one sees that $Z - N(t)$ is the actual number of items in the output buffer. Hence

$$I(t) = \{Z - N(t)\}^+, \quad t \geq 0$$

If $N(t) > Z$, then we need to produce $B'(t) = N(t) - Z$ items to satisfy all the demand for items backlogged at time t. Let $B(t)$ be the number of demanded items backlogged at time t. Because we need to produce a geometric number of items with mean $1/(1-q)$ to meet the demand for one good item,

$$\sum_{k=1}^{B(t)} Y_k \stackrel{d}{=} \{N(t) - Z\}^+$$

Therefore

$$\tilde{F}_{B(t)} \left(\frac{(1-q)z}{1-qz} \right) = \tilde{F}_{B'(t)}(z)$$

where

$$\tilde{F}_{B'(t)}(z) = E[z^{\{N(t)-Z\}^+}]$$

Therefore

$$\tilde{F}_{B(t)}(z) = \tilde{F}_{B'(t)} \left(\frac{z}{1 - q + qz} \right)$$

Observe that

$$E[B(t)] = (1 - q)E[B'(t)]$$

The results from the previous sections can now be applied to this case to obtain the distributions of I and B.

Produce-to-stock systems with machine failures.

So far we have assumed that the machines processing the parts do not fail. In some applications the effect of machine failures on the performance of the system can be significant. Hence in such cases one should incorporate the machine failures in modeling the system and determining the optimal operating policies. As we will see subsequently, if the machine up times, are exponentially distributed we may replace the processing times $\{S_n, n = 1, 2, \ldots\}$ of the parts by appropriately defined processing times $\{\hat{S}_n, n = 1, 2, \ldots\}$ incorporating machine down times and apply the models developed earlier.

Suppose the machine up times are exponentially distributed with mean $1/\zeta$ and the repair (or down) times form an iid sequence $\{R_n, n = 1, 2, \ldots\}$ of random variables with distribution F_R. Then if the number of machine failures during the processing of the rth part is N_r, the total time $\hat{S}$ needed to complete processing of the first part has the same distribution as $S_1 + \sum_{n=1}^{N_1} R_n$. Because $\{N_1 | S_1 = t\}$ has a Poisson distribution with mean ζt, conditioning on S_1 and taking the LST one finds that

$$\tilde{F}_{\hat{S}}(s) = \tilde{F}_S(s + \zeta - \zeta \tilde{F}_R(s))$$

If we define the times $\hat{S}_2, \hat{S}_3, \ldots$ to complete parts 2, 3, $\ldots$ in the same way, it can be seen that by the memoryless property of the exponential machine up times, $\{\hat{S}_n, n = 1, 2, \ldots\}$ is a sequence of iid random variables with distribution $F_{\hat{S}}$ whose LST is given by $\tilde{F}_{\hat{S}}(s)$. Hence all our previous models with no machine failures and their analysis apply to this case as well. When the processing times and repair are also exponentially distributed (say, with means $1/\mu$ and $1/\gamma$, respectively), and the demand arrival process is also Poisson with rate λ, however, explicit modeling of the machine failures can lead to easier computation of system performance measures and hence easier computation of the optimal operating policies. We will illustrate this next by considering a single machine manufacturing facility.

Exponential Machine Up and Down Times.   Consider a single machine produce-to-stock system that operates according to a target level PA mechanism. Let $X(t)$ be the state of the machine at time t, and $I(t)$, $B(t)$, and $N(t)$ be as defined earlier. When the machine is up $X(t)$ takes the value 1 and when the machine is down, $X(t)$ takes the value 0. Then $\{(N(t), X(t)), t \geq 0\}$ is a continuous time Markov process on the state space $S = \{(n, j), j = 0, 1, n = 0, 1, 2, \ldots\}$. The infinitesimal generator $\mathbf{Q}$ of this Markov process is of the block tridiagonal structure. The stationary distribution $(\mathbf{p}_n, n = 0, 1, \ldots)$ with $\mathbf{p}_0 = p_{01}$ and $\mathbf{p}_n = (p_{n0}, p_{n1})$, $n = 1, 2, \ldots$ can be obtained using the results in Section 3.3.3. They are given by

$$\mathbf{p}_0 = 1 - \frac{\lambda}{\mu}\left(1 + \frac{\zeta}{\gamma}\right)$$

$$\mathbf{p}_1 = \left(\frac{\lambda\zeta}{(\lambda + \gamma)\mu}, \frac{\lambda}{\mu}\right)\mathbf{p}_0$$

$$\mathbf{p}_n = \mathbf{p}_1\mathbf{R}^{n-1}, \qquad n = 2, 3, \ldots$$

where

$$\mathbf{R} = \begin{bmatrix} \frac{\lambda}{\lambda+\gamma}\left(1 + \frac{\zeta}{\mu}\right) & \frac{\lambda}{\mu} \\ \frac{\lambda}{\lambda+\gamma}\frac{\zeta}{\mu} & \frac{\lambda}{\mu} \end{bmatrix}$$

As before, the preceding results can now be incorporated in a cost model to calculate the optimal target level Z.

In the preceding model we assumed that the processing times are exponentially distributed and that the demands arrive according to a Poisson process. When the frequency of machine failures is substantially smaller than the production rate and the customer demand rate (i.e., $\zeta << \mu, \lambda$) we may assume in the modeling of machine failures that the manufacturing of parts occurs at a constant rate of μ and that the customer demands arrive at a constant rate of λ as in a deterministic fluid flow. Then treating the inventory level and the back-order level as continuous variables one sees that $\{N(t), t \geq 0\}$ behaves like the water level in an infinite dam where water inflows at a constant rate λ. The water outflows at a rate μ, which is periodically turned on and off according to an alternating renewal process corresponding to the machine up and down times. Particularly when $X(t) = 1$, the output of the dam is open (and thus the level of the dam drops at a rate $\mu - \lambda$ as long as the level is greater than 0) and when $X(t) = 0$, it is closed (and thus the level increases at a rate of λ). A sample path of the process $\{N(t), t \geq 0\}$ is shown in Figure 4.2.

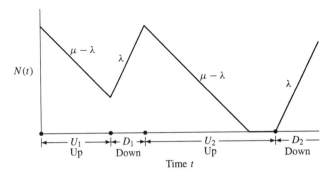

Figure 4.2   Sample Path of $\{N(t), t \geq 0\}$ with Machine Failures

Let $\{U_n, n = 1, 2, \ldots\}$ and $\{D_n, n = 1, 2, \ldots\}$ be the sequence of up and down times of the machine, which we assume is up at time 0. Define

$$W_n = N\left(\sum_{j=1}^{n} U_j + \sum_{j=1}^{n-1} D_j\right)$$

and

$$T_n = N\left(\sum_{j=1}^{n}(U_j + D_j)\right)$$

be the values of $N(t)$ at the nth machine failure epoch, and at the nth repair completion epoch, respectively. It is now easily seen that

$$W_n = (W_{n-1} + \lambda D_{n-1} - (\mu - \lambda)U_n)^+$$

and

$$T_n = W_n + \lambda D_n, \qquad n = 1, 2, \ldots$$

Because λD_{n-1} and $(\mu - \lambda)U_n$ have exponential distributions with mean λ/γ and $(\mu - \lambda)/\zeta$ respectively, W_n has the same distribution as the waiting time in an $M/M/1$ queue with arrival rate $\zeta/(\mu - \lambda)$ and service rate γ/λ. Hence (see Section 3.3.1)

$$F_W(x) = 1 - \frac{\lambda\zeta}{\gamma(\mu - \lambda)}e^{-(\gamma/\lambda - \zeta/(\mu-\lambda))x}$$

and

$$F_T(x) = 1 - e^{-(\gamma/\lambda - \zeta/(\mu-\lambda))x}, \qquad x > 0$$

Because the up and down times are exponentially distributed by the property "Poisson arrivals see time averages,"

$$\lim_{t\to\infty} P\{N(t) \le x | X(t) = 1\} = F_W(x)$$

and

$$\lim_{t\to\infty} P\{N(t) \le x | X(t) = 0\} = F_T(x)$$

Noting that $\lim_{t\to\infty} P\{X(t) = 0\} = \zeta/(\gamma + \zeta)$, and therefore that

$$F_N(x) = \frac{\gamma}{\gamma+\zeta} F_W(x) + \frac{\zeta}{\gamma+\zeta} F_T(x)$$

one has

$$F_N(x) = 1 - \frac{\zeta}{\gamma+\zeta} \left\{ \frac{\mu}{\mu-\lambda} \right\} e^{-(\gamma/\lambda - \zeta/(\mu-\lambda))x}$$

Because $I = (Z - N)^+$, we have

$$F_I(x) = \overline{F}_N(Z - x)$$
$$= \frac{\zeta}{\gamma+\zeta} \left\{ \frac{\mu}{\mu-\lambda} \right\} e^{-(\gamma/\lambda - \zeta/(\mu-\lambda))(Z-x)}, \qquad 0 \le x \le Z$$

Similarly because $B = (N - Z)^+$, we have

$$F_B(x) = F_N(x + Z)$$
$$= 1 - \frac{\zeta}{\gamma+\zeta} \left\{ \frac{\mu}{\mu-\lambda} \right\} e^{-(\gamma/\lambda - \zeta/(\mu-\lambda))(x+Z)}, \qquad x \ge 0$$

These results along with a cost model can be used to obtain the optimal target level Z.

4.3.2 Target-Level PA Mechanism with Threshold for Start-Up

In the target level PA mechanism described in the previous section we see that a PA card is released to an idle machine as soon as the number of items in the output store drops by one from its upper limit Z. Although such a control mechanism is usually good when the machine start-up and shut-down costs are negligible, such a policy can be costly when the machine start-up and shut-down costs are high. To control this frequent start-up and shut-down of the machine, in this section we will introduce a machine start-up policy along with the target-level PA mechanism.

Suppose the output store is full with Z items at time zero. Each item has a tag. As customers arrive their demands are filled from the output store, and the tags on the items are activated as PA cards and moved to the manufacturing facility. If at any time the number of PA cards at the manufacturing facility is greater than or equal to q ($q \geq 1$), the machine operator is authorized to turn on the machine and keep it operational. The PA cards are then used to authorize the production of items. As soon as the production of an item is complete, the PA card is attached to it as a tag and moved to the output store. When there are no PA cards for a machine, the machine is shut off. We call this the (s, Z) threshold–target-level PA mechanism. Here $s = Z - q$.

Now depending on the nature of the customer arrival process, the number of items demanded by the customers, the manufacturing process, and the policy on demand back-logging/lost demands, appropriate stochastic models may be formulated to obtain the performance of a produce-to-stock single-stage manufacturing system with the preceding PA mechanism. We will next consider one such model.

Consider a single-machine single-stage produce-to-stock system operated under the (s, Z) threshold–target-level PA mechanism with Z PA cards. Customers arrive at the system according to a Markov process with arrival rate $\lambda(n)$ depending on the number (n) of finished items in the output store. We assume that $\lambda(n) > 0$, $n = 1, \ldots, Z$ and $\lambda(0) = 0$. Observe that $\lambda(0) = 0$ is equivalent to having a "lost demand" policy for customers who enter the system when the output store is empty. The number of items demanded by each customer is assumed to be one (i.e., $X_n = 1$, $n = 1, 2, \ldots$). The processing times it takes to convert raw material to finished items are assumed to be iid with a phase-type distribution with representation $(\beta, \mathbf{B})$ where β is an m_b-vector, and $\mathbf{B}$ is an $m_b \times m_b$ matrix. Let $I(t)$ be the number of finished items in the output store and $Y(t)$ be the state of processing by the machine at time t. If the machine is shut off at time t, we set $Y(t) = 0$ and otherwise if the machine is on and processing a part and the processing is in phase r, we set $Y(t) = r$, $r = 1, \ldots, m_b$. Then $\{(I(t), Y(t)), t \geq 0\}$ is a continuous time Markov process on the state space $\mathcal{S} = \{((n, r), r = 1, \ldots, m_b); n = 0, \ldots, s; ((n, r), r = 0, \ldots, m_b); n = s + 1, \ldots, Z - 1; (Z, 0)\}$. Let $p(n, r) = \lim_{t \to \infty} P\{I(t) = n, Y(t) = r\}$, $(n, r) \in \mathcal{S}$ be the stationary distribution of $\{(I(t), Y(t)), t \geq 0\}$. Define $\mathbf{p}_n = \{p(n, r) : (n, r) \in \mathcal{S}, n \text{ fixed}\}$, $n = 0, \ldots, Z$. Then the stationary balance equations are

$$\mathbf{pQ} = \mathbf{0},$$

where the infinitesimal generator $\mathbf{Q}$ of the Markov process $\{(I(t), Y(t)), t \geq 0\}$ is given by

$$\mathbf{Q} = \mathbf{R} - (\mathbf{Re})_D$$

and the rate matrix $\mathbf{R}$ in the blocked partition form is

$$\mathbf{R} = \begin{bmatrix} \mathbf{R}_{00} & \mathbf{R}_{01} & & & & \\ \mathbf{R}_{10} & \mathbf{R}_{11} & \mathbf{R}_{12} & & & \\ & \mathbf{R}_{21} & \mathbf{R}_{22} & \mathbf{R}_{23} & & \\ & & & \ddots & & \\ & & & \mathbf{R}_{Z-1,Z-2} & \mathbf{R}_{Z-1,Z-1} & \mathbf{R}_{Z-1,Z} \\ & & & & \mathbf{R}_{Z,Z-1} & \mathbf{R}_{Z,Z} \end{bmatrix}$$

$$R_{n,n+1:i,j} = -\sum_{l=1}^{m_b} B_{il}\beta_j, \quad i,j = 1,\ldots,m_b; n = 0,\ldots,Z-2$$

$$R_{Z-1,Z:i,0} = -\sum_{l=1}^{m_b} B_{il}, \quad i = 1,\ldots,m_b$$

$$R_{n,n:i,j} = B_{ij}, \quad i,j = 1,\ldots,m_b (i \neq j); n = 0,\ldots,Z-1$$

$$R_{n,n-1:i,i} = \lambda(n), \quad i = 1,\ldots,m_b; n = 1,\ldots,Z-1$$

$$R_{n,n-1:0,0} = \lambda(n), \quad n = s+2,\ldots,Z,$$

$$R_{s+1,s:0,j} = \lambda(s+1)\beta_j$$

All other transition rates are zero. The balance equations in the block-partitioned form are

$$\mathbf{p}(0)\mathbf{Q}_{00} + \mathbf{p}(1)\mathbf{Q}_{10} = \mathbf{0}$$

$$\mathbf{p}(n-1)\mathbf{Q}_{n-1,n} + \mathbf{p}(n)\mathbf{Q}_{n,n} + \mathbf{p}(n+1)\mathbf{Q}_{n+1,n} = \mathbf{0}, \quad n = 1,\ldots,Z-1$$

$$\mathbf{p}(Z-1)\mathbf{Q}_{Z-1,Z} + \mathbf{p}(Z)\mathbf{Q}_{Z,Z} = \mathbf{0}$$

Then we have

$$\mathbf{p}(Z) = -\mathbf{p}(Z-1)\mathbf{Q}_{Z-1.Z}\mathbf{Q}_{Z,Z}^{-1} \equiv \mathbf{p}(Z-1)\mathbf{G}(Z)$$

$$\mathbf{p}(n) = \mathbf{p}(n-1)\mathbf{G}(n), \quad n = Z-1,\ldots,1$$

where

$$\mathbf{G}(n) = (-\mathbf{Q}_{n-1,n})[\mathbf{Q}_{n,n} + \mathbf{G}(n+1)\mathbf{Q}_{n+1,n}]^{-1}, \quad n = Z-1,\ldots,1; \mathbf{G}(0) = \mathbf{I}$$

Now $\mathbf{p}(0)$ can be obtained by solving

$$\mathbf{p}(0)\mathbf{Q}_{00} + \mathbf{p}(1)\mathbf{Q}_{10} = \mathbf{0}$$

or equivalently by solving

$$\mathbf{p}(0)[\mathbf{G}(0)\mathbf{Q}_{00} + \mathbf{G}(1)\mathbf{Q}_{10}] = \mathbf{0}$$

and the normalizing equation

$$\mathbf{p}(0)\sum_{n=0}^{Z} \hat{\mathbf{G}}(n)\mathbf{e} = 1,$$

where

$$\hat{\mathbf{G}}(n) = \Pi_{k=0}^{n}\mathbf{G}(k), \quad n = 0,\ldots,Z.$$

Once $\mathbf{p}(0)$ is computed $\mathbf{p}(n)$ can be computed from the relationship $\mathbf{p}(n) = \mathbf{p}(n-1)\mathbf{G}(n)$, $n = 1,\ldots,Z$. The performance measures such as the average inventory $E[I]$ can now be computed.

When $\lambda(n) = \lambda$, $n = 1, \ldots, Z$, the final expression for $\mathbf{p}(n)$ can be simplified (see problems 18 and 19). In addition if the processing times are exponentially distributed, then $S = \{(n, 1), n = 0, \ldots, s; ((n, 0), (n, 1)), n = s + 1, \ldots, Z - 1; (Z, 0)\}$. The stationary probability distribution is given by

$$p(k, 1) = \frac{(1 - \rho)(\rho^{s+1-k} - \rho^{Z+1-k})}{(1 - \rho)(Z - s) - \rho(\rho^{s+1} - \rho^{Z+1})}, \quad k = 0, \ldots, s - 1$$

$$p(k, 1) = \frac{(1 - \rho)(\rho - \rho^{Z+1-k})}{(1 - \rho)(Z - s) - \rho(\rho^{s+1} - \rho^{Z+1})}, \quad k = s, \ldots, Z - 1$$

$$p(k, 0) = \frac{(1 - \rho)^2}{(1 - \rho)(Z - s) - \rho(\rho^{s+1} - \rho^{Z+1})}, \quad k = s + 1, \ldots, Z$$

Suppose the decision to start up and shut down the machines has to be made based on the number of items in the output store alone. Then any (stationary) policy that does not change over time should be of the form of the (s, Z) threshold–target-level PA mechanism. Because there exists a stationary policy that is optimal for the Markovian process, the preceding form of mechanism is optimal when the demand arrival process is Markovian (in the number of items in the output store) and the processing times are exponential.

4.3.3 Target-Level PA Mechanism with Fixed-Batch Size

Let Z be the target inventory level, and suppose a total of Z items with tags are available in the output store. As customers arrive and their demands are met, the tags taken from the items delivered to the customers are activated into PA cards. Whenever q or more PA cards are accumulated at the output store, q PA cards are transmitted to the manufacturing facility. Each of these q PA cards authorizes the operator to produce one item. All q items for a given batch of PA cards will begin manufacture before manufacturing is initiated of items for PA cards in subsequent batches. The actual initiation of the manufacture of items can follow either one of the two policies:

- Policy (1) (nonoverlapping). Complete the manufacture of items belonging to a batch of PA cards before initiating the manufacture of items for a subsequent batch of PA cards,
- Policy (2) (overlapping). As soon as the manufacture of all q items of a batch of PA cards is initiated, the manufacture of items for the subsequent batch of PA cards can be initiated (whenever machines become available).

Note that in the single-machine case or when $Z < 2q - 1$ the preceding two policies will result in the same performance of the system. We will call this the (q, Z) fixed-batch target-level PA mechanism. Either the items produced are shipped along with a PA card as a tag to the output store one at a time as soon as they are ready, or alternatively all q items are shipped along with the q PA cards as tags to the output store. For our analysis we will assume that the latter policy is in effect. Observe that the maximum number of items in the output store at the time of return of a PA card cannot be more than Z. Therefore the maximum number of items in the output store will always be less than or equal to Z. When there is only one machine in the system and $q = 1$, the (q, Z) fixed-batch target-level PA mechanism is the same as the target-level PA mechanism with target level Z. In the general setting this mechanism is essentially the same as the traditional reorder point/order quantity inventory control policy. Unlike in inventory modeling, our approach will explicitly model the process that determines the lead time for replenishment of items at the output store.

Nonoverlapping batch manufacture.   We first consider the case in which $X_n = 1$, $n = 1, 2, \ldots$ and policy (1) is used for initiating manufacture of items. Suppose there are Z items in the output store at time zero. Then at time $\tilde{A}_1 = A_q$, that is, at the qth customer arrival epoch, q PA cards will be transmitted to the machines to produce q items. Let $\tilde{S}_n$ be the total time required to produce q items at the machine for the nth batch of q PA cards. For example, if there is only one machine in the system

$$\tilde{S}_n = \sum_{j=nq+1-q}^{nq} S_j, \quad n = 1, 2, \ldots$$

Conversely, if there are q or more parallel machines $\tilde{S}_n = \max\{S_j, j = nq + 1 - q, \ldots, nq\}$, $n = 1, 2, \ldots$. Then at time $\tilde{D}_1 = \tilde{A}_1 + \tilde{S}_1$, the first q items would have been produced and shipped to the output store. Let $\tilde{A}_n = A_{nq}$, $n = 1, 2, \ldots$. Then if $\tilde{A}_2 > \tilde{D}_1$, the number of PA cards accumulated in the output store at time $\tilde{D}_1$ is less than q. In this case the next batch of q PA cards will be sent to the manufacturing facility only at time $\tilde{A}_2$ and the resulting return of the q PA cards will occur at time $\tilde{D}_2 = \tilde{A}_2 + \tilde{S}_2$. If $\tilde{A}_2 \le \tilde{D}_1$, the number of PA cards accumulated at the output store during $(\tilde{A}_1, \tilde{D}_1)$ is greater than or equal to q. Thus the second batch of q PA cards must have come back to the manufacturing facility at $\tilde{A}_2$. The second return of the batch of q PA cards back to the output store in this case is $\tilde{D}_2 = \tilde{D}_1 + \tilde{S}_2$. Combining the preceding two cases we see that $\tilde{D}_2 = \{\tilde{D}_1 \vee \tilde{A}_2\} + \tilde{S}_2$. Let $\tilde{D}_n$ be the time epoch at which the nth batch of PA cards is returned to the output store. Then

$$\tilde{D}_n = \{\tilde{D}_{n-1} \vee \tilde{A}_n\} + \tilde{S}_n, \quad n = 1, 2, \ldots \tag{4.63}$$

Observe that $\{\tilde{D}_n, n = 1, 2, \ldots\}$ is the departure process of a $\tilde{G}I/\tilde{G}/1$ queueing system with arrival process $\{\tilde{A}_n, n = 1, 2, \ldots\}$ and service times $\{\tilde{S}_n, n = 1, 2, \ldots\}$. Suppose we create a batch tag at each arrival epoch of $\tilde{A}_n = A_{nq}$, $n = 1, 2, \ldots$, (i.e., one tag for every q customers) and destroy one whenever a batch of q PA cards is returned to the output store (i.e., at times $\tilde{D}_n$, $n = 1, 2, \ldots$). The number of batch tags in the manufacturing system is the same as the number of jobs in the $\tilde{G}I/\tilde{G}/1$ queueing system. Next we will relate the number of jobs in the $\tilde{G}I/\tilde{G}/1$ queue, $N(t)$, to the number of items $I(t)$ in the output store, the number of completed items $\overline{C}(t)$ that are in the manufacturing facility waiting to be shipped and the number of customers $B(t)$ backlogged. Let

$$R(t) = A(t) - \left\lfloor \frac{A(t)}{q} \right\rfloor q, \quad t \geq 0 \tag{4.64}$$

be the number of customers arrived at or before time t, but after the last batch tag was created. Then

$$B(t) = \{N(t)q + R(t) - Z\}^+, \quad t \geq 0 \tag{4.65}$$

where $N(t)$ is the number of jobs in the $\tilde{G}I/\tilde{G}/1$ queueing system at time t. Then

$$I(t) = \{Z - N(t)q - R(t)\}^+, \quad t \geq 0 \tag{4.66}$$

Though $N(t)$ and $R(t)$ can be dependent, we will assume that they are independent. From equation 4.64 it can be seen that as t increases (i.e., as $A(t)$ increases), $R(t)$ will uniformly increase from 0 to $q - 1$, drop to 0, increase from 0 to $q - 1$, and so on. Then one sees that

$$P\{R = n\} = \frac{1}{q}, \quad n = 0, 1, \ldots, q - 1 \tag{4.67}$$

Note that this result can be more formally verified by observing that R has the same distribution as the stationary attained age of a discrete renewal process with interarrival time q. Similarly, if there is only a single machine in the system, $P\{\overline{C} = n \mid$ a batch of q PA cards are in the manufacturing system$\} = 1/q$, $n = 0, 1, \ldots, q - 1$. Because the probability there is a batch of q PA cards in the manufacturing facility is $\rho \ (= \lambda E[S] = \tilde{\lambda}E[\tilde{S}])$ we have

$$P\{\overline{C} = n\} = \begin{cases} 1 - \rho + \frac{\rho}{q}, & n = 0 \\ \frac{\rho}{q}, & n = 1, \ldots, q - 1 \end{cases} \tag{4.68}$$

Suppose there are q parallel machines in the manufacturing facility. Then the time needed to produce l items from the time when a batch of q PA cards is released is the lth order statistic $S_{(l)}$ of the q processing times $S_1, \ldots, S_q$. In this case

$$P\{\overline{C} = n\} = \begin{cases} 1 - \tilde{\rho} + \tilde{\rho}\frac{E[S_{(1)}]}{E[S_{(q)}]}, & n = 0 \\ \tilde{\rho}\frac{E[S_{(n+1)}] - E[S_{(n)}]}{E[S_{(q)}]}, & n = 1, \ldots, q - 1 \end{cases} \tag{4.69}$$

where $\tilde{\rho} = \tilde{\lambda}E[\tilde{S}]$, $\tilde{\lambda} = \lambda/q$.

The distribution of N can be approximated by the approximation available for the single-stage queueing model developed in Section 3.3.4. Let $\hat{W}_{GI/G/1}(\lambda, \mu, C_a^2, C_S^2)$ be the approximation selected for a $GI/G/1$ queue. Then

$$w_0 = \hat{W}_{GI/G/1}\left(\frac{\lambda}{q}, \mu q, \frac{C_a^2}{q}, C_{\tilde{S}}^2\right)$$

is the approximate waiting time in the $\tilde{G}I/\tilde{G}/1$ queue. Then $\hat{N} = (\lambda/q)w_0 + \rho$ is the approximation for the number in the system and

$$P\{N = n\} = p_N(n) \approx \begin{cases} 1 - \rho, & n = 0 \\ \rho(1 - \sigma)\sigma^{n-1}, & n = 1, 2, \ldots \end{cases} \tag{4.70}$$

where $\sigma = (\hat{N} - \rho)/\hat{N}$. The squared coefficient of variation of the service time $\tilde{S}$ needs to be computed based on the number of machines in the system. For example, when there is only one machine in the system, recall that $\tilde{S}_n = S_{nq+1-q} + \ldots + S_{nq}$. Therefore in this case $C_{\tilde{S}}^2 = C_S^2/q$. Now assuming that N and R are independent, from equations 4.65, 4.66, and 4.70 it can be shown that

$$P\{B = n\} = \frac{1}{q} p_N \left(\left\lfloor \frac{Z + n}{q} \right\rfloor \right), \quad n = 1, 2, \ldots \tag{4.71}$$

$$P\{I = n\} = \frac{1}{q} p_N \left(\left\lfloor \frac{Z - n}{q} \right\rfloor \right), \quad n = 1, 2, \ldots, Z \tag{4.72}$$

These results can be used to obtain the optimal values of q and Z that minimize the total cost of inventory, backlogging and PA card transmittal, as well as machine set-up costs if the machines are set up for each batch of PA cards.

Overlapped batch manufacture. We will now model and analyze the system when policy (2) is used for initiating manufacture of items. For this analysis we need to explicitly model the manufacturing facility and the output store (unlike in the previous case in which an aggregate service time $\tilde{S}$ was used to model the manufacturing facility implicitly). We restrict our attention to a manufacturing facility with c parallel machines and assume that to guarantee that a machine does not idle when a customer is backlogged we have $Z \geq \lceil (c - 1)/q \rceil q + q$. We model this by a $G\tilde{I}^q/G/c$ queueing system operated as follows:

The nth job arrives at time $\tilde{A}_n = A_{nq}$ and brings q tasks. The service time of the nth task is S_n, $n = 1, 2, \ldots$. Let $\tilde{D}_n$ be the departure time of the nth task from this queueing system. Then $D_n = \tilde{D}_{nq}$ is the time epoch where the nth batch of q PA cards order is filled. To see this equivalence, imagine that each customer brings a tag along with his demand. As soon as q tags are accumulated then q tags are passed along to the service facility as q PA cards. Whenever an item is produced the tag in the manufacturing facility is destroyed. When the qth tag is destroyed q items and the q PA cards are returned to the output storage. Thus, every qth tag departure corresponds to a time at which a batch of q PA card orders is filled.

Let $N(t)$ be the number of tasks in the bulk arrival $G\tilde{I}^q/G/1$ queueing system with arrival process $\{\tilde{A}_n, n = 1, 2, \ldots\}$ and service times $\{S_n, n = 1, 2, \ldots\}$ and bulk size $X_n = q$, $n = 1, 2, \ldots$. Then if $\overline{C}(t)$ is the number of completed items at the manufacturing facility corresponding to the batch of PA cards currently being manufactured, we have

$$\overline{C}(t) = \left\lceil \frac{N(t)}{q} \right\rceil q - N(t), \quad t \geq 0 \tag{4.73}$$

Then the number of customers backlogged at time t is

$$B(t) = \left(\left\lceil \frac{N(t)}{q} \right\rceil q + R(t) - Z \right)^+, \quad t \geq 0 \tag{4.74}$$

The inventory of completed items in the output store is

$$I(t) = \left(Z - \left\lceil \frac{N(t)}{q} \right\rceil q - R(t) \right)^+, \quad t \geq 0 \tag{4.75}$$

We have already shown that R has a uniform distribution on $\{0, 1, \ldots, q-1\}$. Hence if we know the distribution of N, we can use equations 4.73 to 4.75 to obtain the distribution for $\overline{C}$, B, and I. An approximate distribution for N can be obtained from the results outlined in Section 4.3.1 for the $GI^X/G/c$ queue. When q is an integer multiple of c, the approximation suggested there is equivalent to the following. Suppose whenever a job arrives with q tasks we split the tasks into c bundles each consisting of q/c tasks and assign one bundle to each server. If we do this for all jobs, we will have c parallel queueing systems $G\tilde{I}^{q/c}/G/1$ each with an arrival process $\{\tilde{A}_n, n = 1, 2, \ldots\}$ and iid service times equal in distribution to S. Now the approximation given in Section 4.3.1 for $GI^X/G/1$ can be used to obtain approximations for these c parallel queues and use them to approximate the $G\tilde{I}^q/G/1$ system.

In developing the above $G\tilde{I}^q/G/c$ model we implicitly assumed that the items for any given batch of PA cards will be processed one at a time by whichever machine becomes free. In some real manufacturing set-ups all q items associated with a batch of PA cards are produced by the same machine. In such a case the correct model would be a $G\tilde{I}/\tilde{G}/c$ queue for tags generated by every qth customer. The arrival process for this queueing system is $\{\tilde{A}_n, n = 1, 2, \ldots\}$ and the service times $\{\tilde{S}_n, n = 1, 2, \ldots\}$ where $\tilde{S}_n = S_{nq-q+1} + \cdots + S_{nq}, n = 1, 2, \ldots$. In this case we need at least c batches of q PA cards (i.e., $Z \geq cq$) to guarantee that a machine will not be idling when a customer demand is backlogged.

4.4 MULTIPLE-PRODUCT–TYPE PRODUCE-TO-STOCK SYSTEMS

In this section we consider a single-stage manufacturing system that produces r types $\{1, \ldots, r\}$ of products to stock. The nth customer for type i product arrives at time $A_n^{(i)}$, $n = 1, 2, \ldots$. The number of type i items demanded by this customer is $X_n^{(i)}$, $n = 1, 2, \ldots$. As before, if a customer's demand cannot be met from available stock at the output store, the customer may leave and satisfy his demand elsewhere (the lost sales case). Conversely, he may choose to wait until his demand can be satisfied (the backlogging case). The manufacturing process of items involves the transformation of raw material by processing it on a single machine. Items are processed one at a time, and the processing time of the nth type i item is $S_n^{(i)}, n = 1, 2, \ldots; i = 1, \ldots, r$.

4.4.1 Target-Level PA Mechanism

Consider a PA mechanism that assigns Z_i PA cards for type i items, $i = 1, \ldots, r$. Suppose at time 0, there are Z_i type i items in the output store, $i = 1, \ldots, r$. Each type i item is tagged with a type i tag. As soon as a type i item is taken from the output store by a demand, the type i tag attached to it is removed and transmitted to the manufacturing facility as a (type i) PA card authorizing the production of a type i item, $i = 1, \ldots, r$. Therefore if at any time there are Z_i type i items in the output store there will be no type i PA cards in the manufacturing facility. The manufacture of different types of items can be initiated by the corresponding PA cards according to some priority policy. The simplest is to manufacture items in the order of PA card arrivals to the manufacturing facility.

M/M/1 multiclass lost demand FCFS model.   Customers for items of different types arrive according to mutually independent Poisson processes. The demand rate for type i items is λ_i, $i = 1, \ldots, r$. The times needed to manufacture type i items are iid exponential random variables with mean $1/\mu$, $i = 1, \ldots, r$. Items are manufactured according to the order of PA card arrivals at the manufacturing facility. Customer demands not met by items from the output buffer are lost. Let $N_i(t)$ be the number of type i jobs at time t in a $M/M/1/\mathbf{Z}$ multiclass queueing system with finite buffer capacity Z_j for type j jobs ($j = 1, \ldots, r$) and FCFS service protocol, $i = 1, \ldots, r$. Then if $I_i(t)$ is the number of type i items in the output store at time t, one has

$$I_i(t) = Z_i - N_i(t), \quad t \geq 0; i = 1, \ldots, r$$

Unfortunately $\{(N_1(t), \ldots, N_r(t)), t \geq 0\}$ is not a Markov process. To make it a Markov process we need information on the type of jobs in each of the $N(t)$ ($N(t) = \sum_{i=1}^{r} N_i(t)$) positions. So let $X_j(t)$ be the class index of the job in the jth position for service ($j \leq N(t)$). If we define $\mathbf{N}(t) = (N_i(t), i = 1, \ldots, r)$ and $\mathbf{X}(t) = (X_j(t), j = 1, \ldots, N(t))$, then $\{(\mathbf{N}(t), \mathbf{X}(t)), t \geq 0\}$ is a Markov process. Let $q(\mathbf{n}, \mathbf{x}) = \lim_{t \to \infty} P\{\mathbf{N}(t) = \mathbf{n}, \mathbf{X}(t) = \mathbf{x}\}$ be the stationary probability distribution of $\{(\mathbf{N}(t), \mathbf{X}(t)), t \geq 0\}$. The balance equations for the rate of probability inflows and outflows of state $(\mathbf{n}, \mathbf{x})$ are

$$q(\mathbf{n}, \mathbf{x})(\mu I_{\left\{\sum_{i=1}^{r} n_i > 0\right\}} + \sum_{i=1}^{r} \lambda_i I_{\{n_i < Z_i\}})$$

$$= \sum_{i=1}^{r} q(\mathbf{n} - \mathbf{e}_i, x_1, \ldots, x_{n-1}) \lambda_i I_{\{x_n = i\}}$$

$$+ \sum_{i=1}^{r} q(\mathbf{n} + \mathbf{e}_i, i, x_1, \ldots, x_n) \mu I_{\{n_i < Z_i\}}$$

It can be verified, through substitution, that $q(\mathbf{n}, \mathbf{x}) = K \prod_{i=1}^{r} \rho_i^{n_i}$ satisfies these balance equations for all values of $\mathbf{n}$ and $\mathbf{x}$. Hence the marginal distribution $p(\mathbf{n}) = \lim_{t \to \infty} P\{\mathbf{N}(t) = \mathbf{n}\}$ of $\mathbf{N}$ is given by

$$p(\mathbf{n}) = G(\mathbf{Z})^{-1} \begin{pmatrix} \sum_{i=1}^{r} n_i \\ n_1, \cdots, n_r \end{pmatrix} \prod_{i=1}^{r} \rho_i^{n_i}, \quad 0 \leq n_i \leq Z_i; i = 1, \ldots, r \qquad (4.76)$$

where

$$G(\mathbf{Z}) = \sum_{n_1=0}^{Z_1} \cdots \sum_{n_r=0}^{Z_r} \begin{pmatrix} \sum_{i=1}^{r} n_i \\ n_1, \cdots, n_r \end{pmatrix} \prod_{i=1}^{r} \rho_i^{n_i} \qquad (4.77)$$

is the normalizing constant. Because the service level for type i items is $(1 - P\{I_i = 0\}) = (1 - p_i(Z_i))$ one sees from equations 4.76 and 4.77 that

$$SL_i = \left(1 - \sum_{n_1=0}^{Z_1} \cdots \sum_{n_i=Z_i}^{Z_i} \cdots \sum_{n_r=0}^{Z_r} p(\mathbf{n})\right)$$

$$= \frac{G(\mathbf{Z} - \mathbf{e}_i)}{G(\mathbf{Z})}, \quad i = 1, \ldots, r \tag{4.78}$$

It can be verified that (see problem 22) $G(\mathbf{Z})G(\mathbf{Z}) \geq G(\mathbf{Z} + \mathbf{e}_i)G(\mathbf{Z} - \mathbf{e}_i)$ and for $i \neq j$, $G(\mathbf{Z} - \mathbf{e}_i + \mathbf{e}_j)G(\mathbf{Z}) \leq G(\mathbf{Z} + \mathbf{e}_j)G(\mathbf{Z} - \mathbf{e}_i)$. Hence one sees that from equation 4.78 that increasing Z_i will improve the service level for product type i but will decrease the service-level performance for the other product j ($j \neq i$).

M/M/1 multiclass backlogged demand FCFS model. Suppose customers for item i arrive according to a Poisson process with rate λ_i, $i = 1, \ldots, r$. The processing times for type i items are iid random variables with exponential distribution with mean $1/\mu$. Each PA card is given a time stamp corresponding to the arrival time of the demand that generated it. Then the cards are served in chronological order corresponding to the stamped times. Let $N_i(t)$ be the number of type i jobs at time t in an $M/M/1$ multiclass queueing system where jobs are served on a FCFS basis, $i = 1, \ldots, r$. Then if $I_i(t)$ is the number of type i items in the output store and $B_i(t)$ is the number of type i demands backlogged at time t, one has

$$I_i(t) = \{Z_i - N_i(t)\}^+$$

and

$$B_i(t) = \{N_i(t) - Z_i\}^+$$

Let $\mathbf{N}(t) = (N_i(t), i = 1, \ldots, r)$. Because jobs are served on a FCFS basis, without the information on the positions of the different classes of jobs in the queue $\{\mathbf{N}(t), t \geq 0\}$ cannot be a Markov process. With $N(t) = \sum_{i=1}^r N_i(t)$, however, $\{N(t), t \geq 0\}$ is a Markov process, and its stationary distribution is the same as the number of jobs in a $M/M/1$ queue with arrival rate $\lambda = \sum_{i=1}^r \lambda_i$ and service rate μ. That is,

$$p(n) = (1 - \rho)\rho^n, \quad n = 0, 1, \ldots$$

where $\rho = \sum_{i=1}^r \rho_i$ and $\rho_i = \lambda_i/\mu$, $i = 1, \ldots, r$. Because the arrival process of different classes of jobs are mutually independent and Poisson, the stationary probability distribution $p(\mathbf{n}) = \lim_{t\to\infty} P\{\mathbf{N}(t) = \mathbf{n}\}$, $\mathbf{n} \in \mathcal{N}_+^r$ can be obtained from $\mathbf{p}$ by a multinomial thinning with probabilities $(\lambda_i/\lambda, i = 1, \ldots, r)$. Specifically using

$$P\{N_1 = n_1, \ldots, N_r = n_r | N = n\} = \binom{n}{n_1, \cdots, n_r} \prod_{i=1}^r \left(\frac{\lambda_i}{\lambda}\right)^{n_i}, \quad n = \sum_{i=1}^r n_i$$

one obtains

$$p(\mathbf{n}) = \binom{n}{n_1, \cdots, n_r}(1 - \rho)\prod_{i=1}^r \rho_i^{n_i} \tag{4.79}$$

The marginal distribution p_i is then given by

$$p_i(n_i) = P\{N_i = n_i\} = (1 - \hat{\rho}_i)\hat{\rho}_i^{n_i}, \quad n_i = 0, 1, \dots \qquad (4.80)$$

where

$$\hat{\rho}_i = \lambda_i / \left(\mu - \sum_{j \neq i} \lambda_j \right), \quad i = 1, \dots, r$$

$$P\{I_i = 0\} = \hat{\rho}_i^{Z_i}$$

$$P\{I_i = n_i\} = (1 - \hat{\rho}_i)\hat{\rho}_i^{Z_i - n_i}, \quad n_i = 1, \dots, Z_i$$

and

$$P\{B_i = n_i\} = \begin{cases} 1 - \hat{\rho}^{Z_i+1} & n_i = 0 \\ (1 - \hat{\rho}_i)\hat{\rho}_i^{Z_i+n_i}, & n_i = 1, 2, \dots \end{cases}$$

Now the optimal target level can be obtained in exactly the same way as we did for the single-product–type case.

M/G/1 multiple-class backlogged demand FCFS model.

As before we assume that customers requiring different types of items arrive according to a Poisson process. The processing times of type i items are iid random variables with distribution function $F_S^{(i)}$, $i = 1, \dots, r$. Let $N_i(t)$ be the number of type i jobs in an $M/G/1$ multiple-class queueing system with FCFS service discipline, $i = 1, \dots, r$. Then $N(t) = \sum_{i=1}^r N_i(t)$ is the number of jobs in an $M/\hat{G}/1$ (single-class) queueing system with arrival rate $\lambda = \sum_{i=1}^r \lambda_i$ and service time distribution $F_S = \sum_{i=1}^r (\lambda_i/\lambda)F_S^{(i)}$. To see this analogy, pretend that the server at the multiple class queueing system does not observe the class of job it is processing. This is possible because we are using FCFS service discipline. Then the distribution of the nth service time is F_S, and jobs arrive according to a Poisson process with rate λ. This then establishes the analogy. Then from the results in Section 3.3.2 we have

$$\tilde{p}_N(z) = \frac{(1 - \rho)(z - 1)\tilde{F}_S(\lambda - \lambda z)}{z - \tilde{F}_S(\lambda - \lambda z)}$$

Furthermore, the waiting time of an arbitrary job has the LST

$$\tilde{F}_W(s) = \frac{s(1 - \rho)}{s - \lambda(1 - \tilde{F}_S(s))}$$

Then the LST of the total time spent in the system by a class i job is $\tilde{F}_W(s)\tilde{F}_S^{(i)}(s)$. Hence the MGF of the total number of class i jobs arrived during this time is $\tilde{F}_W(\lambda_i - \lambda_i z)\tilde{F}_S^{(i)}(\lambda_i - \lambda_i z)$. Note that this is also the number of type i jobs seen by a type i departure. Because jobs arrive according to a Poisson process, by the property that

"Poisson arrivals see time averages" and the equality between the arrival epoch and departure epoch probabilities, one has

$$
\tilde{p}_{N_i}(z) = \frac{(1-\rho)\lambda_i(1-z)\tilde{F}_S^{(i)}(\lambda_i - \lambda_i z)}{\lambda_i - \lambda_i z - \lambda(1 - \tilde{F}_S(\lambda_i - \lambda_i z))}, \quad i = 1, \ldots, r \tag{4.81}
$$

and

$$
E[N_i] = \frac{\lambda_i \lambda E[S^2]}{2(1-\rho)} + \rho_i, \quad i = 1, \ldots, r \tag{4.82}
$$

Using $P\{N_i = 0\} = \tilde{p}_{N_i}(0)$ we get

$$
P\{N_i = 0\} = \frac{(1-\rho)\lambda_i \tilde{F}_S^{(i)}(\lambda_i)}{\lambda_i - \lambda(1 - \tilde{F}_S(\lambda_i))} \tag{4.83}
$$

Therefore, using an effective utilization $\hat{\rho}_i$ for class i jobs such that $P\{N_i = 0\} = 1 - \hat{\rho}_i$, or equivalently, $\hat{\rho}_i = 1 - P\{N_i = 0\}$, we approximate the marginal distribution of N_i by

$$
p_i(n_i) \approx \begin{cases} 1 - \hat{\rho}_i, & n_i = 0 \\ \hat{\rho}_i(1 - \sigma_i)\sigma_i^{n_i - 1}, & n_i = 1, 2, \ldots, \end{cases}
$$

where

$$
\sigma_i = (E[N_i] - \hat{\rho}_i)/E[N_i], \quad i = 1, \ldots, r
$$

When the processing times are exponentially distributed with the same mean $1/\mu$ for all customer types, it can be verified that the preceding approximation gives the exact result for $p_i(n_i)$ (see equation 4.80).

M/G/1 multiclass NPP service model.

Consider the same produce-to-stock manufacturing system that manufactures multiple types of items described earlier. Suppose instead of manufacturing items in the (time stamped) order of PA cards, we produce type i items as long as there are type i PA cards, and there are no type j ($j < i$) PA cards at the manufacturing facility, $i = 1, \ldots, r$. That is, we give higher NPP to the manufacture of type j item over type i item for $j < i$. Let $N_i(t)$ be the number of type i jobs at time t in an $M/G/1$ multiple-class queueing system with priority service protocol (see Chapter 3). The results given in Section 3.5.2 for this queueing system can be used as in the previous cases to compute the probability distributions of I and B.

4.5 PRODUCE-TO-STOCK SYSTEMS WITH ADVANCE ORDERS

In this section we consider produce-to-stock systems in which orders are received in advance of the time when the items covered by the orders are required to be delivered. Alternatively, and equivalently, we consider the situation in which it is possible to forecast arrival of orders for immediate delivery, and we want to use these forecasts to improve the performance of the produce-to-stock system, either by reducing the amount of inventory carried or improving the service level. Before analyzing these systems, however, we first consider generalized PA systems.

4.5.1 Generalized PA Systems

Up to now we have assumed that there is a tag associated with each item in the output store and hence the number of tags (and the number of PA cards) is the same as the maximum inventory in the store. Another way of viewing the operation of the system is that a customer arrives with an order consisting of a set of tags. The output store consists of items and with each item is associated a material tag. Each order tag is matched with a material tag to create the PA card, and simultaneously the item is issued from the store and delivered to the customer (and could then be accompanied back to the customer by his order tag). The PA card goes to the manufacturing facility, and when the item is produced and delivered to the output store the PA card is converted back to a material tag. If there is no inventory in the store (and hence no material tags) when the customer order arrives then the customer tags would become the queue of backlogs, waiting for items and the associated PA cards to come through the manufacturing facility. Note that with this approach there can be a large number of PA cards in the manufacturing facility, usually more than what is necessary to prevent machines from being idle when there are outstanding orders. Although it is traditionally considered desirable to have a queue of PA cards in the manufacturing facility, this large number of PA cards may create problems in multiple-product systems in ensuring that priority protocols are followed. Thus it is desirable to consider *generalized* PA systems in which the number of PA cards is no longer equal to the maximum inventory in the output store (see Figure 4.3).

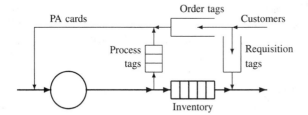

Figure 4.3   Generalized PA System

To implement this we now suppose that the output store, at a time when the manufacturing facility is idle and there are no customer orders, contains Z items and K *process tags* (which will eventually become the PA cards). A customer is now assumed to arrive with two documents, an *order tag* and a *requisition tag*, for each item that he requires. On arrival the requisition tag is matched with a physical item in the output store, and assuming there is an item available, the item (and the requisition tag) are then delivered to the customer. (Note that the customer can take any part in the store.) If no items are available then the waiting requisition tags represent the physical backlog of unmet demands. Also, on arrival, the order tag is matched with a process tag, and when this match occurs the process tag then becomes the PA card. The order tag would then be destroyed. After manufacture of the item authorized by the PA card is completed and the item is delivered to the output store, then the PA card is converted to a process tag, and it waits at the output store. If, when an order tag arrives, there are no process tags available in the output store (irrespective of the existence of physical stock) then the order tags would form a queue, waiting for process tags with which they can be matched.

It is clear that in a single-product single-machine system with immediate transmission of PA cards it is only necessary to have one PA card (or equivalently one process tag) to ensure that the machine is always busy when the store is not full at level Z. Otherwise we would like to find $K_{\min}$, the minimum number of PA cards to ensure that, whenever the total of the inventory in the store and in the manufacturing facility is less than Z, all machines are busy. If we transmit PA cards in batches of size q then

$$K_{\min} = \left\lceil \frac{c-1}{q} \right\rceil q + q.$$

Provided $K \geq K_{\min}$ then all the results and formulas obtained earlier apply irrespective of Z. It is of interest to note that if $Z = 0$ and $K \geq K_{\min}$ then the system behaves like a produce-to-order system.

With multiple products and backlogging we need at least $K^{(i)}_{\min} = \lceil (c-1)/q_i \rceil q_i + q_i$ process tags (or PA cards) for each product i ($i = 1, \ldots, r$). One would never have all available PA cards in process in the facility, however, even if $K^{(i)} = K^{(i)}_{\min}$ for all i.

In a generalized PA system there is a fundamental equality that arises because of the relationship between tags and physical inventory (see Figure 4.4). At time t, let $I(t)$ be the physical inventory in the output store, $B(t)$ be the backlog of unmet demands, $K^+(t)$ be the number of process tags in the output store, and $K^-(t)$ be the number of

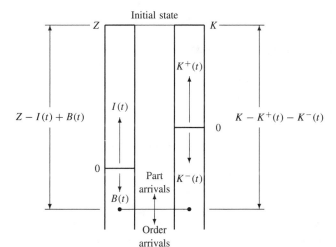

Figure 4.4   Relationship between Inventory and Tags in Generalized PA System

order tags waiting for a match with process tags. Then for all t

$$Z - I(t) + B(t) = K - K^+(t) + K^-(t) \tag{4.84}$$

It follows that with lost sales or stopped arrivals if $K > Z$ there will never be more than Z PA cards in circulation so it follows that there is no point in having $K > Z$. So we choose $K \leq Z$. If $K < Z$ then with lost sales it is necessary to time stamp the arrival of the order tags (and equivalently the instant of filling the order) to ensure that PA cards are served in order of arrival and thus for our model results to remain valid.

4.5.2 Time Lags between Orders and Requisitions

In a generalized PA system there is no inherent reason why the customer should send the order tag and the requisition tag simultaneously. The order tag can be sent in advance of the requisition to inform the producer of the future requirement, but no shipment occurs until the requisition tag is received. Alternatively, if the producer is able to forecast future demands accurately and wants to use this information to improve performance then he could generate a tag that would be used in the same way as a customer-generated order tag. The mechanism of operation is again that on receipt of an order tag it is matched with a process tag to generate a PA card that is then available for transmission to the manufacturing facility. The matched order tag would then be kept until its associated requisition tag arrives, after which it can be destroyed. After the item authorized by the PA card has been made the item goes to the store, and the PA card is converted to a process tag and waits with the other process tags. When the requisition tag arrives, it would be compared with the associated order tag (which could still be waiting for a process tag), the item is then issued, and the item and the requisition tag return to the customer.

Suppose the arrival times of the order tags are $\{A_n, n = 1, 2, \ldots\}$. Then we will assume that the arrival times of the requisition tags are $\{A_n + \tau, n = 1, 2, \ldots\}$, that is, there is a fixed time lag τ between receipt of an order tag and receipt of the corresponding requisition tag. At time 0 assume there are no PA cards in the manufacturing facility and no order tags where the corresponding requisition tag has yet to arrive. The physical inventory in the output store is then Z, and the number of process tags at the output store is K. Because of the lag between receipt of order tags and requisition tags the identity in equation 4.84 no longer applies. Also note that the inventory level in the output store can now exceed Z.

In modeling the system it is necessary to keep track of (1) the processes related to the order tags, the PA cards and the process tags, and (2) the processes related to the requisition tags, the physical inventory, and the backlogged demand. Then considering first the processes related to the order tags, the PA cards and the process tags, let $A(t)$ be the number of order tags arrived in $(0, t]$, $D(t)$ be the number of process tags (and hence items) delivered to the output store in $(0, t]$ and $O(t)$ be the number of PA cards triggered in $(0, t]$. Further, at time t, let $K^+(t)$ be the number of process tags in the output store, $K^-(t)$ be the number of (unmatched) order tags in the output store, and

$C(t)$ be the number of PA cards that are active, that is, have been triggered and are either waiting transmission or at the manufacturing facility. Then

$$K^+(t) = K - C(t) = K + D(t) - O(t) \tag{4.85}$$

$$O(t) = \min\{K + D(t), A(t)\} \tag{4.86}$$

$$K^-(t) = A(t) - O(t) \tag{4.87}$$

$$C(t) = O(t) - D(t) = \min\{K, A(t) - D(t)\} \tag{4.88}$$

Defining $N(t) = A(t) - D(t)$ it can be seen that

$$K^+(t) - K^-(t) = K + D(t) - A(t) = K - N(t) \tag{4.89}$$

and thus

$$K^+(t) = \{K - N(t)\}^+ \tag{4.90}$$

$$K^-(t) = \{N(t) - K\}^+ \tag{4.91}$$

and also

$$C(t) + K^-(t) = N(t) \tag{4.92}$$

So, as before, by studying $N(t)$ it is possible to determine all we need to know about $K^+(t)$, $K^-(t)$, and $C(t)$. With demand backlogging permitted, all the earlier results we have obtained for $N(t)$ apply, and the distributions of $K^+(t)$, $K^-(t)$, and $C(t)$ will be the same as the earlier results for $I(t)$, $B(t)$, and $C(t)$, respectively, except that Z is replaced by K. The condition for all machines to be busy when $K^-(t) > 0$ is now $K \geq \lceil (c-1)/q \rceil q + q$.

Next, consider the processes relating to the requisition tags, physical inventory, and backlogged demand. Let $R(t)$ be the number of requisition tags that arrive in $(0, t]$ and $S(t)$ be the number of items delivered to customers in $(0, t]$. Further, at time t, let $I(t)$ be the inventory in the output store, and $B(t)$ be the number of requisition tags waiting in the output store (i.e., the amount of backlogged demand). Then

$$I(t) = Z + D(t) - S(t) \tag{4.93}$$

$$S(t) = \min\{Z + D(t), R(t)\} \tag{4.94}$$

$$B(t) = R(t) - S(t) \tag{4.95}$$

Now from equations 4.93 and 4.95

$$I(t) - B(t) = Z + D(t) - R(t) \tag{4.96}$$

Next, define $N^*(t) = R(t) - D(t)$ where it must be noted that $N^*(t)$ can take on both positive and negative values. It follows from equations 4.93, 4.94, and 4.96 that

$$I(t) = \{Z - N^*(t)\}^+ \tag{4.97}$$

$$B(t) = \{N^*(t) - Z\}^+ \tag{4.98}$$

Alternatively

$$I(t) = Z + B(t) - N^*(t) \qquad (4.99)$$

Next we relate the two sets of processes by finding a relationship between $N^*(t)$ and $N(t)$. We use the fact that

$$R(t) = A(t - \tau) \qquad (4.100)$$

Hence

$$N^*(t) = R(t) - D(t) = A(t - \tau) - D(t)$$
$$= A(t - \tau) - D(t - \tau) + D(t - \tau) - D(t)$$
$$= N(t - \tau) - D(t - \tau, t) \qquad (4.101)$$

where $D(t - \tau, t)$ denotes the number of customers that departed the queueing system associated with $N(t)$ during the time interval $(t - \tau, t]$. Thus

$$P\{N^*(t) = n\} = \sum_{k=n}^{\infty} P\{D(t - \tau, t) = k - n | N(t - \tau) = k\} P\{N(t - \tau) = k\}, \quad n > 0$$

$$= \sum_{k=0}^{\infty} P\{D(t - \tau, t) = k - n | N(t - \tau) = k\} P\{N(t - \tau) = k\}, \quad n \le 0$$

Now, if $n \ge c$ it can be seen that during $(t - \tau, t]$ all c machines will have been busy all the time. Thus if each machine has exponentially distributed processing time with mean $1/\mu$

$$P\{D(t - \tau, t) = k - n | N(t - \tau) = k\} = \frac{(c\mu\tau)^{k-n}}{(k - n)!} e^{-c\mu\tau}, \quad n \ge c$$

and thus

$$P\{N^*(t) = n\} = \sum_{k=n}^{\infty} \frac{(c\mu\tau)^{k-n}}{(k - n)!} e^{-c\mu\tau} P\{N(t - \tau) = k\}, \quad n \ge c$$

Next, let

$$p^*(n) = \lim_{t \to \infty} P\{N^*(t) = n\}$$

and

$$p(n) = \lim_{t \to \infty} P\{N(t - \tau) = n\}$$

Then with exponential processing times

$$p^*(n) = \sum_{k=n}^{\infty} \frac{(c\mu\tau)^{k-n}}{(k - n)!} e^{-c\mu\tau} p(k), \quad n \ge c$$

Further analysis requires assumptions concerning the pattern of arrival of the order tags and when PA cards are transmitted to the manufacturing facility. Suppose order-tag arrivals are Poisson with parameter λ, and PA cards are transmitted once generated. Then $p(n) = \rho^{n-c} p(c)$ where $\rho = \lambda/c\mu$ and

$$p^*(n) = e^{-c\mu\tau(1-\rho)} p(n), \quad n \geq c$$

Thus, if $Z \geq c - 1$ the following results about $B(t)$ can be obtained

$$\lim_{t\to\infty} P\{B(t) > 0\} = \frac{e^{-c\mu\tau(1-\rho)}}{1-\rho} p(Z+1)$$

and

$$E[B] = \lim_{t\to\infty} E[B(t)] = \frac{e^{-c\mu\tau(1-\rho)}}{(1-\rho)^2} p(Z+1)$$

If $c = 1$ and $Z \geq 0$ then

$$E[B] = e^{-\mu\tau(1-\rho)} \frac{\rho^{Z+1}}{1-\rho}$$

The average inventory level can be found from equations 4.99 and 4.101, which as $t \to \infty$ become

$$E[I] = Z + E[B] - E[N^*]$$

and

$$E[N^*] = E[N] - \lim_{t\to\infty} E[D(t-\tau, t)]$$

Because the item departure rate should equal the demand arrival rate in this system,

$$\lim_{t\to\infty} E[D(t-\tau, t)] = \lambda\tau$$

and thus

$$E[I] = Z + E[B] + \lambda\tau - E[N]$$

For Poisson arrival of order tags, immediate transmission of triggered PA cards and a manufacturing facility consisting of a single machine with exponential processing times, it follows that

$$E[I] = Z + \lambda\tau - \frac{\rho}{1-\rho}(1 - \rho^Z e^{-\mu\tau(1-\rho)}), \quad Z \geq 0.$$

If the inventory carrying cost is k_1 per item per unit time and the backlog cost is k_2 per item per unit time, then the total cost rate $TC(Z, \tau)$ for this system is

$$TC(Z, \tau) = k_1 \left(Z + \lambda\tau - \frac{\rho}{1-\rho} \right) + (k_1 + k_2) \frac{\rho^{Z+1}}{1-\rho} e^{-\mu\tau(1-\rho)}$$

If Z were fixed it would then be possible to find the optimal $\tau^*(Z)$. It would be given by

$$k_1 = (k_1 + k_2)\rho^Z e^{-\mu\tau^*(Z)(1-\rho)}$$

or

$$\mu \tau^*(Z)(1 - \rho) = Z \ln \rho - \ln \left(\frac{k_1}{k_1 + k_2} \right)$$

Note that if $Z > \hat{Z} = \ln(k_1/[k_1 + k_2])/\ln \rho$, where $\hat{Z}$ is the optimal Z obtained for the target level system in Section 4.3.1, then $\tau^*(Z) = 0$ (unless we consider the possibility of a system where orders are sent after requisitions). Provided $\tau^*(Z) > 0$ we have

$$TC(Z, \tau^*(Z)) = k_1(Z + \lambda \tau^*(Z)) \tag{4.102}$$

Note that $\tau^*(Z)$ is a linear function of Z and after substituting for $\tau^*(Z)$ in equation 4.102 the coefficient of Z is $k_1(1 + \ln \rho(\rho/[1 - \rho]))$, which can be shown to be always positive for $\rho < 1$. Hence the minimum of $TC(Z, \tau^*(Z))$ for $Z \geq 0$ is achieved when $Z = 0$ and $\lambda \tau^*(0) = -(\rho/[1 - \rho]) \ln \left(\frac{k_1}{k_1 + k_2} \right)$, or when $\tau = -cE[F]$, where $E[F]$ is the average time from arrival of an order tag until the item authorized by it reach the output store (the average flow time of a job through the manufacturing facility if it were operated as produce-to-order) and $c = \ln(k_1/(k_1 + k_2))$. (Note that $c < 0$.) It also follows that a system where order tags are sent in advance of requisition tags (or equivalently perfect forecasts of the realization of the demand process are available) can be designed so it will have a lower cost than systems that do not have or do not use such information.

It should be noted that although the generalized PA system appears superior to the original PA system the system as we have described it does not allow for two phenomena

1. Unmatched requisition and order tags, that is, the customer takes a different number of items to the number he ordered.
2. Quality problems on delivery to the customer. If the quality problem is detected when the items are delivered to the output store then our models will allow for this because the PA card that arrives with the item found to be defective is immediately returned to the manufacturing facility; however, if the problems show up at delivery to the customer then the system as described would not capture this.

In both cases the assumed relation between $N(t)$ and $N^*(t)$ would no longer be valid. With quality problems it would be necessary to generate an extra order tag every time an item is found to be defective and thus cannot be delivered to the customer. Thus there would be two streams of order tags arriving at the output store: one would be order tags created by customer orders and the other being the order tags generated by defects found in attempting to match a requisition tag with an item for delivery to the customer. (Because of the lack of advance notice of these defects it would no longer be optimal to have zero Z). Alternatively, at the time of receipt of the order tags, extra order tags would then be generated to allow for the expected number of defects. In this case it would not be necessary to create order tags when a defect is found at the time of shipment (key idea is the need to ensure quantity produced is equal to quantity requisitioned plus quantity defective).

In the situation in which the customer order tags and requisition tags do not match (e.g., if the customer decides to take more items than ordered, or he cancels the order after releasing the order tags), then it is necessary to ensure that overall the number of requisition tags is equal to the number of order tags. Thus, if the customer decides to take more than he ordered then for each extra requisition tag it would be necessary to generate an extra order tag as soon as it is known that the customer wants more than he originally ordered. If no requisition tag arrives to match an order tag (i.e., the customer either reduces or cancels his order), however, then the unmatched order tags would be canceled if they are still waiting for process tags. Otherwise if any of his order tags have already generated PA cards, then each such unmatched order tag would be canceled *and* replaced by some other order tag still waiting for a process tag. This replacement would be treated as if it had generated the PA card. If an unmatched order tag has already generated a PA card and there are no order tags waiting for process tags, then either the job authorized by the PA card would be canceled, or alternatively work on it would continue, but as soon as another order tag arrives at the store that order tag would replace the order tag associated with the PA card for the job in process.

4.5.3 Materials Requirement Planning (MRP)–Controlled Systems

MRP is often used to control the release of orders to manufacturing or to a supplier. The basic logic of MRP is described in many texts on operations management; however, the calculations are usually explained using simple numerical examples, and it is rare for the algebraic formulas to be presented. Before we present the formulas it is important to note that, as well as using information about inventory status and forecasts of future requirements, the calculations are based on two parameters that are given in advance and set by management. These parameters are the lead time, τ, and the planned safety stock, Z. The planned safety stock is the amount of inventory that should be on hand in the output store at the end of the lead time if (1) the actual demand in the lead time is equal to the forecast demand over the lead time, and (2) any orders already released to manufacturing or that will be released now (following the MRP calculation) will be delivered to the output store by the end of the lead time.

The calculation of the release quantity at time t, $J(t)$, is based on the following status and forecast information:

1. On hand inventory in the output store $I(t)$ and backlog of unmet demands $B(t)$ with $I(t) \cdot B(t) = 0$ for all t.
2. Outstanding orders, that is, orders released before t and not yet delivered to the output store, $N(t)$.
3. The forecast of demand from the output store over the time interval $(t, t + \tau]$, $\hat{R}(t, t + \tau)$.

Then the MRP calculations project the on-hand inventory at time $t + \tau$ and if $I(t) - B(t) + N(t) - \hat{R}(t, t + \tau) > Z$ then nothing will be released at time $t+$. Otherwise the order quantity released will be given by

$$J(t) = \max\{Z + B(t) - I(t) - N(t) + \hat{R}(t, t + \tau), 0\}. \tag{4.103}$$

Now consider our single-stage manufacturing system where requisition tags arrive at the output store for immediate delivery if inventory is available. As before define $R(t)$ as the total number of requisition tags arrived over $(0, t]$ and $D(t)$ as the total quantity of items delivered to the output store from the manufacturing stage over $(0, t]$. Then $Z + B(t) - I(t) = N^*(t) = R(t) - D(t)$. Suppose order tags are generated using MRP logic, that is, $J(t)$ order tags are generated at time $t+$ according to equation 4.103. Then if $A(t)$ is the number of order tags generated over $(0, t]$, $N(t) = A(t) - D(t)$ and thus

$$J(t) = \max\{R(t) + \hat{R}(t, t + \tau) - A(t), 0\}. \tag{4.104}$$

Usually descriptions of MRP assume that observation of inventory status is made at periodic intervals, and hence opportunity for release occurs only at these observation instants. MRP can also be used with continuous observation of inventory status, however, and thus release will occur whenever an event occurs that results in $R(t) + \hat{R}(t, t + \tau) - A(t) > 0$. This will depend on the nature of the forecasts $\hat{R}(t, t + \tau)$ and also on whether defects are found on delivery to a customer. Consider the following situations:

Perfect Forecasts, No Defects. With perfect forecasts $\hat{R}(t, t + \tau) = R(t + \tau) - R(t)$ and thus $J(t) = \{R(t + \tau) - A(t)\}^+$. That is, the release quantity at time t is equal to the requisition size at time $t + \tau$. That is, the system is identical to the system where order tags arrive at time t and the associated requisition tags arrive at time $t + \tau$. All the results we have developed for this model apply to the MRP-controlled system.

Average Demand Only Known, No Defects. Suppose that it is not possible to forecast the arrival time and magnitude of future orders; however, the average demand level λ is known, that is, $\hat{R}(t, t + \tau) = \lambda\tau$. Then $J(t) = \{R(t) - A(t) + \lambda\tau\}^+$, and this will only differ from zero when a requisition tag arrives at time t. That is, the system is identical to that where order tags and requisition tags arrive simultaneously; however, note that $J(t) = \{Z + \lambda\tau + B(t) - I(t) - N(t)\}^+$. That is, the effective store capacity is not Z but $Z + \lambda\tau$. All values chosen by management for the pair (τ, Z) such that $Z + \lambda\tau = $ constant will have the same performance with respect to average inventory and average back-orders.

Imperfect Forecasts, No Defects. Suppose that forecasts of requisition tag arrivals are made, but these forecasts are not always correct. $J(t) > 0$ at an instant when it is forecast that a requisition will arrive at time $t + \tau$ and thus an order tag will be generated. If a requisition arrives that was not forecast in advance, however, then it will immediately generate an order tag. Conversely, if a requisition forecast to arrive at time t does not materialize at t, then $\{R(t) - A(t) + \hat{R}(t, t + \tau)\} < 0$. This means that either a job in progress should be canceled if it has not been released to the manufacturing stage, or alternatively no order tag will be generated at the next time $t' \geq t$ when it is forecast that a requisition tag will arrive at time $t' + \tau$.

Perfect Forecasts, Defects on Delivery.   If the forecasts of requisition tag arrivals are perfect, but no consideration is given to the possibility of defects, then there will be two streams of order tags generated for each requisition tag that is to arrive at time t: a single order tag at time $t - \tau$, and a number X_t of order tags at time t where X_t is the number of defective items delivered before delivering an acceptable item. Alternatively, the occurrence of defects may be planned for by releasing $f + 1$, $(f \geq 1)$ order tags at time $t - \tau$, but this means that if $X_t > f$ then $X_t - f$ additional order tags will be generated at time t, whereas if $X_t < f$ the surplus $f - X_t$ will be used to either cancel order tags already released or to inhibit generation of order tags at the next time $t' \geq t$ when a requisition tag is forecast to arrive at time $t' + \tau$.

Note that if it is possible to cancel work in process so that $J(t) = R(t) + \hat{R}(t, t + \tau) - A(t)$, then the actual number in process at time $t+$ is $N(t+) = A(t) - D(t) + J(t) = R(t) + \hat{R}(t, t + \tau) - D(t)$. It follows that $I(t) - B(t) = Z - R(t) + D(t) = Z + \hat{R}(t, t + \tau) - N(t)$, and knowing the distributions of $N(t)$ and $R(t, t + \tau)$ enables the distribution of $I(t) - B(t)$ to be found.

4.6 IMPLICATIONS OF MODELS

PA cards transmit information to the manufacturing facility about the occurrence of demands. They enable the manufacturing facility to implement strategies for releasing jobs for processing based on the inventory levels in both the output store and in process at the machines. Using such information significantly enhances the performance of the system with respect to both inventories and service levels. From a modeling perspective the PA card mechanism allows one to understand the dynamics of the system and how information can be used to control material flow. This facilitates a systematic approach to the modeling of the produce-to-stock systems.

It will be noted that we have dealt with many of the situations discussed in classical inventory theory (e.g., see Silver and Peterson [36]), but rather than treating the process of inventory replenishment and the associated lead times as an endogenous feature we have specifically modeled the process by which replenishments occur. The results that we have derived for optimal inventory levels and so on are based on the explicit representation of the replenishment process and any congestion that might occur owing to the variability of processing times or the pattern of arrivals of replenishment orders at the processing facility. In some respects, however, we have been able to go beyond the classical inventory theory. The models described in Section 4.5, because they represent the replenishment process explicitly, provide insight into the respective merits of safety stock versus safety time in MRP systems. Also they emphasize that the "lead time" in MRP systems has to be interpreted as a management set parameter equal to the desired time difference between order and requisition. It is not necessarily the same as the average time for a job to go through the facility. We have shown that the optimal lead time, however, may be a multiple of this average time with the multiple dependent on the costs of inventory and shortages. We also show that the appropriate lead time is critically dependent on the amount of knowledge the manager has about the realization of future demands and that, if the manager cannot get any information, it is better to recommend a zero lead time and hold safety stock.

   With multiple products the models show how the value of the target thresholds of the different products interact, particularly with FCFS service protocol. This interference can have a significant effect on the service level. This emphasizes that producing a variety of products on common facilities brings significant managerial complexities where myopic policies or simplistic approaches to improving performance may not be effective.

4.7 BIBLIOGRAPHICAL NOTE

The recognition that production-inventory systems can be modeled as queueing systems with the number in the queue equal to the amount of empty space in the store is due to Morse [29]. Morse illustrated the use of queueing models to analyze production-inventory systems, with particular emphasis on situations in which the production system (or equivalently the replenishment process) could be regarded as equivalent to an infinite number of parallel servers (i.e., using an $M/G/\infty$ model) if back-orders are permitted or alternatively, with lost sales, by Z parallel servers. Note that queueing effects are not present in these models. Gross and Harris [14] formulate and analyze a $(S-1, S)$ inventory system in which the queueing effects owing to multiple orders present at the manufacturing facility are captured by state-dependent lead times. In the 1950s Moran [28] and Gani [9] developed stochastic models of dams and water storage systems with particular emphasis on understanding policies for control of release of water from the dam. Subsequent work on the control of queueing systems, in particular the work by Bell [4] [5] on when servers should be turned on and off, has provided useful insights into appropriate control policies for production-inventory systems. Nevertheless, despite the enormous literature on inventory models (see, e.g., Silver and Peterson [36]), it is surprising how rarely queueing models are used. This may be due to the preponderance of discrete time, periodic review models in the literature, reflecting batch-processing–based information systems for inventory recording and decision making. The use of tags and cards to control and report job movement and processing on the factory floor probably dates back to around 1920 or earlier in such major U.S. manufacturing organizations as General Electric (see, e.g., Hathaway [15]). The Japanese Kanban system, developed by Toyota in the 1960s [27], sets limits on the number of tags in the system and thus controls work in process. The generalized PA system both as a method of control and as a basis for modeling has been developed by the authors of this book.

PROBLEMS

4.1 Consider a single-product produce-to-stock system with unit demand and backlogging. Demands arrive according to a Poisson process with rate λ, and there is a single machine to process items. Processing times are exponential with mean $1/\mu$. Determine the optimal target inventory Z considering the following costs: a cost of k_1 per item in the output store per unit time, a cost of k_3 per stock-out occasion (i.e., a demand arrives at the output store when $I = 0$ and $B = 0$) and a cost of k_4 per active PA card per unit time (i.e., raw material is acquired as soon as the PA card becomes active).

4.2 Consider a single-product produce-to-stock system with unit demand and lost sales. Suppose the relevant costs are k_1 per item in the output store per unit time and k_3 per stockout occasion (which in the lost sales case means the first demand to arrive after the inventory falls to zero). A gross profit of p is obtained for each demand met. Assuming demands arrive according to a Poisson process, determine the optimal Z if there is a single machine, and processing times are (1) exponential, or (2) Erlang-2, with mean $1/\mu$.

4.3 Consider a single-product produce-to-stock system with unit demand and lost sales. The target stock level is Z, and there are $c = Z$ identical machines. Show that the average inventory and the fraction of demand met from stock depend only on the mean of the processing-time distribution at a machine. Is this also true if the machines have unequal processing times, and an arriving job is allocated to the machine that has been idle longest?

4.4 In a single-product produce-to-stock system with back-ordering and Poisson demand the store knows the mean time between generating a PA card and its return with an item to the store. The store does not know how many machines there are, however. Suppose it assumes there is an infinite number of machines and then estimates the average number of back-orders, and hence decides on the optimal target stock level Z^y. In reality, however, there is only a single machine with average processing time $1/\mu$. Determine what the true optimal Z^* would be as a function of C_S^2, the squared coefficient of variation of the processing time, and $\rho = \lambda/\mu$, and explore the relationship between Z^y and Z^*. Determine the relative magnitude of the cost penalty through using Z^y instead of Z^*.

4.5 In a single-stage produce-to-stock system with back-orders the service level is related to the delay in meeting demand. Show, using sample path arguments, that (1) the service level improves as Z increases, (2) for a given Z the mean delay in meeting a demand increases as the processing times or the demand interarrival times become more variable.

4.6 It is shown in Section 4.3 that a $G/M/1/Z - 1$ lost sales system and a $G/M/1/Z$ system with interruptible demand have identical values of $p^-(n)$, $n = 0, 1, \ldots, Z - 1$. If $p_{LS}(n)$, $n = 0, \ldots, Z - 1$, and $p_{ID}(n)$, $n = 0, \ldots, Z$ are the respective time-average probabilities for lost sales and interruptible demand,
   (a) Find the ratios $p_{LS}(n)/p_{ID}(n)$, $n = 0, \ldots, Z - 1$.
   (b) Show that it is possible to write $p_{LS}(0) = 1 - \rho SL_{LS}$ and $p_{ID}(0) = 1 - \rho SL_{ID}$ where SL_{LS} and SL_{ID} are the service levels with lost sales and interruptible demand, respectively.
   (c) Show that $1 - p_{ID}(0) = \rho(1 - p_{ID}(Z))$.

4.7 Consider a single-product produce-to-stock system with exponential service times and Erlang-k demands. Suppose $Z = 2$. By setting up a Markov process model derive a formula for the service level with lost sales and hence derive a formula for the service level with interruptible demands and $Z = 3$.

4.8 Consider a single-product produce-to-stock system. There is a single machine, and the processing times have an Erlang-2 distribution. The interarrival times between subsequent demands are iid Erlang-2. Compare the exact and approximate service levels for $Z = 1, 2, \ldots, 5$ for both lost sales and interruptible demand. Comment on the adequacy of the approximations.

4.9 The generalized exponential (GE) distribution is a distribution characterized by $F(0) = 1-a$, $F(x) = 1 - ae^{-bx}$, $x > 0$. Consider a single machine with interruptible demand and an output store of size Z. Suppose demand interarrival times have a GE distribution and the processing times have an exponential distribution.

(a) Determine the mean and variance of the GE distribution.

(b) Develop an approximation for the number of jobs in an $M/GE/1$ queue and hence find σ_R and σ. Hence determine the (approximate) service level using equation 4.47.

(c) Develop a Markov model of the $GE/M/1/Z$ stopped arrival model, and compare its predictions of service level with those found in part (b). (It is sufficient to consider only small values of Z).

4.10 Consider a single machine with interruptible demand where both demands and processing times have GE distributions. Assume that these distributions are identical so service level can be approximated by equation 4.49. Develop a Markov model of the $GE/GE/1/Z$ stopped arrival model and compare its predictions of service level with the approximation.

4.11 Repeat problem 10 for $E_2/M/1/Z$ and $E_2/E_2/1/Z$ stopped arrival models.

4.12 Demands arrive according to a Poisson process with rate λ. Rather than using a PA card system jobs are released to the single machine with exponentially distributed interval of mean $\bar{\tau} = 1/\lambda$. Processing times on the machine are exponential with mean $1/\mu < \bar{\tau}$. Initially inventory in the output store is Z. Any demands not met from stock are back-ordered.

(a) Use results from the time dependent solution of the $M/M/1$ queue to characterize the distribution of inventory in the output store at time t. Comment on the asymptotic distribution as $t \to \infty$ and hence the inherent weakness of work-release policies based only on forecasts of demand.

(b) Would your conclusions to part (a) be changed if jobs were released to the machine at a constant interval of τ? For what combination of release interval, processing-time distribution, and interarrival time of demands would a forecast-based release policy work?

4.13 Consider the situation of problem 12 except that demands not met from stock are lost. Is there a job-release rate such that both the mean and variance of the stationary inventory in the output store are finite? If there is a cost of k_1 per item in the output store per unit time and k_3 per lost sale, determine the optimal release rate that minimizes the total cost.

4.14 Now consider the same situation as problem 13, but now impose an upper limit of Z^* on output store inventory. Once inventory reaches Z^* the machine is shut off and job release to the machine is stopped. Determine the service level as a function of Z^*. If, instead of being lost, demand was back-ordered when the output store is empty what work-release rate should be chosen?

4.15 Consider a single machine produce-to-stock system with bulk demand and backlogging. Processing times are iid Erlang-k random variables. If the output store has capacity Z determine the fraction of demand met from stock when bulk sizes are Poisson. Compare with a system where demands arrive one at a time, but the overall demand rate is the same.

4.16 A single-stage produce-to-stock system has an output store with capacity Z. Demand arrivals are Poisson, and any demand not met from stock is lost. The manufacturing facility consists of a single machine that requires an exponential time to process a job. With probability q the processing is defective. If there is a cost of k_1 per item in the output store per unit time and k_3 per lost sale, determine the optimal Z (1) if defects are detected on completion of processing, and (2) if defects are not detected until items are delivered to the customer.

4.17 In a single-machine produce-to-stock system with lost sales the machine is unreliable. Up times, down times, and processing times are exponential with parameters ζ, γ, and μ, respectively. Demands occur according to a Poisson process with rate λ. For each item in the output store a cost of k_1 is incurred per unit time, and for each demand not met a cost of k_3 is incurred. Determine the optimal Z.

4.18 Demands arrive at a single-machine produce-to-stock system with lost sales according to a Poisson process with rate λ. The output store of this system has a capacity Z. Suppose the processing times on the machines are iid phase-type distributions. Find the probability distribution of the number of items in the output store under stationary conditions. Using the ideas from the $M/G/1/Z$ models of Section 4.3 and the matrix geometric solution of the $M/PH/1$ queues (Chapter 3), show that the preceding distribution is a truncated matrix geometric distribution.

4.19 A single-machine produce-to-stock system with lost sales has output store with capacity Z. Starting up the machine costs R_1 and switching it off costs R_2. When the machine is switched off a cost of r_1 is incurred per unit time, and when the machine is switched on a cost of r_2 is incurred per unit time. The machine can only process jobs when it is switched on, but it need not be switched off when it is idle. When it is switched on, however, it will never be idle when there are waiting jobs. The cost of holding one item in the output store is k_1 per unit time, and the cost incurred for every lost sale is k_3. Demands occur according to a Poisson process with rate λ, and processing times on the machine are iid exponential random variables with mean $1/\mu$. Determine the optimal threshold for starting up the machine for given Z and hence determine the optimal Z. Also find under what conditions the machine will never be switched on and under what conditions the machine will never be switched off.

4.20 Consider a single-stage produce-to-stock system with lost sales and with the same costs as problem 19 except that $r_1 = 0$. There are two identical machines available to process jobs. As a result a threshold policy of the following type is to be used. Switch on one machine when the number of PA cards at the manufacturing facility equals q_1, and switch on the other machine once the number of PA cards reaches q_2. If both machines are switched on, switch off one machine once the number of PA cards falls to $q_3 (< q_2)$, and if only one machine is switched on, leave it switched on until the number of PA cards falls to zero. Find the optimal q_1, q_2, and q_3 for a given Z, and show that in general $q_3 < q_1$ (i.e., the optimal policy is a "hysteresis" policy). Find the optimal Z.

4.21 Suppose in the system of problem 19 with just one machine it were possible to speed up or slow down the machine. Assuming that there are no extra costs associated with speed up or slow down, determine how your solution to problem 19 can be improved.

4.22 Consider a single-stage produce-to-stock system with multiple product types and lost sales. PA card arrivals at the single manufacturing facility are served on a FCFS basis. Suppose the target stock in the output store for product j is Z_j. Demands for product j form a Poisson process with rate λ_j, and processing times are iid exponential random variables with mean $1/\mu$, the same for all product types.

(a) Show that for $j \neq i$, the service level of product j decreases in Z_i. Does the service level of product i increase as Z_i increases? Hint: Let

$$g(\mathbf{n}) = \binom{n}{n_1, \cdots, n_r} \prod_{i=1}^{r} \rho_i^{n_i}$$

for $n = \sum_{i=1}^{r} n_i$. Show that $g(\mathbf{n} + \mathbf{e}_j - \mathbf{e}_i)g(\mathbf{n}) \leq g(\mathbf{n} + \mathbf{e}_j)g(\mathbf{n} - \mathbf{e}_i)$. Use this result to show

$$G(\mathbf{Z} - \mathbf{e}_i + \mathbf{e}_j)G(\mathbf{Z}) \leq G(\mathbf{Z} + \mathbf{e}_j)G(\mathbf{Z} - \mathbf{e}_i)$$

for G given by equation 4.77.

(b) If holding one unit of product j in the output store costs k_{1j} per unit time and a cost of k_{3j} is incurred for every demand of product j not met, determine in a two-product system the optimal values of Z_1 and Z_2.

(c) Suppose in a two-product system that the choice of product to produce on the machine whenever the machine becomes free is based on the inventories in the output stores. Consider the following rules: (1) produce the product with the lower inventory, and (2) produce the product where the difference between target inventory and actual inventory is greater. Determine the service levels for each product for each rule, and determine how the service level for product 2 is influenced by increasing Z_1.

(d) Suppose in a two-product system priority is always given to PA cards for product 1 when the machine becomes free. Determine how the service level for product 1 is affected as Z_2 increases. How does the service level of product 2 change if Z_1 increases?

4.23 Consider a single-stage system with time lags between orders and reqisitions. If a requisition arrives and the output store is empty, however, the requisition is not met and the demand is lost. The resulting surplus order tag results in the next arriving order tag not generating a PA card. Determine the performance of the system.

4.24 In a single-stage system with time lags between orders and requisitions the output store inventory is unbounded. Because this might create operational problems, suppose the arrival of order tags is switched off once the output store inventory plus the active PA tags reaches a level Z^*. Determine the performance of the system.

BIBLIOGRAPHY

[1] R. AKELLA and P. R. KUMAR. Optimal control of production rate in a failure prone manufacturing system. *IEEE Trans. on Automatic Control*, AC-31:116–126, 1986.

[2] T. ALTIOK. (R, r) production/inventory systems. *Operations Research*, 37:266–276, 1989.

[3] C. E. BELL. Characterization and computation of optimal policies for operating an $M/G/1$ queue with removable server. *Operations Research*, 19:208–218, 1971.

[4] C. E. BELL. Optimal operation of a $M/M/2$ queue with removable servers. *Operations Research*, 28:1189–1204, 1980.

[5] C. E. BELL. Turning off a server with customers present: is this any way to run an $M/M/c$ queue with removable servers? *Operations Research*, 23:571–574, 1975.

[6] M. L. CHAUDRY and J. G. C. TEMPLETON. *A First Course in Bulk Queues*. John Wiley and Sons, New York, 1983.

[7] A. G. DE KOK, H. C. TIJMS, and F. A. VAN DER DUYN SCHOUTEN. Approximations for the single product production-inventory model with compound Poisson demand and service level constraints. *Adv. Appl. Prob.*, 16:378–401, 1984.

[8] B. T. DOSHI, F. A. VAN DER DUYN SCHOUTEN, and A. J. J. TALMAN. A production-inventory control model with a mixture of back-orders and lost-sales. *Management Science*, 24:1078–1086, 1978.

[9] J. GANI. Problems in the probability theory of storage systems. *J. R. S. S. [B]*, 19:181–206, 1957.

[10] D. P. GAVER, JR. Operating characteristics of a simple production, inventory-control model. *Operations Research*, 9:635–649, 1961.

[11] B. GAVISH and S. C. GRAVES. A one-product production/inventory problem under continuous review. *Operations Research*, 28:1228–1236, 1980.

[12] S. C. GRAVES. The application of queueing theory to continuous perishable inventory systems. *Management Science*, 28:400–406, 1982.

[13] S. C. GRAVES and J. KEILSON. The compensation method applied to a one-product production/inventory model. *Math. of OR*, 6:246–262, 1981.

[14] D. GROSS and C. M. HARRIS. On one-for-one-ordering inventory policies with state-dependent leadtimes. *Operations Research*, 19:735–760, 1971.

[15] H. K. HATHAWAY. Control of shop operations. In S. Person, editor, *Scientific Management in American Industry*. Hive Publishing, Easton, PA, 1929.

[16] D. P. HEYMAN. Optimal disposal policies for a single-item inventory system with returns. *Navel Res. Logistics Q.*, 24:385–405, 1977.

[17] D. P. HEYMAN. Optimal operating policies for $M/G/1$ queueing systems. *Operations Research*, 16:362–382, 1968.

[18] W. J. HOPP, N. PATI, and P. C. JONES. Optimal inventory control in a production flow system with failures. *Int. J. Prod. Res.*, 27:1367–1384, 1989.

[19] H. JONSON and E. A. SILVER. Impact of processing and queueing times on order quantities. *Material Flow*, 2:221–230, 1985.

[20] U. S. KARMARKAR. Lot sizes, lead times and in-process inventories. *Management Science*, 33:409–418, 1989.

[21] H. KASPI and D. PERRY. On a duality between a non-Markovian storage/production process and a Markovian dam process with state-dependent input and output. *J. Appl. Prob.*, 27:835–844, 1989.

[22] H.-S. LEE and M. M. SRINIVASAN. *The continuous review (s, S) policy for production/inventory systems with Poisson demands and arbitrary processing times*. Technical Report 87-33, I&OE, The University of Michigan, Ann Arbor, MI, December 1987.

[23] H.-S. LEE and M. M. SRINIVASAN. *The (s, S) policy for the production/inventory system with compound Poisson demands*. Technical Report 88-2, I&OE University of Michigan, Ann Arbor, MI, April 1988.

[24] R. R. MEYER, M. H. ROTHKOPF, and S. A. SMITH. Reliability and inventory in a production-storage system. *Management Science*, 25:799–807, 1979.

[25] R. G. MILLER, JR. Continuous time stochastic storage processes with random linear inputs and outputs. *J. Math. Mech.*, 12:275–291, 1963.

[26] D. MITRA. Stochastic theory of a fluid model of production and consumers coupled by a buffer. *Adv. Appl. Prob.*, 20:646–676, 1988.

[27] Y. MONDEN. *Toyota Production System*. Industrial Engineering and Management Press, Atlanta, 1983.

[28] P. A. P. MORAN. *The Theory of Storage*. Methuen, London 1959.

[29] P. M. MORSE. *Queues, Inventories and Maintenance: The Analysis of Operational Systems with Variable Demand and Supply*. John Wiley and Sons, New York, 1958.

[30] D. PERRY and B. LEVIKSON. Continuous production/inventory model with analogy to certain queueing and dam models. *Adv. Appl. Prob.*, 21:123–141, 1989.

[31] M. J. M. POSNER and M. BERG. Analysis of a production-inventory system with unreliable production facility. *Operations Research Letters*, 8:339–346, 1989.

[32] N. U. PRABHU. *Queues and Inventories.* John Wiley and Sons, New York, 1964.

[33] H. SAKASEGAWA. *Evaluating the overflow probability using the infinite queue.* Technical Report, Inst. of Socio-Economic Planning, University of Tsukuba, Tsukuba, Ibaraki 305, Japan, 1990.

[34] M. SHAFARALI. On a continuous review production-inventory problem. *Operat. Res. Letters*, 3:199–201, 1984.

[35] A. W. SHOGAN. A single server queue with arrival rate dependent on server breakdown. *Naval Res. Logist. Quart.*, 26:487–497, 1979.

[36] E. A. SILVER and R. PETERSON. *Decision Systems for Inventory Management and Production Planning*, 2nd edition. John Wiley and Sons, New York, 1985.

[37] M. J. SOBEL. Optimal average-cost policy for a queue with start-up and shut-down costs. *Operations Research*, 17:145–162, 1969.

[38] M. M. SRINIVASAN and H.-S. LEE. Random review production/inventory systems with compound Poisson demands and arbitrary processing times. *Management Science*, 37:813–833, 1991.

[39] U. SUMITA and M. KIJIMA. On optimal bulk size of single-server bulk arrival queueing systems with set-up times—numerical exploration via the Laguerre transform. *Selecta Statistica Canadiana*, 7:1986.

[40] H. C. TIJMS and F. A. VAN DER DUYN SCHOUTEN. Inventory control with two switch-over levels for a class of $M/G/1$ queueing systems with variable arrival and service rate. *Stoch. Proc. Appl.*, 6:213–222, 1978.

[41] L. M. WEIN. *Dynamic scheduling of a multiclass make-to-stock queue.* Technical Report, Sloan School of Management, M.I.T., Cambridge, MA, 1990.

[42] T. M. WILLIAMS. Special products and uncertainty in production/inventory systems. *Eur. J. of Opnl. Res.*, 15:46–54, 1984.

[43] M. YADIN and P. NAOR. On queueing systems with variable service capacities. *Naval Res. Log. Quart.*, 14:43–53, 1967.

[44] D. D. YAO, M. L. CHAUDRY, and J. G. C. TEMPLETON. Analyzing the steady-state queue $GI^x/G/1$. *J. Opnl. Res. Soc.*, 35:1027–1030, 1984.

[45] P. ZIPKIN. Models for design and control of stochastic multi-item batch production systems. *Operations Research*, 34:91–104, 1986.

5

Flow Lines

5.1 INTRODUCTION

Ever since Henry Ford achieved dramatic productivity gains and cost savings after introducing the moving belt assembly line into automobile manufacturing in 1913, it has been common belief that the flow line is the ideal way to organize production for products made in sufficient volume to justify the investment in dedicated machines and a dedicated material handling system. In particular, the flow line has been very widely used for assembly operations performed by human operators in situations in which the machines and tooling required have been relatively simple. When higher investment in machinery is required, such as in operations requiring shaping, forming, or cutting of material, flow lines are more difficult to justify and require much higher production volumes to be appropriate.

In this chapter we will focus on the analysis of flow lines with human operators. Such flow lines have been widely used for assembly operations in the automobile, domestic appliance, and high-volume electronic products industries as well as in many other contexts. In the next chapter we will consider automated flow lines or transfer lines where the operations are performed by machines rather than people, and thus the impact of machine breakdown and subsequent repair is the dominant concern. We will begin by summarizing some essential characteristics of human operators working on a flow line.

5.1.1 Characteristics of Human Operators

Task-time variability.   When an operator is required to perform a given task repetitively, it is found that the time required to perform the task varies from one repetition to the next. Much of this variability is not under the control of the operator. Although some variability is due to idle time inserted by the operator, the actual working time also varies, and this component of variability is an innate characteristic of human operators. There is a certain minimum time needed to perform the task determined by attributes of the task and the operator, and the actual time cannot be less than the minimum. The actual shape of the distribution seems to depend on a variety of factors. One is the experience of the operator, and the other relates to the way in which the time available for the task is controlled. It is common to distinguish between two situations: *paced* and *unpaced* work. In unpaced work the operator can take as long as he likes to perform the task. Here it has been observed that for an experienced operator the time to perform the task has the skewed distribution shown in Figure 5.1, in which the frequency of a task time less than the mean is about 65% [27] [54] [70]. For an inexperienced operator the distribution has a greater mean and shows less skewness.

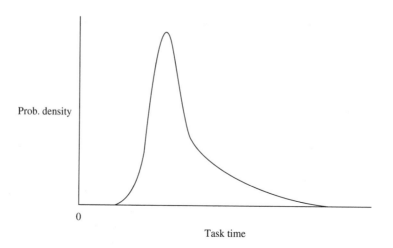

Figure 5.1   Distribution of Unpaced Task Time

When the operator is paced, there is an upper limit on the time available for performing the task. When the upper limit is reached the work will be moved on, so the task may not be complete. Here the consensus is that the distribution of task time is less skewed than in the unpaced case (see [27] [28] [29] [99]).

What is not clear from the literature is whether, in the unpaced situation, there is evidence of serial correlation in the time to perform successive tasks, such as long task times being followed by short task times. In some assembly situations, part of the variability in the task time is due to problems encountered by the operator because of difficulty in fitting together the parts to be assembled. This in turn may be due to part quality problems and these could show some serial correlation in their occurrence.

Variation of working rate.   If the performance of a human operator working in an unpaced situation is monitored, it is found that the working rate, that is, the number of tasks completed per unit time, shows variability over the working day. Typical output curves are as in Figure 5.2. The work rate builds up to a maximum then falls off before lunch break. Then it resumes after the break, builds to a maximum that is not as high as the morning maximum, and then drops off again. It is important to note that the actual time to perform the task does not vary significantly over the day, or rather the distribution of task time is stationary with respect to time of day. The reason for the variability in output rate is that the operator pauses or rests either between successive tasks or while performing a task. In most tasks in modern manufacturing it is rare for the decline in working rate over the day to be due to physical fatigue and exhaustion; the reason for the variation must be sought elsewhere (see [27]). Some inserted idle time is inevitable; for instance, Japanese automobile factories aim at achieving "54-minute" hours. Related to the variation in working rate is a variation of quality of task performance with time of the day and day of the week. The advice not to buy a car built on a Monday or Friday because of quality problems is well known and appears to be supported by evidence of greater quality problems on these days.

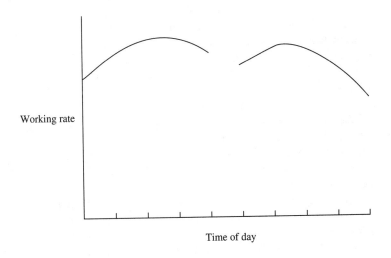

Figure 5.2   Working Rate over Day

Learning effects.   The change in the shape of the task time distribution as the worker acquires experience in the task was mentioned earlier. There have been many studies on the way in which the mean of the task time falls with cumulative experience (see [115] for a review).

There have been a variety of simple models proposed to describe this relationship. One that is frequently used is that the task time t_n on the nth repetition of the task is given by

$$t_n = kn^{-b}, \qquad b \geq 0 \tag{5.1}$$

where b is the learning rate and k is the time required for the first performance of the task. This model implies that if $b > 0$ then $t_n \to 0$ as $n \to \infty$. Another somewhat more complex model which appears to fit the experimental data better [60] [64] is

$$t_n = \frac{a}{n + c} + s \tag{5.2}$$

where s is the asymptotic minimal time below which it is impossible to perform the task, c can be interpreted as a measure of the extent of the worker's prior experience (when $n = 0$) on this or related tasks, and a is the time over and above the asymptotic time for a completely naive subject ($c = 0$) on the first performance of the task ($n = 1$).

In equation 5.1 the learning rate b depends on a variety of factors, in particular the duration of the task and its inherent complexity. In general the learning rate b is greater the shorter the duration of the task.

Job satisfaction and motivation.

Another aspect affecting human operator performance is the satisfaction with the job and motivation to perform at acceptable levels both with respect to the working rate (through inserted idle time) and the quality with which the job is carried out. There are several schools of thought. One claims that longer and more complex tasks are more challenging for the operator and hence will result in better motivation. Another claims that many operators prefer repetitive tasks that can be performed without conscious attention so that the mind is free to occupy itself with other things, daydreaming, or conversation with other workers. Probably both are true (or false) depending on the goals and aspirations of the worker.

There has also been considerable evidence that workers are more satisfied if they can control some aspects of the work environment, for example, if they are responsible for dealing with problems in quality of task performance. That is, some degree of autonomy contributes to job satisfaction. It is desirable that the worker or the work group be able to respond to disturbances and overcome difficulties without always being required to seek outside help.

Apart from the inherent satisfaction with the job there is also the impact of incentives (both financial and otherwise) on task performance. Starting with Taylor and Gilbreth there is an enormous industrial engineering literature on alternative incentive schemes. It is of interest that in [64] it is suggested that there is a close relationship between the mechanism of learning and the impact of incentives, and they suggest that a very similar equation applies, that is, the task time t at external incentive level i is given by

$$t = \frac{a}{i + c} + s$$

where s is the minimum task time (achieved under maximum possible incentive), c is a measure of innate incentive, and a is an indicator of resistance to incentives.

Individual differences.   A final characteristic of human operators is that individuals differ. Even if methods are carefully prescribed, operators are well trained, have had comparable experience on the task, and appear to be equally well motivated, their task performance will not be the same. Some operators will insert more idle time, some will display more variability in actual task performance, and some will learn at a slower pace. Differences of 20% or more seem to be by no means uncommon. In large work groups where speed is measured in terms of performance with respect to standard times it appears that the speed of the fastest operator can be as much as twice the speed of the slowest operator (see [93]).

Implications for modeling.   For our purposes there are only two aspects of operator performance that lend themselves to incorporation in stochastic models of the performance of flow lines with human operators: the task time variability and the impact of individual differences. The variation of working rate over the day and the variation of quality over the day and over the week create a pattern of nonstationarity that is difficult to incorporate in anything other than fairly aggregate models. Learning is also a nonstationary phenomenon that is difficult to model, although differences in cumulative experience of different operators at a given point can be represented in terms of individual differences. The interrelationship between the characteristics of the job, worker satisfaction, and the level of task performance in terms of quality and working rate appears to be very complex, and it does not appear possible to model it explicitly.

5.1.2 Types of Flow Lines

Flow lines can be divided into two broad classes based on their influence on the operator: *paced* or *unpaced* lines. In a paced line the time allowed for an operator to perform the tasks on a job is limited, and once this time is up the job can no longer be worked on; thus it is possible that the tasks are not completed. In an unpaced line there is no maximum limit imposed on the time available for the operator to perform the tasks on a job.

Additionally, flow lines can be classified according to the way in which job movement from one work station to the next is controlled. This control is implemented by means of the material handling system.

- *Indexing lines.* In indexing or synchronous lines transfer of jobs from one station to the next is coordinated so that all jobs begin transfer simultaneously. Indexing lines maintain a fixed number of jobs in the system, usually one job at each work station. Indexing lines can be designed and operated either for the paced mode, where the time available for performing the task is fixed and constant, or the unpaced mode, where transfer does not begin until a signal is received, indicating that all tasks are complete at all stations. There are also various intermediate or partially paced systems that have been proposed, for example, transfer will not occur until a given subset of tasks (the "must do" tasks) are completed at all stations; however, this can result in some deferrable or postponable tasks being incomplete. There may be a minimum time between successive transfers, irrespective of whether all tasks are completed faster.

- *Asynchronous lines.* In asynchronous lines job movement at adjacent stations is not coordinated. The operator starts the next job as soon as it is available, and on completion the job leaves the work station as long as there is space for it. Thus, in asynchronous lines it is possible for the operator to be *starved* (i.e., have no work available) or *blocked* (i.e., have no space to put a completed job), and thus the operator cannot begin the next job because the presence of the completed job in the work station means that the next job cannot be moved into the work station. Usually some work accumulation between stations is possible, although there is almost always a limit on the number of jobs that can be stored between successive work stations. In asynchronous lines the operator is almost always unpaced and thus controls when the job is available to move out of the work station, although when they are used in the automobile industry it is common to have a system of lights that come on to advise the operator (and his supervisor) when the standard time for the job has been exceeded.

If material handling is manual or by roller conveyor the line is usually asynchronous although in the assembly of some very large products such as aircraft (or sometimes ships) all jobs move on from one work area to the next more or less simultaneously, and thus the line behaves like an unpaced indexing line. Power-and-free conveyors result in asynchronous job movement, but with a fixed upper limit on the storage in front of a work station and also a fixed total number of jobs on the line. Similarly systems in which job movement is by means of automated guided vehicles (AGVs) operate asynchronously. If the job remains on the AGV for the tasks at each work station (such as is now common in automobile assembly), then there will also be a fixed number of work carriers (and hence jobs) in the system and a limited number of spaces for AGVs (and jobs) to be stored between work stations.

If material handling is by a moving belt traveling at a constant speed, then the line can be either paced or unpaced, depending on whether the job stays on the belt all the time or not. If items are removed from the line when they reach the next work station and then put into storage until they can be processed, then the operator is unpaced. If the storage area has limited size then jobs may not be able to enter it and thus they will miss the operation. The job may then remain on the belt and recirculate until it is able to enter the storage area for the work station, but in other situations it is removed from the belt and processed in some other way. Sometimes, even though the jobs are removed from the line to perform the required operation, there is some means by which the operator is paced. For example, the belt may be marked into segments, and the operator is required to put the job into a given segment as the belt passes him. This will ensure that the next operator on the line receives successive jobs at uniform time intervals. Lastly, the job may be attached to the moving belt so that the operator can only work on it while it is in a certain zone of accessibility. The traditional drag-chain conveyor that was once used universally in automobile assembly is of this type. Sometimes the spacing of jobs is such that only one job is accessible to the operator at once, in other situations more than one job can be accessed by the operator, and in yet other situations more than one operator may work on the job at once (e.g., the left and right sides of a car).

Another classification of flow lines is by the mix of jobs processed on the line. *Single-model* lines process only one type of job. *Multi-model* lines are designed to be able to process a variety of types of jobs but only process one type at once. That is, the line produces one type for a while, then is shut down to change over to another type that it then produces for a while and so on. *Mixed-model* lines are able to process more than one type of job simultaneously. For example, an automobile assembly line will have both two- and four-door body styles with a wide variety of different colors and options. This means that in a mixed-model line there is an additional source of variability in task times because of the random mix of job types, over and above the variability owing to the characteristics of the human operators.

Most flow lines have a strictly serial arrangement of work stations, although there may be several parallel lines, either with each line dedicated to a different product or with several lines producing the same product. A flow line can have more than one operator at a work station with each operator given different tasks to perform. There may be some paralleling of individual work stations or segments, however, where the same tasks are performed by different operators on different jobs (e.g., in a moving belt line with removable items the overflow from one work station could enter a duplicate work station). With AGVs it is easy to arrange work stations in parallel to form what is called a work cell, but some protocol is required for determining which of the paralleled work stations in the cell the AGV should enter. This will be discussed in Chapter 9.

In reality, many complex manufacturing systems will consist of a variety of different types of flow lines that are in turn linked to form a system that has an overall flow-line–type serial structure. For example, in automobile assembly there may be drag-chain–type flow lines or closed-loop AGV systems in which specific stages of assembly occur. These lines or systems will at a higher level also have a series structure with inventory banks separating the different parts of the overall system. Thus as well as analyzing individual flow lines it is also necessary to model and analyze how the overall system would be coordinated. This is discussed in Chapter 10.

5.2 ISSUES IN FLOW-LINE DESIGN AND OPERATION

The goal in design and operation of a flow line is to meet required levels of both quantity and quality of items produced at least cost, where the significant components of cost are primarily labor, work in process, and space. Indexing lines have no work in process apart from the item at each work station, whereas moving belt lines with the item attached to the belt maintain a fixed number of jobs on the line and a fixed number of jobs between work stations. The other types of lines have fluctuating inventory levels, either at or between work stations, and hence provision must be made for the space required by this inventory. With certain types of material handling systems, such as power-and-free conveyors or AGVs, there may be a limit on storage space at individual stations as well as a limit on the total number of jobs on the line created by the number of work carriers.

The line design and operation problem is usually formulated as that of minimizing the labor cost of meeting a given throughput target, subject to constraints on meeting quality targets and keeping work in process or space less than given maximum levels. Alternatively, one can seek to minimize the sum of labor costs and work-in-progress or space costs subject to quality and throughput constraints. In the past it has been very rare to evaluate the cost of quality and include this cost in the objective. With the increasing importance of quality as a competitive factor, however, the goal becomes to develop line designs and operating strategies that jointly optimize all components of cost.

Of course, the components of labor cost that should be considered are not just the direct cost of time. It is necessary to consider the costs of learning and training of operators and the costs of turnover or absenteeism that might result from the operator's reaction to the nature of the task that he is required to perform. Kilbridge and Wester [50] have made an interesting attempt to model these aspects of labor cost and performance.

In considering the issues involved in flow-line design and operation it is useful to divide them up into three groups.

1. *Configuration and layout.* These are the issues that have to be addressed in designing the line and should be resolved before the line is installed. Once the line is operating it is usually difficult to change the solution adopted without incurring major expense.

2. *Line set-up.* These are design decisions that can be changed after the line is installed, although usually they would be fixed for considerable periods. If there are changes in the product or new members of the product family are introduced then these factors may be changed. Some expense is incurred when they are changed, but this would not be considered major and usually would be considered fine tuning of the design.

3. *Line operation.* These are issues that would be decided on a day-by-day or week-by-week basis and usually would be the responsibility of the line supervisor(s).

We will review each of these groups of issues and then briefly discuss the approaches to production control that will be considered in this chapter.

5.2.1 Configuration and Layout

Before the line is installed the following issues would have to be resolved:

Number of stations.   Suppose the required total production over some long period $(0, t]$ is P. Then the required average throughput rate $TH^* = P/t$ jobs per unit time. Then if the total work content or standard time required for each job at all stations averages W, the minimum number of work stations required m^* will be given by $m^* = W \cdot TH^*$. The actual number of work stations required, m, will be greater than m^* for a variety of reasons:

- *Line balance.* If the work stations are arranged serially as a flow line then each work station will have assigned to it different tasks. Usually there will be precedence relationships between the tasks that limit the way in which they can be assigned to stations. Further, tasks are not infinitely divisible so it may be impossible to come up with an assignment of tasks to m^* work stations so that the standard time required by the tasks assigned to each work station equals W/m^*. If m stations are used in a flow line and the best feasible allocation of tasks to work stations results in the most heavily loaded station having a work load of $w_{max}(m)$ standard time units then it is necessary that m be such that $m/w_{max}(m) \geq TH^*$.

- *Operator variability.* Even if there is a balanced assignment of tasks so that the standard times required at each station are the same, the number of stations required would exceed m^* because of the effect of the variability in the times required by human operators to perform the tasks and the variability between individual operators. The magnitude of this effect will become apparent from the models to be presented in Sections 5.3 and 5.4.

- *Quality and rework.* Not all tasks performed by the operators will be performed correctly. Thus some jobs produced will not pass final inspection, and some tasks may have to be repeated or reworked. This increases the amount of work that has to be performed beyond W and hence increases the number of work stations required.

If m stations are used then the cost per hour of operating the line is $c \cdot m$ where c is the labor cost per hour. The theoretical minimum cost is $c \cdot W \cdot TH^*$ thus the goal is to minimize $c \cdot (m - W \cdot TH^*)$. The quantity $L = (m - W \cdot TH^*)/m$ is known as the balancing loss but in reality it is also measuring loss owing to the impact of operator variability, and quality and rework.

Degree of paralleling.   As the degree of paralleling increases it becomes easier to allocate tasks to stations to minimize balancing losses. This becomes noticeable if the disparity between task times is large, the number of tasks is relatively small, or there are many precedence constraints. If one task is very long and it cannot be divided, then some paralleling may be unavoidable. The advantages of paralleling are primarily in the ease of line balancing and perhaps in the presumed greater job satisfaction from tasks with more work content, the disadvantages are primarily in the more complex material handling system for the jobs, the often greater space requirements at each of the paralleled work stations for the different parts required for many tasks, and hence in some situations the greater likelihood of errors. The time required to learn the larger total set of tasks at a work station is greater, and it is likely that supervision is more demanding.

Inspection and test strategy.   Consideration will have to be given to the inspection and test strategy to be used. Apart from a final inspection or test at the end of the line, it may be appropriate to inspect at other locations to prevent defective work continuing through the system. Also it may be cheaper to repair defects early on in the manufacturing process. Next, decisions will have to be made on what to do with jobs that fail inspection or test. Are they to be repaired or reworked elsewhere? Will they be sent back to the station where the defect occurred and then proceed down the line again? Alternatively, will there be some off-line repair after each inspection followed by reinspection on the line? Although rework and feedback inspection is generally preferable to rework and off-line inspection (see Figure 5.3), the feedback situation is sometimes infeasible from a material handling viewpoint, and if quality problems occur it can result in the on-line inspection station becoming the bottleneck that limits line capacity.

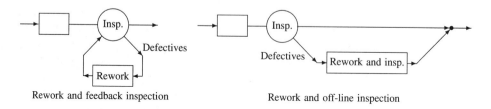

Rework and feedback inspection   Rework and off-line inspection

Figure 5.3   Rework and Inspection

Material handling system.   The selection of the type of material handling system is critical because it determines whether operators will be paced or unpaced, and the extent to which paralleling will be feasible. The material handling system also determines whether the line will be indexing or asynchronous, and this in turn determines the way in which task performance at adjacent stations will interact.

Allocation of storage space for in-process inventory.   In asynchronous lines it will be necessary to allocate the total available storage space for in-process inventory. This requires deciding on the location and size of inventory banks, and how they will be accessed by the material handling system. Banks can be *series* or *shunt* where in a series bank all jobs on the line pass through the bank irrespective of whether they are going to be stored there, whereas jobs only go into a shunt bank for storage (see Figure 5.4). In a series bank the normal travel time of jobs through the bank reduces its effective buffer capacity, whereas in shunt banks there may be time lost in storing and retrieving jobs as required. Series banks may in fact consist of parallel lanes, thus reducing travel times. If off-line repair and rework are used, then storage

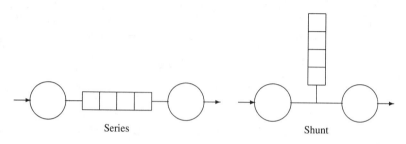

Series   Shunt

Figure 5.4   Series and Shunt Banks

space also has to be provided for jobs waiting rework or, after rework, waiting retest or inspection.

Line layout.   Once the preceding issues have been resolved the layout of the line can be determined. Work stations and inventory banks have to be positioned, and the material handling system must be laid out. Also, any subsidiary material handling systems, such as those that may be required to deliver parts to the line, must be configured. The layout of work stations will also be determined; in particular consideration will be given to the way in which parts and tools required for the operations are to be stored. Rework areas will also have to be located and provision made for storage of defective jobs.

5.2.2 Line Set-Up

Once the line has been laid out it is relatively difficult to change the decisions described earlier. There are several other design parameters, however, that can be more readily changed to improve or modify system performance. These are the following:

Number of work carriers.   In lines where the work carriers circulate in a closed loop an important decision that affects performance is the number of work carriers. Although the total number available would be decided as part of the line layout and configuration it is often desirable to vary the actual number that circulate. This is usually easy to do because there will be a return loop where the work carriers move from the output of the line (after completed jobs have been removed from them) back to the input of the line, and usually carriers can then be moved off the line. Allowing for travel times and operation times the minimum number of work carriers required to meet a given production rate can be estimated, but we will show in Section 5.6 that in asynchronous lines with finite buffers too many work carriers can result in a decline in the maximum throughput achievable from the line.

Allocation of tasks to work stations.   Although in deciding on the line configuration the number of work stations would have been determined by considering the different types of jobs to be produced on the line and how the line would be balanced, the actual mix of jobs would change from time to time. Further it would usually be found that some of the standard times for tasks have to be corrected, particularly because at the design stage they may have not been formally approved; they may also be changed if new methods or tools are introduced. Thus the set of tasks allocated to each work station would be changed from time to time. We will show in Section 5.4.2 that if the variability of the time required to perform the tasks assigned to the work stations differs, even though on the basis of the standard times the stations are balanced, then some adjustment in the assignment of tasks to stations may improve performance.

Level of inspection or test.   During the start-up phase of the line operation it is usual to have quality problems, and thus it would be appropriate to have fairly stringent inspection of in-process jobs. As quality improves it is no longer economical to inspect or test to the same level. Even when the work force has become experienced, however, there will still be variations in quality, either of work done on the line or of parts incorporated in the product. Inspection effort should be allocated to the current quality problems areas to provide essential feedback on the effectiveness of remedial measures, but at the same time it has to ensure that other new problems are not developing and going undetected.

5.2.3 Line Operation

There are several operating parameters of the line that can usually be modified frequently and that are often made the responsibility of the line supervision.

Worker assignment.   Worker absenteeism and turnover means that the set of workers available to man the line changes day by day (and sometimes even hour by hour). Not all workers will be equally experienced, and even if they are there will be individual differences. In some situations the union agreements make the assignment of workers the prerogative of the work group, usually based on seniority giving priority in choice of work station, but in other situations worker assignment is the responsibility of the line supervisor. If he is able to rate the set of workers in some way, for example, by assigning a rating factor r_i to worker i where worker i is considered better than worker j if $r_i > r_j$, then the supervisor would like to develop an assignment that will maximize the performance both with respect to quality and quantity produced.

Cycle time or line speed, feed rate, and zone length.   In a paced line it is also necessary to decide on the time each operator is allowed to perform the tasks at his work station. The hope of the line supervisor is that shortening the allowed time will force the operator to speed up and work faster, but speeding up the line may also result in deterioration in quality because the operator does not have enough time to complete the tasks assigned to him. In a paced indexing line the allowed time is controlled directly. In a moving belt line where the job remains on the belt the allowed time is controlled indirectly by setting the following quantities:

- *Zone length*. This is the length of the zone on the line over which a job is considered accessible to the operator. This will be determined by the requirements of the task and the reach of the operator. In the automobile industry the operator may move with the job while performing his tasks, but it is nevertheless necessary to allow time for the operator to walk back to the appropriate starting point for the next job. Strictly, the zone length is decided as part of the line configuration and layout.
- *Feed rate*. This is the spacing of the items on the line. The minimum would be fixed, but the spacing may be increased as part of the line operation decisions. Greater spacing would be used during the start-up of a line following a change in the job mix or when the work force is inexperienced.

- *Line speed.* The line speed then determines the gross production rate and the time available at each work station for the assigned tasks. The line gross production rate is then

$$\text{Gross production rate} = \frac{\text{line speed}}{\text{job spacing}}$$

and the time available to each operator for performing his set of tasks (the tolerance time) is given by

$$\text{Tolerance time} = \frac{\min\{\text{item spacing, zone length}\}}{\text{line speed}}$$

Job sequencing.   In a mixed-model line it is necessary to decide on the sequence with which jobs will be released to the line. Because the jobs will not have identical time requirements, it is necessary to find sequences that do not overload the operators. In particular a job that requires somewhat more time to process at a work station should be followed by a shorter job so that the operator has time to catch up. For example, in automobile assembly, it would be usual to ensure that cars requiring air conditioning are spaced out in the sequence of job release because they require more work at some stations.

Length of run.   In a multimodel line it is necessary to decide how long the line should run a particular product before being shut down and changed over to the next product. This will depend on change-over and inventory costs, and the demand for the different products.

5.2.4 Production Control

Apart from the issues described earlier, line behavior will also be affected by the approach used for production control. In this chapter we will consider the following approaches:

Produce-to-order.   That is, orders arrive at the beginning of the line and there are always sufficient parts and raw material available. Thus orders (or jobs) queue at the beginning of the line and are processed as soon as possible. When the job completes processing at the last station in the line the order can then be filled by shipping the completed job to the customer. With produce-to-order jobs it is necessary to consider the characteristics of the arrival stream of orders and how this influences the behavior of the line.

Produce-at-line-capacity.   Here the line produces at its maximum capacity. Thus it is assumed that any mismatch between capacity and demand is overcome by varying the number of hours per week that the line operates, such as by adding or removing a second or third shift, or through working overtime. When the line operates, however, it works at its full capacity. In modeling this approach to line operation we will ignore the effect of shutdowns and assume that at the end of a shift all processes cease and then, at the beginning of the next shift, restart where they left off. Nevertheless, we would like to understand what determines the capacity of the line and whether it makes sense to vary the capacity itself rather than adjusting the time that the line is working.

Produce-to-stock.   There is a final output store from which demands are met. The output store has a finite capacity, and the line produces as long as there is space in the output store. An individual work station produces as long as it is not blocked by its downstream inventory bank being full.

Indexing lines almost always use the production control approach of operating at line capacity (for a given task assignment to work stations) if unpaced or operate at a given line speed if paced. Asynchronous lines can use any of the three approaches.

Other approaches.   The whole flow line can be considered equivalent to the single-stage produce-to-stock system discussed in Chapter 4 and thus job arrival to the flow line would be controlled by PA cards. Other approaches to production control which consider the separate control of job flow at each stage of a multistage flow line will be considered in Chapter 10.

5.2.5 Models for Understanding the Issues

In the remainder of this chapter we will describe and develop a variety of different models that provide insight and understanding into many of the issues raised in this section. The work flow coordination in indexing lines is inherently easier to model than the behavior of asynchronous lines so we will begin with models of both unpaced and paced indexing lines, in particular looking at the capacity of unpaced lines and the tradeoff between throughput and quality in paced lines. Several other issues will also be addressed, however. In subsequent sections we will model asynchronous lines with particular emphasis on the impact of limited space for work in process between work stations. It will be shown how the models contribute to our understanding of many of the issues raised earlier, such as the choice of material handling system, the allocation of storage space, inspection, the allocation of tasks to work stations, worker assignment, and selection of line speed along with the impact of different production control strategies.

5.3 MODELS OF INDEXING LINES

5.3.1 Unpaced Indexing Lines

For a given assignment of elemental tasks to work stations, the task durations at the m work stations will be given by a set of random variables $\{T_1, T_2, \ldots, T_m\}$ where each $T_i \geq 0$.

In the unpaced indexing line it follows that the time between release of successive items to the line will be given by the random variable Γ where

$$\Gamma = \max\{T_1, T_2, \ldots, T_m\} \tag{5.3}$$

Let $E[\Gamma] = E[\max\{T_1, T_2, \ldots, T_m\}]$. The utilization ρ_i of station i for $i = 1, \ldots, m$ will be given by

$$\rho_i = \frac{E[T_i]}{E[\Gamma]} \tag{5.4}$$

and the line throughput TH is given by

$$TH = \frac{1}{E[\Gamma]} \tag{5.5}$$

For a given m the maximum production will be attained by that assignment of elemental tasks that maximizes TH.

The classical assembly-line–balancing problem is

Given a set of n_O elemental tasks $\{\sigma_1, \ldots, \sigma_{n_O}\}$ together with precedence constraints specifying for each task σ_i the set of tasks that must be completed before σ_i can start, plus perhaps a variety of other constraints restricting the subset of tasks assigned to the same work station, find an assignment of elemental tasks to stations that minimizes $\max\{E[T_1], \ldots, E[T_m]\}$ where $E[T_i]$ is the expected total duration of all elemental tasks assigned to station i.

Usually, the standard time of each elemental task is assumed to be an unbiased estimator of the time to perform the task, and thus $E[T_i]$ is taken to be the sum of the standard times of the elemental tasks assigned to station i. In general the computational complexity of the classical assembly-line–balancing problem depends exponentially on the size of the problem; however, there are many heuristic procedures for its solution. Empirically it has been found that as n_O/m decreases the balancing loss increases. Because n_O is fixed by the requirements of the job, there is an implication that greater paralleling will result in less balance delay.

Note that the assembly-line–balancing problem does not consider the impact of variability in task times and how the variability at different stations interact. To show the impact of variability in task times suppose that elemental tasks have been assigned to work stations so that the line is balanced with respect to the average task times. Suppose further that the distribution of the T_i's are identical.

Then the distribution of Γ will be the distribution of the maximum of a set of m identically distributed random variables.

$$F_\Gamma(t) = P\{\Gamma < t\} = \prod_{i=1}^{m} P\{T_i < t\} = [F_T(t)]^m \tag{5.6}$$

where $F_T(\cdot)$ is the distribution of T.

An approximate distribution of Γ can be derived in the following way. Let t_m be the value of t such that $F_T(t_m) = 1 - 1/m$. Many distributions are such that the right-hand tail $1 - F_T(t) = ke^{-\alpha t}$ for t sufficiently large. Then we can write

$$1 - F_T(t) = \frac{1}{m}e^{-\alpha(t-t_m)}, \quad t \geq t_m$$

and therefore

$$F_\Gamma(t) = \left(1 - \frac{1}{m}e^{-\alpha(t-t_m)}\right)^m$$

$$ln\, F_\Gamma(t) = m\, ln(1 - (1/m)e^{-\alpha(t-t_m)})$$

$$= m\left\{-\frac{1}{m}e^{-\alpha(t-t_m)} - \frac{1}{2m^2}e^{-2\alpha(t-t_m)} - \cdots\right\}$$

$$\approx -e^{-\alpha(t-t_m)}, \quad t \geq t_m$$

Thus

$$F_\Gamma(t) \approx F_{\hat{\Gamma}}(t) = \exp\{-\exp\{-\alpha(t - t_m)\}\} \tag{5.7}$$

$F_{\hat{\Gamma}}(t)$ is an extreme value distribution, specifically the distribution of the largest extreme. The mode of the distribution of $F_{\hat{\Gamma}}(t)$ is t_m, but because of the skewness of the distribution its mean is given by

$$E[\hat{\Gamma}] = t_m + 0.577/\alpha \tag{5.8}$$

and its standard deviation by

$$std[\hat{\Gamma}] = \frac{\pi}{\alpha}\left(\frac{1}{\sqrt{6}}\right) \tag{5.9}$$

To illustrate the nature of the results, suppose that task times are exponentially distributed with mean $\theta(m)$, that is, $F_T(t) = 1 - e^{-t/\theta(m)}$. Hence, here the exponential assumption on the tail of the distribution applies with $\alpha(m) = 1/\theta(m)$. Table 5.1 shows a comparison of $E[\hat{\Gamma}]$ found using the approximation of equation 5.8 and the exact $E[\hat{\Gamma}]$ using the formula $E[\hat{\Gamma}] = \sum_{i=1}^{m} 1/i$. Note that $E[\Gamma]/\theta(m)$ is the reciprocal of the utilization of a station.

TABLE 5.1 MEAN
INTERDEPARTURE TIMES FOR
EXPONENTIAL TASK TIMES

m	5	10	20
t_m	1.61	2.30	3.00
$E[\hat{\Gamma}]$ approx.	2.19	2.88	3.57
$E[\hat{\Gamma}]$ exact	2.28	2.93	3.60

For a normal distribution of T with mean $\theta(m)$ and variance $\sigma^2(m)$ it is first necessary to determine $\alpha(m)$. If $m = 10$ then $t_m = \theta(m) + 1.28\sigma(m)$. At $t = \theta(m) +$

$1.64\sigma(m)$, $0.5 = e^{-\alpha(m)\sigma(m)(\theta(m)+1.64\sigma(m)-\theta(m)-1.28\sigma(m))}$ and hence $\alpha(m) = 1.93/\sigma(m)$. Therefore the results in Table 5.2 can be obtained.

TABLE 5.2 MEAN INTERDEPARTURE TIMES FOR NORMAL TASK TIMES

m	10	20
t_m	$\theta(10) + 1.28\sigma(10)$	$\theta(20) + 1.64\sigma(20)$
$E[\hat{\Gamma}]$	$\theta(10) + 1.58\sigma(10)$	$\theta(20) + 1.94\sigma(20)$

For either distribution, it can be seen that the variability in task times results in a significant penalty in terms of loss in capacity. The fractional increase in cost owing to task time variability will be

$$L_v(m) = \frac{E[\Gamma] - \max\{E[T_1], E[T_2], \dots, E[T_m]\}}{\max\{E[T_1], \dots, E[T_m]\}} = \frac{1 - \max_{1 \le i \le n} \rho_i}{\max_{1 \le i \le n} \rho_i} \qquad (5.10)$$

which would give, in the case of a balanced line with normally distributed task times

$$L_v(10) = 1.58V(10),$$

$$L_v(20) = 1.94V(20).$$

where $V(m) = \sigma(m)/\theta(m)$, the coefficient of variation of T_m. Because the 10 station line could be designed by combining stations in the 20 station line, it is likely that

$$\theta(10) = 2\theta(20) \text{ and } \sigma(10) = \sqrt{2}\sigma(20)$$

which would result in

$$L_v(10) = 1.58/\sqrt{2}V(20) = \frac{1.58}{1.94\sqrt{2}}L_v(20) \approx 0.6L_v(20)$$

Alternatively, if one were to compare a single line with 20 stations with two parallel lines each of 10 stations and it can be assumed that $\theta(10) = 2\theta(20)$, $\sigma(10) = \sqrt{2}\sigma(20)$, then the cycle time of the 20 station line will be $\theta(20) + 1.94\sigma(20)$, whereas the cycle time of the 10 station line will be $\theta(10) + 1.58\sigma(10)$. Hence

$$\frac{\text{Throughput of two parallel 10 station lines}}{\text{Throughput of 20 station lines}} = \frac{2(\theta(20) + 1.94\sigma(20))}{\theta(10) + 1.58\sigma(10)}$$

$$= 1 + \frac{0.82V(20)}{1 + 1.12V(20)}$$

Of course, the key to this demonstration of the superiority of short paralleled lines to a long single line lies in the assumption that combining more task elements at a station does not result in additional time penalties.

Interchange of stations. Note that in an unpaced indexing line interchange of station i and j, that is, interchanging the tasks performed at i along with the worker at i with the tasks performed at j along with the worker at j (if feasible) has no effect on the cycle time of the line even if the line is unbalanced. Of course, if just the tasks are interchanged and not the workers as well then there could be differences between workers in the speed with which they perform the different tasks and so there could then be advantages in interchanging either just workers or tasks.

5.3.2 Paced Indexing Lines

In a paced line the cycle time will be set and hence the tolerance time, τ, the maximum time available to perform the tasks at any work station. Thus if T_i, a random variable denoting the time required by the operator at station i to perform his required tasks, exceeds τ the tasks will be incomplete, and defective products will result. Hence the probability that a product will not contain any defects when the cycle time is τ, $Q(\tau)$ will be

$$Q(\tau) = P\{T_1 \leq \tau, T_2 \leq \tau, \ldots, T_m \leq \tau\} = \prod_{i=1}^{m} P\{T_i \leq \tau\} \qquad (5.11)$$

assuming independence. Thus the quality of the items produced is given by $Q(\tau) = F_\Gamma(\tau) = \prod_{i=1}^{m} F_i(\tau)$ and $Q(\cdot)$ will have the extreme value distribution discussed earlier. Hence it is possible to determine what $F_i(\tau)$ must be in a line where the F_j's are identically distributed to meet a given quality target Q^*, that is, $F_i(\tau) = (Q^*)^{1/m}$. Table 5.3 shows the required $F_i(\tau)$ for $m = 5$, 10, and 20, and $Q^* = 0.98$ or 0.95. Note that for Q^* close to 1, $F_i(\tau)$ is approximately given by

$$1 - F_i(\tau) \approx \frac{1 - Q^*}{m} \qquad (5.12)$$

In the case of the T_i having identical exponential distributions for $i = 1, \ldots, m$, it follows that the τ required to achieve Q^* values of 0.98 and 0.95 are given by Table 5.3, with $\theta(m)$ the mean of the distribution of T_i in the balanced m station line. Table 5.3 also shows the τ required to achieve Q^* values of 0.98 and 0.95 when the T_i have identical normal distributions for $i = 1, \ldots, m$, and where $\theta(m)$, $\sigma^2(m)$ are the mean and variance of the distribution of T_i in the balanced m station line. Again, a single 20 station line can be compared with paralleled 10 station lines on the assumption that $\theta(10) = 2\theta(20)$ and $\sigma(10) = \sqrt{2}\sigma(20)$. For a quality target of 0.98, we would have

$$\frac{\text{Throughput of parallel 10 station lines}}{\text{Throughput of 20 station lines}} = \frac{2(\theta(20) + k(20)\sigma(20))}{\theta(10) + k(10)\sigma(10)}$$

$$= 1 + \frac{1.05 V(20)}{1 + 2.04 V(20)}$$

Again, as long as the assumption that $\theta(10) \approx 2.\theta(20)$ is true there will be an advantage in the system design with short parallel lines.

TABLE 5.3   TOLERANCE TIME REQUIRED TO MEET QUALITY
TARGETS IN PACED LINES

n	Q^*	Required $F_i(\tau)$	τ (Exponential)	τ (Normal)
5	0.95	0.98979	$4.58\theta(5)$	$\theta(5) + 2.32\sigma(5)$
	0.98	0.99596	$5.51\theta(5)$	$\theta(5) + 2.65\sigma(5)$
10	0.95	0.99488	$5.28\theta(10)$	$\theta(10) + 2.57\sigma(10)$
	0.98	0.99798	$6.21\theta(10)$	$\theta(10) + 2.88\sigma(10)$
20	0.95	0.99744	$5.97\theta(20)$	$\theta(20) + 2.80\sigma(20)$
	0.98	0.99899	$6.90\theta(20)$	$\theta(20) + 3.09\sigma(20)$

Further comparing the paced and unpaced lines, suppose that the tolerance time in the paced line was set at the mean of the unpaced cycle time, that is, set $\tau = E[\Gamma]$. Then for $m = 10$ and identical normal distributions of the T_i for $i = 1, \ldots, m$, $\tau = E[\Gamma] = \theta(10) + 1.6\sigma(10)$. It follows that $F_i(\tau) = 0.945$ and $Q(\tau) = 0.57$. In the unpaced situation, however, there is always sufficient time to complete each task, so Q should equal 1. Hence the unpaced line appears superior to the paced line. Nevertheless, our models have not considered how operators respond to the two situations, in particular, whether the amount of idle time inserted between and within tasks might differ. In the paced situation it may become more evident if the operator inserts idle time, particularly if the idle time exceeds the cycle time and so a task is not even started. Thus paced lines enable managerial control of idle time to be more directly exercised.

The utilization of a work station in a paced indexing line will be given by

$$\rho_i = \frac{\tau \overline{F}_i(\tau) + \int_{t=0}^{\tau} t\, dF_i(t)}{\tau} \tag{5.13}$$

$$= \frac{\int_{t=0}^{\tau} \overline{F}_i(t)\, dt}{\tau}$$

$$\leq \frac{E[T_i]}{\tau}$$

Optimal cycle time.   In a paced line one of the managerial controls is the line speed, or, equivalently, the cycle time or the tolerance time. Too high a line speed reduces the time available to perform the tasks at the work stations and hence quality will be low, whereas too slow a line speed reduces the use of the workers and lowers the productivity. That is, the choice of line speed requires making a tradeoff between quality and productivity. Suppose that the cycle time is equal to τ. Then

$$\text{Gross production rate} = 1/\tau$$

and the throughput, TH, the rate at which nondefective items are produced, is given by

$$TH = \frac{Q(\tau)}{\tau} \tag{5.14}$$

The maximum TH will be achieved when

$$\tau \frac{dQ(\tau)}{d\tau} - Q(\tau) = 0$$

that is,

$$Q'(\tau) = \frac{dQ(\tau)}{d\tau} = \frac{Q(\tau)}{\tau} \qquad (5.15)$$

Thus from Figure 5.5 it can be seen that the maximum throughput will be achieved when $\tau = \tau^*$, the point where the line from the origin is tangent to $Q(\tau)$. In the case in which $Q(\tau)$ has an extreme value distribution, that is,

$$Q(\tau) = \exp\{-\exp\{-\alpha(\tau - t_m)\}\}$$

τ^* will satisfy the equation

$$e^{-\alpha(\tau^* - t_m)} = 1/(\alpha\tau^*)$$

or

$$\alpha(\tau^* - t_m) = \ln(\alpha\tau^*) \qquad (5.16)$$

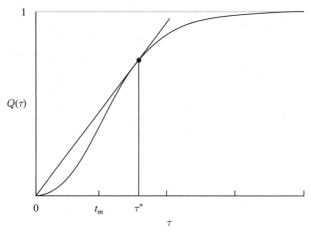

Figure 5.5 $Q(\tau)$ versus τ

In the case in which the T_i has identical exponential distributions for $i = 1, \ldots, m$ with mean $\theta(m)$, Table 5.4 shows the optimal cycle time and throughput.

For a normal distribution $\alpha \approx 2/\sigma$ in the tail area where we are concerned. Therefore

$$\alpha\tau^* - \ln\alpha\tau^* = 2\frac{(\theta(m) + k(m)\sigma(m))}{\sigma(m)}$$

and $k(m)$ is such that for a unit normal $1 - F(k(m)) = 1/m$. Hence if $m = 10$ and $\sigma(m)/\theta(m) = 0.2$, it can be shown that

$$\tau^* = \theta(10) + 2.65\sigma(10)$$

$$Q = 0.96$$

$$TH = \frac{0.96}{1.53\theta(10)} = \frac{0.63}{\theta(10)}$$

TABLE 5.4   OPTIMAL CYCLE TIME
AND THROUGHPUT FOR PACED
LINES

m	τ^*	$Q(\tau^*)$	TH
5	$2.55\theta(5)$	0.66	$0.26/\theta(5)$
10	$3.60\theta(10)$	0.76	$0.21/\theta(10)$
20	$4.50\theta(20)$	0.80	$0.18/\theta(20)$

If the cycle time is increased beyond τ^*, then quality will improve, and throughput will decrease. If the cycle time is reduced below τ^*, however, *both quality and net throughput will decrease.* Thus under no circumstances should the cycle time be set to be less than τ^*. Beyond τ^* it is necessary to make a tradeoff between quality and production rate.

To illustrate, suppose c_l is the labor cost per unit time of all the operators on the line, and suppose c_m is the cost of raw material and parts added to complete a job on the line. Suppose that at the end of the line all jobs are inspected, and if any job is found to be defective it is scrapped. Each scrapped job incurs a further cost of c_d to dispose of it, although if $c_d < 0$ it is implied that there is a credit for each scrapped job. Then the expected cost incurred while y jobs are produced is $c_l y \tau + c_m y + c_d y(1 - Q(\tau))$. Because only $yQ(\tau)$ of these jobs is good, however, the cost per good job $c_g(\tau)$ will be given by

$$c_g(\tau) = \frac{c_l \tau + c_m + c_d(1 - Q(\tau))}{Q(\tau)} \tag{5.17}$$

Now

$$\frac{dc_g(\tau)}{d\tau} = \frac{-(c_m + c_d)Q'(\tau) + c_l(Q(\tau) - \tau Q'(\tau))}{(Q(\tau))^2} \tag{5.18}$$

so it is clear that the optimal cycle time $\tau^{*(s)}$, is such that $\tau^{*(s)} > \tau^*$. Recall that τ^* is the tolerance time maximizing the throughput (except in the somewhat unusual situation that $c_m + c_d < 0$). $\tau^{*(s)}$ will be such that

$$\frac{c_l}{c_m + c_d} = \frac{Q'(\tau^{*(s)})}{Q(\tau^{*(s)}) - \tau^{*(s)} Q'(\tau^{*(s)})} \tag{5.19}$$

and also $dQ'(\tau)/d\tau < 0$ at $\tau = \tau^{*(s)}$, that is, it follows that $\tau^{*(s)}$ is greater than the mode of $Q(\tau)$ if, for example, $Q(\tau)$ is the distribution of the largest extreme.

5.3.3 Off-Line Repair of Defective Jobs

In this section we develop a model for an m-station indexing flow line with imperfect processing. The probability that processing at station i is defective is p_i, $i = 1, \ldots, m$. In between stations k and $k + 1$ is an inspection station where all jobs passing from station k to $k + 1$ are checked for processing defects that may have occurred at stations

1 through k. If at least one defect is identified, the defective job is passed on to an off-line repair station provided storage space is available at the repair facility; otherwise it is allowed to continue its processing by the downstream stations. Because not all jobs needing repair can enter the repair station, at the end of the line, after station m, all the jobs processed by the system will be tested and repaired if necessary. At the repair station the repair time of a defective job is random and depends on the defects that have occurred. A repaired job is transferred to station $k + 1$ as soon as a space at station $k + 1$ becomes available. Observe that in an indexing flow-line such a space will become available only when a job is taken out from the line at or before the inspection station. In our system a job may be taken out only from the inspection station and that happens only when a job is identified as defective. That is, a repaired job may leave the repair station only when a defective job enters it.

Suppose the system is observed just *before indexing occurs at time* n, $n = 1, 2, \ldots$. Let N_n^I and N_n^O be the number of defective jobs (including the one in repair, if any) and the number of repaired jobs at the repair station at time n, $n = 1, 2, \ldots$. If the total buffer space allocated to the repair station is b_R, then the state space of $\{(N_n^I, N_n^O), n = 1, 2, \ldots\}$ is $\mathcal{S} = \{(n_1, n_2); 0 \leq n_1 \leq b_R, 0 \leq n_2 \leq b_R, 0 \leq n_1 + n_2 \leq b_R\}$. The one-step transitions of this process are $(n_1, 0) \to (n_1 + 1, 0)$, $n_1 = 0, \ldots, b_R - 1$; $(n_1, n_2) \to (n_1 + 1, n_2 - 1)$, $n_1 = 0, \ldots, b_R - n_2$, $n_2 = 1, \ldots, b_R$ and $(n_1, n_2) \to (n_1 - 1, n_2 + 1)$, $(n_1, n_2) \in \mathcal{S}$ and $n_1 \geq 1$ (see Figure 5.6). An inspection of these transitions will show that not all states are commutative, and the set of states $\mathcal{S}' = \{(n_1, n_2), 0 \leq n_1 \leq b_R, 0 \leq n_2 \leq b_R, n_1 + n_2 = b_R\}$ is absorbing. Under stationary conditions the process $\{(N_n^I, N_n^O), n = 1, 2, \ldots\}$ will evolve on $\mathcal{S}'$. Hence we may assume without a loss of generality that $N_0^I + N_0^O = b_R$, $N_0^I = 0$ and consider $\{N_n^I, n = 1, 2, \ldots\}$ on the state space $\mathcal{S}' = \{0, \ldots, b_R\}$. Note that if the buffer capacity at the repair facility is unlimited (i.e., $b_R = +\infty$), the total number of jobs at the repair facility will grow without limit. Hence we will assume that $b_R < +\infty$ and focus our attention on the fraction of defectives that will pass from station k to $k + 1$ without being taken to the repair facility.

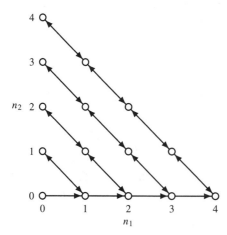

Figure 5.6   State Transition Diagram for Off-Line Repair

Suppose the processing defects occur at each station according to a Bernoulli process and the repair times are iid with distribution G_R. It will be assumed that any repair can only begin immediately after indexing (although it may last over a number of cycles). The probability that a job passing from station k to $k + 1$ is defective is $\hat{p} = 1 - \prod_{i=1}^{k}(1 - p_i)$ and therefore the interarrival time of defective jobs to the repair facility has a geometric distribution with mean $1/\hat{p}$. $\{N_n^I, n = 1, 2, \ldots\}$ is then equal in distribution to the number of jobs in a discrete M/G/1/b_R queueing system with losses. Let $p(k) = \lim_{n \to \infty} P\{N_n^I = k\}$, $k = 0, \ldots, b_R$. Then using an embedded Markov chain model (Section 3.3.2), embedded at observation instants immediately following repair completion but before indexing, it can be seen that if $(q(n)$ $n = 0, \ldots, b_R - 1)$ is the stationary probability vector of this embedded Markov chain, then for $b_R \geq 2$

$$q(n) = \frac{1}{1 - a_1}\left\{q(0)a_n' + \sum_{k=0}^{n-1} q(k)a_{n+1-k} + q(n+1)a_0\right\},$$

$$n = 1, 2, \ldots, b_R - 2 \tag{5.20}$$

$$q(b_R - 1) = \frac{1}{a_0}\left\{q(0)\sum_{l=b_R-1}^{\infty} a_l' + \sum_{k=1}^{b_R-2} q(k)\sum_{l=0}^{\infty} a_{b_R+l-k}\right\}$$

where $a_n = \sum_{l=\max(1,n)}^{\infty}\binom{l}{n}(1 - \hat{p})^{l-n}\hat{p}^n g_R(l)$ is the probability that exactly n defective jobs arrive during a repair duration, $n = 0, 1, \ldots$ and $a_n' = \sum_{l=n}^{\infty}\binom{l}{n}(1 - \hat{p})^{l-n}\hat{p}^n g_R(l+1)$ is the probability n defective jobs arrive during the last $R - 1$ units of the repair time. Recall that $g_R(n) = G_R(n) - G_R(n - 1)$, $n = 1, 2, \ldots$. The mean time between consecutive repair completions is $q(0)\left\{[1 - \hat{p}]/\hat{p}\right\} + E[R]$, and therefore the departure rate of jobs from the repair facility is $\hat{p}/\{(1 - \hat{p})q(0) + \hat{p}E[R]\}$. Because the (defective) job input rate to the repair facility is $\hat{p}p(b_R)$, applying the level crossing argument one obtains

$$p(n) = \left(\frac{1}{(1 - \hat{p})q(0) + \hat{p}E[R]}\right)q(n), \qquad n = 0, \ldots, b_R - 1$$

$$p(b_R) = 1 - \frac{1}{(1 - \hat{p})q(0) + \hat{p}E[R]} \tag{5.21}$$

For $b_R = 1$, $q(0) = 1$ and hence

$$p(0) = \frac{1}{1 - \hat{p} + \hat{p}E[R]} \tag{5.22}$$

and

$$p(1) = \frac{\hat{p}(E[R] - 1)}{1 - \hat{p} + \hat{p}E[R]} \tag{5.23}$$

Because a defective job passing from station k to station $k + 1$ will bypass the repair station if the repair station buffer is full (i.e., $N^I = b_R$) and the arrivals of defective jobs is a Bernoulli process, the fraction of defective jobs bypassing the repair station is $p(b_R)$.

5.4 MODELS OF ASYNCHRONOUS LINES

In an asynchronous line there is no imposed coordination of job transfer from one station to the next. Thus on completion at one station a job moves on to the next station, although it may pass through an intermediate storage. If the storage is series then it is obvious that the flow line and its storages can be modeled by an open tandem queueing network. If the storage is shunt, then from an aggregate perspective the system will also be equivalent to an open tandem queueing network; however, the queue discipline would be changed (e.g., from FCFS to LCFS).

When there are parallel stations it would also be appropriate to model them by parallel servers in a tandem queueing network. It is important, however, to represent adequately the protocol by which arriving jobs are assigned to parallel stations. The predominant protocol used in the queueing literature is that the arriving job goes to the station that has been idle longest. In manufacturing, however, this protocol is often not used because of reasons to do with (1) the material handling system control or (2) the effort to ensure that each operator produces the same number of jobs over a shift. Thus it is not uncommon to use a *cyclic assignment* protocol in which successive jobs are allocated to the parallel stations $1, 2, \ldots, c$ sequentially (e.g., station 1 gets jobs $1, 1 + c, 1 + 2c, \ldots$). Alternatively, in moving-belt systems where items are diverted from the line into parallel stations, the protocol is often equivalent to a *priority* protocol in which jobs first attempt to enter station 1, and if it and any associated buffer space is full, the jobs would then try station 2 and so on. In the remainder of this chapter we will only consider job assignment protocols that do not allow stations to be idle when jobs are waiting for processing by these stations. Other job assignment protocols will be treated in Chapter 9.

5.4.1 Infinite Buffer Tandem Queueing Model:
Exponential Processing Times

In this section we consider a flow-line system with m stages and unlimited buffer capacities. Stage i has c_i parallel stations that are functionally identical, $i = 1, \ldots, m$. The service times of jobs are iid exponential random variables with mean $1/\mu_i$, $i = 1, \ldots, m$. Job orders arrive at the flow line according to a Poisson process with rate λ. We are interested in obtaining the amount of inventory of jobs carried in the system. We model this system by a series of exponential queues. Because the departure process from a $M/M/c_i$ queueing system is Poisson (see Chapter 3), each stage i can be modeled by a $M/M/c_i$ queueing system with arrival rate λ, service rate, μ_i, and c_i parallel stations

$i = 1, \ldots, m$. Then if N_i is the stationary number of jobs at stage i, $i = 1, \ldots, m$, one has for $\rho_i = \lambda/(c_i \mu_i) < 1$,

$$P\{N_i = k_i\} = \begin{cases} p_0 \dfrac{(c\rho_i)^{k_i}}{k_i!}, & 0 \le k_i \le c_i \\[2ex] p_0 \dfrac{\rho_i^{k_i} c_i^{c_i}}{c_i!}, & k_i \ge c_i \end{cases} \tag{5.24}$$

where

$$p_0 = P\{N_i = 0\} = \left[\sum_{k=0}^{c_i-1} \frac{(c_i \rho_i)^k}{k!} + \left(\frac{(c_i \rho_i)^{c_i}}{c_i!} \right) \left(\frac{1}{1 - \rho_i} \right) \right]^{-1}$$

Then the average number of jobs in the system is

$$E[N] = \sum_{i=1}^{m} \hat{N}(\rho_i, c_i) \tag{5.25}$$

where $\hat{N}(\rho, c)$ is the average number of jobs in a $M/M/c$ queueing system with station utilization ρ and c parallel stations.

Work load allocation.   We will now see how the preceding result can be used to allocate optimally a total work load of W units among the m stages. Suppose we have allocated w_i units of work load to stage i. That is, the average processing time of a job at stage i is w_i. The optimization problem is then

$$\min \sum_{i=1}^{m} \hat{N}\left(\frac{\lambda w_i}{c_i}, c_i \right)$$

subject to

$$\sum_{i=1}^{m} w_i = W$$

Because $\hat{N}(\rho, c)$ is increasing and convex in $\rho = \lambda w/c$ from standard optimization theory one sees that the optimal solution to the preceding problem satisfies

$$\frac{\lambda}{c_i} \hat{N}'\left(\frac{\lambda w_i^*}{c_i}, c_i \right) = \text{constant}, \quad i = 1, \ldots, m \tag{5.26}$$

Because $\hat{N}'(\rho, c) = (\partial/\partial\rho)\hat{N}(\rho, c)$ is decreasing in c, from equation 5.26 one has

$$c_i \le c_j \quad \text{implies} \quad \frac{w_i^*}{c_i} \le \frac{w_j^*}{c_j} \text{ for any } j = 1, \ldots, m \tag{5.27}$$

Equation 5.27 is obtained as follows: for $c_i < c_j$

$$\hat{N}'\left(\frac{\lambda w_i^*}{c_i}, c_i \right) = \frac{c_i}{c_j} \hat{N}'\left(\frac{\lambda w_j^*}{c_j}, c_j \right) \le \hat{N}'\left(\frac{\lambda w_j^*}{c_j}, c_j \right)$$

Because $\hat{N}'(\lambda w/c, c)$ is decreasing in c, the preceding can be true only if $w_j^*/c_j \ge w_i^*/c_i$.

From equation 5.27 it is easily seen that if $c_i = c_j$ then $w_i^* = w_j^*$. Therefore if all the stages have the same number of parallel stations it is optimal to allocate the total load W equally (i.e., $w_i^* = W/m$, $i = 1, \ldots, m$) so that the mean number of jobs in the system is minimized. If the number of stations at different stages are different then the work load allocated per station (i.e., w_i^*/c_i) should be larger at a stage that has more stations.

Assignment of workers.   The work load allocation is usually done before the workers are selected for the flow line. Hence the second level of issue faced in the allocation of work load is the assignment of the workers to stations. Suppose we have to allocate one worker per station, and the rating of the worker i is r_i, $i = 1, \ldots, m$; that is, worker i can process r_i units of work load in one unit of real time. If $r_1 = r_2 = \cdots = r_m$, (that is, all workers are equally rated) then we may choose any assignment. In reality this would not be the case. Suppose w_i is the work load allocated to station i, and worker j is assigned to station i. Then the service time for a job at station i is exponentially distributed with mean w_i/r_j. In effect the average number of jobs there will be $\hat{N}(\lambda w_i/r_j, 1)$. Suppose worker $\pi(i)$ is assigned to station i, $i = 1, \ldots, m$. Then for this assignment $\pi = \; < \langle \pi(1), \ldots, \pi(m) \rangle >$ of workers to stations $1, 2, \ldots, m$, the average total number of jobs in the system is

$$\hat{N}(\pi) = \sum_{i=1}^{m} \hat{N}(\lambda w_i/r_{\pi(i)}, 1)$$

Then we wish to find an assignment π^* such that $\hat{N}(\pi^*) = \min\{\hat{N}(\pi), \pi \in \mathcal{P}\}$, where $\mathcal{P}$ is the set of all permutations of $\{1, \ldots, m\}$. Consider two assignments π and π' such that $\pi(i) = \pi'(i)$, $i = 1, \ldots, m; i \neq l, l+1$; $\pi(l) = \pi'(l+1)$ and $\pi(l+1) = \pi'(l)$. Without a loss of generality we assume that $r_{\pi(l)} \geq r_{\pi(l+1)}$. Then if $w_l \geq w_{l+1}$, it can be verified that

$$\frac{w_{l+1}}{r_{\pi(l)}} \leq \min\left\{\frac{w_{l+1}}{r_{\pi(l+1)}}, \frac{w_l}{r_{\pi(l)}}\right\} \leq \max\left\{\frac{w_{l+1}}{r_{\pi(l+1)}}, \frac{w_l}{r_{\pi(l)}}\right\} \leq \frac{w_l}{r_{\pi(l+1)}}$$

and

$$\frac{w_{l+1}}{r_{\pi(l)}} + \frac{w_l}{r_{\pi(l+1)}} \geq \frac{w_l}{r_{\pi(l)}} + \frac{w_{l+1}}{r_{\pi(l+1)}}$$

Because $\hat{N}(\lambda w, 1)$ is increasing and convex in w, it immediately follows that

$$\hat{N}\left(\frac{\lambda w_{l+1}}{r_{\pi(l)}}, 1\right) + \hat{N}\left(\frac{\lambda w_l}{r_{\pi(l+1)}}, 1\right) \geq \hat{N}\left(\frac{\lambda w_l}{r_{\pi(l)}}, 1\right) + \hat{N}\left(\frac{\lambda w_{l+1}}{r_{\pi(l+1)}}, 1\right)$$

and consequently $\hat{N}(\pi') \geq \hat{N}(\pi)$. Thus we can conclude that it is better to assign the higher rated worker ($\pi(l)$ in this case) to the station with the higher work load (w_l in this case). So if $\mathbf{w}_{\hat{\pi}} = (w_{\hat{\pi}(1)}, \ldots, w_{\hat{\pi}(m)})$ is the rearrangement of the work loads in the decreasing order (i.e., work load allocated to station $\hat{\pi}(i)$ is larger than the work load allocated to station $\hat{\pi}(j)$ for $i < j$) and if workers are numbered in the decreasing order of their ratings (i.e., $r_i \geq r_j$ for $i < j$), then $\hat{\pi}$ is an optimal assignment.

5.4.2 Infinite Buffer Tandem Queueing Model: General Processing Times

When the job arrival process or the service process is general (i.e., not Markovian) we may heuristically extend the preceding model. Let C_a^2 be the squared coefficient of variation of the job interarrival times and $C_{S_i}^2$ be the squared coefficient of job service times at stage i, $i = 1, \ldots, m$. Then we model each stage by a $GI_i/GI_i/c_i$ queueing system with renewal input process. The mean and squared coefficient of variation of the interarrival time is $1/\lambda$ and $C_{a_i}^2$. For $i = 1$, $C_{a_1}^2 = C_a^2$. Then choosing the approximation $\hat{N}_{GI/G/c}(\lambda, C_a^2, \mu, C_S^2, c)$ for the mean number of jobs and $C_d^2(\lambda, C_a^2, \mu, C_S^2, c)$ for the squared coefficient of variation of the departure process from a $GI/G/c$ queueing system, we approximate $E[N_i]$ by

$$E[N_i] = \hat{N}_{GI/G/c}(\lambda, C_{a_i}^2, \mu_i, C_{S_i}^2, c_i) \tag{5.28}$$

where

$$C_{a_{i+1}}^2 = C_d^2(\lambda, C_{a_i}^2, \mu_i, C_{S_i}^2, c_i), \quad i = 1, \ldots, m-1; \quad C_{a_1}^2 = C_a^2 \tag{5.29}$$

These approximations can be used as before in the exponential processing time case to address the issues of work-load allocation and station assignment so as to minimize the *approximated* number in the system.

Best order of stations.   Assuming that the work load has been allocated and the workers have been assigned, we will look at the question of the best order of stations. Let $\pi = \langle \pi(1), \ldots, \pi(m) \rangle$ be the order of stations chosen, that is, a job arriving at this system will first go to station $\pi(1)$, next to $\pi(2)$, $\pi(3)$, and up to $\pi(m)$ and leave the system. Then if we choose to approximate the mean flow time at each station by $\hat{T}_1$ (see equation 3.142) and the squared coefficient of variation of the departure process by $\hat{C}_{d:1}^2$ (see Table 3.1), the approximate average flow time $\hat{T}(\pi)$ of an arbitrary job is given by

$$\hat{T}(\pi) = \sum_{i=1}^{m} \frac{\rho_{\pi(i)}^2 \left(1 + C_{S_{\pi(i)}}^2\right)\left(C_{a_{\pi(i)}}^2 + \rho_{\pi(i)}^2 C_{S_{\pi(i)}}^2\right)}{\left(1 + \rho_{\pi(i)}^2 C_{S_{\pi(i)}}^2\right)\left(2\lambda\left(1 - \rho_{\pi(i)}\right)\right)} + \sum_{i=1}^{m} E[S_i] \tag{5.30}$$

where

$$C_{a_{\pi(1)}}^2 = C_a^2 \tag{5.31}$$

and

$$C_{a_{\pi(i+1)}}^2 = (1 - \rho_{\pi(i)}^2)\left\{\frac{C_{a_{\pi(i)}}^2 + \rho_{\pi(i)}^2 C_{S_{\pi(i)}}^2}{1 + \rho_{\pi(i)}^2 C_{S_{\pi(i)}}^2}\right\} + \rho_{\pi(i)}^2 C_{S_{\pi(i)}}^2, \quad i = 1, 2, \ldots, m-1 \tag{5.32}$$

Our objective is then to find an order π^* such that $\hat{T}(\pi^*) = \min\{\hat{T}(\pi), \pi \in \mathcal{P}\}$ where $\mathcal{P}$ is the set of all permutations of $\{1, \ldots, m\}$. We consider a special instance of this problem. Suppose that the loads allocated to the stations are balanced ($w_i = w$, $i = 1, \ldots, m$), and the station ratings are the same (i.e., $r_i = r$, $i = 1, \ldots, m$). Hence the utilization of station i, $\rho_i = \lambda w_i / r_i = \rho$, is the same among all m stations. For this case let π' be an alternate order of stations such that for some $l \in \{1, \ldots, m-1\}$, $\pi(l) = \pi'(l+1)$ and $\pi(l+1) = \pi'(l)$. All other assignments are the same (i.e., $\pi(i) = \pi'(i), i = 1, \ldots, m, i \neq l, l+1$). If $C^2_{S_{\pi(l)}} \leq C^2_{S_{\pi(l+1)}}$ direct computation shows that $\hat{T}(\pi) \leq \hat{T}(\pi')$. Therefore an optimal order of stations that minimizes the average flow time satisfies $C^2_{S_{\pi^*(1)}} \leq C^2_{S_{\pi^*(2)}} \leq \cdots \leq C^2_{S_{\pi^*(m)}}$.

5.4.3 Infinite Buffer Flow Lines with Imperfect Processing

We consider a m-stage flow line with unlimited storage space for raw material and partially processed jobs. The job processing is imperfect, and p_i is the probability that the processing of a job at stage i is defective. Job orders or raw material arrive at a rate λ, and we are interested in investigating the effect an inspection station will have on the work-in-process inventory. Suppose we place the inspection station at the end of the line after stage m. Each job leaving the flow line is inspected by this inspection station. If a job is found to be defective, the job is sent to the stage, say i, that caused the first defect on that job. The job is then reworked by not only stage i, but also by stage $i+1, \ldots, m$ and inspected again. Note that a job arriving at stage $k+1$ for the last time will not have a detectable defect owing to stages $1, \ldots, k$ processing.

Let λ_i be the rate at which jobs are departing from stage i, $i = 1, \ldots, m$. This rate includes both jobs processed for the first time and those jobs that are reworked. Let γ_i be the rate at which jobs are being sent back by the inspection station direct to stage i for rework, $i = 1, \ldots, m$. Also let β_i be the rate at which defective jobs arrive at stage $i+1$ from stage i, $i = 1, \ldots, m$. Let $\lambda_0 = \lambda$ and $\beta_0 = 0$. Then from the conservation of flow at a station (see Figure 5.7) one sees that

$$\lambda_i = \lambda_{i-1} + \gamma_i, \qquad i = 1, \ldots, m \tag{5.33}$$

$$\beta_i = \beta_{i-1} + \gamma_i p_i + (\lambda_{i-1} - \beta_{i-1})p_i, \qquad i = 1, \ldots, m$$

$$\gamma_i = \gamma_i p_i + (\lambda_{i-1} - \beta_{i-1})p_i, \qquad i = 1, \ldots, m$$

From the preceding three equations one sees that $\lambda_i - \beta_i = \lambda_{i-1} - \beta_{i-1} = \lambda$, $i = 1, \ldots, m$ and $\gamma_i = \lambda p_i / (1 - p_i)$. Therefore

$$\lambda_i = \lambda \left(1 + \sum_{j=1}^{i} p_j / (1 - p_j) \right), \qquad i = 1, \ldots, m \tag{5.34}$$

Note the rate at which good jobs pass through any stage i is $\lambda_i - \beta_i = \lambda$. This is to be expected because a job that passes through any given stage i as a good job will not be

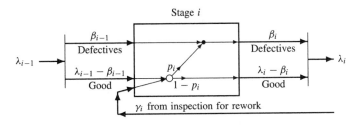

Figure 5.7   Flow of Jobs and Defectives at Station

sent back for rework at any stage j, $j = 1, \ldots, i$. Therefore the rate at which good jobs pass through any stage should be equal to the job arrival rate from outside.

Suppose the number of stations assigned to stage i is c_i and the work load allocated is w_i, $i = 1, \ldots, m$. Then for the system to be stable we need $\lambda_i w_i / c_i < 1$ or equivalently $\lambda < c_i / [w_i (1 + \sum_{j=1}^{i} p_j / (1 - p_j))]$. Let $\eta_i = w_i (1 + \sum_{j=1}^{i} p_j / (1 - p_j)) / c_i$. Then the throughput of this system is

$$TH = \min \left\{ \frac{1}{\eta_i}, i = 1, \ldots, m \right\} \tag{5.35}$$

Now suppose the processing times at stage i are exponentially distributed with mean $w_i = 1/\mu_i$, $i = 1, \ldots, m$. Then the average work in process (i.e., the number of jobs) in the system is

$$E[N] = \sum_{i=1}^{m} \hat{N} (\lambda \eta_i, c_i) \tag{5.36}$$

Optimal order of stages.   We will now look at the order of stages that will maximize the throughput and minimize the average work in process in the system. Note that if $p_i = 0$, $i = 1, \ldots, m$ (i.e., the processing is perfect) the preceding two performance measures are unaffected by the ordering of the stages. Let π be a permutation of $\{1, \ldots, m\}$ and $TH(\pi)$ be the throughput of the system if stage i has $c_{\pi(i)}$ stations, work load $w_{\pi(i)}$, and probability of defective processing $p_{\pi(i)}$, $i = 1, \ldots, m$. For some $l < m$ consider the permutation π' such that $\pi'(i) = \pi(i)$, $i = 1, \ldots, m$; $i \neq l, l + 1$, $\pi'(l) = \pi(l + 1)$ and $\pi'(l + 1) = \pi(l)$. Let

$$\eta_{\pi(i)} = \frac{w_{\pi(i)} (1 + \sum_{j=1}^{i} \frac{p_{\pi(j)}}{1 - p_{\pi(j)}})}{c_{\pi(i)}}, \qquad i = 1, \ldots, m$$

Suppose $w_{\pi(l)} / c_{\pi(l)} \geq w_{\pi(l+1)} / c_{\pi(l+1)}$, then one has

$$\eta_{\pi(l+1)}(\pi) = \frac{w_{\pi(l+1)}}{c_{\pi(l+1)}} \left(1 + \sum_{j=1}^{l+1} \frac{p_{\pi(j)}}{1 - p_{\pi(j)}} \right) \leq \eta_{\pi'(l+1)}(\pi') = \frac{w_{\pi(l)}}{c_{\pi(l)}} \left(1 + \sum_{j=1}^{l+1} \frac{p_{\pi(j)}}{1 - p_{\pi(j)}} \right)$$

and

$$\eta_{\pi'(l+1)}(\pi') = \frac{w_{\pi(l)}}{c_{\pi(l)}}\left(1 + \sum_{j=1}^{l+1} \frac{p_{\pi(j)}}{1 - p_{\pi(j)}}\right)$$

$$\geq \frac{w_{\pi(l+1)}}{c_{\pi(l+1)}}\left(1 + \sum_{j=1}^{l-1} \frac{p_{\pi(j)}}{1 - p_{\pi(j)}} + \frac{p_{\pi(l+1)}}{1 - p_{\pi(l+1)}}\right) = \eta_{\pi'(l)}(\pi')$$

Also

$$\eta_{\pi'(l+1)}(\pi') = \frac{w_{\pi(l)}}{c_{\pi(l)}}\left(1 + \sum_{j=1}^{l+1} \frac{p_{\pi(j)}}{1 - p_{\pi(j)}}\right) \geq \eta_{\pi(l)}$$

Therefore

$$\min\left\{\frac{1}{\eta_{\pi'(l)}(\pi')}, \frac{1}{\eta_{\pi'(l+1)}(\pi')}\right\} = \frac{1}{\eta_{\pi'(l+1)}(\pi')} \leq \min\left\{\frac{1}{\eta_{\pi(l)}(\pi)}, \frac{1}{\eta_{\pi(l+1)}(\pi)}\right\}$$

Hence, from equation 5.35 one sees that if $w_{\pi(l)}/c_{\pi(l)} \geq w_{\pi(l+1)}/c_{\pi(l+1)}$, then $TH(\pi) \geq TH(\pi')$. Therefore we see that a stage with more work load allocated per station should precede a stage with less work load per station to increase the overall throughput of the flow line.

In particular if π^* is a permutation of $\{1, \ldots, m\}$ such that $w_{\pi^*(i)}/c_{\pi^*(i)} \geq w_{\pi^*(i+1)}/c_{\pi^*(i+1)}$, $i = 1, \ldots, m-1$, then

$$TH(\pi^*) \geq TH(\pi)$$

Now we will consider arrangements that will minimize the average work in process in the flow line. For this we consider only the case $c_i = c$, $i = 1, \ldots, m$. Let $\rho_{\pi(i)}(\pi) = \lambda w_{\pi(i)}/\left[c\left(1 + \sum_{j=1}^{i} p_{\pi(j)}/\left(1 - p_{\pi(j)}\right)\right)\right]$, $i = 1, \ldots, m$ for any permutation π of $\{1, \ldots, m\}$. Define π' as before and note that if $w_{\pi(l)} \geq w_{\pi(l+1)}$ and $w_{\pi(l)}(1 - p_{\pi(l)})/p_{\pi(l)} \geq w_{\pi(l+1)}(1 - p_{\pi(l+1)})/p_{\pi(l+1)}$, then

$$\rho_{\pi(l)}(\pi) + \rho_{\pi(l+1)}(\pi) \leq \rho_{\pi'(l)}(\pi') + \rho_{\pi'(l+1)}(\pi')$$

and

$$\max\{\rho_{\pi(l)}(\pi), \rho_{\pi(l+1)}(\pi)\} \leq \max\{\rho_{\pi'(l)}(\pi'), \rho_{\pi'(l+1)}(\pi')\}$$

Because $\hat{N}(\rho, c)$ is increasing and convex in ρ, one sees that

$$\hat{N}(\rho_{\pi(l)}(\pi), c) + \hat{N}(\rho_{\pi(l+1)}(\pi), c) \leq \hat{N}(\rho_{\pi'(l)}(\pi'), c) + \hat{N}(\rho_{\pi'(l+1)}(\pi'), c)$$

Therefore $E[N(\pi)] \leq E[N(\pi')]$. That is, arranging a stage with a larger work load and $w_i(1-p_i)/p_i$ to precede a stage with a smaller work load and $w_j(1-p_j)/p_j$ will decrease the average work in process in the flow line. Particularly if $w_i = w$, $i = 1, \ldots, m$ (i.e., all stages are assigned the same work load), the stage with a better processing reliability $(1 - p_i)$ should precede a stage with a worse processing reliability $(1 - p_j)$ $(\leq (1 - p_i))$ to reduce the average work in process. Particularly, if π^* is a permutation of $\{1, \ldots, m\}$ such that $(1 - p_{\pi^*(i)}) \geq (1 - p_{\pi^*(i+1)})$, $i = 1, \ldots, m - 1$, then

$$E[N(\pi^*)] \leq E[N(\pi)]$$

5.4.4 Finite Buffer Tandem Queueing Models: Single Station per Stage

The open tandem queueing model allows one to compare the relative merits of different work load allocations. In developing that model we have assumed that the storage space available for every stage has unlimited capacity. In many flow lines, the space available for storage of jobs waiting to be processed by the stations of a stage may be limited. Therefore in this section we will develop a finite buffer tandem queueing system model for the flow line incorporating the effects of the finite storage space capacities. Because of the finite storage capacities, the service protocol implemented at each stage should take into consideration the status of the immediate downstream storage space. Let $b_i - c_i$ be the maximum number of jobs that can be stored in the storage space of stage i at any given time. That is, the maximum number of jobs at stage i is restricted to b_i, $i = 1, \ldots, m$. Therefore we need to provide a service protocol to determine when a station that has just completed service on a job and transferred it to the downstream storage should initiate service on the next job. This protocol should take into account the number of jobs in the downstream storage. We will consider two types of service protocols that are employed in flow lines.

- *Production blocking.* Each station will always serve a job as long as there is a job available for processing, and the station is not blocked (i.e., the job it has completed cannot be transferred to the downstream stage because it is full).

- *Communication blocking.* Service to a job at stage $i - 1$ is initiated only if a job is available, and the number of jobs in the downstream stage i and the number of jobs already in process at stage $i - 1$ in total is less than b_i.

Comparison of production and communication blocking. We will next study the job-flow dynamics of flow lines with a single station at each stage that are operated under these two service protocols. Suppose raw jobs arrive according to a stochastic process $\{A(t), t \geq 0\}$, where $A(t)$ is the number of jobs arrived at the flow line during $[0, t]$. Let A_n be the arrival epoch of the nth job to the system, $n = 1, 2, \ldots$, and $S_k^{(i)}$ be the processing time of the kth job to be processed at stage i, $k = 1, 2, \ldots$; $i = 1, \ldots, m$. The buffer capacity of stage i (including the one space available at the station) is b_i, $i = 1, \ldots, m$. Unless otherwise specified we will assume that the storage space of stage 1 has unlimited capacity, that is, $b_1 = \infty$. Suppose there are n_i jobs already at stage i at time zero, $i = 1, \ldots, m$. Let $D_k^{(i)}$ be the time at which the kth movement of a job from the stage i station to stage $i + 1$ occurs. Now, with the production blocking protocol, the job that will depart at time $D_k^{(i)}$ cannot have begun service until

- The job had arrived at the station from stage $i - 1$.
- The previous job had departed stage i.

Now the kth departing job from stage i will have been either the $(k - n_i)$th departing job from stage $i - 1$ if $k > n_i$ or the kth job in the queue at stage i at time 0, if $k \leq n_i$, hence the time at which the job begins service is $\max\{D_{k-n_i}^{(i-1)}, D_{k-1}^{(i)}\}$. The job cannot leave stage i until

- The job has completed service.
- There is a space at stage $i + 1$.

For there to be a space at stage $i + 1$ the number of departures from stage $i + 1$ by time $D_k^{(i)}$, k_{i+1}, must be such that

$$n_{i+1} + k - k_{i+1} \leq b_{i+1}$$

or $k_{i+1} \geq k + n_{i+1} - b_{i+1}$. That is, there will not be space at stage $i + 1$ until time $D_{k+n_{i+1}-b_{i+1}}^{(i+1)}$. Then for production blocking

$$D_k^{(i)} = [\max\{D_{k-n_i}^{(i-1)}, D_{k-1}^{(i)}\} + S_k^{(i)}] \vee D_{k+n_{i+1}-b_{i+1}}^{(i+1)} \tag{5.37}$$

for $k = 1, 2, \ldots$; $i = 1, \ldots, m$. Similarly it can be shown that for communication blocking

$$D_k^{(i)} = \max\{D_{k-n_i}^{(i-1)}, D_{k-1}^{(i)}, D_{k+n_{i+1}-b_{i+1}}^{(i+1)}\} + S_k^{(i)} \tag{5.38}$$

with $D_k^{(i)} = 0$, $k \leq 0$; $i = 1, \ldots, m - 1$; $D_k^{(0)} = A_k$, $k = 1, 2, \ldots$ and $D_k^{(m+1)} = 0$, for all k.

The following conclusions can be immediately made from equations 5.37 and 5.38:

1. The departure times are decreasing functions of b_i, $i = 2, \ldots, m$.
2. The departure times are increasing and convex in the service times $S_k^{(i)}$, $k = 1, 2, \ldots$; $i = 1, \ldots, m$.
3. The departure time $D_k^{(i)}$ under production blocking is smaller than that under communication blocking, $k = 1, 2, \ldots$; $i = 1, \ldots, m$.

Next, we will compare the departure times under production and communication blocking service protocols with different buffer capacities. To explicitly identify the system under consideration we use $D_k^{(i)}(\mathbf{b} : p)$ for the departure time in a flow line with buffer capacities $b_1, \ldots, b_m$ and production blocking and $D_k^{(i)}(\mathbf{b} : c)$ for the departure times in a flow line with buffer capacities $b_1, \ldots, b_m$ and communication blocking service protocol. To simplify the comparison, we will assume that the system is empty at time zero (i.e., $n_i = 0$, $i = 1, \ldots, m$). We will show that $D_k^{(i)}(\mathbf{b} : p) \geq D_k^{(i)}(\mathbf{b} + \mathbf{e} : c)$ where $\mathbf{e} = (1, \ldots, 1)$. As an induction hypothesis we assume that it is true for all $i = 1, \ldots, m$ up to the $(k-1)$th job movements. Then from equations 5.37 and 5.38 one has

$$D_k^{(i)}(\mathbf{b} : p) = \{D_k^{(i-1)}(\mathbf{b} : p) \vee D_{k-1}^{(i)}(\mathbf{b} : p) + S_k^{(i)}\} \vee \{D_{k-b_{i+1}}^{(i+1)}(\mathbf{b} : p)\} \qquad (5.39)$$

and

$$D_k^{(i)}(\mathbf{b}+\mathbf{e} : c) = \{D_k^{(i-1)}(\mathbf{b}+\mathbf{e} : c) \vee D_{k-1}^{(i)}(\mathbf{b}+\mathbf{e} : c) + S_k^{(i)}\} \vee \{D_{k-b_{i+1}-1}^{(i+1)}(\mathbf{b}+\mathbf{e} : c) + S_k^{(i)}\} \qquad (5.40)$$

By the induction assumption

$$D_k^{(i-1)}(\mathbf{b} : p) \vee D_{k-1}^{(i)}(\mathbf{b} : p) + S_k^{(i)} \geq D_k^{(i-1)}(\mathbf{b} + \mathbf{e} : c) \vee D_{k-1}^{(i)}(\mathbf{b} + \mathbf{e} : c) + S_k^{(i)}$$

In addition from equation 5.39 for k replaced by $k - 1$ one has

$$D_{k-1}^{(i)}(\mathbf{b} : p) \geq D_{k-b_{i+1}-1}^{(i+1)}(\mathbf{b} : p) \geq D_{k-b_{i+1}-1}^{(i+1)}(\mathbf{b} + \mathbf{e} : c)$$

Hence $D_k^{(i-1)}(\mathbf{b} : p) \vee D_{k-1}^{(i)}(\mathbf{b} : p) + S_k^{(i)} \geq D_{k-b_{i+1}-1}^{(i+1)}(\mathbf{b} + \mathbf{e} : c) + S_k^{(i)}$. Therefore $D_k^{(i)}(\mathbf{b} : p) \geq D_k^{(i)}(\mathbf{b} + \mathbf{e} : c)$. Combining this to our earlier observation (point 3) we get

$$D_k^{(i)}(\mathbf{b} : c) \geq D_k^{(i)}(\mathbf{b} : p) \geq D_k^{(i)}(\mathbf{b} + \mathbf{e} : c), \quad k = 1, 2, \ldots, i = 1, \ldots, m \qquad (5.41)$$

Let $N^{(i)}(t)$ be the number of jobs at stage i at time t. Clearly $N^{(i)}(t) \leq b_i$, $i = 2, \ldots, m$. If we define $D^{(i)}(t) = \max\{k : D_k^{(i)} \leq t, k = 0, 1, \ldots\}$, one has

$$\sum_{j=1}^{i} N^{(j)}(t) = D^{(i)}(t) - A(t), \quad t \geq 0; i = 1, \ldots, m \qquad (5.42)$$

Therefore the properties we described for $D_k^{(i)}$ can be extended to the number of jobs in the system. Let $N(t) = \sum_{j=1}^{m} N^{(j)}(t)$ be the total number of jobs in the system at time t. Then

1. $N(t)$ decreases in b_i, $i = 2, \ldots, m$.
2. $N(t)$ increases in service times $S_k^{(i)}$, $k = 1, 2, \ldots$; $i = 1, \ldots, m$.
3. $N(t; \mathbf{b} : c) \geq N(t; \mathbf{b} : p) \geq N(t; \mathbf{b} + \mathbf{e} : c)$, using the same notational modification as for $D_k^{(i)}$.

Observe that $N(t)$ includes the number of jobs at the first stage that we assume to have an unlimited buffer capacity. The increase in buffer capacity $b_2, \ldots, b_m$ may cause an increase in the number of jobs at stages $2, \ldots, m$. The decrease in the number of jobs in stage 1 owing to the increase in the buffer capacities, however, is sufficient to guarantee that the total number of jobs in the system decreases. In the single-stage system we have seen that the number of jobs decreases as the service rate increases. Also the service rate is the throughput of a single stage system. This therefore raises the question whether the increase in buffer capacity increases the throughput of the flow line. To answer this we will first need to devise a way to find the throughput of flow lines. As we defined earlier, throughput is the value of the arrival rate above which the number of jobs in the system will no longer be finite. Because $N^{(i)}(t) \leq b_i$, $i = 2, \ldots, m$ we only have to worry about $N^{(1)}(t)$. Let TH be the rate at which jobs leave the system when we have an unlimited number of jobs in front of the first stage (i.e., $A_n = 0$, $n = 1, 2, \ldots$). From equations 5.37 and 5.38 it is easily seen that the departure times $D_k^{(i)}$ increase in $D_k^{(1)} = A_k$, $k = 1, 2, \ldots$. Therefore TH is the largest departure rate that can be realized from this system (note that the departure rate $\lim_{t \to \infty} D^{(i)}(t)/t = \lim_{k \to \infty} k/D_k^{(i)}$, $i = 1, \ldots, m$). Suppose the throughput is $\lambda < TH$. Then with an arrival rate of $\lambda + \epsilon$ there should be infinitely many jobs in front of stage 1 (i.e., $N^{(1)}(t) \to \infty$ as $t \to \infty$). In this case the departure rate should be $TH < \lambda + \epsilon$. Because this should be true for all $\epsilon > 0$, it is easily seen that TH is the throughput of the flow line.

When there are unlimited numbers of new jobs available in front of the first stage, the dynamics of the material flow is given by equations 5.37 and 5.38 with $D_k^{(0)} = 0$, $k = 1, 2, \ldots$. It is easily seen that

1. The throughput increases in the buffer capacities.
2. The throughput decreases in service times $S_k^{(i)}$, $k = 1, 2, \ldots$; $i = 1, \ldots, m$.
3. $TH(\mathbf{b} : c) \leq TH(\mathbf{b} : p) \leq TH(\mathbf{b} + \mathbf{e} : c)$, using the obvious notational modifications.

Effects of Inserted Idleness. In the preceding analysis of the dynamics of the job flow through the flow-line system we assumed that each machine will process a job whenever it is not blocked, and a job is available for processing. That is, no inserted idleness is allowed. Now suppose we allow machine i to idle $I_k^{(i)}$ units of time before initiating a service to the kth job, $k = 1, 2, \ldots$; $i = 1, \ldots, m$. As far as the departure times of the jobs are concerned, we may replace the original service times $S_k^{(i)}$ by $S_k^{(i)} + I_k^{(i)}$ and assume that there is no inserted idleness. Then from equations 5.37 and 5.38 and by an induction argument it can be concluded that the departure times are increased by inserting idleness. Therefore inserted idleness will reduce throughput of a flow line with a single machine at each stage.

Reversibility. *Reversibility* is the property that the production capacity of a flow line in which the sequence of the stations is reversed is the same as the throughput with the original sequence.

Let $A_k^{(i)}$ be the kth time at which service begins at machine i and $D_k^{(i)}$ be the time at which the kth departure occurs from machine i, $k = 1, \ldots, n$; $i = 1, \ldots, m$. Assume that at time 0 there are n jobs waiting for processing at machine 1 and that there are no jobs at the other machines. Then similar to the derivation of equation 5.37 we have for production blocking

$$A_k^{(i)} = \max\{D_k^{(i-1)}, D_{k-1}^{(i)}\}, \quad k = 1, \ldots, n; i = 1, \ldots, m \tag{5.43}$$

and

$$D_k^{(i)} = \max\{A_k^{(i)} + S_k^{(i)}, D_{k-b_{i+1}}^{(i+1)}\}, \quad k = 1, \ldots, n; i = 1, \ldots, m \qquad (5.44)$$

where $D_k^{(0)} = 0$, $k = 1, \ldots, n$ and $D_k^{(m+1)} = 0$. Now consider a reversed flow-line system with m machines in series. The service time of the kth job at machine i is $\hat{S}_k^{(i)} = S_{n+1-k}^{(m+1-i)}$, $k = 1, \ldots, n$, $i = 1, \ldots, m$. The buffer capacity at stage 1 is $b_1 = \infty$ and at stage i is b_{m+2-i}, $i = 2, \ldots, m$. Here also assume that n jobs are available for processing at stage 1, and all other stages are empty. The corresponding dynamics of the system are given by

$$\hat{A}_k^{(i)} = \max\{\hat{D}_k^{(i-1)}, \hat{D}_{k-1}^{(i)}\}, \quad k = 1, \ldots, n; i = 1, \ldots, m \qquad (5.45)$$

and

$$\hat{D}_k^{(i)} = \max\{\hat{A}_k^{(i)} + S_{n+1-k}^{(m+1-i)}, \hat{D}_{k-b_{m+1-i}}^{(i+1)}\}, \quad k = 1, \ldots, n; i = 1, \ldots, m \qquad (5.46)$$

where $\hat{D}_k^{(0)} = 0 = D_k^{(m+1)}$

Subsequently we will show that $\hat{D}_n^{(m)} = D_n^{(m)}$. Therefore if $\{(S_k^{(1)}, \ldots, S_k^{(m)}), k = 1, \ldots, n\}$ are iid random variables the time needed to process n jobs in a reversed system has the same distribution as the time needed to process n jobs in the original system. Therefore the throughputs of both systems are identical.

To show that $\hat{D}_n^{(m)} = D_n^{(m)}$, we will first show that with appropriately defined inserted idleness for the reversed system we will have a departure time say $\hat{D}_n^{*(m)}$ of the nth job from the mth machine that is equal to $D_n^{(m)}$. Because inserted idleness has been shown to increase the departure times one will see that $D_n^{(m)} = \hat{D}_n^{*(m)} \geq \hat{D}_n^{(m)}$. Using a similar construction for the original system we can show that $\hat{D}_n^{(m)} \geq D_n^{(m)}$ thus leading to the conclusion that $\hat{D}_n^{(m)} = D_n^{(m)}$. Consider the reversed system and let us introduce some inserted idleness by restricting the time at which the services at different machines may be initiated. Specifically let $D_n^{(m)} - A_{n+1-k}^{(m+1-i)} - S_{n+1-k}^{(m+1-i)}$ be the earliest time at which the kth job can begin processing at machine i, $k = 1, \ldots, n$; $i = 1, \ldots, m$. If $\hat{A}_k^{*(i)}$ and $\hat{D}_k^{*(i)}$ are the corresponding service initiation and departure epochs of the kth job at machine i in the reversed system with inserted idleness we have

$$\hat{A}_k^{*(i)} = \max\{\hat{D}_k^{*(i-1)}, \hat{D}_{k-1}^{*(i)}, D_n^{(m)} - A_{n+1-k}^{(m+1-i)} - S_{n+1-k}^{(m+1-i)}\}, \quad k = 1, \ldots, n; i = 1, \ldots, m \qquad (5.47)$$

and

$$\hat{D}_k^{*(i)} = \max\{\hat{A}_k^{*(i)} + S_{n+1-k}^{(m+1-i)}, \hat{D}_{k-b_{m+1-i}}^{*(i+1)}\}, \quad k = 1, \ldots, n; i = 1, \ldots, m \qquad (5.48)$$

By induction it is easily verified that $\hat{D}_k^{*(i)} \geq \hat{D}_k^{(i)}$, $k = 1, \ldots, n$; $i = 1, \ldots, m$. We will now show that

$$D_n^{(m)} - \hat{D}_k^{*(i)} = A_{n+1-k}^{(m+1-i)}, \quad k = 1, \ldots, n; i = 1, \ldots, m \qquad (5.49)$$

Observe that for $k = 1$ and $i = 1$ one has $D_n^{(m)} - \hat{D}_1^{*(1)} = D_n^{(m)} - [\max\{D_n^{(m)} - A_n^{(m)} - S_n^{(m)} + S_n^{(m)}, \hat{D}_{1-b_m}^{*(2)}\}] = A_n^{(m)}$. That is, the preceding equality is true for $k = 1$ and $i = 1$.

As an induction hypothesis assume that this equality holds for all $k = 1, \ldots, l - 1$; $i = 1, \ldots, m$ and $k = l$ and $i = 1, \ldots, j$, $j < m$. Then consider

$$D_n^{(m)} - \hat{D}_l^{*(j+1)}$$

$$= \min \left\{ D_n^{(m)} - \hat{D}_l^{*(j)} - S_{n+1-l}^{(m-j)}, D_n^{(m)} - \hat{D}_{l-1}^{*(j+1)} - S_{n+1-l}^{(m-j)}, A_{n+1-l}^{(m-j)}, D_n^{(m)} - \hat{D}_{l-b_{m-j}}^{*(j+2)} \right\}$$

$$= \min \left\{ A_{n+1-l}^{(m+1-j)} - S_{n+1-l}^{(m-j)}, A_{n+2-l}^{(m-j)} - S_{n+1-k}^{(m-j)}, A_{n+1-l}^{(m-j)}, A_{n+1+b_{m-j}}^{(m-j-1)} \right\}$$

$$= A_{n+1-l}^{(m-j)} \tag{5.50}$$

because

$$A_{n+1-l}^{(m+1-j)} \geq A_{n+1-l}^{(m-j)} + S_{n+1-l}^{(m-j)}, \quad A_{n+2-l}^{(m-j)} \geq A_{n+1-l}^{(m-j)} + S_{n+1-l}^{(m-j)}$$

and

$$A_{n+1-l+b_{m-j}}^{(m-j-1)} \geq A_{n+1-l}^{(m-j)}$$

That is, the preceding inequality holds for $k = l$ and $i = j + 1$. Similarly assuming its validity for $k = 1, \ldots, l$; $i = 1, \ldots, m$, it can be shown that $D_n^{(m)} - \hat{D}_{k+1}^{*(1)} = A_{n-k}^{(m)}$. The validity of the preceding equality then follows by induction. Because $A_1^{(1)} = 0$ we have that $D_n^{(m)} = \hat{D}_n^{*(m)} \geq \hat{D}_n^{(m)}$. Now introducing idleness into the original system in a similar manner it can be shown that $\hat{D}_n^{(m)} = D_n^{*(m)} \geq D_n^{(m)}$. Hence $D_n^{(m)} = \hat{D}_n^{(m)}$.

For flow-line systems with communication blocking protocol, one may use a property called job-hole duality to prove the line reversibility (see problem 5.18).

For the remaining portion of this chapter we will assume that the service times $\{S_k^{(i)}, k = 1, 2, \ldots\}$ form an iid sequence mutually independent of each other and the arrival process $\{A_k, k = 1, 2, \ldots\}$.

Impact of Variability of Service Times.   We will now compare two flow lines where one has the generic service time $S^{(i):1}$ at stage i and the other has $S^{(i):2}$, $i = 1, \ldots, m$. All other attributes such as the number of stages and buffer capacities are the same. To carry out this comparison, we need to compare $S^{(i):1}$ and $S^{(i):2}$. The comparison we wish to make is when $S^{(i):2}$ is more variable than $S^{(i):1}$, $i = 1, \ldots, m$. Because $S^{(i):j}$, $j = 1, 2$ are random variables, the notion of "more variable" needs to be formalized for us to carry out the comparison for the flow lines. We denote this by $S^{(i):1} \leq_{cx} S^{(i):2}$. Define $S^{(i):2}$ to be *more variable* than $S^{(i):1}$ if $E[f(S^{(i):1})] \leq E[f(S^{(i):2})]$ for all convex functions f (this ordering is usually called the convex ordering—see Appendix B). Because $f(x) = x$ and $f(x) = -x$ are both convex functions one sees that if $S^{(i):1} \leq_{cx} S^{(i):2}$, then $E[S^{(i):1}] = E[S^{(i):2}]$. Because $f(x) = x^2$ is a convex function, from the preceding observation one sees that $\text{var}[S^{(i):1}] \leq \text{var}[S^{(i):2}]$. This justifies the term *more variable*. Earlier, we have seen that the departure times are increasing convex functions of the service times $S_k^{(i)}$, $k = 1, 2, \ldots$; $i = 1, \ldots, m$. Therefore using the preceding definition of *more variable* service times one sees that

$$S^{(i):1} \leq_{cx} S^{(i):2} \Rightarrow \begin{cases} (i) & E[D_k^{(i):1}] \leq E[D_k^{(i):2}], \ k = 1, 2, \dots; i = 1, \dots, m \\ (ii) & TH^{(1)} \geq TH^{(2)} \end{cases} \qquad (5.51)$$

That is, the throughput with more variable service time is lower.

Three-stage flow lines.

Throughput with Exponential Service Times. In this section we will consider a three-stage flow line with finite buffer storage space. First we will model this system assuming that the service time at each stage is exponentially distributed. That is we assume that $S_k^{(i)}$ is exponentially distributed with mean $1/\mu_i$, $i = 1, \dots, m$. For throughput analysis we assume an infinite number of jobs in front of stage 1. In describing the details of the analysis production blocking service protocol will be assumed.

Let $N_i(t)$ be the number of jobs in the system that are processed by station $i - 1$ but have not yet completed processing by station i, $i = 2, 3$. Then $\{(N_2(t), N_3(t)), t \geq 0\}$ is a Markov process on the state space $S = \{(n_2, n_3) : 0 \leq n_2 \leq b_2 + 1, 0 \leq n_3 \leq b_3 + 1, n_2 + n_3 \leq b_2 + b_3 + 1\}$. Note that when $n_i = b_i + 1$, stage $i - 1$ is blocked by stage i. Let $\mathbf{p} = (p(n_2, n_3), (n_2, n_3) \in S)$ be the stationary probability vector of this Markov process. That is, $p(n_2, n_3) = \lim_{t \to \infty} P\{N_2(t) = n_2, N_3(t) = n_3\}$. The steady-state balance equations for $\mathbf{p}$ are given by

$$\mu_1 p(0, 0) = \mu_3 p(0, 1)$$

$$(\mu_1 + \mu_3) p(0, n_3) = \mu_2 p(1, n_3 - 1) + \mu_3 p(0, n_3 + 1), \qquad 1 \leq n_3 \leq b_3$$

$$(\mu_1 + \mu_3) p(0, b_3 + 1) = \mu_2 p(1, b_3)$$

$$(\mu_1 + \mu_2) p(n_2, 0) = \mu_1 p(n_2 - 1, 0) + \mu_3 p(n_2, 1), \qquad 1 \leq n_2 \leq b_2$$

$$(\mu_1 + \mu_2 + \mu_3) p(n_2, n_3) = \mu_1 p(n_2 - 1, n_3) + \mu_2 p(n_2 + 1, n_3 - 1) + \mu_3 p(n_2, n_3 + 1),$$
$$1 \leq n_2 \leq b_2, 1 \leq n_3 \leq b_3$$

$$(\mu_1 + \mu_3) p(n_2, b_3 + 1) = \mu_1 p(n_2 - 1, b_3 + 1) + \mu_2 p(n_2 + 1, b_3),$$
$$1 \leq n_2 \leq b_2 - 1$$

$$\mu_3 p(b_2, b_3 + 1) = \mu_1 p(b_2 - 1, b_3 + 1) + \mu_2 p(b_2 + 1, b_3)$$

$$\mu_2 p(b_2 + 1, 0) = \mu_1 p(b_2, 0) + \mu_3 p(b_2 + 1, 1)$$

$$(\mu_2 + \mu_3) p(b_2 + 1, n_3) = \mu_1 p(b_2, n_3) + \mu_3 p(b_2 + 1, n_3 + 1), \qquad 1 \leq n_3 \leq b_3 - 1$$

$$(\mu_2 + \mu_3) p(b_2 + 1, b_3) = \mu_1 p(b_2, b_3)$$

These $|S| = (b_2 + 1)(b_3 + 1) - 1$ equations along with the normalizing equation

$$\sum_{(n_2, n_3) \in S} p(n_2, n_3) = 1$$

can be solved for $\mathbf{p}$. Unfortunately they do not posses a formula-type solution.

With present computing facilities direct solution of the equations using a standard procedure is probably as easy a method to use as any other. However, it would be desirable to make use of the fact that the coefficient matrix has a large number of zero entries and thus a sparse matrix solution procedure would be appropriate. It is of interest to note that the state transition matrix has a block tridiagonal structure and the algorithm developed in Section 4.3.2 exploiting this structure can be applied here.

Once **p** has been computed the throughput can be obtained by

$$TH = \mu_3 \left(1 - \sum_{n_2=0}^{b_2+1} p(n_2, 0)\right).$$

For the special case of no (extra) buffer capacity (i.e., $b_2 = 1$, $b_3 = 1$) an explicit formula for the throughput can be obtained. It is

$$TH(\mu_1, \mu_2, \mu_3) =$$

$$1 / \left\{ \frac{1}{\mu_2} + \frac{\mu_2(\mu_1 + \mu_3)}{\mu_1\mu_3(\mu_1 + \mu_2)(\mu_2 + \mu_3)} \left[\frac{(\mu_1^2 + \mu_3^2)(\mu_1 + \mu_2 + \mu_3)^2 - \mu_1^2\mu_3^2}{(\mu_1 + \mu_3)^3 + \mu_2(\mu_1^2 + \mu_1\mu_3 + \mu_3^2)} \right] \right\}$$

For communication blocking protocol the corresponding formula for throughput is

$$TH(\mu_1, \mu_2, \mu_3) = 1 / \left\{ \frac{1}{\mu_2} + \frac{1}{\mu_1} + \frac{1}{\mu_3} - \frac{1}{\mu_1 + \mu_3} \right\}$$

Arrangement of Stations. From the preceding results it can be seen that the throughput $TH(\mu_1, \mu_2, \mu_3) = TH(\mu_3, \mu_2, \mu_1)$. That is, the throughput of this flow line is symmetric in μ_1 and μ_3, which is to be expected because of the reversibility of this system. Let π be the permutation of $\{1, 2, 3\}$ such that $\mu_{\pi(1)} \leq \mu_{\pi(2)}$ and $\mu_{\pi(3)} \leq \mu_{\pi(2)}$. That is, $\pi(2)$ is the fastest station. Then it can be verified that

$$TH(\mu_{\pi(1)}, \mu_{\pi(2)}, \mu_{\pi(3)}) \geq TH(\mu_1, \mu_2, \mu_3)$$

Therefore assigning the fastest station in the middle and the slowest two stations in the first and third stages maximizes the throughput of the system. This result holds even without the assumption of exponential processing times. Let $S_k^{(i)}$ be the processing time of the kth job at stage i, $k = 1, 2, \ldots$; $i = 1, 2, 3$. Assuming production-blocking protocol and that there is exactly one unprocessed job available at each of the three stages at time zero (i.e., $n_i = 1$, $i = 1, 2, 3$), equation 5.37 becomes

$$D_k^{(i)} = \max\left\{\max\left\{D_{k-1}^{(i-1)}, D_{k-1}^{(i)}\right\} + S_k^{(i)}, D_k^{(i+1)}\right\}, \qquad k = 1, 2, \ldots; i = 1, 2, 3 \quad (5.52)$$

where $D_k^{(i)} = 0$, $k \leq 0$, $i = 1, 2, 3$ and $D_k^{(0)} = 0$, $k = 1, 2, \ldots$. From equation 5.52 it can be seen that $D_k^{(i)} \geq D_k^{(i+1)}$, $k = 1, 2, \ldots$, $i = 1, 2, 3$. Hence equation 5.52 can be rewritten as

$$D_k^{(1)} = \max\left\{D_{k-1}^{(1)} + S_k^{(1)}, D_k^{(2)}\right\} \tag{5.53}$$

$$D_k^{(2)} = \max\left\{D_{k-1}^{(1)} + S_k^{(2)}, D_k^{(3)}\right\}$$

$$D_k^{(3)} = D_{k-1}^{(2)} + S_k^{(3)}, \qquad k = 1, 2, \ldots$$

From equation 5.53 one has

$$D_k^{(1)} = \max\left\{D_{k-1}^{(1)} + \max\left\{S_k^{(1)}, S_k^{(2)}\right\}, D_{k-1}^{(2)} + S_k^{(3)}\right\} \tag{5.54}$$

$$D_k^{(2)} = \max\left\{D_{k-1}^{(1)} + S_k^{(2)}, D_{k-1}^{(2)} + S_k^{(3)}\right\}$$

Now let us consider the departure times in another three-stage flow-line system with no buffer capacity and processing times $\{S_k^{(2)}\}$ at the first stage, $\{S_k^{(1)}\}$ at the second stage, and $\{S_k^{(3)}\}$ at the third stage. Following equation 5.54 one gets

$$\hat{D}_k^{(1)} = \max\left\{\hat{D}_{k-1}^{(1)} + \max\left\{S_k^{(1)}, S_k^{(2)}\right\}, \hat{D}_{k-1}^{(2)} + S_k^{(3)}\right\} \tag{5.55}$$

$$\hat{D}_k^{(2)} = \max\left\{\hat{D}_{k-1}^{(1)} + S_k^{(1)}, \hat{D}_{k-1}^{(2)} + S_k^{(3)}\right\}$$

For any increasing function $\phi : R^2 \to R$ consider

$$E\left[\phi\left(D_k^{(1)}, D_k^{(2)}\right) \mid D_{k-1}^{(1)} = \alpha, D_{k-1}^{(2)} = \beta, S_k^{(3)} = \gamma\right] \tag{5.56}$$

$$- E\left[\phi\left(\hat{D}_k^{(1)}, \hat{D}_k^{(2)}\right) \mid \hat{D}_{k-1}^{(1)} = \alpha, \hat{D}_{k-1}^{(2)} = \beta, S_k^{(3)} = \gamma\right]$$

$$= \int_{x=0}^{\infty} \int_{y=0}^{\infty} \psi(x, y) \left[f_{S^{(1)}}(y) f_{S^{(2)}}(x) - f_{S^{(1)}}(x) f_{S^{(2)}}(y)\right] dx\, dy$$

$$= \int_{x=0}^{\infty} \int_{y=0}^{x} \left[\psi(x, y) - \psi(y, x)\right] \left[f_{S^{(1)}}(y) f_{S^{(2)}}(x) - f_{S^{(1)}}(x) f_{S^{(2)}}(y)\right] dx\, dy$$

where $\psi(x, y) = \phi(\max\{\alpha + \max(x, y), \beta + \gamma\}, \max\{\alpha + x, \beta + \gamma\}) (= \{\phi(D_k^{(1)}, D_k^{(2)}) \mid D_{k-1}^{(1)} = \alpha, D_{k-1}^{(2)} = \beta, S_k^{(3)} = \gamma, S_k^{(2)} = x, S_k^{(1)} = y\})$ and $f_{S^{(i)}}$ is the density function of $S^{(i)}$, $i = 1, 2$. It is easily seen that for any increasing function ϕ and $x \geq y$ one has $\psi(x, y) \geq \psi(y, x)$. Hence if $f_{S^{(2)}}(y)/f_{S^{(1)}}(y)$ is increasing in y (i.e., $S^{(2)}$ is larger than $S^{(1)}$ in the stochastic ordering called *likelihood ratio ordering*; we denote this $S^{(2)} \geq_{lr} S^{(1)}$ –see Appendix B), one has from equation 5.56

$$E[\phi(D_k^{(1)}, D_k^{(2)}) \mid D_{k-1}^{(1)} = \alpha, D_{k-1}^{(2)} = \beta, S_k^{(3)} = \gamma]$$

$$\geq E[\phi(\hat{D}_k^{(1)}, \hat{D}_k^{(2)}) \mid \hat{D}_{k-1}^{(1)} = \alpha, \hat{D}_{k-1}^{(2)} = \beta, S_k^{(3)} = \gamma] \tag{5.57}$$

As an induction hypothesis assume that for any increasing function $\phi : R^2 \to R$, $E\phi(D_{k-1}^{(1)}, D_{k-1}^{(2)}) \geq E\phi(\hat{D}_{k-1}^{(1)}, \hat{D}_{k-1}^{(2)})$ for a fixed k. Because $E[\phi(D_k^{(1)}, D_k^{(2)})|D_{k-1}^{(1)} = \alpha, D_{k-1}^{(2)} = \beta, S_k^{(3)} = \gamma]$ is an increasing function of α and β, one sees from equation 5.57 and the induction hypothesis that $E\phi(D_k^{(1)}, D_k^{(2)}) \geq E\phi(\hat{D}_k^{(1)}, \hat{D}_k^{(2)})$ for any increasing function $\phi : R^2 \to R$. Because $D_0^{(i)} = \hat{D}_0^{(i)}$, $i = 1, 2, 3$, by the induction hypothesis $E\phi(D_k^{(1)}, D_k^{(2)}) \geq E\phi(\hat{D}_k^{(1)}, \hat{D}_k^{(2)})$ for any increasing function ϕ and in particular $E[\hat{D}_k^{(1)}] \leq E[D_k^{(1)}]$, $k = 1, 2, \ldots$. Hence the production capacities $TH(1, 2, 3) = \lim_{k\to\infty} 1/E[D_k^{(1)}]$ and $\hat{TH} = TH(2, 1, 3) = \lim_{k\to\infty} 1/E[\hat{D}_k^{(1)}]$ are ordered: that is, $TH \leq \hat{TH}$. Let π be a permutation of $\{1, 2, 3\}$ such that $S_{\pi(1)} \geq_{lr} S_{\pi(2)}$ and $S_{\pi(3)} \geq_{lr} S_{\pi(2)}$. Note that we assume that the service time distributions are such that these comparisons in the likelihood ratio ordering are possible. For example, exponential distributions are always comparable among themselves according to the likelihood ratio ordering. Particularly if $S^{(1)}$ and $S^{(2)}$ are exponentially distributed with means $1/\mu_1$ and $1/\mu_2$, $\mu_2 < \mu_1$ will imply $S^{(2)} \geq_{lr} S^{(1)}$. Now employing the preceding result and the reversibility property of this flow line one sees that

$$TH(\pi) \geq TH(1, 2, 3) \tag{5.58}$$

That is, assigning the fastest station in the middle and the two slowest stations in the first and the last stages will maximize the throughput.

Bounds and approximations to three-stage flow lines with exponential processing times.
We will now look at bounds and approximations for the throughput of a three-stage flow-line system with exponentially distributed processing time and finite buffer storage spaces. Because the throughput is an increasing function of the buffer capacities one sees that

$$TH(\mu_1, \mu_2, \mu_3, b_2, b_3) \leq TH(\mu_1, \mu_2, \mu_3, b_2, +\infty) = \min\{\mu_1(1 - B(\mu_1, \mu_2, b_2+1)), \mu_3\} \tag{5.59}$$

where $B(\lambda, \mu, b) = \rho^b(1 - \rho)/(1 - \rho^{b+1})$, with $\rho = \lambda/\mu$, is the blocking probability (that is, the probability that an arrival sees the system full) in a $M/M/1/b$ queueing system. Then by reversibility

$$TH(\mu_1, \mu_2, \mu_3, b_2, b_3)$$

$$= TH(\mu_3, \mu_2, \mu_1, b_3, b_2) \tag{5.60}$$

$$\leq TH(\mu_3, \mu_2, \mu_1, b_3, +\infty) = \min\{\mu_3(1 - B(\mu_3, \mu_2, b_3 + 1)), \mu_1\}$$

Hence

$$TH \leq \min\{\hat{\mu}_1, \hat{\mu}_3\} \tag{5.61}$$

where

$$\hat{\mu}_1 = \mu_1(1 - B(\mu_1, \mu_2, b_2 + 1))$$

and

$$\hat{\mu}_3 = \mu_3(1 - B(\mu_3, \mu_2, b_3 + 1))$$

Observe that $\hat{\mu}_1$ is the rate at which jobs will be flowing out of station 2, if station 2 is never blocked (i.e., if $b_3 = +\infty$). Hence we may approximate the dynamic behavior of stages 2 and 3 of the flow line by a two-stage flow line with service rates $\hat{\mu}_1$, and μ_3 and intermediate buffer capacity of b_3. The throughput of this approximating system is $\hat{\mu}_1(1 - B(\hat{\mu}_1, \mu_3, b_3 + 1))$. Hence we may choose to use this as an approximation for TH. That is,

$$TH \approx \hat{\mu}_1(1 - B(\hat{\mu}_1, \mu_3, b_3 + 1)) \qquad (5.62)$$

Numerical results comparing the upper bound (equation 5.61) and the preceding approximation (equation 5.62) to the exact throughput are summarized in Table 5.5. It can be seen that the proposed approximation always underestimates the throughput.

TABLE 5.5   TWO-STAGE APPROXIMATIONS FOR THREE-STAGE LINE

$1/\mu_1$	$1/\mu_2$	$1/\mu_3$	b_2	b_2	Throughput		
					Exact	Upper bound	Approx.
1	1	1	1	1	0.5641	0.6667	0.5263
1.2	0.6	1.2	1	1	0.5615	0.7143	0.5181
1.4	0.2	1.4	1	1	0.5263	0.7017	0.4720
0.8	1.4	0.8	1	1	0.5359	0.5914	0.5133
0.8	1.0	1.2	1	1	0.5554	0.6044	0.5221
1	1	1	2	2	0.6705	0.7500	0.6340
1.4	0.2	1.4	2	2	0.5914	0.7125	0.5350
0.8	1.4	0.8	2	2	0.6174	0.6503	0.6029
1	1	1	3	3	0.7263	0.8000	0.6722
1.4	0.2	1.4	3	3	0.6228	0.7140	0.5713
0.8	1.4	0.8	3	3	0.6530	0.6795	0.6511
0.8	1.0	1.2	3	3	0.6928	0.7214	0.6852

Lost arrival queue approximation.   We will now look at ways to improve this approximation. We will consider a three-stage flow line with exponentially distributed service times, finite buffer capacities, and Poisson external arrival process with rate λ.

Observe that if $\lambda < TH$, the job departure rate from the last stage is λ. If jobs were arriving at stage 3 according to a Poisson process with rate $\hat{\lambda}$, the probability that a job is blocked at stage 3 is given by $B(\hat{\lambda}, \mu_3, b_3 + 1)$. Therefore, to achieve a departure rate λ from stage 3, $\hat{\lambda}$ should satisfy

$$\hat{\lambda}[1 - B(\hat{\lambda}, \mu_3, b_3 + 1)] = \lambda \qquad (5.63)$$

Since $\hat{\lambda}[1 - B(\hat{\lambda}, \mu_3, b_3 + 1)]$ is strictly increasing in $\hat{\lambda}$ (see Chapter 4 and Figure 5.8) and $\lim_{\hat{\lambda} \to \infty} \hat{\lambda}[1 - B(\hat{\lambda}, \mu_3, b_3 + 1)] = \mu_3$, it can be seen that equation 5.63 has a unique

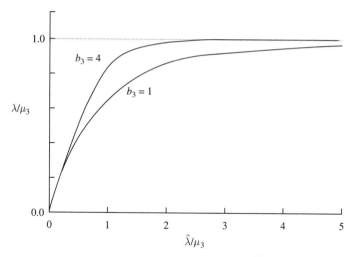

Figure 5.8 Throughput versus $\hat{\lambda}$

solution, say $\hat{\lambda}_3$ for $\hat{\lambda}$ provided λ satisfies $0 \le \lambda < \mu_3$. For $\lambda \ge \mu_3$ we set the solution $\hat{\lambda}_3 = \infty$. Then if $p_I^{(2)}$ is the probability that stage 2 is idle, stage 2 will be servicing jobs at a rate $\mu_2(1 - p_I^{(2)})$. Hence the idle probability $p_I^{(2)}$ should be such that $\mu_2(1 - p_I^{(2)}) = \hat{\lambda}_3$. Hence we impose the condition $\mu_2 > \hat{\lambda}_3$; otherwise we conclude that the system is unstable. Let $p_B^{(i)}$ be the probability that stage i will block stage $i - 1$. Because the rate at which a job is serviced, whenever one is available at stage 2 is $\mu_2(1 - p_B^{(3)})$ we find that the arrival rate $\hat{\lambda}_2$ to stage 2 that will produce the desired idle probability $p_I^{(2)} = 1 - \hat{\lambda}_3/\mu_2$ is given by

$$\hat{\lambda}_2[1 - B(\hat{\lambda}_2, \mu_2(1 - p_B^{(3)}), b_2 + 1)] = \lambda \tag{5.64}$$

Then we set $p_B^{(2)} = B(\hat{\lambda}_2, \mu_2(1 - p_B^{(3)}), b_2 + 1)$. So we have the recursive relations

$$\hat{\lambda}_i[1 - B(\hat{\lambda}_i, \mu_i(1 - p_B^{(i+1)}), b_i + 1)] = \lambda, \qquad i = 3, 2 \tag{5.65}$$

and

$$p_B^{(i)} = B(\hat{\lambda}_i, \mu_i(1 - p_B^{(i+1)}), b_i + 1), \qquad i = 3, 2 \tag{5.66}$$

Starting with $p_B^{(4)} = 0$, equations 5.65 and 5.66 can be recursively solved for $\hat{\lambda}_i$, $i = 3, 2$ and $p_B^{(i)}$, $i = 3, 2$. Then the number of jobs at stage i is approximated by the number of jobs in a $M/M/1/b_i + 1$ queueing system with arrival rate $\hat{\lambda}_i$ and service rate $\mu_i(1 - p_B^{(i+1)})$, $i = 2, 3$. The number of jobs at stage 1 is approximated by the number of jobs in a $M/M/1$ queueing system with arrival rate λ and service rate $\mu_1(1 - p_B^{(2)})$. If the condition $\mu_i > \hat{\lambda}_{i+1}$ is not met for any $i = 1, 2$ we assume that the system is unstable. The actual condition for the stability of the system is $\lambda < TH$. Hence we may use the above condition $\mu_i > \hat{\lambda}_{i+1}$ to obtain an approximation for TH. To see this let

$\hat{\lambda}_i(\lambda)$ be the effective arrival rate computed from equation 5.65 for a given λ. Define $\lambda^* = \min\{\lambda : \mu_i \le \hat{\lambda}_{i+1}(\lambda), i = 1 \text{ or } 2\}$. Then we approximate TH by λ^*.

We will now show that

$$\lambda^* = \mu_1(1 - B(\mu_1, \mu_2(1 - p_B^{(3)}), b_2 + 1) \tag{5.67}$$

Equivalently $\mu_1 = \hat{\lambda}_2(\lambda^*)$. To show this, first observe that $\hat{\lambda}_i(\lambda)$ is increasing in λ for each $i = 2, 3$. Let i^* be the largest value for which $\mu_i \le \lambda_{i+1}(\lambda^*)$. If $i^* = 1$ then it is easily seen that our claim is valid. Therefore suppose $i^* \ge 2$. Then from equation 5.65 it is easily seen that since $\lim_{\mu_i \downarrow \lambda_{i+1}(\lambda^*)} \hat{\lambda}_i(\lambda^*) = +\infty$, there exists a $\lambda < \lambda^*$ such that $\hat{\lambda}_{i^*}(\lambda) > \mu_{i^*-1}$. This contradicts the definition of λ^* and therefore $i^* = 1$. Thus we get equation 5.67.

Substituting our original definition of $\hat{\lambda}_i = \mu_{i-1}(1 - p_I^{(i-1)})$ we see that

$$\lambda^* = \mu_i(1 - p_I^{(i)})(1 - p_B^{(i+1)}), \quad i = 1, 2, 3 \tag{5.68}$$

where $p_I^{(1)} = 0$ and $p_B^{(4)} = 0$.

The preceding approximation procedure does reasonably well when the relative difference between the service rates (i.e., $|\mu_i - \mu_j| \, i, j = 1, 2, 3$) is not large. When this difference is large, the resulting approximation can be very poor. Let $\lambda^*(\mu_1, \mu_2, \mu_3)$ be the approximate production rate obtained by the preceding procedure. Then it can be verified that $\lim_{\mu_2 \to \infty} \lambda^*(\mu_1, \mu_2, \mu_3) = \min(\mu_1, \mu_3)$ independent of the buffer capacities b_2 and b_3. The corresponding limit of the true production capacity is $\mu_3\{1 - (1 - \mu_1/\mu_3)/(1 - (\mu_1/\mu_3)^{b_2+b_3+2})\}$ or $\mu(b_2+b_3+1)/(b_2+b_3+2)$ when $\mu_1 = \mu_3 = \mu$. It is easily verified that the maximum of the error $\min(\mu_1, \mu_3) - \mu_3\{1 - (1 - \mu_1/\mu_3)/(1 - (\mu_1/\mu_3)^{b_2+b_3+2})\}$ occurs at $\mu_1 = \mu_3$ (say $= \mu$). The maximum error is $100/(b_2 + b_3 + 1)\%$. This error decreases with b_2 and b_3. For $b_2 = b_3 = 1$ we have 33% error. This poor performance can be interpreted using equation 5.68. If the service rate μ_i is relatively large compared with μ_{i-1}, $p_I^{(i)}$ will be close to 1. Hence in computing λ^*, the effect of downstream blocking probability $p_B^{(i+1)}$ is largely reduced by this. Because in the original system stage i cannot be simultaneously blocked and be idle the correct form for λ^* should be $\lambda^* = \mu_i(1 - p_I^{(i)} - p_B^{(i+1)})$. For this reason we devise another approximation technique for the multiple stage system that satisfies the correct form for λ^* and is applicable even when the difference in service rate is large.

Multiple-stage flow lines with exponential processing times.   In this section we model the flow line by a tandem queueing system with m stages and with finite buffer capacities $b_2, \ldots, b_m$ and exponentially distributed processing times. That is, we assume the service time $S^{(i)}$ has an exponential distribution with mean $1/\mu_i$, $i = 1, \ldots, m$. The jobs arrive at the flow line according to a Poisson process with rate λ.

We will develop the model for production-blocking service protocol. Because the number of jobs at each of the m stages alone is insufficient to identify whether a station is blocked or not, we let $N_i(t)$ be the number of jobs that have been processed by stage $i - 1$ but not yet completed processing by stage i at time t. Then $\{\mathbf{N}(t), t \geq 0\}$ is a continuous time Markov process. Here $\mathbf{N}(t) = (N_1(t), \ldots, N_m(t))$. Let $p(\mathbf{n}) = \lim_{t \to \infty} P\{\mathbf{N}(t) = \mathbf{n}\}$. Then the balance equations for the probability flow can be written as in the three-stage model.

Unfortunately it is not possible to obtain explicit solutions to these flow balance equations. Because the number of states of this Markov process $\{\mathbf{N}(t), t \geq 0\}$ can be very large, numerical solution to these balance equations may be time consuming. Therefore we will look for an approximate solution. This approximation is essentially the same as the previous one discussed for the three-stage flow line except the service rates $\mu_i(1 - p_B^{(i+1)})$ are replaced by μ_{id}, the reciprocal of the average time it takes stage i to process a job (including the blocking time) when a job is available for processing by stage i, where

$$\frac{1}{\mu_{id}} = \frac{1}{\mu_i} + \frac{1}{\mu_{i+1d}} \hat{p}_B^{(i+1)}, \quad i = 1, \ldots, m - 1 \tag{5.69}$$

with $\mu_{md} = \mu_m$ and $\hat{p}_B^{(i+1)}$ is the probability that at the instant of completion of a service at stage i it is blocked by stage $i + 1$. The rationale for using equation 5.69 is as follows: $1/\mu_{id}$ is the average time it takes stage i to process a job (including the blocking time if any) whenever there is a job available for processing. If stage i is not blocked then it will take on the average $1/\mu_i$ units of time to process a job. When it completes service and is blocked (with probability $\hat{p}_B^{(i+1)}$) then the time to process a job should also include the time to get unblocked. This will take on average $1/\mu_{i+1d}$, the average time required to process a job on stage $i + 1$ and hence equation 5.69 gives the average effective processing time at stage i. Note that because $\hat{p}_B^{(i+1)}$ is the probability that stage i is blocked at the instant of completion of a service it differs from $p_B^{(i+1)}$, the time-average probability that stage i is blocked (see Section 4.3), specifically $\hat{p}_B^{(i+1)} = B(\hat{\lambda}_{i+1}, \mu_{i+1d}, b_{i+1}), i = 1, \ldots, m - 1$.

Now using equation 5.66 with μ_{id} replacing $\mu_i(1 - p_B^{(i+1)})$ we get the following recursive relations to solve for $\hat{\lambda}_i$, $i = 1, \ldots, m - 1$:

$$\hat{\lambda}_i[1 - B(\hat{\lambda}_i, \mu_{id}, b_i + 1)] = \lambda, \quad i = m, \ldots, 2 \tag{5.70}$$

$$\frac{1}{\mu_{i-1d}} = \frac{1}{\mu_{i-1}} + \frac{1}{\mu_{id}} B(\hat{\lambda}_i, \mu_{id}, b_i), \quad i = m, \ldots, 2 \tag{5.71}$$

with the initial condition $\mu_{md} = \mu_m$. Note that for equation 5.70 to have a solution we require $\mu_{id} > \lambda$, $i = 1, \ldots, m - 1$. The number of jobs in the system and at the different stages of the flow line can be approximated by the following algorithm.

Algorithm 5.1   Finite Buffer Open Tandem Queue—Exponential

Step 1: Set $\mu_{md} = \mu_m$.

Step 2: For $i = m, \ldots, 2$

If $\mu_{id} < \lambda$, then the system is unstable, go to step 4.

Otherwise solve $\hat{\lambda}_i(1 - B(\hat{\lambda}_i, \mu_{id}, b_i + 1)) = \lambda$ for $\hat{\lambda}_i$.

Set $1/\mu_{i-1d} = 1/\mu_{i-1} + (1/\mu_{id})B(\hat{\lambda}_i, \mu_{id}, b_i)$.
Compute $N_i = \hat{N}_{M/M/1/b}(\hat{\lambda}_i, \mu_{id}, b_i + 1)$.

Step 3: Compute $N_1 = \hat{N}_{M/M/1}(\lambda, \mu_{1d})$.

Step 4: Stop.

In step 2 we need to solve the nonlinear equation $\hat{\lambda}(1 - B(\hat{\lambda}, \mu, b + 1)) = \lambda$ for the unknown $\hat{\lambda}$ when $\mu > \lambda$. Because $\hat{\lambda}(1 - B(\hat{\lambda}, \mu, b + 1))$ is increasing and concave in $\hat{\lambda}$, the following iterative scheme

$$\hat{\lambda}^{(k)} = \hat{\lambda}^{(k-2)}$$

$$+ \frac{(\hat{\lambda}^{(k-1)} - \hat{\lambda}^{(k-2)})(\lambda - \hat{\lambda}^{(k-2)}(1 - B(\hat{\lambda}^{(k-2)}, \mu, b + 1)))}{\hat{\lambda}^{(k-1)}(1 - B(\hat{\lambda}^{(k-1)}, \mu, b + 1)) - \hat{\lambda}^{(k-2)}(1 - B(\hat{\lambda}^{(k-2)}, \mu, b + 1))},$$

$$k = 2, \ldots$$

starting with $\hat{\lambda}^{(0)} = 0$; $\hat{\lambda}^{(1)} = \lambda$ will monotonically converge to the solution.

From equations 5.70 and 5.71 it is possible to derive an alternative expression for throughput and also a useful relationship between $\hat{\lambda}_i$ and $\hat{\lambda}_{i-1}$. Let $I(\lambda, \mu, b) = (1 - \rho)/(1 - \rho^{b+1})$ be the probability that the server is idle in a stationary $M/M/1/b$ queue with arrival rate λ and service rate μ. We make use of the following identities:

$$(1 - I(\hat{\lambda}_i, \mu_{id}, b_i + 1))B(\hat{\lambda}_i, \mu_{id}, b_i) = B(\hat{\lambda}_i, \mu_{id}, b_i + 1)$$

$$(1 - B(\hat{\lambda}_i, \mu_{id}, b_i + 1))I(\hat{\lambda}_i, \mu_{id}, b_i) = I(\hat{\lambda}_i, \mu_{id}, b_i + 1)$$

and also

$$\lambda = \mu_{id}(1 - I(\hat{\lambda}_i, \mu_{id}, b_i + 1))$$

Thus equation 5.71 can be rewritten as

$$\frac{1 - I(\hat{\lambda}_{i-1}, \mu_{i-1d}, b_{i-1} + 1)}{\lambda} = \frac{1}{\mu_{i-1}} + \frac{B(\hat{\lambda}_i, \mu_{id}, b_i + 1)}{\lambda}, \qquad i = m, \ldots, 2$$

Thus

$$\lambda = \mu_{i-1}(1 - I(\hat{\lambda}_{i-1}, \mu_{i-1d}, b_{i-1} + 1) - B(\hat{\lambda}_i, \mu_{id}, b_i + 1)), \quad i = m, \ldots, 2 \qquad (5.72)$$

and

$$\frac{1}{\hat{\lambda}_i} = \frac{1}{\mu_{i-1}} + \frac{I(\hat{\lambda}_{i-1}, \mu_{i-1d}, b_{i-1} + 1)}{\hat{\lambda}_{i-1}}, \qquad i = m, \ldots, 2 \qquad (5.73)$$

In algorithm 5.1 we see that if $\lambda > \mu_{id}$ for any i, $i = 1, \ldots, m$, the system is unstable. Let $\lambda^* = \max\{\lambda : \mu_{id} \geq \lambda, i = 1, \ldots, m\}$. We will use λ^* as our approximate throughput of the flow line. Similar to equation 5.67 it can be shown that

$$\mu_1(1 - B(\mu_1, \mu_{2d}, b_2 + 1)) = \lambda^* \qquad (5.74)$$

and from equation 5.72 we also have

$$\mu_i(1 - B(\hat{\lambda}_{i+1}^*, \mu_{i+1d}, b_{i+1} + 1) - I(\hat{\lambda}_i^*, \mu_{id}, b_i + 1)) = \lambda^*, \quad i = 2, \ldots, m - 1$$

$$\mu_m(1 - I(\hat{\lambda}_m^*, \mu_m, b_m + 1)) = \lambda^* \tag{5.75}$$

That is, we have $\lambda^* = \mu_i(1 - p_I^{(i)} - p_B^{(i+1)})$, which is the correct form for λ^* (compare this to equation 5.68).

Iterative scheme to compute throughput.   The approximate throughput λ^* we specified earlier cannot be explicitly determined. Here we will describe an iterative procedure to compute this approximation.

Algorithm 5.2   Finite Buffer Tandem Queue Throughput—Exponential
Step 1: Set $k = 1$, $\mu_{md} = \mu_m$ and $\lambda^{(1)} = 1/\sum_{i=1}^m 1/\mu_i$.
Step 2: For $i = m, \ldots, 2$
  If $\mu_{id} < \lambda^{(k)}$, then set $\lambda^{(k)} \leftarrow (1/2)(\lambda^{(k)} + \lambda^{(k-1)})$ and repeat step 2.
  Otherwise solve $\hat{\lambda}_i(1 - B(\hat{\lambda}_i, \mu_{id}, b_i + 1)) = \lambda^{(k)}$.
  Set $1/\mu_{i-1d} = 1/\mu_{i-1} + (1/\mu_{id})B(\hat{\lambda}_i, \mu_{id}, b_i)$.
Step 3: If $|\lambda^{(k)} - \lambda^{(k-1)}| < \epsilon$, then set $\lambda_{app}^* = \lambda^{(k)}$ and stop.
  Otherwise compute $\hat{\lambda}(\lambda^{(k)}) = \mu_1(1 - B(\mu_1, \mu_{2d}, b_2 + 1))$.
  Set $\lambda^{(k+1)} = (1/2)[\hat{\lambda}(\lambda^{(k)}) + \lambda^{(k)}]$, $k \leftarrow k + 1$ and go to step 2.

Some properties of the approximate throughput that we need to establish the convergence of this iterative scheme will be presented next. First observe that the solution $\hat{\lambda}$ of $\hat{\lambda}(1 - B(\hat{\lambda}, \mu, b + 1)) = \lambda$ for given μ and b is increasing in λ ($\lambda < \mu$). Then it is easily seen from equations 5.70 and 5.71 that $\hat{\lambda}_i$ is increasing and μ_{i-1d} is decreasing in λ for $i = 2, \ldots, m$. Suppose we set $\hat{\lambda}(\lambda) = \mu_1(1 - B(\mu_1, \mu_{2d}, b_2 + 1))$. Then we see that $\hat{\lambda}(\lambda)$ is decreasing in λ. The λ^* we are after is the solution to $\hat{\lambda}(\lambda^*) = \lambda^*$. From the monotonicity of $\hat{\lambda}(\lambda)$ in λ the uniqueness of the solution λ^* is immediate. For $\lambda^{(1)} = 1/\sum_{i=1}^m 1/\mu_i$ it can be verified that $\mu_{id} > \lambda^{(1)}$, $i = 1, \ldots, m$. Hence at the end of the first iteration of step 2, one has $\lambda^{(1)} = 1/\sum_{i=1}^m 1/\mu_i \leq \lambda^*$. At the end of the kth iteration of step 2 one has

$$\lambda^{(k)} = \max\{\lambda^{(k-1)} + \left(\frac{1}{2}\right)^l (\hat{\lambda}(\lambda^{(k-1)}) - \lambda^{(k-1)}) \leq \lambda^*, \quad l = 1, 2, \ldots\}, \quad k = 2, 3, \ldots \tag{5.76}$$

Therefore $\lambda^{(k)} \leq \lambda^*$ and because $\hat{\lambda}(\lambda) \geq \lambda$ for $\lambda \leq \lambda^*$, $\lambda^{(k)}$ is increasing in $k = 1, 2, \ldots$ Therefore $\{\lambda^{(k)}\}$ converges. Now we will show that $\lambda^{(k)} \to \lambda^*$ as $k \to \infty$. From equation 5.76 it can be seen that

$$\lambda^* - \lambda^{(k)} = \lambda^* - \lambda^{(k-1)}$$

$$- \max\left\{\left(\frac{1}{2}\right)^l (\hat{\lambda}(\lambda^{(k-1)}) - \lambda^{(k-1)}) \leq \lambda^* - \lambda^{(k-1)}, \quad l = 1, 2, \ldots\right\}$$

$$\leq \frac{1}{2}(\lambda^* - \lambda^{(k-1)}), \quad k = 2, 3, \ldots$$

Therefore $\lambda^* - \lambda^{(k)} \leq (1/2)^{k-1} (\lambda^* - \lambda^{(1)})$ and $\lambda^{(k)} \to \lambda^*$ as $k \to \infty$. Next we will see that algorithm 5.2 will terminate after a finite number of steps and $\lambda^* - \lambda^*_{app} \leq \epsilon$. From equation 5.76 observe that

$$
\lambda^{(k)} - \lambda^{(k+1)} = \max \left\{ \left(\frac{1}{2} \right)^l (\hat{\lambda}(\lambda^{(k-1)}) - \lambda^{(k-1)}) \leq \lambda^* - \lambda^{(k-1)}, \quad l = 1, 2, \ldots \right\}
$$

$$
\geq \min \left\{ \left(\frac{1}{2} \right)^l (\hat{\lambda}(\lambda^{(k-1)}) - \lambda^{(k-1)}) \geq \frac{1}{2}(\lambda^* - \lambda^{(k-1)}), \quad l = 1, 2, \ldots \right\}
$$

$$
\geq \frac{1}{2}(\lambda^* - \lambda^{(k-1)}) \geq \lambda^* - \lambda^{(k)}
$$

Therefore $\lambda^{(k)} - \lambda^{(k-1)} < \epsilon$ implies that $\lambda^* - \lambda^{(k)} = \lambda^* - \lambda^*_{app} < \epsilon$. From the preceding inequalities it can also be seen that $\lambda^{(k)} - \lambda^{(k-1)} \leq \lambda^* - \lambda^{(k-1)} \leq \left(\frac{1}{2} \right)^{k-2} (\lambda^* - \lambda^{(1)})$. Therefore algorithm 5.2 will terminate after at most $2 + \log_2\{(1/\epsilon)(\lambda^* - \lambda^{(1)})\} \leq 2 + \log_2(\mu_1/\epsilon)$ iterations. For example, if we require an order of accuracy of 10^{-3} (i.e., ϵ/μ_1), then the algorithm will terminate after a maximum of 12 iterations.

Simplified Algorithm for Throughput. Next we will look at some additional properties of the solution λ^* that will allow us to develop an alternative algorithm to compute λ^* without the need for an iterative solution of $\hat{\lambda}(1 - B(\hat{\lambda}, \mu, b+1)) = \lambda$. Let $1/\mu_{iu}$ be the average time to process a part by stage i (including any idle time) whenever stage i is not blocked. Then $\mu_{1u} = \mu_1$ and

$$
\frac{1}{\mu_{iu}} = \frac{1}{\mu_i} + \frac{1}{\mu_{i-1u}} I(\mu_{i-1u}, \hat{\mu}_i, b_i), \quad i = 2, \ldots, m \tag{5.77}
$$

Let $\hat{\mu}_i$ be defined by

$$
\mu_{i-1u}(1 - B(\mu_{i-1u}, \hat{\mu}_i, b_i + 1)) = \lambda^*, \quad i = 2, \ldots, m \tag{5.78}
$$

For $i = 2$, from equations 5.78 and 5.74 one sees that $\hat{\mu}_2 = \mu_{2d}$. Also from equation 5.73 one has $\hat{\lambda}^*_2 = \mu_1$. Hence from equations 5.73 and 5.77 one sees that $\mu_{2u} = \hat{\lambda}^*_3$. Hence from equation 5.78 for $i = 3$ one sees that $\hat{\mu}_3 = \mu_{3d}$. Continuing this argument one obtains that $\hat{\mu}_i = \mu_{id}$, $i = 2, \ldots, m - 1$ and $\hat{\lambda}_i = \mu_{i-1u}$, $i = 2, \ldots, m$.

Now suppose for some $\lambda > 0$ we define $\hat{\mu}_i$ to be the solution of

$$
\mu_{i-1u}(1 - B(\mu_{i-1u}, \hat{\mu}_i, b_i + 1)) = \lambda \tag{5.79}
$$

and we define μ_{iu} by

$$
\frac{1}{\mu_{iu}} = \frac{1}{\mu_i} + \frac{1}{\mu_{i-1u}} I(\mu_{i-1u}, \hat{\mu}_i, b_i) \tag{5.80}
$$

for $i = 2, \ldots, m$ starting with $\mu_{1u} = \mu_1$. From equations 5.79 and 5.80 it can be seen that $\hat{\mu}_i$ is an increasing and μ_{iu} is a decreasing function of λ. Suppose for some given $\lambda > 0$, $\hat{\mu}_i(\lambda)$ obtained from equation 5.79 is less than or equal to $\mu_{id}(\lambda)$ computed from equation 5.71. Because $\hat{\mu}_i(\lambda)$ is increasing in λ and $\mu_{id}(\lambda)$ is decreasing in λ and for $\lambda = \lambda^*$, $\mu_{id}(\lambda^*) = \hat{\mu}_i(\lambda)$ one sees that $\lambda < \lambda^*$. Equivalently if $\mu_{iu}(\lambda) > \hat{\lambda}_{i+1}(\lambda)$ then $\lambda < \lambda^*$.

Now one sees from equations 5.74 and 5.78 that finding the solution λ^* is equivalent to finding μ_{iu} and μ_{id} for $i = 1, \ldots, m$ with $\mu_{1u} = \mu_1$, $\mu_{md} = \mu_m$ that satisfy

$$\frac{1}{\mu_{i-1d}} = \frac{1}{\mu_{i-1}} + \frac{1}{\mu_{id}} B(\mu_{i-1u}, \mu_{id}, b_i), \qquad i = 2, \ldots, m \qquad (5.81)$$

and

$$\frac{1}{\mu_{iu}} = \frac{1}{\mu_i} + \frac{1}{\mu_{i-1u}} I(\mu_{i-1u}, \mu_{id}, b_i), \qquad i = 2, \ldots, m \qquad (5.82)$$

From equations 5.81 and 5.82 it is easily seen that for any $\mu_{id}^{(1)} \geq \mu_{id}^{(2)}$ the corresponding solution of equation 5.82 for μ_{iu}'s will satisfy $\mu_{iu}^{(1)} \leq \mu_{iu}^{(2)}$. Similarly if we solve equation 5.81 for μ_{id}'s for $\mu_{iu}^{(1)} \leq \mu_{iu}^{(2)}$, the corresponding μ_{id}'s will satisfy $\mu_{id}^{(1)} \geq \mu_{id}^{(2)}$. Because $0 \leq \mu_{id} \leq \mu_i$ and $0 \leq \mu_{iu} \leq \mu_i$, it is easily verified that the following iterative scheme will converge and provide the solution to equations 5.81 and 5.82, and hence λ^*.

Algorithm 5.3 Finite Buffer Tandem Queue Throughput Direct—Exponential

Step 1: Set $\mu_{1u} = \mu_1$; $\mu_{id} = \mu_i$, $i = 2, \ldots, m$.

Step 2: For $i = 2, \ldots, m$
Compute $1/\mu_{iu} = 1/\mu_i + (1/\mu_{i-1u})I(\mu_{i-1u}\mu_{id}, b_i)$.

Step 3: For $i = m, \ldots, 2$
Compute $1/\mu_{i-1d} = 1/\mu_{i-1} + (1/\mu_{id})B(\mu_{i-1u}, \mu_{id}, b_i)$.

Step 4: If $|\mu_1(1 - B(\mu_1, \mu_{2d}, b_2 + 1)) - \mu_m(1 - I(\mu_{m-1u}, \mu_m, b_m + 1))| < \epsilon$, set $\lambda^* = \mu_1(1 - B(\mu_1, \mu_{2d}, b_2 + 1))$ and stop.
Otherwise go to step 2.

Application to buffer and work-load allocation. Consider the work load allocation problem we studied earlier for a flow line with unlimited buffer capacities. That is, we have a total work load W that needs to be allocated among the m stages. Let $1/\mu_i = w_i$ be the work load allocated to stage i. Then we wish to maximize $TH(\mu)$ subject to $\sum_{i=1}^{m} 1/\mu_i = W$. Because we do not have an easy approach to compute TH or obtain the optimal allocation μ^* we will consider the following: maximize $\lambda^*(\mu)$ subject to $\sum_{i=1}^{m} 1/\mu_i = W$. That is, we will find the allocation μ^* that maximizes the approximate throughput. As we will soon see, the optimal solution to this (modified) allocation problem can be easily characterized.

Now suppose for a given allocation of work load w_i, $i = 1, \ldots, m$ (i.e., for a given $\mu_i = 1/w_i$, $i = 1, \ldots, m$) let $\lambda^*(\mu)$ be the approximate throughput. Then for any i, $(i = 1, \ldots, m - 1)$

$$\mu_{iu}(1 - B(\mu_{iu}, \hat{\mu}_{i+1}, b_{i+1} + 1)) = \lambda^*(\mu) \qquad (5.83)$$

$$\hat{\lambda}_{i+1}(1 - B(\hat{\lambda}_{i+1}, \mu_{i+1d}, b_{i+1} + 1)) = \lambda^*(\mu)$$

Now consider an alternate allocation μ' such that $\mu'_j = \mu_j$, for $j \neq i$ and $j \neq i+1$ and

$$\frac{1}{\mu'_i} + \frac{1}{\mu'_{i+1}} = \frac{1}{\mu_i} + \frac{1}{\mu_{i+1}} \tag{5.84}$$

Then if we computed the quantities μ_{iu} and $\hat{\mu}_i$ using equations 5.78 and 5.77 for $\lambda = \lambda^*(\mu)$ we get

$$\mu'_{iu}(1 - B(\mu'_{iu}, \hat{\mu}'_{i+1}, b_{i+1} + 1)) = \lambda^*(\mu) \tag{5.85}$$

such that

$$\frac{1}{\mu'_{iu}} - \frac{1}{\mu_{iu}} = \frac{1}{\mu'_i} - \frac{1}{\mu_i}$$

Similarly using equations 5.70 and 5.71 one gets

$$\hat{\lambda}'_{i+1}(1 - B(\hat{\lambda}'_{i+1}, \mu'_{i+1d}, b_{i+1} + 1)) = \lambda^*(\mu) \tag{5.86}$$

such that

$$\frac{1}{\mu'_{i+1d}} - \frac{1}{\mu_{i+1d}} = \frac{1}{\mu'_{i+1}} - \frac{1}{\mu_{i+1}}$$

Suppose $\mu_{iu} > \mu_{i+1d}$. Then from equations 5.84, 5.85 and 5.86 one sees that

$$\frac{1}{\mu_{iu}} = \frac{1}{\mu'_{iu}} + \left(\frac{1}{\mu_i} - \frac{1}{\mu'_i}\right) < \frac{1}{\mu'_{i+1d}} - \left(\frac{1}{\mu_{i+1}} - \frac{1}{\mu'_{i+1}}\right) = \frac{1}{\mu_{i+1d}} \tag{5.87}$$

Now suppose we have chosen μ'_i and μ'_{i+1} such that

$$0 < \frac{1}{\mu_i} - \frac{1}{\mu'_i} < \frac{1}{\mu_{i+1d}} - \frac{1}{\mu_{iu}}$$

Then from equations 5.84 and 5.87 one sees that

$$\frac{1}{\mu_{iu}} < \min\left\{\frac{1}{\mu'_{iu}}, \frac{1}{\mu'_{i+1d}}\right\} \leq \max\left\{\frac{1}{\mu'_{iu}}, \frac{1}{\mu'_{i+1d}}\right\} < \frac{1}{\mu_{i+1d}}$$

Now suppose $\mu_{iu} < \mu_{i+1d}$. In this case one can choose as before μ'_i and μ'_{i+1} such that

$$\frac{1}{\mu_{i+1d}} < \min\left\{\frac{1}{\mu'_{iu}}, \frac{1}{\mu'_{i+1d'}}\right\} < \frac{1}{\mu_{i+1d}}$$

Now suppose μ'_i, μ'_{i+1} are chosen such that

$$\min\left\{\frac{1}{\mu_{i+1d}}, \frac{1}{\mu_{iu}}\right\} < \min\left\{\frac{1}{\mu'_{i+1d}}, \frac{1}{\mu'_{iu}}\right\} < \max\left\{\frac{1}{\mu'_{i+1d}}, \frac{1}{\mu'_{iu}}\right\} < \max\left\{\frac{1}{\mu_{i+1d}}, \frac{1}{\mu_{iu}}\right\}$$

Then from the $M/M/1/b$ results it is known that

$$\mu'_{iu}(1 - B(\mu'_{iu}, \mu'_{i+1d}, b_{i+1} + 1)) > \mu_{iu}(1 - B(\mu_{iu}, \mu_{i+1d}, b_{i+1} + 1)) = \lambda^*(\mu) \tag{5.88}$$

Comparing equations 5.85 and 5.88 one sees that $\hat{\mu}'_{i+1} < \mu'_{i+1d}$. Because $\hat{\mu}_i(\lambda)$ obtained from equations 5.79 and 5.80 is increasing in λ and $\mu_{id}(\lambda)$ obtained from equations 5.70 and 5.71 is decreasing in λ we can see that $\hat{\mu}'_{i+1} < \mu'_{i+1d}$ (computed for $\lambda = \lambda^*(\mu)$) implies that the value of λ for which $\hat{\mu}'_{i+1}(\lambda) = \mu'_{i+1d}(\lambda)$ must be larger than $\lambda^*(\mu)$. That is, $\lambda^*(\mu') > \lambda^*(\mu)$.

It is now clear that for an optimal allocation μ^*,

$$\mu_{iu}(\mu^*) = \mu_{i+1d}(\mu^*), \quad 1 = 1, \ldots, m-1 \tag{5.89}$$

Therefore

$$B(\mu_{iu}(\mu^*), \mu_{i+1d}(\mu^*), b_{i+1}+1) = I(\mu_{iu}(\mu^*), \mu_{i+1d}(\mu^*), b_{i+1}+1) = 1/(b_{i+1}+2),$$

$$i = 1, \ldots, m-1$$

Then from equation 5.75 one has

$$\lambda^*(\mu^*) = \mu_i^*(1 - k_i - k_{i+1}), \quad i = 1, \ldots, m \tag{5.90}$$

where we set $k_1 = 0 = k_{m+1}$ and $k_i = 1/(b_i+2)$, $i = 2, \ldots, m$. Because $\sum_{i=1}^m 1/\mu_i^* = W$ one has from equation 5.90

$$w_i^* = \frac{1}{\mu_i^*} = \frac{1 - k_i - k_{i+1}}{\sum_{j=1}^m (1 - k_j - k_{j+1})} W, \quad i = 1, \ldots, m \tag{5.91}$$

The optimal approximated throughput is then

$$\lambda^*(\mu^*) = \sum_{j=1}^m (1 - k_j - k_{j+1})/W \tag{5.92}$$

$$= (m - 2\sum_{j=2}^m \frac{1}{b_j + 2})/W$$

Now suppose we are allowed to allocate the buffer capacities as well. If $(m-1)b$ is the total buffer capacity we are allowed to allocate such that $\sum_{i=2}^m b_i = (m-1)b$, $b_i \geq 1$, $i = 2, \ldots, m$ and $\lambda^*(\mu, b)$ is maximized. From equation 5.92 it can be seen that the optimal allocation $b_i^* = b$, $i = 2, \ldots, m$ and the optimal (approximated) throughput is

$$\lambda(\mu^*, b^*) = \left(\frac{mb + 2}{b + 2}\right)/W \tag{5.93}$$

The corresponding work-load allocation is

$$w_1^* = w_m^* = \frac{(1 - k)W}{2k - 2mk + m} \tag{5.94}$$

$$w_i^* = \frac{(1 - 2k)W}{2k - 2mk + m}, \quad i = 2, \ldots, m-1 \tag{5.95}$$

Models of flow lines with general processing times.   We consider flow-line systems in which the job arrival process cannot be reasonably represented by a Poisson process or the service times are not exponentially distributed. Suppose the jobs arrive at the flow line according to a renewal process. The mean and the squared coefficient of variation of the interarrival times are $1/\lambda$ and C_a^2, respectively. The service times at stage i form a renewal sequence $\{S_k^{(i)}, k = 1, 2, \ldots\}$, $i = 1, \ldots, m$. The service times and the arrival process are all mutually independent. Let $1/\mu_i$ and $C_{S_i}^2$ be the mean and the squared coefficient of variation of the service time at stage i, $i = 1, \ldots, m$. Under certain conditions on the service time distributions and on the arrival process, we can use the results for the exponential case as bounds for this system. Let $\hat{S}^{(i)}$ be an exponential random variable with mean $1/\mu_i$. Then if $S^{(i)}$ has a new better than used (new worse than used) in expectation distribution, one has $S^{(i)} \leq_{cx} (\geq_{cx}) \hat{S}^{(i)}$. Then from the results in Section 5.4.4, one sees that if $S^{(i)}$ is new better than used (new worse than used) in expectation for all $i = 1, \ldots, m$ and the interarrival times are also new better than used (new worse than used) in expectation then

$$\sum_{j=1}^{i} E[N_j] \leq (\geq) \sum_{j=1}^{i} E[N_j^{(e)}] \tag{5.96}$$

and

$$TH \geq (\leq) TH^{(e)} \tag{5.97}$$

where the superscript $^{(e)}$ is used to indicate that the performance measures are for a flow line with Poisson arrival process and exponentially distributed service times.

Because the exact analysis of the general flow line is very difficult, we will develop a decomposition approach to obtain approximate performance measures. In this decomposition approach each stage will be analyzed in isolation as a $GI_i/GI_i/1/b_i + 1$: stopped arrival queueing system. The effective arrival process and the service times will be chosen such that they appropriately reflect upstream and downstream portions of the flow line. Once these two processes are identified (by the mean and squared coefficient of variation) we can use the results in Section 4.3 to find the probability distribution of the number of jobs and the throughput.

Two-Stage Lines.   A two-stage line with a single server per stage is exactly the same as a stopped arrival system with stage 1 corresponding to the arrival process and stage 2 corresponding to the service process. With the production-blocking service protocol the arrival process (service at stage 1) is turned off as soon as there are $b+1$ jobs in the system that have completed service at stage 1, where b is the maximum number of jobs permitted in stage 2. Suppose the mean and squared coefficient of variation of service time at stage i are $1/\mu_i$ and $C_{S_i}^2$ respectively, $i = 1, 2$. Let $\rho = \mu_1/\mu_2$ and $p(n)$ be the steady-state probability that there are n jobs in the system that are processed by stage 1.

Because the throughput of the two-stage line will be given by

$$TH = \mu_1(1 - p(b + 1)) = \mu_2(1 - p(0)) \tag{5.98}$$

substituting the approximations in equation 4.40 for $p(n)$ we get the approximate throughput for $\rho \neq 1$ as

$$TH = \mu_1 \left(\frac{1 - \rho\sigma^b}{1 - \rho^2\sigma^b} \right) \tag{5.99}$$

while from equation 4.43 if $\rho = 1$ and $\mu_1 = \mu_2 = \mu$,

$$TH = \mu \left(\frac{1 + bv}{2 + bv} \right) \tag{5.100}$$

Using the approximation $\hat{N} = \lambda \hat{T}_1$ given by equation 3.142, whence $v = 2/(C_a^2 + C_S^2)$, equation 5.100 becomes

$$TH = \mu \left(\frac{C_{S_1}^2 + C_{S_2}^2 + 2b}{2(C_{S_1}^2 + C_{S_2}^2 + b)} \right) \tag{5.101}$$

Note that for exponential distributions equations 5.99 and 5.101 give the exact result for throughput. Because $TH = \mu_1 SL$, where recall that SL is the service level (Section 4.3), from Tables 4.1 to 4.3 the adequacy of this throughput approximation is clear.

Multiple-Stage Approximate Throughput. Given results for the throughput of a two-stage line, they can be extended to multiple-stage lines in the same way as with exponential service times, provided we can develop results analogous to equations 5.81 and 5.82. We now need two equations corresponding to equations 5.81 and 5.82, however, one for the mean and the other for the squared coefficient of variation.

Consider storage i, $i = 2, \ldots, m$. Suppose $(1/\mu_{i-1u}, C_{S_{i-1u}}^2)$ are the mean and squared coefficient of variation of the input process to stage i when turned on (i.e., not blocked because the stage is full) and $(1/\mu_{id}, C_{S_{id}}^2)$ are the mean and squared coefficient of variation of the output process from stage i when turned on (i.e., not starved because the stage is empty). Let TH_i, $i = 2, \ldots, m$ be the throughput of a two-stage line with stage 1 parameters $(\mu_{i-1u}, C_{S_{i-1u}}^2)$ and stage 2 parameters $(\mu_{id}, C_{S_{id}}^2)$ obtained using either equations 5.99 or 5.100. If there is no blocking by stage $i + 1$, the throughput TH_{i+1} should be equal to μ_{iu}. Because of blocking by stage $i + 1$, however, one would see that $TH_{i+1} \leq \mu_{iu}$. Hence $1/TH_{i+1} - 1/\mu_{iu}$ is the extra delay in the mean interdeparture time from stage i caused by the blocking of stage i by stage $i + 1$. Therefore we set

$$\frac{1}{\mu_{id}} = \frac{1}{\mu_i} + \left(\frac{1}{TH_{i+1}} - \frac{1}{\mu_{iu}} \right), \quad i = 2, \ldots, m - 1 \tag{5.102}$$

with

$$\frac{1}{\mu_{md}} = \frac{1}{\mu_m}$$

Similarly we set

$$\frac{1}{\mu_{1u}} = \frac{1}{\mu_1}$$

$$\frac{1}{\mu_{i+1u}} = \frac{1}{\mu_{i+1}} + \left(\frac{1}{TH_{i+1}} - \frac{1}{\mu_{i+1d}} \right), \quad i = 1, \ldots, m - 2 \tag{5.103}$$

For the recursive equations for the squared coefficient of variation the simplest would be to set

$$C^2_{S_{iu}} = C^2_{S_i}, \quad i = 1, \ldots, m - 1 \tag{5.104}$$

$$C^2_{S_{id}} = C^2_{S_i}, \quad i = 2, \ldots, m \tag{5.105}$$

Numerical results comparing this approximation to the simulated throughputs are given in Table 5.6. For $C^2_{S_i} \leq 1$ this approximation works very well. For $C^2_{S_i} > 1$, this approximation, though it gives good results, can be improved. Such an improved approximation scheme (applicable to the case $C^2_{S_i} > 1$, $i = 1, \ldots, m$) will be presented in Chapter 6. Note that for $C^2_{S_i} = 1$ this approximation is the same as algorithm 5.3.

TABLE 5.6 MULTIPLE STAGE THROUGHPUT WITH
GENERAL SERVICE TIMES

Parameters								TH	
μ_i			$C^2_{S_i}$			b_i			
1	2	3	1	2	3	2	3	Sim.	App.
0.5	0.5	0.5	0.2	0.2	0.2	2	2	0.430	0.438
0.5	0.5	0.5	0.5	0.5	0.5	2	2	0.382	0.384
0.5	0.5	0.5	0.8	0.8	0.8	2	2	0.351	0.347
0.5	0.5	0.5	2	2	2	2	2	0.296	0.272
0.5	0.5	0.5	0.2	2	0.2	2	2	0.360	0.320
0.5	0.5	0.5	2	0.2	2	2	2	0.320	0.331

5.4.5 Finite Buffer Tandem Queueing Models: Multiple Stations per Stage with Exponential Processing Times

We will now analyze flow-line models with multiple stations per stage, finite buffer spaces, and exponentially distributed processing times. There is an infinite supply of raw material available in front of the first stage. There are c_i parallel stations at stage i that are functionally identical with a processing rate of μ_i, $i = 1, \ldots, m$. The buffer capacity at stage i (including the c_i service positions) is b_i (i.e., $b_i \geq c_i$). Let $N_i(t)$ be the number of jobs processed by stage $i - 1$ and that have not yet completed processing by stage i at time t, $i = 2, \ldots, m$. Then $\{\mathbf{N}(t), t \geq 0\}$ is a Markov process, where $\mathbf{N}(t) = (N_2(t), \ldots, N_m(t))$. Therefore in principle the stationary distribution of $\mathbf{N}$ can be computed. Because the number of equations that needs to be solved to obtain this stationary distribution could be large, one may encounter difficulties in the computation of their solution. Consequently we will focus our attention on developing an approximation procedure to compute the throughput of this system. We will use the performance of a two-stage system as a building block for this approximation procedure for computing the throughput of a m-stage flow-line system. Therefore we will first analyze a two-stage flow line.

Two-stage flow lines.   For the two-stage system we only need to consider the number of jobs processed by stage 1 stations that have not yet completed processing by the stations in stage 2 at time t, $t \geq 0$. Let this number be $N(t)$. Then $\{N(t), t \geq 0\}$ is a birth-death process on the state space $\mathcal{S} = \{0, 1, \ldots, b + c_1\}$. The birth rate is $\mu_1 \min\{c_1, b + c_1 - n\}$, and the death rate is $\mu_2 \min\{c_2, n\}$ when $N(t) = n$, $n = 0, \ldots, b + c_1$. Therefore the stationary distribution of N is given by

$$
p(n) = \begin{cases} p(0) \left(\frac{\mu_1}{\mu_2}\right)^n \frac{c_1^n}{n!}, & 0 \leq n \leq c_2 \\ p(0) \left(\frac{\mu_1}{\mu_2}\right)^n \frac{c_1^n}{c_2! c_2^{n-c_2}}, & c_2 + 1 \leq n \leq b \\ p(0) \left(\frac{\mu_1}{\mu_2}\right)^n \frac{c_1^b c_1!}{c_2! c_2^{n-c_2}(b + c_1 - n)!}, & b + 1 \leq n \leq b + c_1 \end{cases} \tag{5.106}
$$

where

$$
p(0) =
$$

$$
\left[\sum_{n=0}^{c_2} \left(\frac{\mu_1}{\mu_2}\right)^n \frac{c_1^n}{n!} + \sum_{n=c_2+1}^{b} \left(\frac{\mu_1}{\mu_2}\right)^n \frac{c_1^n}{c_2! c_2^{n-c_2}} + \sum_{n=b+1}^{b+c_1} \left(\frac{\mu_1}{\mu_2}\right)^n \frac{c_1^b c_1!}{c_2! c_2^{n-c_2}(b + c_1 - n)!} \right]^{-1}
$$

Then the throughput of this system is given by

$$
TH(c_1, c_2; \mu_1, \mu_2, b) = \sum_{n=1}^{b+c_1} \min\{n, c_2\} \mu_2 p(n)
$$

$$
= \sum_{n=0}^{b+c_1} \min\{b + c_1 - n, c_1\} \mu_1 p(n)
$$

If the stations at stage 2 never idle, then the interdeparture time of jobs from the system should be $1/(c_2 \mu_2)$. The actual mean of the interdeparture times will be $1/TH$. The difference $1/TH - 1/(c_2 \mu_2)$ can be interpreted as the average time needed to deliver a job to a station at stage 2 that has just completed processing. If $I(c_1, c_2, \mu_1, \mu_2, b)$ is the fraction of time the delivery of jobs to the station at stage 2 is delayed causing those stations to idle, the average time to delivery is $I(c_1, c_2, \mu_1, \mu_2, b)/(c_1 \mu_1)$. Therefore

$$
I(c_1, c_2, \mu_1, \mu_2, b) = c_1 \mu_1 \left[\frac{1}{TH(c_1, c_2, \mu_1, \mu_2, b)} - \frac{1}{c_2 \mu_2} \right] \tag{5.107}
$$

Similarly the aggregate blocking probability is

$$
B(c_1, c_2, \mu_1, \mu_2, b) = c_2 \mu_2 \left[\frac{1}{TH(c_1, c_2, \mu_1, \mu_2, b)} - \frac{1}{c_1 \mu_1} \right] \tag{5.108}
$$

We will use these two quantities to develop the approximation for a m-stage flow-line system.

Multiple-stage flow lines.   Let μ_{iu} be the effective service rate of a station at stage i given that it will not be blocked and μ_{id} be the effective service rate of a station at stage i if it will not be starved. Then

$$\mu_{1u} = \mu_1 \text{ and } \mu_{md} = \mu_m$$

Following the relationship in equations 5.81 and 5.82 we derived for the single-station per stage flow line system we set

$$\frac{1}{c_i \mu_{iu}} = \frac{1}{c_i \mu_i} + \frac{1}{c_{i-1}\mu_{i-1u}} I(c_{i-1}, c_i, \mu_{i-1u}, \mu_{id}, b_i), \qquad i = 2, \ldots, m \qquad (5.109)$$

and

$$\frac{1}{c_{i-1}\mu_{i-1d}} = \frac{1}{c_{i-1}\mu_{i-1}} + \frac{1}{c_i \mu_{id}} B(c_{i-1}, c_i, \mu_{i-1u}, \mu_{id}, b_i), \qquad i = 2, \ldots, m \qquad (5.110)$$

Note that for $c_i = 1$, $i = 1, \ldots, m$ the preceding equations are exactly the same as equations 5.81 and 5.82. The following iterative algorithm computes the solution to the preceding set of equations much the same way as in the single server case.

Algorithm 5.4   Multiple Server Finite Buffer Flow Line—Exponential

Step 1: Set $\mu_{iu} = \mu_1$; $\mu_{id} = \mu_i$, $i = 2, \ldots, m$.
Step 2: For $i = 2, \ldots, m$
$\quad 1/(c_i \mu_{iu}) = 1/(c_i \mu_i) + I(c_{i-1}, c_i, \mu_{i-1u}, \mu_{id}, b_i)/(c_{i-1}\mu_{i-1u})$.
Step 3: For $i = m, \ldots, 2$
$\quad 1/(c_{i-1}\mu_{i-1d}) = 1/(c_{i-1}\mu_{i-1}) + B(c_{i-1}, c_i, \mu_{i-1u}, \mu_{id}, b_i)/(c_i \mu_{id})$.
Step 4: If $|TH(c_1, c_2; \mu_1; \mu_{2d}, b_2 + 1) - TH(c_{m-1}, c_m; \mu_{m-1u}, \mu_m, b_m + 1)| < \epsilon$, set
$\quad \lambda^* = TH(c_1, c_2; \mu_1, \mu_{2d}, b_2 + 1)$ and stop.
$\quad$ Otherwise go to step 2.

Table 5.7 shows a numerical comparison between this algorithm and simulation.

5.5 PRODUCE-TO-STOCK FLOW LINES

In the flow-line systems considered earlier, it is assumed that jobs are produced at the maximum rate possible. This, although it helped us to identify the throughput of the flow line, may not be true in practice. Particularly when the demand rate for these jobs is less than the throughput, producing jobs at the maximum rate will result in an unlimited number of finished jobs at the output of the flow line. In reality therefore the actual production of jobs will be controlled based on the demand. *Produce-to-order* flow

TABLE 5.7 MULTIPLE-SERVER FINITE BUFFER
APPROXIMATION

System parameters											TH	
μ_i				c_i				$z_i = b_i - c_i$				
1	2	3	4	1	2	3	4	2	3	4	Sim.	App.
1	1	1		2	2	2		5	5		1.705	1.693
1	2	2		2	1	1		5	5		1.685	1.666
1	1	1		4	4	4		2	2		3.291	3.259
1	1	1	1	2	2	2	2	2	2	2	1.499	1.478
1	2	1	2	3	3	3	3	2	2	2	2.707	2.793
1	2	2	1	3	3	3	3	2	2	2	2.781	2.907

lines can be modeled by the open flow line discussed earlier with job or raw material arrivals corresponding to job-order arrivals. In *produce-to-stock* flow lines there is an output buffer (store) from which demands are met. This buffer has capacity b_{m+1}, and in produce-to-stock operation production is stopped at stage m as soon as the number of finished jobs in the output buffer of the flow line reaches b_{m+1}. Production at stage m is resumed as soon as the number of finished jobs in the output buffer becomes $b_{m+1} - 1$. The production at other stages is controlled according to either production or communication-blocking protocol. Thus, in the case of a long time between demand arrivals, the buffer spaces at all stages may get filled up. The choice of the value b_{m+1} of course will depend on the cost of carrying the finished jobs and the cost of backlogging, if backlogging is allowed or otherwise the cost of lost demands. Models for the case of *no backlogging* and the case of *backlogging permitted* will be considered in the next two subsections.

5.5.1 No Backlogging

In this section we assume that demands are not backlogged. That is, if a demand arrives when the output buffer is empty, the demand is lost. Suppose the demand arrives according to a Poisson process with rate λ. Then we may view the output and the demand process as stage $m + 1$ with buffer capacity b_{m+1} and service rate $\mu_{m+1} = \lambda$. The service protocol used at stage m of this model is communication blocking, whereas the service protocols used in the other stages are the same as that in the original flow line. Hence the approximation algorithms given for the flow lines can be adapted for this model. Let $\hat{\lambda}_{m+1}$ and $C_{S_m}^2$ be the rate and squared coefficient of variation of the arrival process to stage $m + 1$ generated by stage m. Then the number of finished jobs in the output buffer of the flow line is determined by a $GI_{m+1}/M/1/b_{m+1}$ queueing system with parameters $\hat{\lambda}_{m+1}$, $C_{S_m}^2$, for arrival process and λ for service rate. Let N be the stationary number of customers in this queueing system. Then the fraction of demand met is $P\{N \geq 1\}$, and the average inventory in the output buffer is $E[N]$. Hence the optimal value of b_{m+1} can be determined from the preceding results and an appropriate cost model.

When the demand arrival process is general we may be tempted to model the output buffer by a $GI_{m+1}/GI_a/1/b_{m+1}$ queueing system. Note that the arrival time of the first customer after the replenishment of one job to an empty output buffer, however, need not be the same as the interarrival time of demands. Hence the first service time after idling in the $GI_{m+1}/GI_a/1/b_{m+1}$ queue must be different from the interarrival time of demands. Thus we could consider a $GI_{m+1}/GI_a/1/b_{m+1}$ queueing system with exceptional service for the first customer to enter an idle server. Alternatively, we may look at the number of empty spaces (i.e., b_{m+1}−number of finished jobs) at the output buffer of the flow line. The number of empty spaces behaves like the number of jobs in $GI/GI/1/b_{m+1}$: lost arrival queue. The arrival parameters are λ and C_a^2 and the service parameters are approximately $\hat{\lambda}_{m+1}$ and $C_{S_m}^2$, where $\hat{\lambda}_{m+1}$ (which is less than the production capacity) needs to be determined. The throughput λ^* is obtained assuming that the final stage m can process jobs at the maximum rate μ_m. Because of the finite output buffer, however, the actual rate will be less than μ_m. Specifically looking at the final buffer by the preceding queueing system, we may approximate the effective service rate $\hat{\mu}_m$ of stage m by

$$\frac{1}{\hat{\mu}_m} = \frac{1}{\mu_m} + \frac{C_a^2 + 1}{2\lambda} I(\lambda, C_a^2, \hat{\lambda}_{m+1}, C_{S_m}^2, b_{m+1}) \tag{5.111}$$

Here $I(\lambda, C_a^2, \hat{\lambda}_{m+1}, C_{S_m}^2, b_{m+1})$ is the probability that the server is idle at an arbitrary time in a $GI/G/1/b_{m+1}$ lost arrival queue. The algorithm for the general flow line can now be modified to include equation 5.111 and solve for $\hat{\lambda}_{m+1}$.

5.5.2 Backlogging Permitted

We will consider the situation in which demands arriving to find an empty output buffer are backlogged and when a finished job becomes available, the backlogged demand is filled. For this case we use the same approximation methods discussed for the no-backlogging case but solve the model on the assumption that the buffer capacity at stage $m + 1$ is unlimited. Let $\hat{\lambda}_{m+1}$ and $\hat{C}_{a_{m+1}}^2$ be the mean and squared coefficient of variation of the finished job arrival process to the output buffer (obtained by either one of the two methods described earlier). If N is the number of jobs in a $GI/GI/1$ queueing system with arrival parameters λ, C_a^2 and service parameters $\hat{\lambda}_{m+1}$ and $\hat{C}_{a_{m+1}}^2$, then $(N - b_{m+1})^+$ is the number of demands backlogged and $(b_{m+1} - N)^+$ is the number of finished jobs that are in the output buffer of the flow line.

5.6 FLOW-LINE SYSTEMS WITH CLOSED-LOOP MATERIAL HANDLING

Consider a flow line in which jobs are moved from one stage to the next by a closed-loop material handling system such as a closed-loop conveyor. In such a system the total number of jobs (or the pallets carrying these jobs) allowed at any given time is limited (say, to n). In addition the number of jobs allowed at any stage is limited (say, b_i at stage i including the job in service, if any, $i = 1, \ldots, m$). A job processed by stage m is removed from the pallet, and raw material is then placed on that pallet. The pallet is then moved on to stage 1.

The behavior of the closed-loop flow line is very much affected by the number of pallets or job carriers provided in the system. This behavior can be very much different from an open flow-line system. To highlight these effects of the number of pallets on the throughput of a closed-loop flow-line system we will first consider a two-stage system.

5.6.1 Two-Stage Closed-Loop Flow Lines with Exponential Processing Times

We consider a two-stage closed-loop flow line system with buffer capacities b_1 and b_2 at stages 1 and 2, respectively, and exponentially distributed processing times with mean $1/\mu_i$ at stage i, $i = 1, 2$. The number of pallets in the system is n and without loss of generality assume that $b_1 \geq b_2$.

Suppose $N(t)$ is the number of jobs in the system that have not yet completed processing by stage 1 at time t. Then $\{N(t), t \geq 0\}$ is a birth-death process on the state space $S = \{0, \ldots, n\}$. When $N(t) = k$, there are $n - k$ jobs at stage 2 (for $k \leq b_1$). If we use production-blocking service protocol in the system, however, when $N(t) = b_1 + 1$, one of these $b_1 + 1$ jobs is at stage 2 and therefore the number of jobs at stage 2, when $N(t) = b_1 + 1$ is $n - b_1$. One of these $n - b_1$ jobs at stage 2 has completed processing at stage 2 and is waiting to be transferred out of the system. (Note that under communication-blocking service protocol for the system $N(t)$ cannot take the value of $b_1 + 1$). The birth and death rates for the birth-death process $\{N(t), t \geq 0\}$ are then given by

$$\lambda(k) = \mu_2, \quad k = 0, \ldots, \min\{n - 1, b_1\};$$

and all other birth rates are zero, and

$$\mu(k) = 0; \quad k = 0, \ldots, \max\{n - b_2 - 1, 0\},$$

$$\mu(k) = \mu_1, \quad k = \max\{n - b_2, 1\}, \ldots, \min\{n, b_1 + 1\}$$

and all other death rates are zero. Therefore under stationary conditions the number of jobs at stage 1 will only take values in $S' = \{\max\{n - b_2, 0\}, \ldots, \min\{n, b_1 + 1\}\}$. If $p(k) = \lim_{t \to \infty} P\{N(t) = k\}$, $k \in S'$, the production capacity of the system is

$$TH(n) = \mu_1 \sum_{k=\max\{n-b_2, 1\}}^{\min\{n, b_1\}} p(k)$$

From the stationary distribution for the birth-death process it can be verified that

$$
TH(n) = \begin{cases}
\mu_1 \left\{ \dfrac{\left(\frac{\mu_2}{\mu_1}\right)-\left(\frac{\mu_2}{\mu_1}\right)^{n+1}}{1-\left(\frac{\mu_2}{\mu_1}\right)^{n+1}} \right\}, & 1 \leq n \leq b_2 + 1 \\[3ex]
\mu_1 \left\{ \dfrac{\left(\frac{\mu_2}{\mu_1}\right)-\left(\frac{\mu_2}{\mu_1}\right)^{b_2+2}}{1-\left(\frac{\mu_2}{\mu_1}\right)^{b_2+2}} \right\}, & b_2 + 1 \leq n \leq b_1 + 1 \\[3ex]
\mu_1 \left\{ \dfrac{\left(\frac{\mu_2}{\mu_1}\right)-\left(\frac{\mu_2}{\mu_1}\right)^{b_1+b_2+3-n}}{1-\left(\frac{\mu_2}{\mu_1}\right)^{b_1+b_2+3-n}} \right\}, & b_1 + 1 \leq n \leq b_1 + b_2
\end{cases}
$$

Observe that the throughput increases as we increase the number of pallets up until b_2+1. Then the throughput remains the same as long as the number of pallets is between b_2+1 and $b_1 + 1$. Beyond that the throughput *decreases* as we increase the number of pallets. Indeed one can observe that $TH(n) = TH(b_1 + b_2 + 2 - n)$, $2 \leq n \leq b_1 + b_2$. If we use communication blocking we will get

$$
TH(n) = \begin{cases}
\mu_1 \left\{ \dfrac{\left(\frac{\mu_2}{\mu_1}\right)-\left(\frac{\mu_2}{\mu_1}\right)^{n+1}}{1-\left(\frac{\mu_2}{\mu_1}\right)^{n+1}} \right\}, & 1 \leq n \leq b_2 \\[3ex]
\mu_1 \left\{ \dfrac{\left(\frac{\mu_2}{\mu_1}\right)-\left(\frac{\mu_2}{\mu_1}\right)^{b_2+1}}{1-\left(\frac{\mu_2}{\mu_1}\right)^{b_2+1}} \right\}, & b_2 \leq n \leq b_1 \\[3ex]
\mu_1 \left\{ \dfrac{\left(\frac{\mu_2}{\mu_1}\right)-\left(\frac{\mu_2}{\mu_1}\right)^{b_1+b_2+1-n}}{1-\left(\frac{\mu_2}{\mu_1}\right)^{b_1+b_2+1-n}} \right\}, & b_1 \leq n \leq b_1 + b_2
\end{cases}
$$

In this case observe that $TH(n) = TH(b_1 + b_2 - n)$, $0 \leq n \leq b_1 + b_2$. Figure 5.9 shows throughput as a function of the number of pallets for both protocols. As a function

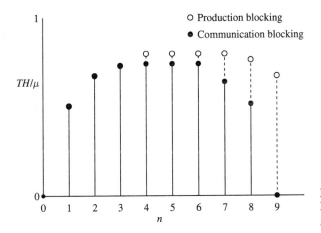

○ Production blocking

● Communication blocking

TH/μ

Figure 5.9 Throughput versus Number of Pallets ($b_1 = 6$, $b_2 = 3$, $\mu_1 = \mu_2 = \mu$)

of the number of pallets the production capacity of this system under communication-blocking service protocol exhibits similar behavior to that under production-blocking service protocol. In particular when the number of pallets in the system is equal to $b_1 + b_2$, the throughput of the system is $TH(b_1 + b_2) = TH(0) = 0$. That is, under communication-blocking service protocol, the system will come to a *deadlock* if we assign $b_1 + b_2$ pallets, that is, if we fill up conveyor space. As we will see in the next section similar behavior is exhibited by the m-stage closed-loop flow-line system as well.

5.6.2 Multiple-Stage Closed-Loop Flow Lines with General Processing Times

Let $b_{max} = \max\{b_i, i = 1, \ldots, m\}$ and without loss of generality assume that $b_1 = b_{max}$. Suppose $\sum_{i=2}^{m} b_i \leq b_{max}$. Then if the total number of jobs n in the system is such that $\sum_{i=2}^{m} b_i \leq n \leq b_1 + 1$ under production blocking neither stage m will be blocked by stage 1 nor will stage 1 be starved by stage m. Therefore the job departure process from each stage will be the same as in an (open) flow line with m stages $1, \ldots, m$ with buffer capacity b_i for stage i, $i = 2, \ldots, m$ and unlimited number of jobs in front of stage 1. Under communication-blocking service protocol, if $\sum_{i=2}^{m} b_i \leq n \leq b_1$, then a similar equivalence exists between the closed and the (open) flow-line systems.

As in the previous flow-line system we focus our attention on only two types of service protocols: production blocking and communication-blocking. We will look at the job-departure process from each of the m stages when the number of servers at each stage is one. Let $S_k^{(i)}$ be the service time of the kth job serviced at stage i and $A_k^{(i)}$ be its arrival time to stage i, $k = 1, 2, \ldots; i = 1, \ldots, m$. Suppose at time zero we have n_i jobs at stage i, $i = 1, \ldots, m$, with $\sum_{i=1}^{m} n_i = n$. Then we set $A_k^{(i)} = 0$, $k = 1, \ldots, n_i \wedge b_i$, $i = 1, \ldots, m$. If $D_k^{(i)}$ is the kth job departure time from stage i we have

$$D_k^{(i)} = \left\{ \max\left\{ A_k^{(i)}, D_{k-1}^{(i)} \right\} + S_k^{(i)} \right\} \vee D_{k-b_{i+1}+n_{i+1}-1}^{(i+1)}, \quad i = 1, \ldots, m \quad (5.112)$$

and

$$A_k^{(i)} = D_{k-(n_i \wedge b_i)}^{(i-1)}, \quad i = 1, \ldots, m \quad (5.113)$$

for production blocking, where the superscript $^{(0)}$ refers to stage m and $(m+1)$ refers to stage 1. In the case of communication blocking we have

$$D_k^{(i)} = \max\left\{ D_{k-n_i}^{(i-1)}, D_{k-1}^{(i)}, D_{k-b_{i+1}+n_{i+1}}^{(i+1)} \right\} + S_k^{(i)}, \quad k = 1, 2, \ldots; i = 1, \ldots, m \quad (5.114)$$

One can use the approach of "inserted idleness" in the reversed system, used in Section 5.4.4, to show the reversibility property of the multistage closed-loop flow line with either production or communication-blocking service protocols. Alternatively, for communication-blocking protocol, the job-hole duality can be used to establish the reversibility property.

Communication blocking and deadlocks. Consider the closed-loop flow line with comunication blocking. Suppose we have a total of $n = \sum_{i=1}^{m} b_i$ jobs in the system. That is stage i has b_i jobs $i = 1, \ldots, m$. Under this condition equation 5.114 for $k = 1$ reduces to

$$D_1^{(i)} = D_1^{(i+1)} + S_1^{(i)}, \quad i = 1, \ldots, m \tag{5.115}$$

Clearly, there is no solution for $D_1^{(i)}$, $i = 1, \ldots, m$, unless $S_1^{(i)} = 0$, $i = 1, \ldots, m$. Therefore $D_1^{(i)} = \infty$, $i = 1, \ldots, m$, indicating that the departures from any stage will never occur. This deadlock situation is easily explained by looking at the communication-blocking service protocol. Here stage i waits for the first service completion of stage $i+1$, which in turn waits for the first service completion at stage $i + 2$ and so forth—waiting forever for the first departure. Consequently we see that the throughput $TH(\sum_{i=1}^{m} b_i)$ is equal to $TH(0)$. Indeed it is easily seen that $TH(n) = TH(\sum_{i=1}^{m} b_i - n)$, $n = 0, \ldots, [(1/2) \sum_{i=1}^{m} b_i]$. The throughput function is symmetric in the total number of jobs allowed in the system, thus reaching a maximum when $n = [(1/2) \sum_{i=1}^{m} b_i]$.

Bounding of production capacity.   From equations 5.112 and 5.114 it can easily be seen that the departure times are increasing and convex in the service times. Then similar to that for the flow line system one has that $S^{(i)}$ is new better than used in expectation, $i = 1, \ldots, m$, then

$$TH(n) \geq TH^{(e)}(n) \tag{5.116}$$

and if $S^{(i)}$ is new worse than used in expectation, $i = 1, \ldots, m$, then

$$TH(n) \leq TH^{(e)}(n). \tag{5.117}$$

5.7 IMPLICATIONS OF MODELS

The overall conclusion from the flow-line models is that variability in processing time can have a significant impact on the performance of flow lines. The variability has several specific implications for their design and operation, and in this section we will summarize these implications for the different types of flow lines.

Indexing lines.   In indexing lines variability results in a loss of productivity and lowered station utilization with unpaced lines. This can, to some extent, be reduced by paralleling provided that combining tasks does not introduce any significant intertask delay. With paced lines variability has an additional impact on quality and the occurrence of defectives because of insufficient time to process a job. Further, it was shown that there is usually a tradeoff between quality and productivity; however, it is possible to have a cycle time that is so short that both quality *and* net production rate would be improved through increasing the cycle time. Aside from the direct impact on quality we also showed that the provision of off-line rework facilities for defects is accompanied by considerable operational difficulties. First of all, any space provided for storing defectives and repaired jobs waiting to get back onto the line will always be full. Further, there will never be enough space), in the sense that some defectives cannot enter the off-line rework area because it is full. Providing more space will reduce the fraction that have to bypass rework but never eliminate it entirely (unless, of course, we provide infinite space), but even then it is still necessary that the cycle time be greater than the average number of defectives arriving during a repair duration.

Asynchronous lines with infinite buffers. Although variability in processing times does not affect the throughput, it does affect the level of work in process and the flow time of jobs. To reduce work-in-process levels it is preferable to assign more work per station to stages where there are parallel stations. Further, the results suggest that if work load is balanced the more variable tasks should be assigned to the end of the line, whereas if all processing times have equal variability then more work should be assigned to the beginning of the line. If work loads have been assigned but the workers have different capability, then the best workers should be assigned to the heavily loaded stations. When there is an inspection station at the end of the line that returns defectives to the first station where a defect occurred, then we showed that throughput is maximized by putting the heaviest loaded stations at the beginning of the line, independent of the defect rates. To minimize the work in process if all stations are equally loaded then the stations with the best processing reliability should be placed at the beginning of the line.

Asynchronous lines with finite buffers. It is possible to prove the very general conclusion that the throughput will increase as any buffer size is increased and also that more variable processing times worsen performance. Further, it is possible to show in the three-station line that it is best to put the fastest station in the middle. Using our approximate throughput calculation procedure we have shown how to determine the optimal work-load allocation, and if it is also possible to allocate a given total buffer space, we have found the jointly optimal work-load and buffer space allocation, with the interesting conclusion that all buffers should have equal size. When all the buffer sizes are equal then the optimal work-load allocation has more work assigned to the first and last station. Models of produce-to-order flow lines can easily be modified to describe produce-to-stock operations.

Closed-loop flow lines. In closed-loop flow lines with finite buffers between stations the major conclusion is that the throughput as a function of the number of work carriers increases to a maximum, corresponding roughly to the number of carriers being equal to half the total number of buffer spaces, then decreases, although there can be a plateau at the maximum where there is a range where throughput is independent of the number of work carriers in that range.

5.8 BIBLIOGRAPHICAL NOTE

The earliest work on modeling flow lines was by R. R. P. Jackson [47] [48] and Hunt [45]. Jackson showed that a multistage multiserver asynchronous line with exponentially distributed service times had a product form solution, whereas Hunt considered asynchronous lines with two or three stages, and a finite buffer or a finite total number of customers in the system. Hillier and Boling [38] then developed methods for solving systems with more than two stages and finite inventory banks, using both numerical techniques and, for exponential processing times, an approximation equivalent to the lost arrival approximation outlined in Section 5.8. They also addressed several design issues and, in particular, recognized the bowl effect [37]. Subsequently, attempts were made to develop approximations for systems with nonexponential processing times, in particular by Knott [52], and to improve on the Hillier and Boling approximation but with little success. Muth [45], however, developed improved numerical techniques for both exponential and Erlang processing times and also in [72] recognized a variety of bounds on performance. The reversibility property was proved (independently) by Yamazaki and Sakasegawa [113], Dattatreya [24], and Muth [73], although it had been conjectured and partly proved by previous authors. The improved approximation for finite buffer multistage lines with exponential processing times is based on approximations used in analyzing transfer lines (see Chapter 6). Systems equivalent to finite buffer flow lines have also been considered in computer system performance modeling, and the recognition of the difference between production blocking and communication blocking is due to Altiok and Stidham [3], and Onvural and Perros [79]. An approximation equivalent to that of Hillier and Boling was developed by Caseau and Pujolle [20] but with the extension to multiple stations per stage. Most attempts at analyzing the performance of closed-loop systems have either relied on the equivalence to open systems when at least one buffer is sufficiently large, or assume that the number of jobs is sufficiently small in relation to buffer capacities that stations are never blocked. An exception to these approximations is the approximation developed in Onvural and Perros [78].

PROBLEMS

5.1 It is required to find the buffer capacity of a two-stage asynchronous flow line that maximizes the net profit per unit time. Assume that there is a gross profit of p per item produced and a cost of k_1 per unit time per unit of buffer capacity and a cost of k_2 per unit time per item in process in the system. If stage i, $i = 1, 2$ has exponential processing time with mean $1/\mu_i$ determine the optimal buffer capacity.

5.2 Consider the off-line repair model of a paced indexing line discussed in Section 5.3.3. Suppose the repair time distribution is geometric with mean $E[R] = 1/r$, that is, $g_R(n) = (1 - r)^{n-1} r$, $n = 1, 2, \ldots$.

    (a) Find the stationary probability distribution of the number of defective jobs in the service facility and the quality (i.e., the fraction of good parts) of the parts going to stage $k+1$.

    (b) Suppose the inventory carrying cost at the repair facility is c_I per unit time, the per unit cost of one unit of buffer space at the repair facility is c_B, and the cost of the defective parts going to stage $k + 1$ is c_D per part. Find the optimal repair facility buffer storage that minimizes the total cost of inventory, buffer storage, and defective parts.

5.3 Consider a m-stage flow line with unlimited buffer capacities and imperfect processing (described in Section 5.4.3). Let p_i be the probability that the processing at stage i is defective, $i = 1, 2, \ldots, m$. Suppose we have two inspection stations assigned to this flow line: one inspection station between stations k and $k+1$ and the other at the output of stage m.

    (a) Obtain an expression for the throughput of this system.

    (b) Find the optimal value of k (i.e., the location of the first inspection station) that will maximize the throughput.

5.4 Consider a multistage asynchronous flow line with m stages, a single station per stage and exponentially distributed processing times with mean $1/\mu$ at all m stages. The buffer capacities of the stages are $b_1 = \infty$, $b_i = b$, $i = 2, \ldots, m$. One approximation that has been suggested for such a flow line is a cyclic queueing model with unlimited buffer capacities, but with a fixed population within the cyclic queue so that the average number of customers at each stage is $b/2$. For two- and three-stage systems compare the throughput given by this approximation with either exact results or the approximations given in Section 5.4.4. Comment on the adequacy of this simple approximation.

5.5 Consider a three-stage flow line with buffer capacities $b_1 = \infty$, $b_i < \infty$, $i = 2, 3$, exponentially distributed processing times with mean $1/\mu_i$ at stage i, $i = 1, 2, 3$, and job flow controlled by communication-blocking service protocol. Let

$$\hat{\mu}_1 = \mu_1(1 - B(\mu_1, \mu_2, b_2))$$

and

$$\hat{TH} = \hat{\mu}_1(1 - B(\hat{\mu}_1, \mu_3, b_3))$$

where $B(\lambda, \mu, b)$ is the blocking probability in a $M/M/1/b$ queueing system with arrival rate λ and service rate μ. $\hat{TH}$ can be used as an approximation for the exact throughput TH of this flow line. Compare this approximation to the exact throughput and show that $\hat{TH} \le TH$.

5.6 Consider a m-stage exponential flow line with buffer capacities $b_i = 1$, $i = 2, \ldots, m$ (i.e., there are no extra storage spaces at any of the stages $2, \ldots, m$) and an unlimited supply of jobs at stage 1. Construct a Markov process that represents the dynamics of the customer flow through the system. Show that the minimal number of states required to construct a Markov process that represents the dynamics of the job flow through this system is given by the Fibonacci number F_{2m-1}, where the Fibonacci numbers satisfy the recursion $F_n = F_{n-1} + F_{n-2}$, $n = 2, 3, \ldots$, and $F_0 = F_1 = 1$.

5.7 In Magazine and Silver [62] an approximation for the throughput of a m-stage exponential flow line with no buffer capacity is proposed. Let $1/\mu_j$, $j = 1, \ldots, m$ be the mean processing time at stage j. Assuming that $1/\mu_j = 1$, $j = 1, \ldots, m$, the approximation is

$$TH_m = \frac{(m-1)TH_{m-1} + F_{2m-3}/F_{2m-1}}{m}$$

with $TH_1 = 1$. Compare the approximation with exact numerical results or with the approximations given by algorithm 5.3.

5.8 Based on extensive numerical studies Knott [51] proposed the following approximation for a two-stage asynchronous flow line with buffer capacity b and where the processing-time distributions are Erlang-k_1 and Erlang-k_2 with identical means equal to $1/\mu$. The approximation is

$$\frac{\mu}{\mu - TH} = 2.5 + (b - 1/2)\frac{2}{1/k_1 + 1/k_2}$$

Compare this formula with the approximation resulting from the use of the $G/G/1$ stopped arrival results (equations 5.99 and 5.100). For $b = 1, 2$ set up and solve the exact equations for predicting throughput and compare with the approximations. (For k_1 and k_2 small this can be done using a spreadsheet program such as Lotus 1-2-3.)

5.9 By modifying some of the state transition equations in a three-exponential–stage flow line Hatcher [34] was able to obtain a closed-form solution for the line throughput. If all stages are identical with mean processing time μ and the buffer capacities are b (including the job on a machine) then his result is that

$$TH = \mu \left\{ 1 - \frac{b + 3 - (1/2)^b}{b^2 + 4b + 5/2} \right\}$$

Compare this result with either exact results or the approximation given in Section 5.4.4.

5.10 Consider a two-stage asynchronous flow line. Stage 1 has deterministic processing time equal to w_1, and stage 2 has exponential processing time with mean w_2. The buffer capacity (including the job on stage 2) is b. Suppose it were technically possible to allocate a given total work content of $W = w_1 + w_2$ between the two stages.

(a) Determine an expression for the throughput for given w_1 and w_2.

(b) Determine for a given W how the throughput depends on the ratio $r = w_1/w_2$. Show the optimal throughput is obtained for $r = r^* > 1$ and show that r^* is decreasing in b. (This is a simple example of variability imbalance with more work being allocated to the less variable station.)

5.11 This problem (based on Rao [87]) is intended to provide insight into the effects of two different types of imbalance in a three-stage asynchronous flow line. Suppose there is no storage space between the stations. Stage processing times are either exponential (M) or deterministic (D). There is a given total work load W that can be allocated to the three stages.

(a) Develop expressions for the throughput for the following arrangements of the stages: MMM, MMD, MDM, DDM, DMD.

(b) For each of the cases considered in part a determine the work-load allocation to the three stages that maximizes the throughput.

(c) In the cases considered in part b characterize the allocations by whether they are examples of the "bowl effect" (i.e., more work is allocated to the end stages than the middle stage), "variability imbalance" (i.e., more work is allocated to the less variable stage), or both. In this example which of the two effects appears to be more significant?

5.12 A fixed cycle flow line is operated as a produce-to-stock system. The output store has capacity z and demand arrives as a Poisson process with rate λ. The line consists of three stations and the processing time distribution of the tasks at each station are identical normal distributions with mean $1/\mu$ and variance σ^2. The line cycle time is set at $\tau = 1/\mu + k\sigma$. The line initially has a job at each station, and it is only operated when there is a job available for input to the line (i.e., the line will always have one job at each station).

(a) Determine the fraction of demand met from stock and the average number of demands backlogged.

(b) If inventory in the output store costs k_1 per item per unit time and back-orders cost k_2 per item back-ordered per unit time determine the optimal inventory target z^*.

(c) Investigate the nature of the relationship between z^* and k (and hence Q the quality of the outgoing product).

(d) Assume that defectives result in the arrival rate of demands being increased from a basic rate of λ^* to $\lambda = \lambda^*/(1 - Q)$. Develop a means of finding the optimal k and z^* combination.

5.13 Suppose manufacture of a product requires three operations. The duration of the tasks required for each operation has identical distributions that are normal with mean $1/\mu$ and variance σ^2. The product is to be produced on a flow line, but it is necessary to compare an unpaced indexing line with a paced indexing line. In the paced indexing line the quality target is $Q = 0.95$. Jobs arrive at the line with mean rate λ and squared coefficient of variation C_a^2. One job is always kept at each station, thus indexing does not occur unless there is a job waiting to begin processing.

(a) Suppose the arrival rate λ can be chosen, but C_a^2 is fixed for all possible λ. Determine the average flow time of a job in each system as a function of λ, and compare the performance of the two systems.

(b) Suppose now it is required that the average flow time in each system be identical. Compare the corresponding arrival rates. Under what circumstances is the unpaced system substantially better than the paced system? Why would one prefer the paced system?

5.14 Consider a paced fixed cycle flow line with m stations. The tasks have been allocated to the stations so that the inherent work content is the same. Suppose that the work content has mean $1/\mu$ and standard deviation σ. The workers for the line, however, are drawn at random from a population in which ability A is a random variable with mean 0 and standard deviation σ_A. This means that the time required by a worker whose ability level is a to perform a task will have mean $1/\mu + a$ and standard deviation σ. If $\sigma_A/\sigma = r$ (typically about 5 for assembly tasks [93]) determine the output quality level if the cycle time is set at $1/\mu + k\sigma$. Consider $m = 2, 3, 4$. Comment on the implications of your results.

5.15 Consider the model of the off-line repair of defective jobs considered in Section 5.3.3. Suppose the repair time is exponential with mean $1/\mu$. Suppose that there is a cost of k_1 per unit of space provided for off-line repair of defective jobs. Defects that are not able to enter the repair station and have to be repaired at the end of the line cost k_2 per defect repaired. Determine the optimal amount of space to provide at the repair station.

5.16 In an m-station asynchronous line with imperfect processing (described in Section 5.4.3) it has been decided to have two inspection stations, one at the end of the line and one at an intermediate location between stations k and $k+1$. At each station there is only one worker and the line is balanced so that $w_i = w$, $i = 1, \ldots, m$. Characterize and develop a procedure to find the optimal location of the inspection station that will minimize the work-in-process inventory.

5.17 Consider the infinite buffer tandem queueing model described in Section 5.4.2. Suppose we choose the following approximation:

$$\hat{T}_{GI/G/1}(\lambda, C_a^2, \mu, C_S^2) = \frac{\rho^2(1 + C_S^2)(C_a^2 + \rho^2 C_S^2)}{(1 + \rho^2 C_S^2)(2\lambda(1 - \rho))} + \frac{1}{\mu}$$

and

$$C_d^2(\lambda, C_a^2, \mu, C_S^2) = (1 - \rho^2)C_a^2 + \rho^2 C_S^2,$$

where $\rho = \lambda/\mu$. Assume that the variability of the processing times at each of the m processing times are the same, that is, $C_{S_i}^2 = C_S^2$, $i = 1, \ldots, m$. The work load w_i and the station rating r_i need not be the same for all values of i. Let $\rho_i = \lambda w_i / r_i$ be the use of station i, $i = 1, \ldots, m$. Show that if $C_a^2 \geq C_S^2$, then an optimal sequence of work stations that minimizes the approximated mean flow time will satisfy $\rho_{\pi^*(1)} \leq \rho_{\pi^*(2)} \leq \cdots \leq \rho_{\pi^*(m)}$. Also show that if $C_a^2 \leq C_S^2$, then an optimal assignment π^* of stations that minimizes the approximated mean flow time will satisfy $\rho_{\pi^*(1)} \geq \cdots \geq \rho_{\pi^*(m)}$.

5.18 Suppose we have a m-stage flow line with buffer capacities $b_1 = \infty$ and $b_i < \infty$, $i = 2, \ldots, m$. Now consider a dual flow line, where a job moves from stage i to stage $i + 1$ whenever a job in the original system moves from stage $m - i$ to $m + 1 - i$, $i = 1, \ldots, m - 1$. A job is taken from the input buffer for processing by stage 1 in the dual system whenever a job is moved from stage m to the output buffer in the original system. Similarly a job from the final stage m is moved to the output buffer of the dual system whenever a job is taken from the input buffer for processing by stage 1 of the original system.

    (a) Show that the sum of the number of jobs in stage i of the original system and the number of jobs in stage $m + 1 - i$ of the dual flow line is equal to b_i, $i = 2, \ldots, m$, provided we start with this number at time zero.

    (b) Consider the original system and imagine that each unoccupied space is occupied by virtual jobs (called *holes*). Each time a job moves forward a hole will move backward. Show that the dynamics of the holes (in the reversed direction) is the same as the dual system.

    (c) When the job flow in the original system is controlled by communication blocking, show that the dynamics of the dual system conforms to the dynamics of the reversed flow line with communication-blocking service protocol. Use this to prove the reversibility of the original flow line.

    (d) When the job flow through the original system is controlled by production-blocking protocol, show that the dynamics of the reversed flow line with production blocking (and hence this duality) cannot be used to prove the reversibility of the original flow line.

5.19 Consider a m-stage flow line with c_j parallel servers, b_j buffer, spaces and exponentially distributed processing times with mean $1/\mu_j$, $j = 1, 2, \ldots, m$. Assume that $b_1 = \infty$ and $b_j < \infty$, $j = 2, \ldots, m$. Keeping $c_j \mu_j = \eta_j$ a constant for each j, compute the throughput of this flow line for increasing values of the c_j's. Is there a trend in the throughput as a function of the c_j's? If there is one explain why.

5.20 Consider the closed-loop flow line discussed in Section 5.6.2. Show the reversibility property of this flow line using (1) the approach of "inserted idleness" when the production-blocking service protocol is used and (2) using job-hole duality when the communication-blocking service protocol is used.

5.21 Consider the closed-loop flow line layout shown in Figure 5.10. When a job is completed at stage i, pallets circulate in the loop containing i in the direction indicated. For $i = 2, \ldots, m - 1$, stage i draws a pallet from the preceding store and places its just completed job in the following store. If the preceding store contains no jobs that have been processed by stage $i - 1$ or the following store contains only jobs that have been processed by stage i then stage i is blocked, and no circulation of pallets occurs. (For example, in Figure 5.10 stage 2 would be blocked if the previous store contained only jobs last processed at stage 4 or if the following store contained only jobs last processed at stage 2). For $i = 1$ and $i = m$ the pallets circulate through only one store, thus stage 1 is blocked once the store

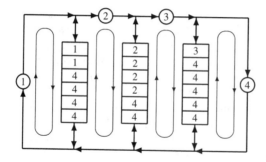

Figure 5.10   Closed-Loop Flow Line with Pallets

is full of jobs that have been processed by stage 1 (i.e., it contains no jobs processed by stage m). Similarly stage m is blocked when the store contains only jobs processed by stage m. Each store can hold $(b-1)$ pallets and the total number of pallets in the system is $(m-1)(b-1)+m$. Processing time at stage i has an exponential distribution with parameter μ_i, $i = 1, \ldots, m$.

     (a) Construct a Markov process for the dynamics of the job flow through the m stages.

     (b) Show that the system has the same behavior as an ordinary series flow line with the maximum number of jobs at stage i equal to b, $i = 2, \ldots, m$.

5.22 **(a)** A two-stage closed-loop flow line has buffers between stages 1 and 2, and between stages 2 and 1. The capacity of each buffer is identical. If the number of pallets in the system is equal to $b + 1$, show that it is possible to locate the buffers so the amount of storage space over and above the job at each station is $b - 1$ and throughput is unchanged.

     (b) Consider now the extension of the idea of part (a) to a m-stage closed-loop system. Suppose the buffer capacities b_i are such that $b_i = b$, $i = 2, \ldots, m$, and $b_1 = \sum_{i=2}^{m} b_i$. Show that it is possible to locate the buffers so that the total storage space other than the space for one job at each station is $b_1 - m$.

     (c) If all m stations in (part b) have identical processing time distributions would you conjecture that the your answer to (part b) would give the optimal throughput for the total storage space of $b_1 - m$?

BIBLIOGRAPHY

[1]   T. ALTIOK. Approximate analysis of exponential tandem queues with blocking. *Eur J of Opnl Res*, 11:390–398, 1982.

[2]   T. ALTIOK. Approximate analysis of queues in series with phase-type service times and blocking. *Opns Res*, 37:601–610, 1989.

[3]   T. ALTIOK and S. STIDHAM, JR. The allocations of interstage buffer capacities in production lines. *IIE Transactions*, 15:292–299, 1983.

[4]   D. R. ANDERSON and C. L. MOODIE. Optimal buffer storage capacity in production line systems. *Int. J. Prod. Res.*, 7:233–240, 1969.

[5]   D. R. ANDERSON and C. L. MOODIE. A quick way to calculate minimum in-process inventory. *Industrial Engineering*, 1:50–51, 1969.

[6]   B. AVI–ITZHAK. A sequence of service stations with arbitrary input and regular service time. *Management Science*, 11:565–571, 1965.

[7]   B. AVI–ITZHAK and M. YADIN. A sequence of two servers with no intermediate queue. *Management Science*, 11:553–564, 1965.

[8]   K. BARTEN. A queueing simulator for determining optimum inventory levels in a sequential process. *Journal of Industrial Engineering*, 13:245–252, 1962.

[9]   R. N. BASU. The interstage buffer storage capacity of non-powered assembly lines: a simple mathematical approach. *Int. J. Prod. Res.*, 15:365–382, 1977.

[10] P. C. BELL. The effect of restricted access to intermediate product inventory in two-stage production lines. *Int. J. Prod. Res.*, 21:75–85, 1983.

[11] P. C. BELL. The use of decomposition techniques for the analysis of open restricted queueing networks. *Operations Research Letters*, 1:230–235, 1982.

[12] D. E. BLUMENFELD. Performance characteristics of assembly systems with fixed and flexible processing times. *Int. J. FMS*, 1:207–222, 1989.

[13] D. E. BLUMENFELD. A simple formula for estimating throughput of serial production lines with variable processing times and limited buffer capacity. *Int. J. Prod. Res.*, 28:1163–1182, 1990.

[14] A. BRANDWAJN and Y. L. JOW. An approximation method for tandem exponential queues with blocking. *Operations Research*, 36:73–83, 1988.

[15] E. BUFFA. Pacing effects in production lines. *J. Ind. Eng.*, 12:383–386, 1961.

[16] G. M. BUXEY and D. SADJADI. Simulation studies of conveyor paced assembly lines with buffer capacity. *Int. J. Prod. Res.*, 14:607–624, 1976.

[17] G. M. BUXEY, N. D. SLACK, and R. WILD. Production flowline systems design—a review. *AIIE Trans.*, 5:37–48, 1973.

[18] J. A. BUZACOTT. The role of inventory banks in flow-line production systems. *Int. J. Prod. Res.*, 9:425–435, 1971.

[19] C. A. CARNALL and R. A. WILD. The location of variable work stations and the performance of production flow lines. *Int. J. Prod. Res.*, 14:703–710, 1976.

[20] P. CASEAU and G. PUJOLLE. Throughput capacity of a sequence of queues with blocking due to finite capacity. *IEEE Trans. on Software Engineering*, SE-5:631–642, 1979.

[21] W.-M. CHOW. Buffer capacity analysis for production lines with variable processing times. *Int. J. Prod. Res.*, 25:1183–1196, 1987.

[22] R. CONWAY, W. MAXWELL, J. O. MCCLAIN, and L. J. THOMAS. The role of work-in-process inventory in serial production lines. *Operations Research*, 36:229–241, 1988.

[23] E. M. DAR-EL and A. MAZE. Predicting the performance of unpaced assembly lines when one station variability may be smaller than the others. *Int. J. Prod. Res.*, 27:2105–2116, 1989.

[24] E. S. DATTATREYA. *Tandem queueing systems with blocking.* PhD thesis, Dept. of Industrial Engineering and Operations Research, University of California, Berkeley, 1978.

[25] J. DING and B. S. GREENBERG. Optimal order for servers in series with no queue capacity. *Prob. Eng. Inf. Sci.*, 5:449–462, 1991.

[26] J. DING and B. S. GREENBERG. Bowl shapes are better in tandem queues with finite buffers. *Prob. Eng. Inf. Sci.*, 5:159–169, 1991.

[27] N. A. DUDLEY. *Work measurement, some research studies.* Macmillan, New York, 1968.

[28] N. A. DUDLEY. Work-time distributions. *Int. J. Prod. Res.*, 2:137–144, 1963.

[29] I. T. FRANKS and R. J. SURY. The performance of operators in conveyor-paced work. *Int. J. Prod. Res.*, 5:97–112, 1966.

[30] H. D. FRIEDMAN. Reduction methods for tandem queueing systems. *Operations Research*, 13:121–131, 1965.

[31] A. GHOSAL and S. C. GUJARIA. Note on the best order for queues in series. *Int. J. Sys. Sci.*, 18:1995–1999, 1987.

[32] H. GOODE and S. SALTZMAN. Estimating inventory limits in a station grouped production line. *J. Ind. Eng.*, 13:484–490, 1962.

[33] B. S. GREENBERG and R. W. WOLFF. Optimal order of servers for tandem queues in light traffic. *Management Science*, 34:500–508, 1988.

[34] J. M. HATCHER. The effect of internal storage on the production rate of a series of stages having exponential service times. *AIIE Trans.*, 1:150–156, 1969.

[35] D. K. HILDEBRAND. On the capacity of tandem server, finite queue, service systems. *Operations Research*, 16:72–82, 1968.

[36] D. K. HILDEBRAND. Stability of finite queue, tandem server systems. *Journal of Applied Probability*, 4:571–583, 1967.

[37] F. S. HILLIER and R. W. BOLING. The effect of some design factors on the efficiency of production lines with variable operation times. *Journal of Industrial Engineering*, 17:651–658, 1966.

[38] F. S. HILLIER and R. W. BOLING. Finite queues in series with exponential or Erlang service times—a numerical approach. *Operations Research*, 15:286–303, 1967.

[39] F. S. HILLIER and R. W. BOLING. On the optimal allocation of work in symmetrically unbalanced production line systems with variable operation times. *Management Sci.*, 25:721–728, 1979.

[40] F. S. HILLIER and K. C. SO. The assignment of extra servers to stations in tandem queueing systems with small or no buffers. *Perf. Evaluation*, 10:219–231, 1989.

[41] F. S. HILLIER and K. C. SO. The effect of the coefficient of variation of operation times on the allocation of storage space in production line systems. *IIE Trans.*, 23:198–206, 1991.

[42] F. S. HILLIER and K. C. SO. *On the simultaneous optimization of server and work allocations in production line systems with variable processing times.* Technical Report, Graduate School of Management, University of California, Irvine, CA, 1991.

[43] K. HITOMI, M. NAKAJIMA, and Y. OSAKA. Analysis of the flow–type manufacturing systems using the cyclic queueing theory. *Transactions of the ASME, Journal of Engineering for Industry*, 100:468–474, 1978.

[44] C. C. HUANG and G. WEISS. On the optimal order of M machines in tandem. *Oper. Res. Letters*, 9:299–303, 1990.

[45] G. C. HUNT. Sequential arrays of waiting lines. *Operations Research*, 4:674–683, 1956.

[46] T. IYAMA and S. ITO. The maximum production rate for an unbalanced multi-server flow line system with finite buffer storage. *Int. J. Prod. Res.*, 25:1157–1170, 1987.

[47] R. R. P. JACKSON. Queueing systems with phase type service. *Operational Research Quarterly*, 5:109–120, 1954.

[48] R. R. P. JACKSON. Random queueing processes with phase type service. *J. R. Stat. Soc. series B*, 18:129–132, 1956.

[49] M. D. KILBRIDGE. Length of task and assembly learning: A model for industrial learning costs. *Management Science*, 8:516–527, 1962.

[50] M. D. KILBRIDGE and L. WESTER. An economic model for the division of labor. *Management Science*, 12:B255–B269, 1966.

[51] A. D. KNOTT. *The efficiency of series production lines*. PhD thesis, The University of New South Wales, Kensington, NSW, 1967.

[52] A. D. KNOTT. The inefficiency of a series of work stations—a simple formula. *Int. J. Prod. Res.*, 8:109–119, 1970.

[53] A. D. KNOTT. Letter to the editor re: Hatcher's "The effect of internal storage on the production rate of a series of stages having exponential service times." *AIIE Trans.*, 2:273, 1970.

[54] K. KNOTT and R. J. SURY. A study of work-time distributions in unpaced tasks. *IIE Trans.*, 19:50–55, 1987.

[55] A. G. KONHEIM and M. REISER. A queueing model with finite waiting room and blocking. *J. ACM*, 23:328–341, 1976.

[56] J. KOTTAS and H.-S. LAU. A total operating cost model for paced lines with stochastic task times. *AIIE Trans.*, 8:234–240, 1976.

[57] S. A. KRAEMER and R. F. LOVE. A model for optimizing the buffer inventory storage size in a sequential production system. *AIIE Trans.*, 2:64–69, 1970.

[58] G. LATOUCHE and M. F. NEUTS. Efficient algorithmic solutions to exponential tandem queues with blocking. *SIAM J. Alg. Disc. Math.*, 1:93–106, 1980.

[59] W. LEE, D. TCHA, and G. YAMAZAKI. *Server assignment for multi-stage production systems with finite buffers*. Technical Report, Department of Management Sciences, Korea Advanced Institute of Science and Technology, Seoul, Korea, 1990.

[60] S. LIPPERT. Accounting for prior practice in skill acquisition. *Int. J. Prod. Res.*, 14:285–293, 1976.

[61] R. F. LOVE. A two–station stochastic inventory model with exact methods of computing optimal policies. *Nav. Res. Log. Quart.*, 14:185–217, 1967.

[62] M. J. MAGAZINE and G. J. SILVER. Heuristics for determining output and work allocations in series flow lines. *Int. J. Prod. Res.*, 16:169–181, 1978.

[63] T. MAKINO. On the mean passage time concerning some queueing problems of the tandem type. *J. Opns. Res. Soc. Japan*, 7:17–47, 1964.

[64] J. E. MAZUR and R. HASTIE. Learning as accumulation: a reexamination of the learning curve. *Psychological Bulletin*, 85:1256–1274, 1978.

[65] G. R. MCGEE and D. B. WEBSTER. An investigation of a two-stage production line with normally distributed interarrival and service time distributions. *Int. J. Prod. Res.*, 14:251–261, 1976.

[66] L. E. MEESTER and J. G. SHANTHIKUMAR. Concavity of the throughput of tandem queueing systems with finite buffer storage. *Adv. Appl. Prob.*, 22:764–767, 1990.

[67] B. MELAMED. A note on the reversibility and duality of some tandem blocking queueing systems. *Management Science*, 32:1648–1650, 1986.

[68] A. MISHRA, D. ACHARYA, N. P. RAO, and G. P. SASTRY. Composite stage effects in unbalancing of series production systems. *Int. J. Prod. Res.*, 23:1–20, 1985.

[69]   R. A. MURPHY. Estimating the output of a series production system. *AIIE Trans.*, 10:139–148, 1978.

[70]   K. F. H. MURRELL. Operator variability and its industrial consequence. *Int. J. Prod. Res.*, 1:39–55, 1962.

[71]   E. J. MUTH. Numerical methods applicable to a production line with stochastic servers. *TIMS Studies in the Management Sciences*, 7:143–159, 1977.

[72]   E. J. MUTH. The production rate of a series of work stations with variable service times. *Int. J. Prod. Res.*, 11:155–169, 1973.

[73]   E. J. MUTH. The reversibility of production lines. *Man. Sci.*, 25:152–158, 1979.

[74]   E. J. MUTH and A. ALKAFF. The bowl phenomenon revisited. *Int. J. Prod. Res.*, 25:161–173, 1987.

[75]   E. J. MUTH and A. ALKAFF. The throughput rate of three-station production lines: a unifying solution. *Int. J. Prod. Res.*, 25:1405–1413, 1987.

[76]   M. F. NEUTS. Two queues in series with a finite, intermediate waitingroom. *J. Applied Probability*, 5:123–142, 1968.

[77]   M. F. NEUTS. Two servers in series, studied in terms of a Markov renewal branching process. *Advances in Applied Probability*, 2:110–149, 1970.

[78]   R. O. ONVURAL and H. G. PERROS. Approximate throughput analysis of cyclic queueing networks with finite buffers. *IEEE Trans. Software Engineering*, SE-15:800–808, 1989.

[79]   R. O. ONVURAL and H. G. PERROS. On equivalencies of blocking mechanisms in queueing networks with blocking. *Operations Research Letters*, 5:293–297, 1986.

[80]   S. S. PANWALKER and M. L. SMITH. A predictive equation for average output of k stage series systems with finite interstage queues. *AIIE Trans.*, 11:136–139, 1979.

[81]   R. L. PATTERSON. Markov processes occurring in the theory of traffic flow through an n-state stochastic flow system. *Journal of Industrial Engineering*, 15:188–193, 1964.

[82]   H. G. PERROS and T. ALTIOK. Approximate analysis of open queueing networks with blocking: tandem configurations. *IEEE Trans. on Software Engineering*, SE-12:450–461, 1986.

[83]   S. M. POLLOCK, J. R. BIRGE, and J. M. ALDEN. *Approximation analysis for open tandem queues with blocking: exponential and general service distribution*. Technical Report 85–30, Department of Industrial and Operations Engineering, University of Michigan, Ann Arbor, 1985.

[84]   B. POURBABAI. Approximating the tandem behavior of a $G/M/1$ queueing system. *Int J. Sys. Sci.*, 20:1035–1051, 1989.

[85]   B. POURBABAI. Tandem behavior of an $M/M/1/N \rightarrow G/M/1$ queueing system. *Comput. Math. Applic.*, 16:215–220, 1988.

[86]   T. O. PRENTING and N. T. THOMOPOULOS, editors. *Humanism and Technology in Assembly Line Systems*. Spartan Books, Rochelle Park, NY, 1974.

[87]   N. P. RAO. A generalization of the "bowl phenomenon" in series production systems. *Int. J. Prod. Res.*, 14:437–443, 1976.

[88]   N. P. RAO. On the mean production rate of a two-stage production system of the tandem type. *Int. J. Prod. Res.*, 13:207–217, 1975.

[89]   N. P. RAO. Two-stage production systems with intermediate storage. *AIIE Trans.*, 7:414–421, 1975.

[90]   N. P. RAO. A viable alternative to the "method of stages" solution of series production systems with Erlang service times. *Int. J. Prod. Res.*, 14:699–702, 1976.

[91]   E. REICH. Waiting times when queues are in tandem. *Ann. Math. Stat.*, 28:768–772, 1957.

[92]   E. RICHMAN and S. ELMAGHRABY. The design of in-process storage facilities. *Journal of Industrial Engineering*, 8:7–9, 1957.

[93]   G. SALVENDY and W. D. SEYMOUR. *Prediction and development of industrial work performance*. John Wiley and Sons, New York, 1973.

[94]   B. R. SARKER. Some comparative and design aspects of series production systems. *IIE Trans.*, 16:229–239, 1984.

[95]   J. G. SHANTHIKUMAR, G. YAMAZAKI, and H. SAKASEGAWA. Characterizing optimal ordering of servers in a tandem queue with blocking. *Oper. Res. Letters*, 10:17–22, 1991.

[96]   L. D. SMITH. Allocating inter-station inventory capacity in unpaced production lines with heteroscedastic processing times. *Int. J. Prod. Res.*, 15:163–172, 1977.

[97]   K. C. SO. On the efficiency of unbalancing production lines. *Int. J. Prod. Res.*, 27:717–729, 1989.

[98]   S. SURESH and W. WHITT. Arranging queues in series: a simulation experiment. *Management Science*, 36:1080–1091, 1990.

[99]   R. J. SURY. An industrial study of paced and unpaced operator performance in a single stage work task. *Int. J. Prod. Res.*, 3:91–102, 1964.

[100]   T. SUZUKI. On a tandem queue with blocking. *J. Operations Res. Soc. Japan*, 6:137–157, 1964.

[101]   Y. TAKAHASHI, H. MIYAHARA, and T. HASEGAWA. An approximation method for open restricted queueing networks. *Operations Research*, 28:594–602, 1980.

[102]   S. V. TEMBE and R. W. WOLFF. The optimal order of service in tandem queues. *Operations Research*, 24:824–832, 1974.

[103]   W. W. THOMPSON, JR., and R. L. BURFORD. Some observations on the bowl phenomenon. *Int. J. Prod. Res.*, 26:1367–1373, 1988.

[104]   N. UDOMKESMALEE. *Work-in-process inventory for assembly systems with variable processing times*. Technical Report, GM Research Laboratories, Warren, MI, September 1987.

[105]   N. VAN DIJK and B. F. LAMOND. Simple bounds for finite single-server exponential queues. *Operations Research*, 36:470–477, 1988.

[106]   F. WANG and R. C. WILSON. Comparative analysis of fixed and removable item mixed model assembly lines. *IIE Trans.*, 18:313–317, 1986.

[107]   R. R. WEBER. The interchangeability of tandem $\cdot/M/1$ queues in series. *J. Appl. Prob.*, 16:690–695, 1979.

[108]   W. WHITT. The best order for queues in series. *Management Science*, 31:475–487, 1985.

[109]   R. WILD. *Mass Production Management: The Design and Operation of Production Flow-Line Systems*. John Wiley and Sons, New York, 1972.

[110]   H. YAMASHITA, H. NAGASAKA, H. NAKAGAWA, and S. SUZUKI. *Modeling and analysis of production line operated according to demand*. Working Paper, Sophia University, Tokyo, Japan, 1988.

[111]   H. YAMASHITA and S. SUZUKI. An approximation method for line production rate of a serial production line with a common buffer. *Computers and Operations Research*, 15:395–402, 1988.

[112]   G. YAMAZAKI, T. KAWASHIMA, and H. SAKASEGAWA. Reversibility of tandem blocking queueing systems. *Management Science*, 31:78–83, 1985.

[113]   G. YAMAZAKI and H. SAKASEGAWA. Properties of duality in tandem queueing systems. *Ann. Inst. Statist. Math.*, 27:201–212, 1975.

[114]   G. YAMAZAKI, H. SAKASEGAWA, and J. G. SHANTHIKUMAR. On optimal arrangement of stations in a tandem queueing system with blocking. *Management Science*, 38:137–153, 1991.

[115]   L. E. YELLE. The learning curve: historical review and comprehensive survey. *Decision Sciences*, 10:302–328, 1979.

[116]   H. H. YOUNG. Optimization models for production lines. *Journal of Industrial Engineering*, 18:70–78, 1967.

6

Transfer Lines

6.1 INTRODUCTION

A transfer line consists of several work stations in series integrated into one system by a common transfer mechanism and a common control system. Each station is a stopping point at which one or more machining, assembly, or inspection operations are performed on the work piece or part. The essential distinguishing features of a transfer line are the following:

- Rather than having all operations performed at one location, the parts processed by the line move between stations specialized to one operation or few closely related operations.
- Movement of the parts between stations is performed automatically by some mechanical means.
- Apart from load/unload stations, operations at all other stations are automated.

The first transfer line, consisting of just three stations, appears to have been installed in 1908 for the purpose of making railroad ties. Then in the 1920s and 1930s rotary transfer lines of 6 to 8 stations were developed, although there was one "unsuccessful" in-line transfer line developed and installed by Morris Motors in the United Kingdom in 1924. Subsequently, integrated transfer lines of 60 to 80 stations were developed and came into widespread use in the automotive and other industries from about 1950. For a fuller account of their development see [66].

Transfer lines can be divided into two broad classes: *rotary* and *in-line*. The in-line type can be further subdivided into *straight lines* or *circular lines*. In straight-line systems the part slides on rails; this is common for heavy castings (e.g., engine blocks) in automotive and farm equipment manufacture. Circular transfer lines have a connection between the end and the beginning of the line, usually because parts are mounted on pallets, and the pallets must return to the beginning of the line. Mounting parts on pallets has several advantages.

- Improved part location because of employing a very accurate and rigid pallet that moves from station to station with the part.
- Improved part rigidity during machining operations,
- Greater ease of movement because of employing a regularly shaped pallet rather than an oddly shaped part.

Conversely, pallets have a high initial cost and require maintenance.

There are various transfer mechanisms for moving parts from one station to the next. For in-line systems it is common to use a transfer bar: a long rod that pushes parts simultaneously from one station to the next using retractable fingers. This bar arrangement requires that each leg or group of stations serviced by a single bar be in a straight line. Other transfer mechanisms use a lift-and-carry device.

To achieve correct sequencing and synchronization of the actual operations at each station and the movement of parts from station to station it is necessary to have a common control system for all stations in each leg or section of the line or even for the whole line. The instructions to activate various components are based on input from a network of sensors throughout the line. The logic of such controllers is now usually performed by microprocessors or programmable controllers. Furthermore, inspection and other work stations can be tied together in such a manner that an inspection station can immediately observe the quality of the work just performed at a preceding machining or assembly station. Based on abnormal observations, corrective action can be taken before producing parts that must be scrapped or repaired. Such corrective action can range from visual or audio annunciation of the condition to shutting down the machine or even automatically correcting the condition through tool compensation. This tie between inspection and production can lead to significant quality improvements.

The nature of the linkage between all stations in a section of a transfer line means that *transfer can only begin when the slowest station has completed its operation*. If no station in the section fails while performing its operation then the mean time between successive transfers of parts between stations in the section is known as the *cycle time* of the section. Thus the cycle time will be determined by the maximum time required to perform operations at the stations plus the time required for transfer. The *gross production rate* of the section is the inverse of the cycle time. Note that if any station in the section fails, and thus its operation is not completed, then no transfer will occur, and all stations in the section will be forced down. The extent to which this section stoppage will affect other sections of the line depends on the degree of integrated linkage and control between sections. If no inventory can be kept between sections then the rest of the line will be forced down almost immediately. In some very complex lines hundreds of machines can be forced down by one machine failure. Because of all the individual machines, and the electronics and hardware for common controllers and transfer mechanisms, there is much equipment that can break down. Thus the *net production rate*, the average number of parts actually completed per unit time, can be substantially less than the gross production rate. The dominant concern in transfer line design and operation is achieving adequate reliability and minimizing the impact of machine failures and the subsequent repair.

One way of increasing the productivity of a transfer line and reducing the impact of stoppages is to insert an in-process storage or bank between any two sections of the line. Banks can be either *shunt* (i.e., last-in first-out), or *series* (i.e., first-in first-out). Control and movement of parts in and out of the banks can be either automatic or manual. Automatic banks have the advantage of not requiring human intervention or initiative to bank parts. A bank has the effect of decoupling the two sections of the line, allowing each section to operate independently. A distinction can be made between *fixed transfer*, where all working sections of the line continue to operate in synchronism so that the cycle time of the line and the cycle times of the sections are always the same, and *free transfer*, where a section of the line begins transfer as soon as all stations in the section have completed their operation. That is, with free transfer two sections separated by an inventory bank do not transfer in synchronism, and they can produce at different rates provided the bank is neither full nor empty. A full bank forces the upstream section to slow to the rate of the downstream section, whereas an empty bank forces the downstream section to slow to the rate of the upstream section.

6.2 ISSUES IN TRANSFER LINE DESIGN AND OPERATION

The overall objective in designing and operating a transfer line is to produce a large volume of parts at the lowest possible cost per piece. The relevant costs that should be considered are the initial costs of acquisition and installation, operating costs such as labor, tools, repair and maintenance, and inventory costs for work in process. The line should also have some flexibility so that it is able to respond to changes in the required product mix and volumes.

Although it is usually quite easy to determine the costs associated with a proposed transfer line design, it is often quite difficult to predict the impact on the net production rate of changes in design and operating practices. In seeking to increase the net production rate there are four general approaches that can be used.

1. *Decrease the cycle time (i.e., increase the gross production rate).* Decreasing the cycle time requires reducing the time to transfer the part from one station to the next and reducing the time to perform the operation. Attention should be paid to the slowest station, whereas other stations need only have a cycle time slightly faster than the slowest station.

2. *Decrease the frequency of occurrence of down time.* Decreasing the occurrence of down times is to a large extent a matter of careful attention to the reliability of machine components and to optimization of tool life. Improved monitoring and on-line control of operations is beneficial.

3. *Decrease the duration of down times.* Decreasing the duration of down times is primarily a matter of the use of appropriate management and operating practices rather than the design of the line. In particular it is necessary to ensure that the common repair actions are planned in advance.

4. *Reduce impact of failures on the line.* The linking of stations by a common transfer mechanism into segments and the overall linking of segments to make up the line mean that the impact of failure of any one component can be very great. It is not just the machining stations but also the transfer, hydraulic, lubrication, coolant, control, and inspection systems that impact on the overall line performance.

Unfortunately, an improvement in one of these four areas may have a negative effect on another. For example, decreasing the cycle time through speeding up the machines will increase the wear rate of most tools and machine components, thus increasing the frequency of occurrence of down time.

When seeking to optimize line performance it is important to consider each of these four approaches. The first three are primarily determined by good engineering and operating management so we will not discuss them further, but the fourth requires a systems' perspective so it will be discussed at length.

To minimize the reduction in production during component failures there are several design alternatives that can be used.

Banking. There are a variety of approaches to incorporating banking between transfer mechanisms of two sections. Automatic banks often employ power and free conveyors, elevators, gravity-fed roller conveyors or silos in some combination. Manually operated banks place parts on roller or monorail conveyors, in containers, or simply on the floor. Palletized parts can be banked either on or off their pallets. Off-pallet banking is not always possible because removing the semifinished parts from one pallet and placing them on another for subsequent operations may lead to location errors because of differences between pallets. The high cost of pallets, however, precludes banking of substantial numbers of palletized parts.

Redundant Stations. Redundant stations can be in the form of standby stations that are idle even when all stations are in working order. This, however, is often too expensive to be justified unless the standby is a less automated or primarily manual backup. Alternatively, one can include two stations that share the production and thus have a longer cycle time than the rest of the line. Some fraction of normal production can then be obtained when one station is down. In either case, however, the layout must be designed so that, when one station is down, parts can get to the up station and maintenance can get to the down station.

Cross Paths. The use of cross-paths is a special case of redundancy in which entire sections of parallel transfer lines act as backup to one another. If the first half of one line and the second half of the identical parallel line were down, then a cross-path between their midpoints would allow the system to operate at one-half capacity. Such a cross-path might be in the form of a power-and-free or a timed conveyor. The former affords a banking capacity, whereas the latter is less expensive.

The design of the line will require consideration of each of these options. Although in principle each is relatively simple, extremely complex alternative systems can easily be generated. Further, it is necessary to consider specifics such as the number, location, and size of inventory banks, or the number and characteristics of redundant stations. Adding the fact that these systems operate in a random environment makes the prediction of the net production rate a difficult task.

6.3 NATURE OF LINE STOPPAGES

To develop realistic models of transfer-line performance it is necessary to study the nature of stoppages of the line. There have been several studies of transfer-line down time reported in the literature. Some have been based on the analysis of records of line down time made by line-operating or maintenance personnel. Others have been based on observations made by industrial engineers. The industrial engineering studies would be expected to be more accurate than those based on the records of the line operator. Because most transfer lines operate on a three-shift, 24-hours-a-day, seven-days-a-week basis, however, the cost of continuous observation of the line has resulted in the duration of the industrial engineering studies being relatively short. Most recently installed lines use a computer for control and monitoring line performance, making it possible to record automatically down-time histories including the location and reasons for failure. So in the future far more information should become available on the nature of line stoppages.

In seeking to understand line stoppages it is necessary to classify them by the following:

Extent. Whether it is the *whole line*, a *group of stations*, or an *individual station* that is directly affected by the stoppage. Examples of stoppages affecting the whole line are power and coolant supply failures. Transfer mechanism failures affect a group of stations, and tool wear or breakage affects an individual station.

Cause. Whether the stoppage is either of the following:

- *Operation dependent.* Can only occur when the station or section is processing a part.
- *Time dependent.* Can occur at any time the line (or section) is operational, even though a part is not in process.

Effect. Whether the stoppage requires the part in process at the instant of stoppage to be removed from the line for rework, repair, or scrap. This is relatively uncommon in machining operations but may be a consequence of failures in assembly operations because of out-of-tolerance parts. Sometimes it is possible that the part in process at the instant of stoppage will be more likely to cause further stoppages as it proceeds along the line.

A further categorization is based on how the task must be completed following repair; whether (1) the task can continue from the point reached when failure occurred or (2) the task must start again from the beginning. Irrespective of which alternative applies, however, it seems reasonable to include any extra processing time beyond the normal processing time in the repair time.

The three-way categorization of stoppages by extent, cause, and effect is required to choose the appropriate structure for a model of the effect of stoppages on line performance.

The most comprehensive study of down-time data reported in the literature is that by Hanifin [47]. He observed several transfer lines that were machining transmission cases within Chrysler Corporation. One line was monitored continuously during a 7-day period during which there were 777 stoppages of the line. (Another study based on line-operator records is that of Law, Baxter, and Massara [63].) The line from which Hanifin's data was collected consisted of 76 stations. Stations 1 to 45 made up the rough machining section, and stations 46 to 76 made up the finish machining section. The stoppages could be classified by whether they were time dependent or operation dependent, and also whether they were whole line stoppages, stoppages of a group of stations owing to failures of one or more transfer bars, or stoppages of just one station. Table 6.1 shows the results of this classification.

TABLE 6.1   CLASSIFICATION OF LINE STOPPAGES
USING HANIFIN'S DATA

	Cause		
Extent	Operation dependent (%)	Time dependent (%)	Total (%)
Line	2	9	11
Station group	21	—	21
Single station	62	6	68
Total	85	15	100

It was found that the up times of the rough section had an exponential distribution, whereas the up times of the finish section appeared to be a mixture of two exponential distributions. Down times appear to have a log-normal distribution. This can be explained by the fact that the observed down times were due to a mixture of different repairs. If one particular repair type, (e.g., changing tools) is examined, then it is usually found to have an exponential distribution. The observed down-time distributions can be quite adequately explained as a mixture of three exponential distributions corresponding to the three general classes of repair types: minor adjustments by the operator, tool changes, and repairs of failed machines.

More sophisticated data analysis seeks to identify and explain patterns in the data. Two particular aspects are of interest.

- *Trends with time (or cumulative production).* It is conceivable that there can be a general improvement or deterioration in performance. Superimposed on this there are likely to be differences owing to the day of the week, the shift, or even between operators. Extensive data would be needed to identify these effects.
- *Nonindependence of successive events.* With Hanifin's data it was found that down times were independent, whereas the up times of the rough section were also independent. For the up times of the finish section, however, the autocorrelations for lags 1 to 6 were all significant.

One of the reasons for the nonindependence of the finish section up times is that the precision of machining requires frequent tool changes and subsequent adjustments until the proper setting is obtained. Further, if performance is judged on production per shift, the operators and maintenance personnel may emphasize restarting the line rather than repairing the line. This in turn can lead to successive restarts of the machine, alternating with short run times, before the gravity of the problem is determined and the proper repair is made. This also explains the finish section up times having a mixture of exponential distributions. One could hypothesize a model of times between stoppages in which there is a basic process generating failures, with times between type 1 failures independent and exponentially distributed. Then, because repair of each type 1 failure is difficult, each type 1 failure is followed by a random number of type 2 failures until effective repair is achieved, with the time between type 2 failures being relatively short.

Further analysis of patterns in the data looks at whether up times and down times are mutually independent. The correlation coefficient between up time and next down time was determined for Hanifin's data and found to be significant for the same reasons as advanced earlier.

6.3.1 Modeling Stoppages

If a simulation model is being developed for an actual line from which data on stoppages is available then the observed down-time history can be used in the model. Even so, it is necessary to distinguish types of failures such as those affecting the whole line or those that are time dependent. Changes in line configuration or operating practices will affect such failures differently from operation-dependent or station group failures. Usually, however, there is no down-time history that can be used because the model is being developed for a hypothetical line. It is then necessary to develop a model of stoppages based on experience and observation of similar transfer lines. Such a model will require the development of a clear set of assumptions about types of failure, the distribution of up and down times, and the pattern of failures.

Unless an actual down-time history is used, which therefore might incorporate complex failure patterns, it is invariably assumed in both simulation and analytical models that the pattern of failures satisfies the following assumptions:

1. All distributions are stationary, that is, there are no trends with time or number of failures.
2. Up times are iid.
3. Down times are iid.
4. Up and down times are mutually independent.

There are then two approaches that can be used to incorporate assumptions concerning distributions of up and down times in the model of stoppages. One is to represent specifically the up and down state of a station in the model. This is the approach we will use in this chapter. Alternatively, one can develop models using the completion time distribution, where the completion time is the total time needed to complete processing a part at a station including any necessary repair times (i.e., completion time = cycle time + repair time if repair of station during processing is necessary).

We will look at two distinct monitoring and repair strategies that lead to different completion time distributions. As we will see, however, suitable modification of repair times will allow us to use models developed for one monitoring and repair strategy to be applied to the other strategy as well. The two strategies are as follows:

1. Stations are continuously monitored, and a repair is initiated as soon as the station fails. At the completion of the repair to the station the processing of the part is resumed. Let F_S be the distribution function of the processing time S and F_R be the distribution function of the iid repair times $\{R_n, n = 1, 2, \ldots\}$. Then if the station up times are exponentially distributed with mean $1/\zeta$, the generic completion time C is then given by

$$C = S + \sum_{n=1}^{N(S)} R_n$$

where $N(S)$ is the number of station failures during the processing time S. Conditioning on $S = x$ and observing that $N(x)$ has a Poisson distribution function with mean ζx, it can be seen that the Laplace-Stieltjes transform of the completion time distribution function is given by

$$\tilde{F}_C(s) = E[e^{-sC}] = \tilde{F}_S(s + \zeta(1 - \tilde{F}_R(s))) \tag{6.1}$$

When the processing time is equal to the cycle time and is fixed, that is, $S = \tau$ is a constant, one has

$$\tilde{F}_C(s) = \exp\{-\tau(s + \zeta(1 - \tilde{F}_R(s)))\}$$
$$= (1 - a)e^{-s\tau} + ae^{-s\tau}\tilde{F}_{\hat{R}}(s), \tag{6.2}$$

where

$$a = 1 - e^{-\zeta\tau}$$

is the probability that at least one failure occurs during a cycle time and

$$\tilde{F}_{\hat{R}}(s) = \frac{1}{1 - e^{-\zeta\tau}} \sum_{n=1}^{\infty} \frac{e^{-\zeta\tau}(\zeta\tau)^n}{n!} \left(\tilde{F}_R(s)\right)^n$$

is the Laplace transform of the conditional distribution function of the total repair times of all failures that occurred during the processing of a part given that at least one failure occurred.

2. Station failures are detected only at the end of a cycle, and a repair is initiated as soon as a failure is detected. At the completion of repair to the station the processing of the part is repeated. Let a be the probability that a station fails during a fixed cycle time τ. Then, if F_R is the distribution function of repair time plus the time to reprocess an item,

$$\tilde{F}_C(s) = (1 - a)e^{-s\tau} + ae^{-s\tau}\tilde{F}_{\hat{R}}(s) \tag{6.3}$$

where

$$\tilde{F}_{\hat{R}}(s) = \sum_{n=1}^{\infty}(1 - a)a^{n-1}(\tilde{F}_R(s))^n$$

The following interpretation can then be given for the completion time in either of these two models: A part is processed for a length of time equal to the cycle time. If the station does not fail during this time the part leaves the station. If the station fails during the cycle time, which can happen with probability a, an additional amount of time with distribution $F_{\hat{R}}$ is needed to repair the station and complete processing that part. After this time, the part leaves the station. If I is an indicator random variable independent of $\hat{R}$ and such that $P\{I = 1\} = a$, we have $C = \tau + I\hat{R}$ for both strategies.

When $a \approx 0$, it can be verified that $\tilde{F}_{\hat{R}}(s) \approx \tilde{F}_R(s)$. Then for $a \approx 0$ we have

$$\tilde{F}_C(s) = (1 - a)e^{-s\tau} + ae^{-s\tau}\tilde{F}_R(s)$$

or, equivalently,

$$F_C(x) = \begin{cases} 0, & x < \tau \\ (1 - a) + aF_R(x - \tau), & x \geq \tau \end{cases} \tag{6.4}$$

where $a = 1 - e^{-\zeta\tau}$ (≈ 0) is the probability that a station failure will occur during one cycle time. In most transfer line systems a is very close to zero, and therefore we may use equation 6.4 to represent the completion time of a part.

It is sufficient to develop models of transfer lines for this situation applied to each of the stations in the line. The potential application of the completion time distribution is in making use of the properties and the results of approximate analytical models of servers in series studied in Chapter 5. The approximate models are often based on using information on just the means and variance of the service time (see Section 5.4.2). The approximate models, however, generally do not give adequate predictions if the coefficient of variation of the service time distribution is greater than one. This is usually a characteristic of the completion time distribution.

Another way of using the completion time distribution is to approximate it by a "Coxian" service station. If the LST $\tilde{F}_C(s)$ of the completion time distribution can be expressed as the ratio of rational polynomials in s, then it is possible to develop a model in which the completion time consists of series-parallel exponential phases. A two-phase version of this approach will be discussed in Section 6.7.

Allowance for whole-line stoppages.   In all the models that follow whole-line stoppages will be ignored. Thus it is necessary to adjust the predictions made by the models to allow for whole-line stoppages. If it is assumed that the repair of whole-line stoppages always preempts repair of individual station stoppages, then if A^W is the fraction of time during which there are no whole-line stoppages the net production rate allowing for whole-line stoppages (NPR) will be A^W times the predicted net production rate ignoring whole-line stoppages (TH), that is,

$$NPR = (A^W)(TH)$$

Similarly, if η^W is the efficiency of the line including whole-line stoppages,

$$\eta^W = A^W \eta$$

where η is the line efficiency ignoring whole line stoppages. Note that if $\bar{\tau}$ is the average cycle time, $NPR = \eta^W/\bar{\tau}$ and $TH = \eta/\bar{\tau}$. In the rest of this chapter we will therefore, without loss of generality, ignore whole-line stoppages, and as performance measures consider either (or both) the production rate TH or the line efficiency η.

6.4 TRANSFER LINES WITH NO INVENTORY BANKS

In this section we model transfer lines in which there are no inventory banks. A variety of different configurations will be considered. In particular we compare different paralleling arrangements as a means of reducing the impact of station failures. First we present, however, models of lines with no paralleling and consider a variety of different assumptions concerning failure mechanism: time-dependent and operation-dependent failures, and whether the part in process at failure is scrapped or not.

6.4.1 No Parallel Stations

Time-dependent failures.   With no inventory banks all stations in the line stop as soon as any station fails. The time-dependent failure assumption means, however, that even though a station is stopped because of failure of some other station in the line the mechanisms generating failure of the station still continue to operate, and thus the stopped station can fail. The line is then equivalent to the series system of reliability theory. That is, if there are m stations in series and station i has availability

$$A_i^T = \frac{\overline{U}_i}{\overline{U}_i + \overline{D}_i} \tag{6.5}$$

where $\overline{U}_i$ is the mean time the station is up (and hence either processing parts or stopped because of failure of some other station) and $\overline{D}_i$ is the mean down time, then the overall line availability A^T, the fraction of time when no station in the line is failed, is given by

$$A^T = \prod_{i=1}^{m} A_i^T \tag{6.6}$$

This formula for line availability assumes independence of up and down times between stations, thus it assumes sufficient repair capability for no station to be idle waiting for repair because all repair facilities are occupied at other stations. Under this assumption of independence the product formula for availability is valid for general distributions of up and down times. Further, it does not require that the up and down times of a station be mutually independent.

Provided that the line is never stopped because of lack of parts at the first station and lack of space after the last station (an assumption common to all models of line performance), then if all parts entering the line eventually leave it as completed parts, the line efficiency $\eta = A^T$ and therefore the production rate of the line will be given by

$$TH = A^T / \overline{\tau} \tag{6.7}$$

where $\overline{\tau}$ is the mean cycle time.

If there were a limited number of repairmen for the whole line, then the availability can be found using the machine interference model (see Chapter 2). Availability is the probability that all servers (repairmen) are idle. For exponentially distributed repair times identical for all stations and a single repairman

$$A^T = \frac{1}{\sum_{k_1=0}^{1} \cdots \sum_{k_m=0}^{1} (\sum_{i=1}^{m} k_i)! \prod_{i=1}^{m} (\frac{\overline{D}}{\overline{U}_i})^{k_i}}. \tag{6.8}$$

Operation-dependent failures.   Consider a line consisting of m stations. Suppose $\overline{T}_i$ is the mean number of cycles station i operates between failures, $\overline{D}_i$ is the mean down time of station i, and $\overline{\tau}$ is the average cycle time when the line is running and $x_i^O = \overline{D}_i / (\overline{T}_i \overline{\tau})$. (Note that the mean operating time between failures is $\overline{T}_i \overline{\tau}_i$.) Then a line consisting of just station i would have efficiency

$$\eta_i^O = \frac{\overline{T}_i}{\overline{T}_i + \overline{D}_i / \overline{\tau}} = \frac{1}{1 + x_i^O} \tag{6.9}$$

If, however, the part in process at the instant of failure must be scrapped, then this formula must be modified. Suppose we observe the line while it operates for n cycles under stationary conditions. The time needed for this is random, and let this time be K. Then the expected number of failures during this time K would be $n/\overline{T}_i$. Hence the average time required to repair these failures is $n\overline{D}_i/\overline{T}_i$, and the average number of parts produced during the observation time K is $n - n/\overline{T}_i$. Note that an average of $n/\overline{T}_i$ parts is scrapped. Therefore the efficiency of the line consisting of just station i is

$$\eta = \frac{n - n/\overline{T}_i}{n + n\overline{D}_i/\overline{T}_i\overline{\tau}} = \frac{1 - 1/\overline{T}_i}{1 + x_i^O} \tag{6.10}$$

and the throughput rate is

$$TH = \left(\frac{1 - 1/\overline{T}_i}{1 + x_i^O}\right)\frac{1}{\overline{\tau}} \tag{6.11}$$

These formulas are valid for general distributions of up and down times.

If there are m stations in the line then the line efficiency can be derived by a simple approach if it is assumed that there is only one repairman, and all parts entering the line leave it as completed parts. Suppose we observe the line while it operates for n cycles under stationary conditions. Then during the observation period the expected number of failures of station i is $n/\overline{T}_i$. The total down time of station i is then $n\overline{D}_i/\overline{T}_i$ and thus, if there is one repairman for the line, the total down time for the line is $\sum_{i=1}^{m} n\overline{D}_i/\overline{T}_i$. Hence

$$TH = \frac{n}{n\overline{\tau} + \sum_{i=1}^{m} n\overline{D}_i/\overline{T}_i} = \frac{1}{\overline{\tau}(1 + \sum_{i=1}^{m} x_i^O)} \tag{6.12}$$

If, however, there is more than one repairman, then in the event of two or more stations failing in the same cycle the line down time will be the maximum of the down time of the failed stations. For example, in a two-station line let $\overline{D}_{12}$ = expected value of the maximum of the down time of stations 1 and 2. Then, if both stations have deterministic repairtime distributions

$$\overline{D}_{12} = \max(\overline{D}_1, \overline{D}_2)$$

If both stations have exponential repair time distributions

$$\overline{D}_{12} = \overline{D}_1 + \overline{D}_2 - \frac{\overline{D}_1\overline{D}_2}{\overline{D}_1 + \overline{D}_2}$$

The probability that both stations will fail in the same cycle is $1/(\overline{T}_1\overline{T}_2)$, whereas the probability of just station 1 failing in a cycle is $(1/\overline{T}_1)(1 - 1/\overline{T}_2)$. Hence

$$TH = \frac{1}{1 + x_1^O + x_2^O + \frac{1}{\overline{T}_1\overline{T}_2}(\overline{D}_{12} - \overline{D}_1 - \overline{D}_2)} \tag{6.13}$$

Similar formulas can be derived for lines with more stations. Note that the distributions of station down times are required to derive the expected value of the line down times when two or more failures occur in the same cycle. With one repairman the efficiency formula applies for general distributions of up and down times.

If the part in process at the instant of failure must be scrapped, then the formulas must be modified. Suppose we observe the first station while it operates for n cycles under stationary conditions. During this time it will have $n/\overline{T}_1$ failures, thus the number of cycles over which the second station will be processing parts will be $n - n/\overline{T}_1$. The second station in turn will have $(n - n/\overline{T}_1)/\overline{T}_2$ failures so the third station will be processing parts for a number of cycles equal to $n(1 - 1/\overline{T}_1)(1 - 1/\overline{T}_2)$. Hence it can be shown that the rth station will be processing parts for $n \prod_{i=1}^{r-1}(1 - 1/\overline{T}_i)$ cycles and will have $n \prod_{i=1}^{r-1}(1 - 1/\overline{T}_i)/\overline{T}_r$ failures. Hence, because the net output of the line is $n \prod_{i=1}^{m}(1 - 1/\overline{T}_i)$

$$TH = \frac{\prod_{i=1}^{m}(1 - 1/\overline{T}_i)}{\overline{\tau}(1 + \sum_{i=1}^{m} x_i^O \prod_{k=1}^{i-1}(1 - 1/\overline{T}_k))} \tag{6.14}$$

This formula applies for general distributions of up and down times.

If it is postulated that not all failures result in scrapping of the part, that is, let q_i be the fraction of failures of station i that result in the part being scrapped, then in the preceding formula it is necessary to replace each $1/\overline{T}_i$ or $1/\overline{T}_k$ by $q_i/\overline{T}_i$ or $q_k/\overline{T}_k$, respectively.

6.4.2 Parallel Stations

The most common configuration with parallel stations is *splitting*: There are two stations in parallel, each of which has a cycle time double that of the rest of the line. If one of the two stations should fail then the rest of the line will be constrained to operate at the speed of the remaining station. Sometimes it is then possible, however, to reduce its cycle time, although this may result in it having a higher failure rate. Another possible configuration is that where there is a standby station that would only be used when the main station is failed. This is rarely used within automated systems because of the resulting low utilization of the standby. There may be some "backup," however, for the whole line in case there were to be a long duration machine failure. This situation will not be analyzed here.

Splitting—no speed up.   Consider a system consisting just of parallel stations. Suppose the cycle time of station j is τ_j with mean $\overline{\tau}_j$, independent of the state of the other stations, and assume repairs of the parallel stations are independent. If there are c stations in parallel and station j has efficiency η_j (failures can be either time dependent or operation dependent),

$$TH_j = \eta_j/\overline{\tau}_j$$

$$TH = \sum_{j=1}^{c} TH_j = \sum_{j=1}^{c} \eta_j/\overline{\tau}_j \tag{6.15}$$

Because gross production rate with no failures is $1/\bar{\tau} = \sum_{j=1}^{c} 1/\bar{\tau}_j$,

$$\eta = \frac{\sum_{j=1}^{c} \eta_j/\bar{\tau}_j}{\sum_{j=1}^{c} 1/\bar{\tau}_j}$$

$$= \sum_{j=1}^{c} f_j \eta_j \qquad (6.16)$$

where $f_j = (1/\bar{\tau}_j)/\sum_{j=1}^{c} 1/\bar{\tau}_j$. Note that if all stations in parallel are identical, such that $\eta_j = \eta$ for all j and $f_j = 1/c$, then the system efficiency is the same as the efficiency of a single station. When the stations are subjected only to time-dependent failures it is reasonable to assume that the failure rates are independent of the processing rates. Therefore it follows that if it is technically possible to design a station to achieve the full production rate then with time-dependent failures splitting offers no advantages.

With operation-dependent failures, however, the time to failure is a function of the number of cycles completed. If the mean number of completed cycles between failures is constant, independent of design speed (i.e., the failure rate in terms of the number of failures per unit time is proportional to the processing rate) then the efficiency of a station working at rate $1/(c\bar{\tau})$ is $1/(1 + x/c)$ where $x = \overline{D}/(\bar{\tau}\overline{T})$. Thus splitting can result in gains in efficiency with operation-dependent failures. Because in many processes such as machining, tool lives are increased by lowering processing speeds (at least up to the point where built-up edge effects occur), one may expect to see an increase in the mean number of cycles between failures when the processing rates are decreased. Combining this observation with the previous result we see that in many machining systems with operation-dependent failures splitting can result in a significant gain in efficiency.

In a system consisting of a group of parallel stations in series with m other stations, if η_p is the efficiency of the parallel group and x_p is defined by

$$\eta_p = \frac{1}{1 + x_p}$$

then with time-dependent failures

$$\eta = \eta_p \prod_{i=1}^{m} \eta_i \qquad (6.17)$$

With operation-dependent failures the system efficiency can be approximated by

$$\eta = \frac{1}{1 + x_p + \sum_{i=1}^{m} x_i} \qquad (6.18)$$

provided that the simultaneous failure of one of the series stations and one of the parallel stations in the same cycle is unlikely.

This result can be shown by considering some period during which each parallel station j, $j = 1, \ldots, c$, produces n_j parts. Then station j will have on average $n_j/\overline{T}_j$

failures and can be down for its repair for a total time with average $n_j \overline{D}_j / \overline{T}_j$. The station will also be forced down by any failure of a series station i, $i = 1, \ldots, m$, and on the average the total time it will be forced down because of their failures is $\sum_{i=1}^{m} \overline{D}_i \sum_{j=1}^{c} n_j / \overline{T}_i$. Thus, during the time $n_j \overline{\tau}_j + n_j \overline{D}_j / \overline{T}_j + \sum_{i=1}^{m} \overline{D}_i (\sum_{j=1}^{c} n_j) / \overline{T}_i$, station j produces n_j parts, and the whole system produces $\sum_{i=1}^{c} n_j$ parts. The system efficiency is then

$$\eta = \frac{\overline{\tau} \sum_{j=1}^{c} n_j}{n_j \overline{\tau}_j + n_j \overline{D}_j / \overline{T}_j + \sum_{i=1}^{m} \overline{D}_i (\sum_{i=1}^{c} n_j) / \overline{T}_i}$$

Now on its own, without the series stations

$$\eta_p = \frac{\overline{\tau} \sum_{j=1}^{c} n_j}{\overline{n}_j \tau_j + n_j \overline{D}_j / \overline{T}_j}$$

$$= \frac{1}{1 + x_p}$$

Hence, it can be seen that

$$\eta = \frac{1}{1 + x_p + \sum_{i=1}^{m} x_i}$$

If, however, there are other groups of parallel stations then it would be necessary to determine the probability of each configuration of working stations and the production rate for that configuration. Let $P(c)$ = probability of configuration c and $TH(c)$ = production rate in configuration c. Then

$$\text{System efficiency} = \overline{\tau} \sum P(c) TH(c)$$

where the summation is over all possible configurations of working stations.

Splitting—speed up possible.   Suppose that there are two identical stations in parallel. Usually each station operates at a rate $1/\overline{\tau}$. When one station is down the other station can speed up to a rate $1/\overline{\tau}'$, however, although when it does so its failure rate increases from γ to γ'.

Assume that there is only one repairman, and suppose that the down time has an exponential distribution with mean $\overline{D}$. Let $N(t)$ be the number of stations working at time t and $p(n) = \lim_{t \to \infty} P\{N(t) = n\}$, $n = 0, 1, 2$. $\{N(t), t \geq 0\}$ is a birth-death process on $\mathcal{S} = \{0, 1, 2\}$ with $p(2) = 1/(1 + 2\gamma \overline{D} + 2\gamma \gamma' \overline{D}^2)$ and $p(1) = 2\gamma \overline{D} p(2)$. Then the production rate of this system is

$$TH' = 2p(2)/\overline{\tau} + p(1)/\overline{\tau}'$$

$$= \frac{2/\overline{\tau} + 2\gamma \overline{D}/\overline{\tau}'}{1 + 2\gamma \overline{D} + 2\gamma \gamma' \overline{D}^2} \tag{6.19}$$

When there is no speed up (i.e., when $\gamma' = \gamma$ and $\overline{\tau}' = \overline{\tau}$), the throughput is

$$TH = \frac{2(1 + \gamma \overline{D})/\overline{\tau}}{1 + 2\gamma \overline{D}(1 + \gamma \overline{D})}$$

Therefore if

$$\gamma' < \frac{\gamma \overline{\tau}}{\overline{\tau}'} + \frac{\overline{\tau} - \overline{\tau}'}{2\overline{D}(1 + \gamma \overline{D})} \tag{6.20}$$

then $TH' > TH$, that is, under the preceding condition speeding up will be beneficial. For example, if the failure rate is proportional to the processing rate (i.e., $\gamma \overline{\tau} = \gamma' \overline{\tau}'$), then this condition is satisfied. Hence if the failure rate is proportional to the processing rate, speedup will increase the production rate of the system. Further, if the relation between γ' and $\overline{\tau}'$ is known then it would be possible to determine the optimum amount of speed up.

6.4.3 General Systems with No Inventory Banks

The results developed for throughput and efficiency of lines consisting of stations in series can also be extended to more complex configurations of stations with no inventory banks (see Figure 6.1). For example, if some stations perform assembly operations and other stations produce several output streams, such as a press producing simultaneously a rotor and a stator lamination for a motor, then when there are no inventory banks all stations are forced to operate with the same cycle time; thus if any one station in the system stops all other stations will be forced down. The results will also apply if some station or section operates at a different rate, but in such a way that its rate is constrained by the cycle time of other stations. For example, a station may only produce a component after a certain number of basic cycles has been completed, such as a motor stator requiring a certain number of laminations before it can be assembled. Suppose $\overline{\tau}_i$ is the mean cycle

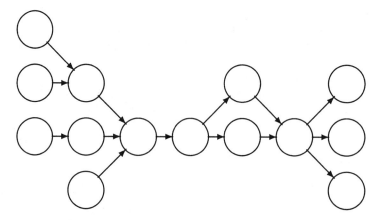

Figure 6.1   General System with No Inventory Banks

time of station i under normal operation when the set of station production rates is such that there is no accumulation of inventory anywhere in the system.

Then, for operation-dependent failures

$$\eta = \frac{1}{1 + \sum_{i=1}^{m} x_i} \tag{6.21}$$

where $x_i = \overline{D}_i / (\overline{T}_i \overline{\tau}_i)$.

6.5 SYSTEMS SEPARATED BY INFINITE INVENTORY BANKS

6.5.1 No Parts Scrapped

If the sections of a system are separated from each other by banks of infinite capacity then these banks will serve to isolate the behavior of different sections. It is possible for each section to operate at a different rate. The overall performance of the system will be determined by the bottleneck section, that is, that section with the lowest net production rate. Thus, if η_k is the efficiency of section k and $\overline{\tau}_k$ its cycle time,

$$TH = \min_k TH_k = \min_k \frac{\eta_k}{\overline{\tau}_k} \tag{6.22}$$

This result does not require that parts flow through sections connected in series. It is valid as long as each section of the system is required to produce for the final product of the system to be produced. That is, it applies to complex systems with branching and merging.

The system efficiency can be determined by observing that, without failures, the mean cycle time of the system $\overline{\tau} = \max_k \overline{\tau}_k$. Thus

$$\eta = \min_k \frac{\eta_k}{\overline{\tau}_k} \cdot \max_k \overline{\tau}_k \tag{6.23}$$

Note that these results assume that each section is independent, that is, no section has to wait repair because other sections are down. If there is a common repair facility for different sections it would be necessary to develop a model that allows for this. The machine interference models discussed in Chapter 2 can be used in some cases to model limited repair facilities.

6.5.2 Scrapping of Parts

Consider an m station series transfer line with unlimited bank capacity in between stations. Suppose when station i fails the part being processed is either scrapped with probability q_i or completed when the station becomes operative with probability $1 - q_i$, $i = 1, \dots, m$. Let I_i and O_i be the intake and production rates of station i if station i is never starved. If parts are not scrapped, conservation of flow implies $O_i = I_i$.

Otherwise $O_i = (1 - q_i/\overline{T}_i)I_i$, because $q_i/\overline{T}_i$ is the fraction of parts scrapped because of station i failures, $i = 1, \ldots, m$. As we have seen earlier $I_i = 1/\{(1 + x_i)\overline{\tau}\}$, $i = 1, \ldots, m$. Because we assume that there is an unlimited supply of raw material available in front of station 1, it will not starve, and hence its production rate of good parts is $TH_1 = (1 - q_1/\overline{T}_1)/\{(1 + x_1)\overline{\tau}\}$. Now suppose TH_i is the production rate of station i, $i = 1, \ldots, m$. If $TH_{i-1} \geq I_i$, then the actual intake rate of station i will be I_i, otherwise the intake rate is TH_{i-1}. Hence

$$TH_i = (1 - q_i/\overline{T}_i) \min\{I_i, TH_{i-1}\} \qquad i = 2, \ldots, m.$$

Then by induction starting with TH_1 it can be shown that

$$TH_i = \min_{1 \leq j \leq i} \left\{ \prod_{k=j}^{i} (1 - q_k/\overline{T}_k)/\{(1 + x_j)\overline{\tau}\} \right\}, \qquad i = 1, \ldots, m$$

Specifically, the production rate of the line is

$$TH = TH_m = \min_{1 \leq j \leq m} \left\{ \prod_{k=j}^{m} (1 - q_k/\overline{T}_k)/\{(1 + x_j)\overline{\tau}\} \right\} \qquad (6.24)$$

6.6 SYNCHRONIZED LINES WITH FINITE CAPACITY BANKS

In this section we consider series transfer lines with m stations and finite intermediate inventory banks. As pointed out earlier we may model the station failures by incorporating them into the part completion times. Then from the results in Chapter 5, it is easily seen that the throughput of the line is increasing in the bank capacities. It is relatively straightforward to develop bounds on the throughput of a transfer line divided into sections separated by finite-capacity inventory banks using the preceding monotonicity. If $TH(\mathbf{z})$ is the throughput of the line with bank capacity z_i between station i and $i + 1$, $i = 1, \ldots, m - 1$ one has

$$TH(\mathbf{0}) \leq TH(\mathbf{z}) \leq TH(\infty). \qquad (6.25)$$

Note that z_i is the capacity of the storage space in between station i and $i + 1$. When there is no bank we set $z_i = 0$. The monotonicity of the production rate with respect to bank capacities holds for more general part flow configurations and is not necessarily restricted to serial part flows. The results from the previous sections can thus be used to compute the upper and lower bounds for lines with relatively complex structure. That is, the sections can have parallel stations, and there can be branching and merging in the part flow. Furthermore, to determine the bounds it is usually not necessary to make any particularly restrictive assumptions concerning the distributions of up and down time. Even when the machine interference models requiring the assumption of exponential failure and repair times have been used, it is known that the performance of such systems is not particularly sensitive to deviations from exponentiality.

The bounds can be relatively wide, however, and, particularly in systems in which the cost of provision of banking capacity is significant, it is desirable to be able to determine the effect of given bank capacities. Nevertheless, the bounds should always be determined as it is sometimes the case that one bottleneck section so dominates the performance that the bounds are very close and the provision of banking capacity would be of very little value. Banks are more useful when there are several different sections that significantly affect performance.

Unfortunately, to analyze transfer lines with finite bank capacities it is necessary to make much more restrictive assumptions concerning the distributions of up and down times. The performance is affected by the nature of these distributions, and, further, exact solutions have only been obtained for models of systems with very simple structure, and either deterministic or "memoryless" (exponential or geometric) distributions of up and down time.

Most solvable analytical models consider a transfer line consisting of two sections in series, separated by a finite-capacity inventory bank. Each section is assumed to consist of a single station unless it is possible to represent several series stations by an equivalent single station. This is what is known as a two-stage transfer line.

The general approach to developing solvable analytical models of two-stage transfer lines will be described first, and then the solutions of specific models will be discussed. This leads to consideration of how to extend the models to more complex systems: either with more stages in series or with sections involving parallel stations.

6.6.1 Two-Stage Lines

There are two approaches that can be used to develop solvable models of two-stage transfer lines with inventory banks. One is based on the use of Markov process models. If the system processes discrete parts and transfer for all sections begins simultaneously then a discrete-state discrete-time model is appropriate. Alternatively, work flow can be viewed as continuous, in which case a continuous-state continuous-time model is used. The other approach that can be used is a semi-Markov process model with observation of the system only at the instant of either a stoppage occurring or a stoppage ending. We will only present the Markov process approach here.

Observation instant.   In a discrete-time model of a continuous process it is necessary to fix on the instant of observation of the system. Because of the cyclical behavior of the transfer line under normal operation there are two possible instants of observation. One is the instant when all stations have completed their required processing and transfer is about to begin. The other is on completion of transfer just before stations beginning their processing. Although both have been used in transfer-line models the simpler appears to be the instant when transfer is about to begin.

Transfer mechanism.   In a discrete-time model it is necessary to define precisely the mechanism of transfer and the conditions under which parts can move out of

TABLE 6.2   CONDITIONS FOR TRANSFER WITH A FULL OR EMPTY BANK

Bank	Station 1 state	Station 2 state	Transfer from station 1	Transfer into station 2
Full	Up	Up	Yes	Yes
	Up	Down	No	No
	Down	Up	No	Yes
Empty	Up	Up	Yes	Yes
	Up	Down	Yes	No
	Down	Up	No	No

a station or out of a store. To avoid excessive complication in the model, it is necessary to assume that both stages have the same cycle time, and the transfer mechanisms of the two stages are synchronized. That is, if both stages are up they will both begin transfer at the same instant. Provided the bank is neither full nor empty, then if one station is down the other station can always transfer on completion of its operation. If the bank is either full or empty, however, Table 6.2 shows the conditions under which transfer will occur.

In reality, the operation times at the two stages would not necessarily be identical. Thus the actual mechanism could be the following:

1. On successful completion of all operations in a stage, transfer, provided a space is available in the output bank and a part is available in the input bank.

2. If no transfer occurs on successful completion of all operations in a stage, monitor continuously the input and the output bank and as soon as the conditions are met (i.e., space becomes available in the output bank and a part is available in the input bank), initiate transfer.

These rules permit stages to operate at different speeds unless the bank is either full or empty. If the bank is empty then the speed will be determined by the upstream station, whereas if the bank is full the speed will be determined by the downstream station.

States and state transitions. In the discrete-time model with the system observed just before the beginning of transfer, then the system state is given by the following:

- States of each stage.
- Level of inventory in bank.

That is, the system state can be written as: (stage 1 state, stage 2 state, inventory level). With a bank of capacity z the inventory can take on values $0, 1, 2, \ldots, z$.

The information necessary to describe the state of a stage depends on the assumptions concerning the distributions of up and down times. The simplest case is when, in a discrete-time model, the up and down times are assumed to be geometric. In that case, assuming ample repair capability so that repair of a station can begin as soon as it fails, the possible states of a station are the following:

- W - Stage working.
- R - Stage under repair.

Sometimes it is convenient, although not necessary, to also use the following states:

- For the upstream stage. B - stage working but unable to transfer because the bank is full.
- For the downstream stage. I - stage working but unable to transfer because the bank is empty.

Thus state $(RW, 0)$ is the same as state $(RI, 0)$, and state (WR, z) is the same as state (BR, z).

The state of the system is represented by (A_n, B_n, X_n) where A_n is the state of stage 1, B_n is the state of stage 2 and X_n is the number of parts in the bank at time n. Then $\{(A_n, B_n, X_n), n = 1, 2, \ldots\}$ is a Markov chain with the state space

$$S = \{(WW, x), (WR, x), (RW, x), (RR, x), x = 0, \ldots, z, \}$$

Two assumptions will be made about the operation of the line.

1. No parts are scrapped, that is, if a station fails while processing a part, the part remains at the station, and repair will include any necessary operations needed to complete processing of the part.
2. If the downstream station completes processing of a part, the bank is empty, and the upstream station is under repair, then the part remains at the downstream station until the next successful transfer can occur. This assumption simplifies the number of states that must be considered and makes no difference to the calculated efficiency.

The state transitions between successive observation instants will be determined by the events occurring in the interval. The change in inventory level is determined by the state at the beginning of the interval in accordance with Table 6.3, whereas the change in station state is determined by whether a failure or repair occurs as in Table 6.4, that is, a_i is the probability of failure of station i in a cycle, and b_i is the probability of completing repair of station i in a cycle, $i = 1, 2$. If the station state is either B or I then the behavior depends on whether failures are assumed time dependent or operation dependent. Table 6.5 shows the respective state transition probabilities. By writing the probabilities as shown in the last column of Table 6.5 it is possible to develop a general model where, through substituting appropriate values of $\hat{a}_1$ and $\hat{a}_2$, both time-dependent and operation-dependent failures can be considered.

TABLE 6.3 CHANGE IN INVENTORY LEVEL OVER CYCLE

Bank level at n	Station 1 state	Station 2 state	Bank level at $n+1$
x $(0 \leq x \leq z)$	W	W	x
x $(0 \leq x < z)$	W	R	$x+1$
x $(0 < x \leq z)$	R	W	$x-1$
x $(0 \leq x \leq z)$	R	R	x
0	R	I	0
z	B	R	z

TABLE 6.4 PROBABILITIES OF CHANGE IN STATION STATE OVER CYCLE

Station i state at time n	Station i state at time $n+1$	Probability
W	W	$1 - a_i$
W	R	a_i
R	W	b_i
R	R	$1 - b_i$

TABLE 6.5 PROBABILITY OF CHANGES IN STATION STATE FROM B OR I STATE

Station state at n	Station state at $n+1$	Time dependent	Operation dependent	General
B	B or W	$1 - a_1$	1	$1 - \hat{a}_1$
B	R	a_1	0	$\hat{a}_1$
I	I or W	$1 - a_2$	1	$1 - \hat{a}_2$
I	R	a_2	0	$\hat{a}_2$

It is possible to combine the information in the tables and draw a system state transition graph. Figure 6.2 shows the diagram in the case of operation dependent failures.

Balance equations.   Let $p(\alpha, \beta, x) = \lim_{n \to \infty} P\{A_n = \alpha, B_n = \beta, X_n = x\}$, $(\alpha, \beta, x) \in S$ be the steady state distribution of $\{(A_n, B_n, X_n), n = 1, 2, \ldots\}$. The balance equations for these steady state probabilities are:
For an empty bank

$$p(WW, 0) = (1 - a_1)(1 - a_2)p(WW, 0) + b_1(1 - a_2)p(RW, 1) + b_1(1 - \hat{a}_2)p(RI, 0)$$
$$+ b_1 b_2 p(RR, 0)$$
$$p(RI, 0) = a_1(1 - a_2)p(WW, 0) + (1 - b_1)(1 - a_2)p(RW, 1)$$
$$+ (1 - b_1)(1 - \hat{a}_2)p(RI, 0) + (1 - b_1)b_2 p(RR, 0)$$

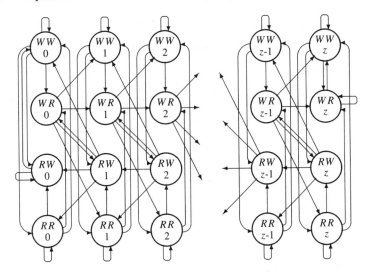

Figure 6.2   System State Transition Graph for Operation-Dependent Failures

$$p(WR,0) = (1 - a_1)a_2 p(WW,0) + b_1 a_2 p(RW,1) + b_1 \hat{a}_2 p(RI,0)$$
$$+ b_1(1 - b_2)p(RR,0)$$
$$p(RR,0) = a_1 a_2 p(WW,0) + (1 - b_1)a_2 p(RW,1) + (1 - b_1)\hat{a}_2 p(RI,0)$$
$$+ (1 - b_1)(1 - b_2)p(RR,0)$$

While for a full bank

$$p(WW,z) = (1 - a_1)(1 - a_2)p(WW,z) + (1 - a_1)b_2 p(WR,z-1)$$
$$+ (1 - \hat{a}_1)b_2 p(BR,z) + b_1 b_2 p(RR,z)$$
$$p(BR,z) = (1 - a_1)a_2 p(WW,z) + (1 - a_1)(1 - b_2)p(WR,z-1)$$
$$+ (1 - \hat{a}_1)(1 - b_2)p(BR,z) + b_1(1 - b_2)p(RR,z)$$
$$p(RW,z) = a_1(1 - a_2)p(WW,z) + a_1 b_2 p(WR,z-1) + \hat{a}_1 b_2 p(BR,z)$$
$$+ (1 - b_1)b_2 p(RR,z)$$
$$p(RR,z) = a_1 a_2 p(WW,z) + a_1(1 - b_2)p(WR,z-1) + \hat{a}_1(1 - b_2)p(BR,z)$$
$$+ (1 - b_1)(1 - b_2)p(RR,z)$$

For intermediate bank inventory levels, that is, $0 < x < z$, the equations are

$$p(WW,x) = (1 - a_1)(1 - a_2)p(WW,x) + (1 - a_1)b_2 p(WR,x-1)$$
$$+ b_1(1 - a_2)p(RW,x+1) + b_1 b_2 p(RR,x)$$

$$p(WR, x) = (1 - a_1)a_2 p(WW, x) + (1 - a_1)(1 - b_2)p(WR, x-1)$$
$$+ b_1 a_2 p(RW, x+1) + b_1(1 - b_2)p(RR, x)$$
$$p(RW, x) = a_1(1 - a_2)p(WW, x) + a_1 b_2 p(WR, x-1)$$
$$+ (1 - b_1)(1 - a_2)p(RW, x+1) + (1 - b_1)b_2 p(RR, x)$$
$$p(RR, x) = a_1 a_2 p(WW, x) + a_1(1 - b_2)p(WR, x-1) + (1 - b_1)a_2 p(RW, x+1)$$
$$+ (1 - b_1)(1 - b_2)p(RR, x).$$

Equation solution.   Solution of the equations is relatively straightforward if it is observed that by summing all equations for inventory levels 0, 1, 2, ..., x ($0 \leq x < z$)

$$p(WR, x) = p(RW, x+1) \tag{6.26}$$

or, directly, the probability of the inventory level increasing from x to $x+1$ must equal the probability of the inventory level decreasing from $x+1$ to x.

This means that in the four equations with left-hand side $p(WW, x)$, $p(WR, x)$, $p(RW, x)$ and $p(RR, x)$ the number of variables reduces from six to four. One of the equations is then redundant, and hence it is possible to express the other state probabilities in terms of say $p(WW, x)$. It is then possible to write for $0 < x < z$

$$p(RW, x) = \frac{\alpha_1}{\beta_1} p(WW, x) \tag{6.27}$$

$$p(WR, x) = \frac{\alpha_2}{\beta_2} p(WW, x) \tag{6.28}$$

$$p(RR, x) = \frac{\alpha_1 \alpha_2}{\beta_1 \beta_2} p(WW, x) \tag{6.29}$$

where

$$\alpha_1 = a_1 + a_2 - a_1 a_2 - b_1 a_2$$
$$\alpha_2 = a_1 + a_2 - a_1 a_2 - a_1 b_2$$
$$\beta_1 = b_1 + b_2 - b_1 b_2 - a_1 b_2$$
$$\beta_2 = b_1 + b_2 - b_1 b_2 - b_1 a_2$$

Further, it follows that

$$p(WW, x) = \frac{\beta_1 \alpha_2}{\alpha_1 \beta_2} p(WW, x-1), \qquad 2 \leq x \leq z-1$$

Define

$$\sigma = \frac{\beta_1 \alpha_2}{\alpha_1 \beta_2} \tag{6.30}$$

where if $a_2/b_2 \leq (\geq) a_1/b_1$, it can be seen that $\sigma \leq (\geq) 1$. Then

$$p(WW, x) = \sigma p(WW, x-1) \qquad 2 \leq x \leq z-1 \tag{6.31}$$

$$p(RW, x) = \sigma p(RW, x-1) \qquad 2 \leq x \leq z \tag{6.32}$$

$$p(WR, x) = \sigma p(WR, x-1) \qquad 1 \leq x \leq z-1 \tag{6.33}$$

$$p(RR, x) = \sigma p(RR, x-1) \qquad 2 \leq x \leq z-1 \tag{6.34}$$

Now defining

$$p(x) = P\{\text{stationary inventory level} = x\} = \lim_{n \to \infty} P\{X_n = x\}$$

$$= p(WW, x) + p(WR, x) + p(RW, x) + p(RR, x)$$

it follows that

$$p(x) = \sigma p(x-1), \qquad 2 \leq x \leq z-1 \tag{6.35}$$

That is, over the range $(1, z-1)$ the inventory-level distribution has a (truncated) geometric distribution. In the case in which $a_2/b_2 = a_1/b_1$ and thus $\sigma = 1$, the inventory-level distribution is uniform $(1 \leq x \leq z-1)$.

At $x = 0$ and $x = z$ there are four equations and four unknowns, with one equation dependent on the other three. Hence it is possible to show that

$$p(RI, 0) = \frac{\alpha_1}{b_1} p(WW, 0)$$

$$p(WR, 0) = \frac{\hat{a}_2 \alpha_1 + a_2 \beta_1}{\beta_2} p(WW, 0)$$

$$p(RR, 0) = \frac{((1-b_1)\hat{a}_2 + b_1 a_2)\alpha_1}{b_1 \beta_2} p(WW, 0)$$

$$p(BR, z) = \frac{\alpha_2}{b_2} p(WW, z)$$

$$p(RW, z) = \frac{\hat{a}_1 \alpha_2 + a_1 \beta_2}{\beta_1} p(WW, z)$$

$$p(RR, z) = \frac{(\hat{a}_1(1-b_2) + a_1 b_2)\alpha_2}{b_2 \beta_1} p(WW, z)$$

These equations make it possible to write all state probabilities in terms of one state probability, for example, $p(WW, 0)$, and then determine the actual state probabilities from the requirement that the sum of all probabilities equals one. Thus

$$p(WW, 1) = \frac{\beta_1}{\alpha_1} p(RW, 1) = \frac{\beta_1}{\alpha_1} p(WR, 0) = \frac{\beta_1(\hat{a}_2 \alpha_1 + a_2 \beta_1)}{\alpha_1 \beta_2} p(WW, 0)$$

$$p(WW, x) = \sigma^{x-1} p(WW, 1) = \sigma^{x-1} \frac{\beta_1(\hat{a}_2 \alpha_1 + a_2 \beta_1)}{\alpha_1 \beta_2} p(WW, 0) \qquad 1 \leq x \leq z-1$$

$$p(WW, z) = \frac{\beta_1}{\hat{a}_1\alpha_2 + a_1\beta_2} p(RW, z) = \frac{\beta_1\sigma^{z-1}}{\hat{a}_1\alpha_2 + a_1\beta_2} p(RW, 1)$$

$$= \frac{\beta_1(\hat{a}_2\alpha_1 + a_2\beta_1)}{\beta_2(\hat{a}_1\alpha_2 + a_1\beta_2)}\sigma^{z-1} p(WW, 0)$$

Further, it is possible to determine $\sum_{x=0}^{z} p(WW, x)$ in terms of $p(WW, 0)$. If $\sigma \neq 1$ then

$$1 - \sigma = 1 - \frac{\alpha_2\beta_1}{\alpha_1\beta_2} = (a_1b_2 - b_1a_2)\frac{(\alpha_1 + \beta_1)}{\alpha_1\beta_2} = (a_1b_2 - b_1a_2)\frac{(\alpha_2 + \beta_2)}{\alpha_1\beta_2}$$

and

$$\sum_{x=0}^{z} p(WW, x) =$$

$$\frac{p(WW, 0)}{1 - \sigma}\left(1 - \sigma + (1 - \sigma^{z-1})\frac{\beta_1(\hat{a}_2\alpha_1 + a_2\beta_1)}{\alpha_1\beta_2} + (1 - \sigma)\sigma^{z-1}\frac{\beta_1(\hat{a}_2\alpha_1 + a_2\beta_1)}{\beta_2(\hat{a}_1\alpha_2 + a_1\beta_2)}\right)$$

This can be further simplified by observing that, for example, the coefficient of terms in σ^{z-1} is given by

$$= -\frac{\beta_1}{\alpha_1\beta_2}(\hat{a}_2\alpha_1 + a_2\beta_1) + (a_1b_2 - b_1a_2)\frac{(\alpha_2 + \beta_2)}{\alpha_1\beta_2}\frac{\beta_1(\hat{a}_2\alpha_1 + a_2\beta_1)}{\beta_2(\hat{a}_1\alpha_2 + a_1\beta_2)}$$

$$= -\frac{\beta_1(\hat{a}_2\alpha_1 + a_2\beta_1)}{\alpha_1\beta_2^2(\hat{a}_1\alpha_2 + a_1\beta_2)}\left(\beta_2(\hat{a}_1\alpha_2 + a_1\beta_2) - (a_1b_2 - b_1a_2)(\alpha_2 + \beta_2)\right)$$

$$= -\frac{\sigma}{\beta_2}\frac{(\hat{a}_2\alpha_1 + a_2\beta_1)}{(\hat{a}_1\alpha_2 + a_1\beta_2)}\left((\hat{a}_1 + b_1)\beta_2 - a_1b_2 + b_1a_2\right)$$

where use is made of the relation

$$a_1\beta_2 - a_1b_2 + b_1a_2 = b_1\alpha_2$$

Hence

$$\sum_{x=0}^{z} p(WW, x) = \frac{p(WW, 0)}{(1 - \sigma)\beta_2}\left((\hat{a}_2 + b_2)\beta_1 + a_1b_2 - b_1a_2\right.$$

$$\left. -\sigma^z\frac{(\hat{a}_2\alpha_1 + a_2\beta_1)}{(\hat{a}_1\alpha_2 + a_1\beta_2)}((\hat{a}_1 + b_1)\beta_2 + b_1a_2 - a_1b_2)\right)$$

Similarly

$$\sum_{x=1}^{z} p(RW, x) = \frac{p(WW, 0)}{(1 - \sigma)\beta_2}(\hat{a}_2\alpha_1 + a_2\beta_1)(1 - \sigma^z)$$

and

$$
\sum_{x=0}^{z} p(WW, x) + \sum_{x=1}^{z} p(RW, x)
$$

$$
= \frac{p(WW, 0)}{(1 - \sigma)\beta_2} \Big((\hat{a}_2 + b_2)\beta_1 + a_1 b_2 - b_1 a_2 + \hat{a}_2 \alpha_1 + a_2 \beta_1
$$

$$
\qquad - \sigma^z \frac{(\hat{a}_2 \alpha_1 + a_2 \beta_1)}{(\hat{a}_1 \alpha_2 + a_1 \beta_2)} ((\hat{a}_1 + b_1)\beta_2 + b_1 a_2 - a_1 b_2 + \hat{a}_1 \alpha_2 + a_1 \beta_2) \Big)
$$

$$
= \frac{p(WW, 0)}{(1 - \sigma)\beta_2} (\alpha_1 + \beta_1) \Big(\hat{a}_2 + b_2 - \sigma^z \frac{(\hat{a}_2 \alpha_1 + a_2 \beta_1)}{(\hat{a}_1 \alpha_2 + a_1 \beta_2)} (\hat{a}_1 + b_1) \Big)
$$

$$
= \frac{p(WW, 0)\alpha_1 (\hat{a}_2 + b_2)}{(a_1 b_2 - b_1 a_2)} \Big(1 - \sigma^z \frac{(\hat{a}_1 + b_1)(\hat{a}_2 \alpha_1 + a_2 \beta_1)}{(\hat{a}_2 + b_2)(\hat{a}_1 \alpha_2 + a_1 \beta_2)} \Big)
$$

$p(WW, 0)$ will be determined by the equation

$$
\sum_{x=0}^{z} p(WW, x) + \sum_{x=1}^{z} p(RW, x) + \sum_{x=0}^{z-1} p(WR, x) + \sum_{x=0}^{z} p(RR, x) + p(BR, z) + P(RI, 0) = 1
$$

After carrying out the necessary algebra this gives

$$
p(WW, 0) = \left[\frac{\alpha_1 (\hat{a}_2 + b_2)}{a_1 b_2 - b_1 a_2} \left(\frac{(a_1 + b_1)}{b_1} - \frac{(a_2 + b_2)}{b_2} r^* \sigma^z \right) \right]^{-1}
$$

where

$$
r^* = \frac{(\hat{a}_1 + b_1)(\hat{a}_2 \alpha_1 + a_2 \beta_1)}{(\hat{a}_2 + b_2)(\hat{a}_1 \alpha_2 + a_1 \beta_2)}
$$

Line efficiency.   Given the state probabilities, it is possible to determine the efficiency of the line. A part will leave the line immediately following an observation instant if the system is observed to be in the following set of states $\{(WW, x) : 0 \leq x \leq z, (RW, x) : 0 < x \leq z\}$. Hence the line efficiency $\eta(z)$ is given by

$$
\eta(z) = \sum_{x=0}^{z} p(WW, x) + \sum_{x=1}^{z} p(RW, x)
$$

Alternatively, a part will enter the line if the system is in the set of states $\{(WW, x) : 0 \leq x \leq z, (WR, x) : 0 \leq x < z\}$. So an alternative expression for efficiency is

$$
\eta(z) = \sum_{x=0}^{z} p(WW, x) + \sum_{x=0}^{z-1} p(WR, x)
$$

Because $p(WR, x) = p(RW, x+1)$, it follows that (as would be expected to maintain conservation of flow) the preceding two expressions for $\eta(z)$ are consistent.

Using the results obtained earlier one can write

$$\eta(z) = \frac{\alpha_1(\hat{a}_2 + b_2)}{a_1 b_2 - b_1 a_2}(1 - r^* \sigma^z)p(WW, 0), \quad a_1 b_2 \neq b_1 a_2$$

Hence, the line efficiency is given by

$$\eta(z) = \frac{1 - r^* \sigma^z}{1 + x_1 - (1 + x_2)r^* \sigma^z} \quad a_1 b_2 \neq b_1 a_2 \tag{6.36}$$

where $x_i = a_i/b_i$, that is, $1/(1 + x_i)$ is the efficiency of station i if it operates on its own (its "isolated efficiency").

From equation 6.36 it can be seen that, because $\lim_{z \to \infty} \sigma^z \to 0$ (∞) if $x_2 < (>)x_1$,

$$\lim_{z \to \infty} \eta(z) = \frac{1}{1 + x_m}$$

where $x_m = \max\{x_1, x_2\}$. That is, the bottleneck stage determines the line efficiency.

Balanced stages. With balanced stages $a_1/b_1 = a_2/b_2$ and $\sigma = 1$. This means that $x_1 = x_2 = \hat{x}$, $\alpha_1 = \alpha_2 = \alpha$, $\beta_1 = \beta_2 = \beta$ and $\hat{x} = \alpha/\beta$. Thus

$$p(WW, x) = p(WW, 1) = (\hat{a}_2 + b_2)p(WW, 0) \quad 1 \leq x \leq z - 1$$

$$p(WW, z) = \frac{\hat{a}_2 + b_2}{\hat{a}_1 + b_1}p(WW, 0)$$

$$p(RW, x) = p(RW, 1) = (\hat{a}_2 + b_2)\hat{x}p(WW, 0) \quad 1 \leq x \leq z$$

$$p(WR, x) = (\hat{a}_2 + b_2)\hat{x}p(WW, 0) \quad 0 \leq x \leq z - 1$$

$$p(RR, x) = (\hat{a}_2 + b_2)\hat{x}^2 p(WW, 0) \quad 1 \leq x \leq z - 1$$

Hence

$$\eta(z)$$

$$= \left(1 + (\hat{a}_2 + b_2)(z - 1) + \frac{\hat{a}_2 + b_2}{\hat{a}_1 + b_1} + (\hat{a}_2 + b_2)z\hat{x}\right)p(WW, 0)$$

$$= \left(\hat{a}_1 + b_1 + \hat{a}_2 + b_2 - (\hat{a}_1 + b_1)(\hat{a}_2 + b_2) + (\hat{a}_1 + b_1)(\hat{a}_2 + b_2)z(1 + \hat{x})\right)\frac{p(WW, 0)}{\hat{a}_1 + b_1}$$

whence it can be shown that

$$\eta(z) =$$

$$\frac{\hat{a}_1 + b_1 + \hat{a}_2 + b_2 - (\hat{a}_1 + b_1)(\hat{a}_2 + b_2) + (\hat{a}_1 + b_1)(\hat{a}_2 + b_2)z(1 + \hat{x})}{(\hat{a}_1 + b_1 + \hat{a}_2 + b_2)(1 + \hat{x}) + (\hat{a}_1 + b_1)(\hat{a}_2 + b_2)(\hat{x}(b_1 + b_2)/b_1 b_2 + (z - 1)(1 + \hat{x})^2)}$$

$$\tag{6.37}$$

Note that if $\hat{a}_1 = \hat{a}_2 = \hat{a}$, $b_1 = b_2 = b$, that is, the stations are identical,

$$\eta(z) = \frac{2 - b - \hat{a} + (b + \hat{a})z(1 + \hat{x})}{2(1 + 2\hat{x} + \hat{x}\hat{a}/b) + (b + \hat{a})(z - 1)(1 + \hat{x})^2} \tag{6.38}$$

Buffer effectiveness.   The buffer effectiveness $g(z)$ can be defined as

$$g(z) = \frac{\eta(z) - \eta(0)}{\eta(\infty) - \eta(0)}$$

Then it can be shown that

$$g(z) = \begin{cases} \frac{(1 - \sigma^z)(1 + x_1)}{(1 + x_1) - (1 + x_2)r^*\sigma^z}, & x_1 > x_2 \\[2mm] \frac{r^*(\sigma^z - 1)(1 + x_2)}{(1 + x_2)r^*\sigma^z - (1 + x_1)}, & x_1 < x_2 \\[2mm] \frac{(\hat{a}_1 + b_1)(\hat{a}_2 + b_2)z(1 + \hat{x})^2}{(\hat{a}_1 + b_1 + \hat{a}_2 + b_2)(1 + \hat{x}) + (\hat{a}_1 + b_1)(\hat{a}_2 + b_2)(\hat{x}(b_1 + b_2)/b_1 b_2 + (z - 1)(1 + \hat{x})^2)}, & x_1 = x_2 \end{cases} \tag{6.39}$$

Operation-dependent and time-dependent failures.   Two limiting cases are of interest.

- Operation-dependent failures: $\hat{a}_1 = \hat{a}_2 = 0$, and thus

$$r^* = \frac{b_1 a_2 \beta_1}{a_1 b_2 \beta_2}$$

- Time-dependent failures: $\hat{a}_1 = a_1$, $\hat{a}_2 = a_2$, and thus

$$r^* = \frac{b_1 a_2 (1 + x_1)}{a_1 b_2 (1 + x_2)}$$

If the stations are identical the line efficiency η and the buffer effectiveness g can be computed.

- Operation-dependent failures

$$\eta(z) = \frac{2 - b + bz(1 + \hat{x})}{2(1 + 2\hat{x}) + b(z - 1)(1 + \hat{x})^2}$$

$$g(z) = \frac{bz(1 + \hat{x})^2}{2(1 + 2\hat{x}) + (z - 1)b(1 + \hat{x})^2}$$

- Time-dependent failures

$$\eta(z) = \frac{2 - b(1 + \hat{x}) + bz(1 + \hat{x})^2}{2(1 + \hat{x})^2 + b(z - 1)(1 + \hat{x})^3}$$

$$g(z) = \frac{bz(1 + \hat{x})}{2 + b(z - 1)(1 + \hat{x})}$$

If $b \ll 1$, the formulas can be written in a way that makes the dependency on bank size more apparent. Define $\hat{z} = bz$.

- Operation-dependent failures

$$\eta(\hat{z}) = \frac{1 + (\hat{z}/2)(1 + \hat{x})}{1 + 2\hat{x} + (\hat{z}/2)(1 + \hat{x})^2} \tag{6.40}$$

$$g(\hat{z}) = \frac{(\hat{z}/2)(1 + \hat{x})^2}{1 + 2\hat{x} + (\hat{z}/2)(1 + \hat{x})^2}. \tag{6.41}$$

- Time-dependent failures

$$\eta(\hat{z}) = \frac{1 + (\hat{z}/2)(1 + \hat{x})^2}{(1 + \hat{x})^2 + (\hat{z}/2)(1 + \hat{x})^3} \tag{6.42}$$

$$g(\hat{z}) = \frac{(\hat{z}/2)(1 + \hat{x})}{1 + (\hat{z}/2)(1 + \hat{x})} \tag{6.43}$$

It can be seen that $(\hat{z}/2)$ determines the effect of the inventory bank on the line efficiency, that is, the principal factor affecting efficiency is the ratio of the bank capacity to the mean repair time. This can be shown in a slightly different way by noting that both equations 6.40 and 6.42 can be written in the form

$$\frac{1}{\eta(\hat{z})} = \frac{1 - w}{\eta(0)} + \frac{w}{\eta(\infty)}$$

with

$$w = \frac{\hat{z}f/2}{1 + \hat{z}f/2}$$

- operation-dependent failures: $f = 1 + \hat{x}$
- time-dependent failures: $f = (1 + \hat{x})^2$

Figure 6.3 shows the effect of bank capacity on line efficiency.

6.6.2 Two-Stage Lines with Operation-Dependent Failures

Operation-dependent failures create a dependence between the behavior of different stations that does not exist with time-dependent failures. Thus it is of interest to analyze two stage lines with stations having operation-dependent failures to characterize various features of their operation.

Output and input patterns. The output (or input) of the line will have a pattern consisting of alternating up and down phases. During an up phase the line will produce parts with the interval between successive parts equal to the (constant) cycle time.

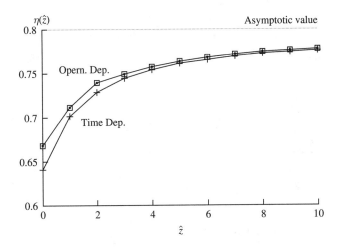

Figure 6.3 Effect of Bank Capacity on Line Efficiency ($\hat{x} = 0.25$).

During a down phase no parts are produced at all. This pattern can be characterized by the up time (the duration of the up phase) and the down time (the duration of the down phase) or the time between stoppages (the total duration of successive down and up phases). If $E[U]$ is the mean up time, $E[D]$ is the mean down time, and $E[T]$ is the mean time between stoppages then

$$E[T] = E[U] + E[D]$$

and the efficiency

$$\eta = \frac{E[U]}{E[T]}$$

thus if $E[U]$ and η are given then $E[D] = E[U](1 - \eta)/\eta$ and $E[T] = E[U]/\eta$.

Mean up time.   Because the system has been described by a finite-state Markov chain there are standard techniques by which the mean and all the moments of the up time can be found. We will only give the results for the mean up time, however. The mean up time can be found from the relationship

$$\frac{1}{E[U]} = P\{\text{line stops during a cycle}|\text{up at beginning of cycle}\}$$

Now the line can be looked at either on the input or output side, and the up-time durations on the input or output side will not necessarily be the same. Denoting U_O and U_I as the up-time durations on the output and input side, respectively, it follows that

$$\frac{1}{E[U_O]} = \frac{a_2(\sum_{x=0}^{z} p(WW, x) + \sum_{x=1}^{z} p(RW, x))}{\sum_{x=0}^{z} p(WW, x) + \sum_{x=1}^{z} p(RW, x)}$$

$$+ \frac{a_1(1 - a_2)p(WW, 0) + (1 - b_1)(1 - a_2)p(RW, 1)}{\sum_{x=0}^{z} p(WW, x) + \sum_{x=1}^{z} p(RW, x)}$$

$$= a_2 + (1 - a_2)\frac{a_1 p(WW, 0) + (1 - b_1)p(RW, 1)}{\sum_{x=0}^{z} p(WW, x) + \sum_{x=1}^{z} p(RW, x)}$$

$$= a_2 + (1 - a_2)a_1 \delta(z) \tag{6.44}$$

where

$$\delta(z) = \frac{1}{a_1}\frac{a_1 p(WW, 0) + (1 - b_1)p(RW, 1)}{\sum_{x=0}^{z} p(WW, x) + \sum_{x=1}^{z} p(RW, x)}$$

$$= \begin{cases} \frac{1 - r^*}{1 - r^* \sigma^z}, & x_1 \neq x_2 \\ \frac{b_1 + b_2 - b_1 b_2}{b_1 + b_2 - b_1 b_2 + z b_1 b_2(1+\hat{x})}, & x_1 = x_2 = \hat{x} \end{cases} \tag{6.45}$$

Similarly

$$\frac{1}{E[U_I]} = a_1 + (1 - a_1)\frac{a_2 p(WW, z) + (1 - b_2)p(WR, z - 1)}{\sum_{x=0}^{z} p(WW, x) + \sum_{x=0}^{z-1} p(WR, x)}$$

$$= a_1 + (1 - a_1)a_2 \delta'(z) \tag{6.46}$$

where

$$\delta'(z) = \frac{1}{a_2}\frac{a_2 p(WW, z) + (1 - b_2)p(WR, z - 1)}{\sum_{x=0}^{z} p(WW, x) + \sum_{x=0}^{z-1} p(WR, x)}$$

$$= \begin{cases} \frac{\sigma^z(1 - r^*)}{1 - r^* \sigma^z}, & x_1 \neq x_2 \\ \frac{b_1 + b_2 - b_1 b_2}{b_1 + b_2 - b_1 b_2 + z b_1 b_2(1+\hat{x})}, & x_1 = x_2 = \hat{x}. \end{cases} \tag{6.47}$$

Mean down time.   Because $E[D] = E[U](1 - \eta)/\eta$

$$\frac{1}{E[D_O]} = \frac{1}{E[U_O](1/\eta - 1)}$$

$$= (a_2 + (1 - a_2)a_1\delta(z))\frac{1 - r^*\sigma^z}{x_1 - x_2 r^*\sigma^z} \qquad x_1 \neq x_2$$

$$= \frac{a_2 + (1 - a_2)a_1\delta(z)}{x_2 + (1 - a_2 b_1/(b_1 + b_2 - b_1 b_2))x_1\delta(z)}$$

$$\approx \frac{a_2 + a_1\delta(z)}{x_2 + x_1\delta(z)}$$

If $x_1 = x_2$ and $b_1 = b_2 = b$ then

$$\frac{1}{E[D_O]} = b\left(1 - \frac{a(1 - b)}{4 - 2b - a + z(a + b)}\right)$$

which would usually be close to b. Even if $x_1 \neq x_2$ but $b_1 = b_2 = b$ then the mean down time is close to $1/b$.

Mean time between stoppages.   Because $E[T] = E[U] + E[D]$ it follows that $E[T] = E[U]/\eta$ thus

$$\frac{1}{E[T_O]} = (a_2 + (1 - a_2)a_1\delta(z))\frac{1 - r^*\sigma^z}{(1 + x_1) - (1 + x_2)r^*\sigma^z}, \quad x_1 \neq x_2$$

Failure transfer coefficients.   An isolated station with parameters a, b would have mean up time $1/E[U] = a$, thus in determining $E[U_O]$, the term $(1 - a_2)a_1\delta(z)$ indicates the additional interruptions to the output owing to the failures of the station 1. Because with no inventory bank the stoppage rate owing to station 1 would be $(1-a_2)a_1$, the effect of the bank is to reduce the apparent stoppage rate of station 1 to $a_1\delta(z)$. The quantities $\delta(z)$ and $\delta'(z)$ are the failure transfer coefficients. They can be expressed in several different ways that provide further insight into their meaning.

Conditional Blocking and Idle Probabilities.   Consider a cycle such that station 1 is working both at the beginning and end of the cycle. Define the *conditional blocking probability* $P'(B)$ by

$$P'(B) = P\{\text{station 1 becomes blocked at end of cycle}|\text{station 1 works}$$

$$\text{at beginning and end of cycle}\}$$

Then

$$P'(B) = \frac{(1 - a_1)(a_2 p(WW, z) + (1 - b_2)p(WR, z-1))}{(1 - a_1)(\sum_{x=0}^{z} p(WW, x) + \sum_{x=0}^{z-1} p(WR, x))}$$

$$= a_2\delta'(z)$$

Similarly the *conditional idle probability* $P'(I)$ is given by

$$P'(I) = P\{\text{station 2 becomes idle at the end of a cycle}|\text{station 2 works}$$

$$\text{at beginning and end of cycle}\}$$

$$= \frac{(1 - a_2)(a_1 p(WW, 0) + (1 - b_1)p(RW, 1))}{(1 - a_2)(\sum_{x=0}^{z} p(WW, x) + \sum_{x=1}^{z} p(RW, x))}$$

$$= a_1\delta(z)$$

Note that the conditional-blocking probability is not the same as the unconditional-blocking probability, the probability that station 1 is blocked at a transfer instant, $p(WR, z)$.

Conditional Empty and Full Bank Probabilities.   Observe that

$$a_1 p(WW, 0) + (1 - b_1)p(RW, 1) = b_1(p(RW, 0) + p(RR, 0))$$

and

$$a_1\left(\sum_{x=0}^{z} p(WW, x) + \sum_{x=1}^{z} p(RW, x)\right) = b_1 \sum_{x=0}^{z}(p(RW, x) + p(RR, x))$$

Hence

$$\delta(z) = \frac{p(RW, 0) + p(RR, 0)}{\sum_{x=0}^{z}(p(RW, x) + p(RR, x))}$$

that is

$$\delta(z) = P\{\text{bank is empty}|\text{station 1 is failed}\}$$

Similarly

$$\delta'(z) = P\{\text{bank is full}|\text{station 2 is failed}\}$$

Note that $\delta(z)$ and $\delta'(z)$ are in fact functions of the five parameters z, a_1, b_1, a_2, b_2 and $\delta(z, a_1, b_1, a_2, b_2) = \delta'(z, a_2, b_2, a_1, b_1)$.

Observe that

$$\frac{p(RW, x) + p(RR, x)}{p(RW, x-1) + p(RR, x-1)} = \begin{cases} \frac{a_2 b_1(\sigma + \beta_1/\alpha_1)}{b_1 + b_2 - b_1 b_2}, & x = 1 \\ \sigma, & 2 \le x \le z. \end{cases}$$

From the preceding result it can be verified that the preceding ratio is increasing in a_2 and a_2/b_2. Hence $1/\delta(z, a_1, b_1, a_2, b_2)$ is increasing and thus $\delta(z, a_1, b_1, a_2, b_2)$ is decreasing in a_2 and a_2/b_2. By symmetry it follows that $\delta'(z, a_1, b_1, a_2, b_2)$ is decreasing in a_1 and a_1/b_1. These properties will be used later to prove the convergence of an approximation algorithm in Section 6.6.5.

Time to next failure. It is of interest to determine the expected time to failure conditional on the line being observed to be up. Suppose the line is observed to be in a state s such that a part will be output at the immediately following transfer, that is, $s \subset W_O = \{(WW, x), x = 0, \ldots, z; (RW, x), x = 1, \ldots, z\}$. Then we would like to determine $E[R_O : s]$, the expected output until the up time ends. $E[R_O : s]$ will be given by the following equations:

$$E[R_O : WW, x]$$
$$= \begin{cases} 1 + (1 - a_2)((1 - a_1)E[R_O : WW, x] + a_1 E[R_O : RW, x]), & 1 \le x \le z \\ 1 + (1 - a_2)(1 - a_1)E[R_O : WW, 0], & x = 0 \end{cases}$$

$$E[R_O : RW, x] = \begin{cases} 1 + (1 - a_2)(b_1 E[R_O : WW, x-1] \\ \quad + (1 - b_1)E[R_O : RW, x-1]), & 2 \le x \le z \\ 1 + (1 - a_2)b_1 E[R_O : WW, 0], & x = 1 \end{cases}$$

These equations can be easily solved recursively, and it can be shown that

$$E[R_O : WW, x] = \frac{1}{a_2}\left(1 - \frac{a_1(1 - a_2)}{a_1 + a_2 - a_1 a_2}\left(\frac{(1 - a_2)\alpha_1}{a_1 + a_2 - a_1 a_2}\right)^x\right), \quad 0 \le x \le z$$

$$E[R_O : RW, x] = \frac{1}{a_2}\left(1 - \left(\frac{(1 - a_2)\alpha_1}{a_1 + a_2 - a_1 a_2}\right)^x\right), \quad 1 \le x \le z$$

It is then possible to determine $E[R_O]$, the mean output before the next stoppage from a randomly chosen instant when the line is observed to be up. $E[R_O]$ will be given by

$$E[R_O] = \frac{\sum_{x=0}^{z} E[R: WW, x]p(WW, x) + \sum_{x=1}^{z} E[R: RW, x]p(RW, x)}{\sum_{x=0}^{z} p(WW, x) + \sum_{x=1}^{z} p(RW, x)}$$

If $a_1 = a_2 = a$ and $b_1 = b_2 = b$ then

$$E[R_O] = \frac{1}{a}\left(1 - \frac{1-a}{2-a}\frac{3-a-b-(1-a)y^z}{2-b+z(a+b)}\right)$$

where $y = (1-a)(2-a-b)/(2-a)$. Given $E[R_O]$ it is possible to determine the second moment of the up time because $E[R_O] = (E[U_O^2] + E[U_O])/2E[U_O]$ using well-known results for the forward recurrence time of a renewal process. Hence the squared coefficient of variation of the up time distribution $C_{U_O}^2$ will be given by $C_{U_O}^2 + 1 = (2E[R_O] - 1)/E[U_O]$. Table 6.6 shows how $C_{U_O}^2$ varies with z if $a/b = 0.2$ and $b = 0.25$. Note that $C_{U_O}^2$ initially decreases, then increases, and finally decreases again. It is, however, always close to 1.

TABLE 6.6 SQUARED
COEFFICIENT OF VARIATION OF
UP TIME AS FUNCTION OF BANK
SIZE ($a = 0.05$, $b = 0.25$)

z	$E[U_O]$	$E[R_O]$	$C_{U_O}^2$
0	10.256	10.256	0.9025
1	11.044	10.907	0.8846
2	11.713	11.507	0.8794
3	12.290	12.057	0.8807
4	12.791	12.558	0.8854
5	13.232	13.015	0.8917
7	13.968	13.809	0.9058
10	14.815	14.757	0.9248
15	15.798	15.878	0.9469
25	16.953	17.165	0.9660
35	17.610	17.853	0.9709
50	18.194	18.429	0.9709
100	19.005	19.171	0.9649
200	19.476	19.574	0.9587
500	19.783	19.827	0.9538
1000	19.890	19.913	0.9520
∞	20	20	0.9500

Distribution of interoutput time.   A part will leave the line immediately following the observation instant if the current state is in $\mathcal{W} = \{(WW, x), 0, \dots, z;$

$(RW, x), 1, \ldots, z\}$. The next output of a part from the line will occur at a random time N later where the distribution of N depends on the stage parameters and the current state. Let $h(k : (\alpha, \beta, x)) = P\{N = k | A_n = \alpha, B_n = \beta, X_n = x\}$, $(\alpha, \beta, x) \in \mathcal{W}$ and let $\hat{N}(\hat{a}, \hat{b})$ be the interoutput times of a single stage that is never starved or blocked and has a probability of failure $\hat{a}$ and probability of repair $\hat{b}$ during one cycle. Then

$$f_{\hat{N}}(k : \hat{a}, \hat{b}) = P\{\hat{N}(\hat{a}, \hat{b}) = k\} = \begin{cases} 1 - \hat{a}, & k = 1 \\ \hat{a}\hat{b}(1 - \hat{b})^{k-2}, & k = 2, 3, \ldots \end{cases}$$

For states (WW, x) $(1 \leq x \leq z)$ and states (RW, x) $(2 \leq x \leq z)$ the time until the next output is determined solely by the characteristics of stage 2 and the distribution is given by

$$h(k : WW, x) = f_{\hat{N}}(k : a_2, b_2), \qquad 1 \leq x \leq z$$

$$h(k : RW, x) = f_{\hat{N}}(k : a_2, b_2), \qquad 2 \leq x \leq z$$

For the other states it can be shown that

$$h(k : WR, x) = b_2(1 - b_2)^{k-1}, \qquad 0 \leq x \leq z, \quad k \geq 1$$

$$h(k : RR, x) = b_2(1 - b_2)^{k-1}, \qquad 1 \leq x \leq z, \quad k \geq 1$$

$$h(k : RR, 0) = b_1(1 - b_1)^{k-1} + b_2(1 - b_2)^{k-1}$$
$$- (b_1 + b_2 - b_1 b_2)(1 - b_1 - b_2 + b_1 b_2)^{k-1}, \quad k \geq 1$$

$$h(k : RW, 0) = b_1(1 - b_1)^{k-1}, \quad k \geq 1$$

$$h(k : WW, 0) = (1 - a_1)f_{\hat{N}}(k : a_2, b_2) + a_1 g(k) \quad k \geq 1$$

$$h(k : RW, 1) = b_1 f_{\hat{N}}(k : a_2, b_2) + (1 - b_1)g(k) \quad k \geq 1$$

where g is the distribution of the time until the output of the next part given that station 1 is failed at the end of the first cycle and the inventory bank is empty. Note that $h(k : RR, 0)$ is the distribution of the maximum of the repair time at station 1, and the repair time at station 2. g can be found by observing that

$$g(1) = 0$$

$$g(k) = (1 - a_2)h(k - 1 : RW, 0) + a_2 h(k - 1 : RR, 0), \qquad k > 1$$

$$= b_1(1 - b_1)^{k-2} + a_2 b_2(1 - b_2)^{k-2} - a_2(b_1 + b_2 - b_1 b_2)(1 - b_1)^{k-2}(1 - b_2)^{k-2}$$

The distribution of the interoutput time $f_N(k) = P\{N = k\}$ is thus given by

$$f_N(k) = \frac{\sum_{x=0}^{z} h(k : WW, x)p(WW, x) + \sum_{x=1}^{z} h(k : RW, x)p(RW, x)}{\sum_{x=0}^{z} p(WW, x) + \sum_{x=1}^{z} p(RW, x)}$$

$$= f_{\hat{N}}(k : a_2, b_2) + \frac{(a_1 p(WW, 0) + (1 - b_1)p(RW, 1))(g(k) - f_{\hat{N}}(k : a_2, b_2))}{\sum_{x=0}^{z} p(WW, x) + \sum_{x=1}^{z} p(RW, x)}$$

$$= f_{\hat{N}}(k : a_2, b_2) + a_1 \delta(z)(g(n) - f_{\hat{N}}(k : a_2, b_2))$$

$$= f_{\hat{N}}(k : a_2, b_2)(1 - a_1 \delta(z)) + g(k)a_1 \delta(z)$$

where $\delta(z)$ is the failure transfer coefficient.

To find the mean and moments of N, the mean and moments of $\hat{N}$ and those corresponding to g are required.

$$E[\hat{N}(a_2, b_2)] = 1 + \frac{a_2}{b_2}$$

$$E[\hat{N}(a_2, b_2)(\hat{N}(a_2, b_2) - 1)] = 2\frac{a_2}{b_2^2}$$

$$\sum_{k=0}^{\infty} kg(k) = 1 + \frac{a_2}{b_2} + \frac{1}{b_1} - \frac{a_2}{b_1 + b_2 - b_1 b_2}$$

$$\sum_{k=1}^{\infty} k(k-1)g(k) = 2\frac{a_2}{b_2^2} + 2\frac{1}{b_1^2} - 2\frac{a_2}{(b_1 + b_2 - b_1 b_2)^2}$$

Hence, the mean and factorial second moment of the interoutput time N are

$$E[N] = 1 + x_2 + a_1\delta(z)\left(\frac{1}{b_1} - \frac{a_2}{b_1 + b_2 - b_1 b_2}\right) \tag{6.48}$$

and

$$E[N(N-1)] = 2\frac{a_2}{b_2^2} + 2a_1\delta(z)\left(\frac{1}{b_1^2} - \frac{a_2}{(b_1 + b_2 - b_1 b_2)^2}\right) \tag{6.49}$$

$E[N]$ can be rewritten as

$$E[N] = 1 + x_2 + \frac{x_1 - x_2}{1 - r^*\sigma^z}, \qquad x_1 \neq x_2$$

$$= 1 + \frac{x_1 - x_2 r^*\sigma^z}{1 - r^*\sigma^z}$$

$$= \frac{1}{\eta(z)}$$

Equivalent zero bank line.   It is of interest to note that f_N can be written as

$$f_N(k) = \begin{cases} 1 - a_2 + 1 - a_1\delta(z) - (1 - a_2 a_1\delta(z)), & k = 1 \\ a_2 b_2(1 - b_2)^{k-2} + a_1\delta(z)b_1(1 - b_1)^{k-2} & \\ \quad - a_2 a_1\delta(z)(b_1 + b_2 - b_1 b_2)(1 - (b_1 + b_2 - b_1 b_2))^{k-2}, & k > 1 \end{cases}$$

$$= f_{\hat{N}}(k : a_2, b_2) + f_{\hat{N}}(k : a_1\delta(z), b_1) - f_{\hat{N}}(k : a_2 a_1\delta(z), b_1 + b_2 - b_1 b_2)$$

That is, the line with finite inventory bank has exactly the same interoutput distribution as a line with zero inventory bank but where station 1 has failure probability $a_1\delta(z)$. Note that the repair probabilities are unchanged.

Distribution of interinput time.   Following a similar argument it can be shown that the distribution f_K of the time K between successive parts entering the line is given by

$$f_K(k) = f_{\hat{N}}(k : a_1, b_1) + f_{\hat{N}}(k : a_2\delta'(z), b_2) - f_{\hat{N}}(k : a_1 a_2\delta'(z), b_1 + b_2 - b_1 b_2)$$

and the mean and factorial second moment are

$$E[K] = 1 + x_1 + a_2\delta'(z)\left(\frac{1}{b_2} - \frac{a_1}{b_1 + b_2 - b_1 b_2}\right)$$

$$= 1 + \frac{x_1 - x_2 r^* \sigma^z}{1 - r^* \sigma^z}, \qquad x_1 \neq x_2,$$

$$E[K(K-1)] = 2\frac{a_1}{b_1^2} + 2a_2\delta'(z)\left(\frac{a_1}{b_1^2} - \frac{a_1}{(b_1 + b_2 - b_1 b_2)^2}\right)$$

As would be expected $E[N] = E[K]$, that is, the mean interoutput time equals the mean interinput time. Also as seen from the input the interinput time has the same distribution as a zero bank line in which the second station has failure probability $a_2\delta'(z)$.

Loss transfer coefficient. Another way of gaining insight into the preceding results is to consider the (compound) states of station i. Station i can be observed to be in one of three states: working ($\mathcal{W}_i$), under repair ($\mathcal{R}_i$), and idle ($\mathcal{I}_2$) for station 2 or blocked ($\mathcal{B}_1$) for station 1. Now

$$p(\mathcal{W}_2) = \sum_{x=0}^{z} p(WW, x) + \sum_{x=1}^{z} p(RW, x)$$

$$p(\mathcal{W}_1) = \sum_{x=0}^{z} p(WW, x) + \sum_{x=0}^{z-1} p(WR, x) = p(\mathcal{W}_2) = p(\mathcal{W})$$

$$p(\mathcal{R}_2) = \sum_{x=0}^{z} (p(WR, x) + p(RR, x))$$

$$p(\mathcal{R}_1) = \sum_{x=0}^{z} (p(RW, x) + p(RR, x))$$

$$p(\mathcal{I}_2) = p(RW, 0)$$

$$p(\mathcal{B}_1) = p(WR, z)$$

Now $p(\mathcal{R}_2) = (a_2/b_2)p(\mathcal{W})$ and $p(\mathcal{R}_1) = (a_1/b_1)p(\mathcal{W})$ and

$$\frac{p(\mathcal{I}_2)}{p(\mathcal{W})} = \frac{p(RW, 0)}{p(RW, 0) + p(RR, 0)} \frac{p(RW, 0) + p(RR, 0)}{p(\mathcal{R}_1)} \frac{p(\mathcal{R}_1)}{p(\mathcal{W})}$$

$$= \left(1 - \frac{b_1 a_2}{b_1 + b_2 - b_1 b_2}\right)\delta(z)\frac{a_1}{b_1}$$

It can be seen that the quantity $\delta(z)(1 - (b_1 a_2)/(b_1 + b_2 - b_1 b_2))$ measures the fraction of the time that station 1 is down during which station 2 is idle. This is the *loss transfer coefficient*, λ_{12}, and it can be seen that the loss transfer coefficient is approximately equal to $\delta(z)$, the failure transfer coefficient. Similarly, the loss transfer coefficient from station

2 to station 1, λ_{21}, is approximately equal to the failure transfer coefficient $\delta'(z)$. Note that one can write

$$E[N] = 1 + x_2 + x_1\lambda_{12}$$

and

$$E[K] = 1 + x_1 + x_2\lambda_{21}$$

6.6.3 Equivalent Single-Machine Approximation for Two-Stage Lines

It is sometimes desirable for modeling purposes to replace two stations separated by a finite inventory bank by a single equivalent station having operation-dependent geometric time to failure and geometric repair time distributions. The parameters of the equivalent single station will depend on whether the two-station system is viewed from its output or input. Denote by a_O, b_O the parameters of the output equivalent single station and by a_I, b_I the parameters of the input equivalent single station. The distribution of the interoutput time of the output equivalent single station will be

$$f(k : a_O, b_O) = \begin{cases} 1 - a_O, & k = 1 \\ a_O b_O (1 - b_O)^{n-2}, & k \geq 1 \end{cases}$$

Now suppose we had a set of n observations of the interoutput time of a system. These observations consist of r observations $k_1, k_2, \ldots, k_r$ where each $k_i > 1$ and $n - r$ observations where the interoutput time is unity (or one cycle). Given these observations we would like to fit the distribution $f(k : a_O, b_O)$ and estimate the parameters of the distribution. Denote by $\hat{a}_O$ and $\hat{b}_O$ the estimates of the parameters. Then there would be two different ways by which these parameters could be estimated, *method of moments* and *maximum likelihood*. For the method of moments we would calculate the first two moments of the observations and then estimate the parameters from the relationships

$$E[N] = 1 + a_O/b_O$$

$$E[N(N - 1)] = 2a_O/b_O^2$$

For the method of maximum likelihood we would form the likelihood function $\mathcal{L}$ of the observations, which would be given by

$$\mathcal{L} = (1 - a_O)^{n-r} \prod_{i=1}^{r} a_O b_O (1 - b_O)^{k_i - 2}$$

$$= (1 - a_O)^{n-r} a_O^r b_O^r (1 - b_O)^{\sum_{i=1}^{r}(k_i - 2)}$$

Hence

$$\ln \mathcal{L} = (n - r)\ln(1 - a_O) + r\ln a_O + r\ln b_O + \left(\sum_{i=1}^{r}(k_i - 2)\right)\ln(1 - b_O)$$

The maximum likelihood estimates of a_O and b_O are obtained by setting $\partial \mathcal{L}/\partial a_O = 0$ and $\partial \mathcal{L}/\partial b_O = 0$. This gives the following formulas for the maximum likelihood estimators

$$\hat{a}_O = \frac{r}{k}$$

$$\hat{b}_O = \frac{r}{\sum_{i=1}^{r}(k_i - 1)}$$

That is, the maximum likelihood estimator of a_O is the fraction of interoutput times that exceed unity, whereas the maximum likelihood estimator of b_O is the reciprocal of the mean duration of the amount by which all non-unit interoutput times exceeds unity.

Thus, given the distribution of the interoutput times of a two-station system, the parameters of the output equivalent single station can be found either by the method of moments or by a method based on maximum likelihood.

Method of moments.   Using equations 6.48 and 6.49, the output equivalent single station will have its parameters determined by

$$E[N] = 1 + \frac{a_O}{b_O} = 1 + \frac{a_2}{b_2} + a_1 \delta(z) \left(\frac{1}{b_1} - \frac{a_2}{b_1 + b_2 - b_1 b_2} \right)$$

$$E[N(N-1)] = 2\frac{a_O}{b_O^2} = 2\frac{a_2}{b_2^2} + 2a_1 \delta(z) \left(\frac{1}{b_1^2} - \frac{a_2}{(b_1 + b_2 - b_1 b_2)^2} \right)$$

hence

$$b_O = \frac{\frac{a_2}{b_2} + a_1 \delta(z) \left(\frac{1}{b_1} - \frac{a_2}{b_1 + b_2 - b_1 b_2} \right)}{\frac{a_2}{b_2^2} + a_1 \delta(z) \left(\frac{1}{b_1^2} - \frac{a_2}{(b_1 + b_2 - b_1 b_2)^2} \right)}$$

$$a_O = \frac{\left(\frac{a_2}{b_2} + a_1 \delta(z) \left(\frac{1}{b_1} - \frac{a_2}{b_1 + b_2 - b_1 b_2} \right) \right)^2}{\frac{a_2}{b_2^2} + a_1 \delta(z) \left(\frac{1}{b_1^2} - \frac{a_2}{(b_1 + b_2 - b_1 b_2)^2} \right)}$$

Note that if $a_2 \ll 1$ and $b_1 = b_2 = b$ then $b_O = b$ and $a_O = a_2 + a_1 \delta(z)$. It can be seen that $a_O > 0$ and $b_O > 0$, and $b_O < 1$. It is not always true that $a_O < 1$.

Maximum likelihood.   The formulas given previously for the maximum likelihood estimation from a set of observations suggest that if a complex distribution is given then the equivalent a_O and b_O, denoted by $a_O^{\mathcal{L}}$ and $b_O^{\mathcal{L}}$, would be given by

$$a_O^{\mathcal{L}} = P\{N > 1\} = 1 - P\{N = 1\}$$

$$b_O^{\mathcal{L}} = \frac{1}{E[N - 1 | N > 1]}$$

Hence for the two stations separated by a finite buffer

$$a_O^{\mathcal{L}} = a_2 + (1 - a_2)a_1\delta(z) = \frac{1}{E[U_O]}$$

$$b_O^{\mathcal{L}} = \frac{P\{N > 1\}}{E[N] - 1}$$

$$= \frac{1}{E[U_O]} \frac{1}{1/\eta(z) - 1}$$

$$= \frac{1}{E[D_O]}$$

Thus the maximum likelihood approach is equivalent to basing the equivalence on the mean up time and the mean down time. Note that $0 < a_O^{\mathcal{L}} < 1$ and $0 < b_O^{\mathcal{L}} < 1$ always, thus this approach is in this respect superior to the method of moments.

6.6.4 Other Solvable Two-Machine Systems

There are several other two-station synchronous systems for which closed-form solutions exist for the state probabilities, and thus formulas for the line efficiency and other performance measures exist. We summarize a selection of these results for stations with operation-dependent failures and comment on the insights that they provide. As a shorthand for presenting these results the following notation will be used. Denote by $A/B : r$ a two-station system in which A denotes the distribution of the operating time between failures of a station, B denotes the distribution of the repair time, and r denotes the number of repairmen ($r = 1$ or $r = 2$). Then the following are some of the situations in which closed-form results apply.

Ge/D: **2.** Special conditions are that the machine up times are geometrically distributed, and the repair times of the stations are the same constant $\overline{D}$ for both stations. Also, either (1) $z = l\overline{D}$ where l is an integer or (2) $0 \leq z \leq \overline{D}$. Let a_1 and a_2 be the failure probability per cycle of stations 1 and 2, respectively.

1. $z = l\overline{D}$, l integer

$$\eta_{Ge/D}(z) = \begin{cases} \dfrac{1 - r_d^* \sigma_d^l}{1 + a_1\overline{D} - r_d^* \sigma_d^l(1 + a_2\overline{D})} & a_1 \neq a_2 \\[2ex] \dfrac{1 + l(1 + a\overline{D})}{1 + 2a\overline{D} - a^2\overline{D} + l(1 + a\overline{D})^2} & a_1 = a_2 = a \end{cases} \tag{6.50}$$

where

$$r_d^* = \frac{a_2(1 - a_1)}{a_1(1 - a_2)}$$

$$\sigma_d = \frac{a_2(1 - a_1)^{\overline{D}+1}}{a_1(1 - a_2)^{\overline{D}+1}}$$

Other parameters

$$\frac{1}{E[U_O]} = a_2 + (1 - a_2)a_1\delta_d(z) \tag{6.51}$$

where

$$\delta_d(z) = \begin{cases} \frac{1 - (1-a_1)/(1-a_2)}{a_1(1-r_d^*\sigma_d^l)} & a_1 \neq a_2 \\ \frac{1}{1+l(1+a\overline{D})} & a_1 = a_2 = a \end{cases}$$

$$\frac{1}{E[U_I]} = a_1 + (1 - a_1)a_2\delta_d'(z) \tag{6.52}$$

where

$$\delta_d'(z) = \sigma_d^l\delta_d(z)$$

$$E[D_O] = E[D_I] = \overline{D} \tag{6.53}$$

and

$$E[N] = 1 + a_2\overline{D} + (1 - a_2)a_1\overline{D}\delta_d(z)$$
$$= \begin{cases} 1 + a_2\overline{D} + \frac{(a_1 - a_2)\overline{D}}{1 - r_d^*\sigma_d^l}, & a_1 \neq a_2 \\ 1 + a\overline{D} + \frac{a\overline{D}(1-a)}{1+l(1+a\overline{D})}, & a_1 = a_2 = a \end{cases} \tag{6.54}$$

If it is required to find a single equivalent station with geometric time to failure and deterministic repair time then the maximum likelihood approach to determining the parameters of the output equivalent station would result in $a_O = 1/E[U_O]$ and $E[D_O] = \overline{D}$, whereas the input equivalent station would have parameters $a_I = 1/E[U_I]$ and $E[D_I] = \overline{D}$.

2. $0 \leq z < \overline{D}, a_1 = a_2 = a$

$$\eta_{Ge/D}(z) = \frac{2 + az}{(2 + az)(1 + a\overline{D}) + a(1 - a)(\overline{D} + (\overline{D} - z)/(1 + az))} \tag{6.55}$$

$$E[U] = a + (1 - a)a\frac{1}{1 + az} \tag{6.56}$$

$$E[D] = \overline{D} - \frac{z(1 - a)}{(1 + az)(2 + a(z - 1))} \tag{6.57}$$

$$E[N] = 1 + a\overline{D} + \frac{a(1 - a)}{1 + az}\left(\overline{D} - \frac{z}{2 + az}\right) \tag{6.58}$$

D/Ge: 2.   If the stations are identical with the time to failure of a station equal to t^* cycles and repair probability per cycle equal to b then the performance depends on whether or not it is possible for the two stations to be simultaneously under repair. If when the bank inventory is zero the downstream station has an age (working time since last failure) that is k cycles greater than the age of the upstream station and $z < k < t^*$

then it is not possible for the stations to be down simultaneously. The efficiency formulas in the two cases are:

$$\eta_{D/Ge}(z) = \begin{cases} \frac{1+(z-1)b}{(1-b)(1+2/t^*b)+zb(1+1/t^*b)} & z < k < t^* \\ \frac{2+(z-1)b}{(2-b)(1+2/t^*b)+zb(1+1/t^*b)-1/t^*b} & 0 \leq k \leq z \end{cases} \tag{6.59}$$

D/D: 2.   If the stations are identical and all times are deterministic then performance depends on whether a station can ever be idle or blocked. This can only occur if when the bank inventory is zero the downstream station has an age that is k cycles greater than the age of the upstream station, $z < k < t^*$ and $z < \overline{D}$ where $\overline{D}$ is the repair time. The efficiency formulas are

$$\eta_{D/D}(z) = \begin{cases} \frac{t^*}{t^*+2\overline{D}-z} & z < k < t^*, \quad z < \overline{D} \\ \frac{t^*}{t^*+\overline{D}} & \text{otherwise} \end{cases} \tag{6.60}$$

Ge/Ge: 1 priority.   If there is only one repairman it is necessary to specify which station the repairman would work on if both stations are failed. A variety of rules can be proposed. It would seem reasonable to consider only those rules where once a repairman begins repair of a station he will continue with that repair until it is completed. The only situation where the repairman would then have a choice would be if both stations failed in the same cycle. Possible rules would then be: *strict priority*—always repair a specific station; *inventory based*—select the station to repair based on the inventory level in the bank. The idea behind the inventory-based rule is that if both stations fail and the bank is empty then it would be better to work on the upstream station as once its repair is completed the station can start working. If the downstream station were repaired first it could not start working until the upstream station is repaired. The inventory-based rule specifies an inventory level y where the repairman works on the upstream stage if the bank level on simultaneous failure is at or below y; otherwise he works on the downstream stage. The formula for the efficiency when $a_1 = a_2 = a$, $b_1 = b_2 = b$ and $\hat{x} = a/b$ is

$$\eta_{Ge/Ge:1} = \frac{1}{1 + 2\hat{x} - g(y, z)} \tag{6.61}$$

where

$$g(y, z) = \frac{a(\sigma^y + \sigma^{y+1} - 1 - \sigma^{2y+1-z})}{(a + b - ab)(\sigma^y + \sigma^{y+1} - 1 - \sigma^{2y+1-z}) + (1 - a)(2 - b)\sigma^y(\sigma - 1)}$$

with $\sigma = 1 + ab/(2 - a - b)$. It can be shown that the optimal y for given z is then $y = z/2$ for z even and $y = (z - 1)/2$ for z odd.

Ge(q)/Ge(1 − q): 2.   A simple model that is occasionally useful is that where the operation of a station is viewed as a sequence of Bernoulli trials. That is, the probability that a station is up during a cycle is set at p and the probability the station is down during a cycle is $q = 1 - p$. This is then equivalent to setting in the geometric time to failure Ge and geometric repair time Ge model $a = q$ and $b = p$ or $b = 1 - a$. It follows that $\alpha_1 = q_1$, $\alpha_2 = q_2$, $\beta_1 = p_1$ and $\beta_2 = p_2$, thus $\sigma = x_2/x_1$ and if the stages are identical

$$\eta(z) = (1 - q)\frac{1 + q + z}{1 + 2q + z} \tag{6.62}$$

Insights.   If the two stations are identical it can be seen that $\eta_{Ge/D}(z) \approx \eta_{Ge/Ge}(2z)$, and $\eta_{D/Ge}(z) \approx \eta_{Ge/Ge}(2z)$, that is, a given bank is twice as effective if either the repair times or the times to failure are deterministic. It is desirable to reduce the variability of either repair times or times to failure to improve the effectiveness of a given bank. Repair time variability can be reduced by preplanning repair activity, but it is also dependent on the mix of failure causes and the consequent repair actions. Thus attempting to reduce the variability of time to failure will also result in a reduction in the variability of the repair times.

6.6.5 Multiple-Stage Transfer Line

In this section we consider an m-stage transfer line with $m - 1$ banks of capacities $z_1, \ldots, z_{m-1}$. The number of cycles of operation before failure and the repair times all have geometric distribution with mean $1/a_i$ and $1/b_i$, respectively, for stage i, $i = 1, \ldots, m$. Then we say that stage i has parameters (a_i, b_i), $i = 1, \ldots, m$. Let $Y_n^{(i)}$ be the state of stage i (i.e., W if working, R if under repair, B if blocked and I if idle) and $X_n^{(i)}$ be the number of parts in bank i (so $0 \leq X_n^{(i)} \leq z_i$) at time n. Then $\{(Y_n^{(i)}, i = 1, \ldots, m; X_n^{(i)}, i = 1, \ldots, m - 1), n = 1, 2, \ldots\}$ is a Markov chain. The number of states of this Markov chain is approximately $2^m \prod_{i=1}^{m-1}(z_i + 1)$ and is very large for even moderately sized transfer lines. Because there are no simple analytical closed-form solutions available for the stationary distribution of this Markov chain and numerical solution tends to be problematic we will look for an approximate solution next.

The basic concept of the approximation approach is the recognition that viewed from any inventory bank the line appears to have two stages, with the arrivals at the bank determined by the upstream stages and the departures from the bank determined by the downstream stages. If the upstream stages are replaced by a single equivalent station and the downstream stages are replaced by a single equivalent station, then, viewed from the bank, the system appears to consist of the bank and two stations. If the equivalent stations have time to failure and repair time distributions for which the two-station system can be solved then the throughput of the system can be determined. The approximation arises because, even when the individual stages have distributions that in a two-station system would be solvable, grouping the upstream or downstream stages results in the grouped stages no longer having the same form of distributions of time to failure and time to repair as their constituent stages. To use the known two-stage results, however, it is necessary to approximate the actual distributions by distributions for which the two-stage system has a solution.

The specifics of the approximation procedure will be described for a system where each stage j, $j = 1, \ldots, m$ has geometric operation-dependent time to failure with failure probability a_j and geometric repair time with repair probability b_j; however, the procedure can easily be adapted to the case in which the stages have geometric time to failure and identical deterministic repair time $\overline{D}$ with all banks having a capacity such that $z_j = l_j \overline{D}$ where l_j is an integer. Viewed from bank j the upstream stages will be assumed to have a geometric time to failure with failure probability a_j^U and geometric repair time with repair probability b_j^U, whereas the downstream stages will be assumed to have geometric time to failure with failure probability a_{j+1}^D and geometric repair time with repair probability b_{j+1}^D. For the system consisting of the equivalent upstream station, bank j and the equivalent downstream station, the failure transfer coefficients $\delta(z_j)$ and $\delta'(z_j)$ can be determined. Write

$$\delta_{j,j+1} = \delta(z_j, a_j^U, b_j^U, a_{j+1}^D, b_{j+1}^D), \qquad 1 \le j < m$$

$$\delta_{j+1,j} = \delta'(z_j, a_j^U, b_j^U, a_{j+1}^D, b_{j+1}^D), \qquad 1 \le j < m$$

Next, the quantities a_j^U, b_j^U, a_{j+1}^D and b_{j+1}^D have to be determined from the parameters of the individual stages and the δ's. This will be done using the maximum likelihood approach discussed earlier for replacing a two-stage line by a single equivalent station. First, consider the two-stage line obtained by viewing the upstream and downstream stages from bank $j - 1$. The (maximum likelihood) parameters of these two stages are (a_{j-1}^U, b_{j-1}^U) and (a_j^D, b_j^D). The parameters of the single station equivalent to the output of this two-stage line are (a_j^*, b_j^*) where $a_j^* = a_j^D + c_1$, and $a_j^*/b_j^* = a_j^D/b_j^D + c_2$. Here c_1 and c_2 are two correction terms given by

$$c_1 = (1 - a_j^D)a_{j-1}^U \delta_{j-1,j}$$

and

$$c_2 = \frac{a_{j-1}^U}{b_{j-1}^U}\left(1 - a_j^D \frac{b_{j-1}^U}{b_j^D + b_{j-1}^U - b_j^D b_{j-1}^U}\right)\delta_{j-1,j}$$

Now if we view the upstream of bank j assuming that stage j is never blocked the corresponding parameters of the aggregated upstream stage are (a_j^U, b_j^U) given by $a_j^U = a_j + c_1$ and $a_j^U/b_j^U = a_j/b_j + c_2$. The rationale behind this representation is that when stage j is never blocked the terms a_j^D and b_j^D in a_j^* and b_j^* should be replaced by a_j and b_j, respectively, to obtain a_j^U and b_j^U. Hence we have

$$a_j^U = a_j + (1 - a_j^D)a_{j-1}^U \delta_{j-1,j}, \quad 2 \le j \le m - 1$$

$$b_j^U = a_j^U \Big/ \left(\frac{a_j}{b_j} + \frac{a_{j-1}^U}{b_{j-1}^U}\delta_{j-1,j}\left(1 - a_j^D \frac{b_{j-1}^U}{b_j^D + b_{j-1}^U - b_j^D b_{j-1}^U}\right)\right), \quad 2 \le j \le m$$

$$a_1^U = a_1$$

$$b_1^U = b_1$$

where it must be pointed out that the $\delta_{j-1,j}$ is calculated using the a_{j-1}^U, b_{j-1}^U, a_j^D and b_j^D. Similarly, considering the two-stage line obtained by viewing the upstream and downstream stages from bank $j+1$, the parameters of the downstream equivalent station will be given by

$$a_{j+1}^D = a_{j+1} + (1 - a_{j+1}^U)a_{j+2}^D \delta_{j+2,j+1}, \quad 0 \le j \le m-2$$

$$b_{j+1}^D = a_{j+1}^D / \left(\frac{a_{j+1}}{b_{j+1}} + \frac{a_{j+2}^D}{b_{j+2}^D} \delta_{j+2,j+1} \left(1 - a_{j+1}^U \frac{b_{j+2}^D}{b_{j+1}^U + b_{j+2}^D - b_{j+1}^U b_{j+2}^D} \right) \right),$$

$$1 \le j \le m-2$$

$$a_m^D = a_m$$
$$b_m^D = b_m$$

There are thus $4(m-1)$ equations in $4(m-1)$ unknowns (the a_j^U, b_j^U, a_{j+1}^D and b_{j+1}^D). Let $\{(a_j^U, b_j^U, a_j^D, b_j^D), j = 1, \ldots, m\}$ be a solution to the preceding equations. Then it can be verified that the efficiency $\eta(z_j, a_j^U, b_j^U, a_{j+1}^D, b_{j+1}^D)$ of the two-stage line is the same for all $j = 1, \ldots, m-1$. Specifically we have

$$\eta^O := \frac{1}{1 + a_m^U / b_m^U} = \eta(z_j, a_j^U, b_j^U, a_{j+1}^D, b_{j+1}^D)$$

$$= \eta^I := \frac{1}{1 + a_1^D / b_1^D}, \quad j = 1, \ldots, m-1$$

Because the equations are nonlinear they are best solved using an iterative technique, where convergence is checked by determining whether $\eta^I = \eta^O$.

Algorithm 6.1   Multistage Transfer Line

Step 0: Set $k = 1$, $a_j^D(0) = a_j$, $b_j^D(0) = b_j$, $j = 2, \ldots, m$

Step 1: Set $a_1^U(k) = a_1$, $b_1^U(k) = b_1$.
For $j = 1, 2, \ldots, m-1$ analyze the two stage transfer line with parameters $(a_j^U(k), b_j^U(k))$ and $(a_{j+1}^D(k-1), b_{j+1}^D(k-1))$ and determine $a_{j+1}^U(k)$ and $b_{j+1}^U(k)$.

Step 2: Set $a_m^D(k) = a_m$, $b_m^D(k) = b_m$. For $j = m-1, m-2, \ldots, 1$ analyze the two-stage transfer line with parameters $(a_{j+1}^D(k), b_{j+1}^D(k))$ and $(a_j^U(k), b_j^U(k))$ and determine $a_j^D(k)$ and $b_j^D(k)$.

Step 3: Calculate $\eta^O = 1/(1 + a_m^U(k)/b_m^U(k))$ and $\eta^I = 1/(1 + a_1^D(k)/b_1^D(k))$. If $|\eta^O - \eta^I| < \epsilon$ stop, otherwise set $k = k+1$ and go to step 1.

We will now establish the convergence of the preceding algorithm. For this we will need the monotonicity properties of $\delta(z, a_1, b_1, a_2, b_2)$ with respect to a_1, b_1, a_2, and b_2. Recall that we have already shown that $\delta(z, a_1, b_1, a_2, b_2)$ is decreasing in a_2 and a_2/b_2. Because σ, $\{a_2(\sigma + \beta_1/\alpha_1)b_1/(b_1 + b_2 - b_1 b_2)\}/a_1$ and $\{a_2(\sigma + \beta_1/\alpha_1)b_1/(b_1 + b_2 - $

$b_1 b_2)\}/\{(a_1/b_1)(1 - b_1 a_2/(b_1 + b_2 - b_1 b_2))\}$ are all decreasing in a_1 and a_1/b_1, it is easily seen that $a_1 \delta(z, a_1, b_1, a_2, b_2)$ and $(a_1/b_1)\delta(z, a_1, b_1, a_2, b_2)(1 - b_1 a_2/(b_1 + b_2 - b_1 b_2))$ are increasing in a_1 and a_1/b_1. Hence from the equations for $a_{j+1}^U(k)$ and $b_{j+1}^U(k)$ it is clear that if

1. $a_j^U(k - 1) \geq a_j^U(k);$ $\dfrac{a_j^U(k-1)}{b_j^U(k-1)} \geq \dfrac{a_j^U(k)}{b_j^U(k)}$ and

2. $a_{j+1}^D(k - 2) \leq a_{j+1}^D(k - 1);$ $\dfrac{a_{j+1}^D(k-2)}{b_{j+1}^D(k-2)} \leq \dfrac{a_{j+1}^D(k-1)}{b_{j+1}^D(k-1)}$

then

$$a_{j+1}^U(k - 1) \geq a_{j+1}^U(k); \quad \frac{a_{j+1}^U(k-1)}{b_{j+1}^U(k-1)} \geq \frac{a_{j+1}^U(k)}{b_{j+1}^U(k)}$$

Observe that $a_j^D(k) \geq a_j^D(0)$ and $a_j^D(k)/b_j^D(k) \geq a_j^D(0)/b_j^D(0)$, $j = 1 \ldots, m - 1$ and $a_1^U(k) = a_1$; $a_1^U(k)/b_1^U(k) = a_1/b_1$. Therefore conditions (1) and (2) are satisfied for $k = 2$ and therefore

$$a_j^U(2) \geq a_j^U(1), \quad j = 1, \ldots, m - 1$$

$$a_j^U(2)/b_j^U(2) \geq a_j^U(1)/b_j^U(1), \quad j = 1, \ldots, m - 1$$

Similarly using the observation $a_m^D(k) = a_m$ and $a_m^D(k)/b_m^D(k) = a_m/b_m$ one gets

$$a_j^D(1) \leq a_j^D(0), \quad j = 2, \ldots, m$$

$$a_j^D(1)/b_j^D(1) \leq a_j^D(0)/b_j^D(0), \quad j = 2, \ldots, m$$

Continuing this argument it can be concluded that $a_j^U(k)$ and $a_j^U(k)/b_j^U(k)$ are decreasing in k and $a_j^D(k)$, and $a_j^D(k)/b_j^D(k)$ are increasing in k. Because these quantities are bounded (for $b_j > 0$, $j = 1, \ldots, m$) the convergence of the recursive procedure is then immediate.

Next we will show that the set of equations defining $(a_j^U, b_j^U, a_j^D, b_j^D)$, $j = 1, \ldots, m$ has a unique solution. To the contrary assume that there are two solutions

$$\{(a_j^{U:i}, b_j^{U:i}, a_j^{D:i}, b_j^{D:i}), j = 1, \ldots, m\}, \quad i = 1, 2$$

that are distinct. Let $\hat{a}_j^D = \max\{a_j^{D:i}, i = 1, 2\}$, $\hat{a}_j^D/\hat{b}_j^D = \max\{a_j^{D:i}/b_j^{D:i}, i = 1, 2\}$, $\hat{a}_j^U = \min\{a_j^{U:i}, i = 1, 2\}$ and $\hat{a}_j^U/\hat{b}_j^U = \min\{a_j^{U:i}/b_j^{U:i}, i = 1, 2\}$, $j = 1, \ldots, m$. Now start the recursion in algorithm 6.1 with $a_j^D(0) = \hat{a}_j^D$ and $b_j^D(0) = \hat{b}_j^D$, $j = 2, \ldots, m$ and continue until it converges to the final solution $\{a_j^{U*}, b_j^{U*}, a_j^{D*}, b_j^{D*}\}$, $j = 1, \ldots, m\}$. Suppose for some i ($i = 1$ or 2) and l ($l = 1, \ldots, m - 1$), either $\hat{a}_l^D > a_l^{D:i}$ or $\hat{a}_l^D/\hat{b}_l^D > a_l^{D:i}/b_l^{D:i}$. Then by the monotonicity properties we discussed for establishing the convergence of the algorithm, we see that $a_j^{D*} > a_j^{D:i}$, $a_j^{D*}/b_j^{D*} > a_j^{D:i}/b_j^{D:i}$, $j = 1, \ldots, m-1$ and $a_j^{U*} < a_j^{U:i}$, $a_j^{U*}/b_j^{U*} < a_j^{U:i}/b_j^{U:i}$, $j = 2, \ldots, m$. Therefore $\eta^{0*} = 1/(1 + a_m^{U*}/b_m^{U*}) > 1/(1 + a_m^{U:i}/b_m^{U:i}) = 1/(1 + a_j^{D:i}/b_j^{D:i}) > 1/(1 + a_1^{D*}/b_1^{D*}) = \eta^{I*}$. Hence $\{(a_j^{U*}, b_j^{U*}, a_j^{D*}, b_j^{D*}), j = 1, \ldots, m\}$ cannot be a solution. This contradicts our assumption. Hence there cannot exist an i and an l such that either $\hat{a}_l^D > a_l^{D:i}$ or $\hat{a}_l^D/\hat{b}_l^D > a_l^{D:i}/b_l^{D:i}$. So $a_j^{D:1} = a_j^{D:2}$ and $b_j^{D:1} = b_j^{D:2}$, $j = 1, \ldots, m$. Similarly it can be argued that $a_j^{U:1} = a_j^{U:2}$ and $b_j^{U:1} = b_j^{U:2}$, $j = 1, \ldots, m$. Hence the uniqueness of the solution.

Convergence to a solution is usually very rapid, and the solution is usually very close to the exact line efficiency. Table 6.7 shows exact and approximate results for a number of three-stage lines, whereas Table 6.8 compares the efficiency given by the approximation algorithm with simulation results for multiple-stage lines with identical stages.

TABLE 6.7 THROUGHPUT OF THREE STATION SYSTEMS

Case	a_1	a_2	a_3	b_1	b_2	b_3	z_1	z_2	η Exact	η Approx.
1	0.03	0.05	0.02	0.20	0.20	0.20	15	15	0.777846	0.777759
2	0.01	0.02	0.005	0.20	0.10	0.15	15	3	0.814949	0.814970
3	0.7	0.9	0.6	0.3	0.4	0.9	7	5	0.285463	0.285463
4	0.9	0.05	0.6	0.4	0.4	0.4	10	10	0.307692	0.307692
5	0.001	0.0003	0.005	0.7	0.02	0.3	8	4	0.970335	0.970337
6	0.6	0.04	0.6	0.8	0.4	0.8	9	9	0.565423	0.571365

TABLE 6.8 THROUGHPUT OF MULTISTAGE SYSTEMS

No. of	Parameters		Bank	Efficiency	
stages	a	b	size	Approx.	Sim. (95% conf. interval)
6	0.1	0.1	8	0.31211	0.31028 ± 0.00501
8	0.1	0.1	8	0.30185	0.29774 ± 0.00292
10	0.1	0.1	8	0.29611	0.29212 ± 0.00497
10	0.01	0.1	10	0.74789	0.75969 ± 0.01448

There are some situations in which convergence of the algorithm is very slow, for example, in some three-station systems in which the failure probability of the middle stage is much less than that of the outer stages (case 6 in Table 6.7). Indeed, in case 6 a better approximation to the efficiency is obtained by ignoring failures of the middle stage. The three-stage line is then equivalent to a two-stage line with an interstage bank capacity of 18. The efficiency of this two-stage line is 0.565916.

If all the b_j are identical and equal to b and the a_j's are sufficiently small that it is unlikely more than one stage is failed at any time, then the preceding $4(m-1)$

equations can be reduced to the following $2(m - 1)$ equations by setting $b_j^U = b_{j+1}^D = b$ for $1 \leq j \leq m - 1$:

$$a_j^U = a_j + a_{j-1}^U \hat{\delta}_{j-1,j}, \quad 2 \leq j \leq m$$

$$a_{j+1}^D = a_{j+1} + a_{j+2}^D \hat{\delta}_{j+2,j+1}, \quad 0 \leq j \leq m - 2$$

$$a_1^U = a_1$$

$$a_m^D = a_m$$

Here

$$\hat{\delta}_{j-1,j} = \hat{\delta}(z_{j-1}, a_{j-1}^U, b, a_j^D, b)$$

$$\hat{\delta}_{j+1,j} = \hat{\delta}'(z_j, a_j^U, b, a_{j+1}^D, b)$$

$$\hat{\delta}'(z, a_1, b, a_2, b) = \hat{\delta}(z, a_2, b, a_1, b)$$

and

$$\hat{\delta}(z, a_1, b, a_2, b) = \frac{1 - a_2/a_1}{1 - \frac{a_2}{a_1} \left(\frac{a_1 + a_2 - a_1 b}{a_1 + a_2 - a_2 b} \right)^z}$$

When $a_1 = a_2$

$$\hat{\delta}(z, a, b, a, b) = \frac{1}{1 + \frac{zb}{2-b}}$$

The efficiency of the line can be found from $\eta^O = 1/(1 + a_m^U/b)$ or $\eta^I = 1/(1 + a_1^D/b)$, or by determining the efficiency of any of the two-stage lines with parameters (a_j^U, b) and (a_{j+1}^D, b) and inventory bank capacity z_j by the efficiency formula

$$\hat{\eta}(z_j, a_j^U, b, a_{j+1}^D, b) = \frac{1}{1 + (a_j^U \hat{\delta}_{j,j+1} + a_{j+1}^D)/b}$$

It is not difficult to show that the same efficiency is obtained for all j. That is

$$\hat{\eta}(z_j, a_j^U, b, a_{j+1}^D, b) = \hat{\eta}(z_{j+1}, a_{j+1}^U, b, a_{j+2}^D, b), \quad j = 1, \ldots, m - 2$$

Therefore the preceding $m - 2$ equations may be used to replace $(m - 2)$ of those specified previously for a_j^U's and a_{j+1}^D's and still solve for $\{(a_j^U, a_{j+1}^D), j = 1, \ldots, m - 1\}$. In this regard for a given efficiency x let $a_j^{U:f}(x)$ and $a_j^{D:f}(x)$ be the solution to

$$x = \eta(z_j, a_j^{U:f}(x), b, a_{j+1}^{D:f}(x), b), \quad j = 1, \ldots, m - 1$$

and

$$a_j^{U:f}(x) = a_j + a_{j-1}^{U:f}(x) \delta(z_{j-1}, a_{j-1}^{U:f}(x), b, a_j^{D:f}(x), b), \quad j = 2, \ldots, m - 1$$

starting with $a_1^{U:f}(x) = a_1$. Observe that $a_j^{D:f}(x)$ is decreasing and $a_j^{U:f}(x)$ is increasing in x. Similarly, for a given efficiency x let $a_j^{U:b}(x)$ and $a_j^{D:b}(x)$ be the solution to

$$x = \eta(z_j, a_j^{U:b}(x), b, a_{j+1}^{D:b}(x), b), \quad j = m-1, \ldots, 1$$

and

$$a_j^{D:b}(x) = a_j + a_{j+1}^{D:b}(x)\delta(z_j, a_j^{U:b}(x), b, a_{j+1}^{D:b}(x), b), \quad j = m-1, \ldots, 2$$

starting with $a_m^{D:b}(x) = a_m$. Observe that $a_j^{U:b}(x)$ is decreasing and $a_j^{D:b}(x)$ is increasing in x. Then the "approximate" efficiency of the line is the value x^* of x that satisfies $a_j^{U:f}(x^*) = a_j^{U:b}(x^*)$ and $a_j^{D:f}(x^*) = a_j^{D:b}(x^*)$. Therefore, if, for some x, we find $a_j^{U:f}(x) < a_j^{U:b}(x)$ or $a_j^{D:f}(x) > a_j^{D:b}(x)$ then $x^* > x$.

Allocation of failure probabilities. In this section we will consider a multi-stage transfer line with equal repair probabilities (b). The total of the failure probabilities $\sum_{i=1}^{m} a_i = A$ is to be allocated to the m stages. Let $\mathbf{a} = (a_1, \ldots, a_m)$ be a given allocation, and let $\{(a_j^U, a_j^D), j = 1, \ldots, m\}$ be the corresponding solution to the set of equations specified earlier. Also let $\hat{\eta}(\mathbf{a})$ be the efficiency of the system for this allocation. Then we are interested in characterizing the allocation $\mathbf{a}^*$ that maximizes this efficiency: $\max\{\eta(\mathbf{a}) : \sum_{i=1}^{m} a_i = A\}$. Suppose for the given allocation $\mathbf{a}$, $a_l^U < a_{l+1}^D$. Now choose an allocation $\mathbf{a}'$ such that $a_j' = a_j, j = 1, \ldots, m, j \neq l, l+1$ and $a_l'+a_{l+1}' = a_l+a_{l+1}$ and compute the corresponding values of the $a_j^{U:f}$'s, $a_j^{D:f}$'s, $a_j^{U:b}$'s and $a_j^{D:b}$'s for $x = \eta(\mathbf{a})$. It is easily seen that

$$a_j^{U:f}(\eta(\mathbf{a})) = a_j^U, \quad a_j^{D:f}(\eta(\mathbf{a})) = a_j^D, \quad j = 1, \ldots, l-1$$

and

$$a_j^{U:b}(\eta(\mathbf{a})) = a_j^U, \quad a_j^{D:b}(\eta(\mathbf{a})) = a_j^D, \quad j = m, \ldots, l+1.$$

Furthermore

$$a_l^{U:f}(\eta(\mathbf{a})) + a_{l+1}^{D:b}(\eta(\mathbf{a})) = a_l^U + a_{l+1}^D$$

We now assume that a_l' and a_{l+1}' are chosen such that

$$a_l^U > \max\{a_l^{U:f}(\eta(\mathbf{a})), a_{l+1}^{D:b}(\eta(\mathbf{a}))\} \geq \min\{a_l^{U:f}(\eta(\mathbf{a})), a_{l+1}^{D:b}(\eta(\mathbf{a}))\} > a_{l+1}^D$$

Then

$$\eta(z_l, a_l^{U:f}(\eta(\mathbf{a})), b, a_{l+1}^{D:f}(\eta(\mathbf{a})), b) = \eta(\mathbf{a}) < \eta(z_l, a_l^{U:f}(\eta(\mathbf{a})), b, a_{l+1}^{D:b}(\eta(\mathbf{a})), b)$$

Therefore $a_{l+1}^{D:b}(\eta(\mathbf{a})) < a_{l+1}^{D:f}(\eta(\mathbf{a}))$ and hence $\eta(\mathbf{a}') > \eta(\mathbf{a})$. Therefore we see that if $a_l^U > a_{l+1}^D$, then a better allocation that balances the upstream and downstream failure probabilities can be obtained. Similarly if $a_l^U > a_{l+1}^D$, it can be shown that a better allocation of failure probabilities exist such that the upstream and downstream failure probabilities are balanced. From this we can conclude that the optimal allocation $\mathbf{a}^*$

will be such that $a_j^{U*} = a_{j+1}^{D*}$, $j = 1, \ldots, m-1$. When $z_j = z$, $j = 1, \ldots, m-1$, the allocation

$$a_1^* = \frac{A(zb + 2 - b)}{mzb + 4 - 2b} = a_m^*$$

and

$$a_j^* = \frac{Azb}{mzb + 4 - 2b}, \quad j = 2, \ldots, m-1$$

satisfies the preceding condition and is therefore optimal. Note that $a_j^* = a_1^*(1 - \delta(z, a, b, a, b))$ for $j = 2, \ldots, m-1$.

It can be shown that if a line is divided into m stages with an optimal allocation of the failure probabilities and if a total bank capacity of Z is divided equally among the $m - 1$ banks then the efficiency will be

$$\eta(Z) = \frac{1}{1 + \frac{A}{b} \frac{Zb + (m-1)(4-2b)}{mZb + (m-1)(4-2b)}} \tag{6.63}$$

Note that for large m the efficiency is independent of m.

Now suppose we have an m-stage transfer line with two repairmen and a single bank. We are interested in finding a location for the bank such that the efficiency of the system is maximized. The placement of the buffer will break the system into two stages where each stage failure will be repaired by one of the two repairmen. The repair probabilities are the same. Furthermore the failure probabilities are very small. Then from the previous results we see that the bank should be located between station j^* and $j^* + 1$ to maximize the "approximate" efficiency of the system, where

$$j^* = \arg\min\{j : |\sum_{i=1}^{j} a_i - \sum_{i=j+1}^{m} a_i|, j = 1, \ldots, m-1\}$$

6.7 ASYNCHRONOUS TRANSFER LINES WITH VARIABLE CYCLE TIMES

Up to this point we have considered transfer lines where the part movement is synchronized, that is, *fixed transfer* systems. It is also common to have *free transfer* lines where the stage operation is not synchronized, and thus movement in and out of the bank is determined solely by whether the downstream stage can accept a part or whether the upstream stage has a part available. The normal time required for the stage to perform a task may be deterministic, typical of lines processing a single part type using machines or robots, or it may be variable, either because the task is performed by people, or the line is a flexible transfer line with the processing requirements of each task varying from task to task.

In this section we consider an m-stage transfer line where the usual task time has an exponential distribution with mean $1/\mu_i$ at stage i, but it can break down, with breakdown not being observed until completion of the usual task time. The probability of breakdown during the task is a constant a_i, independent of the task duration. As soon as breakdown is observed repair is initiated, with the time to repair having an exponential distribution with mean $1/\gamma_i$.

6.7.1 Two-Stage Systems

Suppose the system consists of two stages separated by an inventory bank of capacity z. Then the states of the system are $A(t)$ for stage 1, $B(t)$ for stage 2, and $X(t)$ for the number of parts in the bank at time t. Then $\{(A(t), B(t), X(t)), t \geq 0\}$ is a Markov process on the state space $\mathcal{S} = \{\{(WI, 0); (RI, 0); (WW, x); (RW, x); (WR, x); (RR, x); 0 \leq x \leq z; (BW, z); (BR, z)\}$ where I denotes that the stage is idle and B denotes that the stage is blocked, and thus no processing occurs. It will be assumed that failures are operation dependent. Then if $p(\alpha, \beta, x) = \lim_{t \to \infty} P\{A(t) = \alpha, B(t) = \beta, X(t) = x\}$, $(\alpha, \beta, x) \in \mathcal{S}$ is the steady-state probability of $\{(A(t), B(t), X(t)), t \geq 0\}$, the balance equations describing the system behavior are

$$\mu_1 p(WI, 0) = \mu_2(1 - a_2)p(WW, 0) + \gamma_2 p(WR, 0)$$

$$\gamma_1 p(RI, 0) = \mu_1 a_1 p(WI, 0) + \mu_2(1 - a_2)p(RW, 0) + \gamma_2 p(RR, 0)$$

$$(\mu_1 + \mu_2)p(WW, 0) = \mu_1(1 - a_1)p(WI, 0) + \gamma_1 p(RI, 0) + \mu_2(1 - a_2)p(WW, 1)$$
$$+ \gamma_2 p(WR, 1)$$

$$(\mu_1 + \gamma_2)p(WR, 0) = \mu_2 a_2 p(WW, 0)$$

$$(\mu_1 + \mu_2)p(WW, x) = \mu_1(1 - a_1)p(WW, x-1) + \gamma_1 p(RW, x-1)$$
$$+ \mu_2(1 - a_2)p(WW, x+1) + \gamma_2 p(WR, x+1) \quad 0 < x < z$$

$$(\mu_1 + \gamma_2)p(WR, x) = \mu_1(1 - a_1)p(WR, x-1) + \gamma_1 p(RR, x-1)$$
$$+ \mu_2 a_2 p(WW, x), \qquad\qquad 0 < x \leq z$$

$$(\gamma_1 + \mu_2)p(RW, x) = \mu_1 a_1 p(WW, x) + \mu_2(1 - a_2)p(RW, x+1)$$
$$+ \gamma_2 p(RR, x+1), \qquad\qquad 0 \leq x < z$$

$$(\gamma_1 + \gamma_2)p(RR, x) = \mu_1 a_1 p(WR, x) + \mu_2 a_2 p(RW, x), \qquad 0 \leq x \leq z$$

$$(\mu_1 + \mu_2)p(WW, z) = \mu_1(1 - a_1)p(WW, z-1) + \gamma_1 p(RW, z-1)$$
$$+ \mu_2(1 - a_2)p(BW, z) + \gamma_2 p(BR, z)$$

$$(\gamma_1 + \mu_2)p(RW, z) = \mu_1 a_1 p(WW, z)$$

$$\mu_2 p(BW, z) = \mu_1(1 - a_1)p(WW, z) + \gamma_1 p(RW, z)$$

$$\gamma_2 p(BR, z) = \mu_2 a_2 p(BW, z) + \mu_1(1 - a_1)p(WR, z) + \gamma_1 p(RR, z)$$

Performance Measures. Given the state probabilities, it is then possible to find the performance measures of interest. The main measures required are the following:

$$P\{\mathcal{B}\} = P\{\text{stage 1 blocked}\} = p(BW, z) + p(BR, z)$$

$$P\{\mathcal{I}\} = P\{\text{stage 2 idle}\} = p(WI, 0) + p(RI, 0)$$

$$\eta^O(z) = \gamma_2 p(BR, z) + \mu_2(1 - a_2)p(BW, z)$$

$$+ \sum_{x=0}^{z}(\gamma_2[p(WR, x) + p(RR, x)] + \mu_2(1 - a_2)[p(WW, x) + p(RW, x)])$$

$$= \mu_2 P\{\mathcal{W}_2\}$$

after adding all equations for $p(WR, x)$ and $p(RR, x)$, and using the resulting identity that

$$P\{\mathcal{R}_2\} = \sum_{x=0}^{z}(p(WR, x) + p(RR, x)) + p(BR, z) = \frac{\mu_2 a_2}{\gamma_2}P\{\mathcal{W}_2\}$$

where $P\{\mathcal{W}_2\} = \sum_{x=0}^{z}(p(WW, x) + p(RW, x)) + p(BW, z)$.

Alternatively, because $P\{\mathcal{W}_2\} + P\{\mathcal{R}_2\} + P\{\mathcal{I}\} = 1$ it follows that $P\{\mathcal{W}_2\} = (1 - P\{\mathcal{I}\})/(1 + \mu_2 a_2/\gamma_2)$ and thus

$$\eta^O(z) = \frac{1 - P\{\mathcal{I}\}}{\tau_2}$$

where

$$\tau_2 = \frac{1}{\mu_2} + \frac{a_2}{\gamma_2}$$

is the mean service time at stage 2, that is, the mean time to process a job and carry out any necessary repair if failure occurs.

Similarly, it can be shown that

$$\eta^I(z) = \mu_1(\sum_{x=0}^{z}[p(WW, x) + p(WR, x)] + p(WI, 0))$$

$$= \frac{1 - P\{\mathcal{B}\}}{\tau_1}$$

with $\tau_1 = 1/\mu_1 + a_1/\gamma_1$ being the mean of the stage 1 service time distribution. Obviously, it will be true that $\eta^O(z) = \eta^I(z)$.

Determination of state probabilities.   These equations do not have a closed-form solution. In analyzing multistage systems it is necessary to solve the equations many times so the direct solution of the sparse linear system of $4(z + 2)$ equations becomes unattractive. An alternative approach is to seek a computationally efficient solution technique that exploits the structure of the transition matrix and, in particular, uses the fact that the inventory level only changes by +1, 0 or −1 in any state transition. The following recursive algorithm can be developed to solve the equations efficiently.

Recursive algorithm. The basic idea of this method is the following: There exists a subset of the state probabilities that are defined as boundaries, and if the values of the boundaries are known, the recursive solution of the total system of equations can be carried out efficiently.

The algorithm consists of the following three steps:

1. *Reduction step.* Determine r boundaries and derive a recursive scheme to calculate all other state probabilities from the boundary state probabilities. Then express all state probabilities as linear combinations of the boundary values. To find the coefficient of a particular boundary value in the linear expression, set that boundary value equal to 1 and all other boundary values equal to 0, and then follow through the recursive scheme. This is done r times, corresponding to the r boundaries. There will be r equations not used in calculating the state probabilities in terms of the boundary values. Any $r - 1$ of these r equations together with the normalizing equation requiring the state probabilities to sum to 1 give, after substituting expressions in terms of boundary-state probabilities for nonboundary probabilities, r equations for the r boundary-state probabilities.

2. *Solution step.* Determine the r boundary-state probabilities by solving this set of r equations.

3. *Evaluation step.* From the recursive scheme, determine the remaining state probabilities. Performance measures can now be calculated.

Key to the use of this method is the determination of how many boundaries to use and which specific states should be chosen as the boundary states. By manipulating the equations it is possible to develop a solution approach with $r = 2$. If the equation for $p(WR, x+1)$ is substituted in the equation for $p(WW, x)$ and the equation for $p(RR, x+1)$ is substituted in the equation for $p(RW, x)$ then the original set of equations can be rewritten in the following recursive form, where $\alpha = \mu_2(\mu_1(1 - a_2) + \gamma_2)$ and $\beta = \mu_2(\gamma_1(1 - a_2) + \gamma_2)$:

$$\alpha p(WW, 0) = (\mu_1 + \gamma_2)\mu_1 p(WI, 0)$$

$$(\mu_1 + \gamma_2)p(WR, 0) = \mu_2 a_2 p(WW, 0)$$

$$\beta p(RW, 0) = (\gamma_1 + \gamma_2)(-\mu_1 a_1 p(WI, 0) + \gamma_1 p(RI, 0)) - \mu_1 a_1 \gamma_2 p(WR, 0)$$

$$(\gamma_1 + \gamma_2)p(RR, 0) = \mu_1 a_1 p(WR, 0) + \mu_2 a_2 p(RW, 0)$$

$$\alpha p(WW, 1) = (\mu_1 + \gamma_2)(-\mu_1(1 - a_1)p(WI, 0) - \gamma_1 p(RI, 0)$$

$$+ (\mu_1 + \mu_2)p(WW, 0))$$

$$- \gamma_2(\mu_1(1 - a_1)p(WR, 0) + \gamma_1 p(RR, 0))$$

and then

$$(\mu_1 + \gamma_2)p(WR, x) = \mu_1(1 - a_1)p(WR, x-1) + \gamma_1 p(RR, x-1)$$

$$+ \mu_2 a_2 p(WW, x), \qquad\qquad 1 \leq x \leq z$$

$$\beta p(RW, x) = (\gamma_1 + \gamma_2)(-\mu_1 a_1 p(WW, x-1) + (\gamma_1 + \mu_2)p(RW, x-1))$$
$$- \mu_1 a_1 \gamma_2 p(WR, x), \qquad\qquad 1 \le x \le z$$

$$(\gamma_1 + \gamma_2)p(RR, x) = \mu_1 a_1 p(WR, x) + \mu_2 a_2 p(RW, x), \qquad 1 \le x \le z$$

$$\alpha p(WW, x+1) = (\mu_1 + \mu_2)(-\mu_1(1-a_1)p(WW, x-1) - \gamma_1 p(RW, x-1)$$
$$+ (\mu_1 + \mu_2)p(WW, x)) - \gamma_2(\mu_1(1-a_1)p(WR, x)$$
$$+ \gamma_1 p(RR, x)), \qquad\qquad 1 \le x \le z-1$$

Lastly

$$\mu_2 p(BW, z) = (-\mu_1(1-a_1)p(WW, z-1) - \gamma_1 p(RW, z-1) + (\mu_1 + \mu_2)p(WW, z))$$
$$- (\mu_1(1-a_1)p(WR, z) + \gamma_1 p(RR, z))$$

$$\gamma_2 p(BR, z) = \mu_1(1-a_1)p(WR, z) + \gamma_1 p(RR, z) + \mu_2 a_2 p(BW, z)$$

It can be seen that the preceding equations define a recursive scheme by which all the state probabilities can be expressed in terms of the two boundary state probabilities $p(WI, 0)$ and $p(RI, 0)$. The two equations not used in the recursive scheme are

$$(\gamma_1 + \mu_2)p(RW, z) = \mu_1 a_1 p(WW, z)$$
$$\mu_2 p(BW, z) = \mu_1(1-a_1)p(WW, z) + \gamma_1 p(RW, z)$$

All state probabilities can be expressed as linear combinations of coefficients, that is, the probability of state j is expressed as

$$p(j) = C(1, j)p(WI, 0) + C(2, j)p(RI, 0)$$

The coefficients are then calculated by going through the recursive scheme twice, each time setting one boundary-state probability equal to one and the other equal to zero.

Next, choose one of the two equations not used in the recursive scheme for example, $(\gamma_1 + \mu_2)p(RW, z) = \mu_1 a_1 p(WW, z)$. It can then be written as

$$[(\gamma_1 + \mu_2)C(1, RW, z) - \mu_1 a_1 C(1, WW, z)]p(WI, 0)$$
$$- [(\gamma_1 + \mu_2)C(2, RW, z) - \mu_1 a_1 C(2, WW, z)]p(RI, 0) = 0$$

This and the normalizing equation are the two equations that enable the boundary state probabilities to be determined. The other state probabilities and the performance measures can then be found.

Alternatively one may choose $p(BR, z)$ and $p(BW, z)$ as the boundary-state probabilities and obtain recursive relations for the other state probabilities. We have found that when $\tau_1 > \tau_2$ the first recursion is numerically more stable than the latter recursion, whereas when $\tau_1 < \tau_2$ the opposite is true. For large values of z, however, both recursions will have problems with numerical stability and indicate negative probabilities.

This occurs when the bank is large enough to behave like an infinite bank so the state probabilities no longer depend on the boundary state probabilities (i.e., $C(1, j) = C(2, j)$ for states j where the corresponding x value is large). The throughput is then given by the infinite bank throughput.

Interoutput and interinput time distributions. It is desirable to be able to derive some properties of the distributions of the time between successive output (or input) of the line. This requires deriving some properties of the distribution of the system state at the instant of departure of a job. This distribution is not the same as the distribution at a random instant of time (the $p(XY, x)$ obtained earlier). Define the following:

- $Q(W, x)$ = Probability a departing job leaves x jobs in the bank and stage 1 is in state W, $0 \leq x \leq z$.
- $Q(R, x)$ = Probability a departing job leaves x jobs in the bank and stage 1 is in state R, $0 \leq x \leq z$.
- $Q(B, z)$ = Probability that stage 1 is blocked when a job departs.

Then

$$Q(W, x) = [\mu_2(1 - a_2)p(WW, x) + \gamma_2 p(WR, x)]/\eta(z) \quad 0 \leq x \leq z$$

$$Q(R, x) = [\mu_2(1 - a_2)p(RW, x) + \gamma_2 p(RR, x)]/\eta(z) \quad 0 \leq x \leq z$$

$$Q(B, z) = [\mu_2(1 - a_2)p(BW, z) + \gamma_2 p(BR, x)]/\eta(z)$$

because $\eta(z) = \sum_{x=0}^{z}(Q(W, x) + Q(R, x)) + Q(B, z)$. Note that $Q(W, 0) = \mu_1 p(WI, 0)/\eta(z)$ and $Q(R, 0) = [\gamma_1 p(RI, 0) - \mu_1 a_1 p(WI, 0)]/\eta(z)$.

If the departing job leaves the system in state (W, x) or (R, x), $1 \leq x \leq z$ then the time until the next departure is the service time T_2 of the job at stage 2 (equals processing time plus any necessary repair time). If the state is $(W, 0)$ or $(R, 0)$, however, then as well as the service time T_2 of the job at stage 2 there will also be added idle time at stage 2, corresponding to the remaining repair time S_1' of stage 1 with state $(R, 0)$ or the remaining service time T_1' on stage 1 (processing time on stage 1 plus any necessary repair time) with state $(W, 0)$. Define the following distributions:

- $F_2(t) = P\{T_2 \leq t\}$ where the random variable T_2 is the service time on stage 2 with mean $\tau_2 = 1/\mu_2 + a_2/\gamma_2$;
- $G_{R2}(t) = P\{S_1' + T_2 \leq t\}$ where the random variable S_1' is the remaining repair time on stage 1. (Note that because repair times are exponential the distribution of S_1' is the same as the distribution of S_1, the actual repair time.)
- $G_{W2}(t) = P\{T_1' + T_2 \leq t\}$ where T_1' is the remaining service time at stage 1 (because processing times are exponential the distribution of T_1' is the same as the distribution of T_1, $F_1(t)$, with $E[T_1] = \tau_1 = 1/\mu_1 + a_1/\gamma_1$).
- $H(t) = P\{D \leq t\}$ where D is the time between departures of jobs.

Note that $G_{W2} = F_1 * F_2$, that is, the convolution of the two distributions, and G_{R2} is the convolution of F_2 and the (exponential) repair time distribution G_1. Then

$$H(t) = (1 - Q(W, 0) - Q(R, 0))F_2(t) + Q(W, 0)G_{W2}(t) + Q(R, 0)G_{R2}(t)$$

and the moments of the distribution $H(t)$ are given by

$$E[D] = E[T_2] + Q(W, 0)E[T_1] + Q(R, 0)E[S_1]$$

$$= \tau_2 + \frac{p(WI, 0) + p(RI, 0)}{\eta(z)}$$

$$E[D^2] = E[T_2^2] + Q(W, 0)E[T_1^2] + Q(R, 0)E[S_1^2] + 2E[T_2](Q(W, 0)E[T_1]$$

$$+ Q(R, 0)E[S_1])$$

$$= E[T_2^2] + 2E[D](E[T_1 + T_2]p(WI, 0) + E[S_1 + T_2]p(RI, 0))$$

$$E[D^3] = E[T_2^3] + 3E[T_2^2](E[T_1]Q(W, 0) + E[S_1]Q(R, 0)) + 3E[T_2](E[T_1^2]Q(W, 0)$$

$$+ E[S_1^2]Q(R, 0)) + E[T_1^3]Q(W, 0) + E[S_1^3]Q(R, 0)$$

$$= E[T_2^3] + 3E[D](E[(T_1 + T_2)^2]p(WI, 0) + E[(S_1 + T_2)^2]p(RI, 0))$$

where $E[T_i^2] = 2(1/\mu_i^2 + a_i/(\mu_i\gamma_i) + a_i/\gamma_i^2)$ and $E[T_i^3] = 6(1/\mu_i^3 + a_i/(\mu_i^2\gamma_i) + a_i/(\mu_i\gamma_i^2) + a_i/\gamma_i^3)$. Note that $E[D] = 1/\eta(z)$.

Similarly, it is possible to write down the distribution of the interinput time (or equivalently the time between successive completions on stage 1) and its moments. It is necessary to determine the probabilities at the instant when a job is completed at stage 1, in particular

- $Q(z, W)$ = Probability when a job completes on stage 1 there are z jobs in the bank and stage 2 is in state W.
- $Q(z, R)$ = Probability when a job completes on stage 1 there are z jobs in the bank and stage 2 is in state R.

$$Q(z, W) = [\mu_1(1 - a_1)p(WW, z) + \gamma_1 p(RW, z)]/\eta(z)$$

$$Q(z, R) = [\mu_1(1 - a_1)p(WR, z) + \gamma_1 p(RR, z)]/\eta(z)$$

Hence the first two moments of the interinput time I will be given by

$$E[I] = E[T_1] + Q(z, W)E[T_2] + Q(z, R)E[S_2]$$

$$E[I^2] = E[T_1^2] + Q(z, W)E[T_2^2] + Q(z, R)E[S_2^2] + 2E[T_1](Q(z, W)E[T_2]$$

$$+ Q(z, R)E[S_2])$$

Equivalent single stage.   As a component of approximation schemes it is desirable to replace a two-stage system by a single equivalent stage with either approximately the same interoutput distribution or approximately the same interinput distribution.

We require the output equivalent single stage to have exponential processing time with parameter μ_O, failure probability a_O and exponential repair distribution with parameter γ_O. There are various approaches that can be used. Because there are three parameters for the distribution they can be fitted using the first three moments of the interoutput distribution (moment fitting). That is, we choose μ_O, a_O and γ_O such that

$$\frac{1}{\mu_O} + \frac{a_O}{\gamma_O} = E[D]$$

$$2\left[\frac{1}{\mu_O{}^2} + \frac{a_O}{\mu_O \gamma_O} + \frac{a_O}{\gamma_O{}^2}\right] = E[D^2]$$

$$6\left[\frac{1}{\mu_O{}^3} + \frac{a_O}{\mu_O{}^2 \gamma_O} + \frac{a_O}{\mu_O \gamma_O{}^2} + \frac{a_O}{\gamma_O{}^3}\right] = E[D^3].$$

Another approach is to assume one parameter is given and just fit using the first two moments, for example, assume $\mu_O = \mu_2$ and thus assume that the starving of stage 2 when a departing job leaves the bank empty is equivalent to additional failures. In this case we use the equations for $E[D]$ and $E[D^2]$ to solve for a_O and γ_O. Yet another idea is to assume a relationship between the three parameters of the output equivalent single stage and fit using the first two moments, for example, assume $1/\mu_O = a_O/\gamma_O$ (balanced means) (i.e., the mean processing time is the same as the mean additional service time owing to failures). In this case we find that

$$\mu_O = 2/E[D]$$

$$\gamma_O = E[D]/\text{Var}(D)$$

$$a_O = E[D]^2/2\text{Var}(D)$$

Therefore for this aggregation to be valid we need $C_D^2 = \text{Var}(D)/E[D]^2 \geq 1/2$. Experience suggests that the fitting approaches that use just two moments perform better than approaches that use three moments.

6.7.2 Multiple-Stage Systems

Approximations. The approximate solution to multiple stage lines is based on viewing an m-stage line as a two-stage line in $m - 1$ different ways, corresponding to the $m - 1$ inventory banks of the line. As seen from each inventory bank the line appears as if it is a two-stage line with the upstream stages determining the input process to the bank and the downstream stages determining the output process from the bank. This two-stage line centered on bank j can then be analyzed and the completion instant probabilities $Q(W, 0)$, $Q(R, 0)$, $Q(z, W)$, and $Q(z, R)$ determined. These then can be used to determine the parameters of a single stage that would create the input process to the next downstream bank (bank $j + 1$) and of another single stage that would result in the output process from the previous upstream bank (bank $j - 1$). In particular the equivalent single stages are assumed to have the C_2 distribution with parameters fitted using the techniques described earlier.

The algorithm is as follows:

Algorithm 6.2   Multistage Flexible Transfer Line

Step 0: Set $k = 1$; $(\mu_j^D(0), a_j^D(0), \gamma_j^D(0)) = (\mu_j, a_j, \gamma_j)$, $j = 2, \ldots, m$.

Step 1: Set $(\mu_1^U(k), a_1^U(k), \gamma_1^U(k)) = (\mu_1, a_1, \gamma_1)$. For $j = 1, \ldots, m - 1$ analyze the two-stage system with bank capacity z_j and stage parameters $(\mu_j^U(k), a_j^U(k), \gamma_j^U(k))$, and $(\mu_{j+1}^D(k - 1), a_{j+1}^D(k - 1), \gamma_{j+1}^D(k - 1))$, determine the throughput $\eta_j^O(k)$ and the completion instant idle probabilities $(Q_j(W, 0)(k), Q_j(R, 0)(k))$, and then set $(\mu_{j+1}^U(k), a_{j+1}^U(k), \gamma_{j+1}^U(k))$ to be the parameters of the single-output equivalent stage to a system with stage parameters $(\mu_j^U(k), a_j^U(k), \gamma_j^U(k))$ and $(\mu_{j+1}, a_{j+1}, \gamma_{j+1})$, and idle probabilities $(Q_j(W, 0)(k), Q_j(R, 0)(k))$.

Step 2: Set $(\mu_m^D(k), a_m^D(k), \gamma_m^D(k)) = (\mu_m, a_m, \gamma_m)$. For $j = m - 1, \ldots, 1$ analyze the two-stage system with bank capacity z_j and stage parameters $(\mu_j^U(k), a_j^U(k), \gamma_j^U(k))$, and $(\mu_{j+1}^D(k), a_{j+1}^D(k), \gamma_{j+1}^D(k))$, determine the throughput $\eta_j^I(k)$ and the completion instant blocking probabilities $(Q_j(z, W)(k), Q_j(z, R)(k))$, and then set $(\mu_j^D(k), a_j^D(k), \gamma_j^D(k))$ to be the parameters of the single-input equivalent stage to a two-stage system with stage parameters (μ_j, a_j, γ_j) and $(\mu_{j+1}^D, a_{j+1}^D, \gamma_{j+1}^D)$ and blocking probabilities $(Q_j(z, W)(k), Q_j(z, R)(k))$.

Step 3: If $|\eta_1^I(k) - \eta_{m-1}^O(k)| < \epsilon$ stop, otherwise set $k \leftarrow k + 1$ and go to step 1.

It is easy to show that at convergence $\eta_j^O = \eta_k^I$ for all j, k with $1 \leq j \leq m - 1$ and $1 \leq k \leq m - 1$, that is, the approximation is consistent.

To illustrate the adequacy of the approximation, consider the three stage system with parameters given in Table 6.9. Then if $z_1 = z_2 = z$, Table 6.10 (first reported in [60]) shows for various values of z a comparison of the throughput approximation with simulation results and, for $z \leq 15$, exact results obtained by solving the relevant balance equations for the three-stage asynchronous system. Note that the approximation gives good estimates of the throughput, although as with the synchronous line, if the middle

TABLE 6.9   PARAMETERS OF
THREE-STAGE ASYNCHRONOUS LINE

Stage	μ	γ	a	Squared coefficients of variation
1	1.0	0.20	0.133	2.6
2	1.0	0.25	0.1071	2.08
3	1.0	0.25	0.1000	2.061

TABLE 6.10 COMPARISON OF THROUGHPUT
APPROXIMATION WITH EXACT AND SIMULATION
VALUES

z	Exact	Approximate	Sim. (95% conf. interval)
0	0.3392	0.3325	0.3415 ± 0.0018
1	0.3929	0.4005	0.3961 ± 0.0021
2	0.4275	0.4332	0.4318 ± 0.0046
3	0.4527	0.4573	0.4551 ± 0.0058
4	0.4723	0.4762	0.4784 ± 0.0052
5	0.4880	0.4916	0.4942 ± 0.0056
6	0.5011	0.5043	0.5058 ± 0.0053
8	0.5214	0.5243	0.5261 ± 0.0057
10	0.5365	0.5391	0.5393 ± 0.0043
15	0.5610	0.5630	0.5654 ± 0.0042
20		0.5767	0.5825 ± 0.0052
30		0.5899	0.6001 ± 0.0049
40		0.5955	0.6041 ± 0.0082
50		0.5982	0.6013 ± 0.0068

stage is much faster than the outer stages the approximation would not be expected to perform as well.

Application of flow-line results. Suppose each of the m stations are continuously monitored, and a repair is initiated as soon as a station fails. Let $S^{(i)}$ be the generic processing time at station i, and suppose that station i fails at a constant rate of γ_i, $i = 1, \ldots, m$. The repair times $\{R_n^{(i)}, n = 1, 2, \ldots\}$ of station i are assumed to be IID, $i = 1, \ldots, m$. Then the generic completion time of a part at station i is (see Section 6.3.1)

$$C^{(i)} = S^{(i)} + \sum_{n=1}^{N^{(i)}(S^{(i)})} R_n^{(i)}, \quad i = 1, \ldots, m$$

where $N^{(i)}(S^{(i)})$ is the number of station i failures during the processing time $S^{(i)}$. Observe that $\{N^{(i)}(S^{(i)})|S^{(i)} = \tau\}$ has a Poisson distribution with mean $\gamma_i \tau$, $i = 1, \ldots, m$.

Now suppose $S^{(i):1} \geq_{cx} S^{(i):2}$ and $R_n^{(i):1} \geq_{cx} R_n^{(i):2}$, $i = 1, \ldots, m$. That is, consider two m station transfer lines where the first line has more variable processing and repair times than those of line 2. Then it can be verified that the completion times at line 1 are more variable than the completion times at line 2. That is, $C^{(i):1} \geq_{cx} C^{(i):2}$, $i = 1, \ldots, m$. Therefore, if $TH^{(j)}$ is the throughput of line j, $j = 1, 2$, then from the results for flow lines (see Chapter 5) one has

$$TH^{(1)} \leq TH^{(2)}$$

Observe that the preceding result agrees with those reported for the Ge/D: 2, D/Ge: 2 and Ge/Ge: 2 transfer lines. Note that a geometric distribution is more variable than its deterministic mean.

Now suppose $\gamma_i^{(1)} \geq \gamma_i^{(2)}$, $S^{(i):1} \geq_{st} S^{(i):2}$ and $R_n^{(i):1} \geq_{st} R_n^{(i):2}$, $i = 1, \ldots, m$. That is, each of the m stations in line 1 has a higher failure rate, and longer processing and repair times than those for line 2. Then it is again easily verified that $C^{(i):1} \geq_{st} C^{(i):2}$, $i = 1, \ldots, m$. Then from the results of flow lines it is seen that

$$TH^{(1)} \leq TH^{(2)}$$

Next, consider two transfer lines where the parameters are the same. Processing at transfer line 1, however, is controlled according to a policy that inserts idleness even when a station could process a part and line 2 operates in the following way: An operative station will always process a part as long as there are parts available for processing, and it is not blocked. If we include the idleness into the part completion times, it is easily seen that $\{C_n^{(i):1}, n = 1, 2, \ldots; i = 1, \ldots, m\} \geq_{st} \{C_n^{(i):2}, n = 1, 2, \ldots; i = 1, \ldots, m\}$. Hence the part departure times in line 1 are larger than those in line 2. Hence

$$TH^{(1)} \leq TH^{(2)}$$

That is, we see that any policy that inserts idleness cannot be better than the service policy described earlier. For example, those policies that attempt to keep the banks half full by forcing downstream stations to idle when the number of parts in the bank is low and forcing the upstream stations to idle when the number of parts in the bank is high cannot perform better than the policy described earlier.

6.8 IMPLICATIONS OF MODELS

Role of inventory banks.   The efficiency of a transfer line with no inventory banks can be substantially less than the efficiency of the individual stages. Inventory banks provide a means of improving the line efficiency so that it becomes closer to the efficiency of the worst stage, that is, the stage with the lowest throughput if it were operated on its own. This improvement depends on the quantity zb (i.e., the ratio of the bank capacity to the mean repair time measured in cycles), however, and this should be somewhere in the range of two to five to achieve a significant improvement in the efficiency (at least 50% closer to the efficiency of the worst stage). The required inventory bank capacity z and the associated number of pallets can be quite large so that often the economics of providing bank capacity are not attractive. Spreading the total bank capacity over a number of locations can result in an improvement in the performance. This can be seen from the formula we presented for the efficiency of the line with an optimal failure probability allocation (equation 6.63). Again, however, a relatively small number of banks is likely to be all that would be economically justifiable.

Parallel stations or inventory banks.   If one stage is much worse than the rest of the stages, it is better to improve the performance by paralleling the stage and, if it is technologically possible, having the two parallel stages share the processing task. Parallel standby is unlikely to be economical unless the standby is relatively cheap or needed for technological reasons. Inventory banks are most useful if there are several relatively bad stages so that without banks the line efficiency is substantially worse than the efficiency of a bad stage on its own.

Improving the effectiveness of banks. It has also been shown that any reduction in variability in the times to failure, repair times (or the time to perform usual operations in a flexible transfer line) is beneficial in improving the effectiveness of a given banking arrangement. This implies that there should be management effort devoted to the planning of repairs and maintenance activities, and also in monitoring the effectiveness of repairs so that imperfect repairs do not contribute to the variability in time to failures. In addition it has been shown that any attempt to force idle a stage to balance inventories in banks can only result in the reduction in efficiency.

Limited repair capability. With limited repair capabilities it appears that repair should focus on ensuring that some part of the system is restored to production as soon as possible (i.e., focus repair on the stage whose repair will enable some production soon).

Bank location. Bank placement should be such that the line is divided into stages with approximately equal failure probabilities, although with more than two stages the middle stages should have lower failure probability than the two outside (i.e., the first and the last) stages. The line performance is not very sensitive to deviations from the optimal division, however.

6.9 BIBLIOGRAPHICAL NOTE

The earliest studies of stochastic models of transfer lines were by Vladzievskii in the former Soviet Union in the early 1950s [88] [87]. He introduced the idea of the loss transfer coefficient and developed a simple model to determine it in the case of two equal stations. Vladzievskii's approach then led to the contribution of Sevastyanov [77] who developed the earliest ideas of the approximate algorithm for multistage systems, proved convergence and uniqueness, and found the optimal allocation of failure rates. Subsequent work by Levin and Pasko [65] and Artamonov [4] derived refined versions for the two-stage efficiency for operation and time-dependent failures. Another pioneering paper on transfer lines was by Zimmern [96]. He also recognized the basic idea of the approximation algorithm for analyzing multistage systems, but few subsequent authors were aware of this work until the 1980s. The earliest English language paper on transfer line modeling was by Koenigsberg [59] who described Vladzievskii's results and gave results of unpublished work by Finch [32] for the two-stage transfer line. Apart from the work by Buzacott [10] [13] there were then few publications on this topic until the late 1970s when a variety of improvements or alternative approximate algorithms were developed by Gershwin and co-workers [41] [17] [38]. Some of the transfer-line results were used to validate simulation models of existing or proposed transfer lines starting with Hanifin [47] at Chrysler, and they are believed to have been used by Mercedes and General Motors among others.

A comprehensive review of models developed for transfer lines (and also for flow lines) can be found in Dallery and Gershwin [21].

PROBLEMS

6.1 Consider the two monitoring and repair strategies discussed in Section 6.3.1. Suppose the processing time S has an exponential distribution with mean $1/\mu'$ and the repair times R also have exponential distributions with mean $1/\gamma'$. Show that, for each strategy, station behavior can be represented by an equivalent station that behaves in the same way as the station considered in Section 6.7, that is, exponential processing time with mean $1/\mu$, failure with probability a, and a single (perfect) repair taking exponentially distributed time with mean $1/\gamma$, and derive the formulas by which the parameters (μ, γ, a) of the equivalent station can be found.

6.2 In a multiple-stage manufacturing system with no buffers work movement from one stage to the next is synchronized with a fixed cycle time of $1/h$. Work cannot move from one stage to the next unless the job has been completed satisfactorily at each stage. Processing at a stage is complicated, so at each stage it is necessary to check the job on completion of processing. With probability q_i it is necessary to repeat processing of the job at stage i, $i = 1, \ldots, m$. Show that the throughput of the system is given by

$$TH = \frac{h}{1 + \sum_{i=1}^{m} \frac{q_i}{1-q_i}}$$

6.3 Suppose that the $Ge(q)/Ge(1-q)$: 2 model is to be used as an approximation for the performance of an Ge/Ge: 2 transfer line.
  (a) How should the single parameter q required to describe the behavior of a station be determined from the parameters a and b of the station?
  (b) For a two-stage line with identical stations, compare the efficiency calculated using equation 6.62 and the value of q obtained earlier with the efficiency obtained using equation 6.38. Comment on the adequacy of the Bernoulli approximation.

6.4 **(a)** Write down the efficiency of a two-stage line with unequal stations in the case where operation of each station is viewed as a sequence of Bernoulli trials.
  (b) Use the formula obtained in part (a) as the building block for a three-stage line approximation procedure. Compare this approximation to the exact results given in Table 6.7.

6.5 Performance of the two-stage asynchronous line described in Section 6.7.1 is to be approximated by a two-stage flow line with exponential processing times. Choosing appropriate values for the means of the exponential processing times determine the (approximate) performance of the line and compare with the exact performance. Under what conditions would you expect the exponential approximation to be adequate?

6.6 Now consider the approximation of the performance of a multiple-stage asynchronous transfer line by a series exponential flow line. Compare the throughput predictions using the exponential approximation given in Chapter 5 with the results given in Table 6.10. Comment on the adequacy of the exponential approximation.

6.7 Consider the following approximation for the throughput of a two-stage asynchronous transfer line where $\mu_1 = \mu_2 = \mu$. Its basis is to say that the mean time between parts leaving the line $(=1/\eta(z))$ is equal to $1/\mu$ plus two additional delay components, one owing to the exponential processing times and the other owing to station failures. The delay component owing to the exponential processing times is found using the two-stage exponential flow line model of chapter 5 and is given by $1/TH(z) - 1/\mu$. The component owing to station failures is determined by assuming that the system is a two-stage fixed cycle transfer line with cycle duration $1/\mu$ and hence the equivalent, b_j is given by $b_j = \gamma_j/\mu$. Thus the additional delay owing to station failures is obtained using the fixed-cycle transfer-line formula and is given by $(E[N] - 1)/\mu$. Investigate the adequacy of this approximation. What insight does this approximation provide into the determination of the appropriate size of the inventory bank?

6.8 Suppose in a two-stage fixed-cycle transfer line with finite inventory banks that the part in process when a machine fails must always be scrapped and removed from the machine as part of the repair process. No part can then enter the machine until repair is complete. Write down the state equations describing the operation of the line, and develop a recursive scheme to determine the state probabilities. How many boundary states are required by your recursive scheme?

6.9 Develop a model of a two-stage fixed-cycle transfer line with finite inventory banks in which the part in process when failure of a machine occurs must be reprocessed after completion of repair. Is there a σ such that $p(., x) = \sigma p(., x-1)$, $1 < x \le z-1$ for any combination of machine states "." or must a recursive solution technique be employed?

6.10 Consider a two-stage fixed-cycle transfer line with no inventory bank between the two stages. Each stage, however, consists of a group of n_j, $j = 1, 2$, identical parallel machines. When all machines are up the machine cycle times are $1/h_j$ at stage j, $j = 1, 2$ and $n_1 h_1 = n_2 h_2$. If the failure probability per operating cycle is constant (i.e., failures are operation dependent) while repair times are geometric, develop a formula for the system efficiency if (1) no speed up of a machine is possible to compensate for failed parallel machines, (2) machines can be speeded up so that if at least one of the parallel machines is not failed there is no loss in production from the paralleled group. What is the mean time between parts leaving the line for each condition (1) and (2)?

6.11 Consider the system described in the previous question. What would be the efficiencies under conditions (1) and (2) if there were an infinite capacity bank between the two stages? What is the mean time between parts leaving the line for each condition?

6.12 A possible approximation for the efficiency $\eta(z)$ of a two-stage fixed-cycle transfer line with finite inventory bank, of capacity z and parallel machines at each stage is to set

$$\frac{1}{\eta(z)} = \frac{1 - w(z)}{\eta(0)} + \frac{w(z)}{\eta(\infty)}$$

where $\eta(0)$ is the efficiency of the system with zero inventory bank, and $\eta(\infty)$ is the efficiency of the system with infinite inventory bank. $w(z)$ is then approximated by the corresponding $w(z)$ for a two-stage system with one machine per stage. What would be a reasonable approach to determine the machine parameters required to determine $w(z)$? Try and test the adequacy of the approximation using either simulation or direct solution of the state equations for small z.

6.13 In [9] the following approach is suggested to approximate the efficiency of a multistage fixed-cycle transfer line with failure probability of a_j for stage j, $j = 1, \ldots, m$, and repair time $\overline{D}$, which is a constant and identical for all stages. The approximation assumes (1) only one stage can be failed at a time (i.e., stage failures only occur when *all* stages are working), and (2) whenever a stage failure occurs the inventory in bank j is at its average level x_j, a constant. Assumptions 1 and 2 imply that if stage j fails, stage i ($i < j$) will operate for a time equal to $\min\{\overline{D}, \sum_{k=i}^{j-1}(z_k - x_k)\}$ until it is forced down, that is, the duration of forced down time of i for each failure of j is $\max\{0, \overline{D} - \sum_{k=i}^{j-1}(z_k - x_k)\}$. Develop a set of m equations with equation i, $i = 1, \ldots, m$, giving the total down time of station i owing to failures of stations $j = 1, \ldots, m$. Then because the total down time of each station must be the same, solution of these equations will determine the x_j, $j = 1, \ldots, m-1$, and hence the (approximated) line efficiency. Develop the relevant formulas for $z_j = z$, $j = 1, \ldots, m-1$. Comment on the approximation and compare with the exact two-stage Ge/D: 2 model.

6.14 For the multistage system considered in the previous question, drop assumption 2, and assume that the inventory level in bank j when a stoppage occurs is the level it reached when the previous stoppage ended. Suppose that $z_j = l_j \tau$ where l_j is an integer, $j = 1, \ldots, m-1$. Keeping assumption 1 develop a Markov model to determine the joint distribution of the inventory level in each bank when no station is failed, and hence determine another approximation for the efficiency. Compare the result with the approximation in the previous question and also with the exact two-stage Ge/D: 2 model. Comment on the adequacy of this approximation.

6.15 Suppose a two-stage fixed cycle transfer line is operated using the following policy. Both stations are permitted to be simultaneously working only when the bank is half full. If either station fails when the bank is half full the other station continues to operate until either the failed station is repaired or the operating station is forced down. Once the failed station is repaired, the other station is not permitted to operate until the inventory level is again restored to half full, even if the station restoring the inventory level fails. Determine the efficiency with this operating policy assuming that the two stations are identical and have a constant failure probability per operating cycle and geometric repair time. Alternatively, suppose that if a station fails while the inventory level is being restored then the other station is then permitted to operate until it is either forced down, or repair of the failed station ends. Develop a formula for the line efficiency. Versions of this operating policy are known to have been used in practice, explain why it (superficially) appears appropriate, and give a plausible explanation that does not resort to mathematics as to why the policy of never stopping a station that is not failed or forced down is preferable.

6.16 One method of calculating the failure transfer coefficient in a two-stage fixed-cycle transfer line when the two stages are identical is to assume that the bank is half full when a stage fails and calculate the expected fraction of the down time of the stage that the other stage is forced down, that is, the probability that the repair time exceeds $z/2$. Calculate the resulting approximation for δ if (1) repair times are geometric, and (2) repair times are constant and equal to τ. Explain why in case (1) the approximation is reasonably good, whereas in case (2) it can be poor if $z \geq \overline{D}$.

6.17 Develop a table corresponding to Table 6.6 for the squared coefficient of variation of the interoutput times of a two-stage fixed cycle transfer line with constant (operation-dependent) probability of failure per cycle and deterministic repair time τ when the two stations are identical and the bank capacity $z = l\tau$ where l is an integer.

6.18 Using Lotus 1-2-3 or some other spreadsheet program
   (a) Plot graphs of efficiency versus bank size for various two-stage systems (choosing

different values of the a and b, or different distributions of time to failure or repair time).

(b) For $b_1 = b_2 = b$, plot a graph of efficiency versus a_1 to see how performance depends on line division (i.e., keep $a_1 + a_2 = A$ where A is a constant). Also plot bounds ($z = 0$ and $z = \infty$) for imbalanced systems.

(c) Compare these results to a two-stage system with only one repairman serving the failed stages according to the FCFS protocol.

6.19 In Chapter 4 the stopped arrival model gave a throughput prediction for a $GI/G/1/z$ model. Determine for both fixed-cycle synchronous lines and for asynchronous lines consisting of two identical stations separated by a finite capacity bank the predicted throughput using the $GI/G/1/z$ stopped arrival model and compare with the results obtained using the models in this chapter. Note that the first step is to find the squared coefficient of variation of the completion time at a station.

6.20 In Hanifin's data the completion time at the two stages had the following means: stage 1–1.77 cycles, stage 2–1.33 cycles. He found using simulation the following predictions of the line performance as a function of z:

z	Efficiency %
0	47.3
20	50.2
40	51.5

Use the $Ge/Ge/1/z$ model to predict the throughput and compare the results with the simulation values. Comment on the agreement. Do you think that the completion times in Hanifin's data are more variable or less variable than geometric completion times? Explain.

BIBLIOGRAPHY

[1] T. ALTIOK. Production lines with phase-type operation and repair times and finite buffers. *Int. J. Prod. Res.*, 23:489–498, 1985.

[2] T. ALTIOK and S. STIDHAM JR. A note on transfer lines with unreliable machines, random processing times and finite buffers. *IIE Trans.*, 14:125–127, 1982.

[3] M. H. AMMAR. *Modelling and analysis of unreliable manufacturing assembly networks with finite storages.* Technical Report LIDS–TH–1004, MIT Laboratory for Information and Decision Systems, Cambridge, MA, 1980.

[4] G. T. ARTAMONOV. Productivity of a two instrument discrete processing line in the presence of failures. *Cybernetics*, 12:464–468, 1976.

[5] M. ATHANS, N. H. COOK, et al. *Complex Materials Handling and Assembly Systems— Second Interim Progress Report.* Technical Report ESL–IR–771, MIT Electronic Systems Laboratory, Cambridge, MA, 1977.

[6] B. AVI-ITZHAK and P. NAOR. Some queueing problems with the service stations subject to breakdowns. *Operations Research*, 11:303–320, 1963.

[7]   O. BERMAN. Efficiency and productivity of a transfer line with two machines and a finite storage buffer. *Eur. J. Opnl. Res.*, 9:295–308, 1982.

[8]   G. BOOTHROYD and A. H. REDFORD. Free transfer can improve assembly machine economics. *Metalworking Production*, 67–70, June 1, 1966.

[9]   G. BOOTHROYD and A. R. REDFORD. *Mechanized Assembly*. McGraw-Hill, New York, 1968.

[10] J. A. BUZACOTT. Automatic transfer lines with buffer stocks. *Int. J. Prod. Res.*, 5:183–200, 1967.

[11] J. A. BUZACOTT. The effect of station breakdowns and random processing times on the capacity of flow-lines with in-process storage. *AIIE Trans*, 4:308–312, 1972.

[12] J. A. BUZACOTT. Methods of reliability analysis of production systems subject to breakdowns. In D. Grouchko, editor, *Operations research and reliability*, pages 211–232, Gordon and Breach, New York, 1971.

[13] J. A. BUZACOTT. Prediction of the efficiency of production systems without internal storage. *Int. J. Prod. Res.*, 6:173–188, 1968.

[14] J. A. BUZACOTT. *Reliability of systems with in–service repair*. PhD thesis, University of Birmingham, 1967.

[15] J. A. BUZACOTT and L. E. HANIFIN. Models of automatic transfer lines with inventory banks: a review and comparison. *AIIE Trans.*, 10:197–207, 1978.

[16] J. A. BUZACOTT and L. E. HANIFIN. Transfer line design and analysis—an overview. In *Proceedings 1978 Fall Industrial Engineering Conference, Atlanta, Georgia*, pages 277–286, AIIE, American Institute of Industrial Engineers, Inc., 1978.

[17] Y. F. CHOONG and S. B. GERSHWIN. A decomposition method for the approximate evaluation of capacitated transfer lines with unreliable machines and random processing times. *IIE Trans.*, 19:150–159, 1987.

[18] C. COMMAULT and Y. DALLERY. Production rate of transfer lines without buffer storage. *IIE Trans.*, 22:315–329, 1990.

[19] Y. DALLERY, R. DAVID, and X.-L. XIE. Approximate analysis of transfer lines with unreliable machines and finite buffers. *IEEE Trans. on Automatic Control*, 34:943–953, 1989.

[20] Y. DALLERY, R. DAVID, and X.-L. XIE. An efficient algorithm for analysis of transfer lines with unreliable machines and finite buffers. *IIE Trans.*, 20:280–283, 1988.

[21] Y. DALLERY and S. B. GERSHWIN. Manufacturing flow line systems: a review of models and analytical results. *Queueing Systems: Theory and Applications*, in press.

[22] A. J. DE KOK. Computationally efficient approximations for balanced flow lines with finite intermediate buffers. *Int. J. Prod. Res.*, 28:401–419, 1990.

[23] M. B. M. DE KOSTER. Estimation of line efficiency by aggregation. *Int. J. Prod. Res.*, 25:615–626, 1987.

[24] M. B. M. DE KOSTER. An improved algorithm to approximate the behaviour of flow lines. *Int. J. Prod. Res.*, 26:691–700, 1988.

[25] M. B. M. DE KOSTER and J. WIJNGAARD. On the equivalence of multi-stage production lines and two-stage lines. *IIE Trans.*, 19:351–353, 1987.

[26] D. DUBOIS and J. P. FORESTIER. Productivité et en-cours moyen d'un ensemble de deux machines séparées par une zone de stockage. *RAIRO Automatique*, 16:105–132, 1982.

[27] A. DUDICK. *Fixed-cycle production systems with in-line inventory and limited repair capability*. PhD thesis, Columbia University, New York, 1979.

[28] E. A. ELSAYED and C. C. HWANG. Analysis of two-stage manufacturing systems with buffer storage and redundant machines. *Int. J. Prod. Res.*, 24:187–201, 1986.

[29] E. A. ELSAYED and R. E. TURLEY. Reliability analysis of production systems with buffer storage. *Int. J. Prod. Res.*, 18:637–645, 1980.

[30] YU. B. ERPSHER. Losses of working time and division of automatic lines into sections. *Stanki i Instrument*, 23(7):7–16, 1952. English Translation DSIR Ref. CTS 631 and CTS 634.

[31] P. D. FINCH. A storage problem along a production line of discrete flow. *Aust. J. Statist.*, 13:165–167, 1971.

[32] P. D. FINCH. Storage problems along a production line. 1956, manuscript.

[33] P. D. FINCH. Storage problems along a production line of continuous flow. *Annales Univ. de Eötvös, Sectio Mathematika*, III–IV:67–84, 1961.

[34] R. J. FOX and D. R. ZERBE. Some practical system availability calculations. *AIIE Trans.*, 6:228–234, 1974.

[35] M. C. FREEMAN. The effects of breakdowns and interstage storage on production line capacity. *Journal of Industrial Engineering*, 15:194–200, 1964.

[36] D. P. GAVER. Time to failure and availability of paralleled systems with repair. *IEEE Trans. of Reliability*, R-12:30–38, 1963.

[37] S. B. GERSHWIN. An efficient decomposition algorithm for unreliable tandem queueing systems with finite buffers. In H. G. Perros and T. Altiok, editors, *Queueing Networks with Blocking*, pages 127–146, North-Holland, Amsterdam, 1989.

[38] S. B. GERSHWIN. An efficient decomposition method for the approximate evaluation of tandem queues with finite storage space and blocking. *Operations Research*, 35:291–305, 1987.

[39] S. B. GERSHWIN. Representation and analysis of transfer lines with machines that have different failure rates. *Annals of Operations Research*, 9:511–530, 1987.

[40] S. B. GERSHWIN and O. BERMAN. Analysis of transfer lines consisting of two unreliable machines with random processing times and finite storage buffers. *AIIE Trans.*, 13:2–11, 1981.

[41] S. B. GERSHWIN and I. C. SCHICK. Modelling and analysis of three-stage transfer lines with unreliable machines and finite buffers. *Operations Research*, 31:354–380, 1983.

[42] V. N. GLUKHOV. Execution time and other reliability characteristics. *Automation and Remote Control*, 33:1726–1733, 1972.

[43] L. A. GOLEMANOV. On the problem of storage and control strategy optimization. *Int. J. Syst. Sci.*, 4:197–210, 1973.

[44] M. P. GROOVER. Analyzing automatic transfer lines. *Industrial Engineering*, 7:26–31, 1975.

[45] Y. P. GUPTA, T. M. SOMERS, and L. GRAU. Modelling the interrelationship between downtimes and uptimes of CNC machines. *Eur. J. Opnl. Res.*, 37:254–271, 1988.

[46] R. HAHN. *Produktionsplanung bei Linienfertigung*. Walter de Gruyter, Berlin, 1972.

[47] L. E. HANIFIN. *Increased transfer line productivity using systems simulation*. PhD thesis, University of Detroit, 1975.

[48] L. E. HANIFIN, J. A. BUZACOTT, and K. S. TARAMAN. *A comparison of analytical and simulation models of transfer lines*. Technical Report EM75-374, SME, 1975.

[49] L. E. HANIFIN, S. G. LIBERTY, and K. TARAMAN. *Improved transfer line efficiency utilizing systems simulation*. Technical Report MR75–169, SME, 1975. Presented at the International Engineering Conference, Detroit.

[50] Y. C. HO, M. A. EYLER, and T. T. CHIEN. A gradient technique for general buffer storage design in a production line. *Int. J. Prod. Res.*, 17:557–580, 1979.

[51] Y. C. HO, M. A. EYLER, and T. T. CHIEN. A new approach to determine parameter sensitivities of transfer lines. *Management Science*, 29:700–714, 1983.

[52] E. IGNALL and A. SILVER. The output of a two stage system with unreliable machines and limited storage. *AIIE Trans*, 9:183–188, 1977.

[53] M. A. JAFARI. Effect of scrapping on the performance of multistage transfer lines. *Int. J. Prod. Res.*, 25:525–530, 1987.

[54] M. A. JAFARI and J. G. SHANTHIKUMAR. An approximate model of multistage automatic transfer lines with possible scrapping of workpieces. *IIE Trans.*, 19:252–265, 1987.

[55] M. A. JAFARI and J. G. SHANTHIKUMAR. Determination of optimal buffer storage capacities and optimal allocation in multistage automatic transfer lines. *IIE Trans.*, 21:130–135, 1989.

[56] M. A. JAFARI and J. G. SHANTHIKUMAR. Exact and approximate solutions to two-stage flow lines with general uptime and downtime distributions. *IIE Trans.*, 19:412–420, 1987.

[57] E. KAY. Buffer stocks in automatic transfer lines. *Int. J. Prod. Res.*, 10:155–165, 1972.

[58] K.-P. KISTNER. Betriebstörungen bei Fliessbändern (breakdowns in flowlines). *Zeitschrift für Operations Research*, 17:B47–B65, 1973.

[59] E. KOENIGSBERG. Production lines and internal storage—a review. *Management Science*, 5:410–433, 1959.

[60] D. KOSTELSKI. *A study of models for automatic transfer lines with unreliable stations and limited capacity in-process inventory buffers*. Master's thesis, University of Waterloo, Department of Management Sciences, Waterloo, Ontario, 1985.

[61] P. KUBAT and U. SUMITA. Buffers and backup machines in automatic transfer lines. *Int. J. Prod. Res.*, 23:1259–1270, 1985.

[62] S. S. LAW. A factorial analysis of automatic transfer line systems. *Int. J. Prod. Res.*, 21:827–834, 1983.

[63] S. S. LAW, R. J. BAXTER, and G. M. MASSARA. Analysis of in–process buffers for multi-input manufacturing systems. *Transactions of the ASME Journal of Engineering for Industry*, 97:1079–1086, 1975.

[64] B. LEV and D. I. TOOF. The role of internal storage capacity in fixed cycle production systems. *Naval Research Logistics Quarterly*, 27:477–487, 1980.

[65] A. A. LEVIN and N. I. PASKO. Calculating the output of transfer lines. *Stanki i Instrument*, 40:12–16, 1969.

[66] E. D. LLOYD. *Transfer and unit machines*. Industrial Press, New York, 1969.

[67] J. MASSO and M. L. SMITH. Interstage storage for three stage lines subject to stochastic failures. *AIIE Trans.*, 6:354–358, 1974.

[68] G. J. MILTENBURG. Variance of the number of units produced on a transfer line with buffer inventories during a period of length t. *Naval Research Logistics*, 34:811–822, 1987.

[69] R. A. MURPHY. The effect of surge on system availability. *AIIE Trans.*, 7:439–443, 1975.

[70] R. A. MURPHY. Estimating the output of a series production system. *AIIE Trans.*, 10:139–148, 1978.

[71] R. A. MURPHY. Examining the distribution of buffer protection. *AIIE Trans.*, 11:113–120, 1979.

[72] T. OHMI. An approximation for the production efficiency of automatic transfer lines with in–process storages. *AIIE Trans.*, 13:22–28, 1981.

[73] K. OKAMURA and H. YAMASHINA. Analysis of the effect of buffer storage capacity in transfer line systems. *AIIE Trans.*, 9:127–135, 1977.

[74] K. OKAMURA and H. YAMASHINA. Justification for installing buffer stocks in unbalanced two stage automatic transfer lines. *AIIE Trans.*, 11:308–312, 1979.

[75] N. I. PASKO. A probabilistic model of a system with intermediate accumulation of production. *Engineering Cybernetics*, 12:75–82, 1974.

[76] M. SAVSAR and W. E. BILES. Two-stage production lines with a single repair crew. *Int. J. Prod. Res.*, 22:499–514, 1984.

[77] B. A. SEVASTYANOV. Influence of storage bin capacity on the average standstill time of a production line. *Theory of Probability and Its Applications*, 7:429–438, 1962.

[78] J. G. SHANTHIKUMAR. On the production capacity of automatic transfer lines with unlimited buffer space. *AIIE Trans.*, 12:273–274, 1980.

[79] J. G. SHANTHIKUMAR and C. C. TIEN. An algorithm solution to two stage transfer lines with possible scrapping of units. *Management Science*, 29:1069–1086, 1983.

[80] T. J. SHESKIN. Allocation of interstage storage along an automatic production line. *AIIE Trans.*, 8:146–152, 1976.

[81] T. J. SHESKIN. Letter to the editor. *AIIE Trans.*, 9:120, 1977.

[82] A. L. SOYSTER, J. W. SCHMIDT, and M. W. ROHRER. Allocation of buffer capacities for a class of fixed cycle production lines. *AIIE Trans.*, 11:140–146, 1979.

[83] A. S. SOYSTER and D. I. TOOF. Some comparative and design aspects of fixed cycle production systems. *Naval Research Logistics Quarterly*, 23:437–454, 1976.

[84] R. V. STETTEN. *Auslegung von Störungspuffern in Kapitalintensiven Fertigungslinien [Locating Buffers in Capital Intensive Production Lines]*. Krausskopf, Mainz, 1977.

[85] H. C. TOWN. *Automatic machine tools*. Iliffe Books, 1968.

[86] P. VANDERHENST, F. V. VAN STEELANDT, and L. F. GELDERS. Efficiency improvement of a transfer line via simulation. *J. Opl. Res. Soc.*, 32:555–562, 1981.

[87] A. P. VLADZIEVSKII. Losses of working time and the division of automatic lines into sections. *Stanki i Instrument* 24(10):9–15, 1953.

[88] A. P. VLADZIEVSKII. The probability law of operation of automatic lines and internal storage in them. *Avtomatika i Telemekhanika*, 13:227–281, 1952.

[89] A. P. VLADZIEVSKII. Problems of the theory of production lines. In A. N. Gavrilov, editor, *Automation and Mechanization of Production Processes in the Instrument Industry*, pages 19–44, Pergamon Press, Oxford, 1967.

[90] A. P. VLADZIEVSKII. The theory of internal stock and their influence on the output of automatic lines. *Stanki i Instrument*, 21(12):4–7, 1950 and 22(1):16–17, 1951.

[91] J. WIJNGAARD. The effect of interstage buffer storage on the output of two unreliable production units in series, with different production rates. *AIIE Trans.*, 11:42–47, 1979.

[92] I. YA RETSKER and A. A. BUNIN. Determining the main parameters of transfer lines. *Stanki i Instrument*, 35(6):17, 1964.

[93] H. YAMASHINA and K. OKAMURA. Analysis of in-process buffers for multi-stage transfer line systems. *Int. J. Prod. Res.*, 21:183–195, 1983.

[94] S. YERALAN, W. E. FRANCK, JR., and M. A. QUASEM. A continuous materials flow production line model with station breakdown. *Eur. J. Opnl. Res.*, 27:289–300, 1986.

[95] S. YERALAN and E. J. MUTH. A general model of a production line with intermediate buffer and station breakdown. *IIE Trans.*, 19:130–139, 1987.

[96] B. ZIMMERN. Études de la propagation des arrêts aléatoires dans les chaines de production. *Revue de statistique appliquée*, 4:85–104, 1956.

7

Dynamic Job Shops

7.1 INTRODUCTION

A job shop consists of several different types of machines with each type capable of performing a specific set of production operations, for example drilling, milling, or forming. Usually, the layout of a job shop is such that all machines of the same type are located together, that is, all drilling machines are in one area of the shop, whereas all milling machines are located together in another area of the shop. The different groups of machines may even be located in different buildings and indeed the buildings may not be all on the same site. Thus associated with the job shop there will be material handling or transport facilities to move jobs between machine groups if they require operations on more than one type of machine. Apart from the machines and the material handling facilities there will also be space required for the storage of jobs in process. This may be at or near machines, but it is often away from machines and sometimes just in the factory yard. Job shop operation is almost always based on the premise that space to store work in process can always be found somewhere so as a result machines are never blocked.

The key distinguishing feature of a job shop is that different types of jobs with different routings (i.e., sequence of machines to be visited by a job) can be manufactured. Jobs can be introduced to the shop at almost any machine, and the routing of a job may require it to return to a given machine several times.

Job shops have been used for organizing manufacture ever since some form of division of labor and specialization of function developed. In particular with the advent of the factory system of manufacture at the time of the Industrial Revolution and the development of powered machinery to perform production tasks the job shop became the predominant form of production organization. Job shops represent a natural form of evolution of manufacturing capability. As the required production volume increases, additional machines are acquired one at a time to alleviate bottlenecks, and it is natural to locate the new machine alongside the old to share the new capacity with the old and exploit the existing skills and experience with the machine type. This also makes it possible for one operator to look after more than one machine. This is particularly true with NC machines where the operator's role is primarily to identify and respond to any processing problems, even though this might mean that much of the time he is just watching the machines follow their programs.

Even though we will consider that the job shop processes individual jobs, in practice each job is usually a batch of identical parts. The batch size is usually small and in some cases can be of size one. All the parts in the batch are processed together and moved together. Thus the time to process a job on a machine is made up of the time to set up the machine for the batch, the time to process all parts in the batch, and also perhaps the time to tear down or remove the set-up and tools. In traditional job shops the set-up of the machine was the responsibility of a skilled machinist or setter, and the setter had to ensure that the first part produced of the batch met the required process specifications. Subsequent parts were produced by semiskilled operators whose main function was to load and unload parts on the machine, and either the setter or inspectors periodically checked that the parts produced were still meeting specifications. Thus for the semiskilled operators job shop production entailed a substantial number of repetitive tasks. To a large extent, however, the time required to perform the operations on a part was determined by the set-up of the machine, and hence there was little reduction of processing time because of learning by the operators as the batch size increased. In modern job shops machines are often numerically controlled, thus much of the skill of the setter has been transferred to the NC parts programmer. As a result there is minimal learning during processing of a batch, although there is the opportunity to revise the program between successive batches of a given part. In this chapter we will not consider the determination of the optimal batch size or whether it is sometimes appropriate to split batches, but it should be noted that very large batch sizes can create scheduling problems because they occupy a machine for too long and thus delay all other jobs. Also, there may be natural maximum batch sizes created by tool or die life considerations. The only issue relating to batch sizing we will consider is the combining of batches (jobs) for bulk material movement.

7.1.1 Advantages of Job Shops

Ease of supervision.   Probably the major advantage of job shops is the ease of expert supervision or operation by skilled people owing to grouping similar machines. This is particularly true if the machines or processes are not fully understood, making it difficult to achieve consistent quality and predictable performance. Firms that are able to develop reliable processes and learn how to achieve high quality have a considerable competitive advantage. Close attention to the technical problems of a group of similar machines tends to result in faster learning. At later stages of development, once the machines or processes are better understood it becomes advantageous to use one highly skilled person to set up all the machines in a group and deal with any processing problems and use only semiskilled labor to load, unload, and monitor performance of the individual machines. For example, this division of labor was widespread in the munitions plants of World Wars I and II.

Machine utilization.   Another advantage of grouping similar machines in one location, particularly when there is uncertainty about the time required to process a job, is that it reduces the likelihood that there will simultaneously be an idle machine and a job that is available for processing. With grouping, all available jobs will be waiting at the machines so jobs can be assigned to machines as the machines become available. Thus, utilization of expensive machines can be improved. Even if the machines in the group have different capabilities such as reliability or process capability (inherent variability in quality of operation performance) it is easier for supervision to select the appropriate job for the appropriate machine.

Flexibility.   The other principal advantage of job shops is that they enable a wide variety of different products to be manufactured, that is, they have broad scope. Changes in designs and customer requirements can often be accommodated without major expense, even when the changes may require different operation sequences or different processing steps. Within conventional cost-accounting systems there appears to be no penalty for increasing the variety of jobs in a job shop. Job shops have a high degree of flexibility in that they can accommodate changes in product mix, changes in processing requirements, and changes in production volumes (through varying the number of operating machines in a group). Also if there are problems with a machine or process it is possible to bypass the machine and still meet the desired production volume.

7.1.2 Disadvantages of Job Shops

Work in progress and flow time.   The major disadvantages of job shops are usually in the high level of work in progress and long flow times. It is quite common to have flow times that are 20 or more times the actual time that the job is being processed at machines. This is because the job has to wait in front of machines for processing or has to wait to be moved from one machine group to another, either because of limited material handling facilities or because the information that the job is complete and ready to move has not reached the production scheduler.

Priorities.   Because each group of machines is usually the responsibility of a skilled machinist or setter, the selection of the next job to be processed on a machine is often determined by the setter on the basis of such factors as ease of setting up the machine, the natural desire to put off difficult jobs, or the bonus to be earned from the job. In traditional job shops it usually requires some negotiation between the scheduler and the setter to establish the priorities of jobs, and it is thus often the case that the priorities are not consistent with customer desires or promises made by sales departments. Higher-level plant management objectives can also influence priorities. For example, management may be evaluated on tonnage or monetary value of work completed and the desire to meet targets of this type may be given precedence over customer-related priorities.

Meeting promise dates.   The combination of high levels of work in process and the inability to control priorities means that job shop manufacture rarely results in promises on delivery dates to customers being met, even if there are no processing or quality problems with the job and the promise date appeared realistic. It is easy for jobs to get "lost" and thus traditional job shops require many expediters or progress chasers to ensure that jobs move through the shop, to answer customer's queries on the progress of their order, and to urge setters or supervisors to process next the jobs for the important customer.

Variety.   Even though conventional cost-accounting systems do not identify costs associated with the variety of jobs processed by a job shop, these costs can be significant. First of all, variety reduces the opportunity to benefit from learning when a job is repeated many times. As mentioned in Section 5.1.1 these reductions in time or cost can be significant. Next, the variety makes the scheduling and control of production much more difficult, in particular it greatly increases the amount of information that has to be gathered and processed about the jobs in the shop.

7.2 ISSUES IN PLANNING, CONTROL, AND SCHEDULING

Job shops should be considered as consisting not only of the physical facilities, machines, and material handling, but also the production planning, control, and scheduling system. Job shops can be operated as either *produce-to-order* or *produce-to-stock* systems. Produce to order is typical of situations in which every job is different and will not ever be repeated, or, alternatively, where the firm has a set of standard designs or catalog from which customers select, but no manufacture is begun until there is a firm order from the customer. In this case, identical items for different customers are treated as totally distinct jobs. In produce-to-stock systems there is an inventory of completed items from which customer orders are met, and this inventory is replenished as required. Many job shops have both produce-to-stock jobs ("standards") and produce-to-order jobs ("specials"). Sometimes a job shop may be predominantly produce to stock in periods of low demand and produce to order in periods of high demand (or vice versa as in a McDonald's outlet). In this chapter we will assume that with a produce-to-stock operation work release to the job shop is controlled in a manner similar to the single-stage manufacturing system discussed in Chapter 4, that is, when PA cards are generated and transmitted to the input store of the job shop the PA card (or, more generally, the batch of q PA cards) becomes the "job" to be processed by the shop. It will be assumed that, apart from the output store, there are no other planned stocks within the job shop. In the systems described in Section 4.5 in which the order tags are received separately from the requisition tags, it is again the PA cards that create the jobs to be processed by the job shop. All the results that relate the actual number of items in the store to the number of PA cards in the manufacturing stage will continue to apply. In produce-to-order operation of a job shop there is often a promised delivery date associated with the job that would have been agreed on at the time the order was accepted and became a job, and that may be renegotiated from time to time as the job moves through the shop. In a produce-to-stock operation there is no promise date for a job, although it is possible to develop a priority index associated with the job depending on the inventory level in the output store. This means that if promise dates or priorities are relevant to the issue being considered there can be substantial differences between how these issues are addressed in produce-to-order and produce-to-stock operations (and it is still more complicated if a given job shop processes some jobs generated on a produce-to-stock basis and some jobs generated on a produce-to-order basis). We will, however, see that there are many issues where promise dates and priorities are not relevant and the same models will apply to both produce-to-stock and produce-to-order operations. Because of this we will describe and develop most job shop models as if the job shop were operated on a produce-to-order basis.

In planning, control, and scheduling of produce-to-order job shops there are always two perspectives: a customer orientation and a resource orientation. The customer orientation focuses on the customer orders and the resulting jobs that have to be processed by the shop, whereas the resource orientation focuses on the resources required for processing jobs, the machines and labor requirements, and how available time on these resources can be created and managed. Although to a large extent effective management of resources will improve customer-related performance, and meeting customer requirements will demand effective utilization of resources, there can be a conflict between the two perspectives. This makes job-shop management difficult and complex, particularly because customer requirements are always changing, and unexpected problems with machines or workers can often occur.

In discussing the issues in job-shop design, planning, control, and scheduling we will subdivide them into four groups: (1) job-shop design and layout (i.e., those issues that arise at the time when the job-shop design or layout is determined), (2) order acceptance and planning (i.e., those issues that arise at the time the customer negotiates with the firm on placing an order), (3) loading and resource commitment (i.e., those issues that arise when the order has been accepted and a production authorization has been generated), and (4) scheduling and resource assignment (i.e., those issues that arise in processing a job at a particular machine group).

Job-shop design and layout.   In contrast to flow lines and transfer lines it is rare for there to be much in the way of formal design activity when a job shop is first created apart from identifying the types of machines required. The predominant concern initially is getting products of acceptable and consistent quality out the door. It is only when this has been achieved and a customer base has been developed that capacity and bottlenecks become of concern. Job shops tend to evolve over time as the result of one-at-a-time decisions on machine acquisition, installation, or disposal that depend on the current bottlenecks in either capacity or quality. Similarly, material handling or transport facilities are acquired and changed in response to perceived bottlenecks or capacity shortfalls, and generally these decisions are made without much of an overall design or systems analysis.

Job shop layout also evolves over time with machines being located where space can be found or in such a way that the minimum number of other machines has to be moved. Occasionally new machine types are required, and, if they are particularly expensive or novel, additional space may be acquired. A group of part types may have sufficient volume to justify acquiring a flow line or transfer line, and creating the space for this new facility may require rearrangement of existing machines in the remaining job-shop section of the plant.

The major issue that models can help resolve is the determination of the capacity of the job shop for a given mix of jobs and the identification of the bottleneck machines or material handling facilities. Models are also useful in analyzing the impact of changes in layout on the load on the material handling system and seeing whether this has any impact on the overall capacity of the job shop. For design and layout decisions it is usually immaterial whether the job shop is operated on a produce-to-order or produce-to-stock basis, and the models apply to either situation.

Order acceptance and resource planning.   When a job shop is approached by a potential customer it is usually necessary for the job shop to develop a quotation covering such aspects as price and delivery. For each quote the job shop has to estimate the resources needed to meet the customer's requirements. Often, apart from the actual manufacture of the product substantial resources will be required in design and manufacturing engineering, and the elapsed time for these activities can represent a substantial part of the total time required between order acceptance and delivery. Apart from the resources required by this job, it is also necessary to consider other jobs already in process and other outstanding quotes that might become firm orders. Thus there are two crucial issues at this stage.

1. *Promised delivery date.* Apart from such factors as the product design, and reputation for quality and price, a critical factor in determining whether the quote will result in a firm order will be the promised delivery date and the confidence the customer has in that date. Deciding on a realistic promise date involves not only considering the uncertainties in processing existing jobs but also the uncertainties associated with other potential jobs. Further, in seeking to get the order, the job shop may be under some pressure to modify the promised delivery date by giving this job higher priority, yet it cannot give almost all jobs higher priority and still meet all promises. Thus a crucial issue for modeling is characterizing the distribution of flow times and determining how it is influenced by resource levels and priority rules.

2. *Resource levels.* One way of trying to improve the delivery performance is to reduce resource utilization by increasing the capacity of key resources (e.g., hiring more skilled machinists or acquiring more machines). There is often a lengthy lead time associated with acquiring new machines or training skilled workers. Further, the arrival rate of orders and the consequent loading on resources will vary with business conditions, so the firm will often be reluctant to add resources until it is convinced that the increase in resource loading will last a sufficient time to recover the investment associated with acquisitions of new capacity. If there are decreases in the level of demand then it may be necessary for the firm to lay off skilled people or dispose of apparently surplus machines to match capacity to available work. Determination of the required resource levels is complicated by the need to make forecasts of future orders when order arrivals are both random and nonstationary over the period required to justify changes in levels.

In a job shop that produces to stock there are no promise dates, so it is only resource levels that need be planned. This requires a tradeoff to be made between the costs of inventory in the output store and in process, the costs of back-orders, and the costs associated with the resource level such as the cost of staffing machines. Thus, to make the resource level decision it is necessary to determine the impact of resource levels on inventory and back-orders for given demand rates for finished products.

Loading and resource commitment.   Once the order has been accepted, along with a promised delivery date, the job proceeds through design and manufacturing engineering where the processing steps required, the sequence in which the operations should be performed, and the time required for each operation on each machine are determined. Also any necessary tools or special materials required by the operation are identified. Once this has been done then the decision of when to release the job to the job shop can be made. To make this decision two approaches are used. In one approach there are a small number (one or two) of critical machine groups and when a job arrives at the job shop it is allocated to time slots (typically in weeks) on these groups in accordance with the job priority and available uncommitted capacity. The sequence of jobs on these critical groups then determines the sequence in which these jobs will be processed at all other machines and the timing of release of the job to the shop. The other approach is periodically (typically once a week) to "load" all jobs onto all the machines where an operation has to be performed on the jobs, where loading means developing a plan that "reserves" time for each operation required by each job on the appropriate machine. Conventionally there are two methods of loading: *forward loading* in which time is reserved forward from the present, perhaps using promise date as the priority index to decide which of two available jobs goes first on a machine, and *backward loading* in which jobs are loaded backward from their promise date. With forward loading it can happen that some jobs may not meet their promise date, whereas with backward loading it may be found that no feasible loading exists. Jobs generated by produce-to-stock operation do not have promise dates set by an external customer, and, particularly for backward loading, it is necessary to determine an equivalent priority index for these jobs.

The loading plan, usually called the production schedule, is used to identify resource conflicts, control the release of jobs to the job shop, ensure that the required raw materials are ordered in time, and identify whether delivery promises can be met. It is usually developed assuming that the time required for operations is precisely known and the only allowance for machine failures or processing problems is to assume that the available time on machines is somewhat less than the actual time that it is planned for the machine to be operated. In reality the actual time required for an operation may differ, sometimes substantially, from the planned time because of processing and quality problems. Further, machines may break down, operators may get sick, essential tools or raw materials may not be available, and customers may change their requirements, causing the size of the order or the promised delivery date to vary. Even in make-to-stock systems unexpectedly high demand may exhaust available stock, thus increasing the urgency for the replenishment order, or demand may be less so that the priority can be lowered. All this means that the schedule is almost never adhered to and it is almost always wrong and out of date within a very short time after being prepared. Experienced schedulers seem to recognize this from the beginning in preparing the schedule and tend to build into the schedule the ability, should the need arise, to make changes that do not disrupt the whole plan. For example, they may ensure that there is spare capacity on machines that can be used for a variety of different operations. An issue that arises is whether there is an alternative approach to loading that takes into account the stochastic characteristics of job processing from the beginning. One approach would be to formulate this as a stochastic control problem, but this is outside the scope of this book.

Allocation and resource assignment.   Once the job is released to the shop it will move from one machine group to the next in accordance with its operation sequence. However, this movement is by no means automatic. Information that the operation is complete has to be collected and communicated to the material handling facilities. Then, once job movement has occurred and the job is waiting in front of the group of machines for its next operation, decisions must be made as to which job is to be processed next on an available machine and, usually, which worker will perform the operation. The schedule may provide some guidance, but it is rare for the schedule itself to provide complete instructions because the decision should consider differences in skill between individual workers and differences in capability of individual machines within the group. Further the fact that the schedule is usually not quite up to date will result in expediters and schedulers trying to influence the allocation decision. If the machine perspective is dominant then the decision on which job to be done next may be influenced by the amount of set-up or change-over time required or the desire to maximize machine utilization or short-term output. In this case, promise dates may not be considered at all in making the allocation decision.

7.2.1 Dealing with Variety, Changes, and Disturbances

The dominant concern in managing job shops and ensuring satisfactory performance in meeting promised delivery dates is almost always trying to deal with the variety of jobs, in particular their differing operation sequences and their different processing times (and also their different batch sizes). Changes are always happening: changes in the mix of jobs, changes in the volume of work, changes in the work force and their skills. Disturbances such as machine failures, processing problems, worker absences, and customer order changes are frequent. Stochastic models are essential to describe these changes and disturbances. Until very recently, however, most job shops had very inadequate data collection systems, and it was almost impossible to have reliable information about the status of jobs and machines. Thus models have been difficult to apply and to validate. Now, with modern data collection and information processing systems, it is possible to have up-to-date information on the status of the job shop. Thus, it should be possible to use models to aid scheduling, but it is also likely that models will suggest alternative approaches to scheduling, particularly in loading and job release, where it will be recognized that the continual change makes detailed plans of future activities somewhat pointless. Apart from the hard information that could be captured by an information system, experienced schedulers acquire and use soft information through gossip, hunch, and intuition. This cannot be incorporated readily into formal models, and probably limits their applicability in allocation and resource assignment.

7.3 REPRESENTATION OF JOB FLOW

In this chapter we will develop models to identify policies that could be used to eliminate some of the problems causing excessive flow times and in-process jobs. To do this, we will first consider an abstract version of job shops and use it to represent the job flow through the system, and determine the capacity of the job shop.

7.3.1 Abstract Description of Job Shop

A job shop in abstraction is a set of machine centers between which jobs can be moved from any machine center to another. This job movement may be carried out by a central transport center, by a dedicated transporter between machine centers, or by a combination of these. Jobs of different types arrive at the job shop over time. Each job has a specific sequence of machine or machine centers it will visit before it leaves the job shop. We will assume that the jobs, machine centers, and the operating policies conform to the following assumptions:

Job-based assumptions

J1. A job arriving in the system goes directly to a machine center for its first operation (this assumption will be relaxed when we introduce two levels of control to improve the performance of the job shops).

J2. The characteristics of each job are statistically independent of those of all other jobs.

J3. Each job has a specified sequence of machine centers it should visit.

J4. Each job requires a finite processing time for each of its operations. The processing time of all jobs of a specific type at any machine center are iid with a known distribution function (in some instances we will also assume that the processing times can be determined before the jobs are processed).

J5. Each job may have to wait between processing at different machine centers, and thus in-process inventory is allowed.

Machine-based assumptions

M1. Each machine center consists of one or several identical machines.

M2. Each machine in the shop operates independently of other machines and thus is capable of operation at its own maximum output rate.

M3. Each machine is continuously available for processing jobs, and there are no interruptions owing to breakdowns, maintenance, or other such causes.

Operating policies

O1. Each job is considered as an indivisible entity even though it may be composed of several individual units.

O2. Each job, once accepted, is processed to completion, that is, no cancellation or interruption of jobs is permitted.

O3. Each job (operation), once started on a machine, is completed before another job is started on that machine.

O4. Each job is processed on no more than one machine at a time.

O5. Each machine center is provided with adequate waiting space for allowing jobs to wait before starting their processing.

O6. Each machine center is provided with adequate output space for allowing completed jobs to wait until they are moved out of the machine center.

Our focus on the job shop is to study the effect of control and scheduling policies on the job flow through the system. We are particularly interested in the number and flow times of jobs in the system, at machine centers, and in transition between machine centers.

In abstraction, a job shop can therefore be viewed as a multiple-class open queueing network. Because the number of types of jobs may be very large in a job shop, it is convenient to develop aggregate models of job flow through the machine centers by aggregating different types of jobs into a single or several classes of jobs. We will next describe this aggregation procedure. It should be noted that even if aggregation is not desired in a specific case of interest, the models and the solution procedures to be developed in this chapter are still applicable, simply by treating each job type as a distinct job class.

7.3.2 Modeling Job Shops by Open Queueing Networks

The job shop consists of a set $M = \{1, \ldots, m\}$ of m machine centers with c_i identical machines at machine center i, $i = 1, \ldots, m$. There is also a set of $H = \{1, \ldots, h\}$ of h material handling devices that are used to transport jobs from one machine center to another. The set of different job types is $R = \{1, \ldots, r\}$. Let $N_l(t)$ be the number of type l jobs that arrived during $(0, t]$, $l = 1, \ldots, r$. Each job of type l requires a total of n_l operations at machine centers $C_1^{(l)}, \ldots, C_{n_l}^{(l)}$, in that order, $l = 1, \ldots, r$. If two or more consecutive operations are required on the same machine center we will view these operations as one big operation, and therefore in the machine sequence $C_1^{(l)}, \ldots, C_{n_l}^l$ of a type l job we will not have $C_k^{(l)} = C_{k+1}^{(l)}$ for any $k = 1, \ldots, n_l - 1$; and $l = 1, \ldots, r$. The processing time of the jth operation of the nth type l job at machine center $C_j^{(l)}$ is $S_{j,n}^{(l)}$: $S_{j,n}^{(l)}$ could be random or deterministic, $j = 1, \ldots, n_l; l = 1, \ldots, r; n = 1, 2, \ldots$. The time to transport a job from machine center i to j is $T_{ij,n}$ where this time could be random or deterministic, $i, j = 1, \ldots, m; n = 1, 2, \ldots$. We will first see how we can aggregate the job flow such that the aggregated job flow will mimic the actual job flow through the system.

Let

$$\lambda(l) = \lim_{t \to \infty} \frac{1}{t} N_l(t)$$

be the arrival rate of type l jobs to the job shop. Then the arrival rate of the aggregated job is $\lambda = \sum_{l=1}^{r} \lambda(l)$. The corresponding arrival process is $\{N(t), t \geq 0\}$, where $N(t) = \sum_{l=1}^{r} N_l(t), t \geq 0$.

The first machine center visited by the aggregated job cannot be identified with certainty unless all the r types of jobs visit the same machine center for their first operation. Therefore we need to represent the first machine center visited by the aggregate job by a probability distribution, say γ_i, $i = 1, \ldots, m$, where γ_i is the probability that the aggregate job will visit machine center i for its first operation. The rate at which aggregate jobs visit machine center i for their first operation is $\lambda \gamma_i$, $i = 1, \ldots, m$. We should therefore choose γ_i such that these rates are the same as the rates at which jobs enter machine center i from outside, $i = 1, \ldots, m$. The true arrival rate of jobs from outside to machine center i is $\sum_{l=1}^{r} \lambda(l) I_{\{C_1^{(l)} = i\}}$, $i = 1, \ldots, m$, where $I_{\{C_1^{(l)} = i\}}$ is the indicator function that takes on the value of 1 if the first operation of type l job is on machine center i (i.e., $C_1^{(l)} = i$) and 0 otherwise. Therefore we set

$$\gamma_i = \frac{1}{\lambda} \sum_{l=1}^{r} \lambda(l) I_{\{C_1^{(l)} = i\}}, \quad i = 1, \ldots, m \tag{7.1}$$

Let p_{ij} be the fraction of jobs leaving machine center i that will go next to machine center j. Observe that the number of times a type l job will go from machine center i to j is $\sum_{k=1}^{n_l-1} I_{\{C_k^{(l)} = i, C_{k+1}^{(l)} = j\}}$. Therefore the rate of flow of jobs from machine center i to machine center j is $\lambda_{ij} = \sum_{l=1}^{r} \lambda(l) \sum_{k=1}^{n_l-1} I_{\{C_k^{(l)} = i, C_{k+1}^{(l)} = j\}}, i, j = 1, \ldots, m$. $\lambda_{ii} = 0$,

$i = 1, \ldots, m$ because $I_{\{C_k^{(l)}=i, C_{k+1}^{(l)}=i\}} = 0$. The rate of flow of jobs into (and out of) machine i is $\lambda_i = \sum_{l=1}^{r} \lambda(l) \sum_{k=1}^{n_l} I_{\{C_k^{(l)}=i\}}$. Therefore $p_{ii} = 0$, $i = 1, \ldots, m$ and

$$p_{ij} = \frac{\lambda_{ij}}{\lambda_i}, \quad i, j = 1, \ldots, m \tag{7.2}$$

Now suppose we model the job transfer behavior of the aggregate jobs by a transfer probability matrix $\mathbf{P} = (p_{ij})_{i,j=1,\ldots,m}$. That is, an aggregate job that completes its service at machine center i goes next to machine center j with probability p_{ij}, independent of its past history of visits. Let $\hat{\lambda}_i$ be the rate at which aggregate jobs enter machine center i when aggregate jobs are transferred from one machine center to another according to the transfer probability matrix $\mathbf{P}$. Then, for our aggregate job representation to be a reasonable one we need $\hat{\lambda}_i = \lambda_i$, $i = 1, \ldots, m$. When this is true, the aggregate job flow rate $\hat{\lambda}_i p_{ij}$ from machine center i to j will be the same as the actual job flow rate from machine center i to j, λ_{ij}. We will next show that indeed $\hat{\lambda}_i = \lambda_i$, $i = 1, \ldots, m$.

The rate at which aggregate jobs arrive from outside to machine center i is $\lambda \gamma_i$, and the rate at which aggregate jobs arrive at machine center i from machine center j is $\hat{\lambda}_j p_{ji}$, $j = 1, \ldots, m$. Therefore it is easy to observe that the rate at which the aggregate jobs flow into machine center i is given by

$$\hat{\lambda}_i = \lambda \gamma_i + \sum_{j=1}^{m} \hat{\lambda}_j p_{ji}, \quad i = 1, \ldots, m \tag{7.3}$$

Substituting $\hat{\lambda}_j = \lambda_j$, $j = 1, \ldots, m$ on the right-hand side of equation 7.3 one sees from equations 7.1 and 7.2 that

$$\hat{\lambda}_i = \sum_{l=1}^{r} \lambda(l) I_{\{C_1^{(l)}=i\}} + \sum_{j=1}^{m} \sum_{l=1}^{r} \lambda(l) \sum_{k=1}^{n_l-1} I_{\{C_k^{(l)}=j, C_{k+1}^{(l)}=i\}} \tag{7.4}$$

because $\sum_{j=1}^{m} I_{\{C_k^{(l)}=j, C_{k+1}^{(l)}=i\}} = I_{\{C_{k+1}^{(l)}=i\}}$. Since $\mathbf{P}$ is substochastic (i.e., for at least one i, $i = 1, \ldots, m$, we have $\sum_{j=1}^{m} p_{ij} < 1$), the solution $\hat{\lambda}_i$, $i = 1, \ldots, m$ of equation 7.3 is unique and therefore $\hat{\lambda}_i = \lambda_i$, $i = 1, \ldots, m$ as desired. So we have seen that by representing the flow of the aggregate jobs by the transfer probability matrix $\mathbf{P}$ for inside transfers and by the probability vector γ for the first operation we have captured the job flow rates of the job shop exactly. Under certain assumptions on the service time distributions, we will see later on in Section 7.5 that this aggregation does give the exact system performance with respect to the number of jobs in the system and at the machine centers. Now to complete the aggregation procedure we need to obtain the service time and the transport time of the aggregate jobs.

Suppose we randomly sample a job that arrived at a machine center i. The probability that this job is of type l and it came to machine center i for its kth operation is $(1/\lambda_i)\lambda(l) I_{\{C_k^{(l)}=i\}}$. Therefore the service time of a randomly chosen job that arrived at machine center i is

$$S_i = S_k^{(l)} \text{ with probability } \frac{1}{\lambda_i}\lambda(l)I_{\{C_k^{(l)}=i\}}, \quad k=1,\ldots,n_l; l=1,\ldots,r; i=1,\ldots,m$$

(7.5)

where $S_k^{(l)}$ is a generic random variable representing the service time of the kth operation of a type l job.

We will assume that the service times of the aggregate jobs at machine center i are iid with distribution the same as that of S_i, $i=1,\ldots,m$. Note that

$$E[S_i] = \frac{1}{\lambda_i}\sum_{l=1}^{r}\sum_{k=1}^{n_l}\lambda(l)I_{\{C_k^{(l)}=i\}}E[S_k^{(l)}], \quad i=1,\ldots,m$$

(7.6)

and

$$E[S_i^2] = \frac{1}{\lambda_i}\sum_{l=1}^{r}\sum_{k=1}^{n_l}\lambda(l)I_{\{C_k^{(l)}=i\}}E[(S_k^{(l)})^2], \quad i=1,\ldots,m$$

(7.7)

The average work load brought to work center i by all types of jobs is

$$w_i = \sum_{l=1}^{r}\lambda(l)\sum_{k=1}^{n_l}I_{\{C_k^{(l)}=i\}}E[S_k^{(l)}], \quad i=1,\ldots,m$$

This is exactly the same as the average workload brought to machine center i by the aggregate job (i.e., $\lambda_i E[S_i] = w_i$, $i=1,\ldots,m$). Thus the aggregate job behavior captures the exact job flow rates and work load at each machine center of the job shop under consideration.

All that is left for us to characterize is the transport time. To do this we need more information about the structure of the transport system. For example, if the transport system is such that there are transporters dedicated to each link (i,j) connecting machine center i to j, we may regard these transporters as an equivalent machine center, say (i,j). The service times at this machine center are iid with distribution the same as that of T_{ij}, $i,j=1,\ldots,m$. The modified transfer probabilities are then $p'_{i,(i,j)} = p_{ij}$ and $p'_{(i,j),j} = 1$, $i,j=1,\ldots,m$. In addition $\gamma'_{(0,i)} = \gamma_i$, $p'_{(0,i),i} = 1$ and $p'_{i,m+1} = 1 - \sum_{j=1}^{m}p_{ij}$, $p'_{(i,m+1),m+1} = 1$. Here $(m+1)$ represents the outside and $(i,m+1)$ represents the transport system that moves jobs from machine center i to the outside. Therefore in this system we assume that there is a total of $2m+m(m-1) = m(m+1)$ transporters. In this case, even with transporters included in the model we can view it as a job-shop model with $m' = m(m+2)$ machine centers, job-transfer probabilities characterized by the modified transfer probability matrix $\mathbf{P}'$, a job's first operation given by the probability vector γ', and no job transporters in the system.

When a central transport system is used to handle job movements from one machine to another, we need an intermediate step first to identify the destination of an aggregate job that is leaving any machine center i, $i=1,\ldots,m$ or an aggregate job that is entering the machine center j from outside. We will do this by tagging each aggregate job in transit by a class index. Specifically a class (i, j) job is in transit from machine center i toward machine center j. Jobs that are either waiting or in process at a machine center are tagged by the class index 1. Then $p_{i0}^{(1)(i,j)} = p_{ij}$ is the probability that a job (class 1) completing service at machine center i moves to the central transport system as class (i, j), requesting its movement from machine center i to machine center j. Here 0 represents the central transport system and T_{ij} is the generic service time of a class (i, j) job at the central transport center 0. Then $p_{0j}^{(i,j)(1)} = 1$. A class $(0, j)$ job from outside is an aggregate job that requires its first operation at machine center j. Then $\gamma_0^{(0,j)} = \gamma_j$ and $p_{0,j}^{(0,j)(1)} = 1$, $j = 1, \ldots, m$. Now to represent the flow of aggregate jobs to the outside, $(m + 1)$, we set $p_{i0}^{(1)(i,m+1)} = 1 - \sum_{j=1}^{m} p_{ij}$ and $p_{0,m+1}^{(i,m+1)(m+1)} = 1$. Thus we have represented the aggregate job flow in the central transport system by a multiple-class aggregate job flow. If we wish to aggregate these (artificially created) job classes into a single (new) aggregate job class, we need an aggregation procedure for classes of jobs with probabilistic job transfers and class changes. We will next describe this aggregation procedure in a general context.

7.3.3 Aggregation of Job Flow with Probabilistic Routing

In the initial discussion of the job shop we specified the flow of jobs of each type through the machine centers by a fixed sequence of machine centers. Because of rework and reject requirements it may happen that the actual sequence of machine centers visited by any specific job may be random. The randomness in a machine sequence, however, evolves in a systematic way such that it can be characterized by a set of deterministic machine sequences and probabilistic switch-over from one machine sequence to another. The details of this switch-over is as follows. Suppose a job type l arrives at the job shop requiring processing at machine centers $C_k^{(l)}$, $k = 1, \ldots, n_l$ in that order. After it receives its kth service at machine center $C_k^{(l)}$, suppose it has been detected that the job processing of the job is defective. To rectify this the job may need to be processed through the machine centers according to a machine center sequence that is different from $\{C_{k+1}^{(l)}, \ldots, C_{n_l}^{(l)}\}$. For example, if all we need is to reprocess the last operation, the new machine center sequence is $\{C_k^{(l)}, C_{k+1}^{(l)}, \ldots, C_{n_l}^{(l)}\}$. So depending on the type of defect detected the job will be assigned one of many possible new machine center sequences. Because the type of defect that may be encountered is random, we represent the switch-over of machine center sequences by probability distributions. More specifically $q_{kk'}^{(l)(l')}$ is the probability that a job after receiving service at machine center $C_k^{(l)}$ as a type l job is transferred to machine center $C_{k'}^{(l')}$ as a type l' job. Note that if the processing of the job is found not to be defective after receiving service at machine center $C_k^{(l)}$ as a type l job it then proceeds to machine center $C_{k+1}^{(l)}$ as a type l job (provided $k + 1 \leq n_l$); otherwise it leaves the system. Conversely, if the processing was defective, but requires only reprocessing, the job is then routed back to machine center $C_k^{(l)}$ as a type l job. We will next see how we can represent the job flow through the machine centers by a transfer probability matrix.

If the machine center sequence of a type l job is such that no machine center is visited more than once we may represent the machine center sequence by a transfer probability matrix, say $\mathbf{A}^{(l)}$ as follows:

$$a_{ij}^{(l)} = \sum_{k=1}^{n_l-1} I_{\{C_k^{(l)}=i, C_{k+1}^{(l)}=j\}}, \quad i, j = 1, \ldots, m$$

The probability that the job joins machine center i for its first operation is $\gamma_i^{(l)} = I_{\{C_1^{(l)}=i\}}$, $i = 1, \ldots, m$. If it happens that the machine center sequence of a type l job requires processing by the same machine center more than once, then we will introduce artificial job types such that when the job visits the same machine center for the second time it will change its job type. Suppose $C_1^{(l)}, \ldots, C_{k-1}^{(l)}$ are all different machine centers, but $C_k^{(l)} \in \{C_1^{(l)}, \ldots, C_{k-1}^{(l)}\}$. That is a type l job for its kth operation requires the machine center $C_k^{(l)}$ that it has already visited before. In this case we replace the type l job by two jobs with different types (indexed l' and l''). The type l' job has a machine center sequence $\{C_1^{(l)}, \ldots, C_{k-1}^{(l)}\}$ and the type l'' job has a machine center sequence $\{C_k^{(l)}, \ldots, C_{n_l}^{(l)}\}$. In addition $q_{k-1,1}^{(l')(l'')} = q_{k-1,k}^{(l)(l)}$, $q_{i'i''}^{(l')(j)} = q_{i'i''}^{(l)(j)}$, $q_{i''i'}^{(j)(l')} = q_{i''i'}^{(j)(l)}$, $q_{ii'}^{(l')(l')} = q_{ii'}^{(l)(l)}$, $i, i' = 1, \ldots, k-1$; $i'' = 1, \ldots, n_j$; $j = 1, \ldots, r$ and $q_{i'i''}^{(l'')(j)} = q_{k-1+i',i''}^{(l)(j)}$, $i' = 1, \ldots, n_l + 1 - k$; $i'' = 1, \ldots, n_j$; $j = 1, \ldots, r$, $j \neq l$. Observe that type l' jobs do not require the same machine center twice. Now decomposing type l'' jobs in this way, if necessary, and continuing we will end up with a set of job types that does not require the same machine center twice. Therefore without loss of generality we will assume that this is true for the given set R of job types. Then we can combine $\gamma^{(l)}$, $\mathbf{A}^{(l)}$ and $q_{kk'}^{(l)(l')}$, $k = 1, \ldots, n_l$, $k' = 1, \ldots, n_{l'}$, $l, l' = 1, \ldots, r$ to create a job transfer probability matrix $\mathbf{P} = \left(p_{ij}^{(l)(l')}\right)_{i,j=1,\ldots,m}^{l,l'=1,\ldots,r}$, where $p_{ij}^{(l)(l')}$ is the probability that a job leaving machine center i as a type l job will join machine center j as a type l' job. This probability is easily seen to be given by

$$p_{ij}^{(l)(l')} = \begin{cases} a_{ij}^{(l)} - \sum_{l'' \neq l} \sum_{k=1}^{n_l} \sum_{k''=1}^{n_{l''}} q_{kk''}^{(l)(l'')} I_{\{C_k^{(l)}=i, C_{k''}^{(l'')}=j\}}, \\ \quad\quad\quad\quad\quad\quad i, j = 1, \ldots, m; l = l' = 1, \ldots, r \\ \sum_{k=1}^{n_l} \sum_{k'=1}^{n_{l'}} q_{kk'}^{(l)(l')} I_{\{C_k^{(l)}=i, C_{k'}^{(l')}=j\}}, \\ \quad\quad\quad\quad\quad\quad i, j = 1, \ldots, m; l \neq l' = 1, \ldots, r \end{cases} \quad (7.8)$$

Note that $\mathbf{P}$ represents the job flow exactly.

Now we will carry out the aggregation procedure such that the aggregate job flow will have the same job-flow rates and machine center work loads as in the original system. Let $\lambda_{ij}(l)$ be the rate of type l job flow into machine center j directly from machine center i independent of the job type when it entered machine center i. Then $\lambda_i(l) = \sum_{j=1}^{m} \lambda_{ji}(l)$ is the total job flow of type l jobs to machine center i. It is then not difficult to see that

$$\lambda_i(l) = \lambda(l)\gamma_i + \sum_{j=1}^{m}\sum_{l'=1}^{r}\lambda_j(l')p_{ij}^{(l')(l)}, \quad i = 1, \ldots, m, l = 1, \ldots, r \quad (7.9)$$

The set of linear equations in equation 7.9 can be solved to obtain $\lambda_i(l)$, $i = 1, \ldots, m$; $l = 1, \ldots, r$ once **P** is obtained. Let **P**$'$ be defined so that

$$p'_{ij} = \lambda_{ij}/\lambda_i \quad , \quad i, j = 1, \ldots, m \quad (7.10)$$

where $\lambda_i = \sum_{l=1}^{r}\lambda_i(l)$; $\lambda_{ij} = \sum_{l=1}^{r}\lambda_{ij}(l)$ and $\lambda_{ij}(l) = \sum_{l'=1}^{r}\lambda_i^{(l')}p_{ij}^{(l')(l)}$, $i, j = 1, \ldots, m$. The probability vector γ' corresponding to the first operation of the aggregate job is given by

$$\gamma'_i = \frac{1}{\lambda}\sum_{l=1}^{r}\lambda(l)\gamma_i^{(l)} \quad (7.11)$$

where $\lambda = \sum_{l=1}^{r}\lambda(l)$ is the total job arrival rate. Remember that because some or many of the job types are artificially created to satisfy the assumption of not requiring the same machine center twice or more by the same type of job, the external arrival rates of these job types will be zero. It can be verified routinely by substitution that the job flow rates induced by γ' and **P**$'$ are the same as the original rates λ_{ij}, $i, j = 1, \ldots, m$. Because of feedback, however, it may be possible that $\lambda_{ii} > 0$ for some or all $i = 1, \ldots, m$. As before the generic service times of the aggregate jobs at machine center i is

$$S_i = S_i^{(l)} \text{ with probability } \lambda_i(l)/\lambda_i, \quad l = 1, \ldots, r; i = 1, \ldots, m \quad (7.12)$$

When we have transporters dedicated to each link (i, j) from machine center i to j, the transporters can be incorporated into this model in exactly the same way as before.

We will now apply this aggregation procedure for the job shop with a central transporter system. For this we will carry out an aggregation procedure for job *classes* as we did for job *types* in Section 7.3.2. Then we will end up with a new aggregate job flow with $m + 1$ machine centers consisting of the original m machine centers and one central transport center. The corresponding job transfer probability matrix **P**$'$ is given by

$$p'_{0i} = \frac{1}{\Lambda}\lambda_i, \quad i = 1, \ldots, m$$
$$p'_{i0} = 1 \text{ and } p_{ij} = 0, \quad j \neq 0 \quad (7.13)$$

Here

$$\Lambda = \lambda + \sum_{i=1}^{m}\lambda_i \quad (7.14)$$

is the job-flow rate at the transporter. In addition the probability γ'_i that an external aggregate job arrival joins machine center i is given by

$$\gamma'_0 = 1, \quad \gamma'_i = 0, \quad i = 1, \ldots, m \quad (7.15)$$

It is easily verified that the flow rate and the work load brought in by this new aggregate job at each machine center are the same (i.e., λ_i and w_i, respectively) as in the original system. To have the same effect on the transport center, we set the generic transport time of a (new) aggregate job by

$$T_0 = T_{i,j} \text{ with probability } \frac{\lambda_{ij}}{\Lambda}, \quad i, j = 0, \ldots, m \tag{7.16}$$

where $\lambda_{0j} = \gamma_j \lambda$ and $\lambda_{i0} = \lambda_i(1 - \sum_{j=1}^{m} p_{ij})$.

In many instances the single-class aggregation is sufficient to study the aggregate behavior of a dynamic job shop. In some cases, however, a natural classification of the set of job types $\{1, \ldots, r\}$ into a number of job classes $\{1, \ldots, p\}$ may be required. This need may arise because jobs belonging to different classes receive different priorities when selected for service or their flow paths within the job shop are drastically different. We will now see how a given set R of r job types can be aggregated into a set C of p classes of jobs. Let R_l be the set of job types that belong to class l, $l = 1, \ldots, p$. Then $\{R_1, \ldots, R_p\}$ is a partition of R. At this stage we will assume that we have carried out the exact representation of the job flows of the job types by the transfer probability matrix $\mathbf{P} = (p_{ij}^{(l)(l')})_{i,j=1,\ldots,m}^{l,l'=1,\ldots,r}$ as discussed earlier. Let $\hat{\mathbf{P}} = (\hat{p}_{ij}^{(l)(l')})_{i,j=1,\ldots,m}^{l,l'=1,\ldots,p}$ be the transfer probability matrix for the p aggregated classes of jobs. We wish to select $\hat{\mathbf{P}}$ such that the aggregate flow rates, say $\hat{\lambda}_{ij}(l)$, of class l jobs into machine center j from machine center i is the same as that in the original system. If $\lambda_{ij}(l)$ is the flow rate of type l jobs into machine center j from machine center i, we need $\hat{\lambda}_{ij}(l) = \sum_{l' \in R_l} \lambda_{ij}(l')$, $l = 1, \ldots, p; i, j = 1, \ldots, m$. Define $\hat{\mathbf{P}}$ by

$$\hat{p}_{ij}^{(l)(l')} = \frac{\sum_{k \in R_l} \sum_{k' \in R_{l'}} \lambda_i(k) p_{ij}^{(k)(k')}}{\sum_{k \in R_l} \lambda_i(k)}, \quad l, l' = 1, \ldots, p; i, j = 1, \ldots, m \tag{7.17}$$

It can be easily verified that the job flow rates $\hat{\lambda}_{ij}(l)$ induced by $\hat{\mathbf{P}}$ are equal to $\sum_{l' \in R_l} \lambda_{ij}(l')$, $l = 1, \ldots, p; i, j = 1, \ldots, m$. The generic service time $S_i^{(l)}$ of a class l job at machine center i is then given by

$$S_i^{(l)} = S_i^{(l')} \text{ with probability } \lambda_i(l')/\hat{\lambda}_i(l), \quad l' \in R_l \tag{7.18}$$

where $\hat{\lambda}_i(l) = \sum_{l' \in R_l} \lambda_i(l')$, $l = 1, \ldots, p; i = 1, \ldots, m$. The transport center can now be incorporated into this model exactly as we did before.

7.3.4 Capacity of Job Shops

In this section we will see how, without making additional assumptions, the aggregate model can be used to obtain the capacity of a job shop. Recall that the capacity of a job shop is the minimum value of the external job arrival rate such that for any job arrival rate larger than this value the number of jobs in the system will grow without limit. Let $\mathbf{P}$ be the job transfer probability matrix of the aggregate job and γ be the probability vector of the machine center visited by an aggregate job for its first operation. If λ is the external job arrival rate we have (see equation 7.3)

$$\lambda_i = \lambda \gamma_i + \sum_{j=1}^{m} \lambda_j p_{ji}, \quad i = 1, \ldots, m \tag{7.19}$$

for the aggregate job arrival rate to machine center i, $i = 1, \ldots, m$. Then the work load offered to machine center i is $\lambda_i E[S_i]$. From Section 3.2.1 we know that if $\lambda_i E[S_i] < c_i$, then the number of jobs in machine center i will be finite. Therefore as long as $\lambda_i E[S_i] < c_i$, $i = 1, \ldots, m$, the number of jobs in the system is finite. The capacity of the job shop is therefore the minimum value of λ for which at least for one i, $\lambda_i E[S_i] = c_i$. To obtain this capacity λ^* first note that λ_i's are linear in λ. More specifically, if we set v_i, $i = 1, \ldots, m$, to be the solution of

$$v_i = \gamma_i + \sum_{j=1}^{m} v_j p_{ji}, \quad i = 1, \ldots, m \tag{7.20}$$

then $\lambda_i = \lambda v_i$. Note that v_i is the average number of times an arbitrary aggregate job visits machine center i during its stay in the system. Therefore the capacity of the job shop is given by

$$\lambda^* = \min_{1 \leq i \leq m} \left\{ \frac{c_i}{v_i E[S_i]} \right\} \tag{7.21}$$

7.4 JACKSON OPEN QUEUEING NETWORK MODEL

A single (probably aggregated) class of jobs arrive at the job shop according to a Poisson process with arrival rate λ. The fraction of jobs that will join machine center i on their arrival is γ_i, $i = 1, \ldots, m$. ($\sum_{i=1}^{m} \gamma_i = 1$). The fraction of jobs that complete service at machine center i that will directly go to machine center j is p_{ij}. Then $1 - \sum_{i=1}^{m} p_{ij}$ is the fraction of jobs among those completing service at machine center i, that will directly leave the system. Of course, at least for one $i = 1, \ldots, m$, $1 - \sum_{j=1}^{m} p_{ij} > 0$ so that the jobs will eventually leave the system. The service times of jobs at machine center i are iid exponential random variables with mean $1/\mu_i$, $i = 1, \ldots, m$. All the service times and the arrival times are mutually independent.

The rate at which a job is processed at machine center i when there are n jobs is assumed to be $\mu_i r_i(n)$, $n = 0, 1, \ldots$. This allows us to represent, as special cases, single or multiple machines in parallel at the machine center i. Specifically if there are c_i machines in parallel at machine center i, we set $r_i(n) = \min\{n, c_i\}$, $n = 0, 1, \ldots$; $i = 1, \ldots, m$. Leaving this representation as a general function r_i allows one to model the effect of the number of jobs in a machine center on the worker efficiency.

In this section we concentrate on job shops that use service protocols such as FCFS or LCFS that are independent of the job service time requirements.

We model this system by a Markovian open queueing network (i.e., Jackson open queueing network) where jobs are routed from one service center to another according to a transfer probability matrix $\mathbf{P} = (p_{ij})_{i,j=1,\ldots,m}$.

Let $N_i(t)$ be the number of jobs at machine center i at time t; $i = 1, \ldots, m$ and $\mathbf{N}(t) = (N_1(t), \ldots, N_m(t))$. Then $\mathbf{N}(t)$ is a continuous time Markov process on $\mathcal{N}_+^m$. Define the stationary distribution of $\mathbf{N}$ by $p(\mathbf{n}) = \lim_{t \to \infty} P\{\mathbf{N}(t) = \mathbf{n}\}$, $\mathbf{n} \in \mathcal{N}_+^m$ assuming its existence. The inflow into state $\mathbf{n}$ can occur from state $\mathbf{n} - \mathbf{e}_i$ owing to an external job arrival joining machine center i, or from state $\mathbf{n} + \mathbf{e}_i$ owing to a service completion of a job at machine center i that leaves the system, or from state $\mathbf{n} + \mathbf{e}_j - \mathbf{e}_i$ owing to a service completion at machine center j ($j = 1, \ldots, m; j \neq i$) that joins machine center i directly; $i = 1, \ldots, m$. Here $\mathbf{e}_i$ is the ith unit vector. Outflow from state $\mathbf{n}$ can occur because of an external arrival or a service completion that is not immediately fed back to the same machine center. Now equating the rates of probability inflow and outflow of state $\mathbf{n}$ one gets the following balance equations

$$\sum_{i=1}^{m} \lambda \gamma_i \, p(\mathbf{n} - \mathbf{e}_i) + \sum_{i=1}^{m} \mu_i r_i (n_i + 1) \left(1 - \sum_{j=1}^{m} p_{ij}\right) p(\mathbf{n} + \mathbf{e}_i)$$

$$+ \sum_{j=1}^{m} \sum_{i=1, i \neq j}^{m} \mu_j r_j (n_j + 1) p_{ji} \, p(\mathbf{n} + \mathbf{e}_j - \mathbf{e}_i)$$

$$= \sum_{i=1}^{m} (\lambda \gamma_i + \mu_i r_i (n_i)(1 - p_{ii})) p(\mathbf{n}), \quad \mathbf{n} \in \mathcal{N}_+^m \qquad (7.22)$$

where $p(\mathbf{n}) = 0$, $\mathbf{n} \notin \mathcal{N}_+^m$. Equation 7.22 along with the normalizing equation $\sum_{\mathbf{n} \in \mathcal{N}_+^m} p(\mathbf{n}) = 1$ can be used to solve for p.

Observe that p is a joint distribution of $(N_1, \ldots, N_m)$. It is therefore a good strategy to see whether a product form solution of the form $p(\mathbf{n}) = \prod_{i=1}^{m} p_i(n_i)$ exists for p. If it did exist, then it may be easy to solve for p. Therefore we will first assume that a product form solution indeed exists and solve for p_i, $i = 1, \ldots, m$. Later we will check whether this solution is consistent with equation 7.22. Substituting the product form for p in equation 7.22 and dividing by $p(\mathbf{n})$ we get

$$\sum_{i=1}^{m} \lambda \gamma_i \frac{p_i(n_i - 1)}{p_i(n_i)} + \sum_{j=1}^{m} \mu_j r_j (n_j + 1) \left(1 - \sum_{i=1}^{m} p_{ji}\right) \frac{p_j(n_j + 1)}{p_j(n_j)}$$

$$+ \sum_{j=1}^{m} \mu_j r_j (n_j + 1) \sum_{i=1; i \neq j}^{m} p_{ji} \frac{p_j(n_j + 1)}{p_j(n_j)} \frac{p_i(n_i - 1)}{p_i(n_i)}$$

$$= \sum_{i=1}^{m} (\lambda \gamma_i + \mu_i r_i (n_i)(1 - p_{ii})), \quad \mathbf{n} \in \mathcal{N}_+^m \qquad (7.23)$$

Rewriting equation 7.23 we have

$$\sum_{i=1}^{m} \left\{ \lambda\gamma_i + \sum_{j=1; j\neq i}^{m} \mu_j r_j(n_j+1) \frac{p_j(n_j+1)}{p_j(n_j)} p_{ji} \right\} \frac{p_i(n_i-1)}{p_i(n_i)}$$

$$+ \sum_{i=1}^{m} \left\{ \mu_i r_i(n_i+1)(1-p_{ii}) \frac{p_i(n_i+1)}{p_i(n_i)} - \sum_{j=1; j\neq i}^{m} \mu_j r_j(n_j+1) \frac{p_j(n_j+1)}{p_j(n_j)} p_{ji} \right\}$$

$$= \sum_{i=1}^{m} (\lambda\gamma_i + \mu_i r_i(n_i)(1-p_{ii})) \tag{7.24}$$

Suppose the solution p_i for

$$\left\{ \lambda\gamma_i + \sum_{j=1; j\neq i}^{m} \hat{\lambda}_j p_{ji} \right\} \frac{p_i(n_i-1)}{p_i(n_i)} + \mu_i r_i(n_i+1)(1-p_{ii}) \frac{p_i(n_i+1)}{p_i(n_i)} - \sum_{j=1; j\neq i}^{m} \hat{\lambda}_j p_{ji}$$

$$= \lambda\gamma_i + \mu_i r_i(n_i)(1-p_{ii}), \quad i = 1, \ldots, m \tag{7.25}$$

is consistent such that $\mu_j r_j(n_j+1)p_j(n_j+1)/p_j(n_j) = \hat{\lambda}_j$, for all $n_j = 0, 1, \ldots$; $j = 1, 2, \ldots, m$. Then p_i, $i = 1, \ldots, m$ is the solution to equation 7.24. To see this, substitute $\mu_j r_j(n_j+1)p_j(n_j+1)/p_j(n_j)$ for $\hat{\lambda}_j$ in (7.25) and sum the resulting equation over all $j = 1, \ldots, m$. Observe that this transformation gives an equation the same as equation 7.24. Therefore the solution p_i, $i = 1, \ldots, m$ satisfies equation 7.24 and hence it is the solution to equation 7.24 as well.

Let $(\hat{\lambda}_i, i = 1, \ldots, m)$ be the solution to the set of linear equations

$$\hat{\lambda}_i = \lambda\gamma_i + \sum_{j=1}^{m} \hat{\lambda}_j p_{ji}, \quad i = 1, \ldots, m \tag{7.26}$$

Then using the observation that $p_i(-1) = 0$ and solving equation 7.25 sequentially for $n_i = 0, 1, 2, \ldots$, one obtains

$$\hat{\lambda}_i p_i(n_i) = \mu_i r(n_i+1)p_i(n_i+1), \quad n_i = 0, 1, \ldots; i = 1, \ldots, m \tag{7.27}$$

$$p_i(n_i) = p_i(0) f_i(n_i), \quad n_i = 0, 1, \ldots; i = 1, \ldots, m \tag{7.28}$$

where

$$f_i(0) = 1; \ f_i(n_i) = f_i(n_i-1)\lambda_i/(\mu_i r_i(n_i)), \quad n_i = 1, 2, \ldots; i = 1, \ldots, m \tag{7.29}$$

$$p_i(0) = 1/\sum_{n_i=0}^{\infty} f_i(n_i), \quad i = 1, \ldots, m \tag{7.30}$$

Observe that the preceding solution is consistent with the requirement

$$\hat{\lambda}_j p_j(n_j) = \mu_j r_j(n_j+1)p_j(n_j+1)$$

and therefore the solution p for equation 7.24 is

$$p(\mathbf{n}) = \prod_{i=1}^{m} p_i(n_i), \quad \mathbf{n} \in \mathcal{N}_+^m \tag{7.31}$$

In addition note that $\hat{\lambda}_j = \lambda_j$ is the rate of job arrivals (including both internal and external arrivals) to the machine center j, $j = 1, \ldots, m$. Furthermore it is important to observe that $p_i(\cdot)$ is the stationary distribution of the number of jobs in a $M/M(n)/1$ queueing system with arrival rate λ_i and state-dependent service rate $\mu_i r_i(n_i)$ when there are n_i jobs in it, $n_i = 0, 1, \ldots$. Therefore the results given in Sections 3.3 and 3.4 for the $M/M/1$ and $M/M/c$ queueing systems can be directly applied to this queueing network model. Particularly, when there is only a single machine at each machine center (i.e., $c_i = 1$, $i = 1, \ldots, m$) we have

$$p(\mathbf{n}) = \prod_{i=1}^{m} (1 - \rho_i) \rho_i^{n_i}, \quad \mathbf{n} \in \mathcal{N}_+^m \tag{7.32}$$

where $\rho_i = \lambda_i / \mu_i < 1$, $i = 1, \ldots, m$. In this case the average number of jobs in machine center i is

$$E[N_i] = \frac{\rho_i}{1 - \rho_i}, \quad i = 1, \ldots, m \tag{7.33}$$

and the total number of jobs in the job shop is

$$E[N] = \sum_{i=1}^{m} \frac{\rho_i}{1 - \rho_i} \tag{7.34}$$

Applying Little's formula, one then obtains the average flow time of an arbitrary job as

$$E[T] = \frac{1}{\lambda} \sum_{i=1}^{m} \frac{\rho_i}{1 - \rho_i} \tag{7.35}$$

Because $\lambda_i = \lambda v_i$, where v_i is the expected number of visits made to machine center i by an arbitrary job before it leaves the system, equation 7.35 can be rewritten as

$$E[T] = \sum_{i=1}^{m} v_i E[T_i] \tag{7.36}$$

where

$$E[T_i] = \frac{1}{\lambda_i} \frac{\rho_i}{1 - \rho_i} \tag{7.37}$$

is the average flow time of an arbitrary job in machine center i each time it visits machine center i, $i = 1, \ldots, m$. Now equation 7.36 is self-explanatory.

The independence of $N_1, \ldots, N_m$ allows one to compute the variance of the total number of jobs in the job shop. Particularly because $\text{var}(N_i) = \rho_i/(1 - \rho_i)^2$, $i = 1, \ldots, m$, we have

$$\text{var}(N) = \sum_{i=1}^{m} \frac{\rho_i}{(1 - \rho_i)^2} \tag{7.38}$$

Even though the stationary number of jobs in the machine centers are statistically independent, which we used in obtaining equation 7.38, the stationary flow time of arbitrary jobs through the different machine centers are in general not independent. Consequently we cannot obtain a simple expression for the variance of the flow time of an arbitrary job. Later in this chapter we will see how a good approximation for this variance can be obtained.

7.4.1 Optimal Assignment of Tasks to Machine Centers

We will illustrate how the results obtained previously can be applied to control the assignments of tasks to machines or workers, and to control the assignment of workers to machine centers. First consider the problem of allocating the different tasks to workers preassigned to machine centers. For ease of illustration we will assume that there is only one worker and one machine at each machine center. The way in which tasks are assigned to the machine centers essentially determines the expected number of visits v_i made to machine center i ($i = 1, \ldots, m$) by an arbitrary job. Therefore no matter how we allocate the tasks $\sum_{i=1}^{m} v_i = K$ remains a constant because K is the average number of tasks (i.e., operations) required by an arbitrary job. We may therefore wish to obtain an optimal allocation of tasks among work centers such that the mean flow time of an arbitrary job or the average inventory carrying cost is minimized.

Because $E[N] = \lambda E[T]$, it is obvious that an allocation that minimizes $E[T]$ will also minimize $E[N]$. Therefore we will choose to minimize $E[N]$. Before we obtain the task allocation, we will identify the "ideal" values of v_i, $i = 1, \ldots, m$ that will minimize $E[N]$. These ideal values of v_i, $i = 1, \ldots, m$ may be such that no task allocation can be found in which each task is assigned to only one machine center. For such a case we will provide a simple allocation method to identify the tasks that need to be allocated to more than one machine center. To facilitate our analysis we will assume that a task if assigned to machine center i will take an operation time that is exponentially distributed with mean $1/\mu_i$. With this assumption it can be seen that for a task allocation that results in an expected number of visits v_i to machine center i, $i = 1, \ldots, m$, the average number of jobs in the job shop is $\sum_{i=1}^{m} \lambda v_i/(\mu_i - \lambda v_i)$. The ideal values of v_i, $i = 1, \ldots, m$ are then obtained from

$$\min \left\{ \sum_{i=1}^{m} \frac{\lambda v_i}{\mu_i - \lambda v_i} \mid \sum_{i=1}^{m} v_i = K, v_i \geq 0, i = 1, \ldots, m \right\} \tag{7.39}$$

It can be easily verified that the optimal values of v_i, $i = 1, \ldots, m$ for equation 7.39 are given by

$$v_i^* = \frac{1}{\lambda} \left(\mu_i - \left(\sum_{j=1}^{m} \mu_j - \lambda K \right) \frac{\sqrt{\mu_i}}{\sum_{j=1}^{m} \sqrt{\mu_j}} \right), \quad i = 1, \ldots, m \qquad (7.40)$$

We assume that $\lambda K < \sum_{j=1}^{m} \mu_j$. Otherwise there is no allocation of tasks that will lead to a stable system. $\sum_{j=1}^{m} \mu_j - \lambda K$ can be viewed as a measure of excess capacity available for this job shop. As λ increases this excess capacity decreases, and it is easily seen from equation 7.40 that a balanced task allocation such that v_i/μ_i is a constant becomes preferable.

Let $\alpha = \{1, \ldots, L\}$ be the set of L tasks that needs to be assigned to the m machine centers. Define w_i to be the expected number of times task i needs to be performed on an arbitrary job. Clearly $\sum_{i=1}^{L} w_i = K$. Suppose that there exists a partition $\{S_1, \ldots, S_m\}$ of α such that $\sum_{j \in S_i} w_j = v_i^*$, $i = 1, \ldots, m$. Then assigning the tasks $j \in S_i$ to machine center i will give us the optimal task allocation. Identifying such a partition, even if it exists, will take an exponential amount of time with respect to L. Therefore we suggest the following simple allocation rule that will provide us with the optimal v_i^*, but at the expense of allocating some (at most m) of the L tasks to more than one machine center. Find $l(k)$, $k = 1, \ldots, m$ such that $l(k)$ is the largest value that satisfies $\sum_{i=1}^{l(k)} w_i \leq \sum_{j=1}^{k} v_j^*$, $k = 1, \ldots, m$. Note that $l(m) = L$. Then if $\sum_{j=1}^{k} v_j^* - \sum_{i=1}^{l(k)} w_i > 0$, task $l(k) + 1$ is assigned to both machine centers k and $k + 1$. The fraction of jobs requiring the task $l(k) + 1$ that will be routed to machine center k is

$$\frac{\sum_{j=1}^{k} v_j^* - \sum_{i=1}^{l(k)} w_i}{w_{l(k)+1}}$$

The remaining fraction is routed to machine center $k + 1$ for processing task $l(k) + 1$, $k = 1, \ldots, m - 1$.

7.4.2 Optimal Allocation of Workers to Machine Centers

The next problem we consider concerns the allocation of workers to machine centers. Suppose we have a total of W workers that we need to allocate among the m machine centers. Each worker can operate any one of the $\sum_{i=1}^{m} k_i$ machines and at most one machine may be assigned to a worker. There are k_i identical machines available in machine center i, $i = 1, \ldots, m$. If $\sum_{i=1}^{m} k_i \leq W$, then we can allocate the workers on a one-to-one basis for all $\sum_{i=1}^{m} k_i$ machines. In a real job shop it is usually the case that $\sum_{i=1}^{m} k_i > W$. In such a situation it is important that the limited number of workers we have are assigned to the machine centers in the right numbers. Suppose we have assigned c_i workers to machine center i, $i = 1, \ldots, m$. If the job arrival rate to machine center i is λ_i, $i = 1, \ldots, m$, assuming that the open Jackson queueing network model is applicable here, the expected number of jobs in the system is $E[N] = \sum_{i=1}^{m} g_i(c_i)$, where $g_i(c_i)$ is the expected number of jobs in a $M/M/c_i$ queue with arrival rate λ_i, service rate μ_i and c_i parallel servers, $i = 1, \ldots, m$. The exact form of $g_i(c_i)$ can be obtained from Section 3.4. The optimal allocation of workers that minimize the average number of jobs in the job shop (and the average flow time of an arbitrary job) is given by

$$\min \left\{ \sum_{i=1}^{m} g_i(c_i) \mid \sum_{i=1}^{m} c_i = W, c_i \geq 1 \right\} \tag{7.41}$$

It can be verified from the results given in Section 3.4 that $g_i(c_i)$ is a decreasing and convex function of c_i, $i = 1, \ldots, m$. Therefore equation 7.41 can be solved using a marginal allocation approach. Suppose $(c_i^*(k), i = 1, \ldots, m)$ is an optimal solution to equation 7.41 when $W = k$. Then if $j^* = \arg \max \{-g_i(c_i^*(k) + 1) + g_i(c_i^*(k)), \quad i = 1, \ldots, m\}$ is the machine center where the maximum reduction in the average number of jobs is achieved by an additional server, we set $c_{j^*}^*(k + 1) = c_{j^*}^*(k) + 1$ and $c_i^*(k + 1) = c_i^*(k)$, $i = 1, \ldots, m$, $i \neq j^*$. These results, apart from giving us an optimal allocation of workers to machine centers, also allow us to identify the reallocation of available workers during worker absenteeism. Specifically if we had an allocation $(c_i^*(k+1), i = 1, \ldots, m)$ with $k + 1$ workers and one of them (say one of those assigned to machine center j) is absent. Let j^* be as defined before, then the optimal allocation of workers is obtained by keeping the previous allocation with one worker from machine center j^* switched to machine center j. This apart from being an optimal allocation of the k workers also guarantees to have minimal disruption owing to absenteeism, because we transfer at most one in the reallocation procedure.

We may also use the preceding formulation to determine the optimal number of workers to hire. Suppose h is the unit time cost of hiring one extra worker. Then we wish to solve

$$\min \left\{ h.W + \sum_{i=1}^{m} g_i(c_i) \right\} \tag{7.42}$$

subject to

$$\sum_{i=1}^{m} c_i = W$$

$$c_i \geq 1$$

$$W \geq m$$

This problem too can be solved using a marginal allocation approach. Let $(c_i^*(k), i = 1, \ldots, m)$ be an optimal allocation of workers when $W = k$. Then the optimal number of workers is given by

$$W^* = \min\{W : h \geq \{g_i(c_i^*(W)) - g_i(c_i^*(W) + 1), i = 1, \ldots, m\} : W \geq m\} \tag{7.43}$$

Equation 7.43 implies that we keep hiring workers one at a time until the cost per worker per unit time is at least as large as the marginal benefit that can be realized by adding one more worker.

7.5 MULTIPLE-JOB-CLASS OPEN JACKSON QUEUEING NETWORK MODEL

In this section we will look at queueing network models for job shop systems where aggregating all job types into a single class is unacceptable. As pointed out earlier, this may be because the service protocols are such that the classification of the jobs by different classes is essential (e.g., the service protocol is the priority index protocol discussed in Section 3.5), or the operation/machine sequence of the jobs belonging to different classes of jobs and their processing requirements are vastly different. In this section, however, we will concentrate only on situations in which the operation/machine sequence is different for different classes of jobs, but the service requirements of the different classes at any machine center are probabilistically almost the same. In addition we will assume that jobs are selected for service according to a FCFS service protocol.

Class l jobs arrive at the job shop according to Poisson process with rate $\lambda^{(l)}, l = 1, \ldots, r$. All these r arrival processes are mutually independent. The fraction of class l jobs that join machine center i is $\gamma_i^{(l)}$ $\left(\sum_{i=1}^{m} \gamma_i^{(l)} = 1 \right)$ and the fraction of class l jobs among those class l jobs that complete service at machine center i, that proceed directly to machine center j as a class k job is $p_{ij}^{(l)(k)}$, $i, j = 1, 2, \ldots ; l, k = 1, 2, \ldots, r$. We assume that these probabilities are such that each job will eventually leave the system. The service time of a class l job at machine center i is exponentially distributed with mean $1/\mu_i$, $i = 1, \ldots, m$. Jobs are served at a rate $r_i(n_i)$ at machine center i when there are n_i jobs, $r_i(0) = 0, r_i(n_i) > 0, n_i = 1, 2, \ldots$. Note that this service rate is independent of the number of individual classes of jobs but depends only on the total number. The arrival, service and job transfers from one machine center to another are mutually independent. Assuming that each job is transferred from one machine center to another and from one class to another according to a transfer probability matrix $\left(p_{ij}^{(l)(k)} \right)_{i,j=1,\ldots,m}^{l,k=1,\ldots,r}$ we model this job shop by a multiple–job-class open Jackson queueing network. Let $N_i(t)$ be the number of jobs at machine center i at time t and $X_{i:j}(t)$ be the class index of the job in the jth position of the queue at machine center i, $j = 1, \ldots, N_i(t)$; $i = 1, \ldots, m$. We assume that the job in the first position is in service, being served at a rate $r_i(n_i)$ when $N_i(t) = n_i$, $i = 1, \ldots, m$. Then $\{\mathbf{X}(t), t \geq 0\}$ is a continuous time Markov process where $\mathbf{X}(t) = (X_{i:j}(t), j = 1, \ldots, N_i(t), i = 1, \ldots, m)$. Let $q(\mathbf{x}) = \lim_{t \to \infty} P\{\mathbf{X}(t) = \mathbf{x}\}$ be the stationary probability distribution of $\mathbf{X}$. Suppose the state of the process $\{\mathbf{X}_i(t) = (X_{i:j}(t), j = 1, \ldots, N_i(t)), t \geq 0\}$ is $\mathbf{x}_i = (x_{i:1}, \ldots, x_{i:n_i})$ at time t. If an arrival of a class l job to machine center i occurs, the state will immediately change to $(\mathbf{x}_i, l) = (x_{i:1}, \ldots, x_{i:n_i}, l)$ and $N_i(t)$ will take the value $n_i + 1$. Conversely, if a service completion occurs at time t, the state will change immediately to $(x_{i:2}, \ldots, x_{i:n_i})$ and $N_i(t)$ will take on the value $n_i - 1$. Now, as before, equating the rates of probability inflow to probability outflow, accounting for the arrivals and service completions one gets

$$
\sum_{i=1}^{m} \lambda^{(x_{i:n_i})} \gamma_i^{(x_{i:n_i})} q(\mathbf{x}_1, \ldots, \mathbf{x}_i^-, \ldots, \mathbf{x}_m)
$$

$$
+ \sum_{i=1}^{m} \sum_{l=1}^{r} \mu_i r_i (n_i + 1) \left(1 - \sum_{k=1}^{r} \sum_{j=1}^{m} p_{ij}^{(l)(k)} \right) q(\mathbf{x}_1, \ldots, (l, \mathbf{x}_i), \ldots, \mathbf{x}_m)
$$

$$
+ \sum_{i=1}^{m} \sum_{j=1, j \neq i}^{m} \sum_{l=1}^{r} \mu_i r_i (n_i + 1) p_{ij}^{(l)(x_{j:n_j})} q(\mathbf{x}_1, \ldots, (l, \mathbf{x}_i), \ldots, \mathbf{x}_j^-, \ldots, \mathbf{x}_m)
$$

$$
+ \sum_{i=1}^{m} \sum_{l=1}^{r} \mu_i r_i (n_i) p_{ii}^{(l)(x_{i:n_i})} q(\mathbf{x}_1, \ldots, (l, \mathbf{x}_i^-) \ldots, \mathbf{x}_m)
$$

$$
= \left\{ \sum_{l=1}^{r} \lambda^{(l)} + \sum_{i=1}^{m} \mu_i r_i (n_i) \right\} q(\mathbf{x}) \tag{7.44}
$$

where $\mathbf{x}_i^- = (x_{i:1}, \ldots, x_{i:n_i-1})$ is the vector $\mathbf{x}_i$ without its last element. In comparison with balance equations 7.22 it should be noted that in equation 7.44 we are accounting for the self-transition rates from state $\mathbf{x}$ to state $\mathbf{x}$, both as an inflow and as an outflow.

As suggested earlier, let us see whether the joint distribution of $\mathbf{X}$ can be represented as the product of its marginal distribution. Suppose $q(\mathbf{x}) = \prod_{i=1}^{m} \prod_{j=1}^{n_i} q_{i:j}(x_{i:j})$. Substituting this in equation 7.44 and dividing by $q(\mathbf{x})$ on both sides one gets

$$
\sum_{i=1}^{m} \lambda^{(x_{i:n_i})} \gamma_i^{(x_{i:n_i})} \frac{1}{q_{i:n_i}(x_{i:n_i})}
$$

$$
+ \sum_{i=1}^{m} \sum_{l=1}^{r} \mu_i r_i (n_i + 1) \left(1 - \sum_{k=1}^{r} \sum_{j=1}^{m} p_{ij}^{(l)(k)} \right) \frac{q_{i:1}(l) \prod_{s=2}^{n_i+1} q_{i:s}(x_{i;s-1})}{\prod_{s=1}^{n_i} q_{i:s}(x_{i:s})}
$$

$$
+ \sum_{i=1}^{m} \sum_{j=1, j \neq i}^{m} \sum_{l=1}^{r} \mu_i r_i (n_i + 1) p_{ij}^{(l)(x_{j:n_j})} \frac{q_{i:1}(l) \prod_{s=2}^{n_i+1} q_{i:s}(x_{i;s-1})}{\prod_{s=1}^{n_i} q_{i:s}(x_{i:s}) q_{j:n_j}(x_{j:n_j})}
$$

$$
+ \sum_{i=1}^{m} \sum_{l=1}^{r} \mu_i r_i (n_i) p_{ii}^{(l)(x_{i:n_i})} q_{i:1}(l) \frac{\prod_{s=2}^{n_i} q_{i:s}(x_{i;s-1})}{\prod_{s=1}^{n_i} q_{i:s}(x_{i:s})}
$$

$$
= \sum_{l=1}^{r} \lambda^{(l)} + \sum_{i=1}^{m} \mu_i r_i (n_i) \tag{7.45}
$$

Suppose the solution to

$$\left\{ \lambda^{(x_{i:n_i})} \gamma_i^{(x_{i:n_i})} + \sum_{j=1}^{m} \sum_{l=1}^{r} \hat{\lambda}_j^{(l)} p_{ji}^{(l)(x_{i:n_i})} \right\} \prod_{s=1}^{n_i-1} q_{i:s}(x_{i:s})$$

$$+ \sum_{l=1}^{r} \left\{ \lambda_i^{(l)} - \sum_{k=1}^{r} \sum_{j=1}^{m} \hat{\lambda}_j^{(k)} p_{ji}^{(k)(l)} \right\} \prod_{s=1}^{n_i} q_{i:s}(x_{i:s})$$

$$= \left\{ \sum_{l=1}^{r} \lambda^{(l)} \gamma_i^{(l)} + \mu_i r_i(n_i) \right\} \prod_{s=1}^{n_i} q_{i:s}(x_{i:s}), \quad i = 1, \ldots, m \qquad (7.46)$$

is consistent such that for any $n_j = 0, 1, \ldots$,

$$\hat{\lambda}_j^{(l)} = \mu_j r_j(n_j + 1) \frac{q_{j:1}(l) \prod_{s=2}^{n_j+1} q_{j:s}(x_{j:s-1})}{\prod_{s=1}^{n_j} q_{j:s}(x_{j:s})}, \quad j = 1, \ldots, m; l = 1, \ldots, r \qquad (7.47)$$

If such a solution exists it is then clear that it will satisfy equation 7.45 as well. To see this substitute equation 7.47 in equation 7.46 and sum the resulting equations for all $i = 1, \ldots, m$ and note that it is exactly the same as equation 7.45. Then the product form solution is valid for equation 7.44 also. Now to solve equation 7.46, let $\hat{\lambda}_j^{(l)}$, $l = 1, \ldots, r$; $j = 1, \ldots, m$ be the solution to

$$\hat{\lambda}_i^{(l)} = \lambda^{(l)} \gamma_i^{(l)} + \sum_{j=1}^{m} \sum_{k=1}^{r} \hat{\lambda}_j^{(k)} p_{ji}^{(k)(l)}, \quad l = 1, \ldots, r; i = 1, \ldots, m \qquad (7.48)$$

Rewriting equation 7.46 and substituting equation 7.48 one obtains

$$\hat{\lambda}_i^{(x_{i:n_i})} \prod_{s=1}^{n_i-1} q_{i:s}(x_{i:s}) = \mu_i r_i(n_i) \prod_{s=1}^{n_i} q_{i:s}(x_{i:s}) \qquad (7.49)$$

Solving equation 7.49 sequentially starting with $n_i = 1$, $x_{i:n_i} = 1, \ldots, r$; $n_i = 1, 2, \ldots$, one obtains

$$\prod_{s=1}^{n_i} q_{i:s}(x_{i:s}) = \prod_{s=1}^{n_i} \frac{\hat{\lambda}_i^{(x_{i:s})}}{\mu_i r_i(s)}, \quad n_i = 0, 1, \ldots \qquad (7.50)$$

Observe that $q_{i:j}(\cdot)$ presented in equation 7.50 is not a probability distribution function because it does not add up to one. Particularly observe that

$$\sum_{n_i=0}^{\infty} \sum_{x_{i:1}=1}^{r} \sum_{x_{i:2}=1}^{r} \cdots \sum_{x_{i:n_i}=1}^{r} q_i(\mathbf{x}) = \sum_{n_i=0}^{\infty} \prod_{s=1}^{n_i} \frac{\hat{\lambda}_i}{\mu_i r_i(s)}$$

$$= \sum_{n_i=0}^{\infty} f_i(n_i)$$

where $\hat{\lambda}_i = \sum_{l=1}^{r} \lambda_i^{(l)}$, $f_i(0) = 1$ and $f_i(n_i) = f_i(n_i - 1)\hat{\lambda}_i/(\mu_i r_i(n_i))$; $n_i = 1, 2, \ldots$. Then

$$q_i(\mathbf{x}_i) = \frac{1}{\sum_{n=0}^{\infty} f_i(n)} \prod_{s=1}^{n_i} \frac{\hat{\lambda}_i^{(x_{i:s})}}{\mu_i r_i(s)}, \quad i = 1, \ldots, m \tag{7.51}$$

are proper probability distribution functions and

$$q(\mathbf{x}) = \prod_{i=1}^{m} q_i(\mathbf{x}_i) = \prod_{i=1}^{m} \frac{1}{\sum_{n=0}^{\infty} f_i(n)} \prod_{s=1}^{n_i} \frac{\hat{\lambda}_i^{(x_{i:s})}}{\mu_i r_i(s)} \tag{7.52}$$

is the solution to equation 7.44, that is the joint probability distribution of $\mathbf{X}$.

We can now use equation 7.52 to obtain the joint probability distribution of $\mathbf{N}$, the number of jobs of each class at each machine center. Let $p(\mathbf{n}) = P\{\mathbf{N} = \mathbf{n}\}$ and $p_i(\mathbf{n}_i) = P\{\mathbf{N}_i = \mathbf{n}_i\}$. Here $\mathbf{n}_i = (n_i^{(1)}, n_i^{(2)}, \ldots, n_i^{(r)})$ and $n_i^{(l)}$ is the number of class l jobs at machine center i. Then from equation 7.51 we have

$$p_i(\mathbf{n}_i) = \binom{n_i}{n_i^{(1)}, \ldots, n_i^{(r)}} \prod_{l=1}^{r} \left(\frac{\hat{\lambda}_i^{(l)}}{\hat{\lambda}_i}\right)^{n_i^{(l)}} \frac{f_i(n_i)}{\sum_{n=0}^{\infty} f_i(n)}, \quad \mathbf{n}_i \in \mathcal{N}_+^r \tag{7.53}$$

where $n_i = \sum_{l=1}^{r} n_i^{(l)}$.

Therefore the joint distribution of $\mathbf{N}$ is

$$p(\mathbf{n}) = \prod_{i=1}^{m} \binom{n_i}{n_i^{(1)}, \ldots, n_i^{(r)}} \prod_{l=1}^{r} \left(\frac{\hat{\lambda}_i^{(l)}}{\hat{\lambda}_i}\right)^{n_i^{(l)}} \left\{\frac{f_i(n_i)}{\sum_{n=0}^{\infty} f_i(n)}\right\}, \quad \mathbf{n} \in \mathcal{N}_+^{r \times m} \tag{7.54}$$

Observe that $f_i(n_i)/\sum_{n=0}^{\infty} f_i(n)$, $n_i = 0, 1, \ldots$. is the probability distribution of the number of customers in a $M/M(n)/1$ queueing system with arrival rate $\hat{\lambda}_i$ and state dependent service rate $\mu_i r_i(\cdot)$. Conversely

$$\binom{n_i}{n_i^{(1)}, \ldots, n_i^{(r)}} \prod_{l=1}^{r} \left(\frac{\hat{\lambda}_i^{(l)}}{\lambda_i}\right)^{n_i^{(l)}}$$

is the multinominal probability of choosing $n_i^{(l)}$ of type l items according to a probability $\hat{\lambda}_i^{(l)}/\hat{\lambda}_i$, $l = 1, \ldots, r$. Therefore the joint distribution of the number of jobs at the different machine centers is the same as that in a single–job-class queueing network with arrival rate $\hat{\lambda}_i$ to machine center $i = 1, \ldots, m$. We may therefore solve the open queueing network model with single class and obtain the joint distribution of the number of jobs of different classes at different machines using a multinominal sampling with probability $\hat{\lambda}_i^{(l)}/\hat{\lambda}_i$ for class l jobs ($l = 1, \ldots, r$) at machine center i ($i = 1, \ldots, m$).

For a machine center that has only one machine we have

$$p_i(\mathbf{n}_i) = \binom{n_i}{n_i^{(1)}, \ldots, n_i^{(r)}} \prod_{l=1}^{r} \left(\frac{\hat{\lambda}_i^{(l)}}{\hat{\lambda}_i}\right)^{n_i^{(l)}} (1 - \rho_i)\rho_i^{n_i}, \quad \mathbf{n}_i \in \mathcal{N}_+^r \tag{7.55}$$

where $\rho_i = \hat{\lambda}_i/\mu_i < 1$. For a machine center, say i that has $c_i > 1$ parallel machines, however, our analysis is not valid because we assumed that at each service completion only the job at the head of queue leaves that machine center. This is not the case when we have parallel machines. In any event a modified analysis will indeed show that our results for $p_i(\cdot)$ is valid even in this case. For $c_i = +\infty$ we have

$$p_i(\mathbf{n}_i) = \prod_{l=1}^{r} \left(\rho_i^{(l)}\right)^{n_i^{(l)}} \frac{e^{-\rho_i^{(l)}}}{n_i^{(l)}!}, \quad \mathbf{n}_i \in \mathcal{N}_+^r \tag{7.56}$$

where $\rho_i^{(l)} = \hat{\lambda}_i^{(l)}/\mu_i$, $l = 1, \ldots, r$. The number of jobs of different classes at a machine center with infinitely many machines are independent. In the other cases this need not be true. As we will see next the marginal distribution of any class of job, say l, at any machine center with a single machine, say i, is exactly the same as that in an $M/M/1$ queue with arrival rate $\hat{\lambda}_i^{(l)}$ and service rate $\mu_i - \sum_{s=1;s\neq l}^{r} \hat{\lambda}_i^{(s)}$. From equation 7.55 one sees that

$$P\left\{N_i^{(l)} = n_i^{(l)}\right\}$$

$$= \sum_{n_i=n_i^{(l)}}^{\infty} \sum_{n_i^{(1)}=0}^{n_i} \cdots \sum_{n_i^{(s)}=0}^{n_i - \sum_{z=1}^{s-1} n_i^{(z)}} \cdots \sum_{n_i^{(r)}=0}^{n_i - \sum_{z=0}^{r-1} n_i^{(z)}} \binom{n_i}{n_i^{(1)}, \ldots, n_i^{(r)}}$$

$$\prod_{k=1}^{r} \left(\frac{\hat{\lambda}_i^{(k)}}{\hat{\lambda}_i}\right)^{n_i^{(k)}} (1 - \rho_i)\rho_i^{n_i}$$

$$= \sum_{n_i=n_i^{(l)}}^{\infty} \binom{n_i}{n_i^{(l)}} \left(\frac{\hat{\lambda}_i^{(l)}}{\hat{\lambda}_i}\right)^{n_i^{(l)}} \left(1 - \frac{\hat{\lambda}_i^{(l)}}{\hat{\lambda}_i}\right)^{n_i - n_i^{(l)}} (1 - \rho_i)\rho_i^{n_i}$$

$$= \left\{\frac{1 - \rho_i}{1 - \rho_i + \rho_i^{(l)}}\right\} \left\{\frac{\rho_i^{(l)}}{1 - \rho_i + \rho_i^{(l)}}\right\}^{n_i^{(l)}}$$

$$= \left(1 - \hat{\rho}_i^{(l)}\right) \left(\hat{\rho}_i^{(l)}\right)^{n_i^{(l)}}, \quad n_i^{(l)} = 0, 1, \ldots \tag{7.57}$$

where

$$\hat{\rho}_i^{(l)} = \frac{\rho_i^{(l)}}{1 - \rho_i + \rho_i^{(l)}} = \frac{\hat{\lambda}_i^{(l)}}{\mu_i - \sum_{k=1,k\neq l}^{r} \hat{\lambda}_i^{(k)}}$$

and $\rho_i^{(l)} = \hat{\lambda}_i^{(l)}/\mu_i$. Therefore if we are interested only in studying the flow performance of a particular class of job, (say l) in a job shop with single or infinite machine centers, all we have to do is redefine the service rates of the single-machine machine center (say i) by $\mu_i - \sum_{k=1,k\neq l}^{r} \hat{\lambda}_i^{(k)}$ and analyze a single open queueing network with only class l

jobs and the modified service rates. We cannot, however, do this *job-class decomposition* when there are machine centers with a limited number of parallel machines.

We have seen how the number of jobs of different classes in single- and infinite-machine machine centers decomposes. Next we will make an observation about the effect of aggregating the number of jobs at different infinite-machine machine centers. Let $I \subset \{1, \ldots, m\}$ be the set of machine centers that have infinitely many machines. Then from equations 7.52 and 7.56 one sees that

$$P\{N_i^{(l)} = n_i^{(l)}, l = 1, \ldots, r; i \in I\} = \prod_{i \in I} \prod_{l=1}^{r} \frac{e^{-\rho_i^{(l)}} (\rho_i^{(l)})^{n_i^{(l)}}}{n_i^{(l)}!}.$$

Then it is easily verified that

$$P\left\{ \sum_{i \in I} N_i^{(l)} = n^{(l)}, l = 1, \ldots, r \right\} = \prod_{l=1}^{r} \frac{e^{-\rho_I^{(l)}} (\rho_I^{(l)})^{n^{(l)}}}{n^{(l)}!} \tag{7.58}$$

where $\rho_I^{(l)} = \sum_{i \in I} \rho_i^{(l)}$, $l = 1, \ldots, r$. Therefore if we are interested only in the total number of jobs of different classes in the set I of machine centers we may aggregate all these $|I|$ machine centers into one infinite-machine machine center.

It should be noted that the results discussed here for the infinite-machine machine centers hold true even if the distributions of the service times are general.

7.6 INCORPORATION OF MATERIAL HANDLING

In this section we will pay explicit attention to material handling systems in modeling the job shop. We will consider two material handling configurations. In the first configuration we assume that for each link (i, j) that connects machine centers i and j $(i, j = 1, \ldots, m; i \neq j)$ we have sufficient number of transporters or a conveyor system such that when a job completes processing at machine center i it can immediately begin movement from machine center i to its destination. The average time taken to transfer a job from machine center i to machine center j is $1/\mu_{(i,j)}$, $i, j = 1, \ldots, m$; $i \neq j$. For the second configuration we have a central transportation system with c_0 transporters. The average transportation time is $1/\mu_0$. It is appropriate to assume that the transport time in a centralized configuration is random. The randomness is created by the response time for the central control system and the random positioning of the transporters at any given time. In our model we will assume that this transport time is exponentially distributed.

Dedicated transport system.   Consider the first configuration. Suppose after aggregation of all jobs into a single class we can model the job transfers from one machine center to another by a job transfer probability matrix $\mathbf{P} = (p_{ij})_{i,j=1,\ldots,m}$. If we now incorporate the transporters on each link (i, j) as a service center, the job transfer probabilities are $p'_{ii} = p_{ii}$, $p'_{i,(i,j)} = p_{ij}$ and $p'_{(i,j),j} = 1$, $i, j = 1, \ldots, m$;

$i \neq j$. All other transfer probabilities are zero. Here (i, j) represents the service center corresponding to the transporters on link (i, j), $i, j = 1, \ldots, m$; $i \neq j$. If we model this system by an open Jackson queueing network as described in Section 7.4 we find that the stationary distribution of the number of jobs in the machine centers ($\mathbf{n}$) and the number of jobs in the transporters ($\mathbf{l}$) is given by

$$p(\mathbf{n}, \mathbf{l}) = \left\{ \prod_{i=1}^{m} p_i(n_i) \right\} \prod_{i=1}^{m} \prod_{j=1; j \neq i}^{m} p_{(i,j)}(l_{(i,j)}), \quad \mathbf{n} \in \mathcal{N}_+^m, \mathbf{l} \in \mathcal{N}_+^{m^2 - m} \tag{7.59}$$

where $p_i(n_i) = f_i(n_i) p_i(0)$, $p_i(0) = 1 / \sum_{n_i=0}^{\infty} f_i(n_i)$, and $f_i(n_i) = \lambda_i f_i(n_i - 1)/\mu_i r_i(n_i)$, $f_i(0) = 1$; $i = 1, \ldots, m$,

$$p_{(i,j)}(l_{(i,j)}) = \frac{e^{-\rho_{(i,j)}} \rho_{(i,j)}^{l_{(i,j)}}}{l_{(i,j)}!}, \quad i, j = 0, 1, \ldots; i \neq j \tag{7.60}$$

and $\rho_{(i,j)} = \lambda_{(i,j)}/\mu_{(i,j)} = \lambda_i p_{ij}/\mu_{(i,j)}$, $i \neq j, i, j = 1, \ldots, m$. Here n_i is the number of jobs in machine center i and $l_{(i,j)}$ is the number of jobs in transit from machine center i to j, $i \neq j, i, j = 1, \ldots, m$. Observe that if $p_{ij} = 0$ then $\rho_{(i,j)} = 0$ and the number of jobs in transit on link (i, j) is also zero, as it should be. If we are only interested in the total number of jobs in transit, we may aggregate all the jobs in transit (see equation 7.58) to obtain, from equation 7.60

$$P\{l \text{ jobs are in transit}\} = e^{-\hat{\rho}_H} \hat{\rho}_H^l / l!, \quad l = 0, 1, \ldots \tag{7.61}$$

where

$$\hat{\rho}_H = \sum_{i=1}^{m} \lambda_i \sum_{j=1; j \neq i}^{m} p_{ij}/\mu_{(i,j)}$$

and $\hat{\rho}_H$ is the mean number of jobs in transit. Applying Little's law one sees that the additional average time spent by an arbitrary job in the shop owing to material handling is

$$E[\hat{T}_H] = \frac{1}{\lambda} \sum_{i=1}^{m} \lambda_i \sum_{j=1; j \neq i}^{m} p_{ij}/\mu_{(i,j)} = \sum_{i=1}^{m} \sum_{j=1; j \neq i}^{m} \frac{v_{(i,j)}}{\mu_{(i,j)}} \tag{7.62}$$

where $v_{(i,j)}$ is the expected number of times an arbitrary job is moved along the link (i, j), $i \neq j$; $i, j = 1, \ldots, m$. Now observe that the distribution of the number of jobs at different machine centers given in equation 7.59 is independent of the transportation times. Hence we see that the number of jobs at different machine centers is unaffected by the actual transportation time. All it does is to add an additional amount $E[\hat{T}_H]$ to the average flow time of an arbitrary job.

Central transport system.   Next, consider the second configuration. Suppose job transfers from one machine center to another are modeled by a job transfer probability matrix $\mathbf{P} = (p_{ij})$, $i, j = 1, \ldots, m$. Because we use a central transporter, any job that is being transported, say from machine center i to j, should be explicitly tagged as a class (i, j) job at the central transport service center (indexed by 0). For $i = 0$, a class $(0, j)$ job is a job that is being transported from the input location to machine center j, and for $j = 0$, a class $(i, 0)$ job is a job that is being transported from machine center i to the output location. In some job shops the input and output locations may be the same. Therefore, to incorporate the central transporter in our job shop model we need to model the job transfers from one machine center to another, and from the input and output locations to and from the machine centers using a multiple-class job-transfer probability matrix as follows:

$$p_{i0}^{(1)(i,j)} = p_{ij}, \quad i \neq j, i, j = 1, \ldots, m$$

$$p_{i0}^{(1)(i,0)} = 1 - \sum_{j=1}^{m} p_{ij}, \quad i = 1, \ldots, m$$

$$p_{0j}^{(i,j)(1)} = 1, \quad i \neq j, i, j = 1, \ldots, m$$

$$\gamma_{0}^{(0,i)} = \gamma_i, \quad i = 1, \ldots, m$$

where class 1 is the tag of all jobs at the machine centers. If we model the job shop with a central transporter by a multiple-class open Jackson queueing network (see Section 7.5) we get the stationary distribution

$$p(\mathbf{n}, \mathbf{l}) = \left\{ \prod_{i=1}^{m} p_i(n_i) \right\} \hat{p}(\mathbf{l}) \tag{7.63}$$

where $p_i(n_i)$ is as defined in equation 7.59,

$$\hat{p}(\mathbf{l}) = \left(\begin{matrix} |\mathbf{l}| \\ l_{(i,j)}, i \neq j, i, j = 0, \ldots, m \end{matrix} \right) \prod_{i=0}^{m} \prod_{j=0; j\neq i}^{m} \left\{ \frac{\lambda_{(i,j)}}{\lambda + \sum_{i=1}^{m} \lambda_i (1 - p_{ii})} \right\}^{l_{(i,j)}} q(|\mathbf{l}|), \tag{7.64}$$

$\lambda_{(i,j)} = \lambda_i p_{ij}$ is the job-flow rate on the link (i, j), $i, j = 1, \ldots, m$, $(i \neq j)$, $\lambda_{(0,j)} = \lambda \gamma_i$ is the job-flow rate from the input location to machine center j, $j = 1, \ldots, m$, $\lambda_{(i,0)} = \lambda_i (1 - \sum_{j=1}^{m} p_{ij})$ is the job-flow rate from machine center i to the output location, $i = 1, \ldots, m$, and $q(\cdot)$ is the stationary distribution of the number of jobs in a $M/M/c_0$ queueing system with arrival rate $\lambda + \sum_{i=1}^{m} \lambda_i (1 - p_{ii})$, service rate μ_0, and c_0 parallel servers. Therefore for stationarity of the dynamic job shop we need $c_0 \mu_0 > \lambda + \sum_{i=1}^{m} \lambda_i (1 - p_{ii})$. Here $l_{(i,j)}$ is the number of jobs in transit from machine center i to machine center j, $i, j = 1, \ldots, m$, $i \neq j$; $l_{(0,j)}$ is the number of jobs in transit from the input location to machine center j, $j = 1, \ldots, m$ and $l_{(i,0)}$ is the number of jobs in transit from machine center i to the output location, $i = 1, \ldots, m$. If we are interested only in the total number of jobs in transit, from equation 7.64 it can be seen that the distribution of this is $q(.)$. Therefore if we model the job transfers by a single-class job-transfer probability matrix $\mathbf{P}'$ where

$$p'_{ii} = p_{ii}$$

$$p'_{i0} = 1 - p_{ii}, \quad i = 1, \ldots, m$$

$$p'_{0j} = \frac{\lambda_j}{\lambda + \sum_{i=1}^{m} \lambda_i (1 - p_{ii})}, \quad j = 1, \ldots, m$$

and $\gamma_0 = 1$

it can be verified that the resulting stationary distribution for the number of jobs at different machine centers and in transit is the same as that obtained from the multiple-class open Jackson queueing network model. This demonstrates that the multiple classes in a Jackson queueing network can be aggregated into one class of jobs without violating the validity of the distribution for the number of jobs at different machine centers and in transit.

Suppose we have c_i machines in machine center i, $i = 1, \ldots, m$. Then the machine utilization at machine center i is $\rho_i = \lambda_i / (c_i \mu_i)$, $i = 1, \ldots, m$. Conversely we have observed that the use of the transporter is $\rho_0 = \left\{ \lambda + \sum_{i=1}^{m} \lambda_i (1 - p_{ii}) \right\} / (c_0 \mu_0)$. Because the central transporters are handling all the job flow (with a total rate of $\lambda + \sum_{i=1}^{m} \lambda_i (1 - p_{ii})$) it is possible that the transporter system itself may become the bottleneck station in the system. This happens when $\rho_0 > \rho_i$, $i = 1, \ldots, m$. Rewriting ρ_0 we have

$$\rho_0 = \frac{\lambda}{c_0 \mu_0} \left\{ 1 + \sum_{i=1}^{m} v_i (1 - p_{ii}) \right\} \tag{7.65}$$

where v_i is the average number of operations required by an arbitrary job at machine center i, $i = 1, \ldots, m$. This is independent of the routing sequence of the operations and is fixed in advance. The only way to reduce the loading of the transporter is therefore to increase p_{ii}. That is the number of sequential operations at any given machine center should be maximized as much as possible.

7.7 GENERAL JOB SHOP WITH LOCAL SERVICE PROTOCOLS

In this section we develop an open queueing network model for job shops where (1) the service times or interarrival times cannot be approximated by exponential distributions, and (2) the jobs are serviced at each machine center according to a service protocol (such as SPT) that uses only the information about the jobs at that machine center. To model this job shop we will first use the lesson learned in the Jackson network model: Each machine center can be analyzed in isolation with the appropriate arrival process and then the results of this analysis can be combined together to yield the system performance measures. Therefore, suppose the interarrival times of jobs to machine center i have a mean $1/\lambda_i$ and squared coefficient of variation $C_{a_i}^2$. Then machine center i will be modeled by a $GI/G/c_i$ queueing system with renewal input process $\{A_{i:n}, n = 1, 2, \ldots\}$, iid service times $\{S_{i:n}, n = 1, 2, \ldots\}$, c_i parallel servers, and the appropriate service protocol. The mean and squared coefficient of variation of the interarrival times are $1/\lambda_i$, $C_{a_i}^2$, and the mean and squared coefficient of variation of the service times are $1/\mu_i$ and $C_{S_i}^2$. Again c_i is the number of machines at machine center i, $i = 1, \ldots, m$.

7.7.1 Number in System and Mean Flow Time

Let $\hat{N}(\lambda_i, C_{a_i}^2, \mu_i, C_{S_i}^2, c_i)$ and $\hat{T}(\lambda_i, C_{a_i}^2, \mu_i, C_{S_i}^2, c_i)$ be the approximate number of jobs in the system and the flow time of an arbitrary job in a $GI/GI/c_i$ queueing system with the appropriate service protocol (see Section 3.5). For notational convenience, we will use $\hat{N}_i$ and $\hat{T}_i$ instead of $\hat{N}(\ldots)$ and $\hat{T}(\ldots)$, respectively. Then the system performance measures are approximated by

$$P\{N_i = n_i, i = 1, \ldots, m\} = \prod_{i=1}^{m} P\{\hat{N}_i = n_i\}, \quad \mathbf{n} \in \mathcal{N}_+^m \tag{7.66}$$

$$E[N] = \sum_{i=1}^{m} E[\hat{N}_i] \tag{7.67}$$

and

$$E[T] = \sum_{i=1}^{m} v_i E[\hat{T}_i] \tag{7.68}$$

where N_i is the number of jobs at machine center i. $N = \sum_{i=1}^{m} N_i$ is the total number of jobs in the system, and T is the flow time of an arbitrary job. As defined before v_i is the average number of visits made to machine center i by an arbitrary job.

Therefore, once the parameters $\lambda_i, C_{a_i}^2, \mu_i, C_{S_i}^2, c_i$ and v_i, $i = 1, \ldots, m$ are known we can obtain approximate performance measures for the job shop. These parameters may be obtained from actual data as described in Section 7.3. Suppose we have access only to the following parameters λ, C_a^2, γ_i, $\mathbf{P}$, μ_i, $C_{S_i}^2$, and c_i, $i = 1, \ldots, m$. Here $1/\lambda$ and C_a^2 are the mean and squared coefficient of variation of the external job arrival process, γ_i is the fraction of external jobs that join machine center i for its first operation, $i = 1, \ldots, m$, $\mathbf{P}$ is the job transfer probability matrix, and the others are as defined earlier. Then λ_i, $i = 1, \ldots, m$ is the solution to the set of linear equations

$$\lambda_i = \lambda \gamma_i + \sum_{j=1}^{m} \lambda_j p_{ji}, \quad i = 1, \ldots, m$$

and $v_i = \lambda_i / \lambda$, $i = 1, \ldots, m$. Note that all instantaneous transitions of a job from a machine center to the same machine center are accounted as a single operation with a larger processing time. All that remains is to find $C_{a_i}^2$, $i = 1, \ldots, m$. To do this we will first look at the squared coefficient of variation $C_{d_i}^2$ of the job interdeparture times from machine center i, $i = 1, \ldots, m$. Corresponding to approximation 3, Table 3.1 of

Chapter 3, independent of the service protocol we approximate this by

$$C_{d_i}^2 = 1 + (1 - \rho_i)\left(1 + \rho_i + \rho_i C_{a_i}^2\right)\left(C_{a_i}^2 - 1\right) + \frac{\rho_i^2}{c_i}\left(C_{S_i}^2 - 1\right), \quad i = 1, \ldots, m \quad (7.69)$$

Observe that the preceding set of equations are nonlinear in $C_{a_i}^2$, $i = 1, \ldots, m$. If $C_{a_i}^2 < 1$, then $|(1 - \rho_i)\rho_i C_{a_i}^2\left(C_{a_i}^2 - 1\right)| \leq 1/16$. Therefore if $C_{S_i}^2 \leq 1$, $i = 1, \ldots, m$ and $C_a^2 \leq 1$, then we may use the following set of linear equations on $C_{a_i}^2$, $i = 1, \ldots, m$:

$$C_{d_i}^2 = 1 + \left(1 - \rho_i^2\right)\left(C_{a_i}^2 - 1\right) + \frac{\rho_i^2}{c_i}\left(C_{S_i}^2 - 1\right), \quad i = 1, \ldots, m \quad (7.70)$$

This gives us m equations for $C_{d_i}^2$, $C_{a_i}^2$, $i = 1, \ldots, m$, the $2m$ unknowns. To construct m more equations we need to study how the departure processes from the machine centers are converted into arrival processes to the machine centers. This conversion involves two steps. First the departure process from each machine center is *split* into several streams depending on where the jobs are being routed. The second step involves the *composition* of the split streams that are directed to each machine center. Observe that a binomial sampling with probability p of a renewal process with squared coefficient of variation C^2 for interevent times leads to a renewal process with squared coefficient of variation $pC^2 + 1 - p$ for its interevent times. Then it is easy to see that the squared coefficient of variation for the job stream on link (i, j) can be approximated by $1 - p_{ij} + p_{ij}C_{d_i}^2$, $j \neq i, i, j = 1, \ldots, m$.

For the second step observe that the limit $\lim_{t \to \infty} \text{Var}(N(t))/t = \hat{\lambda}\hat{C}^2$ for a renewal counting process $\{N(t), t \geq 0\}$ with interevent time with mean $1/\hat{\lambda}$ and squared coefficient of variation $\hat{C}^2$. Because the variance of the sum of independent random variables is the sum of their variances, it is reasonable to approximate the squared coefficient of variation of the composition of renewal processes arriving at machine center i by

$$C_{a_i}^2 = \frac{1}{\hat{\lambda}_i}\sum_{j=1; j\neq i}^{m}\hat{\lambda}_j p_{ji}\left[p_{ji}C_{d_j}^2 + (1 - p_{ji})\right] + \frac{\lambda\gamma_i}{\hat{\lambda}_i}\left[\gamma_i C_a^2 + (1 - \gamma_i)\right], \quad i = 1, \ldots, m$$

$$(7.71)$$

where $\hat{\lambda}_i = \lambda\gamma_i + \sum_{j=1; j\neq i}^{m}\hat{\lambda}_j p_{ji}$. Now equations 7.70 and 7.71 can be used together to solve for $C_{a_i}^2$ and $C_{d_i}^2$, $i = 1, \ldots, m$.

We will now apply the preceding approximation technique to a multiple-center dynamic job shop with single machines at each machine center and Poisson external arrival process. At each machine center jobs are processed according to the shortest processing time first service protocol. Then if $1/\lambda_i$ and $C_{a_i}^2$ are the mean and squared coefficient of variation of the interarrival time to machine center i, then the mean time $E[\hat{T}_i]$ spent at machine center i by an arbitrary job is approximated by (see Section 3.5.4)

$$E[\hat{T}_i] \approx \hat{W}(\lambda_i, \mu_i, C_{a_i}^2, C_{S_i}^2)(1 - \rho_i)\int_{x=0}^{\infty}\frac{dF_{Si}(x)}{(1 - \rho_i(x))^2} + \frac{1}{\mu_i} \quad (7.72)$$

where $\hat{W}(\lambda_i, \mu_i, C_{a_i}^2, C_{S_i}^2)$ is a chosen approximation for the mean waiting time in a $GI/G/1$ queue and $\rho_i(x) = \lambda_i\int_0^x y\, dF_{S_i}(y)$, $x \geq 0$. From equations 7.70 and 7.71 one obtains λ_i and $C_{a_i}^2$, $i = 1, \ldots, m$ by solving

$$\lambda_i = \lambda \gamma_i + \sum_{j=1}^{m} \lambda_j p_{ji}, \qquad i = 1, \ldots, m \tag{7.73}$$

$$C_{d_i}^2 = \left(1 - \rho_i^2\right) C_{a_i}^2 + \rho_i^2 C_{S_i}^2, \qquad i = 1, \ldots, m \tag{7.74}$$

$$\text{and } C_{a_i}^2 = \sum_{j=1}^{m} \frac{\lambda_j p_{ji}}{\lambda_i} \left(p_{ji} C_{d_j}^2 + (1 - p_{ji}) \right) + \frac{\gamma_i}{\lambda_i} \tag{7.75}$$

Now equations 7.73 to 7.75 can be used to obtain the approximate flow time $E[T] = \sum_{i=1}^{m} v_i E[\hat{T}_i]$ and the approximate mean number of jobs $E[N] = \lambda E[T]$ in the system. We will consider two extreme types of job shops: (1) symmetric and (2) uniform flow.

Symmetric job shop.   In a symmetric job shop a job leaving a machine center is equally likely to go to one of the other $m - 1$ machine centers or leave the system. That is, $p_{ij} = 1/m$; $i \neq j$, $i, j = 1, \ldots, m$; $p_{ii} = 0$, $i = 1, \ldots, m$. Then it can be verified that

$$\lambda_i = \lambda, \qquad i = 1, \ldots, m \tag{7.76}$$

It can be seen from equation 7.71 that $C_{a_i}^2 \to 1$ as $m \to \infty$. It is therefore possible to simplify the earlier approximation by replacing $\hat{W}\left(\lambda_i, \mu_i, C_{a_i}^2, C_{S_i}^2\right)$ in equation 7.72 by the mean waiting time $\rho_i^2(1 + C_{S_i}^2)/[2\lambda_i(1 - \rho_i)]$ in a $M/G/1$ queue. That is, we approximate $E[T]$ by

$$E[T] = \sum_{i=1}^{m} \left\{ \frac{\rho_i^2 \left(1 + C_{S_i}^2\right)(1 - \rho_i)}{2\lambda_i} \int_{x=0}^{\infty} \frac{dF_{S_i}(x)}{(1 - \rho_i(x))^2} + \frac{1}{\mu_i} \right\} \tag{7.77}$$

We will call the above approximation an M-arrival approximation and the former (obtained from equation 7.72) a GI-arrival approximation. In Tables 7.1 to 7.3 we present these approximations along with the simulation results for various values of m, λ and k. In all the cases reported it is assumed that the service times at all machines are iid with Erlang-k distributions with mean 1.

From these results we can see that both the M-arrival and the GI-arrival approximations give good approximations for the mean flow time.

Recall that among all nonpreemptive service protocols, SPT minimizes the mean flow time in a $M/G/1$ queue. Therefore when M-arrival approximations are reasonable, one may conclude that SPT service protocol will minimize the mean flow time in a dynamic job shop among all local service protocols. Therefore it is reasonable to expect the SPT service protocol to produce a small mean flow time in a large job shop with random job routing.

Uniform flow job shop.   Next we consider a uniform-flow job shop where all jobs follow the same sequence of machine centers. That is, without a loss of generality,

TABLE 7.1 MEAN FLOW TIME IN
TWO-MACHINE–CENTER SYMMETRIC JOB
SHOP

		Simulation	Approximation	
k	λ	95% C.I.	GI-arrival	M-arrival
1	0.4	2.98 ± 0.12	3.04	3.04
	0.6	3.97 ± 0.16	3.92	3.92
	0.8	5.94 ± 0.27	5.76	5.76
2	0.4	2.82 ± 0.10	2.81	2.83
	0.6	3.62 ± 0.14	3.53	3.61
	0.8	5.52 ± 0.24	5.16	5.40
5	0.4	2.72 ± 0.09	2.68	2.71
	0.6	3.40 ± 0.13	3.32	3.44
	0.8	5.52 ± 0.24	4.87	5.29
∞	0.4	2.61 ± 0.09	2.63	2.67
	0.6	3.41 ± 0.13	3.33	3.50
	0.8	6.23 ± 0.28	5.28	6.00

TABLE 7.2 MEAN FLOW TIME IN
FOUR-MACHINE–CENTER SYMMETRIC JOB
SHOP

		Simulation	Approximation	
k	λ	95% C.I.	GI-arrival	M-arrival
1	0.4	6.11 ± 0.23	6.07	6.07
	0.6	8.16 ± 0.34	7.85	7.85
	0.8	11.90 ± 0.47	11.53	11.53
2	0.4	5.64 ± 0.21	5.63	5.66
	0.6	7.43 ± 0.27	7.11	7.22
	0.8	11.21 ± 0.48	10.45	10.80
5	0.4	5.43 ± 0.19	5.37	5.41
	0.6	7.22 ± 0.28	6.71	6.89
	0.8	10.98 ± 0.48	9.96	10.59
∞	0.4	5.21 ± 0.19	5.28	5.33
	0.6	7.19 ± 0.29	6.76	7.00
	0.8	12.97 ± 0.58	10.95	12.00

TABLE 7.3   MEAN FLOW TIME IN
NINE-MACHINE–CENTER SYMMETRIC JOB
SHOP

		Simulation	Approximation	
k	λ	95% C.I.	GI-arrival	M-arrival
1	0.4	13.58 ± 0.51	13.67	13.67
	0.6	18.09 ± 0.81	17.66	17.66
	0.8	27.18 ± 1.63	25.94	25.94
2	0.4	12.86 ± 0.42	12.69	12.73
	0.6	16.51 ± 1.26	16.12	16.25
	0.8	25.68 ± 2.43	23.90	24.30
5	0.4	12.33 ± 0.44	12.13	12.18
	0.6	15.64 ± 0.99	15.30	15.50
	0.8	25.01 ± 1.94	23.11	25.82
∞	0.4	11.15 ± 0.51	11.95	12.00
	0.6	16.03 ± 1.07	15.49	15.75
	0.8	28.07 ± 2.62	25.81	27.00

$\gamma_1 = 1$, $p_{i,i+1} = 1$, $i = 1, \ldots, m - 1$ and the jobs leaving machine center m leave the system immediately. Here again

$$\lambda_i = \lambda, \qquad i = 1, \ldots, m$$

But observe, however, that $C_{a_i}^2$ for $i \leq m$ is unaffected by the value of m. Approximate mean flow time using both the M-arrival and GI-arrival approximation methods are presented in Tables 7.4 and 7.5, along with the simulation results. In the cases considered it is assumed that the service times are iid Erlang-k random variables with mean 1 and results are reported for various values of λ, m, and k.

It is clear from these results that the M-arrival approximation is not appropriate for a uniform-flow job shop. The GI-arrival approximation, however, provides good estimates of the mean flow time. Therefore, overall we recommend the GI-arrival approximation for any type of job shop, whereas we recommend the use of M-arrival approximations for large randomly routed job shops.

7.7.2 Job Routing Diversity

Next we will compare the relative performance of the symmetric job shops and uniform flow shops considered previously, and define and study *job routing diversity*. Observe that in the above examples the mean flow time in the four-machine uniform-flow job shop with a given work load is smaller than that in the symmetric job shop with the same work load. We will now formalize this for job shops with high work loads (i.e., $\rho_i \approx 1$, $i = 1, \ldots, m$). Consider an m machine center job shop with the same number

TABLE 7.4 MEAN FLOW TIME IN TWO-MACHINE–CENTER UNIFORM-FLOW SHOP

k	λ	Simulation 95% C.I.	Approximation	
			GI-arrival	M-arrival
1	0.4	3.06 ± 0.10	3.04	3.04
	0.6	3.87 ± 0.14	3.92	3.92
	0.8	5.82 ± 0.25	5.76	5.76
2	0.4	2.70 ± 0.08	2.79	2.83
	0.6	3.53 ± 0.12	3.48	3.61
	0.8	5.18 ± 0.21	4.97	5.40
5	0.4	2.56 ± 0.06	2.66	2.71
	0.6	3.26 ± 0.10	3.24	3.44
	0.8	4.68 ± 0.19	4.51	5.29
∞	0.4	2.33 ± 0.04	2.60	2.67
	0.6	2.76 ± 0.08	3.19	3.50
	0.8	4.03 ± 0.16	4.60	6.00

TABLE 7.5 MEAN FLOW TIME IN FOUR-MACHINE–CENTER UNIFORM FLOW SHOP

k	λ	Simulation 95% C.I.	Approximation	
			GI-arrival	M-arrival
1	0.4	6.14 ± 0.18	6.07	6.07
	0.6	7.77 ± 0.25	7.85	7.85
	0.8	11.34 ± 0.38	11.53	11.53
2	0.4	5.37 ± 0.13	5.47	5.66
	0.6	6.74 ± 0.19	6.59	7.22
	0.8	9.83 ± 0.37	9.14	10.80
5	0.4	4.93 ± 0.10	5.14	5.41
	0.6	5.99 ± 0.15	5.84	6.89
	0.8	8.44 ± 0.30	7.50	10.59
∞	0.4	4.32 ± 0.04	4.96	5.33
	0.6	4.72 ± 0.08	5.48	7.00
	0.8	6.13 ± 0.22	6.70	12.00

of machines at all m machine centers (i.e., $c_i = c$, $i = 1, \ldots, m$), the same service time distributions (i.e., $S_i \overset{d}{=} S$, $i = 1, \ldots, m$) and the same utilization (i.e., $\rho_i = \rho = \lambda E[S]$,

$i = 1, \ldots, m$). It is further assumed that the average number of times a job visits machine center i is one (i.e., $v_i = 1, i = 1, \ldots, m$). Observing that the job transfer probability matrix **P** and the first operation vector $\gamma = (\gamma_i, i = 1, \ldots, m)$ reflect the job routing diversity we wish to study the mean flow time of an arbitrary job with respect to **P** and γ. First we will look at the "job routing diversity" that will give us the minimum mean flow time. That is, we wish to

$$\min E[T]$$

subject to

$$\sum_{i=1}^{m} p_{ji} \leq 1, \qquad j = 1, \ldots, m$$

$$\sum_{i=1}^{m} \gamma_i = 1$$

$$\gamma_i + \sum_{j=1}^{m} p_{ji} = 1, \qquad i = 1, \ldots, m$$

$$\gamma_i \geq 0, \qquad i = 1, \ldots, m$$

$$p_{ij} \geq 0, \qquad i, j = 1, \ldots, m$$

$$p_{ii} = 0, \qquad i = 1, \ldots, m$$

Note that the constraint $\gamma_i + \sum_{j=1}^{m} p_{ji} = 1$ ensures that the average number of visits made to machine center i by an arbitrary job is $1, i = 1, \ldots, m$. Because the exact mean flow time $E[T]$ is difficult to obtain we will use the approximation described earlier. Let $C_{a_i}^2$ be the squared coefficient of variation of the arrival process to machine center i. Then from equations 3.142, 3.155 and 7.68 we get

$$E[T] \approx \hat{T} = \frac{E[W]_{M/M/c}}{E[W]_{M/M/1}} \cdot \left\{ \frac{\rho^2 \left(1 + C_S^2\right)}{1 + \rho^2 C_S^2} \right\} \left\{ \frac{\left(\sum_{i=1}^{m} C_{a_i}^2 + m\rho^2 C_S^2\right)}{2\lambda(1 - \rho)} \right\} + m E[S]$$

Therefore minimizing $\hat{T}$ is equivalent to minimizing $\sum_{i=1}^{m} C_{a_i}^2$. From equations 7.70 and 7.71 one sees that for $\rho_i \approx 1$,

$$\sum_{i=1}^{m} C_{a_i}^2 = \sum_{i=1}^{m} \left\{ \sum_{j=1}^{m} \left[p_{ji} \left(p_{ji} C_S^2 + (1 - p_{ji}) \right) \right] + \gamma_i \left(\gamma_i C_a^2 + (1 - \gamma_i) \right) \right\}$$

$$= m - \left\{ \sum_{i=1}^{m} (1 - C_S^2) \left(\sum_{j=1}^{m} p_{ji}^2 + \gamma_i^2 \right) + \gamma_i^2 \left(C_a^2 - C_S^2 \right) \right\}$$

Suppose $C_a^2 \geq C_S^2$ and $C_S^2 \leq 1$. Then

$$\left(1 - C_S^2\right) + \frac{C_a^2 - C_S^2}{m}$$

$$\leq \sum_{i=1}^{m} \left\{ \left(1 - C_S^2\right) \left(\sum_{j=1}^{m} p_{ji}^2 + \gamma_i^2 \right) + \gamma_i^2 (C_a^2 - C_S^2) \right\} \leq m \left(1 - C_S^2\right) + C_a^2 - C_S^2$$

Conversely if $C_a^2 \leq C_S^2$ and $C_S^2 \geq 1$, the preceding inequalities are reversed. The lower limit of the preceding inequality is attained by a symmetric job routing (i.e., $p_{ji} = 1/m$, $i \neq j, i, j = 1, \ldots, m$; $\gamma_i = 1/m$, $i = 1, \ldots, m$) and the upper bound is attained by a uniform flow job routing (i.e., $\gamma_1 = 1$, $p_{i,i+1} = 1$, $i = 1, \ldots, m - 1$, and all other $p_{ij} = 0$). Therefore we see that

1. If $C_a^2 \geq C_S^2$ and $C_S^2 \leq 1$, then uniform-flow job routing will minimize the mean flow time and symmetric job routing will maximize the mean flow time.
2. If $C_a^2 \leq C_S^2$ and $C_S^2 \geq 1$, then symmetric job routing will minimize the mean flow time, and uniform-flow job routing will maximize the mean flow time.

7.7.3 Flow Time

Now we will develop an approximation for the flow time T of an arbitrary job. Let T_{ij} be the time spent in machine center i by a tagged arbitrary job during its jth visit to machine center i. If K_i is the number of visits to machine center i made by the tagged job, the flow time is then given by

$$T = \sum_{i=1}^{m} \sum_{j=1}^{K_i} T_{ij} \tag{7.78}$$

In general the T_{ij}'s and K_i's are all dependent random variables. To make the analysis tractable we will, however, assume that all these random variables are independent and that for each i, T_{ij}, $j = 1, 2, \ldots$ are identical. Then conditioning on K_i, $i = 1, \ldots, m$, one obtains

$$E[T|K_i, i = 1, \ldots, m] = \sum_{i=1}^{m} K_i E[T_i] \tag{7.79}$$

and

$$E[T^2|K_i, i = 1, \ldots, m] = \sum_{i=1}^{m} K_i \text{var}[T_i] + \sum_{i=1}^{m} \sum_{j=1}^{m} K_i K_j E[T_i] E[T_j] \tag{7.80}$$

Now taking expectations of equations 7.79 and 7.80 with respect to K_i, $i = 1, \ldots, m$ one gets

$$E[T] = \sum_{i=1}^{m} E[K_i] E[T_i] \tag{7.81}$$

and

$$\text{var}(T) = \sum_{i=1}^{m} E[K_i]\text{var}(T_i) + \sum_{i=1}^{m}\sum_{j=1}^{m} \text{cov}(K_i, K_j)E[T_i]E[T_j]. \qquad (7.82)$$

Because $E[K_i] = v_i$, and $E[T_i]$ is the average time spent in machine center i by an arbitrary job during a single visit to machine center i, it is clear from Little's formula that equation 7.81 is indeed an exact result. For this, note that $\lambda E[T] = \sum_{i=1}^{m} \lambda v_i E[T_i] = \sum_{i=1}^{m} E[N_i]$ is the total number of jobs in the system as it should be. To obtain the variance of the flow time we will need $\text{cov}(K_i, K_j)$, $i, j = 1, \ldots, m$. We will obtain these quantities using a Markov chain model $\{X_k, k = 1, 2, \ldots\}$ for the sequence of machines visited by the tagged job. If the tagged job is at machine center i for its kth operation, then $X_k = i$, $i = 1, \ldots, m$. Conversely if the number of operations required by the tagged job is less than k, we set $X_k = 0$, indicating that it has left the job shop. Then $\{X_k, k = 1, 2, \ldots\}$ is an absorbing Markov chain on $\{0, 1, \ldots, m\}$ with the absorption state 0. Furthermore

$$P\{X_k = j | X_{k-1} = i\} = p_{ij}, \quad i, j = 1, \ldots, m$$

$$P\{X_k = 0 | X_{k-1} = i\} = 1 - \sum_{j=1}^{m} p_{ij}$$

$$P\{X_k = 0 | X_{k-1} = 0\} = 1$$

and $P\{X_1 = i\} = \gamma_i$, $i = 1, \ldots, m$. Then K_i is the number of times this Markov chain visits state i before its absorption to state 0. Let K_{ji} be the number of times this Markov chain visits state i before its absorption, given that $X_1 = j$, $i, j = 1, \ldots, m$.

Then from the Markov property of $\{X_k\}$ it is easily seen that by accounting for the number of visits (either zero or one) to state i in the first step and the state of $\{X_k\}$ at its second step

$$K_{ji} = \begin{cases} \delta_{ji}, & \text{with probability } 1 - \sum_{l=1}^{m} p_{jl} \\ K_{li} + \delta_{ji}, & \text{with probability } p_{jl}, \quad l = 1, \ldots, m \end{cases} \qquad (7.83)$$

where $\delta_{ji} = 1$ if $j = i$ and 0 otherwise. Similarly

$$K_{ji}K_{jr} = \begin{cases} \delta_{ji}\delta_{jr}, & \text{with probability } 1 - \sum_{l=1}^{m} p_{jl} \\ (K_{li} + \delta_{ji})(K_{lr} + \delta_{jr}), & \text{with probability } p_{jl}, \quad l = 1, \ldots, m \end{cases} \qquad (7.84)$$

Taking expectations on both sides of equations 7.83 and solving for $E[K_{ji}]$ one obtains

$$(E[K_{ji}])_{j,i=1,\ldots,m} = (\mathbf{I} - \mathbf{P})^{-1} \qquad (7.85)$$

Taking expectations of both sides of equation 7.84 and solving for $E[K_{ji}K_{jr}]$ one obtains

$$E[K_{ji}K_{jr}] = E[K_{jr}]E[K_{ri}] + E[K_{ji}]E[K_{ir}], \quad i \neq r \qquad (7.86)$$

and

$$E[K_{ji}^2] = E[K_{ji}]\{2E[K_{ii}] - 1\} \tag{7.87}$$

Because $E[K_i K_r] = \sum_{i=1}^{m} \gamma_j E[K_{ji} K_{jr}]$ and $E[K_i^2] = \sum_{j=1}^{m} \gamma_j E[K_{ji}^2]$ from equations 7.86 and 7.87 we obtain

$$E[K_i K_r] = E[K_i]E[K_{ir}] + E[K_r]E[K_{ri}], \quad i \neq r \tag{7.88}$$

and

$$E[K_i^2] = E[K_i]\{2E[K_{ii}] - 1\} \tag{7.89}$$

where $E[K_i] = \sum_{j=1}^{m} \gamma_j E[K_{ji}]$, $i = 1, \ldots, m$. Therefore

$$\text{cov}(K_i, K_j) = E[K_i]E[K_{ij}] + E[K_j]E[K_{ji}] - E[K_i]E[K_j], \quad i \neq j \tag{7.90}$$

and

$$\text{cov}(K_i, K_i) = \text{var}(K_i) = E[K_i]\{2E[K_{ii}] - E[K_i] - 1\} \tag{7.91}$$

The covariance of K_i and K_j evaluated by equation 7.90 is for the aggregate job that is transferred from one machine center to another according to the transfer probability matrix $\mathbf{P}$. Covariance of K_i and K_j corresponding to the actual data that we used to obtain $\mathbf{P}$ (see Section 7.3.2) can be different than that obtained by equation 7.90. To see this consider a job shop with two machine centers ($m = 2$) and two types of jobs ($r = 2$) with equal arrival rates $\lambda(1) = \lambda(2)$. The machine center sequence for type 1 jobs is $\langle 1, 2 \rangle$ and for type 2 jobs is $\langle 2, 1 \rangle$. Then $\text{cov}(K_1, K_2)$ evaluated by equation 7.90 is 1/3 and the actual covariance of K_1 and K_2 is 0. This would therefore introduce an error in the evaluation of var (T) by equation 7.82.

The mean and variance of T can be used to obtain an approximate distribution for the flow time. If $\text{var}(T) \leq (E[T])^2$, we suggest the use of a generalized Erlang approximation, if $\text{var}(T) \simeq (E[T])^2$, an exponential distribution with mean $E[T]$ should be used, and if $\text{var}(T) > (E[T])^2$, a hyperexponential distribution could be used. When the job transfers are symmetric, that is, $p_{ij} = 1/m$, $i \neq j$, $i, j = 1, \ldots, m$, we have $E[K_i] = 1$ and $\text{cov}(K_i, K_j) = (m-1)/(m+1)$, $i \neq j$ and $\text{var}(K_i) = 2(m-1)/(m+1)$. If all the m machine centers are identical we have $E[T_i] = E[T_j]$ for all $i, j = 1, \ldots, m$. Therefore $E[T] = mE[T_i]$ and $\text{var}[T] = m\text{var}(T_i) + m(m-1)E[T_i]^2$. For large m, $\text{var}(T) \simeq E[T]^2$ and therefore an exponential distribution with mean $mE[T_i]$ can be a good approximation for the distribution of T. In addition it can be expected that any service protocol that reduces the mean flow time at a machine will stochastically reduce T (i.e., reduce second and higher moments of T as well). For a uniform-flow job shop with $p_{i,i+1} = 1$, $i = 1, \ldots, m-1$ one has $E[K_i] = 1$ and $\text{cov}(K_i, K_j) = 0$ $i, j = 1, \ldots, m$. Then $E[T] = \sum_{i=1}^{m} E[T_i]$ and $\text{var}(T) = \sum_{i=1}^{m} \text{var}(T_i)$.

7.7.4 Due Date Assignment

We can use the distribution of the flow time to set due dates for job shipments. The following simple model illustrates this. Suppose we may set a due date t_d such that if

$t_d \leq A$, where A is an acceptable due date limit, no cost is incurred for setting the due date. Otherwise if $t_d > A$, a linear cost of $K_1(t_d - A)$ is incurred. A job completed after its due date incurs a cost of $K_2(> K_1)$ per unit time of tardiness and a job completed before its due date incurs a cost of K_3 per unit time of earliness. Thus, if T is the flow time of a job, the expected cost of setting a due date t_d is

$$C(t_d) = K_1(t_d - A)^+ + K_2 E\left[(T - t_d)^+\right] + K_3 E\left[(t_d - T)^+\right]$$

Since

$$\frac{dC(t_d)}{dt_d} = \begin{cases} K_1 - K_2 \bar{F}_T(t_d) + K_3(1 - \bar{F}_T(t_d)), & t_d \geq A \\ -K_2 \bar{F}_T(t_d) + K_3(1 - \bar{F}_T(t_d)), & t_d < A \end{cases} \tag{7.92}$$

the optimal due date t_d^* is given by

$$t_d^* = \begin{cases} \bar{F}_T^{-1}\left(\frac{K_1 + K_3}{K_2 + K_3}\right), & \text{if} \geq A \\ \bar{F}_T^{-1}\left(\frac{K_3}{K_2 + K_3}\right), & \text{if} < A \\ A, & \text{otherwise} \end{cases} \tag{7.93}$$

7.7.5 Job Shops with Finite Storage Space

So far we have assumed that there is sufficient space in the job shop such that all jobs arriving at the job shop can be dispatched to the shop. In this section we look at systems that consist of a dispatch area and a job shop with a limited storage space of Z. Any job that arrives when there are Z jobs in the job shop (i.e., when there are Z or more jobs in the system) is kept in the dispatch area until a space in the job shop becomes available. If, however, the total number of jobs in the system is B ($\geq Z$) any external arrival is lost. For a system where no loss occurs we will set $B = +\infty$, and if jobs are not allowed to be in the dispatch area we set $B = Z$. Suppose $p(\mathbf{n})$, $\mathbf{n} \in \mathcal{N}_+^m$, is an approximation (or the exact solution, depending on the model assumptions) for the stationary distribution of the number of jobs at the machine centers for such a system when $B = Z = +\infty$. Then we approximate the total number of jobs in the system by $q_{B,Z}(.)$ where

$$q_{B,Z}(k) = \alpha q(k), \quad k = 0, \ldots, Z$$

$$q_{B,Z}(k) = \alpha q(Z)\left\{\frac{q(Z)}{q(Z-1)}\right\}^{k-Z}, \quad k = Z+1, \ldots, B$$

$$q(k) = \sum_{\mathbf{n} \in S_k} p(\mathbf{n}), \quad k = 0, 1, \ldots, Z$$

$S_k = \{\mathbf{n} : |\mathbf{n}| = k, \mathbf{n} \in \mathcal{N}_+^m\}$ and

$$\alpha = 1/\left\{\sum_{k=0}^{Z} q(k) + q(Z) \sum_{k=Z+1}^{B} \left\{\frac{q(Z)}{q(Z-1)}\right\}^{k-Z}\right\}$$

It is worth noting that when $p(\mathbf{n})$ is obtained using the results in Section 7.4 for the single-class open Jackson queueing network, this result is exact for $B = Z$.

7.7.6 Job Shops with Bulk Job Transfers

As we have illustrated earlier in section 7.6, it may happen that the material handling system may become the bottleneck station, or it may contribute to a major portion of the flow time of an arbitrary job. To reduce the work load on the material handling system (i.e., the number of movements) it is possible to move jobs in batches. We will now develop an open queueing network model of a job shop where jobs are moved in batches of size k and show how the optimal batch size can be computed such that the overall mean flow time, or, equivalently, the number of jobs in the system is minimized.

When jobs are moved in batches of size k we can view each of these batches as a "superjob." The movement of these superjobs through the job shop can be modeled by an open queueing network as shown previously. Let λ, C_a^2, μ_i, $C_{S_i}^2$, $i = 1, \ldots, m$ be as defined earlier. Here we will assume that the service center m is the central transport center. When jobs are moved one at a time we have seen that the squared coefficient of variation of the arrival process to each machine center is approximated by the solution to the system of equations given by equations 7.70 and 7.71. When jobs are moved in batches the arrival process to each machine center will be affected.

Let $C_{a_i}^{2(k)}$ be the squared coefficient of variation of the arrival process of superjobs to machine center i, $i = 1, 2, \ldots, m$. Because the squared coefficient of variation of the sum of k iid random variables $X_1, \ldots, X_k$ is C_X^2/k, we approximate $C_{a_i}^{2(k)}$ by $C_{a_i}^2/k$, $i = 1, \ldots, m$. The service time of a superjob is the sum of k iid service times. Therefore the squared coefficient of variation of the service time of a superjob at machine center i is $C_{S_i}^{2(k)} = C_{S_i}^2/k$, $i = 1, \ldots, m-1$. The transport times, however, are unaffected by the batch size. Therefore $C_{S_m}^{2(k)} = C_{S_m}^2$. A similar treatment shows that the arrival rates and the service rates for the superjobs are $\lambda_i^{(k)} = \lambda_i/k$ and $\mu_i^{(k)} = \mu_i/k$, respectively, at machine center i, $i = 1, \ldots, m-1$. The arrival rate to the transport center is $\lambda_m^{(k)} = \lambda_m/k$, but the service rate is $\mu_m^{(k)} = \mu_m$, because the service time is unaffected by the batch size. The mean number of superjobs in the system is then approximated by

$$E[\hat{N}^{(k)}] = \sum_{i=1}^{m} \hat{N}_i \left(\lambda_i^{(k)}, \mu_i^{(k)}, C_{a_i}^{2(k)}, C_{S_i}^{2(k)}, c_i \right)$$

where $\hat{N}_i(\cdot)$ is a suitably chosen approximation for the average number in a $GI/G/c_i$ queue.

Because the total number of jobs in the system is $E[N^{(k)}] = kE[\hat{N}^{(k)}] + (k-1)/2$, using the approximation

$$\hat{N}\left(\lambda, \mu, C_a^2, C_S^2, c \right) = \left\{ \frac{\rho^2 \left(1 + C_S^2 \right)}{1 + \rho^2 C_S^2} \right\} \left\{ \frac{\left(C_a^2 + \rho^2 C_S^2 \right)}{2(1 - \rho)} \right\} \frac{\hat{L}_{M/M/c}(\rho)}{\hat{L}_{M/M/1}(\rho)} + \rho$$

we get

$$
E\left[N^{(k)}\right] = \sum_{i=1}^{m-1} \left[\frac{\rho_i^2 \left(k + C_{S_i}^2\right) \left(C_{a_i}^2 + \rho^2 C_{S_i}^2\right) \hat{L}_{M/M/c_i}(\rho_i)}{2 \left(k + \rho_i^2 C_{S_i}^2\right) (1 - \rho_i) \hat{L}_{M/M/1}(\rho_i)} + k\rho_i \right]
$$
$$
+ \left[\frac{\rho_m^2 \left(1 + C_{S_m}^2\right) \left(k C_{a_m}^2 + \rho_m^2 C_{S_m}^2\right)}{2 \left(k^2 + \rho_m^2 C_{S_m}^2\right) (k - \rho_m)} + \rho_m \right] + \frac{k-1}{2} \qquad (7.94)
$$

Note that we assume availability of only one transporter (i.e., $c_m = 1$). The optimal value of k that minimizes $E[N^{(k)}]$ can be easily computed from equation 7.94. The computational effort can be reduced by observing that when $C_{a_m}^2 \le 1$ and $C_{S_m}^2 \le 1$, $E[N^{(k)}]$ given by equation 7.94 is convex in k. Therefore in this case the local optimum k is also globally optimum.

For example, consider the symmetric job shop with exponentially distributed processing times and Poisson external job arrival process. Job arrival rate is λ and the mean processing time at each of the m single-machine centers is one. The central transport center 0 has a single transporter with exponentially distributed service time with mean $1/\mu$. Then the utilization of machine center i is $\rho_i = \lambda$, $i = 1, \ldots, m$ and the utilization of the transporter is $\rho_0 = (m+1)\lambda/\mu_0$. Furthermore $C_{a_i}^2 = 1$, $i = 0, \ldots, m$. Then using the approximation given by equation 7.94 one obtains

$$
E\left[N^{(k)}\right] \approx m \left\{ \frac{\rho^2(k+1)\left(1+\rho^2\right)}{2\left(k+\rho^2\right)(1-\rho)} + k\rho \right\}
$$
$$
+ \left\{ \frac{\rho_0^2 \left(k + \rho_0^2\right)}{\left(k^2 + \rho_0^2\right)(k - \rho_0)} + \rho_0 \right\} + \frac{k-1}{2} \qquad (7.95)
$$

From equation 7.95 it can be easily verified that $E[N^{(k)}]$ is convex in k. In addition the batch size k^* that minimizes $E[N^{(k)}]$ is decreasing in ρ (because the increase in the quantity in the first braces in equation 7.94 with respect to ρ is increasing in k) and decreasing in ρ_0 (because the increase in the quantity in the second braces in equation 7.94 with respect to ρ_0 is decreasing in k).

7.8 MULTIPLE-CLASS GENERAL JOB SHOP

We will now consider a dynamic job shop that cannot be modeled by a single–job-class queueing network. This may be because (1) the service protocols used at different machine centers use information on the "type" of job, (2) because there is a set-up time (or change-over time) that depends on the sequence of the class indices of the jobs being serviced, or (3) because the machine center sequence of different job types are very different and the number of job classes is not very large. In all cases we assume that the job types are aggregated into r job classes. Class l jobs arrive at the job shop according to a renewal process with interarrival times that have a mean $1/\lambda^{(l)}$ and a squared coefficient of variation $C_a^{2(l)}$, $l = 1, \ldots, r$. A class l job will go through the job shop without changing its class index according to a job transfer probability matrix $\mathbf{P}^{(l)} = (p_{ij}^{(l)})_{i, j=1, \ldots, m}$. The processing times of class l jobs at machine center i form a sequence of iid random variables with mean $1/\mu_i^{(l)}$ and squared coefficient of variation $C_{S_i}^{2(l)}$, $i = 1, \ldots, m$; $l = 1, \ldots, r$. Jobs at machine center i are served according to the service protocol SP_i (such as FCFS or NPP) that uses only the information on jobs available at that machine center ($i = 1, \ldots, m$). Exact analysis of such a queueing network is extremely difficult. Therefore we will develop two approximations (1) *aggregation* and (2) *job-class decomposition* to analyze this queueing system.

7.8.1 Aggregation

Suppose we could not model the job shop by a single-class queueing network because (1) the service protocols used at different machine centers utilize information on the class index of jobs present at that center, or (2) there is a set-up time (or change-over time) that depends on the sequence of the class indices of the jobs being serviced. Suppose, however, the routing of jobs from one machine center to another are similar for all job classes. Furthermore the processing requirements of all job classes at each machine center are similar (i.e., $|\mu_i^{(l)} - \mu_i^{(l')}|$ and $|C_{S_i}^{2(l)} - C_{S_i}^{2(l')}|$ are not large for all $l, l' = 1, \ldots, r$). Then we may aggregate all job classes into a single job class (see Section 7.5) and obtain the mean $1/\lambda_i$ and squared coefficient of variation $C_{a_i}^2$ of the aggregated job interarrival time to machine center i, $i = 1, \ldots, m$ (see Section 7.7). That is, if $1/\mu_i$ and $C_{S_i}^2$ are the mean and squared coefficient of variation of the service time of an aggregated job at machine center i, then

$$\frac{1}{\mu_i} = \frac{1}{\lambda_i} \sum_{l=1}^{r} \frac{\lambda_i^{(l)}}{\mu_i^{(l)}}, \qquad i = 1, \ldots, m$$

and

$$C_{S_i}^2 = \left(\frac{\mu_i^2}{\lambda_i} \right) \sum_{l=1}^{r} \frac{\lambda_i^{(l)}}{(\mu_i^{(l)})^2} (1 + C_{S_i}^{2(l)}) - 1, \qquad i = 1, \ldots, m$$

Here $\lambda_i^{(l)}$ is the class l job arrival rate to machine center i, $l = 1, \ldots, r$ and $\lambda_i = \sum_{l=1}^{r} \lambda_i^{(l)}$ is the aggregate job arrival rate to machine center i, $i = 1, \ldots, m$. If $\mathbf{P}$ is the job transfer probability matrix of the aggregate jobs

$$p_{ij} = \frac{1}{\lambda_i} \sum_{l=1}^{r} \lambda_i^{(l)} p_{ij}^{(l)}, \qquad i, j = 1, \ldots, m$$

These parameters of the aggregated jobs are then used along with equations 7.69 (or 7.70) and 7.71 to obtain $(\lambda_i, C_{a_i}^2)$, $i = 1, \ldots, m$. Now each machine center is modeled by a multiple-class $GI/G/c_i$ (SP_i) queueing system with $\lambda_i^{(l)}/\lambda_i$ fraction of class l jobs, $l = 1, \ldots, r$. Let $\hat{N}^{(l)} \left(\lambda_i, \mu_i^{(l')}, C_{a_i}^2, C_{S_i}^{2(l')}, \lambda_i^{(l')}/\lambda_i, l' = 1, \ldots, r \right)$ be an approximate number of class l jobs in this queueing system (see Section 3.5). Then we approximate the total number of class l jobs in the system by

$$E[N^{(l)}] \approx \hat{N}(l) = \sum_{i=1}^{m} \hat{N}^{(l)}\left(\lambda_i, \mu_i^{(l')}, C_{a_i}^2, C_{S_i}^{2(l')}, \lambda_i^{(l')}/\lambda_i, l' = 1, \ldots, r\right), \qquad l = 1, \ldots, r$$

and the average flow time of an arbitrary class l job by

$$E[T^{(l)}] \approx \hat{N}(l)/\lambda^{(l)}, \qquad l = 1, \ldots, r$$

In some cases we may simplify the preceding approximation by assuming that $C_{a_i}^2 = 1$, $i = 1, \ldots, m$ (this is the M-arrival approximation, see Section 7.7).

We will now apply this approximation procedure to a three-machine-center job shop with 12 job classes. There is only one machine at each machine center and the processing times of all jobs at machine center i are iid exponential random variables with mean 0.5 (i.e., $\mu_i^{(l)} = 2$ and $C_{S_i}^{2(l)} = 1$, $l = 1, \ldots, 12$; $i = 1, 2, 3$). Job class l and $6+l$ has a uniform-flow job-transfer matrix according to the permutation π_l of $\{1, 2, 3\}$, $l = 1, \ldots, 6$ (note that $\pi = \{\pi_l, l = 1, \ldots, 6\}$ is the collection of 6 distinct permutations of $\{1, 2, 3\}$). Class l jobs arrive at the job shop from outside according to a Poisson process with rate $\lambda^{(l)}$, $l = 1, \ldots, 12$. We assume that $\lambda^{(l)} = p/6$, $l = 1, \ldots, 6$ and $\lambda^{(l)} = (1 - p)/6$, $l = 7, \ldots, 12$. At each machine center jobs within job class $1, \ldots, 6$ (and within job classes $7, \ldots, 12$) are served on a FCFS basis. Jobs between these two sets $\mathcal{S}_1 = \{1, \ldots, 6\}$ and $\mathcal{S}_2 = \{7, \ldots, 12\}$ of job classes, however, are served according to either (1) FCFS, (2) NPP, or (3) AP service protocols (see Sections 3.5 and 4.2). There is a switch-over time between the two sets of classes of jobs that we assume to be a constant equal to $1/\mu_{(1,2)}$. The approximated mean flow time and the simulation results are given in Table 7.6.

Note that because $C_{S_i}^{2(l)} = 1$, $\mu_i^{(l)} = 2$, $l = 1, \ldots, r$; $i = 1, \ldots, m$ and $C_a^2 = 1$, one finds that $C_{a_i}^2 = 1$, $i = 1, \ldots, m$. Hence in this case both the GI-arrival and M-arrival approximations are the same. The $M/G/1$ results needed for these approximations are taken from Buzacott and Gupta [5]. As can be seen from the Table 7.6 the proposed approximation provides a good estimate of the mean flow time. In this example observe that the routings of different classes of jobs are very different. Yet because the processing time of all these classes are probabilistically identical at each machine center the proposed aggregation method works. We will next look at an approach that could be used when the processing requirements of different classes of jobs are different.

7.8.2 Job-Class Decomposition

In Section 7.5 we saw that the performance of a single class of jobs in a multiple-class open Jackson queueing network can be obtained by analyzing the single-class open Jackson queueing network with appropriately modified service times. Specifically if we are interested in the performance of class l jobs, then we consider a single-class open Jackson queueing network with mean processing time $1/\mu_i$ at machine center i where

TABLE 7.6 MEAN FLOW TIME IN THREE-MACHINE–CENTER JOB SHOP WITH
TWELVE JOB CLASSES AND CHANGE-OVER TIMES

Parameters		Service protocol					
		FCFS		NPP		AP	
$1/\mu_{(1,2)}$	p	App.	Sim.	App.	Sim.	App.	Sim.
0.1	0.1	3.174	3.15 ± 0.062	3.136	3.16 ± 0.062	3.120	3.09 ± 0.056
	0.2	3.312	3.28 ± 0.064	3.316	3.28 ± 0.060	3.120	3.17 ± 0.054
	0.3	3.411	3.37 ± 0.065	3.396	3.35 ± 0.062	3.264	3.23 ± 0.051
	0.4	3.480	3.43 ± 0.062	3.415	3.37 ± 0.062	3.312	3.26 ± 0.059
	0.5	3.501	3.44 ± 0.060	3.426	3.37 ± 0.062	3.312	3.27 ± 0.058
0.5	0.1	4.161	4.02 ± 0.079	4.200	4.06 ± 0.079	3.720	3.65 ± 0.056
	0.2	5.331	5.04 ± 0.086	5.016	4.88 ± 0.091	4.230	4.11 ± 0.064
	0.3	6.405	5.88 ± 0.119	5.838	5.44 ± 0.084	4.539	4.40 ± 0.063
	0.4	7.200	6.48 ± 0.116	6.159	5.70 ± 0.090	4.710	4.56 ± 0.077
	0.5	7.500	6.71 ± 0.127	6.135	5.70 ± 0.076	4.776	4.62 ± 0.080

$\mu_i = \mu_i^{(l)} - \sum_{k=1, k \neq l}^{r} \lambda_i^{(k)}$, $i = 1, \ldots, m$. The job-transfer probability matrix of this single class of job is the same as that of class l jobs (i.e., equal to $\mathbf{P}^{(l)}$). We will adopt this job class decomposition to obtain the approximate performance of the multiple-class general network queueing model. To this end consider a multiple class single server $M/G/1$ queueing system where jobs are served according to a local service protocol SP. Let $\hat{N}^{(l)}(SP)$ be the average number of class l jobs in the system. Then if we were to represent the performance of class l jobs by that of a single-class $M/G/1$ queue with service time having a mean $1/\mu$ and squared coefficient of variation C_S^2 we need

$$E[N^{(l)}(SP)] = \frac{\lambda^{(l)}\left(1 + C_S^2\right)}{2(1 - \lambda^{(l)}/\mu)} + \frac{\lambda^{(l)}}{\mu} \tag{7.96}$$

We have two unknowns, μ and C_S^2, and only one equation. To obtain the second equation observe that the server in the multiple class $M/G/1$ queue is available only for the fraction of time equal to $\left(1 - \sum_{k=1; k \neq l}^{r} \lambda^{(k)}/\mu^{(k)}\right)$. Therefore the mean time needed to process a class l job with processing requirement x is $x/(1 - \sum_{k=1; k \neq l}^{r} \lambda^{(k)}/\mu_{(k)})$. Therefore we set

$$\mu = \mu^{(l)}\left(1 - \sum_{k=1; k \neq l}^{r} \frac{\lambda^{(k)}}{\mu^{(k)}}\right) = \mu^{(l)}\left(1 - \rho + \rho^{(l)}\right) \tag{7.97}$$

where $\rho^{(l)} = \lambda^{(l)}/\mu^{(l)}$ and $\rho = \sum_{k=1}^{r} \rho^{(k)}$. Based on this equivalent representation, the job class decomposition for class l jobs works as follows: The service parameters for machine center i in the single-class general job shop model are set by

$$\mu_i = \mu_i^{(l)}(1 - \rho_i + \rho_i^{(l)}), \qquad i = 1, \ldots, m$$

and

$$C_{S_i}^2 = 2 \left(\frac{1-\rho_i}{1-\rho_i+\rho_i^{(l)}} \right) \frac{1}{\lambda_i^{(l)}} \left[E\left[N_i^{(l)}(SP_i) \right] - \frac{\lambda_i^{(l)}}{\mu_i} \right] - 1, \qquad i = 1, \ldots, m \qquad (7.98)$$

where $E\left[N_i^{(l)}(SP_i) \right]$ is the mean number of class l jobs in a $M/G/1(SP_i)$ queue with class k arrival rate $\lambda_i^{(k)}$, generic service distribution $S_i^{(k)}$, $k = 1, \ldots, r$ and SP_i service protocol; $i = 1, \ldots, m$. Note that with this selection of parameter values the mean waiting time obtained from equation 7.96 is $E\left[N_i^{(l)}(SP_i) \right]$. For example if $SP_i = $ FCFS then (see Section 3.5.2)

$$E\left[N_i^{(l)}(\text{FCFS}) \right] = \frac{\lambda_i^{(l)} \sum_{k=1}^r \lambda_i^{(k)} \left(1 + C_{S_i}^{2(k)} \right) / \left(\mu_i^{(k)} \right)^2}{2(1-\rho_i)} + \rho_i^{(l)}, \qquad l = 1, \ldots, r$$
$$(7.99)$$

Once these parameters are obtained we may estimate the performance of class l jobs by analyzing a single–job-class queueing network as was done in Section 7.7. Therefore from equations 7.98 and 7.99 one finds that the squared coefficient of variation of the equivalent service time is

$$C_{S_i}^2 = 2 \left\{ \frac{1-\rho_i}{1-\rho_i+\rho_i^{(l)}} \right\} \left[\frac{\sum_{k=1}^r \lambda_i^{(k)}(1+C_{S_i}^{2(k)})/(\mu_i^{(k)})^2}{2(1-\rho_i)} - \left(\frac{\rho_i - \rho_i^{(l)}}{1-\rho_i+\rho_i^{(l)}} \right) \frac{1}{\mu_i^{(l)}} \right] - 1,$$
$$i = 1, \ldots, m \qquad (7.100)$$

7.8.3 Job Diversity in Processing Time

We will now study the effect of having a collection of job classes that have diversified processing times in a job shop operated under a FCFS service protocol. Let J be a collection of r job classes with class l job processing times at machine center i having a mean $E[S_i^{(l)}]$ and squared coefficient of variation $C_{S_i}^{2(l)}$, $l = 1, \ldots, r$; $i = 1, \ldots, m$. Also let $\hat{J}$ be another collection of r job classes with class l job processing times at machine centers having a mean $E[\hat{S}_i^{(l)}]$ and squared coefficient of variation $\hat{C}_{S_i}^{2(l)}$, $l = 1, \ldots, r$; $i = 1, \ldots, m$. Because the variability in processing times is caused by the machine or the worker in a machine center we may assume that

$$C_{S_i}^{2(l)} = \hat{C}_{S_i}^{2(l)} = \tilde{C}_{S_i}^2, \qquad l = 1, \ldots, r; i = 1, \ldots, m \qquad (7.101)$$

Suppose the job routing matrix and the first operation vectors are the same for both collections of classes of jobs. Let $\lambda_i^{(l)}$ be the class l job arrival rate to the machine center i, $l = 1, \ldots, r$; $i = 1, \ldots, m$. Given that the work load brought in to each machine center i is the same for both collections J and $\hat{J}$ for all $i = 1, \ldots, m$ we wish to study the effect of the mean job processing times on the mean flow time. The constraints that we have for these processing times is therefore

$$\sum_{l=1}^{r} \lambda_i^{(l)} E\left[S_i^{(l)}\right] = \sum_{l=1}^{r} \lambda_i^{(l)} E\left[\hat{S}_i^{(l)}\right], \qquad i = 1, \ldots, m \tag{7.102}$$

If we aggregate these job classes, then the aggregated jobs will have a processing time with mean

$$\frac{1}{\mu_i} = \frac{1}{\lambda_i} \sum_{l=1}^{r} \lambda_i^{(l)} E\left[S_i^{(l)}\right] \tag{7.103}$$

the same for both collection of job classes, and squared coefficient of variation

$$C_{S_i}^2 = \left(\frac{\mu_i^2}{\lambda_i}\right)\left(1 + \tilde{C}_{S_i}^2\right) \sum_{l=1}^{r} \lambda_i^{(l)} E\left[S_i^{(l)}\right]^2 - 1 \tag{7.104}$$

for the collection J, and

$$\hat{C}_{S_i}^2 = \left(\frac{\mu_i^2}{\lambda_i}\right)\left(1 + \tilde{C}_{S_i}^2\right) \sum_{l=1}^{r} \lambda_i^{(l)} E\left[\hat{S}_i^{(l)}\right]^2 - 1 \tag{7.105}$$

for the collection $\hat{J}$ (for $i = 1, \ldots, m$). If

$$\left(E\left[S_i^{(l)}\right], l = 1, \ldots, r\right) \leq_m \left(E\left[\hat{S}_i^{(l)}\right], l = 1, \ldots, r\right) \tag{7.106}$$

then from equations 7.104 and 7.105 it can be verified that

$$C_{S_i}^2 \leq \hat{C}_{S_i}^2$$

Here, for two vectors $\mathbf{x}, \hat{\mathbf{x}} \in \mathcal{R}^r$, we write $\mathbf{x} \leq_m \hat{\mathbf{x}}$, if $\sum_{l=1}^{k} x_{[l]} \leq \sum_{l=1}^{r} \hat{x}_{[l]}$, $k = 1, \ldots, r - 1$, and $\sum_{l=1}^{r} x_l = \sum_{l=1}^{r} \hat{x}_l$. In this case we say that $\hat{\mathbf{x}}$ *majorizes* $\mathbf{x}$. The two extreme values in this ordering are given by $(|\mathbf{x}|/r, \ldots, |\mathbf{x}|/r) \leq_m \mathbf{x} \leq_m (|\mathbf{x}|, 0, \ldots, 0)$. If equation 7.106 is satisfied for all $i = 1, \ldots, m$ we will say that the collection of job classes $\hat{J}$ is *more diversified in processing times* than the collection of job classes J. Because the mean flow time of a job in a single class job shop is increasing in $C_{S_i}^2$ (see Section 7.7) we have the following conclusion: The mean flow time of an arbitrary job in a job shop that uses the FCFS service protocol at all machine centers is increasing in the "processing time job diversity."

Next we will consider the effect of processing time job diversity on the mean flow time of an arbitrary class l job when the mean processing time of class l jobs are the same in both collections J and $\hat{J}$ (i.e., $E\left[S_i^{(l)}\right] = E\left[\hat{S}_i^{(l)}\right]$, $i = 1, \ldots, m$). Under job-class decomposition from equation 7.100 we see that

$$
C_{S_i}^2 = 2 \left(\frac{1 - \rho_i}{1 - \rho_i + \rho_i^{(l)}} \right) \frac{1}{\lambda_i^{(l)}} \times
$$

$$
\left[\frac{\lambda_i^{(l)} \left(1 + \tilde{C}_{S_i}^2 \right) \sum_{k=1}^{r} \lambda_i^{(k)} E \left[S_i^{(k)} \right]^2}{2 \lambda_i (1 - \rho_i)} - \left(\frac{\rho_i - \rho_i^{(l)}}{1 - \rho_i + \rho_i^{(l)}} \right) \frac{1}{\mu_i^{(l)}} \right] - 1 \qquad (7.107)
$$

for class l jobs in collection J and

$$
\hat{C}_{S_i}^2 = 2 \left(\frac{1 - \rho_i}{1 - \rho_i + \rho_i^{(l)}} \right) \frac{1}{\lambda_i^{(l)}} \times
$$

$$
\left[\frac{\lambda_i^{(l)} \left(1 + \tilde{C}_{Si}^2 \right) \sum_{k=1}^{r} \lambda_i^{(k)} E \left[\hat{S}_i^{(k)} \right]^2}{2 \lambda_i (1 - \rho_i)} - \left(\frac{\rho_i - \rho_i^{(l)}}{1 - \rho_i + \rho_i^{(l)}} \right) \frac{1}{\mu_i^{(l)}} \right] - 1 \qquad (7.108)
$$

for class l jobs in collection $\hat{J}$. Then as before we have the following conclusion: The mean flow time of an arbitrary class l job in a job shop that uses FCFS service protocol at all machine centers is increasing in the "processing time job diversity."

7.9 IMPLICATIONS OF MODELS

Capacity.   The capacity of the job shop and the identity of the bottleneck machine center are solely determined by the rate of job flows and the average work load at each machine center and are unaffected by diversity in processing time requirements and routing.

Impact of diversity in processing time and routing.   We see from the single-class general job shop model that the mean flow time of a job is minimized by a uniform-flow job routing if the arrival process is more variable than the service times and the service times are more regular than exponential random variables. That is, diversity of job routing degrades the performance of a job shop in this case. In the same case, symmetric job routing maximizes the mean flow time. Conversely if the arrival process is more regular than the service times and the service times are more variable than exponential random variables, then symmetric job routing minimizes the mean flow time. In this case we see that diversity in job routing improves the performance of a job shop. The multiple-class queueing network models demonstrate that the average number of jobs in the system and the mean flow time are increased as the processing time diversity of jobs handled by the job shop increases. Diversity in processing time requirements impairs performance of a job shop that uses FCFS service protocol.

Worker allocation.   It is observed that when the worker efficiencies are identical, the performance of the job shop can be improved by allocating equal work load among all machines. Conversely if the worker efficiencies are not identical, it is better to assign the most unreliable worker to the machine center with the least work load per machine.

Choice of local service protocol.   Models also demonstrate that if the job routing is random and the number of machine centers is large, then the flow time of an arbitrary job is minimized by the SPT service protocol among all local service protocols.

Impact of delays in material handling.   The transport times in a dedicated transport system do not affect the number of jobs at the machine centers or the flow time through the machine centers. Any increase in transport time just increases the job flow time through the system by the *same* amount. That is, there is a *linear* relationship between the job-flow time and the transport times in a dedicated transport system. In contrast the flow time in a job shop with a central transport system can exponentially increase as a function of the increase in transport time.

7.10 BIBLIOGRAPHICAL NOTE

A queueing network model of a dynamic job shop under the assumption of Poisson external arrival process, exponentially distributed processing times, and Markovian job transfers between machines was first developed by Jackson [13] [14]. The simplicity of the result derived by Jackson has raised the curiosity of several researchers. As a result several explanations and approaches such as quasi-reversibility, local balance, station balance, and so on have been devised by many researchers (e.g., see Kelly [35], Walrand [36]). In our presentation we have chosen to give a straightforward algebraic derivation that avoids the need to be familiar with any advanced topics such as quasi-reversibility. The approximate extension of this model to systems with general service times was first presented by Kuehn [20] for FCFS service protocol. Shanthikumar [26] and Shanthikumar and Buzacott [28] extended this approach to job shops with service time–dependent protocols such as SPT, SPTT-α and 2C-NP-α. This decomposition approach employed by Kuehn, and Shanthikumar and Buzacott was refined by Whitt [38], who in addition developed a commonly available software: QNA. Approximations for the second moment of the sojourn time were reported in Shanthikumar and Buzacott [29], and extension to approximate its distribution and application to due-date setting is described in Shanthikumar and Sumita [30]. Further refinement and extensions for multiple-class job-shop systems are reported in Bitran and Tirupati [3] [4]. The exact job-class decomposition of multiclass open Jackson queueing networks was part of Huang's [12] contribution. Karmarkar, Kekre, and Kekre [15], Kekre [16], and Kekre [17] have studied the effects of batch size on job performance. For a comprehensive review of job-shop modeling by open queueing networks, see Bitran and Dasu [2].

PROBLEMS

7.1 Consider a three-machine center job shop where four types of jobs are processed. The arrival rates $(\lambda^{(l)})$, number of operations (n_l), and the machine-center sequence $\left(C_1^{(l)}, \ldots, C_{n_l}^{(l)}\right)$ are given subsequently.

Job type (l)	$\lambda^{(l)}$	n_l	$C_1^{(l)}, \ldots, C_{n_l}^{(l)}$
1	0.2	4	(2, 3, 2, 1)
2	0.1	1	(3)
3	0.3	3	(1, 3, 1)
4	0.2	1	(2)

(a) Find the job arrival rate λ_i to machine center i, $i = 1, 2, 3$.

(b) Obtain a job-transfer probability matrix that will give the same job-arrival rate as in part a. Verify your answer by solving for the job-arrival rate induced by the job-transfer probability you derived.

7.2 Consider the example given in problem 1. Suppose the mean processing times for the kth operation of a type l job, $E\left[S_k^{(l)}\right]$, is given by

Job type l	Mean processing times $(E\left[S_k^{(l)}\right], k = 1, \ldots, n_l)$
1	(2, 1, 1, 2)
2	(2)
3	(0.5, 1, 0.5)
4	(2)

(a) With the current arrival rate is the system stable?

(b) Suppose the ratio of the job-type mix is to be maintained at 2:1:3:2. Find the capacity of this job shop, assuming each machine center has only one machine.

7.3 Consider the example of problems 1 and 2. Suppose the revenue per a type l job processed is $v^{(l)}$ given subsequently.

Job type l	1	2	3	4
Revenue $v^{(l)}$	4	2	3	1

(a) Neglecting any queueing effects formulate a linear programming model and find the optimal arrival rate of each type of job subject to the constraint that the job-shop capacity is not exceeded.

(b) Discuss the shortcomings of the preceding Linear Programming formulation for this job mix problem.

7.4 A new job shop is being designed. It is estimated that there are six types of jobs that will be handled by this job shop. Three different machine centers are to be formed to cover all the processing requirements of the six types of jobs. The details of the estimated job flow characteristics are shown subsequently.

Job type l	$\lambda^{(l)}$	n_l	$C_1^{(l)}, \ldots, C_{n_l}^{(l)}$	$E\left[S_1^{(l)}\right], \ldots, E\left[S_{n_l}^{(l)}\right]$
1	0.3	4	(1, 2, 1, 3)	(2, 1, 1, 1)
2	0.2	2	(2, 1)	(1, 4)
3	0.4	3	(2, 1, 3)	(3, 2, 2)
4	0.1	5	(1, 2, 1, 2, 3)	(1, 1, 2, 2, 1)
5	0.6	2	(1, 3)	(2, 2)
6	0.2	1	(3)	(4)

(a) Find the minimum number of machines needed in each of the three machine centers.

(b) Suppose we wish to keep the use of any machine below 90%. What is the minimum number of machines needed in each of the machine centers so that this constraint is satisfied?

7.5 An initial study of a job shop has resulted in aggregating the different types of jobs into three classes of jobs. The job flow of class l jobs between the three machine centers of the job shop is represented by the job-transfer probability matrix $\mathbf{P}^{(l)}$ given by

$$\mathbf{P}^{(1)} = \begin{bmatrix} 0 & 0.2 & 0.6 \\ 0.9 & 0 & 0.1 \\ 0 & 0 & 0 \end{bmatrix}$$

$$\mathbf{P}^{(2)} = \begin{bmatrix} 0 & 0.4 & 0.4 \\ 0.8 & 0 & 0.1 \\ 0 & 0.6 & 0 \end{bmatrix}$$

and

$$\mathbf{P}^{(3)} = \begin{bmatrix} 0 & 1 & 0 \\ 0 & 0 & 1 \\ 0 & 0 & 0 \end{bmatrix}$$

If the arrival rate of class l jobs is $\lambda^{(l)}$ given by

$$\lambda^{(1)} = 0.4; \quad \lambda^{(2)} = 0.2; \quad \lambda^{(3)} = 0.2$$

then
  (a) Find the aggregate flow of all three classes of jobs and represent it by a single–job-transfer probability matrix.
  (b) Find the capacity of this job shop if the mix of different job classes is to be kept fixed at 4:2:2.

7.6 Suppose the total processing time needed by an aggregate job in a job shop is W. This work W needs to be allocated among the m machine centers of the job shop. Let c_i be the number of machines available in machine center i, $i = 1, \ldots, m$. Using an open Jackson queueing network model (for this job shop) and assuming that the work load can be allocated to the different machine centers in any amount (without altering the processing time requirement) show that an optimal work-load allocation (w_i^*, $i = 1, \ldots, m$) that minimizes the mean flow time satisfies

$$\text{(a)} \qquad c_i = c_j \quad \Rightarrow \quad w_i^* = w_j^*$$

and

$$\text{(b)} \qquad c_i > c_j \quad \Rightarrow \quad \frac{w_i^*}{c_i} > \frac{w_j^*}{c_j}$$

7.7 A dynamic job shop consists of m machine centers. Machine center i has c_i machines. An aggregate class of jobs arrive at this job shop according to a Poisson process with rate λ. The processing time at machine center i is exponentially distributed with mean $1/\mu_i$, $i = 1, \ldots, m$. The expected number of visits made by an arbitrary job to machine center i is v_i, $i = 1, \ldots, m$.
  (a) Specify the condition under which this system will be stable.
  (b) Suppose a total of S operators are available, and they are to be allocated to these machine centers. Each operator can operate only one machine. Find the allocation (s_i^*, $i = 1, \ldots, m$) that will minimize the mean flow time and show that

$$\frac{v_i}{\mu_i} > \frac{v_j}{\mu_j} \quad \Rightarrow \quad s_i^* \geq s_j^*$$

7.8 Consider a job shop consisting of three machine centers. Each machine center has three machines of the same kind. Jobs are received at an input-output store (center 0), from which they are then transferred to the appropriate machine centers. The three machine centers are located separately in three separate buildings. Consequently the material handling delay from one machine center to another is high. The material handling delay within a machine center is negligible, however. The management is thinking of forming three cells, each set up within a building with one machine of each of the three kinds. The incoming jobs will then be equally divided among the three cells. Assuming that there are always dedicated transporters available between the input-output store and the three machine centers, formulate an open Jackson queueing network model to study the effects of management's idea. Under what condition on the transportation times will it be beneficial to form the three cells?

7.9 Consider the job shop discussed in Section 7.5. Suppose machine center i has c_i identical servers, $i = 1, \ldots, m$. Formulate a Markov process and find the steady-state joint probability distribution of the different classes of jobs at different machine centers. (Note that the analysis given in Section 7.5 needs to be modified, because there it is assumed that at each service completion only the job at the head of the queue leaves.)

7.10 Consider a job shop with a central transport service center (described in Section 7.6). Suppose the per unit time cost of a transporter is v, and the per unit time cost of waiting per job is w. Find the optimal number of transporters that will minimize the total transporter and job waiting cost per unit time.

7.11 Consider the job-shop system described in Section 7.6. Suppose a subset $\mathcal{P}'$ of paths (or links) between machine centers has dedicated transporters. All job transfers on other links are handled by the central transport service center.

(a) Find the mean flow time of an arbitrary job in this system.

(b) Suppose we wish to choose the subset $\mathcal{P}'$ of paths between machine centers that are to be provided with dedicated transporters subject to the constraint that $|\mathcal{P}'| = p$ (fixed). Assuming that the travel times are unaffected whether we use a central transporter or a dedicated transporter, find the set of links $\mathcal{P}'$ to be provided with dedicated transporters so that the mean flow time is minimized.

7.12 Consider the work-load allocation considered in problem 7. Suppose the squared coefficient of variation of the processing time at machine i is $C^2_{S_i}$ ($i = 1, \ldots, m$) and that it is sufficiently different from one so that the open Jackson queueing network model is not applicable. Assuming that $c_i = 1$, $i = 1, \ldots, m$, and that the M-arrival approximation is applicable here, find the optimal work-load allocation and show that

$$C^2_{S_i} > C^2_{S_j} \quad \Rightarrow \quad w^*_i < w^*_j$$

7.13 Using the flow-time approximation presented in Section 7.7.3 show that the flow time of an arbitrary job in a large job shop (i.e., m is large) with symmetric job transfers can be approximated by the exponential distribution. Use the due-date model given in Section 7.7.4 to find the optimal due date in a large job shop with symmetric job transfers.

7.14 Consider a large job shop with symmetric job transfers. Show that an M-arrival approximation is acceptable for such a system. Using this result, show that SPT service protocol will stochastically minimize the flow time of an arbitrary job among all local service protocols.

7.15 Find the optimal batch size for the example of a job shop with bulk job transfers described in Section 7.7.6 (equation 7.95). Show that the optimal batch size k^* is decreasing in the machine utilization and increasing in the transporter utilization.

BIBLIOGRAPHY

[1] F. BASKETT, K. M. CHANDY, R. R. MUNTZ, and PALACIOS. Open, closed and mixed networks of queues with different classes of customers. *J. ACM*, 22:248–260, 1975.

[2] G. R. BITRAN and S. DASU. A review of open queueing network models of manufacturing systems. *Queueing Systems: Theory and Applications* (in press).

[3] G. R. BITRAN and D. TIRUPATI. Multiproduct queueing networks with deterministic routing. *Management Science*, 34:75–100, 1988.

[4] G. R. BITRAN and D. TIRUPATI. Tradeoff curves, targeting and balancing in manufacturing queueing networks. *Operations Research*, 37:547–564, 1989.

[5] J. A. BUZACOTT and D. GUPTA. Impact of flexible machines on automated manufacturing systems. *Annals of Operations Research*, 15:169–205, 1988.

[6] J. A. BUZACOTT and J. G. SHANTHIKUMAR. On approximate queueing models of dynamic job shops. *Management Science*, 31:870–888, 1985.

[7] R. W. CONWAY, W. L. MAXWELL, and L. W. MILLER. *Theory of Scheduling*. Addison-Wesley, Reading, MA, 1967.

[8]   P. J. DENNING and J. P. BUZEN. The operational analysis of queueing network models. *Computing Surveys*, 10:225–261, 1978.

[9]   R. L. DISNEY. Random flow in queueing networks: a review and critique. *AIIE Trans.*, 7:268–288, 1975.

[10] C. C. GALLAGHER. The history of batch production and functional factory layout. *CME*, 73–76, April 1980.

[11] R. H. HOLLIER. Two studies of work flow control. *Int. J. Prod. Res.*, 3:253–283, 1964.

[12] P.-Y. HUANG. *Job class decomposition of multiclass queueing systems*. PhD thesis, School of Business Administration, University of California, 1987.

[13] J. R. JACKSON. Job–shop–like queueing systems. *Management Science*, 10:131–142, 1963.

[14] J. R. JACKSON. Networks of waiting lines. *Operations Research*, 5:518–521, 1957.

[15] U. KARMARKAR, S. KEKRE, and S. KEKRE. Lotsizing in multi-item multi-machine job shops. *IIE Trans.*, 17:290–297, 1985.

[16] S. KEKRE. *Management of job shops*. PhD thesis, Graduate School of Management, University of Rochester, 1984.

[17] S. KEKRE. *Some issues in job shop design*. PhD thesis, Graduate School of Management, University of Rochester, 1984.

[18] F. P. KELLY. *Reversibility and Stochastic Networks*. John Wiley and Sons, New York, 1979.

[19] P. C. KIESSLER and R. L. DISNEY. Further remarks on queueing network theory. *Eur. J. Opnl. Res.*, 36:285–296, 1988.

[20] P. J. KUEHN. Analysis of complex queueing networks by decomposition. In *Proceedings 8th ITC, Melbourne*, pages 236.1–236.8, 1976.

[21] P. J. KUEHN. Approximate analysis of general queueing networks by decomposition. *IEEE Trans. Commun.*, 27:113–126, 1979.

[22] J. C. LEIGH. A model for predicting the performance of a job shop. *Opl. Res. Q.*, 25:131–142, 1974.

[23] R. A. MARIE. An approximate analytical method for general queueing networks. *IEEE Trans. on Software Eng.*, SE-5:530–538, 1979.

[24] M. REISER and S. S. LAVENBERG. Mean-value analysis of closed multichain queueing networks. *J. ACM*, 27:313–322, 1980.

[25] P. J. SCHWEITZER. Maximum throughput in finite-capacity open queueing networks with product-form solutions. *Management Science*, 24:217–223, 1977.

[26] J. G. SHANTHIKUMAR. *Approximate queueing models of dynamic job shops*. PhD thesis, Department of Industrial Engineering, University of Toronto, 1979.

[27] J. G. SHANTHIKUMAR. Stochastic majorization of random variables with proportional equilibrium rates. *Adv. Appl. Prob.*, 19:1440–1452, 1987.

[28] J. G. SHANTHIKUMAR and J. A. BUZACOTT. Open queueing network models of dynamic job shops. *Int. J. Prod. Res.*, 19:255–266, 1981.

[29] J. G. SHANTHIKUMAR and J. A. BUZACOTT. The time spent in a dynamic job shop. *Eur. J. Opnl. Res.*, 17:215–226, 1984.

[30] J. G. SHANTHIKUMAR and U. SUMITA. Approximations for the time spent in a dynamic job shop with applications to due-date assignment. *Int. J. Prod. Res.*, 26:1329–1352, 1988.

[31] J. G. SHANTHIKUMAR and D. D. YAO. Optimal server allocation in a system of multi-server stations. *Management Science*, 33:1173–1180, 1987.

[32] J. SPRAGINS. Analytical queueing models. *Computer*, 4:9–11, 1980.

[33] H. J. STEUDEL and S. M. WU. A time series approach to queueing systems with applications for modeling job-shop in-process inventories. *Management Science*, 23:745–755, 1977.

[34] S. M. STEUDEL, H. J. PANDIT, and S. M. WU. A multiple time series approach to modelling the manufacturing job-shop as a network of queues. *Management Science*, 24:456–463, 1977.

[35] R. SURI. Robustness of queueing network formulae. *J. ACM*, 30:564–594, 1983.

[36] J. WALRAND. *An Introduction to Queueing Networks*. Prentice Hall, Englewood Cliffs, NJ, 1988.

[37] L. M. WEIN. Capacity allocation in generalized Jackson Networks. *Oper. Res. Letters*, 8:143–146, 1990.

[38] W. WHITT. The queueing network analyser. *Bell Syst. Tech. J.*, 62:2779–2815, 1983.

[39] D. D. YAO. Majorization and arrangement orderings in open networks of queues. *Annals of Operations Research*, 9:531–543, 1987.

[40] D. D. YAO and S. C. KIM. Reducing the congestion in a class of job shops. *Management Science*, 34:1165–1172, 1987.

8

Flexible Machining Systems

8.1 INTRODUCTION

8.1.1 Components of Flexible Machining Systems (FMS)

A *FMS* consists of several computer-controlled machines where machining operations such as milling, drilling, turning, or shaping are carried out, along with other work stations where related operations such as washing, inspection, or measurement are performed. The machines and work stations are linked by an automated material handling system that enables work pieces to move from any one machine or work station to any other machine or work station in the system, and it also enables work pieces to move to and from the loading/unloading station(s) and perhaps also to and from locations for storage of work in process. The system is under coordinated computer control, either one computer or several computers and programmed controllers that can exchange data and commands.

The number of machines can range from 2 to 30 or more. If the number of machines is 2 to 4 then the facility would usually be called a *flexible machining cell*. Often material handling would then be performed by a single robot that can reach all machines. Some writers define a *flexible machining module* as a single computer-controlled machine equipped with automatic means of loading and unloading work pieces. For modeling purposes, however, such a module is indistinguishable from the single-stage manufacturing systems considered in Chapters 3 and 4. In this chapter we will consider FMSs where work pieces may be required to visit any of the machines in the system in any sequence. Sometimes large systems may consist of more or less independent subsystems or cells where a given work piece typically only visits one of the cells, or, alternatively, all work pieces visit the same sequence of cells. Such systems will be considered in Chapter 10.

In a FMS all machines and work stations except perhaps for the load/unload work station(s) would be computer controlled. Machines are provided with tool storage and automatic tool changing capability so that tools can be swapped between the machine and its local tool store or carousel. Because this carousel has limited capacity and enlarging it increases the tool changing time, some systems also have a more or less automated tool management system that enables either tool carousels or individual tools on it to be exchanged with tools kept at a central location. Nevertheless, because of the limited capacity of the tool carousel and the difficulty of exchanging tools between the carousel and the central tool store, many FMSs are such that although several machines are functionally identical, different assignments of tools to functionally identical machines makes these machines dissimilar in their capability to perform operations on work pieces; thus a given work piece may have to visit more than one of these "identical" machines to have required operations performed.

A variety of material handling systems are used. Some systems use AGVs, others use rail mounted carts, and still others use roller conveyors typically arranged in a loop layout. Early systems used rail mounted shuttle carts, but AGVs are now more commonly used. Work pieces are usually mounted on pallets and secured by fixtures in a specific orientation. To perform machining operations on all required surfaces it may be necessary to send a work piece to a work station for reorientation and refixturing, but as far as possible it is desirable only to have to fixture a work piece once. Because the pallets are used to create precise location of work pieces at machines, they have to be made to close tolerances and thus are expensive. Further, different work pieces may require different pallets.

Machines are provided with pallet changers so that a waiting pallet and a pallet where the work piece has completed processing by the machine can be exchanged automatically. Other than the waiting pallet or the just completed pallet, however, it is not usual to provide local storage at machines. In AGV systems the AGV can hold a pallet (and thus act as a storage location), whereas in other systems a few pallet storage locations accessible by the work transporter system may be provided. Empty or full pallets can also wait at the load/unload station.

For modeling purposes the key components of a flexible machining system are as follows:

- *Machines or work stations*, where only if the tool assignment is such that a group of machines is capable of performing the same operations on all types of jobs processed by the system can the group be considered as forming a parallel group of machines.
- *Material handling system*, where it is necessary to distinguish between unit load systems (such as AGVs and shuttle cars) that have a few material handling devices and hence may cause delays while jobs wait for a material handling device to

become available, and systems such as roller conveyors in which jobs will not be delayed by lack of material handling capability.

- *Central and local storage*, where local storage will be at the input or output of individual machines or machine groups, whereas central storage can be used by jobs irrespective of the next machine they are to visit.

- *Pallets and fixtures*, where it is possible that there may be several different types of pallets or fixtures with pallets of a given type only being usable for a specific subset of jobs.

- *Information and control system*, where the system has to acquire information on the status of jobs, machines, and material handling facilities, and control the processing of jobs and their movement into and through the system.

8.1.2 Evolution

The development of the concept of FMSs occurred during the 1960s. At that time it appeared to be a natural evolution of numerically controlled (NC) machine tools. These had developed in the late 1940s using the concepts of automatic feedback control and servomechanisms developed during World War II. The NC machines use digital control to guide precisely the path of a tool during a metal cutting operation, thus eliminating the need to create complex cams and fixtures. Then in the late 1950s NC machines began to incorporate the idea of automatic tool changing and tool selection from a tool magazine at the machine so that several different machining tasks could be carried out on a work piece while it was at the machine. The machine, at least for nonrotational parts, became a machining center that could perform milling, shaping, or drilling operations. Shortly after, the machining center was provided with automatic pallet changers so that work could be taken in and out of the machine automatically with the fixturing of the work piece on the pallet being carried out at another work station. Further, in the same period computer numerical control (CNC) and direct numerical control (DNC) systems became common, eventually meaning that each machine could be provided with its own computer or microprocessor to control tool paths, tool changing, pallet input and output, and essential monitoring functions (such as detecting tool breakage). The machine's computer was linked to a central computer and thus could exchange information on job progress, and receive programs and data for each new job as it arrived at the machine.

When there was a family of related parts produced in medium total volume and requiring fairly complex machining operations, stand-alone machining centers had short-comings. Many of the parts in the family might require similar operations using the same tools, making it reasonable to perform these operations at the same machine. The machine might not have enough capacity, however, either in available hours or in terms of the size of its tool magazine. Thus, to process the family, jobs would have to visit several machines. Some of these machines could be functionally identical (e.g., all horizontal spindle machining centers) but assigned different tools. Also, certain operations required by the family might be better carried out at specialized machines (e.g., multiple-spindle drilling machines). To process the family efficiently and move parts between machines, the machines were linked by an automated material handling system, thus creating a flexible machining system. The first such systems were installed in the late 1960s (see Talavage and Hannan [95]). A key feature is that the FMS can process a variety of related parts that can usually share tooling, fixtures, and pallets. It was the ability to manufacture simultaneously a mix of parts in medium volume that made the system preferable to dedicated transfer lines.

Another concept that motivated some of the early developments of FMSs was the ideal of unmanned manufacturing, that is, the system would be manned for one shift and left unmanned for the other one or two shifts. All set-up and work preparation such as mounting parts on pallets would be done on one shift while the machines could continue their operation the rest of the time. This goal appears to have been somewhat elusive because of the difficulty of achieving sufficient reliability, however, and there are very few systems operated in this way at present.

In almost all systems, until very recently, tool delivery has not been automated, and thus tools have had to be assigned to machines in a batch mode. The system has had to be set up to make a given part mix by assigning specific tools to specific locations in the machine tool magazines, and this assignment could not be changed easily without shutting down the system. Recently automated tool management systems have been developed that enable the tool assignments to machines to be changed for every job even while a job is in process at a machine (see Graver and McGinnis [21]). If the processing times of jobs at a machine are significantly longer than the time to change tools then this capability has the potential to make all functionally identical machines, such as all horizontal machining centers, behave as if they were identical parallel machines. The consequent reduction in the amount of job movement probably makes the system behave more like a system of loosely linked cells whose modeling issues will be discussed in Chapter 10. Our interest in this chapter is in modeling "traditional" flexible machining systems where all machines are tightly linked by the material handling system and the different jobs processed by the system may have to visit different machines to have all their required operations performed.

8.1.3 Flexibility

There has been a lot of discussion of the meaning of "flexible" in the context of FMSs (see, e.g., Browne et al. [4] and Sethi and Sethi [64]). Essentially there are two key respects in which these systems are flexible.

Routing. In contrast to transfer lines, the material handling system of FMSs permits work pieces to visit machines in any sequence and imposes no constraint on the number of visits of work pieces to machines.

Processing. Numerical control of tool paths and automatic tool changing enable a wide variety of different operations to be carried out on a work piece at a machine. This flexibility may be limited by the size of the tool magazine and hence by the set of tools available at the machine. Also some machine types are inherently more flexible than others (e.g., horizontal spindle machining centers can access four sides of a prismatic work piece, whereas vertical spindle machining centers can usually access only one side).

These two flexibilities mean that a variety of other benefits can be achieved. For example, if a machine fails it may be possible to divert work to another machine. Sometimes this can be done because the machines share common tooling; in other cases if the failure is major then the tools can be transferred to another machine thus making it functionally identical to the failed machine. Also, if minor changes in the design of a part are made, the numerical control programs can be modified rapidly and easily.

Because of the high investment in material handling and computer control, FMSs are not particularly flexible with respect to production volume. If required production volume drops off so that the system becomes underutilized the economics deteriorate because of the high fixed costs. In such a case it is usual to try and find other parts that can be produced on the system so that utilization can be improved. If this cannot be done, the system may then be dismantled. The individual machines still have value, however, in contrast to a transfer line where the individual stations cannot normally be used on their own (see Jablonski [31] for examples of these ideas at Avco-Lycoming, International Harvester/Detroit Diesel, and Massey-Ferguson).

Further, "traditional" FMSs are also not particularly flexible with respect to changes in the set of parts that are to be produced on the system. Each new part may require new pallets and fixtures that may require a substantial investment and also may require time to procure. Also the NC programs have to be tested to ensure that the operations can be performed correctly and reliably. Most U.S. FMSs have as a result a relatively low product mix (5–30 different parts) (see Jaikumar [34]).

8.2 DESIGN, PLANNING AND SCHEDULING OF FMS

We will first outline the major decisions involved in design, planning, and scheduling, and then comment on the issues that can be addressed using stochastic models.

8.2.1 Design

In designing a FMS there are several key decisions that have to be made. These decisions are interrelated, thus the order in which we present them does not mean that the designer can go from one to the next in sequence. The decisions, however, do break down into what have been called *initial specification decisions* and *subsequent implementation decisions* (see Stecke [83]).

Initial specification decisions

1. *Selection of parts to be produced on system.* The parts should have similar dimensions, or at least fall within a similar envelope, be made from the same material and have similar processing requirements so that they can share common tooling, pallets, and fixtures. The economics of the system are usually improved if it is anticipated that the volumes of each part selected will be such that the relative part mix is reasonably constant. This is of course likely to be the case if all parts are required by the same end product.

2. *Identification of machine types required.* The required operations on each part have to be identified as well as the various options for each operation (and also their sequence). As mentioned earlier it is desirable to choose machine types that offer the greatest flexibility in processing unless specialized machines give significant productivity and cost advantages.

3. *Determination of number of each machine type needed.* Sometimes this number can be reduced if use is made of the opportunity to process a given part in more than one way. Also this will be influenced by the tool magazine capacities.

4. *Choice of material handling system.* As mentioned earlier the choice is usually among shuttle car systems, AGVs, and roller conveyors. The choice will be influenced by the size and weight of the parts, the volume of material handling, and also the relative times of typical move and processing tasks.

5. *Tool management system.* This can be automated to permit rapid exchange of tool magazines or individual tools at the machine, or it can be slow relative to the processing times of parts.

6. *Provision of central storage.* Most systems provide some central storage within the system for work in process, so its location and capacity have to be determined. Some local storage at machines may be provided but usually this is limited to one or two jobs so it is not normally a major design issue.

7. *Computers and control structure.* The structure of the control hierarchy has to be decided, along with the type of computers required to implement the control software. Typically there are local computers at machines and on some material handling components such as AGVs along with a central control computer for the whole system.

Subsequent implementation decisions

1. *Layout of system.* This requires locating machines, storage, and material handling paths to provide adequate access to all equipment for maintenance and to minimize the time to move jobs between machines, avoiding bottlenecks and other sources of delays.

2. *Number of pallets and fixtures.* It is necessary to decide how many pallets of each type, along with the number of fixtures, are required. The total number of pallets determines the maximum number of jobs in the system, whereas the number of pallets of each type can influence the relative throughput of different job types. Similar considerations apply for the number of fixtures.

3. *Number of AGVs or shuttle cars.* In unit-load material handling systems it is necessary to decide on the number of material handling vehicles or cars. The same considerations arise as were discussed in the context of flow lines with closed-loop material handling systems: Too few vehicles will result in machines being starved; too many vehicles can result in delays owing to blocking.

4. *Specification of planning, scheduling, and control software.* It is necessary to develop clear functional specifications of the software needed to run the system. Further, it is necessary to ensure that the software is designed so that it can easily be modified as experience in running the system is accumulated or changes are made in the system.

5. *Produce-to-stock versus produce-to-order.* Like any other manufacturing system it is necessary to decide whether the jobs to be made on the system will be determined directly by customer orders (where a customer could be a downstream assembly plant) or by stock levels of the different parts downstream of the system. Because of the need to maintain high utilization of the system, it would be expected that a mixture of the two would be used (i.e., any outstanding orders would be produced, and any unused capacity would be filled up by making to stock).

8.2.2 Planning

Planning decisions are those decisions that must be made before jobs are released to the system. The context of these decisions is influenced by the system design. In particular we assume that the system does not have a tool management system that permits tools to be changed while jobs are in process. Thus the key planning decisions are as follows:

1. *Choice of set of part types for simultaneous manufacture.* That is, the system is to be set up to make the parts in this set, which may be smaller than the set of all parts that the system could possibly make. This choice will be based on considering either due dates or stock levels combined with considerations of how frequently the system should be changed over from one tool assignment to another.

2. *Decision on production targets.* Once the set of part types has been decided then production targets can be set for each type for the period until the next system is set up.

3. *Assignment of tools to machines.* If machines are identically tooled then the system is more flexible, however, the limited number of tools that can be held in the tool magazines means that this will often not be the case. Tools have to be assigned to meet the production targets and ensure a reasonably balanced work load on the machines.

4. *Allocation of pallets and fixtures.* Even though in some systems pallets and fixtures could be used for a variety of part types, sometimes pallets and fixtures are allocated to specific part types to ensure that all the part types are made simultaneously. It is not clear, however, that this is necessary to achieve the production targets.

5. *Decisions on work release strategy.* If work release is not determined by the allocation of parts and fixtures then it is necessary to have rules for deciding which job should be released next. This could be based on due dates, inventory levels, unmet targets, or such attributes as the total processing time or number of machines the job has to visit.

8.2.3 Scheduling

Often the scheduling of jobs once they are released to the system is relatively straightforward because the limited work in process means that there are not many options. The main tasks are as follows:

1. *Material handling vehicle scheduling.* If more than one job requires transport then it is necessary to decide which job should be moved first. This might depend on the closeness of the material handling vehicle to the job locations, the transport time, or the job priorities.

2. *Job sequencing.* When a machine becomes free it is necessary to choose which job should be processed first on the machine. This may depend on the job priority, the duration of the operation, or the current location of the job.

3. *Refixturing.* Sometimes a job may have to be refixtured to complete all operations. The refixturing may be a manual operation so it is necessary to provide information to the worker on which fixture to use, and it may also be advisable to inform him which job should be refixtured next when there are several jobs waiting.

8.2.4 Issues

The major issues for which stochastic models are required are as follows:

1. *Capacity evaluation.* Because of the mix of jobs being simultaneously produced in a FMS and the random processing times (along with the possibility of tool or machine failure) stochastic models are required to determine the capacity. Capacity will be influenced by such design decisions as the type of material handling system, the number of pallets and fixtures, the number of material handling vehicles, and the amount of storage, as well as by planning decisions such as the tool assignments to machines. Further, capacity will be influenced by the mix of jobs, and the work-release and scheduling rules.

2. *Design optimization.* Once appropriate models for capacity evaluation have been developed then they can be used to optimize the system design and determine such quantities as the optimal number of machines of a given type and the optimal number of pallets. They can also be used to decide whether a new part type should be added to the system.

3. *Tool allocation and machine grouping.* The ability to allocate tools to machines in different ways, thereby creating varying degrees of pooling of machine capability, is one of the distinctive features of traditional FMSs. Stochastic models can be used to gain insight into how this capability should be used to maximize performance.

4. *Release and scheduling.* Stochastic models can also be used to identify good rules for the timing of work release, deciding between alternative operation sequences, and arriving at guidelines for detailed scheduling and control of work flow.

8.3 MODELING FMS BY QUEUEING NETWORKS

FMSs, in abstract, are a set of processors such as NC machines and inspection stations, interconnected by an automated material handling system. A given set of different types of parts is to be processed through this system. Because of the flexibility of the system, the set of parts that *could* be processed through the FMS is usually much larger than the given set of parts. Each part type requires a certain set of operations: These operations may have precedence requirements resulting in a partial or total order of processing. Hence there may be several different sequences of operations that a part type could follow. The routing flexibility of a FMS enables this flexibility in operation sequence to be exploited. Each operation requires a processor and in addition may require a set of tools, programs, and special fixtures. There may be several identical processors, but whether an operation can be performed on any one of these processors will depend on whether the tools and programs needed for that operation are available (i.e., assigned or loaded) at that processor. Therefore, once these assignments are made, the operation sequence of any part type can be converted into one or many sequences of processors that the part type needs to follow. It is possible to reassign the tools and program even when there are parts being processed in the system. For example, such a reassignment may be carried out because of a processor failure, a change in part mix, or because of the introduction of a new part type.

Queueing models of FMSs are often formulated to capture the aggregate behavior of material flow through the system. The FMS that we model here can be an entire system or a part of a larger manufacturing system. The influence of the rest of the actual system will be represented by appropriate material inflow and demands for output. In abstract we view the system to be a set $M = \{1, 2, \ldots, m\}$ of m service centers with c_i identical processors at each service center. There is also a set $H = \{1, 2, \ldots, h\}$ of h material handling devices that are used to transport parts from one service center to another. Note that a service center with more than one processor is formed only if all processors in that service center are assigned identical tools and programs. Therefore if only some of the tools and programs assigned to two processors are identical, we will not group them into the same service center.

The set of part types to be processed by the system is $R = \{1, 2, \ldots, r\}$. Let $A_i(t)$ be the number of parts of type i that arrived at the system during $(0, t]$. The nature of the input process $\{A_i(t), t \geq 0\}$ is determined by the demand pattern and by the type of input control policies exercised. As we will see later, depending on the input control policies we may choose to formulate either open or closed queueing network models for the FMS.

For each part type there is a set of service center sequences corresponding to the sequences of processors that the part type needs to visit. Each part of type j, if processed according to service center sequence s requires a total of $v_j(s)$ operations at service centers $C_1^{(j)}(s), \ldots, C_{v_j(s)}^{(j)}(s)$ in that order (i.e., $s = \langle C_1^{(j)}(s), \ldots, C_{v_j(s)}^{(j)}(s) \rangle$). Let $S_l^{(j)}(s)$ be the (generic) processing time of the lth operation of a type j part when processed according to the service center sequence s. The (generic) time needed to transport a part from service center i to service center j is T_{ij}. These service and transport times can be deterministic or random.

A detailed model of the nature described can be used in the development of input control and operational control (such as part type selection, assignment of tools and programs, routing of parts to service centers, and scheduling). In its entirety, however, this model is too detailed to be used in strategic and medium range production planning. Specifically, when one needs to evaluate the production capacity (for capacity planning) or study the effect of different system configurations (for facility design) on system performance measures, it is adequate that we further aggregate the preceding model for material flow.

The aggregation procedure for material flow with part routing flexibility involves two steps.

Step 1: Predefined splitting. Earlier we pointed out that a part type can have several different service center sequences (owing to operation or machine flexibility). This flexibility induces a control problem and in reality the choice of service center sequence that any specific part of a part type will take depends on the state of the system. In this section, however, we will assume that the fraction of parts (of a part type that has several service center sequences) that are to follow each of its possible service center sequences is predefined. Then we can split that part type into several new part types, each with a single service center sequence. The input processes of these artificially created part types can be obtained by appropriate splitting of the input processes $\{A_i(t), t \geq 0\}$, $i = 1, \ldots, r$. The queueing models can indeed be used to obtain optimal values for these fractions so that some performance measure of interest such as the production capacity is optimized. These "optimal" fractions can then be used as a guideline in the optimal dynamic sequence selection of parts. Therefore in our aggregation model we will without loss of generality assume that each of the part types has a fixed service center sequence. Because we have restricted certain types of control owing to predefined splitting, the aggregate model can be expected to provide pessimistic estimates of the system performance.

Step 2: Aggregation. With predefined splitting the internal part flow in the system is modeled in exactly the same way as we modeled the job flow inside the job shop. To complete the aggregation procedure, we need to look at the nature of the part input processes $\{A_i(t), t \geq 0\}$, $i = 1, \ldots, r$. As we will soon see, the nature of the input process (and control) not only dictates the type of queueing network models (i.e, open or closed) to use but also determines the level of aggregation (i.e., single versus multiple classes) that we may use. At one extreme we may have input processes that are totally independent of the inside dynamics of the FMS. That is, the input processes of parts are similar in nature to that of job flows into a job shop. Then we may use probabilistic aggregation and develop single– or multiple–job-class open queueing network models as we did in Chapter 7 for job shops. Specifically the open queueing network model may be used to represent an FMS if the following conditions are satisfied:

- Free inflow of raw parts, that is, the system at hand is a job-shop–like FMS where outflow from a preceding stage, inflow of raw material, and the receipt, processing, and dispatch of an order do not use any information on the status of the parts and processors within the system.
- Sufficient number of pallets (of each kind if needed) are available to load all the incoming parts.
- Each part type has a fixed service center sequence (after the tools and programs assignment), or we carry out "predefined splitting."

The other extreme case in comparison with the free inflow process is where input is triggered only when a part departs the system. Here, whenever a part completes its processing through the system, the part is removed from the pallet, and a raw part is loaded onto that pallet and immediately fed back into the system. The type of part fed back of course will depend on the part type that completed service, the pallet type required by the parts, and the part input control policy. In a simplistic setting, if each part type needs its own type of pallet, one would then replace a departing part type by the same part type. Alternatively, if all the part types require the same type of pallet, one may then replace a departing part type by any type of part. For the purpose of modeling we will assume that such a replacement will be carried out (1) to maintain a fixed ratio of production of different part types, (2) to maintain a fixed number of different types of parts in the system, or (3) to maintain a fixed ratio of production of the different part types within each class when the set of part types R is partitioned into p classes of part types $\{R_1, \ldots, R_p\}$ while the total number of each class of part types in the system is kept a constant. It should be noted that even when the total numbers of each part type or each class of part types is fixed, these numbers should be selected such that the desired production ratio of the different part types is achieved.

First, consider the scenario in which the production ratios $D_1 : \cdots : D_r$ of the r types of parts are prespecified. In addition, all these part types can be loaded onto the same type of pallets. The total number of pallets available (or allowed) in the system is n. Because the ratio at which different part types flow through the system is known, we may use it to obtain the flow pattern of an aggregate part that will produce the same average work load and flows as in the original system. For this, note that in the aggregation procedure used in the job shop we needed only the relative ratios of the input rates of different types of jobs rather than their actual rates. The flow pattern of the aggregate part type is represented by a transfer probability matrix and the probability vector of the service center needed for the first operation. Let $\mathbf{Q}$ and γ be this matrix and vector, obtained by the aggregation procedure described for the job shop with D_i replacing $\lambda(i)$, $i = 1, \ldots, r$ (see Section 7.3.2). Unlike in the job-shop case, however, we cannot model this system by an open queueing network because of the type of input process. Here the input process seeks to maintain the number of aggregate parts at a constant value of n. Each departing aggregate part is replaced by a new (raw) aggregate part. Hence we model this by a single-class closed queueing network with a total of n parts in it. The parts are routed among service centers according to a (stochastic) transfer probability matrix $\mathbf{P} = (p_{ij})_{i,j=0,1,...,m}$, where

$$p_{ij} = q_{ij}, \qquad i, j = 1, \ldots, m \qquad\qquad (8.1)$$

$$p_{0j} = \gamma_j, \qquad j = 1, \ldots, m$$

$$p_{i0} = 1 - \sum_{j=1}^{m} q_{ij}, \qquad i = 1, \ldots, m$$

and service center 0 acts as the load/unload server. If required, we may add the model of the material handling system to this model as we did before in Section 7.6 for the job shop model. It should be noted that the expected number of times service center i visited by an aggregated part v_i, $i = 1, \ldots, m$ is given by

$$v_i = \gamma_i + \sum_{j=1}^{m} v_j p_{ji}, \qquad i = 1, \ldots, m; \qquad v_0 = 1 \qquad (8.2)$$

The second scenario we consider is the one in which the number of pallets and parts of type i in the system is maintained at a constant level of n_i, $i = 1, \ldots, r$. It is then immediately clear that we can use a multiple class closed queueing network (with r classes of customers and a population size of n_i for class i customers, $i = 1, \ldots, r$).

The third scenario we consider is a combination of the first and the second scenarios. Let $\{R_1, \ldots, R_p\}$ be a partition of R. A type j pallet can hold any one of the part types in R_j, $j = 1, \ldots, p$. The part types in R_j are to be produced according to the ratio $D_1^{(j)} : \cdots : D_{|R_j|}^{(j)}$, $j = 1, \ldots, p$. The total number of class j parts (i.e., the type of parts in R_j) in the system is to be kept constant at a level of n_j, $j = 1, \ldots, p$. Part types belonging to each class can be aggregated in the same way as we did for the first case. Let $\mathbf{P}^{(j)}$ be the resulting transfer probability matrix incorporating the load/unload service center. The model we use for this third scenario is then a multiple-class closed queueing network with p customer classes and a population size n_j for class j (aggregate) parts. Class j customers are moved from one service center to another according to a transfer probability matrix $\mathbf{P}^{(j)}$, $j = 1, \ldots, p$.

Closed queueing network models of these three scenarios are considered in Sections 8.4, 8.6.1, and 8.6.2, respectively.

8.4 SINGLE-CLASS CLOSED JACKSON QUEUEING NETWORK MODEL

Suppose n parts of a single (probably aggregated) class of parts circulate from one service center to another according to a transfer probability matrix $\mathbf{P} = (p_{ij})_{i,j=0,...,m}$. We will assume that all self-transitions have been eliminated so that $p_{ii} = 0$, $i = 1, \ldots, m$. The service requirements of parts at service center i are iid exponential random variables with mean $1/\mu_i$, $i = 0, \ldots, m$. The sequence of service requirements at the different service centers are mutually independent. The rate at which unit service requirement of a part is processed at a machine center i when there are k parts is assumed to be $r_i(k)$, $k = 1, \ldots, n$; $i = 0, \ldots, m$. This allows us to represent, as special cases, single or multiple servers in parallel at each service center. The parts at each service center are served according to a FCFS service protocol.

Let $N_i(t)$ be the number of parts at service center i at time t, $i = 0, 1, \ldots, m$ and let $\mathbf{N}(t) = (N_0(t), \ldots, N_m(t))$. Then $\{\mathbf{N}(t), t \geq 0\}$ is a continuous time Markov process on the state space $\mathcal{S}_n$, where

$$\mathcal{S}_l = \{\mathbf{k} : \mathbf{k} \in \mathcal{N}_+^{m+1}, |\mathbf{k}| = l\}, \qquad l = 1, 2, \ldots$$

$$\mathcal{S}_0 = \{0, \ldots, 0\} \tag{8.3}$$

Define the stationary distribution of $\mathbf{N}$ by $p(\mathbf{k}) = \lim_{t \to \infty} P\{\mathbf{N}(t) = \mathbf{k}\}$, $\mathbf{k} \in \mathcal{S}_n$. Equating the rate of probability inflow and outflow of state $\mathbf{k}$, one gets

$$\sum_{j=0}^{m} \sum_{i=0}^{m} \mu_j r_j(k_j + 1) p_{ji} \, p(\mathbf{k} + \mathbf{e}_j - \mathbf{e}_i) = \sum_{i=0}^{m} \mu_i r_i(k_i) p(\mathbf{k}), \qquad \mathbf{k} \in \mathcal{S}_n \tag{8.4}$$

Equation 8.4 along with the normalizing equation $\sum_{\mathbf{k} \in \mathcal{S}_n} p(\mathbf{k}) = 1$ can be used to obtain p. As with the open Jackson queueing network model, we will try to see whether a product form solution of the form $p(\mathbf{k}) = K \prod_{i=0}^{m} p_i(k_i)$ will lead to a consistent solution for equation 8.4. Substituting this product form in equation 8.4 and dividing by $p(\mathbf{k})$ we get

$$\sum_{j=0}^{m} \sum_{i=0}^{m} \mu_j r_j(k_j + 1) p_{ji} \frac{p_j(k_j + 1)}{p_j(k_j)} \frac{p_i(k_i - 1)}{p_i(k_i)} = \sum_{i=0}^{m} \mu_i r_i(k_i), \quad \mathbf{k} \in \mathcal{S}_n \tag{8.5}$$

Equating term by term in equation 8.5 for each i we get

$$p_i(k_i - 1) \sum_{j=0}^{m} \mu_j r_j(k_j + 1) \frac{p_j(k_j + 1)}{p_j(k_j)} p_{ji} = \mu_i r_i(k_i) p_i(k_i), \quad i = 1, \ldots, m; \mathbf{k} \in \mathcal{S}_n$$

$$\tag{8.6}$$

Therefore, if equation 8.6 has a consistent solution for p_i, $i = 0, 1, \ldots, m$, it is also a solution to equation 8.5. For this to be true the term

$$\sum_{j=0}^{m} \mu_j r_j(k_j + 1) \frac{p_j(k_j + 1)}{p_j(k_j)} p_{ji}$$

should be independent of k_j. So, if we set the preceding term equal to $\hat{v}_i$, from equation 8.6 we get

$$p_i(k_i - 1)\hat{v}_i = \mu_i r_i(k_i) p_i(k_i), \quad k_i = 1, \ldots, n; \quad i = 0, \ldots, m \qquad (8.7)$$

Substituting equation 8.7 in equation 8.6 we obtain

$$p_i(k_i - 1) \sum_{j=0}^{m} \hat{v}_j p_{ji} = \mu_i r_i(k_i) p_i(k_i), \quad i = 0, \ldots, m \qquad (8.8)$$

Because v_i, $i = 0, \ldots, m$ with $v_0 = 1$ defined earlier is the solution to $\hat{v}_i = \sum_{j=0}^{m} \hat{v}_j p_{ji}$, $i = 0, \ldots, m$, setting $\hat{v}_i = v_i$ we see that the solution to

$$p_i(k_i - 1)v_i = \mu_i r_i(k_i) p_i(k_i), \quad k_i = 1, \ldots, n \qquad (8.9)$$

is consistent with equations 8.7 and 8.8 and therefore $p(\mathbf{k}) = K \prod_{i=0}^{m} p_i(k_i)$, $\mathbf{k} \in \mathcal{S}_n$ is a solution to (8.4). Because

$$p_i(k_i) = \frac{v_i^{k_i}}{\prod_{j=1}^{k_i} \mu_i r_i(j)}, \quad k_i = 0, \ldots, n; \quad i = 0, \ldots, m \qquad (8.10)$$

satisfies equation 8.7 we see that

$$p(\mathbf{k}) = K \prod_{i=0}^{m} \frac{v_i^{k_i}}{\prod_{j=1}^{k_i} \mu_i r_i(j)}, \quad \mathbf{k} \in \mathcal{S}_n \qquad (8.11)$$

where

$$K = \frac{1}{\sum_{\mathbf{k} \in \mathcal{S}_n} \prod_{i=0}^{m} \frac{v_i^{k_i}}{\prod_{j=1}^{k_i} \mu_i r_i(j)}} \qquad (8.12)$$

Let Y_i be a generic random variable representing the stationary distribution of the number of parts in a $M/M(n)/1$ queueing system with arrival rate $\lambda_i = \lambda v_i$ and state dependent service rate $\mu_i r_i(k_i)$ when there are k_i parts in the system ($k_i = 1, 2, \ldots$). λ can be any value that guarantees the existence of a stationary distribution for the $M/M(n)/1$ queue. For example, if $r_i(k_i) = \min\{k_i, c_i\}$, that is, we have c_i parallel servers at service center i, then we require that $\lambda_i < \mu_i c_i$ or equivalently $\lambda < \mu_i c_i / v_i$. The level crossing rate balance equation for the stationary distribution $P\{Y_i = k_i\}$, $k_i = 0, 1, \ldots$, is $\lambda v_i P\{Y_i = k_i\} = \mu_i r_i(k_i + 1) P\{Y_i = k_i + 1\}$, $k_i = 0, 1, \ldots$ (see Section 3.4.1). Therefore

$$P\{Y_i = k_i\} = \frac{\lambda^{k_i} v_i^{k_i}}{\prod_{j=1}^{k_i} \mu_i r_i(j)} P\{Y_i = 0\}, \quad k_i = 0, 1, 2, \ldots \qquad (8.13)$$

From equation 8.13 and assuming independence of $Y_0, \ldots, Y_m$, one obtains,

$$P\{\mathbf{N} = \mathbf{k}\} = P\{\mathbf{Y} = \mathbf{k} \mid |\mathbf{Y}| = n\}, \quad \mathbf{k} \in \mathcal{S}_n \tag{8.14}$$

It is also clear from the results in Section 7.4 that the distribution of $\mathbf{Y}$ is the stationary distribution of an open Jackson queueing network with a set of service centers $\{0, \ldots, m\}$, and an arbitrary part visiting service center i on the average v_i number of times before it leaves the system $(i = 1, \ldots, m)$ and the load/unload center twice $(= 2v_0)$. The service rate at service center i is $\mu_i r_i(k_i)$ when there are k_i parts for $i = 1, \ldots, m$ and $2\mu_0 r_0(k_0)$ when there are k_0 parts at the load/unload service center. Note that the average load or unload times are $1/(2\mu_0)$, whereas the combined load/unload operation (incorporated as a single operation) in the closed queueing network model requires an average of $1/\mu_0$ units of time. The external part arrival rate is λ. Note that this open Jackson network could be our model for the FMS if we had free inflow of raw parts into the system at rate λ. Therefore the distribution of the number of parts in the closed queueing network model is the same as that in an open queueing network model provided we observe the open queueing network only when there are a total of n parts in it. Hence the probability distribution of the number of parts in an open queueing network model can be used to compute the probability distribution of the number of parts in the closed queueing network. To do this we need to compute the convolution of the probability distributions $P\{Y_i = k_i\} = P\{Y_i = 0\}(\lambda v_i/\mu_i)^{k_i} / \prod_{j=1}^{k_i} r_i(j), i = 0, \ldots, m$. The following algorithm will compute this convolution and the stationary distribution of the number of parts in the closed queueing network.

Algorithm 8.1   Convolution Algorithm

Step 1: Set $p_i(0) = 1$, $i = 0, \ldots, m$.

Step 2: For $i = 0, \ldots, m$
  For $k = 0, \ldots, n - 1$, set $p_i(k + 1) = p_i(k)v_i/(\mu_i r_i(k + 1))$.

Step 3: Set $\hat{p}(k) = p_0(k)$, $k = 0, \ldots, n$.
  For $i = 1, \ldots, m$
  For $k = 0, \ldots, n$, set $\hat{q}(k) = \sum_{l=0}^{k} \hat{p}(l) p_i(k - l)$.
  For $k = 0, \ldots, n$, set $\hat{p}(k) = \hat{q}(k)$.

Step 4: $p(\mathbf{k}) = (1/\hat{q}(n)) \prod_{i=0}^{m} p_i(k_i)$, $\mathbf{k} \in \mathcal{S}_n$.

Step 5: Stop.

Let $TH(n)$ be the throughput rate of this closed queueing network. This is the rate at which parts are loaded/unloaded at the load/unload service center, that is, $TH(n) = E[\mu_0 r_0(N_0)]$. Using equation 8.14 we see that

$$\begin{aligned} TH(n) &= E[\mu_0 r_0(Y_0) \mid |\mathbf{Y}| = n] \\ &= \sum_{\mathbf{k} \in \mathcal{S}_n} \mu_0 r_0(k_0) P\{\mathbf{Y} = \mathbf{k}\}/P\{|\mathbf{Y}| = n\} \end{aligned} \tag{8.15}$$

Because $\mu_0 r_0(k_0) P\{Y_0 = k_0\} = \lambda P\{Y_0 = k_0 - 1\}$, $k_0 \geq 1$ and $r_0(0) = 0$, we get from equation 8.15

$$TH(n) = \lambda \sum_{\mathbf{k}' \in \mathcal{S}_{n-1}} P\{\mathbf{Y} = \mathbf{k}'\}/P\{|\mathbf{Y}| = n\}$$

$$= \lambda P\{|\mathbf{Y}| = n - 1\}/P\{|\mathbf{Y}| = n\} \tag{8.16}$$

Equivalently

$$\lambda P\{|\mathbf{Y}| = n - 1\} = TH(n) P\{|\mathbf{Y}| = n\}, \quad n = 1, 2, \ldots \tag{8.17}$$

Hence we see that the total number of parts in the open queueing network is the same in distribution as that in a $M/M(n)/1$ queueing system with arrival rates λ and state-dependent service rates $TH(n)$, $n = 1, 2, \ldots$. For the purpose of analyzing the aggregate behavior of the total number of parts in the system, we can aggregate the whole FMS and replace it by an equivalent single-stage server with state-dependent service rates equal to the throughput rates. We will use this property later in this chapter to analyze FMSs that have controlled part input flow and in Chapter 10 to analyze multiple-cell manufacturing systems.

Next we will look at another property of the closed queueing network process $\{\mathbf{N}(t), t \geq 0\}$ that will help us to compute efficiently the production capacity of the flexible machining system. Let $\tau_n^{(i)}$ be the nth time epoch at which an arrival occurs at service center i. We are interested in

$$p^{(i)}(\mathbf{k}) = \lim_{n \to \infty} P\{\mathbf{N}(\tau_n^{(i)}) = \mathbf{k} + \mathbf{e}_i\}, \quad \mathbf{k} \in \mathcal{S}_{n-1}$$

the probability distribution of number of parts at different service centers seen by an arbitrary arrival at its arrival epoch to service center i. $\mathcal{S}_{n-1}$ contains all possible states that could be seen by any part excluding itself.

The rate of occurrence of a part seeing the system state $\mathbf{k}$ at its arrival epoch to service center i is $\sum_{j=0}^{m} p(\mathbf{k}+\mathbf{e}_j)\mu_j r_j(k_j+1)p_{ji}$. The rate at which parts arrive at service center i is $\sum_{\mathbf{k} \in \mathcal{S}_{n-1}} \sum_{j=0}^{m} p(\mathbf{k}+\mathbf{e}_j)\mu_j r_j(k_j+1)p_{ji}$. Therefore

$$p^{(i)}(\mathbf{k}) = \frac{\sum_{j=0}^{m} p(\mathbf{k}+\mathbf{e}_j)\mu_j r_j(k_j+1)p_{ji}}{\sum_{\mathbf{k}' \in \mathcal{S}_{n-1}} \sum_{j=0}^{m} p(\mathbf{k}'+\mathbf{e}_j)\mu_j r_j(k_j'+1)p_{ji}}, \quad \mathbf{k} \in \mathcal{S}_{n-1} \tag{8.18}$$

Substituting equation 8.11 into equation 8.18 and using equation 8.9 we get

$$p^{(i)}(\mathbf{k}) = \frac{\prod_{l=0}^{m} \frac{v_l^{k_l}}{\prod_{j=1}^{k_l} \mu_l r_l(j)}}{\sum_{\mathbf{k}' \in \mathcal{S}_{n-1}} \prod_{l=0}^{m} \frac{v_l^{k_l}}{\prod_{j=1}^{k_l} \mu_l r_l(j)}}, \quad \mathbf{k} \in \mathcal{S}_{n-1} \tag{8.19}$$

Observe that $\{p^{(i)}(\mathbf{k}), \mathbf{k} \in \mathcal{S}_{n-1}\}$ is exactly the same as the stationary distribution of the number of parts in the closed queueing network with a total of $n - 1$ parts.

Furthermore, this probability distribution is independent of the service center i. Because we will be using this relationship between the closed queueing network with n and $n-1$ parts in it to compute the performance measures we will use an index n to associate the total number of parts to its performance measures. Earlier we saw how to obtain the joint probability distribution $p(\mathbf{k})$, $\mathbf{k} \in \mathcal{S}_n$. In most applications, however, it may be sufficient to compute the marginal probability distribution $p_i(k_i; n)$, $k_i = 0, \dots, n$; its average (i.e., the average number of parts at service center i) $E[N_i(n)]$, the average sojourn time of an arbitrary part in service center i, $E[T_i(n)]$, $i = 0, \dots, m$, and the throughput rate $TH(n)$. Then $TH(n) = \mu_0 E[r_0(N_0(n))]$ is the rate at which parts are being loaded/unloaded at the load/unload service center. Using a simple recursive algorithm over the values of n we can compute these performance measures with considerably less computational effort than required by the convolution algorithm described earlier.

The rate at which parts arrive at service center i is $v_i TH(n)$. Because the probability that an arrival at service center i sees $k_i - 1$ parts in service center i is $p_i(k_i - 1; n - 1)$, the rate of upcrossings from the compound state $\{\mathbf{k}' : \mathbf{k}' \in \mathcal{S}_n, k_i' = k_i - 1\}$ is $v_i TH(n) p_i(k_i - 1; n - 1)$. Equating this to the rate of downcrossings $(\mu_i r_i(k_i) p_i(k_i; n))$ into this compound state we get

$$p_i(k_i; n) = \frac{v_i}{\mu_i r_i(k_i)} TH(n) p_i(k_i - 1; n - 1), \quad k_i = 1, \dots, n$$

$$p_i(0; n) = 1 - \sum_{k_i=1}^{n} p_i(k_i; n), \quad n = 1, 2, \dots \tag{8.20}$$

From equation 8.20 and Little's results we find that

$$E[N_i(n)] = TH(n) \sum_{k_i=0}^{n-1} \frac{(k_i + 1) v_i}{\mu_i r_i(k_i + 1)} p_i(k_i; n - 1) \tag{8.21}$$

and

$$E[T_i(n)] = \sum_{k_i=0}^{n-1} \frac{k_i + 1}{\mu_i r_i(k_i + 1)} p_i(k_i; n - 1) \tag{8.22}$$

Because $\sum_{i=0}^{m} E[N_i(n)] = n$, from Little's result, we obtain

$$TH(n) = \frac{n}{\sum_{i=0}^{m} v_i E[T_i(n)]} \tag{8.23}$$

Equations 8.20 to 8.23 provide a recursive relationship for $p_i(.; n)$ and $TH(n)$. As an initial value for this recursion observe that $p_i(0; 0) = 1$, $i = 0, \dots, m$. Then $TH(1) = 1 / \sum_{i=0}^{m} \frac{v_i}{\mu_i r_i(1)}$. Algorithm 8.2 computes these performance measures using this recursion.

Algorithm 8.2   Marginal Distribution Analysis (MDA)

Step 1: Set $p_i(0; 0) = 1$, $i = 0, \dots, m$.

Step 2: For $l = 1, \ldots, n$, compute

$$E[T_i(l)] = \sum_{k_i=0}^{l-1} \frac{k_i + 1}{\mu_i r_i(k_i + 1)} p_i(k_i; l - 1); \quad i = 0, \ldots, m$$

$$TH(l) = l / \left\{ \sum_{i=0}^{m} v_i E[T_i(l)] \right\}$$

$$E[N_i(l)] = v_i TH(l) E[T_i(l)], \quad i = 0, \ldots, m$$

$$p_i(k_i; l) = \frac{v_i TH(l)}{\mu_i r_i(k_i)} p_i(k_i - 1; l - 1), \quad k_i = 1, \ldots, l; i = 0, \ldots, m$$

$$p_i(0; l) = 1 - \sum_{k_i=1}^{l} p_i(k_i; l), \quad i = 0, \ldots, m$$

Step 3: Stop.

When we have only a single server at a service center i (i.e., $c_i = 1$) the computational effort required for the MDA can be reduced by the following observation $E[T_i(n)] = \{E[N_i(n - 1)] + 1\}/\mu_i$. Particularly if each service center has only a single server (i.e., $c_i = 1$, $i = 0, \ldots, m$) the following mean value analysis algorithm provides an efficient way to compute the system performance measures.

Algorithm 8.3   Mean Value Analysis (MVA)

Step 1: Set $E[N_i(0)] = 0$, $i = 0, \ldots, m$.

Step 2: For $l = 1, \ldots, n$, compute

$$E[T_i(l)] = \{E[N_i(l - 1)] + 1\}/\mu_i, \quad i = 0, \ldots, m$$

$$TH(l) = l / \left\{ \sum_{i=0}^{m} v_i E[T_i(l)] \right\}$$

$$E[N_i(l)] = v_i TH(l) E[T_i(l)], \quad i = 0, \ldots, m$$

Step 3: Stop.

8.4.1 Properties of FMS Throughput Rate

We will next list some of the useful properties of the throughput rate $TH(n)$. Because these properties are in relation to the number of servers c_i at service center i (i.e., $r_i(k_i) = \min\{k_i, c_i\}$) and the work load $\hat{\rho}_i = v_i/\mu_i$ assigned to service center i, $i = 0, \ldots, m$, we will use $TH(\hat{\rho}, \mathbf{c}, n)$ to represent the throughput rate of this system, where $\mathbf{c} = (c_0, \ldots, c_m)$.

Property 1: $TH(\hat{\rho}, \mathbf{c}, n)$ is increasing and concave in n.

Property 2: Suppose $c_0 = c_1 = \cdots = c_m$. Then for any two relative work allocations $\hat{\rho}^{(j)}$, $j = 1, 2$, if $\hat{\rho}^{(1)} \geq_{wm} \hat{\rho}^{(2)}$ then $TH(\hat{\rho}^{(1)}, \mathbf{c}, n) \leq TH(\hat{\rho}^{(2)}, \mathbf{c}, n)$, where $\hat{\rho}^{(1)} \geq_{wm} \hat{\rho}^{(2)}$ means $\sum_{i=0}^{l} \hat{\rho}_{[i]}^{(1)} \geq \sum_{i=0}^{l} \hat{\rho}_{[i]}^{(2)}$, $l = 0, \ldots, m$ and $\{\hat{\rho}_{[i]}, i = 0, \ldots, m\}$ is the decreasing rearrangement of $\{\hat{\rho}_i, i = 0, \ldots, m\}$. Observe that for any $\hat{\rho}$, $\sum_{i=0}^{l} \hat{\rho}_{[i]} \geq \sum_{i=0}^{l} \rho_i^*$, $l = 0, 1, \ldots, m$ where $\rho_i^* = (1/(m + 1)) \sum_{l=0}^{m} \hat{\rho}_l$, $l = 0, \ldots, m$. Therefore $TH(\hat{\rho}, \mathbf{c}, n) \leq TH(\rho^*, \mathbf{c}, n)$. That is, having an equal relative work load on service centers when they have the same number of servers maximizes the throughput rate.

Property 3: Suppose $\hat{\rho}_0 \geq \hat{\rho}_1 \geq \cdots \geq \hat{\rho}_m$. Then $TH(\hat{\rho}, \mathbf{c}, n) \leq TH(\hat{\rho}, \mathbf{c}^*, n)$ where $\mathbf{c}^* = (c_{[0]}, \ldots, c_{[m]})$ is a decreasing rearrangement of $(c_0, \ldots, c_m)$. That is giving more servers to those service centers with higher relative work load increases the throughput rate.

Property 4: $TH(\hat{\rho}, \mathbf{c}, n)$ is increasing and concave in c_i, $i = 0, \ldots, m$.

Property 5: $TH(\hat{\rho}, \mathbf{c}, n)$ is decreasing and componentwise convex in $\hat{\rho}_i$, $i = 0, \ldots, m$.

Property 6: Let $\{S_k, k = 1, \ldots, r\}$ be a partition of the set of service centers and define

$$\rho_0' = \hat{\rho}_0$$

$$\rho_k' = \sum_{i \in S_k} \hat{\rho}_i, \quad k = 1, \ldots, r$$

$$c_0' = c_0$$

$$c_k' = \sum_{i \in S_k} c_i, \quad k = 1, \ldots, r$$

Then

$$TH(\rho', \mathbf{c}', n) \geq TH(\hat{\rho}, \mathbf{c}, n)$$

Effects of pooling service centers. Property 6 states that if we pool service centers to form a larger service center, the throughput of the system will increase. Hence if we pool all service centers together we will get the maximum throughput. Particularly if $1/\mu_0 = 0$ we have

$$TH(\hat{\rho}, \mathbf{c}, n) \leq \min \left\{ n, \sum_{i=1}^{m} c_i \right\} / \sum_{i=1}^{m} \hat{\rho}_i$$

Therefore one sees that the system throughput can be improved by providing more duplicates of part programs and tools and increasing tool magazine capacity.

Property 7: Suppose $c_i' \leq c_i$ and $\rho_i'/c_i' = \rho_i/c_i$, $i = 0, \ldots, m$. Then $TH(\rho, \mathbf{c}, n') \geq TH(\rho', \mathbf{c}', n) \geq TH(\rho, \mathbf{c}, n)$, where $n' = n + \sum_{i=0}^{m} (c_i - c_i')$.

Few fast servers are better than more slower servers. Property 7 states that if we have fewer (i.e., $c_i' \leq c_i$) servers that are faster (so that the work load is $\rho_i' = \rho_i c_i'/c_i \leq \rho_i$) we may obtain a higher throughput rate than having a larger number of servers who are slow. This decrease in throughput, however, can be overcome by providing $\sum_{i=0}^{m}(c_i - c_i')$ extra pallets in the system that has a larger number of slower servers.

8.4.2 Modeling the Effects of Dedicated Material Handling Systems

Suppose we have material handling systems available for each link connecting any two service centers such that no queueing delays occur during part movement. The only time expended on moving parts is the travel time. Let $1/\mu_{(i,j)}$ be the average travel time of a part on the link (i, j), $i, j = 0, \ldots, m$. Now treating such a link (i, j) as a service center with infinitely many servers one sees that $v_{(i,j)} = v_i p_{ij}$ and $E[T_{ij}(n)] = 1/\mu_{(i,j)}$, $i, j = 0, \ldots, m$. Therefore $TH(n) = n/(\sum_{i=0}^{m} v_i E[T_i(n)] + \sum_{i=0}^{m} \sum_{j=0}^{m} v_i p_{ij}/\mu_{(i,j)})$ (compare with equation 8.23). Incorporating this in the MVA algorithm we get the following algorithm to compute the system performance with material handling effects when $c_i = 1$, $i = 0, \ldots, m$.

Algorithm 8.4   MVA with Material Handling

Step 1: Let $E[N_i(0)] = 0$, $i = 0, \ldots, m$.

Step 2: For $l = 1, \ldots, n$, compute

$$E[T_i(l)] = \{E[N_i(l - 1)] + 1\}/\mu_i, \quad i = 0, \ldots, m$$

$$TH(l) = l/\left\{\sum_{i=0}^{m} v_i E[T_i(l)] + \sum_{i=0}^{m} \sum_{j=0}^{m} v_i p_{ij}/\mu_{(i,j)}\right\}$$

$$E[N_i(l)] = v_i TH(l) E[T_i(l)], \quad i = 0, \ldots, m$$

Step 3: Stop.

8.4.3 Server Allocation

We will now illustrate how the results obtained so far can be used to tackle one of the production planning issues in FMSs. As we pointed out in Section 8.2.2, at the beginning of a production set-up we need to allocate the tools and programs to the processors. The allocation can be done in several ways. We will describe only one of such approaches. Here we first partition the set of tools and programs into several, say m sets. Set i contains a limited number of replicates r_i of tools and programs so that all or some of them can be assigned to c_i ($c_i \leq r_i$) processors such that all these c_i processors are made functionally identical, ($i = 1, \ldots, m$). The relative work load required to be handled by the tools and programs, and hence the processors assigned with these tools and programs are $\hat{\rho}_i = v_i/\mu_i$, $i = 1, \ldots, m$ and are known in advance. Therefore we wish to maximize the throughput rate $TH(\hat{\rho}, \mathbf{c}, n)$ by properly choosing the number of processors c_i to be allocated to set i of tools and programs, $i = 1, \ldots, m$. Observe that because 0 is a load/unload service center we do not consider it in our allocation problem. Formally the optimization problem is

$$\max\{TH(\hat{\rho}, \mathbf{c}, n)\}$$

subject to

$$1 \leq c_i \leq r_i, \qquad i = 1, \ldots, m$$

$$\sum_{i=1}^{m} c_i \leq C$$

where C is the total number of processors available for tool and program allocation, and $\hat{\rho}$ and n are fixed. Property 4 states that $TH(\hat{\rho}, \mathbf{c}, n)$ is increasing and concave in c_i, $i = 1, \ldots, m$. If the objective function is a separable function in c_i, $i = 1, \ldots, m$ (as in our job-shop example) we know that a marginal allocation approach will provide an optimal solution. In this case we will use marginal allocation as a heuristic approach to obtaining the allocation. Numerical results presented in Table 8.1 for a sample problem show that the proposed heuristic approach gives optimal solutions in all cases considered.

TABLE 8.1 THROUGHPUT AND GREEDY SERVER ALLOCATION IN A FOUR-MACHINE FMS WITH WORK-LOAD ALLOCATION (1/4, 1/8, 1/12, 1/16) AND 20 PALLETS

Number of servers C	(Greedy) Allocation of servers $\mathbf{c}$	Throughput $TH(\hat{\rho}, \mathbf{c}, n)$
5	(2,1,1,1)	7.543
6	(2,2,1,1)	7.995
7	(3,2,1,1)	11.133
8	(3,2,2,1)	11.889
9	(3,2,2,2)	11.965
10	(4,2,2,2)	14.934
11	(4,3,2,2)	15.931
12	(5,3,2,2)	19.131
13	(6,3,2,2)	20.761
14	(6,3,3,2)	21.787

8.5 GENERAL SINGLE-CLASS CLOSED QUEUEING NETWORK MODEL OF FMS

In this section we will look at an FMS where the aggregated processing times of parts do not have exponential distributions. When the deviation of these distribution of the part processing times from exponential is not significant, the closed Jackson queueing network model can still be used to approximate the performance measures of the FMS. Even if the deviations are large, in some cases it is possible to relate the performance measures obtained from the closed Jackson queueing network model to the actual performance measures. For example, if the logs of the density functions of the part processing times at all service centers are concave (convex), then the throughput rate evaluated by the closed Jackson queueing network is smaller (larger) than the actual throughput rate.

We will next develop methods to obtain approximations for the performance measures of the general closed queueing network model. In this model we assume that there are c_i ($c_i \geq 1$) parallel servers at service center i, and the part service times at service center i form an iid sequence of random variables with distribution function F_{S_i}, mean $E[S_i]$, and second moment $E[S_i^2]$, $i = 0, \ldots, m$. Therefore each service center behaves as a $\cdot / G_i / c_i$ queueing system with the input process "$\cdot$" determined by the rest of the service centers. In the case of exponential service times, through the marginal distributional analysis algorithm we see that we could represent service center i by a $M_i(n)/M_i(n)/1$ queueing system with state-dependent arrival rate $\lambda_i(k_i - 1) = \mu_i r_i(k_i) p_i(k_i)/p_i(k_i - 1)$, $k_i = 1, \ldots, n$; $\lambda_i(n) = 0$ and state-dependent service rates $\mu_i r_i(k_i)$, $k_i = 1, \ldots, n$. Mimicking this, we therefore represent the marginal behavior of service center i by a $M_i(n)/M_i(n)/1$ queueing system with a state-dependent arrival rate $\lambda_i(k_i)$, $k_i = 0, \ldots, n$ with $\lambda_i(n) = 0$, $i = 0, \ldots, m$ and state-dependent service rates $\mu_i r_i(k_i)$, $k_i = 0, \ldots, n$ such that the stationary distribution of the number of parts in the $M_i(n)/M_i(n)/1$ queueing system is the same as that in a $M_i(n)/G_i/c_i$ queueing system. In addition we require that the marginal distributions obtained by the MDA algorithm with these state-dependent service rates $\mu_i(k_i)$, $k_i = 0, \ldots, n$; $i = 0, \ldots, m$ are the same as the stationary distributions of the number of parts in the $M_i(n)/G_i/c_i$ queueing system with state-dependent arrival rates $\lambda_i(k_i)$, $k_i = 0, \ldots, n$; $i = 0, \ldots, m$. To formalize this requirement, let $\mathbf{f}(\lambda, G, c)$ be the stationary probability vector of the number of parts in a $M(n)/G/c$ queueing system with state-dependent arrival rates $\lambda(k)$, $k = 0, \ldots, n$; let $\mathbf{g}(\lambda, \mu)$ be the stationary probability vector of the number of parts in a $M(n)/M(n)/1$ queueing system with state-dependent arrival rate $\lambda(k)$, $k = 0, \ldots, n$ and state-dependent service rate $\mu(k)$, $k = 1, \ldots, n$; and let $\mathbf{h}_i(\mu_i \mathbf{r}_i, i = 0, \ldots, m)$ be the marginal stationary probability vector of the number of parts at service center i in the closed Jackson queueing network with state-dependent service rates $\mu_i r_i(k_i)$, $k_i = 0, \ldots, n$. Then we wish to find a λ_i and $\mu_i \mathbf{r}_i$, $i = 0, \ldots, m$ such that

$$\mathbf{h}_i(\mu_i \mathbf{r}_i, i = 0, \ldots, m) = \mathbf{g}(\lambda_i, \mu_i \mathbf{r}_i)$$

$$= \mathbf{f}(\lambda_i, G_i, c_i), \qquad i = 0, \ldots, m \qquad (8.24)$$

It is not possible to obtain an explicit solution for equation 8.24. Note, however, that when $G_i = M_i$ (i.e., the service times have exponential distributions) $\mu_i r_i(k_i) = \mu_i \min\{k_i, c_i\}$, $k_i = 0, \ldots, n$ and $\lambda_i(k_i - 1) = \mu_i r_i(k_i) h_{i:k_i}(\mu_j \mathbf{r}_j, j = 0, \ldots, m)/h_{i:k_i-1}$ $(\mu_j \mathbf{r}_j, j = 0, \ldots, m)$, $k_i = 1, \ldots, n$; $\lambda_i(n) = 0$; $i = 0, \ldots, m$ satisfies equation 8.24. For general service time distributions, the following recursive algorithm computes a solution $(\lambda_i, \mu_i \mathbf{r}_i, i = 0, \ldots, m)$ for equation 8.24.

Algorithm 8.5   Extended Marginal Distribution Analysis (EMDA).

Step 1: Set $l = 1$, $r_i^{(l)}(k_i) = \min\{k_i, c_i\}$, $k_i = 1, \ldots, n$; $r_i^{(l)}(0) = 0$, $i = 0, \ldots, m$ and $\mu_i = 1/E[S_i]$, $i = 0, \ldots, m$.

Step 2: Apply the MDA algorithm with $\mathbf{r}_i = \mathbf{r}_i^{(l)}$, $i = 0, \ldots, m$; and obtain $h_{i:k_i}(\mu_j \mathbf{r}_j, j = 0, \ldots, m) = p_i(k_i; n)$, $k_i = 0, \ldots, n$; $i = 0, \ldots, m$.

Step 3: Compute

$$\lambda_i(k_i - 1) = \mu_i r_i(k_i) h_{i:k_i}(\mu_j \mathbf{r}_j, j = 0, \ldots, m) / h_{i:k_i-1}(\mu_j \mathbf{r}_j, j = 0, \ldots, m)$$

$$k_i = 1, \ldots, n; \lambda_i(n) = 0; i = 0, \ldots, m$$

Obtain the stationary probability distribution $\{g_{i:k_i}(\lambda_i, G_i, c_i), k_i = 0, \ldots, n\}$ of the number of parts in a $M_i(n)/G_i/c_i$ queueing system with state-dependent arrival rate $\lambda_i(k_i)$, $k_i = 0, \ldots, n$.

Step 4: If the sum of the absolute differences $\sum_{i=0}^{m} \sum_{k_i=0}^{n} |p_i(k_i; n) - g_{i:k_i}(\lambda_i, G_i, c_i)| \leq \epsilon$, an acceptable error, go to step 5. Otherwise set $l = l + 1$ and $r_i^{(l)}(k_i) = \lambda_i(k_i - 1) g_{i:k_i-1}(\lambda_i, G_i, c_i) / (\mu_i g_{i:k_i}(\lambda_i, G_i, c_i))$, $k_i = 1, \ldots, n$, $r_i^{(l)}(0) = 0$; $i = 0, \ldots, m$ and go to step 2.

Step 5: The marginal probability distribution $p_i(.; n)$, $i = 0, \ldots, m$ and the throughput rate $TH(n)$ obtained by the MDA algorithm during the last iteration are the approximate performance measures of the general closed queueing network delivered by the EMDA algorithm.

When we have only a single server (i.e., $c_i = 1$) at each of the service centers a simple approximation algorithm extending the MVA algorithm can be developed. This extended MVA is as follows: Each arrival to service center i on its arrival will see the part in service, if any, requiring an average of $E[S_i^2]/2E[S_i]$ additional processing time to complete its service. This mimics the property of a $M/G/1$ queueing system. With this and the assumption that an arrival with n parts in the system sees the time average behavior of a system with $n - 1$ parts, we have

$$E[T_i(n)] = \{E[N_i(n-1)] - v_i TH(n-1)E[S_i] + 1\}E[S_i] + v_i TH(n-1)E[S_i^2]/2,$$

$$i = 0, \ldots, m \quad (8.25)$$

The preceding relationship is then applied in the following algorithm:

Algorithm 8.6   Extended Mean Value Analysis (EMVA)

Step 1: Set $E[N_i(0)] = 0$, $i = 0, \ldots, m$; $TH(0) = 0$.

Step 2: For $l = 1, \ldots, n$, compute

$$E[T_i(l)] = \{E[N_i(l-1)] - v_i TH(l-1)E[S_i] + 1\}E[S_i]$$

$$+ v_i TH(l-1)E[S_i^2]/2, \quad i = 0, \ldots, m$$

$$TH(l) = l / \left\{ \sum_{i=0}^{m} v_i E[T_i(l)] \right\}$$

$$E[N_i(l)] = v_i TH(l)E[T_i(l)], \quad i = 0, \ldots, m$$

Step 3: Stop.

The following numerical examples test the accuracy of the EMVA approximation relative to exact and simulation results, for various system parameters. In all cases considered $c_i = 1$, $i = 1, \ldots, m$.

Example 8.1

$m = 2$; $E[S_0] = 0.5$, $E[S_1] = 1.0$, $E[S_2] = 0.5$; $C_{S_0}^2 = 0.5$, $C_{S_1}^2 = 1$, $C_{S_2}^2 = 1$, and $n = 6$.

$$\mathbf{P} = \begin{bmatrix} 0 & 0.3 & 0.7 \\ 1 & 0 & 0 \\ 1 & 0 & 0 \end{bmatrix}$$

EMVA approximate throughput $= 1.931$

Exact throughput $= 1.885$

Table 8.2 gives the comparison of exact and EMVA approximate mean queue length and mean waiting time for this example.

TABLE 8.2 A COMPARISON OF EMVA APPROXIMATION AND EXACT PERFORMANCE—EXAMPLE 8.1.

Machine center i	Mean queue length $E[N_i]$		Mean waiting time $E[T_i]$	
	EMVA approx.	Exact	EMVA approx.	Exact
0	3.167	3.481	1.640	1.847
1	1.198	1.102	2.069	1.948
2	1.634	1.417	1.209	1.074

Example 8.2

$m = 2$; $E[S_0] = 0.63$, $E[S_1] = 1$, $E[S_2] = 2$; $C_{S_0}^2 = 1.32$, $C_{S_1}^2 = 1$, $C_{S_2}^2 = 1$, and $n = 5$.

$$\mathbf{P} = \begin{bmatrix} 0 & 0.7 & 0.3 \\ 1 & 0 & 0 \\ 1 & 0 & 0 \end{bmatrix}$$

EMVA approximate throughput $= 1.088$

Exact throughput $= 1.089$

Table 8.3 gives the comparison of exact and EMVA approximate mean queue length and mean waiting time for this example.

TABLE 8.3 A COMPARISON OF EMVA APPROXIMATION
AND EXACT PERFORMANCE—EXAMPLE 8.2

Machine center i	Mean queue length $E[N_i]$		Mean waiting time $E[T_i]$	
	EMVA approx.	Exact	EMVA approx.	Exact
0	1.699	1.635	1.562	1.502
1	1.902	1.949	2.498	2.558
2	1.400	1.416	4.289	4.336

Example 8.3

$m = 2$; $E[S_0] = 0.32$, $E[S_1] = 0.7$, $E[S_2] = 0.5$; $C_{S_0}^2 = 1.11$, $C_{S_1}^2 = 1.24$, $C_{S_2}^2 = 1$, and $n = 4$.

$$\mathbf{P} = \begin{bmatrix} 0 & 0.4 & 0.6 \\ 1 & 0 & 0 \\ 1 & 0 & 0 \end{bmatrix}$$

EMVA approximate throughput $= 2.169$

Exact throughput $= 2.176$

Table 8.4 gives the comparison of exact and EMVA approximate mean queue length and mean waiting time for this example.

TABLE 8.4 COMPARISON OF EMVA APPROXIMATION
AND EXACT PERFORMANCE—EXAMPLE 8.3

Machine center i	Mean queue length $E[N_i]$		Mean waiting time $E[T_i]$	
	EMVA approx.	Exact	EMVA approx.	Exact
0	1.516	1.515	0.699	0.696
1	1.203	1.180	1.387	1.355
2	1.281	1.305	0.984	1.000

Example 8.4

$m = 3$; $E[S_0] = 1.25$, $E[S_1] = 2$, $E[S_2] = 1.6$, $E[S_3] = 1.25$; $C_{S_0}^2 = 0.25$, $C_{S_1}^2 = 0.33$, $C_{S_2}^2 = 1$, $C_{S_3}^2 = 0.5$ and $n = 10$.

$$\mathbf{P} = \begin{bmatrix} 0 & 0 & 0 & 1 \\ 1 & 0 & 0 & 0 \\ 1 & 0 & 0 & 0 \\ 0 & 0.45 & 0.55 & 0 \end{bmatrix}$$

EMVA approximate throughput $= 0.736$

Simulation (95% C.I.) $= 0.743 \pm 0.003$

Table 8.5 gives the comparison of the 95% confidence interval of a simulated and EMVA approximate mean queue length and mean waiting time for this example.

TABLE 8.5 COMPARISON OF EMVA APPROXIMATION AND EXACT PERFORMANCE—EXAMPLE 8.4

Machine center i	Mean queue length $E[N_i]$		Mean waiting time $E[T_i]$	
	EMVA approx.	Sim. (±95% C.I.)	EMVA approx.	Sim. (±95% C.I.)
0	3.249	3.630 ± 0.086	4.417	4.890 ± 0.106
1	1.384	1.290 ± 0.026	4.181	3.860 ± 0.068
2	1.653	1.550 ± 0.053	4.086	3.790 ± 0.118
3	3.714	3.530 ± 0.116	5.050	4.760 ± 0.169

Example 8.5

$m = 3$; $E[S_0] = 1.25$, $E[S_1] = 1.35$, $E[S_2] = 1.45$, $E[S_3] = 1.25$; $C_{S_0}^2 = 0.25$, $C_{S_1}^2 = 1$, $C_{S_2}^2 = 1$, $C_{S_3}^2 = 0.5$, and $n = 15$.

$$P = \begin{bmatrix} 0 & 1/3 & 1/3 & 1/3 \\ 1/3 & 0 & 1/3 & 1/3 \\ 1/3 & 1/3 & 0 & 1/3 \\ 1/3 & 1/3 & 1/3 & 0 \end{bmatrix}$$

EMVA approximate throughput $= 0.634$

Simulation (95% C.I.) $= 0.634 \pm 0.005$

Table 8.6 gives the comparison of the 95% confidence interval of a simulated and EMVA approximate mean queue length and mean waiting time for this example.

TABLE 8.6 COMPARISON OF EMVA APPROXIMATION AND EXACT PERFORMANCE—EXAMPLE 8.5

Machine center i	Mean queue length $E[N_i]$		Mean waiting time $E[T_i]$	
	EMVA approx.	Sim. (±95% C.I.)	EMVA approx.	Sim. (±95% C.I.)
0	2.289	2.630 ± 0.096	3.610	4.150 ± 0.135
1	4.272	4.080 ± 0.171	6.738	6.420 ± 0.300
2	5.852	5.570 ± 0.153	9.231	8.780 ± 0.240
3	2.588	2.910 ± 0.103	4.081	4.280 ± 0.150

Example 8.6

$m = 3$; $E[S_0] = 1$, $E[S_1] = 2$, $E[S_2] = 1.6$, $E[S_3] = 1.25$; $C_{S_0}^2 = 2$, $C_{S_1}^2 = 1.5$, $C_{S_2}^2 = 1$, $C_{S_3}^2 = 0.5$, and $n = 15$.

$$\mathbf{P} = \begin{bmatrix} 0 & 0 & 0 & 1 \\ 1 & 0 & 0 & 0 \\ 1 & 0 & 0 & 0 \\ 0 & 0.45 & 0.55 & 0 \end{bmatrix}$$

EMVA approximate throughput $= 0.781$

Simulation (95% C.I.) $= 0.765 \pm 0.005$

Table 8.7 gives the comparison of the 95% confidence interval of a simulated and EMVA approximate mean queue length and mean waiting time for this example.

TABLE 8.7 COMPARISON OF EMVA APPROXIMATION AND EXACT PERFORMANCE—EXAMPLE 8.6

Machine center i	Mean queue length $E[N_i]$		Mean waiting time $E[T_i]$	
	EMVA approx.	Sim. ($\pm$95% C.I.)	EMVA approx.	Sim. ($\pm$95% C.I.)
0	4.174	3.430 ± 0.103	5.341	4.490 ± 0.128
1	2.549	2.110 ± 0.147	7.248	6.130 ± 0.406
2	2.044	1.790 ± 0.083	4.757	4.250 ± 0.159
3	6.233	7.670 ± 0.151	7.976	10.000 ± 0.252

8.5.1 Approximation of Closed Queueing Networks by Open Queueing Networks

Conditional distribution with fixed population size.   Earlier we saw how the open and closed Jackson networks are related to one another. We may use this relationship to approximately extend the results we obtained for the general open queueing networks (in Section 7.7) to general closed queueing networks. For example, let $\mathbf{Y}$ be the generic random variable representing the approximate probability distribution of the number of parts in an open queueing network. Then for the general closed queueing network we set

$$P\{\mathbf{N} = \mathbf{k}\} \simeq P\{\mathbf{Y} = \mathbf{k} \mid |\mathbf{Y}| = n\}, \qquad \mathbf{k} \in \mathcal{S}_n \tag{8.26}$$

and

$$TH(n) = \lambda P\{|\mathbf{Y}| = n - 1\}/P\{|\mathbf{Y}| = n\} \tag{8.27}$$

Previously discussed algorithms (convolutional and MDA) can be used to compute the performance measures of the general closed queueing network (with $r_i(k_i) = \lambda_i P\{Y_i = k_i - 1\}/\mu_i P\{Y_i = k_i\}$, $k_i = 1, \ldots, n$; $r_i(0) = 0$, $i = 0, \ldots, m$). The following algorithm implements this idea:

Algorithm 8.7   Extended Conditional Distribution Approach

Step 1: Set $\lambda^{(1)} = 1/\sum_{i=0}^{m} E[S_i]$; $k = 1$.

Step 2: Obtain the marginal probability distribution $(p_i^{(k)}(n_i), n_i = 0, 1, \ldots)$ of the number of jobs in machine center i of the open queueing network model of the FMS with Poisson external arrival process with a rate $\lambda^{(k)}$ to the Load/Unload (L/U) station 0 for $i = 0, \ldots, m$ (approximate procedures described in Section 7.7 may be used).

Step 3: Compute $\mu_i^{(k)}(n_i) = \lambda^{(k)} v_i p_i^{(k)}(n_i - 1)/p_i^{(k)}(n_i)$, for $n_i = 1, 2, \ldots, n$; $i = 0, \ldots, m$.

Step 4: Compute (using the MDA algorithm 8.2) the throughput $TH(n)$ of a single-class closed Jackson queueing network with visit ratios $v_0 : \cdots : v_m$ and state-dependent service rates $(\mu_i^{(k)}(n_i), n_i = 0, 1, \ldots, n)$ at machine center i, $i = 0, \ldots, m$.

Step 5: If $|TH(n) - \lambda^{(k)}| < \epsilon$, then stop; else set $k \leftarrow k+1$, $\lambda^{(k)} = (1/2)\left(\lambda^{(k-1)} + TH(n)\right)$ and go to step 2.

Average population size equivalent arrival-rate method. An alternative way to relate an open queueing network to a closed queueing network is as follows. Suppose the average number of parts in an open queueing network with arrival rate λ is n. That is, when we have free inflow of parts at rate λ we see on the average n parts in the system. It is then possible that λ is very close to the throughput rate of the closed queueing network with n parts in it. In the case of Jackson networks we know that the total number in the open queueing network is the same in distribution as that in a $M/M(n)/1$ queueing system with arrival rate λ and state-dependent service rate $TH(k)$, $k = 0, 1, \ldots$. Because the departure rate from this $M/M(n)/1$ queue is $E[TH(|\mathbf{Y}|)] = \lambda$ and $TH(k)$ is increasing and concave (property 1) we see that $\lambda = E[TH(|\mathbf{Y}|)] \leq TH(E[|\mathbf{Y}|]) = TH(n)$. The arrival rate λ that will give an average number of parts in the system equal to n cannot be explicitly specified. A bisection search procedure combined with the approximations developed in Section 7.7 for open queueing networks can be used to obtain λ. The following algorithm implements this idea:

Algorithm 8.8   Extended Mean Population Approach

Step 1: Set $\lambda^{(1)} = 1/\sum_{i=0}^{m} E[S_i]$; $\beta = 1/\max\{v_i E[S_i], i = 0, \ldots, m\}$; $k = 1$.

Step 2: Obtain the average number of jobs, $E[N]$, in the open queueing network model of the FMS with a Poisson external arrival rate $\lambda^{(k)}$ to the L/U station 0.

Step 3: If $|E[N] - n| < \epsilon$, then set $TH(n) = \lambda^{(k)}$ and stop; else set $k \leftarrow k + 1$ and if $E[N] > n$, then set $\beta = \lambda^{(k-1)}$, $\lambda^{(k)} = (1/2)(\alpha + \beta)$ and go to step 2; else (i.e., if $E[N] < n$) set $\alpha = \lambda^{(k-1)}$, $\lambda^{(k)} = (1/2)(\alpha + \beta)$ and go to step 2.

Approximate throughputs of the six sample FMSs considered earlier in this section (examples 8.1–8.6) are computed using the extended conditional distribution approach (ECD) and the extended mean population approach (EMP). Comparison of these approximations to the exact results is shown in Table 8.8. From these results it is clear that the ECD approach is better than the EMP approach. This is to be expected because even in the closed Jackson queueing network case, the ECD approach will give the exact results, whereas the EMP approach will only give a bound.

TABLE 8.8 APPROXIMATE THROUGHPUT
COMPUTED BY ECD AND EMP
APPROACHES

Example	Throughput		
	ECD	EMP	Exact/simulation
1	1.881	1.656	1.885
2	1.092	0.945	1.089
3	2.186	1.857	2.176
4	0.733	0.690	0.742 ± 0.003
5	0.636	0.600	0.634 ± 0.005
6	0.766	0.711	0.763 ± 0.004

8.6 MULTIPLE-CLASS CLOSED JACKSON QUEUEING NETWORK

8.6.1 Single-Chain Multiple-Class Model

Earlier, in the modeling of job shops, we needed a multiple-class queueing network to incorporate a central material handling system (see Section 7.6). Similarly, if we wish to include a central material handling system in the closed queueing network model of the FMS, we should consider a multiple-class closed queueing network. We will first consider a general multiple-class closed Jackson queueing network model with p classes where a part type can change from one class to another whenever it is moved from one service center to another. There are altogether n parts of all classes circulating in the system. As before we have $m + 1$ service centers $0, \ldots, m$. Service center 0 is the load/unload station. $\mathbf{P} = (p_{ij}^{(l)(l')})_{i,j=0,\ldots,m}^{l,l'=1,\ldots,p}$ is the part transfer probability matrix, that is, $p_{ij}^{(l)(l')}$ is the probability that a job completing service at service center i as a class l part next goes to service center j as a class l' part. Then the expected number of times a part visits service center i as a class l part is $v_i^{(l)}$, $i = 0, \ldots, m; l = 1, \ldots, p$: the solution to

$$v_i^{(l)} = \sum_{j=0}^{m} \sum_{l'=1}^{p} v_j^{(l')} p_{ji}^{(l')(l)}, \qquad i = 0, \ldots, m; l = 1, \ldots, p \qquad (8.28)$$

Because each part visits the load/unload station only once before it is replaced by a new (raw) part we also have

$$\sum_{l=1}^{p} v_0^{(l)} = 1 \qquad (8.29)$$

The parts are serviced according to the FCFS service protocol and the service times are iid exponential random variables with mean $1/\mu_i$ at service center i $(i = 0, \ldots, m)$

independent of the part class. The service rate at service center i is $\mu_i r_i(k)$ when there are k parts, $i = 0, \ldots, m$. Let $N_i(t)$ be the number of parts at service center i at time t and $X_{ij}(t)$ be the class index of the part in the jth position of the queue at service center i at time t; $\mathbf{X}_i(t) = (X_{ij}(t), j = 1, \ldots, N_i(t))$. Note that $\sum_{i=0}^{m} N_i(t) = n$. We assume that the part, if any, in the first position is in service being served at a rate $\mu_i(k_i)$ when $N_i(t) = k_i$, $i = 0, \ldots, m$.

Then $\{\mathbf{X}(t), t \geq 0\}$ with $\mathbf{X}(t) = (\mathbf{X}_0(t), \ldots, \mathbf{X}_m(t))$ is a continuous time Markov process on the state space $\mathcal{S}_n^{(p)}$, where $\mathcal{S}_n^{(p)} = \{\mathbf{x} = (\mathbf{x}_0, \ldots, \mathbf{x}_m) : \mathbf{x}_i \in \{1, \ldots, p\}^{k_i}, i = 0, \ldots, m; \mathbf{k} \in \mathcal{S}_n\}$. Let $q(\mathbf{x}) = \lim_{t \to \infty} P\{\mathbf{X}(t) = \mathbf{x}\}$ be the stationary probability distribution of $\{\mathbf{X}(t), t \geq 0\}$ and $\mathbf{x} = (x_{01}, \ldots, x_{0k_0}, \ldots, x_{m1}, \ldots, x_{mk_m})$ such that $x_{i,j} \in \{1, \ldots, p\}$ and $\sum_{i=0}^{m} k_i = n$. The balance equation for probability flow into and out of state $\mathbf{x}$ is

$$\sum_{i=0}^{m} \sum_{j=0}^{m} \sum_{l=1}^{p} \mu_i r_i(k_i + 1) p_{ij}^{(l)(x_{jk_j})} q(\mathbf{x}_1, \ldots, (l, \mathbf{x}_i), \ldots, \mathbf{x}_j^-, \ldots, \mathbf{x}_m) = \sum_{i=1}^{m} \mu_i r_i(k_i) q(\mathbf{x})$$

(8.30)

where $\mathbf{x}_j^- = (x_{j1}, \ldots, x_{jk_j-1})$ is the vector $\mathbf{x}_j$ without the last element. It can be verified (either by substitution or by deriving the result following the approach given in Section 7.5) that

$$q(\mathbf{x}) = \frac{\prod_{i=0}^{m} \prod_{j=1}^{k_i} \frac{v_i^{(x_{ij})}}{\mu_i r_i(j)}}{\sum_{\mathbf{y} \in \mathcal{S}_n^{(p)}} \prod_{i=0}^{m} \prod_{j=1}^{k_i'} \frac{v_i^{(y_{ij})}}{\mu_i r_i(j)}}, \qquad \mathbf{x} \in \mathcal{S}_n^{(p)}$$

(8.31)

Then it is easily shown that the probability $\hat{p}(\hat{\mathbf{k}}) = \hat{p}(\hat{k}_{il}, l = 1, \ldots, p; i = 0, \ldots, m)$ that there are $\hat{k}_{il}$ parts of class l at service center i is given by

$$\hat{p}(\hat{\mathbf{k}}) = \frac{\prod_{i=0}^{m} \frac{\hat{k}_i!}{\hat{k}_{i1}!, \ldots, \hat{k}_{ip}!} \prod_{l=1}^{p} (v_i^{(l)})^{\hat{k}_{il}} \big/ \prod_{j=1}^{\hat{k}_i} \mu_i r_i(j)}{\sum_{\hat{\mathbf{k}}' \in \mathcal{S}_n^{[p]}} \prod_{i=0}^{m} \frac{\hat{k}_i!}{\hat{k}_{i1}'!, \ldots, \hat{k}_{ip}'!} \prod_{l=1}^{p} (v_i^{(l)})^{\hat{k}_{i,l}'} \big/ \prod_{j=1}^{\hat{k}_i'} \mu_j r_j(j)}, \qquad \hat{\mathbf{k}} \in \mathcal{S}_n^{[p]}$$

(8.32)

where $\mathcal{S}_n^{[p]} = \{\hat{\mathbf{k}} = (\hat{\mathbf{k}}_0, \ldots, \hat{\mathbf{k}}_m) : \hat{\mathbf{k}}_i \in \mathcal{N}_+^p, |\hat{\mathbf{k}}_i| = k_i, i = 0, 1, \ldots, m; \mathbf{k} \in \mathcal{S}_n\}$. Letting $p(\mathbf{n}) = \lim_{t \to \infty} P\{\mathbf{N}(t) = \mathbf{n}\}$, we find after some algebra

$$p(\mathbf{k}) = \frac{\prod_{i=0}^{m} v_i^{k_i} \big/ \prod_{j=1}^{k_i} \mu_i r_i(j)}{\sum_{\mathbf{k}' \in \mathcal{S}_n} \left\{ \prod_{i=0}^{m} v_i^{k_i'} \big/ \prod_{j=1}^{k_i'} \mu_i r_i(j) \right\}}, \qquad \mathbf{k} \in \mathcal{S}_n$$

(8.33)

where $v_i = \sum_{l=1}^{p} v_i^{(l)}$, $i = 0, \ldots, m$. Observe that equation 8.33 is exactly the same result as in the single-class closed Jackson queueing network. Therefore we can aggregate the p classes of parts into a single-class of parts and use the single-class closed Jackson queueing network model to characterize the performance of the aggregated material flow. In particular the overall throughput rate of the single-chain multiple-class closed Jackson queueing network is the same as the aggregated single-class closed Jackson queueing network. Consequently we see that the single-class Jackson queueing network can be used even to incorporate the central material handling system into the flexible machining system model.

8.6.2 Multiple-Chain Multiple-Class Model

In the preceding model we assumed that each part can be in any one of the p classes. If different part types require different types of pallets we need to model the part types by different classes of parts. To be specific suppose the set of part types, $R = \{1, \ldots, r\}$, is partitioned into p subsets $R_1, \ldots, R_p$ such that any part type in R_s will need a type s pallet, $s = 1, \ldots, p$. Suppose we model the part flows of those part types in R_s by p_s classes $\{(s, 1), \ldots, (s, p_s)\}$, $s = 1, \ldots, p$. Then once an arbitrary part enters a class, say (s, l), it cannot become a class (s', l') for any $s' \neq s$ during its sojourn through the system. Hence the multiple-class closed queueing network model presented earlier cannot be used here. Instead we develop an alternative multiple-class closed Jackson queueing network model for this scenario. Let $p_{ij}^{(s,l)(s,l')}$ be the probability that a part completing a service at service center i as a class (s, l) part next goes to service center j as a class (s, l') part, $i, j = 0, \ldots, m$; $l, l' = 1, \ldots, p_s$; $s = 1, \ldots, p$. Let $v_i^{(s,l)}$, $i = 0, \ldots, m$; $l = 1, \ldots, p_s$ be the solution to

$$v_i^{(s,l)} = \sum_{j=0}^{m} \sum_{l'=1}^{p_s} v_j^{(s,l')} p_{ji}^{(s,l')(s,l)} \tag{8.34}$$

with $\sum_{l=1}^{p_s} v_0^{(s,l)} = 1$ for each $s = 1, \ldots, p$. Then $v_i^{(s,l)}$ is the expected number of times an arbitrary part belonging to the types in R_s visits service center i as a class (s, l) part during its sojourn through the system. Parts at each service center are served according to the FCFS service protocol, and the service times are iid exponential random variables with mean $1/\mu_i$ independent of part class. The part at the head of the queue receives service at rate $r_i(k_i)$ when there are k_i parts in it. Let $N_i(t)$ be the number of parts at service center i at time t and $X_{ij}(t)$ be the class index of the part in the jth position of the queue at service center i at time t. Then $\{X(t), t \geq 0\}$ is a continuous time Markov process. Note $X(t) = (X_i(t), i = 0, \ldots, m)$ and $X_i(t) = (X_{i1}(t), \ldots, X_{iN_i(t)}(t))$. The balance equation for the stationary distribution of X is given by

$$\sum_{i=0}^{m} \sum_{j=0}^{m} \sum_{l=1}^{p_{x_{jn_j}:1}} \mu_i r_i(k_i + 1) p_{ij}^{(x_{jn_j}:1, l)(x_{jn_j})} \, q(x_0, \ldots, ((x_{jn_j}:1, l), x_i), \ldots, x_j^-, \ldots, x_m)$$

$$= \sum_{i=0}^{m} \mu_i r_i(k_i) q(x) \tag{8.35}$$

where $x_{jn_j:1}$ is the first element of $\mathbf{x}_{jn_j}$. The solution to equation 8.35 is

$$q(\mathbf{x}) = \frac{\prod_{i=0}^{m} \prod_{j=1}^{k_i} \frac{v_i^{(\mathbf{x}_{ij})}}{\mu_i r_i(j)}}{\sum_{\mathbf{x}' \in \mathcal{S}_n^{(p)}} \prod_{i=0}^{m} \prod_{j=1}^{k_i'} \frac{v_i^{(\mathbf{x}_{ij}')}}{\mu_i r_i(j)}}, \quad \mathbf{x} \in \mathcal{S}_{\mathbf{n}}^{(p)} \tag{8.36}$$

where

$$\mathcal{S}_{\mathbf{n}}^{(p)} = \left\{ \mathbf{x} = (\mathbf{x}_0, \dots, \mathbf{x}_m) : \mathbf{x}_{ij} \in \mathcal{C}, \sum_{i=0}^{m} \sum_{j=1}^{k_i} I_{\{x_{ij:1}=s\}} = n_s, s = 1, \dots, p; \quad \mathbf{k} \in \mathcal{S}_n \right\}$$

and

$$\mathcal{C} = \{(s, l), l = 1, \dots, p_s; s = 1, \dots, p\}$$

is the set of classes of parts. For any $\mathbf{k} \in \mathcal{S}_{\mathbf{n}}^{[p]} = \{\mathbf{k}' = (\mathbf{k}_0', \dots, \mathbf{k}_m') : \mathbf{k}_i' \in \mathcal{N}_+^p$ and $\sum_{i=0}^{m} k_{is} = n_s, s = 1, \dots, p\}$ let $\hat{p}(\mathbf{k})$ be the probability that there are k_{is} class $s = \{(s, l), l = 1, \dots, p_s\}$ parts at service center i, $s = 1, \dots, p$; $i = 0, \dots, m$. Then from equation 8.36 one obtains

$$\hat{p}(\mathbf{k}) = \frac{\prod_{i=0}^{m} \left\{ \frac{k_i!}{k_{i1}!, \dots, k_{ip}!} \prod_{s=1}^{p} (v_i^{(s)})^{k_{is}} \Big/ \prod_{j=1}^{k_i} \mu_i r_i(j) \right\}}{\sum_{\mathbf{k}' \in \mathcal{S}_n^{[p]}} \prod_{i=0}^{m} \left\{ \frac{k_i'!}{k_{i1}'!, \dots, k_{ip}'!} \prod_{s=1}^{p} (v_i^{(s)})^{k_{is}'} \Big/ \prod_{j=1}^{k_i'} \mu_i r_i(j) \right\}}, \quad \mathbf{k} \in \mathcal{S}_n^{[p]} \tag{8.37}$$

where $v_i^{(s)} = \sum_{l=1}^{p_s} v_i^{(s,l)}$ is the expected number of visits made to service center i by an arbitrary class s part during its sojourn through the system.

Now consider an open analog of the preceding closed queueing network. Let λ be the overall arrival rate of parts, and each part will visit service center i as a class s part on the average $v_i^{(s)}$ number of times during its sojourn through the system. Let $\lambda_i^{(s)} = \lambda v_i^{(s)}$, $s = 1, \dots, p$; $i = 0, \dots, m$ and let $\{Y_{is}, s = 1, \dots, p\}$ be the generic random variable representing the stationary distribution of the number of parts of different classes at service center $i = 0, \dots, m$. The stationary distribution of the multiple class $M/M(n)/1$ queueing system is

$$P\{Y_{i1} = k_{i1}, \dots, Y_{ip} = k_{ip}\} = \binom{k_i}{k_{i1}, \dots, k_{ip}} \prod_{s=1}^{p} \left(\frac{\lambda_i^{(s)}}{\lambda_i} \right)^{k_{is}} \frac{\lambda_i^{k_i}}{\prod_{j=1}^{k_i} \mu_i r_i(j)} P\{Y_i = 0\} \tag{8.38}$$

where $k_i = |\mathbf{k}|$, $\lambda_i = \sum_{s=1}^{p} \lambda_i^{(s)}$ and $Y_i = \sum_{s=1}^{p} Y_{is}$. This is obtained by (1) observing that through multinomial thinning, $\binom{k_i}{k_{i1}, \dots, k_{ip}} \prod_{s=1}^{p} \left(\lambda_i^{(s)}/\lambda_i \right)^{k_{is}}$ is the conditional probability that the number of type s parts in the $M/M(n)/1$ system is k_{is}, $s = 1, \dots, p$, given the total number in the system is k_i (i.e., equal to $P\{Y_{i1} = k_{i1}, \dots, Y_{ip} = k_{ip}\mid$

$\sum_{s=1}^{p} Y_{is} = k_i\}$; and (2) observing that the total number of parts, $\sum_{s=1}^{p} Y_{is}$, has a distribution $P\{\sum_{s=1}^{p} Y_{is} = k_i\} = \lambda_i^{k_i} / \left\{ \prod_{j=1}^{k_i} \mu_i r_i(j) \right\}$ (see Section 4.4). From equations 8.38 and 8.37 it can be verified that

$$\hat{p}(\mathbf{k}) = P\left\{ \mathbf{Y}_i = \mathbf{k}_i, i = 0, \dots, m \mid \sum_{i=0}^{m} Y_{is} = n_s, s = 1, \dots, p \right\}, \quad \mathbf{k} \in \mathcal{S}_n^{[p]} \quad (8.39)$$

Because each class of part visits the load/unload service center only once, the throughput rate TH_s of class s parts is given by

$$TH_s = \sum_{\mathbf{x} \in \mathcal{S}_n^{(p)}} q(\mathbf{x}) I_{\{x_{01:1}=s\}} \mu_0 r_0(k_0)$$

$$= \frac{v_0^{(s)} \sum_{\mathbf{x} \in \mathcal{S}_{n-e_s}^{(p)}} \left\{ \prod_{i=0}^{m} \prod_{j=1}^{k_i} \frac{v_i^{(x_{ij})}}{\mu_i r_i(j)} \right\}}{\sum_{\mathbf{x}' \in \mathcal{S}_n^{(p)}} \left\{ \prod_{i=0}^{m} \prod_{j=1}^{k_i'} \frac{v_i^{(x_{ij}')}}{\mu_i r_i(j)} \right\}}$$

$$= \lambda \frac{P\{(\sum_{i=0}^{m} Y_{il}, l = 1, \dots, p) = \mathbf{n} - \mathbf{e}_s\}}{P\{(\sum_{i=0}^{m} Y_{il}, l = 1, \dots, p) = \mathbf{n}\}}, \quad s = 1, \dots, p \quad (8.40)$$

Recall that $I_{\{x_{01:1}=s\}}$ takes on the value one if a class s part is at the head of the load/unload service center queue and takes the value zero otherwise.

Now in principle we can compute the performance measures of the multiple-class closed Jackson queueing network using equations 8.36 and 8.37. Explicit computation of $q(\mathbf{x})$ and $\hat{p}(\mathbf{k})$ can be very cumbersome. Therefore we will restrict our attention to some more useful performance measures such as the throughput rate $TH_s(n_s)$ of each class $(s = 1, \dots, p)$ of part, average number $E[N_{is}]$ of parts of each class $(s = 1, \dots, p)$ at each service center $(i = 0, \dots, m)$, average flow time $E[T_{is}]$ of each class $(s = 1, \dots, p)$ of part through each service center i $(i = 0, \dots, m)$, and the marginal probability $p_{i,s}(k_{is})$ that there are k_{is} class s parts at service center i $(s = 1, \dots, p; i = 0, \dots, m)$. As before in the single-class model we will devise efficient algorithms to compute these quantities using recursive relation of these quantities on the values of $n_1, \dots, n_p$. Hence we will use the index $\mathbf{n}$ explicitly sometimes to identify the number of pallets of each type available in the system.

We will next look at the conditional probability distribution, $\hat{q}_{i,s}$, of the system state seen by a class s part on its arrival to service center i. For $\mathbf{x}' \in \mathcal{S}_{n-e_s}^{(p)}$ the probability $\hat{q}_{i,s}(\mathbf{x}')$ can be shown to be given by

$$\hat{q}_{i,s}(\mathbf{x}')$$

$$= \frac{\sum_{j=0}^{m} \sum_{l=1}^{p_s} \sum_{l'=1}^{p_s} \mu_j r_j(k_j' + 1) p_{ji}^{(s,l)(s,l')} q(\mathbf{x}_1', \dots, ((s,l), \mathbf{x}_j'), \dots, \mathbf{x}_i', \dots, \mathbf{x}_m')}{\sum_{\mathbf{x} \in \mathcal{S}_{n-e_s}^{(p)}} \sum_{j=0}^{m} \sum_{l=1}^{p_s} \sum_{l'=1}^{p_s} \mu_j r_j(k_j + 1) p_{ji}^{(s,l)(s,l')} q(\mathbf{x}_1, \dots, ((s,l), \mathbf{x}_j), \dots, \mathbf{x}_i, \dots, \mathbf{x}_m)}$$

$$(8.41)$$

Then from equation 8.36 one finds that

$$\hat{q}_{i,s}(\mathbf{x}) = q(\mathbf{x}; \mathbf{n} - \mathbf{e}_s), \quad \mathbf{x} \in \mathcal{S}_{\mathbf{n}-\mathbf{e}_s}; \quad i = 0, \dots, m; s = 1, \dots, p \tag{8.42}$$

Therefore if $\hat{p}_{i,s}(k_i)$ is the marginal probability that a class s part on its arrival to service center i sees k_i parts in it then

$$\hat{p}_{i,s}(k_i) = p_i(k_i; \mathbf{n} - \mathbf{e}_s), \quad i = 0, \dots, m; s = 1, \dots, p \tag{8.43}$$

Let $TH_s(\mathbf{n})$ be the throughput rate of class s parts. Then $v_i^{(s)} TH_s(\mathbf{n})$ is the rate at which class s parts arrive at service center i. Then equating the rate of upcrossings and downcrossings into the compound state $\{\mathbf{k}' : \mathbf{k}' \in \mathcal{S}_n^{(p)}, k_i' \le k_i\}$ one obtains

$$\sum_{s=1}^{p} v_i^{(s)} TH_s(\mathbf{n}) p_i(k_i; \mathbf{n}-\mathbf{e}_s) = \mu_i r_i(k_i+1) p_i(k_i+1; \mathbf{n}), \quad k_i = 0, \dots, n-1; i = 0, \dots, m \tag{8.44}$$

Therefore

$$p_i(k_i; \mathbf{n}) = \frac{1}{\mu_i r_i(k_i)} \sum_{s=1}^{p} v_i^{(s)} TH_s(\mathbf{n}) p_i(k_i - 1; \mathbf{n} - \mathbf{e}_s), \quad k_i = 1, \dots, n \tag{8.45}$$

and

$$p_i(0; \mathbf{n}) = 1 - \sum_{k_i=1}^{n} p_i(k_i; \mathbf{n}), \quad i = 0, \dots, m \tag{8.46}$$

for $s = 1, \dots, p$. Equations 8.45 and 8.46 provide a recursive relation that can be used to compute these quantities. To do so we need $TH_s(\mathbf{n})$, $s = 1, \dots, p$.

The following recursive relation on $p_i(\mathbf{k}_i)$ when the total number of class s parts in the system is n_s and $n_s - 1$ can be obtained from the preceding result and will be used later to develop an algorithm to compute TH_s. From equations 8.38, 8.39, and 8.40 one gets,

$$
\begin{aligned}
p_i(\mathbf{k}_i; \mathbf{n}) &= P\{\mathbf{Y}_i = \mathbf{k}_i\}/P\left\{\sum_{i=0}^{m} Y_{i,l} = n_l, l = 1, \dots, p\right\} \\
&= \frac{\lambda k_i v_i^{(s)}}{k_{i,s} \mu_i r_i(k_i)} P\{\mathbf{Y}_i = \mathbf{k}_i - \mathbf{e}_s\}/P\left\{\sum_{i=0}^{m} Y_{i,l} = n_l, l = 1, \dots, p\right\} \\
&= \frac{k_i v_i^{(s)} TH_s(\mathbf{n})}{k_{i,s} \mu_i r_i(k_i)} p_i(\mathbf{k}_i - \mathbf{e}_s; \mathbf{n} - \mathbf{e}_s) \text{ for any } \mathbf{k}_i \text{ with } k_{i,s} \ge 1
\end{aligned}
$$

Now multiplying both sides by $k_{i,s}$ and summing over all values of $0 \le k_{i,l} \le n_l$, $l = 1, \dots, p \ne s$ and $1 \le k_{i,s} \le n_s$ we get

$$E[N_{i,s}(\mathbf{n})] = v_i^{(s)} TH_s(\mathbf{n}) \sum_{k_i=1}^{n} \frac{k_i}{\mu_i r_i(k_i)} p_i(k_i - 1; \mathbf{n} - \mathbf{e}_s) \tag{8.47}$$

Using $\sum_{i=0}^{m} E[N_{i,s}(\mathbf{n})] = n_s$ one gets from equation 8.47

$$TH_s(\mathbf{n}) = n_s \Big/ \left\{ \sum_{i=0}^{m} v_i^{(s)} E[T_{i,s}(\mathbf{n})] \right\} \tag{8.48}$$

where

$$E[T_{i,s}(\mathbf{n})] = \frac{E[N_{i,s}(\mathbf{n})]}{v_i^{(s)} TH_s(\mathbf{n})}$$

$$= \sum_{k_i=1}^{n} \frac{k_i}{\mu_i r_i(k_i)} p_i(k_i - 1; \mathbf{n} - \mathbf{e}_s), \quad i = 0, \ldots, m \tag{8.49}$$

The following algorithm computes these performance measures recursively over different values of $\mathbf{n}$.

Algorithm 8.9   Multiclass MDA

Step 1: Set $p_i(0; \mathbf{0}) = 1$, $i = 0, \ldots, m$.

Step 2: For $l_1 = 0, \ldots, n_1; l_2 = 0, \ldots, n_2; \ldots; l_p = 0, \ldots, n_p$ except $\mathbf{l} = \mathbf{0}$ repeat steps 3 and 4.

Step 3: Set $l = \sum_{s=1}^{p} l_s$. For $s = 1, \ldots, p$, if $l_s \geq 1$, then for $i = 0, \ldots, m$, compute

$$E[T_{i,s}(\mathbf{l})] = \sum_{k_i=1}^{l} \frac{k_i}{\mu_i r_i(k_i)} p_i(k_i - 1; \mathbf{l} - \mathbf{e}_s)$$

$$TH_s(\mathbf{l}) = l_s \Big/ \left\{ \sum_{i=0}^{m} v_i^{(s)} E[T_{i,s}(\mathbf{l})] \right\}$$

else

$$E[T_{i,s}(\mathbf{l})] = 0$$

$$TH_s(\mathbf{l}) = 0$$

Step 4: For $i = 0, \ldots, m$, compute

$$p_i(k_i; \mathbf{l}) = \frac{1}{\mu_i r_i(k_i)} \sum_{s=1}^{p} v_i^{(s)} TH_s(\mathbf{l}) p_i(k_i - 1; \mathbf{l} - \mathbf{e}_s), \quad k_i = 1, \ldots, l,$$

$$p_i(0; \mathbf{l}) = 1 - \sum_{k_i=1}^{l} p_i(k_i; \mathbf{l})$$

Step 5: Stop.

This algorithm can be simplified if $r_i(k_i) = 1$, $k_i = 1, 2, \ldots$; $i = 0, \ldots, m$ (i.e., if we have only a single server at each of the $m + 1$ service centers.) In this case we have $E[T_{i,s}(\mathbf{n})] = \{E[N_{i,s}(\mathbf{n} - \mathbf{e}_s)] + 1\}/\mu_i$. Therefore we can bypass the computation of the marginal probability distribution $p_i(\cdot)$, $i = 0, \ldots, m$ and work only with the mean values $E[T_{i,s}(\cdot)]$ and $E[N_{i,s}(\cdot)]$. The corresponding algorithm is

Algorithm 8.10   Multiclass MVA

Step 1: Set $E[N_{i,s}(\mathbf{l})] = 0$, $l_s \leq 0$; $s = 1, \ldots, p$; $i = 0, \ldots, m$

Step 2: For $l_1 = 0, \ldots, n_1; \ldots; l_p = 0, \ldots, n_p$ repeat step 3.

Step 3: For $i = 0, \ldots, m$ and $s = 1, \ldots, p$, compute

$$E[T_{i,s}(\mathbf{l})] = \{E[N_{i,s}(\mathbf{l} - \mathbf{e}_s)] + 1\}/\mu_i$$

$$TH_s(\mathbf{l}) = l_s / \left\{ \sum_{i=0}^{m} v_i^{(s)} E[T_{i,s}(\mathbf{l})] \right\}$$

$$E[N_{i,s}(\mathbf{l})] = v_i^{(s)} TH_s(\mathbf{l}) E[T_{i,s}(\mathbf{l})]$$

Step 4: Stop.

Properties of throughput rate.   Unlike in the single-class closed Jackson queueing network, the throughput function of the multiple-class closed Jackson queueing network does not possess many desirable properties. We will summarize some of these properties (or lack thereof) and illustrate them through examples.

Property 1. $TH_s(\mathbf{n})$ is increasing in n_s for each s ($s = 1, \ldots, p$).

Property 2. $TH_l(\mathbf{n})$ *need not* increase (and may *decrease*) in n_s for $s \neq l$, $s, l = 1, \ldots, p$.

Property 3. $TH(\mathbf{n}) = \sum_{s=1}^{p} TH_s(\mathbf{n})$ *need not* increase (and may *decrease*) in n_s, $s = 1, \ldots, p$.

Example 8.7

Consider a two-chain two-class closed Jackson queueing network with $m + 1$ service centers $\{0, \ldots, m\}$. The transition probability matrix of class 1 parts is $\mathbf{P}$ and that of class 2 parts is $\hat{\mathbf{P}}$ given by

$$\hat{p}_{ij} = p_{ij} + (1 - q)p_{i0}p_{0j}, \quad i, j = 1, \ldots, m$$

$$\hat{p}_{i0} = qp_{i0}, \quad i = 1, \ldots, m$$

$$\hat{p}_{0j} = p_{0j}, \quad j = 1, \ldots, m$$

for some $0 < q < 1$. For $q = 1$ we get $\hat{\mathbf{P}} = \mathbf{P}$. It is now easily verified that with $v_0^{(1)} = v_0^{(2)} = 1$ we have

$$v_i^{(2)} = \frac{1}{q} v_i^{(1)}, \quad i = 1, \ldots, m \tag{8.50}$$

Now let $TH_s(n_1, n_2)$ be the throughput rate of class s parts when there are n_s class s parts in the system ($s = 1, 2$). Also let $\widehat{TH}(n_1 + n_2)$ be the throughput of a single-class closed Jackson queueing network with job-transfer probability matrix $\mathbf{P}$ and the same service rates as the two-class network. Then if $1/\mu_0 = 0$, it can be verified using equation 8.50 that

$$TH_1(n_1, n_2) = \frac{n_1}{n_1 + n_2} \widehat{TH}(n_1 + n_2)$$

and

$$TH_2(n_1, n_2) = \frac{qn_2}{n_1 + n_2} \widehat{TH}(n_1 + n_2)$$

Because $\widehat{TH}(n)$ is increasing in n, clearly $TH_1(n_1, n_2)$ is increasing in n_1 and $TH_2(n_1, n_2)$ is increasing in n_2. Because $\widehat{TH}(n)$ is concave in n, however, we have $\widehat{TH}(n)/n$ is decreasing in n. Hence $TH_1(n_1, n_2)$ is decreasing in n_2 and $TH_2(n_1, n_2)$ is decreasing in n_1. Now consider the total throughput

$$TH(n_1, n_2) = \left(\frac{n_1 + qn_2}{n_1 + n_2} \right) \widehat{TH}(n_1 + n_2)$$

$$= q\widehat{TH}(n_1 + n_2) + \frac{(1-q)n_1}{n_1 + n_2} \widehat{TH}(n_1 + n_2).$$

Therefore we see that the total throughput $TH(n_1 + n_2)$ is increasing in n_1. But if

$$q < \frac{1}{1 + \frac{1}{n_1} \left(\dfrac{\widehat{TH}(n_1+n_2+1) - \widehat{TH}(n_1+n_2)}{\dfrac{\widehat{TH}(n_1+n_2)}{n_1+n_2} - \dfrac{\widehat{TH}(n_1+n_2+1)}{n_1+n_2+1}} \right)} := \hat{q}$$

then

$$TH(n_1, n_2 + 1) < TH(n_1, n_2)$$

Because $\widehat{TH}(n_1 + n_2 + 1) - \widehat{TH}(n_1 + n_2) \geq 0$ and $\left(\widehat{TH}(n_1 + n_2)/n_1 + n_2 \right) - \left(\widehat{TH}(n_1 + n_2 + 1)/n_1 + n_2 + 1 \right) \geq 0$ we see that $0 \leq \hat{q} \leq 1$. Hence it is possible to choose a q such that the overall throughput decreases when the number of class 2 parts circulating in the system increases. For example, if $c_i = 1$, $\mu_i = 1$ and $v_i^{(1)} = 1$, $i = 1, \ldots, m$, then $\widehat{TH}(n) = n/(n + m + 1)$, and $\hat{q} = n_1/(m + n_1 + 1)$. Specifically for $n_1 = 20$ and $m = 4$ we have $\hat{q} = 0.8$. Hence if we choose $q < 0.8$, the condition is satisfied.

It is surprising to see that the overall throughput of the system goes down when the number of pallets of one kind is increased. One is therefore forced to ask whether this multiple-chain multiple-class model captures the real system behavior. Observe that in this modeling we have assumed that the parts are served on a FCFS basis, and a departing class l part is *always* replaced by a class l part ($l = 1, \ldots, p$). If the real system is indeed operated under such a *blind* policy, the preceding behavior can be expected to occur. With proper control, however, the overall performance (i.e., the individual class and overall throughput of the system) can be improved as the number of pallets is increased. One way is to use a proper feedback control when replacing a removed part (see Section 8.6.3). The other is to use a proper service protocol. For example, suppose we tag the $(n_2 + 1)$th type 2 pallet. At each of the $m + 1$ service centers, the part mounted on this pallet is given the lowest priority. That is, this part will not be processed by a service center if any other part is waiting at that service center (and its service will be interrupted if another part should arrive). This way the dynamics of the previously existing n_1 type 1 pallets and n_2 type 2 pallets are unaffected by the extra type 2 pallet. This simple policy thus guarantees that by adding a type 2 pallet we can increase the throughput of class 2 parts without affecting the throughput of class 1 parts. This does, however, require proper control of the system. Therefore we recommend not using the multiple-chain multiple-class model in the design of a FMS unless it seems that the FMS will be operated blindly without proper control.

Property 4.   Pooling of service centers need not increase the total throughput.

Example 8.8

Consider a two-chain two-class closed queueing network with four service centers. Service center 0 is the load/unload service center and we assume that $1/\mu_0 = 0$. There are ample servers at service center 1 (i.e., $c_1 = +\infty$) and only one server each at service centers 2 and 3 (i.e., $c_2 = c_3 = 1$). Furthermore $\mu_1 = \mu_2 = \mu_3 = \mu$. The job-transfer probabilities of class 1 parts are given by

$$p_{01}^{(1)} = 1; \quad p_{12}^{(1)} = 1; \quad p_{21}^{(1)} = 1 - q; \quad p_{20}^{(1)} = q$$

and all other transfer probabilities are zero. For class 2 we have

$$p_{01}^{(2)} = 1; \quad p_{13}^{(2)} = 1; \quad p_{30}^{(2)} = 1$$

and all other transfer probabilities are zero (see Figure 8.1). Suppose there are a total of n_l class l parts ($l = 1, 2$) circulating in this system. Because service centers 0 and 1 are the only service centers shared by the two classes, $c_1 = +\infty$ and $1/\mu_0 = 0$, we may use a single-class closed queueing network to compute the throughput of each class in this system.

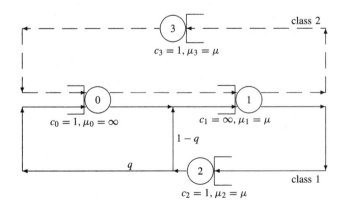

Figure 8.1   Two-Chain Two-Class Network with Four-Service Centers

The throughputs are

$$TH_1(n_1, n_2) = \mu \left\{ \frac{\sum_{n=0}^{n_1-1} \frac{1}{n!}}{\sum_{n=0}^{n_1} \frac{1}{n!}} \right\} q$$

$$TH_2(n_1, n_2) = \mu \left\{ \frac{\sum_{n=0}^{n_2-1} \frac{1}{n!}}{\sum_{n=0}^{n_2} \frac{1}{n!}} \right\}$$

$$TH(n_1, n_2) = \mu \left\{ \frac{\sum_{n=0}^{n_1-1} \frac{1}{n!}}{\sum_{n=0}^{n_1} \frac{1}{n!}} q + \frac{\sum_{n=0}^{n_2-1} \frac{1}{n!}}{\sum_{n=0}^{n_2} \frac{1}{n!}} \right\}$$

Observe that as $n_1 \to \infty$ we have the first and the third of the preceding three quantities go to μq and $\mu q + \mu \sum_{n=0}^{n_2-1} \frac{1}{n!} / \sum_{n=0}^{n_2} \frac{1}{n!}$. Suppose we pool centers 2 and 3. Then we get a queueing network with three service centers $\{0, 1, 2\}$ with $c_1 = +\infty$; $c_2 = 2$; $1/\mu_0 = 0$; $\mu_1 = \mu_2 = \mu$. The job-transfer probabilities for this pooled system are

$$\hat{p}_{01}^{(l)} = 1; \quad \hat{p}_{12}^{(l)} = 1, \quad l = 1, 2$$

$$\hat{p}_{20}^{(1)} = q; \quad \hat{p}_{21}^{(1)} = 1 - q$$

and $\hat{p}_{20}^{(2)} = 1$ (see Figure 8.2). Because the time spent by class 1 and 2 jobs is zero at the load/unload stations, excluding this station in the job routing will result in a network that has the same job routing probabilities for both classes of customers. Hence we may use a single-class queueing network with two servers 1 and 2 and job routing $\tilde{p}_{12} = 1$, $\tilde{p}_{21} = 1$, $c_1 = +\infty$ and $c_2 = 2$. Let $\widehat{TH}(n_1 + n_2)$ be the throughput of this network. Using property 7 of the throughput function of a single class closed queueing network (i.e., it is better to have $c_2' = 1$ with $\mu_2' = 2\mu$) we have that the throughput of this single class network is bounded by

$$\widehat{TH} \leq \mu \left\{ \frac{\sum_{n=0}^{n_1+n_2-1} \frac{1}{n!} 2^n}{\sum_{n=0}^{n_1+n_2} \frac{1}{n!} 2^n} \right\}$$

Hence the throughput of the pooled system is bounded by

$$\widehat{TH}_1(n_1, n_2) \leq \mu \left\{ \frac{\sum_{n=0}^{n_1+n_2-1} \frac{1}{n!} 2^n}{\sum_{n=0}^{n_1+n_2} \frac{1}{n!} 2^n} \right\} \frac{n_1}{n_1 + n_2} q$$

$$\widehat{TH}_2(n_1, n_2) \leq \mu \left\{ \frac{\sum_{n=0}^{n_1+n_2-1} \frac{1}{n!} 2^n}{\sum_{n=0}^{n_1+n_2} \frac{1}{n!} 2^n} \right\} \frac{n_2}{n_1 + n_2}$$

$$\widehat{TH}(n_1, n_2) \leq \mu \left\{ \frac{\sum_{n=0}^{n_1+n_2-1} \frac{1}{n!} 2^n}{\sum_{n=0}^{n_1+n_2} \frac{1}{n!} 2^n} \right\} \frac{(n_1 q + n_2)}{n_1 + n_2}$$

Observe that as $n_1 \to \infty$, the right-hand sides of the above inequalities goes to μq, 0 and μq, respectively. Comparing these limits to the limits of $TH_2(n_1, n_2)$, and $TH(n_1, n_2)$ one concludes that there exists a $n_1 < \infty$ such that

$$\widehat{TH}_2(n_1, n_2) < TH_2(n_1, n_2)$$

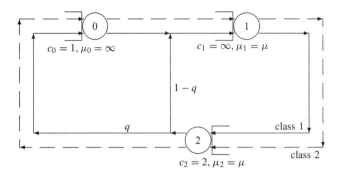

Figure 8.2   Two-Class Two-Chain
Network with Servers 2 and 3 Pooled

and

$$\widehat{TH}(n_1, n_2) < TH(n_1, n_2)$$

That is, in this example, pooling decreases the class 2 parts throughput and the overall throughput. This result is counterintuitive. When there is no overhead one would always expect to get better performance by pooling resources. One may, however, improve the throughput of all classes of parts by pooling provided proper controls such as service protocols, feedback, and part input controls are exercised. This example supports our contention that the multiple-chain multiple-class model should not be used in design and policy decisions in an FMS unless it is expected that the FMS will be operated blindly with no or inadequate control.

We see for this model that because of interference, actions that would intuitively be expected to improve the system may in fact have a deleterious effect. This underscores the usefulness of modeling to tell us the impact of decisions and the necessity of careful control to obtain expected behavior.

8.6.3 Multiple-Chain Multiple-Class Model with Mixed Feedback

Consider an FMS that produces multiple classes of parts. A part while in the system will not change class. When a class l part is removed from the system (at the load/unload service center), however, it is replaced by a class l' part with probability $d_{l'}$ (> 0), $l' = 1, \ldots, p$. Observe that in the multiple-chain multiple-class model we assumed that a class l part removed is *always* replaced by a class l part, $l = 1, \ldots, p$. As we will see the mixed feedback will allow us to achieve the desired production ratios $d_1 : \cdots : d_p$ of different classes of parts. When we model the system by a closed queueing network it will appear that a class l part can change into a class l' part, $l' = 1, \ldots, p$. Therefore in modeling this system we will use a single-chain multiple-class model (of Section 8.6.1).

Let $\mathbf{P}^{(l)}$ be the job-transfer probability matrix of a class l part, $l = 1, \ldots, p$. Then the job transfer probability matrix of the single chain model is $\hat{\mathbf{P}}$ where

$$\hat{p}_{ij}^{(l)(l)} = p_{ij}^{(l)}, \quad i = 1, \ldots, m; j = 0, \ldots, m; l = 1, \ldots, p$$

$$\hat{p}_{0i}^{(l)(l')} = d_{l'} p_{0i}^{(l')}, \quad i = 1, \ldots, m; l, l' = 1, \ldots, p$$

and all other transfer probabilities are zero. Let $\hat{v}_i^{(l)}$ be the expected number of times a part visits service center i as a class l part, that is, $\hat{v}_i^{(l)}$ is the solution to

$$\hat{v}_i^{(l)} = \sum_{l'=1}^{p} \sum_{j=0}^{m} \hat{v}_j^{(l')} \hat{p}_{ji}^{(l')(l)}, \quad \sum_{l=1}^{p} \hat{v}_0^{(l)} = 1$$

Then it can be verified that $\hat{v}_i^{(l)} = d_l v_i^{(l)}, i = 0, \ldots, m; l = 1, \ldots, p$ where $v_i^{(l)}$ is the expected number of times a class l part visits machine center i before it leaves the system. Let $v_i = \sum_{l=1}^{p} v_i^{(l)}, i = 0, \ldots, m$ and TH be the throughput rate of a single-class closed Jackson network with visit ratios $v_i, i = 0, \ldots, m$. Then TH is the total throughput rate of the single-chain multiple-class closed Jackson network (see Section 8.6.1). Therefore it follows that the rate at which class l parts are fed into the system is $d_l TH$. Hence the throughput rate of class l parts is

$$TH_l = d_l TH, \quad l = 1, \ldots, p$$

Observe that the ratio of the throughput rate of different classes of parts is $d_1 : \cdots : d_l$ and the ratio is independent of the number of pallets (n) used in the system. Indeed the preceding result holds even if the feedback (i.e., part reloading) mechanism is arbitrary but independent of the identity of the departing part types. For example, we could use a predefined loading sequence of part types (which could be random) such that the long-run ratio of the different part types loaded is $d_1 : \cdots : d_p$. We can now use this result to obtain the optimal number of pallets that will maximize the profit.

Optimal number of pallets.   Consider a FMS that produces p classes of parts. The profit (without accounting for the pallet costs) is w_l per class l part produced ($l = 1, \ldots, p$). The per unit time pallet cost is k. We wish to find the optimal number of pallets that will maximize the per unit time profit. That is, we wish to

$$\max \left\{ \sum_{l=1}^{p} w_l d_l TH(n) - nk | n \in \mathcal{N}_+ \right\}$$

Observe that in the preceding formulation we have incorporated the constraint of fixed ratios $d_1 : \cdots : d_l$ of throughput rates for different classes of parts. Alternatively, one may leave these variables also as decision variables, subject to the obvious constraints $\left\{ \sum_{l=1}^{p} d_l = 1, d_l \geq 0, l = 1, \ldots, p \right\}$. Because $TH(n)$ is increasing and concave in n, this optimization problem can be easily solved.

8.7 MULTIPLE-CLASS GENERAL QUEUEING NETWORK

In our earlier model of multiple-class closed Jackson queueing networks we assumed that the service times of all parts at service center i, independent of the class of parts, are iid exponential random variables with mean $1/\mu_i, i = 0, \ldots, m$. In many cases this assumption may not be valid. For example, the processing times may be deterministic, or the processing rates of different classes of parts may be different. Therefore, in this section, we develop approximation methods to evaluate the performance measures of a closed queueing network with class-dependent processing times.

8.7.1 Single-Chain Multiple-Class Model

Consider a multiple-class closed queueing network in which a part can change its class index from one to another whenever it is moved from one service center to another in such a way that a part can belong to any one of the p classes circulating in the system. There are $m + 1$ service centers $\{0, \ldots, m\}$ in the system with c_i parallel servers at service center i, $i = 0, \ldots, m$. A total of n parts circulate in the system. $\mathbf{P}$ is the part transfer probability matrix with $p_{ij}^{(l)(l')}$ being the probability that a part completing service at service center i as a class l part moves next to service center j as a class l' part. Then $v_i^{(l)}$ obtained from equation 8.28 is the expected number of times a part visits service center i as a class l part, $l = 1, \ldots, p$; $i = 0, \ldots, m$. The generic processing time of a class l part at service center i is $S_i^{(l)}$, $l = 1, \ldots, p$; $i = 0, \ldots, m$. At each service center the parts are serviced according to a FCFS service protocol. We will use an aggregation procedure to obtain approximate performance measures of this system.

Aggregation.   Let TH be the throughput of this system (through the load/unload service center 0). Then $\lambda_i^{(l)} = v_i^{(l)} TH$ is the rate at which class l parts arrive at service center i, $l = 1, \ldots, p$; $i = 0, \ldots, m$. Then applying the aggregation approach described in Section 7.8.1 we get a single-class general closed queueing network with job-transfer probability matrix $\hat{\mathbf{P}}$ given by

$$\hat{p}_{ij} = \frac{1}{\sum_{l=1}^{p} v_i^{(l)}} \sum_{l=1}^{p} v_i^{(l)} \sum_{l'=1}^{p} p_{ij}^{(l)(l')}, \quad i, j = 0, \ldots, m$$

The aggregate processing times at service center i have a mean

$$E[S_i] = \frac{1}{v_i} \sum_{l=1}^{p} v_i^{(l)} E\left[S_i^{(l)}\right]$$

and second moment

$$E\left[S_i^2\right] = \frac{1}{v_i} \sum_{l=1}^{p} v_i^{(l)} E\left[(S_i^{(l)})^2\right]$$

for $i = 0, \ldots, m$. The approximation procedures described in Section 8.5 for single-class general closed queueing networks can be used to obtain approximations for the performance measures of the single-chain multiple-class closed queueing network. Specifically if TH is the throughput of the aggregate single-class closed queueing network, then $TH_s = v_0^{(s)} TH$ is the throughput of class s parts, $s = 1, \ldots, p$.

8.7.2 Multiple-Chain Multiple-Class Model

Consider a closed queueing network where r types of parts circulate. This set $\mathcal{R} = \{1, \ldots, r\}$ of part types are partitioned into p subsets $\mathcal{R}_1, \ldots, \mathcal{R}_p$ such that if a part belongs to a part type in $\mathcal{R}_l$ it can never become a part with a part type in $\mathcal{R}_{l'}$ for any $l \neq l'$. If a part belongs to a part type in $\mathcal{R}_l$, however, then it has a positive probability that it will switch to any one of the other part types belonging to $\mathcal{R}_l$. Therefore, we assume that all part types belonging to $\mathcal{R}_l$ are aggregated (following the procedure given in Section 8.7.1) into a single class l of parts ($l = 1, \ldots, p$). Let $\mathbf{P}^{(l)}$ be the job transfer probability matrix of a class l part and $S_i^{(l)}$ be the generic processing time of a class l part at machine center i, $l = 1, \ldots, p$; $i = 0, \ldots, m$. There are a total of n_l class l parts circulating in the system, $l = 1, \ldots, p$. At each machine center, parts are served according to a FCFS service protocol. We will now heuristically adopt the multiclass MVA algorithm presented earlier for the multiclass closed Jackson queueing network. Let $E[N_{i,s}(\mathbf{l})]$ be the average number of class s parts at service center i when the total number of class s parts in the network is l_s, $s = 1, \ldots, p$. Also let $E[T_{i,s}(\mathbf{l})]$ be the average flow time of a class s part through service center i, and $TH_s(\mathbf{l})$ be the throughput rate of class s parts. Then the rate at which parts arrive at service center i is

$$\lambda_i(\mathbf{l}) = \sum_{s=1}^{p} v_i^{(s)} TH_s(\mathbf{l}), \quad i = 0, \ldots, m$$

The average number of busy servers at service center i is then $\lambda_i(\mathbf{l})E[S_i(\mathbf{l})]$, where $E[S_i(\mathbf{l})]$ is the average service time of a part at service center i. Therefore the probability that a randomly chosen server at service center i is busy is $(1/c_i)\lambda_i(\mathbf{l})E[S_i(\mathbf{l})]$. We may therefore approximate the probability that all c_i servers are busy at service center i by $((1/c_i)\lambda_i(\mathbf{l})E[S_i(\mathbf{l})])^{c_i}$, $i = 0, \ldots, m$. Suppose an imaginary class s part arrives at service center i when the system is under stationary conditions. Then this imaginary class s part will see $E[N_{i,\hat{s}}(\mathbf{l})]$ class $\hat{s}$ parts at the service center, $\hat{s} = 1, \ldots, p$, an average of $\lambda_i(\mathbf{l})E[S_i(\mathbf{l})]$ at the servers and see all servers busy with probability $((1/c_i)\lambda_i(\mathbf{l})E[S_i(\mathbf{l})])^{c_i}$. Hence the mean flow time of this "imaginary" class s part can be approximated by

$$E[T_{i,s}(\mathbf{l} + \mathbf{e}_s)] = E\left[S_i^{(s)}\right] +$$

$$\frac{1}{c_i} \sum_{\hat{s}=1}^{p} (E[N_{i,\hat{s}}(\mathbf{l})] - \lambda_i^{(\hat{s})} E\left[S_i^{(\hat{s})}(\mathbf{l})\right]) E\left[S_i^{(\hat{s})}\right]$$

$$+ \left(\frac{1}{c_i}\lambda_i(\mathbf{l})E[S_i(\mathbf{l})]\right)^{c_i} \left\{ \frac{1}{c_i} \frac{E[S_i^2(\mathbf{l})]}{2E[S_i(\mathbf{l})]} \right\}, \quad i = 0, \ldots, m \qquad (8.51)$$

where

$$E[S_i(\mathbf{l})] = \frac{1}{\lambda_i(\mathbf{l})} \sum_{\hat{s}=1}^{p} \lambda_i^{(\hat{s})}(\mathbf{l})E[S_i^{(\hat{s})}], \quad i = 0, \ldots, m$$

and

$$E[S_i^2(\mathbf{l})] = \frac{1}{\lambda_i(\mathbf{l})} \sum_{\hat{s}=1}^{p} \lambda_i^{(\hat{s})}(\mathbf{l}) E[(S_i^{(\hat{s})})^2], \quad i = 0, \ldots, m$$

are the first and second moments of the service time of an arbitrary part at service center i. Because $\sum_{i=0}^{m} E[N_{i,s}(\mathbf{l} + \mathbf{e}_s)] = TH_s(\mathbf{l} + \mathbf{e}_s) \sum_{i=0}^{m} v_i^{(s)} E[T_{i,s}(\mathbf{l} + \mathbf{e}_s)] = l_s + 1$, one has

$$TH_s(\mathbf{l} + \mathbf{e}_s) = \frac{l_s + 1}{\sum_{i=0}^{m} v_i^{(s)} E[T_{i,s}(\mathbf{l} + \mathbf{e}_s)]} \tag{8.52}$$

These equations now provide a recursive scheme to compute $TH_s(\mathbf{n})$, $s = 1, \ldots, p$ starting with $TH_s(\mathbf{0}) = 0$, $s = 1, \ldots, p$. The following algorithm computes the approximate performance measures.

Algorithm 8.11   Extended Multiclass MVA

Step 1: Set $E[N_{i,s}(\mathbf{l})] = 0$, $TH_s(\mathbf{l}) = 0$, $-1 \le l_{\hat{s}} \le 0$, $s, \hat{s} = 1, \ldots, p$; $i = 0, \ldots, m$.
Step 2: For $l_1 = 0, \ldots, n_1$; $l_2 = 0, \ldots, n_2$; $\ldots$; $l_p = 0, \ldots, n_p$ repeat step 3.
Step 3: For $i = 0, \ldots, m$; $s = 1, \ldots, p$ and $l_s \le n_s - 1$, compute

$$\lambda_i^{(s)}(\mathbf{l}) = v_i^{(s)} TH_s(\mathbf{l})$$

$$\lambda_i(\mathbf{l}) = \sum_{s=1}^{p} \lambda_i^{(s)}(\mathbf{l})$$

and $E[T_{i,s}(\mathbf{l} + \mathbf{e}_s)]$ by equation 8.51. Then compute $TH_s(\mathbf{l} + \mathbf{e}_s)$ by equation 8.52 and

$$E[N_{i,s}(\mathbf{l} + \mathbf{e}_s)] = v_i^{(s)} TH_s(\mathbf{l} + \mathbf{e}_s) E[T_{i,s}(\mathbf{l} + \mathbf{e}_s)]$$

Step 4: Stop.

Approximate throughput obtained by this extended mean value analysis (EMVA) for two class FMS with four (example 8.9) and three (example 8.10) single-machine machine centers are compared with the 95% confidence interval of the throughput obtained by simulation. The service time parameters used for these examples are given in Tables 8.9 and 8.10.

The job-transfer probability matrices for example 8.9 are

$$\mathbf{P}^{(1)} = \begin{pmatrix} 0 & 1/3 & 1/3 & 1/3 \\ 1/3 & 0 & 1/3 & 1/3 \\ 1/3 & 1/3 & 0 & 1/3 \\ 1/3 & 1/3 & 1/3 & 0 \end{pmatrix}$$

$$\mathbf{P}^{(2)} = \begin{pmatrix} 0 & 0 & 0 & 1 \\ 1 & 0 & 0 & 0 \\ 1 & 0 & 0 & 0 \\ 0 & 0.45 & 0.55 & 0 \end{pmatrix}$$

TABLE 8.9 SERVICE TIME PARAMETERS FOR
EXAMPLE 8.9

Machine center i	Class l	Mean service time $E\left[S_i^{(l)}\right]$	Squared coefficient of variation $C_{S_i^{(l)}}^2$
1	1	1.25	0.25
	2	1.00	2.00
2	1	1.35	1.00
	2	2.00	1.50
3	1	1.45	1.00
	2	1.60	1.00
4	1	1.25	0.50
	2	1.25	0.50

TABLE 8.10 SERVICE TIME PARAMETERS FOR
EXAMPLE 8.10

Machine center i	Class l	Mean service time $E\left[S_i^{(l)}\right]$	Squared coefficient of variation $C_{S_i^{(l)}}^2$
1	1	0.9335	1.3151
	2	0.2225	1.1065
2	1	2.0000	1.0000
	2	1.1000	1.2449
3	1	8.0000	1.0000
	2	0.5000	1.0000

and for example 8.10 are

$$\mathbf{P}^{(1)} = \begin{pmatrix} 0 & 0.7 & 0.3 \\ 1 & 0 & 0 \\ 1 & 0 & 0 \end{pmatrix}$$

$$\mathbf{P}^{(2)} = \begin{pmatrix} 0 & 0.4 & 0.6 \\ 1 & 0 & 0 \\ 1 & 0 & 0 \end{pmatrix}$$

The corresponding comparisons are given in Tables 8.11 to 8.14. In these tables n_i is equal to the number of type i pallets, $i = 1, 2$.

TABLE 8.11   THROUGHPUT OF CLASS 1 PARTS IN EXAMPLE 8.9

n_2	n_1				
	2	4	6	8	10
2	0.220 (0.214 ± 0.004)	0.352 (0.334 ± 0.006)	0.435 (0.414 ± 0.005)	0.491 (0.468 ± 0.003)	0.529 (0.505 ± 0.005)
4	0.170 (0.164 ± 0.003)	0.283 (0.271 ± 0.004)	0.361 (0.345 ± 0.003)	0.417 (0.400 ± 0.004)	0.459 (0.442 ± 0.004)
6	0.138 (0.132 ± 0.002)	0.236 (0.230 ± 0.004)	0.309 (0.298 ± 0.004)	0.364 (0.354 ± 0.005)	0.407 (0.394 ± 0.004)
8	0.117 (0.113 ± 0.003)	0.204 (0.197 ± 0.003)	0.271 (0.264 ± 0.006)	0.323 (0.315 ± 0.003)	0.365 (0.357 ± 0.006)
10	0.101 (0.100 ± 0.002)	0.179 (0.175 ± 0.003)	0.241 (0.237 ± 0.003)	0.291 (0.285 ± 0.005)	0.332 (0.325 ± 0.003)

TABLE 8.12   THROUGHPUT OF CLASS 2 PARTS IN EXAMPLE 8.9

n_2	n_1				
	2	4	6	8	10
2	0.309 (0.297 ± 0.002)	0.248 (0.236 ± 0.001)	0.206 (0.195 ± 0.001)	0.176 (0.166 ± 0.001)	0.153 (0.146 ± 0.002)
4	0.471 (0.446 ± 0.004)	0.394 (0.375 ± 0.004)	0.338 (0.320 ± 0.002)	0.296 (0.279 ± 0.001)	0.262 (0.250 ± 0.003)
6	0.569 (0.538 ± 0.002)	0.490 (0.463 ± 0.003)	0.429 (0.407 ± 0.002)	0.382 (0.361 ± 0.003)	0.344 (0.327 ± 0.002)
8	0.633 (0.597 ± 0.004)	0.556 (0.525 ± 0.004)	0.495 (0.469 ± 0.002)	0.447 (0.423 ± 0.003)	0.407 (0.386 ± 0.002)
10	0.677 (0.637 ± 0.004)	0.604 (0.570 ± 0.004)	0.545 (0.517 ± 0.003)	0.476 (30.470 ± 0.004)	0.456 (0.432 ± 0.003)

TABLE 8.13   THROUGHPUT OF CLASS 1 PARTS IN EXAMPLE 8.10

n_2	\multicolumn{4}{c}{n_1}
	2	4	6	8
2	0.645 (0.641 ± 0.007)	0.920 (0.900 ± 0.010)	1.067 (1.040 ± 0.012)	1.156 (1.130 ± 0.014)
4	.541 (0.541 ± 0.007)	.810 (0.796 ± 0.012)	.966 (0.946 ± 0.011)	1.065 (1.040 ± 0.011)
6	0.467 (0.468 ± 0.006)	0.724 (0.715 ± 0.013)	0.882 (0.864 ± 0.008)	0.987 (0.967 ± 0.017)
8	0.410 (0.412 ± 0.004)	0.653 (0.647 ± 0.008)	0.811 (0.799 ± 0.012)	0.920 (0.907 ± 0.015)

TABLE 8.14   THROUGHPUT OF CLASS 2 PARTS IN EXAMPLE 8.10

n_2	\multicolumn{4}{c}{n_1}
	2	4	6	8
2	0.828 (0.748 ± 0.009)	0.552 (0.495 ± 0.004)	0.407 (0.373 ± 0.006)	0.322 (0.304 ± 0.006)
4	1.364 (1.220 ± 0.016)	0.956 (0.867 ± 0.006)	0.730 (0.673 ± 0.008)	0.590 (0.555 ± 0.008)
6	1.723 (1.550 ± 0.011)	1.262 (1.160 ± 0.008)	0.991 (0.922 ± 0.014)	0.816 (0.767 ± 0.011)
8	1.977 (1.790 ± 0.013)	1.501 (1.370 ± 0.019)	1.207 (1.120 ± 0.013)	1.009 (0.966 ± 0.011)

8.8 SEMI-OPEN QUEUEING NETWORK MODELS

As we pointed out earlier, if (raw) parts or orders arrive at the FMS freely, independent of the state of the system, then we may use an open queueing network model. On the contrary, if the input of (raw) parts is such that each departing part is immediately replaced by a new (raw) part, then a closed queueing network model can be used. This requires that we have an infinite supply of raw parts. In reality the true situation may be somewhat in between the preceding two. The raw parts arrive at the system (say from an upstream manufacturing facility) in such a way that we may not have a raw part available whenever a completed part departs the system. Yet the raw parts may not follow a free inflow process, because it is likely that the production at the upstream manufacturing facility is controlled based on the state of the system we are considering. Suppose $\lambda(n)$ is the rate at which the upstream facility delivers a raw part to the FMS given that there are n parts in the system. For example, if the delivery policy at the upstream facility is to continue delivery at a constant rate, say λ, as long as the number of parts in the system is less than n and stop delivery as soon as the number of parts in the system reaches n we have $\lambda(k) = \lambda$, $k = 0, \ldots, n - 1$; $\lambda(k) = 0$, $k = n, n + 1, \ldots$. So it appears that neither the open nor the closed queueing network model is applicable here. The system in this case should be modeled by a semi-open queueing network where the input process is controlled by the state of the system. When $n = \inf\{k : \lambda(k) = 0, k = 0, 1, \ldots\}$ is finite (i.e., the delivery stops as soon as the number of parts in the system reaches n), we can use a closed queueing network model. We do this by introducing an imaginary service center "$m + 1$" that represents the delivery process. The service rate of this service center is $\mu_{m+1}(k) = \lambda(n - k)$, $k = 0, \ldots, n$. The total number of parts in the closed queueing network is n. Each part departing from this service center will go direct to the load/unload service center, and each part unloaded from the load/unload service center will go directly to this service center. It is not difficult to see that this mimics the behavior of the delivery process appropriately (i.e., the delivery rate when there are k parts in the system is $\mu_{m+1}(n - k) = \lambda(k)$). Therefore all the results derived earlier for the single- and multiple-class closed queueing networks are applicable here.

8.9 IMPLICATIONS OF MODELS

Capacity evaluation. The models clearly demonstrate that the throughput of a FMS is not easy to predict without using models that recognize the stochastic effects owing to the mix of jobs and their random processing times. In particular, constraints on the allocation of tools and programs that make it necessary for different part types to visit different machines and follow different routes through the system introduce the possibility of some surprising behavior. Thus it is essential for the designer to have available software that implements the various algorithms discussed in this chapter (for a review of available packages see [72]) so that he can try out a wide variety of system configurations and acquire an understanding of the influence of various design factors on system performance. Most simulation models are too complex to enable the designer to try out a wide variety of parameters.

Design optimization. Selection of part types to be processed by the FMS should take account of (1) their diversity in processing times and (2) the extent to which they can share the same pallet types.

1. From the open/closed queueing network relationship and the implications derived from the open queueing network models of job shops it is clear that job diversity in processing times degrades the performance of a FMS. The reduction in set-up and transport times achieved in a FMS, however, means that the FMS will usually perform better than an ordinary job shop processing the same set of parts.

2. In general, our models suggest that when part types use the same pallets the system performance will not be critically dependent on how work is released into the system. For example, selecting the next part type to be released with a probability proportional to the production ratios of the part types is an adequate approach. When different part types require different pallets, however, then it is easy to end up with degraded performance because of the use of what may appear to be "sensible" release rules (e.g., release a part whenever a pallet is available for it). If different pallets must be used for the different part types the designer should develop appropriate guidelines for the system control at an early stage so that system performance targets will be met, and economic viability can be demonstrated. The discussion on pooling indicates that if different pallets are used for different part types the designer should be aware that pooling is not necessarily always advisable, and some specialization of machines to part types can be advisable under normal operating conditions. Indeed it can be inferred that a collection of small flexible machining cells is easier to control than a large FMS, so the designer should consider such options. (For some further insights on this see Ranta and Tchijov [54].)

Tool allocation and machine grouping.   The models clearly demonstrate that throughput is improved by proper allocation of work load among the service centers. If all service centers have the same number of parallel processors then work load should be balanced. Service centers that have higher work load should have more processors assigned to them, or equivalently more tools and programs should be assigned to the set of tasks that require higher processing times. Not just designers but system management should also be aware of the potential negative effects of pooling in a system that processes multiple classes of parts and the need for a careful analysis of the impact of any proposed change in pooling.

Release and scheduling.   Internal control of job flow and service protocol within an FMS is usually difficult to exercise. Although performance can be improved by dynamic routing (i.e., deciding which operation should be performed next by which processor depending on current machine work loads), it is difficult to exercise such control once jobs are released to the system except when machine breakdown makes the need for a change in routing obvious. The models show that a certain amount of improvement, however, can be achieved by effective part loading and release (and also that inappropriate release rules can impair performance). This is particularly important in systems with multiple pallet types where it is sometimes appropriate *not* to release work even though a pallet is available.

8.10 BIBLIOGRAPHICAL NOTE

The first application of closed queueing networks to model an FMS was by Solberg [75] (using the name CAN-Q for the software implementation). Subsequently Hildebrant and Suri [29] [92] applied the MVA algorithm to FMS, and Suri implemented this and other queueing network algorithms in the MANUPLAN software [90]. Several other software packages implementing various closed queueing network algorithms have been developed. For a review see [72] and [63].

Buzacott and Shanthikumar [10] were the first to show the desirability of balancing work loads in a FMS with one machine per service center. They also showed the need for careful consideration of work-release policies (see also [8] [9]). Stecke and Solberg [88], however, showed that with an unequal number of machines per service center it was no longer necessarily optimal to balance work loads at the different service centers. Yao [115] developed a comprehensive set of closed queueing network models of FMS and approximations for general service times (for an overview see [12]).

PROBLEMS

8.1 A flexible machining system consists of three identical NC machines. The tools and part programs are allocated such that the average processing time per part per visit to machine i is $1/\mu_i$, $i = 1, 2, 3$.

   (a) Machine i is visited on the average v_i times by a part, $i = 1, 2, 3$. Find the throughput, $TH(n)$, of this FMS as a function of the number of pallets, n. What is the throughput when $n = 1$ and $n \to \infty$?

   (b) Suppose we have a sufficient number of tools so that all three machines can be pooled (i.e., identically tooled). What is the throughput of the FMS with this pooled configuration? Find this throughput, $TH'(n)$, as a function of the number of pallets, n, and compare it to $TH(n)$, the throughput of the previous configuration.

   (c) How many pallets should be provided in the first configuration to match the throughput of the second configuration with three pallets?

8.2 Consider a FMS consisting of two NC machines, each with an unlimited tool magazine capacity and n pallets. A part requires a total of k operations. Associated with each operation is a distinct set of tools and a processing time. Let $\mathcal{T}_i$ be the set of tools needed for operation i and let x_i be the average processing time of operation i, $i = 1, \ldots, k$.

   (a) Find an allocation of the set of tools $\{\mathcal{T}_i, i = 1, \ldots, k\}$ among the two machines such that the throughput of the FMS is maximized.

   (b) Does your answer depend on the value of n?

   (c) If there is a spare set of tools for all k operations, what would be the maximum throughput of this FMS?

8.3 Consider an FMS consisting of two machine centers and n pallets. Machine center 1 has two NC machines, and machine center 2 has one NC machine. Each part requires a total of w units of processing. Suppose it is possible to tool these two machine centers in such a way that the total processing requirement w can be split into w_1 and w_2 (such that $w_1 + w_2 = 1$) and assigned to machine centers 1 and 2, respectively.

   (a) Find the work-load allocation (w_1, w_2) that will maximize the throughput of this FMS.

   (b) Does your answer to part a depend on n?

   (c) What is the limit of the optimal work-load allocation as $n \to \infty$?

8.4 The operations of a FMS have been grouped into two sets S_1 and S_2. The total processing time needed by all those operations in set S_i is w_i, $i = 1, 2$. Suppose we have n pallets and a total of five NC machines that need to be assigned to perform these sets of operations. Suppose we tool c_i machines to perform the operations in set S_i, $i = 1, 2$. Let $TH(c_1, c_2)$ be the throughput of this FMS.

(a) Find the optimal values of (c_1, c_2) that will maximize the throughput $TH(c_1, c_2)$.

(b) Does your answer to part a depend on the number of pallets n available for the system?

(c) What is the limit of the optimal allocation of (c_1, c_2) as $n \to \infty$?

8.5 A FMS consists of two machine centers. Machine center i has c_i identically tooled NC machines, $i = 1, 2$. The work load allocated to machine center i is w_i, $i = 1, 2$. There are n pallets available for loading the parts. Let $TH(w_1, w_2, c_1, c_2, n)$ be the throughput of this FMS. Show that the throughput is

(a) Increasing and concave in n.

(b) Increasing and concave in c_1 (and in c_2).

(c) Decreasing and convex in w_1 (and in w_2).

8.6 Consider a FMS with m machine centers, each with a single machine. A single type of part is loaded onto n pallets and processed by this FMS. Each part visits machine center i on the average v_i times and the average processing requirement is $1/\mu_i$ per part per visit to machine center i, $i = 1, \ldots, m$. Suppose the increase in the number of parts at the different machine centers owing to an increase in the number of pallets is approximately proportional to the average workloads $\hat{\rho}_i = v_i/\mu_i$, $i = 1, \ldots, m$, allocated to these machine centers. That is,

$$E[N_i(l + 1)] - E[N_i(l)] \approx \hat{\rho}_i / \left\{ \sum_{j=1}^{m} \hat{\rho}_j \right\}, \quad i = 1, \ldots, m$$

(a) Using the mean value analysis (see Section 8.4), show that the throughput $TH(n)$ of this system can be approximated by

$$TH(n) \approx \frac{n}{\sum_{j=1}^{m} x_j}$$

where x_i, $i = 1, 2, \ldots, m$, is the solution to

$$x_i = \frac{x_i \hat{\rho}_i}{\sum_{j=1}^{m} x_j} n - (1 - \delta_i)\hat{\rho}_i, \quad i = 1, \ldots, m$$

and where $\delta_i = \hat{\rho}_i / \{\sum_{j=1}^{m} \hat{\rho}_j\}$, $i = 1, 2, \ldots, m$.

(b) Show that when $\hat{\rho}_i = \rho$, $i = 1, \ldots, m$, the preceding approximation gives the exact throughput.

8.7 Consider the FMS described in problem 6. Show that

$$TH(n) \leq \left\{ \frac{n}{m + n - 1} \right\} \left\{ \frac{1}{m} \sum_{j=1}^{m} \hat{\rho}_j \right\}$$

8.8 Two alternative configurations for a FMS with two machining centers are being considered. In the first, there will be two machines in machine center 1 and three machines in machine center 2. The processing rate of the machines in machine center i is μ_i, $i = 1, 2$. In the second configuration there will be one machine each at the two machine centers, but the processing rate of the machine in machine center i is $c_i \mu_i$, $i = 1, 2$, where $c_1 = 2$ and $c_2 = 3$.

(a) Which configuration has a higher throughput rate?

(b) Suppose it costs $b_i^{(j)}$ per unit time for machine i in configuration j, $i = 1, 2$; $j = 1, 2$, and it costs a per pallet per unit time. Then show that if

$$ a \leq \frac{1}{3} \left\{ \sum_{i=1}^{2} (b_i^{(2)} - c_i b_i^{(1)}) \right\} $$

then configuration 1 has a better profit rate.

8.9 Let $TH(n)$ be the throughput rate of an FMS with n pallets and an average work load of $\hat{\rho}_i$ per part allocated to machine center i, $i = 1, \ldots, m$.

(a) Show that $TH(n) \leq n / \sum_{i=1}^{m} \hat{\rho}_i$, $n = 1, 2, \ldots$.

(b) Show that the preceding bound is the actual throughput for $n = 1$ and as $n \to \infty$.

(c) Find an idealized FMS (with m machine centers and with work load $\hat{\rho}_i$ assigned to machine center i, $i = 1, \ldots, m$) that will attain the upper bound throughput.

8.10 Consider the "average population size equivalent method" described in Section 8.5.1. Apply this approximation to a FMS with balanced work-load allocation. Compare this approximation to the exact throughput.

(a) For what values of the number of pallets, n, is the percentage error in the approximation a maximum? Explain your observation.

(b) What is the percentage error as $n \to \infty$? Explain your observation.

8.11 Two types of parts, $\{1, 2\}$, requiring the same type of pallets, are loaded into a FMS with three machine centers. The parts are loaded into the FMS according to the cyclic loading sequence $\langle 1, 2, 1, 2, 1, 2, 2, 2, 2, 1, 2, 1, 2, 1, 2, 2, 2, 2, \ldots \rangle$ while maintaining the total number of pallets in the system always equal to 6. The cycle length of this sequence is 9. Suppose type 1 (2) parts' machine center sequence is $\langle 1, 2, 3 \rangle$ $(\langle 2, 3, 1, 3 \rangle)$. There is only one NC machine assigned to machine center i, $i = 1, 2, 3$, and the part processing times are exponentially distributed with mean 1 at machine center i, $i = 1, 2, 3$.

(a) Formulate a single-chain multiple-class closed Jackson queueing network model of this FMS and find the throughput of each of the two types of parts.

(b) Suppose the part loading sequence is changed to $\langle 1, 1, 1, 2, 2, 2, 2, 2, 2, 1, 1, 1, 2, 2, 2, 2, 2, 2, \ldots \rangle$. Find the new throughput rates of each of the two types of parts.

8.12 Consider the FMS described in problem 11, but assume that the type of pallets needed for the different types of part are different. Suppose a total of four type 1 pallets loaded with type 1 parts and two type 2 pallets loaded with type 2 parts are always maintained in the system.

(a) Formulate a multiple-chain multiclass closed Jackson queueing network model of this FMS, and obtain the throughput of each type of part.

(b) Suppose it is desired that the ratio of the two types of parts produced be 1:2 for types 1 and 2, respectively. Subject to the limitation of the number of pallets of type 1 to be 4 and type 2 to be 2, find a loading policy that will result in throughputs with the required ratio.

(c) Show that the throughputs obtained in part b for each part type are less than those obtained in problem 11.

8.13 FMS(1) is being replicated with identical processing capabilities but with faster NC machines. Suppose the machines in FMS(2) are α ($\alpha > 1$) times faster than the corresponding machines in FMS(1). If a total of n pallets are to be shared among these two FMSs,

  (a) Should all the pallets be allocated to FMS(2)?

  (b) If not, how many should be allocated to FMS(1)? (Give this answer as a function of the throughput function of FMS(1)).

  (c) Under what condition would you allocate all n pallets to FMS(2)?

8.14 A supplier is under contract to keep the n pallets in an FMS loaded with parts. Therefore, as soon as a part is processed by the FMS, the supplier needs to provide a raw part to replace the removed finished part. The supplier itself needs to process raw material to produce raw parts for the FMS. Therefore to provide such service to the FMS, the supplier needs to keep an inventory of raw parts.

  (a) Assuming that the supplier has an instantaneous supply of unlimited raw material, formulate a semiopen queueing network model to study the system.

  (b) Suppose the inventory carrying cost of raw parts for the supplier is h per unit time per unit and the cost per unit time of delay in delivery of a raw part to the FMS is d. Formulate a cost model to find the optimal inventory level that the supplier should carry.

BIBLIOGRAPHY

[1] M. M. BARASH. The future of numerical controls. *Mechanical Engineering*, 101:26–31, September 1979.

[2] P. P. T. BOLWIJN et al. *Flexible manufacturing: integrating technological and social innovation*. Elsevier, New York, 1986.

[3] R. BONETTO. *FMS in Practice*. Hemisphere Publishing, New York, 1988.

[4] J. BROWN, D. DUBOIS, K. RATHMILL, S. P. SETHI, and K. E. STECKE. Classification of flexible manufacturing systems. *The FMS Magazine*, 2:114–117, April 1984.

[5] S. C. BRUELL and G. BALBO. *Computational Algorithms for Closed Queueing Networks*. North-Holland, New York, 1980.

[6] J. A. BUZACOTT. The fundamental principles of flexibility in manufacturing systems. In *Proc. 1st Int. Conf. on FMS,* Brighton, pages 13–22, IFS, Bradford, UK, 1982.

[7] J. A. BUZACOTT. Modelling flexible manufacturing systems. In J. P. Brans, editor, *Operational Research '84*, pages 546–560, IFORS, North-Holland, New York, 1984.

[8] J. A. BUZACOTT. "Optimal" operating rules for automated manufacturing systems. *IEEE Trans. on Automatic Control*, AC-27:80–86, 1982.

[9] J. A. BUZACOTT. The production capacity of job shops with limited storage space. *Int. J. Prod. Res.*, 14:597–606, 1976.

[10] J. A. BUZACOTT and J. G. SHANTHIKUMAR. Models for understanding flexible manufacturing systems. *AIIE Trans.*, 12:339–349, 1980.

[11] J. A. BUZACOTT and D. D. YAO. Flexible manufacturing systems: a review of analytical models. *Management Science*, 32:890–905, 1986.

[12] J. A. BUZACOTT and D. D. YAO. On queueing network models of flexible manufacturing systems. *Queueing Systems: Theory and Applications*, 1:5–27, 1986.

[13] F. CHOOBINCH and R. SURI, editors. *Flexible Manufacturing Systems: Current Issues and Models*. Industrial Engineering and Management Press, Norcross, GA, 1986.

[14] H. C. CO and R. A. WYSK. The robustness of CAN-Q in modelling automated manufacturing systems. *Int. J. Prod. Res.*, 24:1485–1503, 1986.

[15] W. D. COMPTON, editor. *Design and Analysis of Integrated Manufacturing Systems*. National Academy Press, Washington, DC, 1988.

[16] Y. DALLERY. On modelling flexible manufacturing systems using closed queueing networks. *Large Scale Systems*, 11:109–119, 1986.

[17] Y. DALLERY and Y. FREIN. An efficient method to determine the optimal configuration of a flexible manufacturing system. *Annals of Operations Research*, 15:207–225, 1988.

[18] Y. DALLERY and K. STECKE. On the optimal allocation of servers and workloads in closed queueing networks. *Operations Research*, 38:694–703, 1990.

[19] J. D. GOLDHAR and M. JELINEK. Plan for economics of scope. *Harvard Business Review*, 141–148, Nov.–Dec. 1983.

[20] W. J. GORDON and G. F. NEWELL. Closed queueing systems with exponential servers. *Operations Research*, 15:254–265, 1967.

[21] T. W. GRAVER and L. F. MCGINNIS. A tool provisioning problem in an FMS. *Int. J. FMS*, 1:239–254, 1989.

[22] N. R. GREENWOOD. *Implementing Flexible Manufacturing Systems*. IFS, Bradford, UK, 1984.

[23] Y. P. GUPTA and S. GOYAL. Flexibility of manufacturing systems: concepts and measurements. *Eur. J. Opnl. Res.*, 43:119–135, 1989.

[24] S. GUSTAVSSON. Flexibility and productivity in complex production processes. *Int. J. Prod. Res.*, 22:801–808, 1984.

[25] E. H. HAHNE. *Dynamic routing in an unreliable manufacturing network with limited storage*. Technical Report LIDS-TH-1063, Laboratory for Information and Decision Systems, Massachusetts Institute of Technology, 1981.

[26] J. HARRINGTON, JR. *Computer integrated manufacturing*. Industrial Press, New York, 1973.

[27] J. HARTLEY. *FMS at Work*. IFS, Bradford, UK, 1984.

[28] D. E. HEGLAND. Flexible manufacturing—your balance between productivity and adaptability. *Production Engineering*, 39–43, May 1981.

[29] R. R. HILDEBRANT. *Scheduling flexible machining systems when machines are prone to failure*. PhD thesis, Department of Aeronautics and Astronautics, Massachusetts Institute of Technology, 1980.

[30] G. K. HUTCHINSON and B. E. WYNNE. A flexible manufacturing system. *Industrial Engineering*, 5:10, 1973.

[31] J. JABLONSKI. Reexamining FMS: American Machinist special report 774. *American Machinist*, 129:125–140, March 1985.

[32] M. A. JAFARI. Performance modelling of a manufacturing cell with a single material transporter: an approximation. *Int. J. FMS*, 2:63–86, 1989.

[33] R. JAIKUMAR. *Flexible manufacturing systems: a management perspective*. Technical Report WP 1-784-078, Division of Research, Harvard Business School, Jan. 1984.

[34] R. JAIKUMAR. Postindustrial manufacturing. *Harvard Business Review*, 301–308, Nov.–Dec. 1986.

[35] F. P. KELLY. *Reversibility and Stochastic Networks*. John Wiley and Sons, New York, 1979.

[36] J. G. KIMEMIA. *Hierarchical control of production in flexible manufacturing systems*. PhD thesis LIDS-TH-1215, Laboratory for Information and Decision Systems, Massachusetts Institute of Technology, 1982.

[37] J. G. KIMEMIA and S. B. GERSHWIN. An algorithm for the computer control of production in flexible manufacturing systems. *IIE Trans.*, 15:353–362, 1983.

[38] J. G. KIMEMIA and S. B. GERSHWIN. Flow optimization in flexible manufacturing systems. *Int. J. Prod. Res.*, 23:81–96, 1985.

[39] H. T. KLAHORST. Flexible manufacturing systems: combining elements to lower costs, add flexibility. *Industrial Engineering*, 13:112–117, 1981.

[40] E. KOENIGSBERG. Twenty-five years of cyclic queues and closed queue networks: a review. *J. Opl. Res. Soc.*, 33:605–619, 1982.

[41] A. KUSIAK, editor. *Flexible Manufacturing Systems: Methods and Studies*. North-Holland, New York, 1986.

[42] A. KUSIAK, editor. *Modelling and Design of FMS*. Elsevier, New York, 1986.

[43] M. MANDELBAUM. *Flexibility in decision making: an exploration and unification*. PhD thesis, University of Toronto, 1978.

[44] R. MARIE. An approximate analytical method for general queueing networks. *IEEE Trans. on Software Engineering*, SE-5:530–538, 1979.

[45] M. E. MERCHANT. Future of computer-automated machine tool systems. *Carbide and Tool*, 10:7–11, 1978.

[46] H. NAGASAWA. *A design for a separated functional FMS with balanced loading*. Technical Report, Department of Ind. Eng., University of Osaka Prefecture, Osaka 591, Japan, 1990.

[47] S. Y. NOF, M. M. BARASH, and J. J. SOLBERG. Operational control of item flow in versatile manufacturing systems. *Int. J. Prod. Res.*, 17:479–489, 1979.

[48] G. J. OLSDER and R. SURI. Time-optimal control of parts-routing in a manufacturing system with failure-prone machines. In *Proc. 19th IEEE Conference on Decision and Control*, pages 722–727, 1980.

[49] J. OSBORNE. Direct on–line computer control of machine tools and material handling. In M. A. De Vries, editor, *The Expanding World of NC*, pages 260–268, Numerical Control Society, Chicago, 1971.

[50] C. B. PERRY. Variable-mission manufacturing systems. In M. A. De Vries, editor, *NC: 1971 the opening door to productivity and profits*, pages 409–433, Numerical Control Society, 1971.

[51] J. POND. On the road to CAD/CAM—Part I: Japan. *Iron Age*, 219:37–44, 28 Mar 1977.

[52] P. G. RANKY. *Computer integrated manufacturing*. Prentice Hall, Englewood Cliffs, NJ, 1986.

[53] P. G. RANKY. *The Design and Operation of FMS*. IFS, Bradford, UK, 1983.

[54] J. RANTA and I. TCHIJOV. Economics and success factors of flexible manufacturing systems: the conventional explanation revisited. *Int. J. of FMS*, 2:169–190, 1990.

[55] A. RAOUF, editor. *Flexible Manufacturing*. Elsevier, New York, 1985.

[56] M. REISER. Mean-value analysis and convolution method for queue-dependent servers in closed queueing networks. *Performance Evaluation*, 1:7–18, 1981.

[57] M. REISER and S. S. LAVENBERG. Mean-value analysis of closed multichain queueing networks. *J. ACM*, 27:313–322, 1980.

[58] R. RIGHTER and J. G. SHANTHIKUMAR. Extremal properties of the FIFO discipline in queueing networks. *J. Appl. Prob.*, in press.

[59] P. J. SCHWEITZER, A. SEIDMANN, and P. B. GOES. Performance management in flexible manufacturing systems. *Int. J. FMS*, 4:17–50, 1991.

[60] G. SECCO-SUARDO. *Optimization of a closed network of queues*, vol. 3 of *Complex Materials Handling and Assembly Systems*. Technical Report ESL-FR-834-3, Electronic Systems Laboratory, Massachusetts Institute of Technology, 1978.

[61] A. SEIDMANN and S. Y. NOF. Unitary manufacturing cell design with random product feedback flow. *IIE Trans.*, 17:188–193, 1985.

[62] A. SEIDMANN and P. J. SCHWEITZER. Part selection policies for a flexible manufacturing cell feeding several production lines. *IIE Trans.*, 16:355–362, 1984.

[63] A. SEIDMANN, P. J. SCHWEITZER, and S. SHALEV-OREN. Computerized closed queueing network models of flexible manufacturing systems: A comparative evaluation. *Large Scale Systems*, 12:91–107, 1987.

[64] A. K. SETHI and S. P. SETHI. Flexibility in manufacturing: a survey. *Int. J. FMS*, 2:289–328, 1990.

[65] S. SHALEV-OREN, A. SEIDMANN, and P. J. SCHWEITZER. Analysis of flexible manufacturing systems with priority scheduling: PMVA. *Annals of Operations Research*, 3:115–139, 1985.

[66] J. G. SHANTHIKUMAR and M. GOCMEN. Heuristic analysis of closed queueing networks. *Int. J. Prod. Res.*, 21:675–690, 1983.

[67] J. G. SHANTHIKUMAR and R. G. SARGENT. A hybrid simulation/analytic model of a computerized manufacturing system. In *Proc. 9th IFORS Conference,* Hamburg, pages 901–915, 1981.

[68] J. G. SHANTHIKUMAR and K. E. STECKE. Reducing the work-in-process inventory in certain classes of flexible manufacturing systems. *Eur. J. Opnl. Res.*, 26:266–271, 1986.

[69] J. G. SHANTHIKUMAR and D. D. YAO. The effect of increasing service rates in a closed queueing network. *J. Appl. Prob.*, 23:474–483, 1986.

[70] J. G. SHANTHIKUMAR and D. D. YAO. On server allocation in multiple center manufacturing systems. *Operations Research*, 36:333–342, 1988.

[71] J. G. SHANTHIKUMAR and D. D. YAO. Optimal buffer allocation in a multicell system. *Int. J. FMS*, 1:347–356, 1989.

[72] J. SNOWDON and J. C. AMMONS. A survey of queueing network packages for the analysis of manufacturing systems. *Manufacturing Review*, 1:14–25, 1988.

[73] K. C. SO. Allocating buffer storages in a flexible manufacturing system. *Int. J. FMS*, 1:223–238, 1989.

[74] J. J. SOLBERG. Capacity planning with a stochastic workflow model. *AIIE Trans.*, 13:116–122, 1981.

[75] J. J. SOLBERG. A mathematical model of computerized manufacturing systems. In *Proc. 4th Intl. Conf. on Production Research,* Tokyo, pages 22–30, 1977.

[76]   J. J. SOLBERG and S. Y. NOF. Analysis of flow control in alternative manufacturing config-urations. *Journal of Dynamic Systems, Measurement and Control*, 102:141–147, September 1980.

[77]   P. SOLOT. A heuristic method to determine the number of pallets in a flexible manufacturing system with several pallet types. *Int. J. FMS*, 2:191–216, 1990.

[78]   P. SOLOT and J. M. Bastos. MULTIQ: a queueing model for FMSs with several pallet types. *J. Opl. Res. Soc.*, 39:811–821, 1988.

[79]   P. SOLOT and MARIO VAN VLIET. *Analytical models for FMS design optimization*. Techni-cal Report ORWP 90/16, Department of Mathematics, Ecole Polytechnique Federale de Lausanne, Lausanne, Switzerland, 1990.

[80]   G. SPUR, K. MARTINS, and B. VIEHWEGER. Flexible manufacturing systems in Europe. *Robotics and Computer Integrated Manufacturing*, 1:355–364, 1984.

[81]   DRAPER LABS. Staff. *FMS Handbook*. Noyes, Park Ridge, NJ, 1984.

[82]   K. STECKE. On the nonconcavity of throughput in certain closed queueing networks. *Per-formance Evaluation*, 6:293–305, 1986.

[83]   K. E. STECKE. Design, planning, scheduling, and control problems of flexible manufactur-ing systems. *Annals of Operations Research*, 3:3–12, 1985.

[84]   K. E. STECKE. Formulation and solution of nonlinear integer production planning problems for flexible manufacturing systems. *Management Science*, 29:273–288, 1983.

[85]   K. E. STECKE and I. KIM. Performance evaluation for systems of pooled machines of unequal sizes: unbalancing vs. blocking. *Eur. J. Opnl. Res.*, 42:22–38, 1989.

[86]   K. E. STECKE and T. L. MORIN. The optimality of balancing workloads in certain types of flexible manufacturing systems. *Eur. J. Opnl. Res.*, 20:68–82, 1985.

[87]   K. E. STECKE and J. J. SOLBERG. *The CMS loading problem: The optimal planning of com-puterized manufacturing systems*. Technical Report 20, School of Industrial Engineering, Purdue University, West Lafayette, IN, 1981.

[88]   K. E. STECKE and J. J. SOLBERG. The optimality of unbalancing both workloads and ma-chine group size in closed queueing networks of multi-server queues. *Operations Resarch*, 33:882–910, 1985.

[89]   R. SURI. An overview of evaluative models of FMS. *Annals of Operations Research*, 3:13–21, 1985.

[90]   R. SURI. RMT puts manufacturing at the helm. *Manufacturing Engineering*, 100:41–44, 1988.

[91]   R. SURI. Robustness of queuing network formulas. *J. ACM*, 30:564–594, 1983.

[92]   R. SURI and R. R. HILDEBRANT. Modelling flexible manufacturing systems using mean-value analysis. *Journal of Manufacturing Systems*, 3:27–38, 1984.

[93]   R. SURI and C. K. WHITNEY. Decision support requirements in flexible manufacturing. *Journal of Manufacturing Systems*, 3:61–69, 1984.

[94]   E. SZELKE and J. BROWNE, editors. *Advances in Production Management Systems '85*. North-Holland, New York, 1986.

[95]   J. TALAVAGE and R. G. HANNAN. *Flexible Manufacturing Systems in Practice*. Marcel Dekker, New York, 1988.

[96]   E. TEICHOLZ and J. NORR. *CIM Handbook*. McGraw-Hill, New York, 1987.

[97]   H. TEMPLEMEIER, H. KUHN, and U. TETZLAFF. Performance evaluation of flexible manufacturing systems with blocking. *Int. J. Prod. Res.*, 27:1963–1979, 1989.

[98]   A. TENENBAUM and A. SEIDMAN. Dynamic load control policies for a flexible manufacturing system with stochastic processing rates. *Int. J. FMS*, 2:93–120, 1989.

[99]   U. TETZLAFF. *Optimal Design of Flexible Manufacturing Systems*. Physica-Verlag, Heidelberg, Germany, 1990.

[100]   B. VINOD and M. SABBAGH. Optimal performance analysis of manufacturing systems subject to tool unavailability. *Eur. J. Opnl. Res.*, 24:398–409, 1986.

[101]   B. VINOD and J. J. SOLBERG. The optimal design of flexible manufacturing systems. *Int. J. Prod. Res.*, 23:1141–1151, 1985.

[102]   B. VINOD and J. J. SOLBERG. Performance evaluation models for unreliable flexible manufacturing systems. *Omega*, 12:299–308, 1984.

[103]   W. WHITT. Open and closed models for networks of queues. *AT&T Bell Lab. Tech J.*, 63:1911–1979, 1984.

[104]   D. T. N. WILLIAMSON. A better way of making things. *Science Journal*, 4:53–59, June 1968.

[105]   D. T. N. WILLIAMSON. The pattern of batch manufacture and its influence on machine tool design. *Proc. Instn. Mech. Engrs.*, 182, pt. 1:870–895, 1968.

[106]   D. D. YAO. Material and information flows in flexible manufacturing systems. *Material Flow*, 2:143–149, 1985.

[107]   D. D. YAO. An optimal storage model for a flexible manufacturing system. In A. Kusiak, editor, *Flexible Manufacturing Systems: Methods and Studies*, pages 113–125, North-Holland, New York, 1986.

[108]   D. D. YAO. Some properties of the throughput function of closed networks of queues. *Oper. Res. Letters*, 3:313–317, 1984.

[109]   D. D. YAO and J. A. BUZACOTT. The exponentialization approach to flexible manufacturing system models with general processing times. *Eur. J. Opnl. Res.*, 24:410–416, 1986.

[110]   D. D. YAO and J. A. BUZACOTT. Modeling a class of flexible manufacturing systems with reversible routing. *Operations Research*, 35:87–93, 1987.

[111]   D. D. YAO and J. A. BUZACOTT. Modeling a class of state-dependent routing in flexible manufacturing systems. *Annals of Operations Research*, 3:153–167, 1985.

[112]   D. D. YAO and J. A. BUZACOTT. Modelling the performance of flexible manufacturing systems. *Int. J. Prod. Res.*, 23:945–959, 1985.

[113]   D. D. YAO and J. A. BUZACOTT. Models of flexible manufacturing systems with limited local buffers. *Int. J. Prod. Res.*, 24:107–118, 1986.

[114]   D. D. YAO and F. F. PEI. Flexible parts routing in manufacturing systems. *IIE Trans.*, 22:48–55, 1990.

[115]   D. D. W. YAO. *Queueing models of flexible manufacturing systems*. PhD thesis, Department of Industrial Engineering, University of Toronto, September 1983.

[116]   D. D. W. YAO and J. A. BUZACOTT. Queueing models of a flexible machining station. Part I: the diffusion approximation. *Eur. J. Opnl. Res.*, 19:223–240, 1985.

[117]   D. D. W. YAO and J. A. BUZACOTT. Queueing models of a flexible machining station. Part II: the method of Coxian phases. *Eur. J. Opnl. Res.*, 19:241–252, 1985.

[118] C. YOUNG and A. GREENE. *Flexible menufacturing systems*. AMA Membership Publications Division, American Management Association, 1986.

[119] D. M. ZELENOVIC. Flexibility—a condition for effective production systems. *Int. J. Prod. Res.*, 20:319–337, 1982.

[120] L. ZHUANG and K. S. HINDI. Mean value analysis for multiclass closed queueing network models of flexible manufacturing systems. *Eur. J. Opnl. Res.*, 46:366–379, 1990.

[121] L. ZHUANG and K. S. HINDI. Convolution algorithm for closed queueing network models of flexible manufacturing systems with limited buffers. *Information and Decision Technologies*, 17:83–90, 1991.

9

Flexible Assembly Systems

9.1 INTRODUCTION

The assembly of discrete parts into a final product is typically organized around either flow lines or job shops. With low-volume or large and bulky products, it is common to use fixed position assembly where the product does not move, and all parts are brought to the assembly location (i.e., a single-stage manufacturing system). Starting in the 1960s, however, interest began to develop in alternative manufacturing systems that would overcome what were perceived to be the inherent weaknesses of conventional flow lines or job shops. The primary goals were to achieve greater flexibility and improved quality, where flexibility was understood to mean the ability to produce a variety of different products without losing productivity. The earliest application of these new ideas would appear to be in the clothing and footwear industries, although they later spread to other industries, in particular in the early 1980s to electronics assembly.

Until recently it was quite rare for assembly operations to be highly automated, apart from the very high volume assembly of such products as electric lamps. Most assembly operations were carried out by human operators. Further, except for welding operations, most assembly operations usually did not require much beyond fairly simple power tools. The rapid evolution of robotics, however, has meant that in designing assembly systems it has become desirable to allow for the eventual replacement of people by robots on some tasks. Also, development of automated data collection systems using bar codes and bar code readers plus the advent of cheap computing facilities (PCs) for data storage and processing have enabled the control of work flow to be automated. Inspection and test are often extensively automated, and the information on the nature of any defects in the product can be used to determine work flow. Nevertheless, the key concepts of the new flexible assembly systems are not related to sophisticated automation of individual work stations (except perhaps for information provided by automated inspection and test stations), but rather to the way in which jobs are delivered to work stations and the way in which job flow through the system is controlled.

9.1.1 Background—Problems with Flow Lines and Job Shops

To understand the features of flexible assembly systems it is desirable to review the problems with traditional flow lines and job shops when used for assembly.

Problems with flow lines. As shown in Chapter 5 conventional flow lines have several shortcomings if they are used to produce a variety of different products or when task times at stations are variable. This variability could be due to the innate characteristics of human operators or the inability to control task times at the station because either the process is poorly understood, or it is inherently variable (e.g., if the task involves finding and repairing defects). Because of the variability the following problems occur:

- *Imbalance with job variety.* Flow lines require that the work load per station be approximately balanced. If the line makes only one product this can usually be achieved, however, when the line makes a variety of products and the mix of products changes continually, it is difficult to achieve and maintain balance. As a result, the utilization of some stations may be relatively poor. If only one product is produced at a time a higher utilization can be achieved, but this means that there must then be a substantial inventory of finished products in a produce-to-stock environment, or long lead times in a produce-to-order situation.

- *Paced: Quality versus productivity.* In paced lines we have shown that variability of task times, whether owing to innate human variability or the unequal processing requirements of different jobs, means that there is a tradeoff between quality and productivity that worsens as variability increases or as the duration of tasks assigned to a work station diminishes.

- *Unpaced: Inventory (and space) versus productivity.* With unpaced lines we showed that for given in process inventory limits (and thus, equivalently, given space assignment for in process inventory) the throughput worsens with increasing variability of task times. Alternatively more in-process inventory space is required to maintain the same level of production as variability increases.

Problems with job shops. If job-shop organization is used for assembly there are a different set of problems, in particular the following:

- *Work flow progress and control.* The major difficulty in traditional job shops lies in controlling the flow of work and, in particular, finding out that operations are complete, issuing instructions to move jobs from one work center to another, and ensuring that the job selection for processing meets required priorities. Expediters and progress chasers are needed to follow up on jobs, and there is a divided responsibility between expediters, foremen, and operators on what job is to be done next. This can make it difficult to produce rush jobs.

- *Impact of diversity in processing time.* Diversity in processing times arising from different work content per job or different batch sizes can result in long flow times if work release is aimed at achieving high work station utilization. Job diversity also makes it difficult to make good estimates of flow times and to meet delivery promises.

- *Impact of diversity in routing.* Diversity in routing in assembly can arise either because of different processing requirements or because of the need for rework or reprocessing of jobs found to contain defects at inspection and test. This can result in a loss of capacity and, depending on the variability of the arrival and service processes and the effectiveness of control of work flow, an increase in work in process and flow times.

9.1.2 Attributes of Solutions

The solutions that have been developed for improving assembly system design have attempted (1) to use concepts from flow lines and job shops that are known to be beneficial, and (2) to adopt new concepts to overcome the shortcomings of flow lines and job shops.

Beneficial concepts from flow lines

- *Unpaced.* We showed that removing pacing effects generally enabled both productivity and quality to be improved as long as workers were allowed the time necessary to complete the job properly, and their performance did not degrade once given this responsibility.

- *Paralleling.* We also showed that paralleling work stations and thus increasing the assigned work content resulted in improved performance, at least as long as the increased work content did not result in too long a learning time (relative to the turnover rate that existed in the organization).

- *Groups rather than individuals.* Another way of achieving similar benefits to paralleling is to have groups of workers responsible for performing a set of tasks. Sometimes this will look like paralleling, but if the workers in parallel are not equally skilled then paralleling unequal workers can result in some production loss. The idea of group work is that the faster workers will help the slower workers, thus making the paralleled individuals more equal. Also with group incentive schemes the better workers are motivated to train the less skilled and competent workers.

Beneficial concepts from job shops

- *Unit load movement.* Unequal batch sizes and hence significant differences in work content per job (i.e., batch) worsen performance. This implies that performance would be improved by having all batch sizes more or less the same. Having the work content per batch the same for different batches would be better still for performance. In practice, if the system assembles jobs of the same family this can be achieved by designing standard-sized containers or totes, and always moving jobs in these containers. In many cases the preferable batch size will be one.

- *Flexible routing (totally or partially).* We also mentioned that one of the advantages of job shops is that, if there is a problem with a machine or process, flexible routing enables production to continue, even if the production of some particular job is halted because of the down machine. Some other job that does not use the down machine can continue to be produced, and thus all production is not halted because of the problem. Flexible routing also enables job routing to be changed as the job goes through the shop. For example, a job may be diverted to a repair or rework area if a problem is found on inspection. Of course, the manufacturing process may be such that totally flexible routing is not required because once certain stages have been successfully completed a job may never have to return to them. Thus the routing only needs to be sufficiently flexible to permit all technologically feasible routings.

New concepts to overcome flow-line and job-shop problems

Material Handling System Structure.   To overcome the problems with work-flow progress and control that characterize traditional job shops, flexible assembly systems limit the amount of work that is stored at work stations, and try and keep other work in process at a central location. Initially flexible assembly systems used (and many still use) the concept of *central storage, central dispatch, all jobs routed via central store* (see Figure 9.1). That is, no jobs are stored at work stations except the job in process and perhaps the next job to be processed. Otherwise all jobs are kept in a central store and are dispatched from this store to a work station only when that work station is ready to accept a new job. On completion of a job at a work station it is returned directly to the central store. Thus all local control over what to do next is removed from the individual workers. The responsibility for deciding what to do next is transferred to the central dispatcher who can thus make the decision on the basis of the latest priorities and with all possible jobs available to him. For this concept to be implemented there has to be a material handling system (usually a transporter) by which jobs can be sent from the dispatcher to specific work stations automatically, and there should also be an automatic

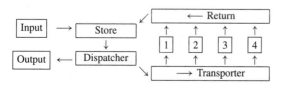

Figure 9.1   Central Storage, Central Dispatch System with All Jobs Routed via Central Store

return conveyor from the work stations back to the central store. Such systems have been widely used in the clothing industry (see [24]) and also in electronics assembly (see [6]). Note that with this implementation the return of a job to the central store is sufficient information to notify the dispatcher that a work station has completed a job. Hence the information needs of the dispatcher are restricted to information about the processing requirements and priorities of the jobs in the central store. In less sophisticated systems the dispatcher may dispatch a job to a set of parallel work stations with the exact work station that processes the dispatched job being determined locally at the paralleled work stations, depending on the availability of space at the work stations.

With improvements in information collection systems, such as the advent of bar code readers, this system can be modified to *central storage, central dispatch, random routing* (see Figure 9.2). That is, work flow is still controlled centrally (implying an automated materials handling system), but jobs do not always return to the central store. On completion of a job at a work station, the job then looks to see if the next station on its route is free and if so it moves to that station directly. Otherwise, if the station is not free the job would then return to the central store. Usually, there would be a small number (one or two) of storage spaces at each work station on its input and output side so that jobs can be held for short periods if either the next work station, the material handling system, or the information processing system (including the dispatcher) are busy. Nevertheless, the dispatcher still retains full control over work allocation and job movement, although dispatching may be automated. (See Manning [37] for an excellent description of such a system in the shoe industry. Its use in electronics assembly is illustrated in [35]). Because all jobs do not return to the central store the load on the dispatcher is reduced, and the likelihood that he will be the bottleneck is less.

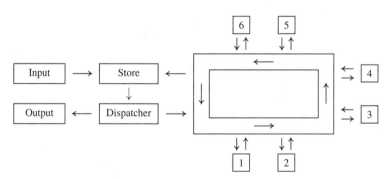

Figure 9.2 Central Storage, Central Dispatch System with Random Routing

For large and bulky items (such as automobile bodies) another possibility is *feedback storage, central dispatch* (see Figure 9.3). As jobs arrive they are dispatched to work stations and on successful completion of the task they usually proceed on to the next set of work stations. If quality problems arise so that the job requires repair and then reprocessing, however, the job is diverted to repair stations and then returns to the

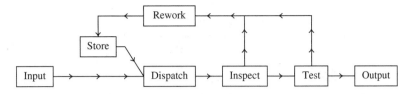

Figure 9.3   Feedback Storage, Central Dispatch System

dispatch point where it waits in a separate queue from the jobs arriving from outside the system. The dispatcher then has to decide whether to release a fresh job or a reprocessed job next, but usually it is only in the feedback loop that a significant amount of storage is provided.

Grouping of Stations.   Another new concept is *logical grouping rather than locational grouping* of stations. With central storage and central dispatch it is no longer necessary to put stations performing the same tasks (effectively parallel stations) together in the same area of the plant, because the central dispatch enables jobs to be routed to the stations wherever they are. This has the added advantage that, by changing the tools or moving people with different skills from one station to another, the set of stations assigned to a particular task (thus in effect paralleled) can be changed. Of course, this ability may be constrained by the material handling system if it does not possess total flexibility. A consequence of the ability to group stations logically is *modularity for expansion or contraction.* Modules, or sets of stations, can be added to the system as the demand for its products increases. Each module can be provided with a material handling system that can be linked to the overall material handling system, and functions or capabilities can be assigned to stations within the module to alleviate bottlenecks elsewhere in the system. Then if demand should decrease modules can be removed from the system. Thus overall capacity can be matched to demand. The actual distances traveled by jobs in such a system may be greater than in a comparable flow line or a comparable flexible system designed for a specific production rate and with the stations optimally located to minimize material handling distances. The ease of adapting the system to other production rates, however, compensates for this slightly higher material movement cost. In one approach to implementing the modular system there is a fixed "spine" that connects all modules, and each module has its own material handling system that connects stations within the module and also to the spine. As far as possible stations are provided with standard interfaces with the material handling system so as to make it easy to change the function of the station within the module. (See [23] and [15] for a description of such a system.)

Job and Part Delivery.   The material handling systems have to deliver jobs *and* parts to a work station. To retain flexibility and permit easy change-over of work stations, particularly those where human operators perform assembly tasks, it is desirable to try and deliver the job and associated parts as a kit. That is, when the job leaves the dispatcher it would have all the required parts with it. Another alternative is to arrange work stations around a carousel (automated storage and retrieval system [AS/RS]) and have the carousel deliver or present the parts to the station as required. If the station is automated or it assembles the same parts on all jobs then it would be more common to have the parts delivered directly to the station, perhaps using a different material handling system. When, because of size or weight, parts have to be delivered separately to work stations and different jobs require different parts then difficulties can arise in matching jobs and parts. Limited space for storage of such parts at a station means that it is desirable to try and match as closely as possible the sequence in which the parts are sent to the station and the sequence in which the jobs are processed at the station. Because of time delays in the part delivery system and the way in which job sequences can be affected by random events, it is difficult to match sequences exactly. Thus it is necessary to choose between either storing parts for several different jobs at the station so as to accommodate changes in job sequence or forcing jobs to follow a prescribed sequence. The latter may mean that jobs are blocked at the central store or at other machines resulting in a loss of throughput for the system.

The early flexible assembly systems used transporters that delivered jobs (and parts) to a station and then either had a separate system to return jobs to the central store, or used the transporter to move jobs from the station to another station or back to the central store. The latter is called "drop off/pick up." This is still used in many flexible assembly systems, although the means of delivering the job (and parts) may be an AGV. An alternative that is used frequently in the automotive industry is "docking," that is, having the material handling device or vehicle stay with the job all the time, thus enabling the material handling vehicle to provide an adjustable platform on which to carry out the assembly tasks on the job. Docking is attractive if the job is heavy or difficult to maneuver, and it eliminates the drop-off or pick-up time, at the expense of requiring more vehicles. (See [3] for a description of one of the first such systems in the automobile industry.)

9.1.3 "Putting-Out" System

The putting-out system was in widespread use before the advent of the factory system at the time of the Industrial Revolution. Materials were supplied to the workers by a contractor, and the work was performed at home, perhaps while also running a small farm or some other part-time occupation. The contractor oversaw both quality and the meeting of deadlines, although with the transport systems of the times the latter could not be closely controlled. The factory system supplanted putting out for a variety of reasons, some to do with the advent of powered machinery that was not appropriate for individual workers, and some to do with the need for tighter control over quality and task performance (and perhaps also reducing in process inventories). There is evidence that, at least in high value added assembly operations the putting-out system is becoming attractive again. There are several reasons for this. One is the advent of much-improved communication of information and the development of highly efficient package handling and delivery networks. Another is the development of PC-controlled machinery in such activities as weaving and knitting, which give very high-quality automatic production for low capital cost per unit of output and without significant economies of scale. Workers either form cooperatives or arrange for an association or service center to market the products, provide accounting services, coordinate purchase of raw materials, arrange credit, and monitor quality. Product design can either be the responsibility of individual workers, the cooperative as a whole, or the customers. The best developed examples of this appear to be in central and northeastern Italy where the industry sectors cover products with a high design or fashion component such as knitwear, pottery, ceramic tiles, and engineering products such as production equipment and farm machinery (see [1], [7], [11], [12]).

From a system viewpoint the putting-out system is essentially the same as a central dispatch flexible assembly system, particularly if design is done centrally or provided by customers. Thus, key issues are the assignment of jobs to individual workers and the monitoring of quality. The modern cooperative form of organization has, however, somewhat different objectives, in particular balancing individual worker goals with the goals of the cooperative. For example, the cooperative may prefer to give more work to the more productive or higher-quality worker, thereby benefiting the cooperative as a whole but perhaps reducing the income of some other less productive worker. The same issues can arise in conventional flexible assembly systems, however, where a faster worker may object to being assigned more work than a slower worker and thus may deliberately slow his pace of working. Also, in both systems it is sometimes appropriate to have workers who are kept idle even though there is work they could be doing.

9.2 DESIGN, PLANNING, AND CONTROL OF FLEXIBLE ASSEMBLY SYSTEMS

Flexible assembly systems require careful design and planning for successful implementation. If this is done then the scheduling and control issues are relatively straightforward from a systems perspective, although the specification, development, and testing of the software that implements the scheduling and control is a major management concern as the software has to provide for the changing configuration and product mix that the system will experience over its life.

9.2.1 Design

Flexible assembly systems achieve their performance because of the combination of the material handling system and the information system that they implement. By and large at the design stage the idea is to ensure that at the work station level there is the freedom to use varying degrees of automation. In general this should not be fixed in advance except perhaps for some specialized operations like soldering or oven drying. In many cases it is also desirable to allow for the number of work stations to change over time, although thought must be given as to how the material handling and information system will adapt to such a change. The key design issues that have to be resolved are the following:

- *Material handling system structure.* This is closely linked with the decision on the type of material handling system to be used. Essentially the options are (1) central storage, central dispatch, with all jobs routed via the central store; (2) central storage, central dispatch, with random routing (i.e., jobs do not always have to return to the central store but the system has the capability for this); and (3) feedback storage, central dispatch where any input storage would not have the random access usual with the central store in alternatives 1 and 2. In option 3 many jobs would only go through the system once, whereas some would be fed back and go through the system again. The flexible assembly system can be part of a larger system, consisting of flexible assembly cells that are linked together in series and share material handling facilities.

- *Material handling devices.* Transporter systems are common for systems of type 1 and can also be used for systems of type 2. If a transporter is used with systems of type 1 then it may be preferable to use a separate return conveyor to take jobs from work stations back to the central store. Alternatively, AGVs, power and free (normal or inverted) or car-on-track systems can be used, the essential feature being unit loads moved by a work carrier (see [4] for a description of these material handling systems). For each of these alternatives it would be necessary to decide whether the unit load is to remain on the work carrier at all times or whether a drop-off/pick-up system would be used.

- *Storage.* Another key issue concerns the provision of storage, both centrally and at work stations. In flexible assembly systems work station storage is usually restricted to the next job to be processed and the job just completed, although in random routing (option 2) there may sometimes be a need for a greater local storage capacity. For central storage systems it is desirable that the store permit any job to be selected (i.e., a carousel or AS/RS). The feedback systems may have just a queue for the new arrivals and a queue for the feedback jobs, however, and only the first job in either queue can be selected. Sometimes when such systems are used in automobile assembly it may be necessary to provide some output storage to either reestablish the input sequence or to modify the output sequence to meet certain sequence constraints for a subsequent assembly system.

- *Parts delivery.* If all jobs use the same parts then parts can be held at work stations, but this is rare in flexible assembly systems. In systems with central storage it is preferable for the dispatcher to send all the parts along with the job to the work station as a kit. Preparing the kits may be an additional task for the dispatcher. In systems without central storage parts would be delivered to the dispatch point, but it is then necessary to prespecify the sequence of jobs or maintain a sufficient stock of parts at the dispatch location.

- *Information system structure.* It is necessary to decide on the structure of the information system and whether dispatching will be automated or otherwise.

- *Modularity.* Another issue is the extent to which the system is designed to permit modular expansion (or contraction) and changes in the mix of different types of work stations. The system should be designed so that different types of work stations can be easily interfaced with the material handling system and so that work stations can be changed from one function to another. Related to this is the development of standard approaches to hold jobs so that it is possible to accommodate variations in product design.

- *Capacity and bottlenecks in the material handling or information system.* One issue of particular importance in both designing and planning flexible assembly systems is the avoidance of bottlenecks in either the material handling system or the information processing system. Because flexible assembly systems tend to result in more material and information movement it is essential that the system design involve a careful check for possible bottlenecks. There can often be intersecting, recirculating, and overlapping paths in the material flow, so bottlenecks are not always at machines but can be at other points in the system. Thus there has to be a careful and comprehensive evaluation of capacity, and how it is influenced by the layout of the material handling system. Also, systems with central dispatch and random routing tend to have a high volume of information to be processed, and thus it is also desirable to ensure that the system response remains acceptable.

9.2.2 Planning

Once the basic system has been designed there are several other issues that arise and that can also require reexamination as the mix of jobs changes. These are the following:

- *Number of material handling units.* The number of work carriers or units allowed in the system has to be decided. Often the work carriers are quite expensive so the goal is to minimize the number required. In drop-off/pick-up systems the number has to be sufficient to ensure that material handling will not be a bottleneck and stations are not blocked waiting for completed jobs to be moved. In docking systems the number of work carriers determines the number of jobs at work stations and in the central store. The permissible number of work carriers has to be related to the available space for storage of work carriers. Too many can sometimes result in a deterioration in performance because jobs are blocked, whereas too few may result in work stations being starved.

- *Central store size.* The central store has to be sized. Again there is a compromise between a store that is too small to give the dispatcher enough choice versus a store that is too large so that some "undesirable" jobs get delayed inordinately because of the other jobs in the system getting preference. In feedback systems the feedback store also has to be sized. If it is too small it may cause blocking. Further, if there is an output store for resequencing of jobs it too must be sized.

- *Control rules for work allocation.* Also at the planning stage it is necessary to identify the sorts of rules that will be used to allocate jobs to work stations. This is particularly important if automatic dispatch is going to be used, but in any case the general structure of the permitted rules has to be determined to avoid potential problems and loss of productivity. Because of the inherent complexity of flexible assembly systems from a systems viewpoint it is possible for people with a background in traditional manufacturing to propose rules that do not work well and result in a substantial deterioration in performance.

9.2.3 Control

Once the system is operating there are several issues that arise on a frequent basis. Some of these issues are the following:

- *Motivation of work force.* In a flexible assembly system there is no inherent pacing of the operator. Achieving the desired benefits, however, is critically dependent on the continuing motivation of the work force. In some systems workers who differ considerably in motivation and performance can create major problems because of the nature of the linkages that exist. Indeed in some cases it is better to close down a station with a poor worker than to allow him to remain in the system. Flexible assembly systems in some rather subtle respects give the worker considerable control over the system, and thus it is important to ensure that workers remain motivated, poor performance is identified, and appropriate training is provided.

- *Monitoring of performance (quality and quantity).* Usually quality problems create more job movements and increase the likelihood of bottlenecks being created. Thus it is important to monitor continually the performance of processes, robots, and workers so that any incipient quality problems can be identified. Also, because individual differences in skill and the time to perform a task can sometimes create problems, it is also important to monitor individual performance.

- *Local buffer size and occupancy.* It is usually possible to vary the use of local buffers. Sometimes it is desirable to use them fully to avoid overloading the material handling system. At other times, however, sending a job to a station too soon can result in a deterioration in performance because unexpected problems may develop with the job in process. This will delay the waiting job when it could have been sent to another station.

- *Regrouping of stations.* The dynamic capability of flexible assembly systems to change the nature of work stations, and the assignment of people or tasks to them means that this capability has to be used and appropriate rules developed.

- *Addition or deletion of stations.* In a flexible assembly system it is also possible to vary the number of stations that are in operation. Thus it is also necessary to develop rules for deciding when stations should be opened or closed. This decision is also influenced by the individual capability of the human or robot and the "balance" of the system.

9.2.4 Issues for Modeling

There are several issues in design, planning, and control for which stochastic models can provide insight.

Why flexible assembly systems? Perhaps the most important issue is why flexible assembly systems should be considered in the first place. In particular, what are the advantages of central storage and central dispatch, and in what circumstances are they particularly attractive in comparison with traditional assembly systems?

Material handling bottlenecks. Because job routing in a flexible assembly system is not fixed in advance but can depend on the need for rework, inspection, or test, it is necessary to have stochastic models to identify possible bottlenecks. These bottlenecks need not be at work stations or test facilities. It is possible that they can be at points where material handling paths cross.

Number of material handling units. If material handling is by discrete work carriers such as AGVs then stochastic models are needed to find how many AGVs are needed and how the number is influenced by whether pick up/drop off or docking is used. The closed queueing network models of Chapter 8 can be used to address this issue, although they may need to be modified to allow for the impact of limited storage space at work stations.

Central and local storage space. Throughput and other performance measures are influenced by the amount of space for jobs in the central store and at individual work stations. In particular it is necessary to understand how a system with no space at work stations apart from the space for the job in process will perform.

Incentives and equity. Although we will not propose any models that will suggest appropriate worker (or management) incentive structures, nevertheless existing manufacturing operations often place a high weight on equity (i.e., ensuring that all workers have the same amount of work expected from them over the day). That is, there should not be more work assigned to the faster worker just because he or she is faster. Stochastic models are required to understand the extent of lost throughput because of equity requirements.

Part delivery sequence. If part delivery systems require that job sequence be maintained throughout, then stochastic models are needed to determine how much throughput is lost because of resequencing delays.

Assignment of workers and tasks. In systems in which task assignment is determined by layout (e.g., in a conveyor system), it is necessary to decide which worker should be put at which station to optimize performance. In systems in which task assignment is determined by decisions made by a central dispatcher then it is necessary to have rules to guide the dispatcher in choosing which available worker should be given which available task.

9.3 MATERIAL HANDLING BOTTLENECKS IN FLEXIBLE ASSEMBLY SYSTEMS

One of the crucial design issues in flexible assembly systems is ensuring that the material handling system has sufficient capacity. Usually it is automated material handling that makes the obvious difference to traditional manufacturing, and the incremental investment in the material handling system has to be weighed against the benefits achieved through the new system structure and control. Because of the "nonproductive" nature of material handling it may well be designed with minimum surplus capacity, and as a result it can easily become the bottleneck. In FMSs or transfer lines, the relatively high cost of the work stations and hence the desire to ensure maximum utilization means that material handling usually has ample surplus capacity. Conversely, however, in flexible assembly systems in which work station investment is low, material handling often is designed with little margin.

The two crucial components of any material handling system are (1) the transporters and (2) the storage/retrieval system. For transport systems, because jobs are often randomly routed, traditional approaches to sizing the system do not give adequate answers. The modeling approach needed for designing transport systems has essentially already been described in Chapters 7 and 8, so we will present an example of an actual system to illustrate the application of the approach and also to indicate how it can be used to address some related issues in system design and operation.

9.3.1 Analysis of AGV System

The example is based on the AGV system installed in the paint shop of a leading automobile manufacturer. Car bodies are transported via conveyor through the automatic sealer line, the prime and interior color line, the wet sand booth, and then to one of two parallel first color booths, booth A and booth B (with 50% of bodies going to A and 50% to B) (see Figure 9.4). After leaving either first color booth A or B a body is transferred from conveyor to AGV, and this is the beginning of the AGV system. The AGV system enables the bodies to follow a variety of different routings depending on the quality of the painted body, and whether the body will be monotone or tu-tone (requiring another color to be applied). If the body is tu-tone, then the AGV system delivers it to the input conveyor and storage area for the tu-tone color booth. After leaving the tu-tone color booth on a conveyor the body is transferred back to an AGV, and then subsequent routings depend on quality. When either monotone or tu-tone bodies have completed color application and are on an AGV they are given an inspection. Four different possibilities then exist.

- If the quality is quite inadequate, the body is delivered by AGV to the beginning of the wet-sand booth. The AGV drops it off there, and the body will go through wet sand and then back to one of the color booths again. We will subsequently use p_{WS} to denote the fraction of bodies inspected that fall into this category.

- The body may be sent to the repair booth. It would be dropped off onto the repair booth conveyor, freeing up the AGV. After repair the body is picked up by an AGV and goes through a "finesse" lane for final polishing. Then the AGV delivers the body to the load-bar system, which delivers it to subsequent manufacturing stages. The fraction of bodies inspected that are sent to the repair booth will be denoted by p_R.

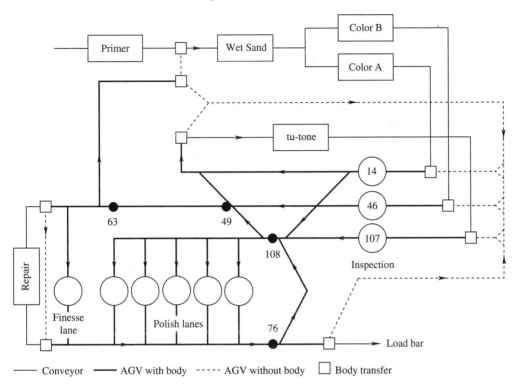

Figure 9.4   Paint Shop AGV System

- Most bodies are sent to the polishing hall where there are several parallel lanes at which any minor blemishes are removed while the body remains on the AGV. The fraction of bodies sent direct to polish is p_P. Sometimes during polish it is found that the defects are more serious so the body may be sent on to either the wet-sand booth or the repair booth. Otherwise the body is delivered to the load-bar system after polish. Denote by $p_{P,WS}$, $p_{P,R}$ and $p_{P,P}$ the relative frequencies of these three eventualities.

- Sometimes the body is free of any apparent blemishes or defects so it is then sent to the finesse lane and then to the load-bar system. The fraction of bodies inspected in this category will be denoted by p_F.

Although bodies coming from the tu-tone booth could also be sent to any of these four destinations, in practice, because of surplus capacity in the tu-tone booth and limited capacity in the repair booth, any necessary repairs would be carried out before the body leaves the tu-tone booth. Thus it can be assumed that all bodies coming from tu-tone will only go to polish. Also, we will assume that no tu-tone bodies sent to polish need subsequent repair or wet sand.

The AGV system uses AGVs that follow a guide path embedded in the floor. For control purposes, the guide path is divided up into segments, and the movement of the AGVs is controlled so that only one AGV is allowed to be in any segment at the same time. If guide paths merge or cross there is the possibility of conflict so the segment boundaries are defined so that there is no risk of collision while an AGV takes any of the possible paths through the intersection. Because it takes a finite time for the AGV to move through the intersection from one boundary to another, it is necessary to treat each intersection as equivalent to a work station where the task time is the time during which it is occupied by an AGV, and no other AGV can use it. It is possible that an intersection could be the system bottleneck. Figure 9.4 shows the structure of the AGV system with all intersection stations marked by numbers corresponding to their reference number in the AGV system control segment structure.

Before analyzing the system we have to define its boundaries. We choose the boundary shown on the figure so that we can represent the feedback of jobs through wet sand and the flow of jobs through the tu-tone and repair booths. The first step in analyzing the system is to represent the different possible routings of jobs through the system. To begin with we consider that the system processes two types of jobs: monotone and tu-tone. The normal routing of a monotone job is prime → wet sand → {A paint → 14} or {B paint → 46} → 108 → polish → 76 → load bar, whereas the normal routing of a tu-tone job is prime → wet sand → {A paint → 14} or {B paint → 46} → tu-tone → 107 → 108 → polish → 76 → load bar. Next we identify other possible routings.

    i. *Diversion to wet sand.* This routing is 14 or 46 → 49 → wet sand and then subsequent routing as if a new job arriving at wet sand is from prime.

    ii. *Diversion to repair.* This routing is 14 or 46 → 49 → 63 → repair → 76 → 108 → 49 → 63 → finesse → 76 → load bar.

    iii. *Diversion to finesse.* This routing is 14 or 46 → 49 → 63 → finesse → 76 → load bar.

    iv. *Polish diversion to wet sand.* This routing is polish → 76 → 108 → 49 → wet sand and then routing as if a new job arriving at wet sand is from prime.

    v. *Polish diversion to repair.* This routing is polish → 76 → 108 → 49 → 63 → repair and from then on as in point 2.

It is now possible to define the set of job classes. Because the time taken by an AGV to go through some of the intersection stations depends on source and destination, for example, a job following the route {107 → 108 → polish} takes a shorter time to clear 108 than a job going on the route {14 → 108 → polish}, it is necessary to define distinct classes for jobs that go through color booth A or color booth B. Thus the following 12 classes of jobs can be defined:

    1. Prime → wet sand → A paint → 14 → 108 → polish → 76 → load bar.

    2. B paint → 46 → 108 → polish.

3. Prime $\rightarrow$ wet sand $\rightarrow$ A paint $\rightarrow$ 14 $\rightarrow$ tu-tone $\rightarrow$ 107 $\rightarrow$ 108 $\rightarrow$ polish $\rightarrow$ 76 $\rightarrow$ load bar.

4. B paint $\rightarrow$ 46 $\rightarrow$ tu-tone,

5. 14 $\rightarrow$ 49 $\rightarrow$ wet sand.

6. 46 $\rightarrow$ 49 $\rightarrow$ wet sand.

7. 14 $\rightarrow$ 49 $\rightarrow$ 63 $\rightarrow$ repair $\rightarrow$ 76 $\rightarrow$ 108 $\rightarrow$ 49.

8. 46 $\rightarrow$ 49 $\rightarrow$ 63 $\rightarrow$ repair.

9. 14 $\rightarrow$ 49 $\rightarrow$ 63 $\rightarrow$ finesse $\rightarrow$ 76 $\rightarrow$ load bar.

10. 46 $\rightarrow$ 49 $\rightarrow$ 63 $\rightarrow$ finesse.

11. Polish $\rightarrow$ 76 $\rightarrow$ 108 $\rightarrow$ 49 $\rightarrow$ wet sand.

12. Polish $\rightarrow$ 76 $\rightarrow$ 108 $\rightarrow$ 49 $\rightarrow$ 63 $\rightarrow$ repair.

By design it is possible to define several transition probabilities between classes $p^{(l)(l')}$ at specific stations.

- Before wet sand $p^{(5)(1)} = 1$, $p^{(6)(1)} = 1$, and $p^{(11)(1)} = 1$.
- After wet sand $p^{(1)(1)} = 0.5$, $p^{(1)(2)} = 0.5$, $p^{(3)(3)} = 0.5$, and $p^{(3)(4)} = 0.5$.
- Before polish $p^{(2)(1)} = 1$.
- Before tu-tone $p^{(4)(3)} = 1$.
- Before 49 $p^{(7)(9)} = 1$.
- Before repair $p^{(8)(7)} = 1$, $p^{(12)(7)} = 1$.
- Before finesse $p^{(10)(9)} = 1$.

The other transition probabilities depend on the quality levels, in particular the fraction of bodies found to be in the various categories at inspection or at polish. Specifically, we have the following probabilities:

- After inspection point 14: $p^{(1)(1)} = p_P$, $p^{(1)(5)} = p_{WS}$, $p^{(1)(7)} = p_R$ and $p^{(1)(9)} = p_F$.
- Similarly, after inspection point 46: $p^{(2)(2)} = p_P$, $p^{(2)(6)} = p_{WS}$, $p^{(2)(8)} = p_R$ and $p^{(2)(10)} = p_F$.
- After polish $p^{(1)(1)} = p_{P,P}$, $p^{(1)(11)} = p_{P,WS}$ and $p^{(1)(12)} = p_{P,R}$.

For all other stations the probabilities $p^{(l)(l)} = 1$ for all l. Lastly, it is necessary to know p_{TT}, the fraction of jobs leaving prime that are tu-tone, and p_{MT}, the fraction of jobs leaving prime that are monotone.

It is then straightforward to determine the $v_i^{(l)}$, the number of visits to station i as a class l job for each job arriving at prime (see Section 7.3). This requires setting up a system of equations that are of very simple structure, for example,

$$v^{(1)}_{wet\ sand} = p_{MT} + v^{(5)}_{wet\ sand} + v^{(6)}_{wet\ sand} + v^{(11)}_{wet\ sand}$$

$$v^{(5)}_{wet\ sand} = v^{(6)}_{wet\ sand} = 0.5 p_{WS} v^{(1)}_{wet\ sand}$$

$$v^{(11)}_{wet\ sand} = p_{P,WS} v^{(1)}_{polish} = p_P p_{P,WS} v^{(1)}_{wet\ sand}$$

Hence

$$v^{(1)}_{wet\ sand} = \frac{p_{MT}}{1 - p_{WS} - p_P p_{P,WS}}$$

Similarly it can be shown

$$v^{(3)}_{wet\ sand} = p_{TT}$$

$$v^{(1)}_{polish} = \frac{p_P p_{MT}}{1 - p_{WS} - p_P p_{P,WS}}$$

$$v^{(3)}_{polish} = p_{TT}$$

$$v^{(7)}_{repair} = \frac{(p_R + p_P p_{P,R}) p_{MT}}{1 - p_{WS} - p_P p_{P,WS}}$$

$$v^{(9)}_{finesse} = \frac{(p_R + p_P p_{P,R} + p_F) p_{MT}}{1 - p_{WS} - p_P p_{P,WS}}$$

Hence it is then possible to determine the $v_i = \sum_l v^{(l)}_i$. For example,

$$v_{repair} = v^{(7)}_{repair}$$

$$v_{polish} = v^{(1)}_{polish} + v^{(3)}_{polish}$$

$$v_{finesse} = v^{(9)}_{finesse}$$

Further, it is possible to determine the $v^{(l)}_i$ for the key intersection on the AGV path, 108. We have

$$v^{(1)}_{108} = v^{(2)}_{108} = \frac{0.5 p_{MT}}{1 - p_{WS} - p_P p_{P,WS}}$$

$$v^{(3)}_{108} = p_{TT}$$

$$v^{(7)}_{108} = v_{repair}$$

$$v^{(11)}_{108} = p_{P,WS} v^{(1)}_{polish}$$

$$v^{(12)}_{108} = p_{P,R} v^{(1)}_{polish}$$

Because of the different routes taken by jobs through 108 the times during which the different classes of jobs occupy 108 differ. The following times apply:

- *Class 1.* 63 s.
- *Class 2.* 56 s.
- *Class 3.* 49 s.
- *Classes 7, 11, and 12.* 59 s.

Hence the capacity of 108 measured in terms of the maximum rate in jobs per hour at which jobs can leave prime without 108 becoming the bottleneck is given by

$$\lambda^*_{108} = \frac{3600}{63v^{(1)}_{108} + 56v^{(2)}_{108} + 49v^{(3)}_{108} + 59(v^{(7)}_{108} + v^{(11)}_{108} + v^{(12)}_{108})}$$

Example 9.1

Suppose that the quality-related probabilities are as shown in Table 9.1. Also suppose $p_{MT} = 0.86$ and $P_{TT} = 0.14$. Then the $v^{(l)}_i$'s are given by Table 9.2. Hence $\lambda^*_{108} = 45.9$ jobs per hour.

TABLE 9.1   QUALITY-RELATED PROBABILITIES

Probability	p_P	p_{WS}	p_R	p_F	$p_{P,P}$	$p_{P,WS}$	$p_{P,R}$
Value	0.88	0.03	0.088	0.002	0.86	0.07	0.07

TABLE 9.2   VALUES OF $v^{(l)}_i$

$v^{(l)}_i$	$v^{(1)}_{wet\ sand}$	$v^{(3)}_{wet\ sand}$	$v^{(1)}_{polish}$	$v^{(3)}_{polish}$	$v^{(7)}_{repair}$	$v^{(9)}_{finesse}$
Value	0.9467	0.14	0.8331	0.14	0.1416	0.1435

$v^{(l)}_{108}$	$v^{(1)}_{108}$	$v^{(2)}_{108}$	$v^{(3)}_{108}$	$v^{(7)}_{108}$	$v^{(11)}_{108}$	$v^{(12)}_{108}$
Value	0.4734	0.4734	0.14	0.1416	0.0583	0.0583

Similarly it is possible to determine the conditions under which other stations become the bottleneck on system throughput. For example with c_{polish} parallel polishing lanes and mean polish time $E[S^{(1)}_{polish}]$ for monotone bodies and $E[S^{(3)}_{polish}]$ for tu-tone bodies the capacity of polish is given by

$$\lambda^*_{polish} = \frac{c_{polish}}{v^{(1)}_{polish}E[S^{(1)}_{polish}] + v^{(3)}_{polish}E[S^{(3)}_{polish}]}$$

The contribution of the repair booth to determining system capacity is complicated by the fact that AGVs drop off the jobs to be repaired and then pick up a job that has completed repair. Thus, while the repair booth might be operating at capacity, this does not mean that the repair lane is necessarily going to limit AGV system throughput. In practice, because of tu-tone surplus capacity, the tu-tone booth functions as a backup repair facility to which jobs requiring repair are diverted when the repair booth and associated spaces for jobs waiting repair are full. The preceding analysis should be modified to reflect this and additional job classes introduced. The maximum throughput such that no diversion of bodies to tu-tone booth is necessary will be

$$\lambda^*_{repair} = \frac{1}{v^{(7)}_{repair} E[S^{(7)}_{repair}]}$$

It is of interest to note that although the number of AGVs in the system is much below the number of sections into which the AGV system is divided and thus lock-up does not appear possible, the number of AGVs that the system can accommodate in the loop $108 \rightarrow 49 \rightarrow 63 \rightarrow$ repair $\rightarrow 76 \rightarrow 108$ is less than the total number of AGVs in the system. Thus it is possible for all spaces in the loop to become occupied, and thus system lock-up will occur. This in fact happened when the system was first installed but, through changing the control logic to ensure that there is always an empty space in the loop, lock-up was eliminated.

The preceding analysis can be extended to investigate other operating procedures (which might change the definitions of the classes). For example, the finesse lane could be eliminated so that after repair all jobs go to a polish lane. Also some jobs coming from the tu-tone booth may be sent to repair or wet sand.

9.3.2 Access Times in AS/RSs

Very often in flexible assembly systems AS/RSs are used to store partially processed items. AS/RS are mostly fully (or at least partially) automated devices that perform storage and retrieval of material under the control of real-time computer systems. The state of the art in the measurement, positioning, and speed control and in electronic controls and computers is such that there are few problems in reliability in a well-designed AS/RS system. To design an AS/RS one should estimate the throughput capacities of different configurations of such systems. The basic component that determines the throughput of an AS/RS is the access time needed for order picking. We will next evaluate the access times for different configurations of AS/RSs.

We will first look at a one-dimensional AS/RS where there are n storage locations (or pigeon holes) available in an array as shown in Figure 9.5. First assume that the I/O station is located at the beginning of the array, and the crane that picks up the orders is always positioned at the I/O station when it is not in use. Let the unit of time be chosen such that the time needed for the crane to move from one location i to $i + 1$ is one unit of time. Let q_j be the probability that an arriving I/O (i.e., storage/retrieval) request is for the item in location j, $j = 1, \ldots, n$. Then if T_k is the access time to fill the kth I/O request then $\{T_k, k = 1, 2, \ldots\}$ are iid random variables with

$$P\{T_k = 2j\} = q_j, \qquad j = 1, \ldots, n$$

$$E[T_k] = 2 \sum_{j=1}^{n} j q_j,$$

$$E[T_k^2] = 2 \sum_{j=1}^{n} j^2 q_j$$

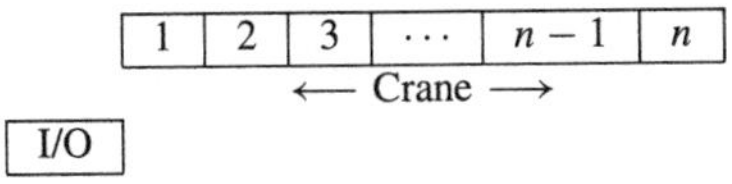

1	2	3	$\cdots$	$n-1$	n

$\longleftarrow$ Crane $\longrightarrow$

I/O

Figure 9.5   One-Dimensional AS/RS

Observe that we may position the different items at different locations according to their demand frequencies q_j, $j = 1, \ldots, n$. Clearly, if we position the most demanded item $j^* = \arg\{\max\{q_j, j = 1, \ldots, m\}\}$ at position 1, the access time will be reduced. It is indeed easily verified by pairwise interchanges that placing the items in such a way that position j has the item with demand frequency $q_{[j]}$, $j = 1, \ldots, n$ will minimize the access time in the sense of usual stochastic ordering. Note that $q_{[j]}$, $j = 1, \ldots, n$ is the decreasing rearrangement of q_j, $j = 1, \ldots, n$. The idea of keeping the most frequently demanded items closer to the I/O station is always a good rule in not only this but other AS/RS configurations as well.

The throughput of this system is

$$TH = 1/E[T_k].$$

When the demand frequencies are uniform among different items (i.e., $q_j = 1/n$) the mean access time is

$$E[T_k] = n + 1$$

and hence the throughput is

$$TH = 1/(n + 1)$$

Now suppose the I/O station is placed in the middle of the array (i.e., at location $(n + 1)/2$). Locations j and $n - j + 1$ are then equidistant from the I/O station. If n is even, that is, $n = 2l$ where l is an integer, the access time is given by

$$P\{T_k = n + 1 - 2j\} = q_j + q_{n-j+1} \qquad j = 1, \ldots, l.$$

When the demand frequencies are uniform among different items (i.e., $q_j = 1/n$) the mean access time is

$$E[T_k] = n/2$$

and hence the throughput is

$$TH = 2/n$$

If n is odd then $E[T_k] = n/2 - 1/2n$. It is worth noting that the throughput is doubled simply by placing the I/O station at the middle of the array as opposed to placing it at the beginning of the array. This underscores the importance of proper layout design of AS/RS to realize the full potential.

It is much commoner to have a two-dimensional AS/RS system in which several vertically arranged racks of the one-dimensional arrays discussed earlier are serviced by a crane that has the capability of simultaneous horizontal and vertical movement (see Figure 9.6). Let h be the time required to move horizontally from horizontal location i

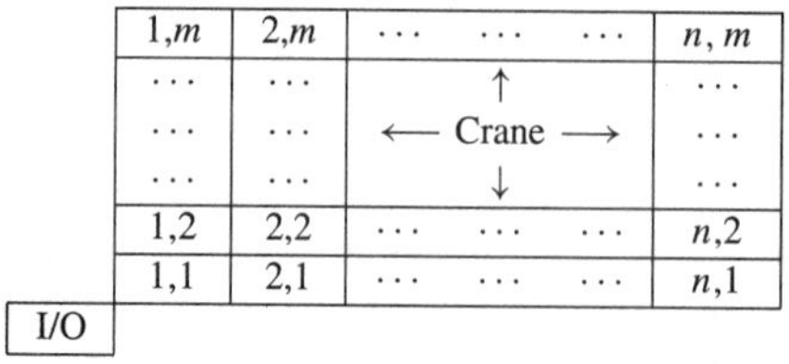

1,m	2,m	...	...	...	n, m
...	...		↑		...
...	...		← Crane →		...
...	...		↓		...
1,2	2,2	...	...	...	n,2
1,1	2,1	...	...	...	n,1

I/O

Figure 9.6   Two-Dimensional AS/RS

to $i + 1$ and v be the time required to move the crane vertically from vertical location j to $j + 1$. Ignoring the acceleration/deceleration times for start/stop stages, the travel time from location (i, j) to (i', j') is then

$$T_{(i,j)(i',j')} = \max\{|i' - i|h, |j' - j|v\}, \qquad i = 1, \ldots, n; j = 1, \ldots, m$$

Suppose we have the I/O station at (0,0), then the access time for order picking is given by

$$P\{T_k = 2\max\{ih, jv\}\} = q_{ij}, \qquad i = 1, \ldots, n; j = 1, \ldots, m$$

$$= 0, \qquad \text{otherwise}$$

where q_{ij} is the demand frequency of the item stored in location (i, j). If $h = v$ and $m = n$, the number of locations that can be accessed in time $2jh$ is $2j - 1$ (i.e., locations $(x, j), 1 \le x \le j$ and locations $(j, y), 1 \le y < j$). If the demand frequency is uniform (i.e., $q_{ij} = 1/n^2$),

$$P\{T_k = 2jh\} = \frac{2j - 1}{n^2}, \qquad j = 1, \ldots, n$$

$$= 0, \qquad \text{otherwise}$$

Hence

$$E[T_k] = \frac{(n + 1)(4n - 1)}{3n}h$$

Observe that when n is very large and we set $nh = 1$ unit of time, we have (as $n \to \infty$ and $nh \to 1$), $E[T_k] = 4/3$. The throughput of this system is

$$TH = \frac{3n}{(n + 1)(4n - 1)h}$$

As before the throughput of the system can be doubled by placing the I/O station at the middle (i.e., at location $((n + 1)/2, (n + 1)/2)$).

An alternative AS/RS that has improved throughput capabilities is the carousel system. The carousel AS/RS is a circular conveyor having many carriers with each carrier consisting of several levels of storage locations (see Figure 9.7). The order picking (i.e., L/U unit is stationary and has the capability to move up and down to access the different levels of the storage locations. The carousel rotates and aligns the correct carrier into the L/U device for removing or loading an item. Suppose the carousel

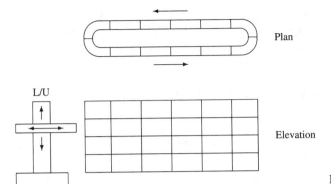

Figure 9.7   Carousel AS/RS

can move in only one direction, and the vertical movement of the L/U device is very fast. Then if the demand frequencies are uniform over all locations the access time and the throughput of this system has the same form as in a one-dimensional AS/RS described earlier with I/O device located at the beginning. Conversely if the carousel can move in both directions the access time and the throughput have the same form as that in a one-dimensional AS/RS described earlier with I/O device located in the middle.

9.4 FUNCTIONALLY IDENTICAL ASSEMBLY STATIONS WITH NO INFORMATION FEEDBACK

We will first consider models of assembly systems in which there is no information feedback to the dispatcher on the completion of work. These models will serve as a base case for evaluating the role of the dispatcher and the value of information feedback. Kits are delivered (probably from a central dispatch area) to m functionally identical work stations by a conveyor. The stations are arranged along the path of the conveyor in the order $1, 2, \ldots, s$ (see Figure 9.8). The work stations may or may not have storage space to store kits received for processing. Kits that cannot be withdrawn from the conveyor by station i will move to station $i + 1$ (i.e., the kit is said to overflow from station i to station $i + 1$), $i = 1, \ldots, m - 1$; if a kit on the conveyor is not withdrawn by any of the m stations it will be considered an overflow from the system and stored in an overflow storage area. The kits stored in the overflow storage may be processed by a separate worker or may be used as a backup supply whenever the conveyor breaks down or there is an interruption in kits supply at the input of the conveyor. Some of the design issues that arise in such a setting are (1) how much storage space, if any, should be assigned to each work station; (2) how much space is needed at the overflow

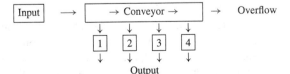

Figure 9.8   Conveyor System with Functionally Identical Stations

storage; and (3) if the workers have different efficiencies how should they be assigned to the work stations. We will next develop a stochastic model of this system to answer these questions. First, we will consider the conveyor system in which none of the work stations has any extra storage space beyond the space for the kit in process.

9.4.1 Overflow Conveyor with No Buffer Space at Stations

Suppose the kits arrive at the first station according to a renewal process $\{A_n, n = 1, 2, \ldots\}$ with interarrival times $\{\tau_n, n = 1, 2, \ldots\}$. We will assume that the processing times at station i form an iid sequence of exponential random variables with mean $1/\mu_i$, $i = 1, \ldots, m$. Because there is no extra buffer space at the first station, the number of kits in it behaves like the number of jobs in a $GI/M/1/1$ queue.

Let $N(t)$ be the number of kits at station 1. Then $\{N(t), t \geq 0\}$ is a semi-Markov process on the state space $\{0, 1\}$ associated with the (Markov) renewal times $\{A_n, n = 1, 2, \ldots\}$. Now define N_n to be the number of kits at station 1 when the nth kit arrives at the system at time A_n (i.e., $N_n = N(A_n-)$, $n = 1, 2, \ldots$). Suppose there was an arrival at time zero and that it was rejected (i.e., there is one kit in process at station 1 at time zero; $N_0 = 1$). Then the probability that the first kit arriving at time A_1 will also be rejected is given by

$$q_1 = \int_0^\infty e^{-\mu_1 x} dF_\tau(x) = \tilde{F}_\tau(\mu_1) \tag{9.1}$$

Therefore the number of kits K_1 arrived before the first rejection by station 1 after time zero has a geometric distribution with mean $1/q_1$. Because the time epochs at which kits are rejected from station 1 forms a regeneration epoch for the process $\{N_n, n = 1, 2, \ldots\}$, the stationary probability that an arrival sees one kit at station 1 is $1/E[K_1]$. Hence the fraction of kits rejected by (i.e., overflowed to station 2 from) station 1 is $1/E[K_1]$. This is called the blocking probability, say B_1, of station 1. Then

$$B_1 = \tilde{F}_\tau(\mu_1) \tag{9.2}$$

Because $\lambda = 1/E[\tau]$ is the kits arrival rate to station 1, one sees that the throughput rate of station 1 is

$$TH_1 = \lambda(1 - B_1) \tag{9.3}$$

and that the overflow rate from station 1 is

$$\lambda_2 = \lambda B_1 \tag{9.4}$$

Observe that the kits arrival rate to station 2 is λ_2. To analyze station 2, however, we need more than just the arrival rate. Actually we need the characterization of the arrival process. Let $\{A_n^{(2)}, n = 1, 2, \ldots\}$ be the kits arrival process to station 2. Clearly $\{A_n^{(2)}, n = 1, 2, \ldots\}$ is a subsequence of $\{A_n, n = 1, 2, \ldots\}$. Because the time epochs at which overflows occur are regeneration epochs for the semi-Markov process $\{N(t), t \geq 0\}$

associated with the (Markov) renewal times $\{A_n, n = 1, 2, \ldots\}$, it is clear that $\{A_n^{(2)}, n = 1, 2, \ldots\}$ is a renewal process. Let $\{\tau_n^{(2)}, n = 1, 2, \ldots\}$ be the corresponding interarrival times. Now conditioning on $A_1 = \tau_1$ and using the renewal argument one sees that the LST $\tilde{F}_{\tau^{(2)}}(s)$ of the distribution $F_{\tau^{(2)}}$ of $\tau_1^{(2)} = A_1^{(2)}$ is given by

$$
\begin{aligned}
\tilde{F}_{\tau^{(2)}}(s) &= \int_0^\infty e^{-\mu_1 x} e^{-sx} dF_\tau(x) + \int_0^\infty (1 - e^{-\mu_1 x}) e^{-sx} \tilde{F}_{\tau^{(2)}}(s) dF_\tau(x) \\
&= \frac{\tilde{F}_\tau(\mu_1 + s)}{1 - \tilde{F}_\tau(s) + \tilde{F}_\tau(\mu_1 + s)}
\end{aligned} \tag{9.5}
$$

Then station 2 will also behave like a $GI/M/1/1$ queue with interarrival time distribution $F_{\tau^{(2)}}$ and service rate μ_2. Continuing this way to stations $3, \ldots, m$ one has

$$
\begin{aligned}
B_i &= \tilde{F}_{\tau^{(i)}}(\mu_i) \\
TH_i &= \lambda_i (1 - B_i) \\
\lambda_{i+1} &= \lambda_i B_i, \quad i = 1, \ldots, m
\end{aligned} \tag{9.6}
$$

where

$$
\tilde{F}_{\tau^{(i+1)}}(s) = \frac{\tilde{F}_{\tau^{(i)}}(\mu_i + s)}{1 - \tilde{F}_{\tau^{(i)}}(s) + \tilde{F}_{\tau^{(i)}}(\mu_i + s)}, \quad i = 1, \ldots, m
$$

Here B_i is the blocking probability at station i, TH_i is the throughput of station i, λ_{i+1} is the kits overflow rate from station i, and $\{\tau_n^{(i+1)}, n = 1, 2, \ldots\}$ is the renewal interoverflow times of kits from station i whose distribution has LST $\tilde{F}_{\tau^{(i+1)}}(s)$, $i = 1, \ldots, m$. As initial conditions we have $\lambda_1 = \lambda$ and $\tau_n^{(1)} = \tau_n, n = 1, 2, \ldots$.

Observe that $LR = \lambda_{m+1}$ is the overflow rate from the system, and $TH = \sum_{i=1}^m TH_i$ is the throughput of the system. Then

$$
LR = \lambda \prod_{i=1}^m \tilde{F}_{\tau^{(i)}}(\mu_i)
$$

$$
TH = \lambda \left\{ 1 - \prod_{i=1}^m \tilde{F}_{\tau^{(i)}}(\mu_i) \right\} \tag{9.7}
$$

Arrangement of stations/assignment of workers.   Suppose $\mu_j < \mu_{j+1}$, and we consider a new system where worker $j + 1$ is assigned to station j, and worker j is assigned to station $j + 1$. Let $\tilde{F}'_{\tau^{(i+1)}}(s)$ be the Laplace transform of the interoverflow times of kits from station i of the new system $i = 1, \ldots, m$. Then using the observation that $\tilde{F}_\tau(s)$ is decreasing and convex in s it can be verified that

$$
\begin{aligned}
\tilde{F}'_{\tau^{(i+1)}}(s) &= \tilde{F}_{\tau^{(i+1)}}(s), \quad i = 1, \ldots, j - 1 \\
\tilde{F}'_{\tau^{(i+1)}}(s) &\le \tilde{F}_{\tau^{(i+1)}}(s), \quad i = j, \ldots, m
\end{aligned} \tag{9.8}
$$

Therefore comparing the mean interoverflow times in the new system and the original system one sees that

$$E[\tau^{(i+1)'}] = E[\tau^{(i+1)}], \qquad i = 1, \ldots, j-1$$

and

$$E[\tau^{(i+1)'}] \geq E[\tau^{(i+1)}], \qquad i = j, \ldots, m$$

Therefore we see that assigning the faster worker ahead of the slower worker increases the throughput of the system and reduces the overflow rate. Continuing this interchange one finds that a worker assignment such that $\mu_1 \geq \mu_2 \geq \cdots \geq \mu_m$ will maximize the throughput and minimize the overflow rate.

Sizing of overflow buffer storage. Observe that kits arrive at the overflow buffer according to a renewal process, say $\{A^{(m+1)}, t \geq 0\}$ with interrenewal times $\{\tau_n^{(m+1)}, n = 1, \ldots\}$. The LST of the distribution of $A_n^{(m+1)}$ is $\tilde{F}_{\tau^{(m+1)}}(s)$. Suppose the size of the overflow buffer storage is b_{m+1}. Then the kits in the overflow buffer will have to be cleared after every b_{m+1}th arrival. The clearing rate is therefore $1/\{b_{m+1}.E[\tau_m^{(m+1)}]\} = LR/b_{m+1}$. Now suppose the clearing cost is $f(b_{m+1})$ for each clearing. Then the clearing cost is $LRf(b_{m+1})/b_{m+1}$ per unit time. Therefore if $g(b_{m+1})$ is the per unit time cost of providing an overflow buffer storage of size b_{m+1}, we wish to

$$\min\{LRf(b_{m+1})/b_{m+1} + g(b_{m+1})|b_{m+1} \geq 1\}$$

Suppose $f(k)/k$ is decreasing and convex and $g(k)$ is increasing and convex. The optimal buffer capacity can then be easily obtained once LR is determined.

9.4.2 Overflow Conveyors with Finite Buffer Space at Stations

Consider an overflow conveyor at which kits arrive at times $\{A_n, n = 1, 2, \ldots\}$ according to a renewal process. The interarrival times are $\{\tau_n = A_n - A_{n-1}, n = 1, 2, \ldots\}$ with $A_0 = 0$. Station i has a buffer capacity b_i that can store kits, including the kit in service, if any. We will assume that the service times of kits at station i form an iid sequence of exponential random variables with mean $1/\mu_i$, $i = 1, \ldots, m$. We will assume that all m buffers are full at time zero and let $A_n^{(i)}$ be the time at which the nth kit overflow from station $i - 1$ occurs (that is, arrives at station i), $i = 1, \ldots, m$. Note that $A_n^{(1)} = A_n$, $n = 1, 2, \ldots$. Also let $\{\tau_n^{(i+1)} = A_n^{(i+1)} - A_{n-1}^{(i+1)}, n = 1, 2, \ldots\}$ be the sequence of interoverflow times from station i, $i = 1, \ldots, m$ ($A_0^{(i)} = 0, i = 1, \ldots, m$). Suppose $N^{(i)}(t)$ is the number of kits at station i ($i = 1, \ldots, m$) at time t. Define $N_n^{(i)} = N^{(i)}(A_n-)$ to be the number of kits in the station i when the nth kit arrives at the system. Then $\{N_n^{(1)}, \ldots, N_n^{(m)}\}$ is a Markov process and $\{\mathbf{N}(t), t \geq 0\}$ is a semi-Markov process with Markov renewal times $\{A_n, n = 1, 2, \ldots\}$. Furthermore

$$A_n^{(i)} = \inf\{A_k; A_k > A_{n-1}^{(i)}, N_k^{(j)} = b_j, j = 1, \ldots, i-1; k = 1, 2, \ldots\}$$

$$i = 2, \ldots, m+1; n = 1, 2, \ldots \qquad (9.9)$$

Then it is also clear that $\{A_n^{(i)}, n = 1, 2, \ldots\}$ is a renewal process and $\{\tau_n^{(i+1)}, n = 1, 2, \ldots\}$ is a renewal sequence for each $i = 1, 2, \ldots, m$. Therefore individually each station will behave like a $GI/M/1/b$ queueing system. So consider a $GI/M/1/b$ queue with interarrival times $\{\tau_n, n = 1, 2, \ldots\}$, service rate μ and finite buffer capacity b. Suppose an arriving customer finds k jobs in the system at time 0. Let T_k be the time at which an overflow occurs (i.e., an arriving customer sees b in the system). Let K be the number of kits serviced during the first interarrival time τ_1. Then $\{K|\tau_1\}$ has a truncated Poisson distribution. Then conditioning on τ_1 one has

$$\{T_k|\tau_1\} \stackrel{\mathrm{d}}{=} \tau_1 + T_{k+1-\{K|\tau_1\}} \tag{9.10}$$

If $\tilde{F}_k(s)$ is the Laplace transform of T_k and $\tilde{F}_b(s) = 1$, then taking the Laplace transform on both sides of the preceding equation one finds that for $k = 0, \ldots, b-1$

$$E\{e^{-sT_k}|\tau_1\} = e^{-s\tau_1} \left[\sum_{l=0}^{k} E\{e^{-sT_{k+1-l}}\} \frac{e^{-\mu\tau_1}(\mu\tau_1)^l}{l!} + \sum_{l=k+1}^{\infty} E\{e^{-sT_0}\} \frac{e^{-\mu\tau_1}(\mu\tau_1)^l}{l!} \right] \tag{9.11}$$

and

$$\tilde{F}_k(s) = \sum_{l=0}^{k} \frac{\mu^l}{l!} \tilde{F}_\tau^{(l)}(\mu+s) \tilde{F}_{k+1-l}(s) + \sum_{l=k+1}^{\infty} \frac{\mu^l}{l!} \tilde{F}_\tau^{(l)}(\mu+s) \tilde{F}_0(s) \tag{9.12}$$

where $\tilde{F}_\tau^{(l)}$ is the lth derivative of $\tilde{F}_\tau$, $l = 1, 2, \ldots$ and $\tilde{F}_\tau^{(0)} = \tilde{F}_\tau$. The preceding set of equations can be recursively solved for the Laplace transform $\tilde{F}_0(s), \tilde{F}_1(s), \ldots$ in that order. For example, we have for $b = 1$

$$\tilde{F}_0(s) = \frac{\tilde{F}_\tau(\mu+s)}{1 - \tilde{F}_\tau(s) + \tilde{F}_\tau(\mu+s)}$$

and for $b = 2$,

$$\tilde{F}_1(s) = \frac{\tilde{F}_\tau(\mu+s)}{1 - \mu\tilde{F}_\tau^{(1)}(\mu+s) - \tilde{F}_\tau(\mu+s) \left\{ \frac{\tilde{F}_\tau(s) - \tilde{F}_\tau(\mu+s) - \tilde{F}_\tau^{(1)}(\mu+s)\mu}{1 - \tilde{F}_\tau(s) + \tilde{F}_\tau(\mu+s)} \right\}}$$

Because the state just after an arrival is the same whether an arrival sees b or $b-1$ jobs in the system, the interoverflow times in a $GI/M/1/b$ system have a LST $\tilde{F}_{b-1}(s)$. Hence for an overflow system with a finite buffer capacity of two (i.e., $b_i = 2$, $i = 1, \ldots, m$), we have

$$\tilde{F}_{\tau^{(i+1)}}(s) =$$

$$\frac{\tilde{F}_{\tau^{(i)}}(\mu_i+s)}{1 - \mu_i\tilde{F}_{\tau^{(i)}}^{(1)}(\mu_i+s) - \tilde{F}_{\tau^{(i)}}(\mu_i+s) \left\{ \frac{\tilde{F}_{\tau^{(i)}}(s) - \tilde{F}_{\tau^{(i)}}(\mu_i+s) - \tilde{F}_{\tau^{(i)}}^{(1)}(\mu_i+s)\mu_i}{1 - \tilde{F}_{\tau^{(i)}}(s) + \tilde{F}_{\tau^{(i)}}(\mu_i+s)} \right\}},$$

$$i = 1, \ldots, m \tag{9.13}$$

Hence the loss rate for this system is

$$LR = 1/E[\tau^{(m+1)}] = -1/\tilde{F}^{(1)}_{\tau^{(m+1)}}(0) \qquad (9.14)$$

and the throughput rate of this system is

$$TH = 1/E[\tau^{(1)}] - 1/E[\tau^{(m+1)}]$$
$$= \lambda - LR \qquad (9.15)$$

9.4.3 Recirculating Conveyor with Finite Common Buffer

Consider a recirculating conveyor that transports kits to the stations $1, 2, \ldots, m$. These m stations are arranged in the same order along the path of the conveyor. There is a limited common buffer at the input to the conveyor where kits arrive according to a Poisson process with rate λ (see Figure 9.9). If the buffer space is full the arriving kit is rejected. Kits from the central buffer are loaded onto the conveyor whenever space is available there. If a server becomes free it will wait for the next kit on the conveyor to pass by and pick it for processing. There is no storage space available for extra kits at any of the work stations. The processing times of kits at station i form an iid sequence of exponential random variables with mean $1/\mu_i$, $i = 1, \ldots, m$. We will assume that the conveyor speed is such that the transportation delays can be ignored. The total space available for storing kits at the central buffer, conveyor, and work stations is b. When there are more than m kits in the system the system as a whole will behave like a $M/M/1/b$ queue with arrival rate λ and service rate $\hat{\mu} = \sum_{i=1}^{m} \mu_i$. When the number of kits in the system is less than or equal to m, the system will behave like the loss system studied earlier. The performance of this system can be obtained by appropriate convex combination of the preceding two cases. The mean duration of the interval during which there are more than m kits in the system will be the same as the mean busy cycle in a $M/M/1/b - m$ queue, that is, $(1 - \hat{\rho}^{b-m})/[\hat{\mu}(1 - \hat{\rho})]$. The mean interoverflow time in the "overflow" system is $1/\{\lambda \prod_{i=1}^{m} \tilde{F}_{\tau^{(i)}}(\mu_i)\}$, and this will be the same as the mean duration of intervals when there are no customers in the $M/M/1/b - m$ queue. Then if a is the fraction of time the number of kits in the system is less than or equal to m, we have

$$a = \frac{1 - \hat{\rho}}{1 - \hat{\rho} + \hat{\rho}(1 - \hat{\rho}^{b-m}) \prod_{i=1}^{m} \tilde{F}_{\tau^{(i)}}(\mu_i)} \qquad (9.16)$$

Then the probability $p(n)$ that there are n kits waiting in the buffer or recirculating on the conveyor is

$$p(n) = (1 - a) \left\{ \frac{1 - \hat{\rho}}{\hat{\rho} - \hat{\rho}^{b-m+1}} \right\} \hat{\rho}^n, \qquad n = 1, \ldots, b - m$$

$$p(0) = a \qquad (9.17)$$

One may improve the performance of an assembly system by having an allocation policy such as dispatching an item only when a station is available, thereby avoiding

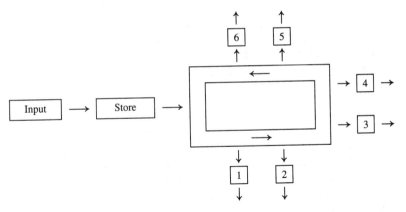

Figure 9.9   Recirculating conveyor system

overflows. Alternatively, it may be necessary to carry out an equitable assignment of kits to workers so as to avoid any possibility of slowdown of workers. Therefore we will next consider assembly systems with allocations performed by central dispatch.

9.5 FUNCTIONALLY IDENTICAL ASSEMBLY STATIONS WITH CENTRAL DISPATCH

We consider a flexible assembly system consisting of m parallel stations. These m assembly stations have the same functional capabilities but may have different efficiencies. Let $1/\mu_i$ be the mean time needed to assemble one kit at station i, $i = 1, \ldots, m$. Components for assembly (i.e., kits) are dispatched from a central dispatch area. A dispatcher stationed at the dispatch area has information on the status of all m stations and dispatches components to assembly stations based on this information. Usually the processing time at the dispatch area and the transportation times are negligible in comparison to the assembly time. Hence ignoring these delays, it is easily seen that the maximum achievable throughput from such an assembly system is

$$TH^* = \sum_{i=1}^{m} \mu_i \tag{9.18}$$

To achieve such a throughput, the dispatcher needs, over the long run, to assign μ_i kits per unit time to assembly station i, $i = 1, \ldots, m$. This implies that a faster station will be assigned more kits per unit time than a slower station. Because of worker dissatisfaction with the inequity of such a work assignment, in a manual assembly system this may lead to slowdown among stations to an extent that the overall assembly rate of any individual station will fall below $\mu_{\min} = \min\{\mu_i, i = 1, \ldots, m\}$. To avoid this, the dispatch policy as a precondition should be equitable to all m stations. Suppose we have at least m kits available all the time. Then we can keep the slowest station busy all the time by assigning one kit each time it completes assembly of a kit. This way the assembly rate at the slowest station can be kept at $\mu_{\min}$. Now dispatching kits to the other stations so as to match the number of dispatches to the slowest station, one can achieve an assembly rate of $\mu_{\min}$ from the other stations as well. Consequently the maximum throughput that can be achieved with a constraint on an equitable assignment of kits is

$$TH^e(m) = m\mu_{\min} \tag{9.19}$$

Observe here that the throughput of the system is very tightly controlled by the slowest (i.e., the bottleneck) station even though there is *no* material flow between the m stations. Thus it is questionable whether the overall throughput of such a system can be improved by adding additional stations. To answer this let us look at the optimal number of stations to operate when m stations with assembly rates μ_i, $i = 1, \ldots, m$ are available for inclusion. Without a loss of generality assume that $\mu_1 \geq \mu_2 \geq \cdots \geq \mu_m$. Then the maximum throughput obtainable with k stations and equitable kit allocation is

$$TH^e(k) = k\mu_k, \qquad k = 1, \ldots, m$$

Therefore the optimal number of stations to operate (given a pool of m available assembly stations) is

$$k^* = \arg\max\{TH^e(k) : k = 1, \ldots, m\}$$

It is not difficult to see that k^* need not be equal to m.

As pointed out earlier we can achieve the maximum throughput TH^e with m assembly stations as long as at least m kits are always available. Next we will see what will happen if we have only $n(< m)$ kits available all the time and see whether we can achieve the same maximum throughput. Suppose $n = 1$. Then an equitable assignment of kits among the m stations implies that we cyclically assign the kits whenever the preceding kits are assembled. Therefore the mean time between the assignment of consecutive kits to station k is $\sum_{i=1}^m 1/\mu_i$. Therefore the overall throughput rate of assemblies is

$$TH^e(m, 1) = m / \sum_{i=1}^m 1/\mu_i$$

Clearly $TH^e(m, m) > TH^e(m, 1)$ for $m > 1$. That is, the availability of just one kit is insufficient to achieve the maximum throughput $TH^e(m)$ with m stations. Next we will construct a lower bound on the minimum number of kits we need to have to achieve the maximum throughput $TH^e(m)$. Suppose we have k kits available to us always. Let $\hat{T}H(k)$ be the maximum throughput achievable under an equitable kit allocation policy. Let $\bar{T}_i$ be the average flow time and $\bar{N}_i$ be the average number of kits available at station i, $i = 1, \ldots, m$. The rate at which kits are allocated to station i is $\hat{T}H(k)/m$, $i = 1, \ldots, m$. Then by Little's result $(\hat{T}H(k)/m)\bar{T}_i = \bar{N}_i$, and hence

$$\frac{\hat{T}H(k)}{m} \sum_{i=1}^{m} \bar{T}_i = k$$

Because $\bar{T}_i \geq 1/\mu_i$, $i = 1, \ldots, m$, one has

$$\hat{T}H(k) \leq mk / \sum_{i=1}^{m} 1/\mu_i$$

Therefore for $\hat{T}H(k)$ to be greater than or equal to $TH^e(m) = m\mu_{\min}$, we need

$$k \geq \sum_{i=1}^{m} \frac{\mu_{\min}}{\mu_i}$$

9.5.1 Equitable Kit Allocations with Exponential Assembly Times

Two obvious ways to carry out equitable kit allocations are

- Cyclic assignment.
- Random assignment with equal probability $(1/m)$ for all stations.

We will next illustrate this for an assembly system with two stations ($m = 2$) and exponential assembly times. Let k be the number of positions at the dispatch area that can be accessed by the dispatcher for assigning kits to the stations. As kits arrive at the dispatch area they are tagged with their destination station number in accordance with the assignment policy so all the kits in these k positions would be tagged with a destination station number. We will also assume that there is always at least $k + 2$ kits in the system available for the system (i.e., all k positions will be filled all the time).

Cyclic assignment.   First we consider the cyclic assignment policy. Let $Z(t)$ be the station index of the kit that will enter the access area next and $N_i(t)$ be the number of kits assigned (or tagged) with station index i, $i = 1, 2$, at time t. Note that $N_i(t)$ includes the kit being assembled at work station i at time t, $i = 1, 2$. Then $\{(Z(t), N_1(t), N_2(t)), t \geq 0\}$ is a continuous time Markov process on the state space $S = \{(1, 0, k + 1), (j, n_1, n_2), j = 1, 2; n_1 + n_2 = k + 2, (2, k + 1, 0)\}$. The transition rate diagram of this Markov process is shown in Figure 9.10. Let $p(j, n_1, n_2) = \lim_{t \to \infty} P\{Z(t) = j, N_i(t) = n_i, i = 1, 2\}$, $(j, n_1, n_2) \in S$ be the stationary distribution of this Markov process. Then from the transition rate diagram (which is the same as that

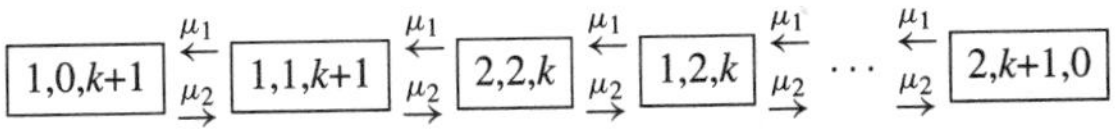

Figure 9.10   Transition Rate Diagram for Cyclic Assignment

for a birth-death process) it is easily derived that

$$p(1, 0, k+1) = \frac{1 - \left(\frac{\mu_2}{\mu_1}\right)}{1 - \left(\frac{\mu_2}{\mu_1}\right)^{2k+2}}$$

$$p(1, n_1, k+2-n_1) = \left\{ \frac{1 - \left(\frac{\mu_2}{\mu_1}\right)}{1 - \left(\frac{\mu_2}{\mu_1}\right)^{2k+2}} \right\} \left(\frac{\mu_2}{\mu_1}\right)^{2n_1-1}, \qquad n_1 = 1, 2, \ldots, k$$

$$p(2, n_1, k+2-n_1) = \left\{ \frac{1 - \left(\frac{\mu_2}{\mu_1}\right)}{1 - \left(\frac{\mu_2}{\mu_1}\right)^{2k+2}} \right\} \left(\frac{\mu_2}{\mu_1}\right)^{2n_1-2}, \qquad n_1 = 2, \ldots, k+1$$

$$p(2, k+1, 0) = \left\{ \frac{1 - \left(\frac{\mu_2}{\mu_1}\right)}{1 - \left(\frac{\mu_2}{\mu_1}\right)^{2k+2}} \right\} \left(\frac{\mu_2}{\mu_1}\right)^{2k+1} \tag{9.20}$$

Therefore the throughput of this system with k accessible positions is

$$TH^c(2, k) = \mu_1(1 - p(1, 0, k+1)) + \mu_2(1 - p(2, k+1, 0))$$

$$= 2\mu_1\mu_2 \left\{ \frac{\mu_1^{2k+1} - \mu_2^{2k+1}}{\mu_1^{2k+2} - \mu_2^{2k+2}} \right\} \tag{9.21}$$

It is easily verified that for $k > 0$, $TH^c(2, k) < TH^e(2)$ and $\lim_{k\to\infty} TH^c(2, k) = 2\mu_{\min} = TH^e(2)$. When $\mu_1 = \mu_2 = \mu$,

$$TH^c(2, k) = 2\mu \left\{ 1 - \frac{1}{2k+2} \right\} \tag{9.22}$$

Thus we see the effectiveness of increasing the size (k) of the access area on the system throughput very rapidly decreases.

Random assignment.   Now we will consider the random assignment policy where each kit transferred into the access area is randomly assigned one of the m station indices with equal probability. Let $N_i(t)$ be the number of kits in the access area assigned to station i including any kit that may be present at the station i at time t, $i = 1, \ldots, m$. Then $\{(N_1(t), \ldots, N_m(t)), t \geq 0\}$ is a Markov process on $S = \{\mathbf{n} : |\mathbf{n}| = k + m\}$. This process is probabilistically identical to the number of jobs in a closed Jackson queueing network with $(k+m)$ jobs and symmetric routing (i.e., each job departing a server selects its destination among all the m servers with equal probability). Then (see Section 8.4) the stationary distribution $\{p(\mathbf{n}), \mathbf{n} \in S\}$ of this process is given by

$$p(\mathbf{n}) = \frac{\prod_{i=1}^m 1/\mu_i^{n_i}}{\sum_{\mathbf{n}' \in S} \prod_{i=1}^m 1/\mu_i^{n_i'}}, \qquad \mathbf{n} \in S \tag{9.23}$$

If $TH^r(m, k)$ is the throughput of the system with random assignment one has

$$TH^r(m, k) = m \left\{ \sum_{\mathbf{n}' \in \mathcal{S}'} \prod_{i=1}^{m} 1/\mu_i^{n'_i} \right\} / \left\{ \sum_{\mathbf{n} \in \mathcal{S}} \prod_{i=1}^{m} 1/\mu_i^{n_i} \right\} \tag{9.24}$$

where $\mathcal{S}' = \{\mathbf{n}' : |\mathbf{n}'| = k + m - 1\}$. It can be easily verified that (see Section 8.4) $TH^r(m, k)$ is increasing in k, $\lim_{k \to \infty} TH^r(m, k) = m.\mu_{\min}$ and therefore $TH^r(m, k) < m.\mu_{\min} = TH^e(m)$ for $k \geq 0$. When $\mu_i = \mu$, $i = 1, \ldots, m$ one finds that

$$TH^r(m, k) = m.\frac{m + k}{2m + k - 1}\mu \tag{9.25}$$

For $m = 2$, $TH^r(2, k) = 2\mu \{1 - 1/(k + 3)\}$.

Observe that for $k = 1$, $TH^r(2, 1) = TH^c(2, 1)$ and for $k \geq 2$, $TH^c(2, k)$ is slightly better than $TH^r(2, k)$. For $k > 1$, none of these two assignments will achieve the maximum throughput that is achievable under the constraint of equitable kit allocation. For either $k = 1$ or as $k \to \infty$, however, these two policies would result in the maximum throughput. Because the implementation of such policies is straightforward it would not be surprising to come across these policies in real systems.

9.5.2 Kit Allocation with External Delivery

Next we will focus our attention on systems where kits for assembly arrive from outside and are received at the dispatch area. As before, the kits received are dispatched to the assembly station according to some dispatch policy. In such a scenario one is usually interested in choosing the dispatch policy so as to minimize the mean number of kits in the assembly system.

Suppose the only restriction we have on the dispatch policy is that we cannot intentionally keep a station idle when a kit is available for dispatch to that station. Then in such a case, the only times at which allocation decisions have to be made are the arrival times of kits that see two or more idling stations. In such an instance the arriving kit should be allocated to one of the idling stations. An obvious policy that could minimize the number of kits in the system allocates the arriving kit to the fastest station idling at that time. When the processing times are exponentially distributed, this policy is indeed optimal.

The dynamic behavior of the number of kits in the assembly system with this policy is probabilistically the same as the number of jobs in a recirculating conveyor system with infinite common buffer space (see Section 9.4.3). Then if N is the total number of kits in the dispatch area under steady-state conditions, we have

$$P\{N = 0\} = a \tag{9.26}$$

and

$$P\{N = n\} = (1 - a)(1 - \hat{\rho})\hat{\rho}^{n-1}, \quad n = 1, 2, \ldots \tag{9.27}$$

where

$$\hat{\rho} = \lambda / \sum_{i=1}^{m} \mu_i$$

and a is given by equation 9.16. When the transportation times are comparable with the assembly times, it may be no longer advisable to wait until a station becomes idle before allocating the next available kit to that station. In such a situation it is advisable to preassign at least one kit to a station selected according to some allocation policy. That is, we are required to assign the kit as soon as it arrives at the head of the dispatch line. Even in this case it seems desirable to allocate this kit to the station that has the least mean remaining work. When the processing times are exponentially distributed the preceding policy will always assign the kit at the head of the dispatch line to the fastest station. (In general if we have to preassign the first k kits in the dispatch line, they should be preassigned to the k fastest stations.)

We will now analyze the performance of the two equitable dispatch policies: (1) cyclic assignment and (2) probabilistic assignment. In this analysis we will assume that the kits arrive at the dispatch area according to a Poisson process with rate λ, that the assembly times are exponentially distributed, and that the dispatcher can access all kits stored in the dispatch area.

Cyclic assignment.   Suppose the first arrival of a kit is assigned to station 1. Then $(mk + j)$th arrival will be assigned to station j, and this will be the $(k + 1)$th kit to be assigned to station j, $j = 1, \ldots, m$; $k = 0, 1, \ldots$. The interarrival times of kits to station j are iid Erlang-m random variables with mean m/λ. Hence the number of kits N_i at station i has the same distribution as the number of jobs in an $E_m/M/1$ queueing system with arrival rate λ/m and service rate μ_i, $i = 1, \ldots, m$. Then (see Section 4.3)

$$E[N_i] = \frac{\sigma_i}{1 - \sigma_i}, \qquad i = 1, \ldots, m \tag{9.28}$$

where σ_i $(0 < \sigma_i < 1)$ is the solution to

$$\sigma = \left(\frac{\lambda}{\lambda + \mu_i - \mu_i \sigma} \right)^m, \qquad i = 1, \ldots, m$$

Random assignment.   Because the arrival stream of kits are randomly split, the kits will arrive at the assembly station according to a Poisson process with rate λ/m. Therefore each station i will behave like a $M/M/1$ queue with arrival rate λ/m and service rate μ_i, $i = 1, \ldots, m$. Therefore in this case (see Section 3.3)

$$E[N_i] = \frac{\rho_i}{1 - \rho_i}, \qquad i = 1, \ldots, m \tag{9.29}$$

where $\rho_i = \lambda/(m\mu_i)$, $i = 1, \ldots, m$. It is easily verified that $\sigma_i < \rho_i$ for $m > 1$. Hence it can be seen that the mean number of kits in the system under the random assignment policy is larger than that under the cyclic assignment policy. Particularly when m is large, the difference between these two quantities can be very large.

9.6 SINGLE-SEQUENCE FLEXIBLE ASSEMBLY SYSTEM

We consider a flexible assembly system consisting of m stations. Several different products are assembled in this system. The sequence of stations visited by all the product types are the same, however, say $\{1, \ldots, m\}$. We will next compare two configurations of such an assembly system. In system 1 each station has its local buffer space for incoming kits. Let b_i be the buffer capacity for storing incoming kits including the one in assembly, if any, at station i, $i = 1, \ldots, m$. In system 2, there is a common buffer with a total capacity of $\sum_{i=1}^{m} b_i - m$. All kits processed at station i will go to the common buffer from which kits are dispatched to station $i + 1$ whenever it becomes idle, $i = 1, \ldots, m - 1$. Kits departing from station m are completely assembled and thus leave the system (see Figure 9.11). In both systems the total number of kits in the system is kept constant at n ($\leq \sum_{i=1}^{m} b_i$). We will next show that the throughput attained from the common buffer system 2 is better than that obtained from the local buffer system 1.

Suppose at time 0 there are n_i kits waiting for assembly at station i ($n_i \leq b_i$), $i = 1, \ldots, m$ in both systems. Let $S_k^{(i)}$ be the assembly time of the kth kit to be assembled at station i, $i = 1, \ldots, m$. Then if $D_k^{(i):1}$ ($D_k^{(i):2}$) is the time at which the kth movement of kit from station i to the station $(i + 1)$ local buffer (to the central buffer) occurs (see Section 5.6.2)

$$D_k^{(i):1} = \max \left\{ \left[\max \left\{ D_{k-n_i}^{(i-1):1}, D_{k-1}^{(i):1} \right\} + S_k^{(i)} \right], D_{k+n_{i+1}-b_{i+1}}^{(i+1):1} \right\}, \qquad i = 1, \ldots, m \quad (9.30)$$

and

$$D_k^{(i):2} = \max \left\{ D_{k-n_i}^{(i-1):2}, D_{k-1}^{(i):2} \right\} + S_k^{(i)}, \qquad i = 1, \ldots, m \quad (9.31)$$

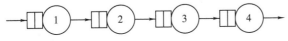

System 1: Local Buffers

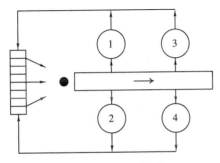

System 2: Common Buffer

Figure 9.11 Local Buffer and Common Buffer Assembly Systems.

where $b_{m+1} = b_1$, $D_k^{(m+1):j} = D_k^{(1):j}$, $j = 1, 2$, and $D_k^{(i):j} = 0$, $i = 1, \ldots, m + 1$; $j = 1, 2$; $k \leq 0$. Now using induction it can be verified that

$$D_k^{(i):1} \geq D_k^{(i):2}, \qquad i = 1, \ldots, m; k = 1, 2, \ldots \tag{9.32}$$

Because the throughput $TH^{(j)}$ of system j is given by $TH^{(j)} = \lim_{k \to \infty} k/E[D_k^{(m):j}]$, $j = 1, 2$, one has

$$TH^{(1)} \leq TH^{(2)} \tag{9.33}$$

It is not difficult to see that when $n \leq \min\{b_i, i = 1, \ldots, m\}$ there will be no blocking and then $D_k^{(i):1} = D_k^{(i):2}$, $i = 1, \ldots, m$; $k = 1, 2, \ldots$, and $TH^{(1)} = TH^{(2)}$.

As was the case in the flow-line system the departure times are convex in the service times for both systems. Therefore if the service times all have NBUE (NWUE) (see Appendix C) distributions, then the throughput evaluated assuming that the assembly times are exponentially distributed will be a lower (upper) bound on the actual throughput.

9.6.1 Central Buffer Assembly Systems with Exponential Assembly Times

Here we consider an assembly system that has a central buffer of size b and assume that the processing times are exponentially distributed (i.e., $\{S_k^{(i)}, k = 1, 2, \ldots\}$ are exponential random variables with mean $1/\mu_i$). It is then easily seen that such an assembly system with n kits ($n \leq b$), behaves like a Jackson closed queueing network. Let $p(\mathbf{n})$ be the stationary distribution of n_i kits waiting for assembly at station i, $i = 1, \ldots, m$. Then (see Section 8.4)

$$p(\mathbf{n}) = \frac{\prod_{i=1}^{m}(1/\mu_i)^{n_i}}{\sum_{\mathbf{n}' \in S_n} \prod_{i=1}^{m}(1/\mu_i)^{n_i'}}, \qquad \mathbf{n} \in S_n, \tag{9.34}$$

where $S_n = \{\mathbf{n} : |\mathbf{n}| = n, \mathbf{n} \in \mathcal{N}_+^m\}$. Also

$$TH(n) = \sum_{\mathbf{n} \in S_{n-1}} \prod_{i=1}^{m}(1/\mu_i)^{n_i} \Big/ \sum_{\mathbf{n}' \in S_n} \prod_{i=1}^{m}(1/\mu_i)^{n_i'} \tag{9.35}$$

From the properties of $TH(n)$ (see Section 8.4) we have the following:

1. Balancing the assembly times among all the m stations would maximize the throughput (i.e., if $m/\mu = \sum_{i=1}^{m} 1/\mu_i$, then

$$TH(n) \leq \widehat{TH}(n) = \frac{n}{m + n - 1}\mu$$

where $\widehat{TH}(n)$ is the throughput of the assembly system when the assembly times at all m stations are exponentially distributed with mean $1/\mu$).

2. The throughput as a function of the number of kits (n) assigned to the system is increasing and concave in n.

3. The maximum throughput achievable by this system is $\mu_{\min} = \min\{\mu_i, i = 1, \ldots, m\}$.

9.6.2 Local Buffer Assembly Systems with Exponential Assembly Times

Consider an assembly system consisting of m stations $\{1, \ldots, m\}$. Station i has a local buffer of size b_i where the kits needing assembly at station i and the kit being assembled at station i are stored, $i = 1, \ldots, m$. The assembly times $\{S_k^{(i)}, k = 1, 2, \ldots\}$ at station i form a sequence of exponentially distributed iid random variables with mean $1/\mu_i$, $i = 1, \ldots, m$. Whenever station i finishes its assembly on a kit, it is passed along to station $(i+1)$'s buffer. If no space is available there, the kit is kept at station i forcing it to idle. We are interested in obtaining the performance of this system. One may formulate a multidimensional Markov process for the number of kits in the buffer spaces. Solving it for the stationary distribution cannot be done explicitly, however. Therefore we will look at bounds and approximations for the performance measures of this system. We will do this by first looking at a system with a slightly modified kit transfer policy.

Consider an assembly system with finite local buffers. Whenever a kit arrives at station i to see the buffer at station i full, the kit will be immediately transferred to station $i + 1$. If the buffer at station $i + 1$ is also full it will be immediately transferred to station $i + 2$ and so on. Otherwise, it will remain at station $i + 1$ until it completes its assembly there, and it is then transferred to station $i + 2$. One way to view this kit transfer mechanism is to imagine that as soon as the number of kits in station i buffer reaches $b_i + 1$, the service rate at station i becomes infinite, thus causing an immediate transfer of a kit from station i to station $i + 1$. Then taking the appropriate limit (i.e., $\mu_i(b_i + 1) \to \infty$) of the closed queueing network result we have in Section 8.4 we get

$$p(\mathbf{n}) = \frac{\prod_{i=1}^{m}(1/\mu_i)^{n_i}}{\sum_{\mathbf{n}' \in \mathcal{S}_{n:\mathbf{b}}} \prod_{i=1}^{m}(1/\mu_i)^{n_i'}}, \qquad \mathbf{n} \in \mathcal{S}_{n:\mathbf{b}}, \tag{9.36}$$

where $\mathcal{S}_{n:\mathbf{b}} = \{\mathbf{n} : |\mathbf{n}| = n, 0 \le n_i \le b_i, i = 1, \ldots, m\}$ and $p(\mathbf{n})$ is the stationary probability distribution that there are n_i kits waiting at station i buffer, $i = 1, \ldots, m$. The corresponding throughput of the system is

$$TH^a(n) = \sum_{\mathbf{n} \in \mathcal{S}_{n-1:\mathbf{b}}} \prod_{i=1}^{m}(1/\mu_i)^{n_i} \Big/ \sum_{\mathbf{n}' \in \mathcal{S}_{n:\mathbf{b}}} \prod_{i=1}^{m}(1/\mu_i)^{n_i'} \tag{9.37}$$

Note that all the transfers from station i to $i + 1$ (including those not processed by station i) are included in this throughput. Because the departure times $D_k^{(i):1}$ are increasing in the service times it is not difficult to see that

$$TH^a(n) \ge TH(n) \tag{9.38}$$

where $TH(n)$ is the throughput of the assembly system with local buffer and proper kit transfer policy. Let B_i be the probability that a kit arriving at station i sees the station i

buffer full. Using Bayes's formula for conditional probabilities (see Section 8.4)

$$B_i = \frac{\sum_{\mathbf{n} \in S_{n-1:\mathbf{b}}, n_i = b_i} \prod_{j=1}^{m} (1/\mu_j)^{n_j}}{\sum_{\mathbf{n}' \in S_{n-1:\mathbf{b}}} \prod_{j=1}^{m} (1/\mu_j)^{n'_j}}, \qquad i = 1, \ldots, m \tag{9.39}$$

When the B_is are small one may expect $TH^a(n)$ to be a good approximation. To obtain a better approximation, however, we should not count those kits that are not processed by stations i, $i = 1, \ldots, m$. The approximate probability that a kit entering station 1 will proceed all the way through to station m without getting blocked is $\prod_{i=1}^{m}(1 - B_i)$. Hence we choose to approximate $TH(n)$ by

$$TH(n) \approx TH^a(n) \prod_{i=1}^{m}(1 - B_i) \tag{9.40}$$

Table 9.3 compares the approximation of equation 9.40 with $TH^a(n)$ and the exact throughputs with local buffers and a common buffer. It appears that the common buffer throughput is a better approximation than equations 9.37 and 9.40.

TABLE 9.3 LOCAL AND COMMON
BUFFER THROUGHPUT, $\mu_1 = \mu_3 = 2$,
$\mu_2 = 1$, $b_1 = b_3 = 2$, $b_2 = 4$

n	Exact $TH^{(1)}$	Exact $TH^{(2)}$	$TH^a(n)$	Approx. TH
3	0.846	0.846	0.917	0.758
4	0.898	0.912	0.980	0.750
5	0.916	0.950	1.485	0.735
6	0.910	0.972	1.941	0.621
7	0.886	0.984	3.400	0.346

9.7 ASSEMBLY SYSTEMS WITH FLEXIBLE ROUTING

Consider an assembly system with m stations that assembles several types of products. The assembly sequence of these different product types are different. Therefore kits that are assembled at station i are sent to a central store from which they are dispatched to the next station on the kits assembly sequence. There may or may not be local buffer space available at the stations.

9.7.1 Exponential Assembly Times

If the dispatcher is not a bottleneck (i.e., the dispatch delay is significantly smaller than the assembly times) and the transportation times are negligible, there is no real need to have more than a single storage space at each station. In this case the system will behave like a multiple class closed queueing network with m stations and no blocking. Suppose there are s_i assembly workers available at station i, and the fraction of kits routed to station i is r_i, $i = 1, \ldots, m$.

Suppose the assembly times are exponentially distributed. Then the throughput of the flexible assembly system with n kits always in the system is (see Section 8.4)

$$TH(n) = \frac{\sum_{\mathbf{n} \in \mathcal{S}_{n-1}} \prod_{i=1}^{m} \left(\frac{r_i}{\mu_i}\right)^{n_i} f_i(n_i)}{\sum_{\mathbf{n'} \in \mathcal{S}_n} \prod_{i=1}^{m} \left(\frac{r_i}{\mu_i}\right)^{n_i'} f_i(n_i')}, \tag{9.41}$$

where $f_i(n_i) = f_i(n_i - 1)/\min\{s_i, n_i\}$, $n_i = 1, 2, \ldots$; $f_i(0) = 1$; $i = 1, \ldots, m$. From the results given in Section 8.4 we know that if $r_i/\mu_i \geq r_j/\mu_j$, then the number of assembly workers assigned to station i should be at least as many as those assigned to station j so that a better throughput is achieved. A nonlinear optimization problem can be formulated to obtain the optimal allocation of the $S = \sum_{i=1}^{m} s_i$ workers among the m stations (see Section 8.4).

Therefore, let us consider the case where the dispatcher is a bottleneck. Let b_i be the local buffer space at station i, $i = 1, \ldots, m$ and assume that the central buffer has enough storage space to accommodate all n kits if necessary. Assume that the dispatch times are iid exponential random variables with mean $1/\mu_0$. Even with these assumptions the exact analysis of this system is difficult. Therefore we will look at bounds and approximations for this system performance. First suppose $b_i \geq n$, $i = 1, \ldots, m$ so that no blocking occurs at any of the stations. Then the system will behave like a closed Jackson queueing network with no blocking. The corresponding throughput $TH^{\infty}(n)$ is given by

$$TH^{\infty}(n) = \frac{\sum_{\mathbf{n} \in \hat{\mathcal{S}}_{n-1}} \prod_{i=0}^{m} \left(\frac{r_i}{\mu_i}\right)^{n_i} f_i(n_i)}{\sum_{\mathbf{n'} \in \hat{\mathcal{S}}_n} \prod_{i=0}^{m} \left(\frac{r_i}{\mu_i}\right)^{n_i'} f_i(n_i')} \tag{9.42}$$

where $\hat{\mathcal{S}}_n = \{\mathbf{n} : |\mathbf{n}| = n, \mathbf{n} \in \mathcal{N}_+^{m+1}\}$ and $r_0 = 1$, $f_0(n_0) = f_0(n_0 - 1)/\min\{n_0, s_0\}$, $n_0 = 1, 2, \ldots$; $f_0(0) = 1$. If $TH(n)$ is the actual throughput of the assembly system with finite buffer space,

$$TH(n) \leq TH^{\infty}(n) \tag{9.43}$$

$TH^{\infty}(n)$ is an increasing and concave function of the number of dispatchers s_0. A cost model can be therefore formulated to find the optimal number of dispatchers. The local buffer capacity needed to achieve at least a throughput of $TH^{\infty}(n) - \epsilon$ can be obtained by requiring $\sum_{\mathbf{n} \in s_n} p(\mathbf{n}) \left(\sum_{i=1}^{m} r_i I\{n_i \geq b_{i+1}\}\right) I\{n_0 \geq 1\}\mu_0 < \epsilon$. Note that the term on the left hand side of the above inequality is the rate of dispatch to any station that has more kits than the buffer capacity we choose. Conversely if the system local buffer capacity b_i is such that the preceding quantity is large, then an alternate analysis is needed to compute the throughput of the system. To do this we will first consider a system with a slightly modified kit transfer mechanism.

Consider an assembly system with both common and local buffers. If a kit dispatched from the central buffer to station i arrives at station i to find its buffer full, that kit is removed from the system and replaced at the end of the central buffer queue by a new kit that requires assembly at station i with probability l_i, $i = 1, \ldots, m$. We will later, however, choose the probabilities in such a way that the actual fraction of kits assembled at station i is r_i, $i = 1, \ldots, m$. Note that if there is no blocking at any of the stations then we should choose $l_i = r_i$, $i = 1, \ldots, m$. Observe that when a blocked kit is replaced by a new kit its future dynamics is probabilistically the same as that of a kit returning from a station. Hence we may as before in Section 9.6.2 pretend that the service rate of station i goes to infinity as the number of kits there reaches $b_i + 1$, $i = 1, \ldots, m$. Then modifying the results of Section 8.4 (see equations 9.36 and 9.37) we have the throughput of this system given by

$$\widehat{TH}(n) = \sum_{\mathbf{n} \in \mathcal{S}_{n-1:\mathbf{b}}} \prod_{i=0}^{m} \left(\frac{l_i}{\mu_i} \right)^{n_i} f_i(n_i) / \sum_{\mathbf{n}' \in \mathcal{S}_{n:\mathbf{b}}} \prod_{i=0}^{m} \left(\frac{l_i}{\mu_i} \right)^{n_i'} f_i(n_i') \qquad (9.44)$$

The fraction of arrivals at station i that sees a full buffer is

$$B_i = \sum_{\mathbf{n} \in \mathcal{S}_{n-1:\mathbf{b}}, n_i = b_i} \prod_{j=0}^{m} \left(\frac{l_j}{\mu_j} \right)^{n_j} f_j(n_j) / \sum_{\mathbf{n}' \in \mathcal{S}_{n-1:\mathbf{b}}} \prod_{j=0}^{m} \left(\frac{l_j}{\mu_j} \right)^{n_j'} f_j(n_j'), \qquad i = 1, \ldots, m \qquad (9.45)$$

Therefore the rate at which kits are assembled at station i is

$$TH_i(n) = l_i \widehat{TH}(n)(1 - B_i), \qquad i = 1, \ldots, m \qquad (9.46)$$

and the fraction of kits assembled at station i is $TH_i(n) / \sum_{j=1}^{m} l_j \widehat{TH}(n)(1 - B_j)$, $i = 1, \ldots, m$. Hence we choose l_i, $i = 1, \ldots, m$, such that $r_i = TH_i(n) / \sum_{j=1}^{m} TH_j(n)$ or equivalently

$$r_i = l_i(1 - B_i) / \sum_{j=1}^{m} l_j(1 - B_j), \qquad i = 1, \ldots, m \qquad (9.47)$$

Because the modified network with this selection of l_i, $i = 1, \ldots, m$, processes the kits at the different stations according to the correct ratios $r_1 : r_2 : \cdots : r_m$ we approximate the throughput of the assembly system described earlier by

$$TH(n) \approx \sum_{i=1}^{m} l_i \widehat{TH}(n)(1 - B_i) \qquad (9.48)$$

Table 9.4 lists numerical results that show the required l_i and compare the throughput for this model of block-and-recirculate (BAR) and the throughput with block-and-hold (BAH), that is, the kit waits at the dispatcher until its destination is free, thus blocking the dispatcher. The number of kits in the system is the value of b_0.

9.7.2 General Assembly Times

We will next show that the throughput $\widehat{TH}(n)$ obtained for the modified kit transfer mechanism remains valid even if the assembly times at the assembly stations have general distributions provided the buffer space at station i is s_i, $i = 1, \ldots, m$ (i.e., there

TABLE 9.4 COMPARISON OF BAR AND BAH
THROUGHPUT

| | Parameters | | | | BAR | | BAH |
Station	μ_i	s_i	b_i	r_i	l_i	TH_i	TH_i
0	3.00	1	10			1.989	1.866
1	0.50	1	4	0.45	0.50	0.901	0.840
2	0.40	1	6	0.55	0.50	1.086	1.027
0	2.00	1	10			1.405	1.235
1	0.50	1	3	0.3	0.32	0.417	0.371
2	0.80	1	5	0.5	0.44	0.709	0.618
3	0.40	1	2	0.2	0.22	0.279	0.247
0	3.50	1	20			2.384	2.057
1	0.40	2	5	0.3	0.29	0.720	0.617
2	0.55	1	3	0.2	0.22	0.473	0.411
3	0.30	3	6	0.35	0.37	0.845	0.720
4	0.20	2	6	0.15	0.12	0.346	0.309

are no extra buffer spaces available at any of the assembly stations). To simplify this derivation we will assume that the number of kits in the system is larger than $\sum_{i=1}^{m} s_i$ (i.e., $n \geq \sum_{i=1}^{m} s_i + 1$) and that the assembly times $\{S_k^{(i)}, k = 1, 2, \ldots\}$ at station i are iid random variables with generalized phase type distribution with representation $(\hat{\mu}_i, \hat{\mathbf{p}}_i)$, $i = 1, \ldots, m$. That is (see Appendix A),

$$f_{S^{(i)}}(x) = \sum_{l=0}^{\infty} \frac{e^{-\hat{\mu}_i x}(\hat{\mu}_i x)^l}{l!} \hat{\mu}_i \hat{p}_i(l+1), \qquad x > 0 \qquad (9.49)$$

The generalized phase-type distribution is such that if the service is in phase l, the completion of service occurs at a rate of $\hat{\mu}_i h_i(l)$, and the continuation to phase $l+1$ occurs at rate $\hat{\mu}_i (1 - h_i(l))$, where

$$h_i(l) = \hat{p}_i(l) / \sum_{k=l}^{\infty} \hat{p}_i(k), \qquad l = 1, 2, \ldots; h(0) = 0$$

is the hazard rate corresponding to the probability distribution $\hat{p}_i$. Let $N_{ij}(t)$ be the status of the jth server at station i at time t. If $N_{ij}(t) = 0$ then there are no kits assigned to that server, and otherwise there is one kit in assembly at that server and the assembly has progressed into its $N_{ij}(t)$th phase. We will assume that the kit dispatched to station i will be assigned to one of the idling servers, if any, at random with equal probability. Then $\{\mathbf{N}(t), t \geq 0\}$, where $\mathbf{N}(t) = (\mathbf{N}_1(t), \ldots, \mathbf{N}_m(t))$ and $\mathbf{N}_i(t) = (N_{i1}(t), \ldots, N_{is_i}(t))$, is a continuous time Markov process on $\mathcal{S} = N_{+}^{\sum_{i=1}^{m} s_i}$. Let $q(\mathbf{n}) = \lim_{t \to \infty} P\{\mathbf{N}(t) = \mathbf{n}\}$, $\mathbf{n} \in \mathcal{S}$ be the stationary probability distribution of $\{\mathbf{N}(t), t \geq 0\}$. Then the balance equations for **q** are

$$q(\mathbf{n}_1, \ldots, \mathbf{n}_m) \left\{ \sum_{i=1}^{m} \hat{\mu}_i (\sum_{j=1}^{s_i} I\{n_{ij} \geq 1\}) + l_i \mu_0 I\{\sum_{j=1}^{s_i} I\{n_{ij} = 0\} > 0\} \right\}$$

$$= \sum_{i=1}^{m} \hat{\mu}_i \sum_{j=1}^{s_i} q(\mathbf{n}_1, \ldots, \mathbf{n}_i - \mathbf{e}_j, \ldots, \mathbf{n}_m)(1 - h_i(n_{ij} - 1))I\{n_{ij} \geq 2\}$$

$$+ \sum_{i=1}^{m} l_i \mu_0 \frac{1}{1 + \sum_{l=1}^{s_i} I\{n_{il} = 0\}} \sum_{j=1}^{s_i} q(\mathbf{n}_1, \ldots, \mathbf{n}_i - \mathbf{e}_j, \ldots, \mathbf{n}_m) I\{n_{ij} = 1\}$$

$$+ \sum_{i=1}^{m} \hat{\mu}_i h_i(k) \sum_{j=1}^{s_i} q(\mathbf{n}_1, \ldots, \mathbf{n}_i - \mathbf{e}_j, \ldots, \mathbf{n}_m) I\{n_{ij} = 0\}, \qquad \mathbf{n} \in \mathcal{S} \qquad (9.50)$$

By substitution it can be verified that for any $\mathbf{n} \in \mathcal{S}$,

$$q(\mathbf{n}_1, \ldots, \mathbf{n}_m) = K \prod_{i=1}^{m} \left(\frac{l_i \mu_0}{\hat{\mu}_i} \right)^{\sum_{j=1}^{s_i} I\{n_{ij} \geq 1\}} \left[\frac{1}{(\sum_{j=1}^{s_i} I\{n_{ij} \geq 1\})!} \right] \prod_{j=1}^{s_i} \prod_{l=0}^{n_{ij}} (1 - h_j(l))$$
$$(9.51)$$

is the solution to the preceding set of balance equations. Here K is a normalizing constant to be chosen such that $\sum_{\mathbf{n} \in \mathcal{S}} q(\mathbf{n}) = 1$.

Let $\hat{N}_i(t) = \sum_{j=1}^{s_i} I\{N_{ij}(t) \geq 1\}$ be the number of kits at station i at time t. Then if $p(\mathbf{n})$ is the stationary distribution of $(\hat{N}_1(t), \ldots, \hat{N}_m(t))$ for $0 \leq n_i \leq s_i, i = 1, \ldots, m$, one has

$$p(\mathbf{n}) = \frac{\prod_{i=1}^{m} \left(\frac{l_i \mu_0}{\hat{\mu}_i} \right)^{n_i} \frac{1}{n_i!}}{\sum_{i=1}^{m} \sum_{0 \leq n_i' \leq s_i} \prod_{i=1}^{m} \left(\frac{l_i \mu_0}{\hat{\mu}_i} \right)^{n_i'} \frac{1}{n_i'!}}, \qquad 0 \leq n_i \leq s_i, i = 1, \ldots, m \qquad (9.52)$$

It can be easily verified that this marginal distribution of the number of kits at the different service stations is the same as that obtained for the exponential service time case.

9.8 ASSEMBLY LINES WITH STRICT KIT SEQUENCE REQUIREMENTS

In this section we consider an assembly system where kits are moved from one assembly station to another sequentially (from station 1 to 2 to 3 to ... etc.). Kits of different types are dispatched to station 1 in a *given sequence*. Because of preassignment of instructions and components for assembly at different stations, it is required that kits arrive at stations 2 up to m in the *same* order as they arrived at station 1. If we have only one server at each of the first $m - 1$ stations, then using FCFS service policy at these stations will satisfy the preceding strict sequence requirement. Balancing work loads among assembly stations, however, may require that some stations be subdivided into parallel lanes with a kit being processed on one (and only one) of the parallel lanes. Sometimes there is more than one worker per lane, for example, in car assembly there could be a worker who performs tasks on the left side of the car and a worker who performs tasks on the right side. Instructions and components for assembly at the station are typically added to the kit at the input to the station before assigning the kit to one of the parallel lanes. If the assembly times at each lane of a station are fixed constants that are the same for all kits irrespective of which of the s_i lanes at station i a kit takes, then again using FCFS at each assembly station ensures that the strict sequence requirement will be satisfied. Unfortunately, however, there will usually be variability in assembly times owing to either variations in the specific assembly requirements of each kit (e.g., owing to variations in option content of different cars on the line) or the natural variability in task times characteristic of human operators (see Section 5.1). This variability will result in the sequence of kit departures from the stations differing from their input sequence. To resequence the departures to match the input sequence, one uses an output buffer. This output buffer space may be common to all s_i lanes (see Figure 9.12) or dedicated to each of the s_i lanes (see Figure 9.13). All locations of the common output buffer are accessible by the output store. Kits departing from station i that are out-of-sequence are stored in its output buffer until the kit that is ahead in sequence of all those in the output buffer or at the other $s_i - 1$ lanes at station i completes processing. At this time all kits in buffer that are in sequence behind the just completed kit are dispatched to station $i + 1$. This system is very susceptible to assembly-time variations. As we will soon see, when the assembly times on different kits at different lanes vary widely, it may be possible to improve the throughput of the system by closing one or more lanes at some of those work stations that have more than one lane. To illustrate this we will consider a single assembly station with two lanes and a common output buffer (see Figure 9.14).

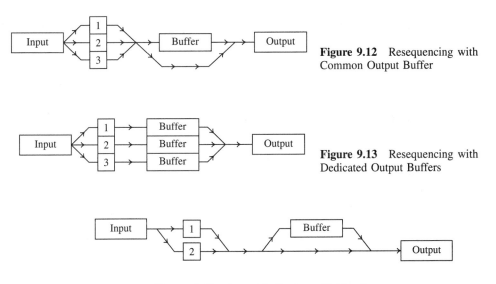

Figure 9.12   Resequencing with Common Output Buffer

Figure 9.13   Resequencing with Dedicated Output Buffers

Figure 9.14   Assembly Station with Two Lanes

We assume that the assembly times of kits at lane i are iid exponential random variables with mean $1/\mu_i$, $i = 1, 2$, and that kits in preassigned order are always available for input to this station. Suppose the kits are indexed $1, 2, \ldots$, in the order of their input sequence. Let $K_i(t)$ be the index of the kit being processed at lane i at time t, $i = 1, 2$. Also let $X_i(t)$ be the indicator that takes the value 1 if the workers at lane i are working at time t or takes the value 0 if the workers at lane i are blocked (because the buffer is full) at time t, $i = 1, 2$. Observe that if lane $3 - i$ completes assembling a kit while lane i is blocked, all the kits processed at these two lanes, including those in the output buffer, will be dispatched to the subsequent station. At such an instant both lanes will be ready to accept kits for assembly. We should therefore provide a policy that determines to which lane we will assign the kit at the head of the sequence. Because there are only two lanes, we may without a loss of generality assume that the kit at the head of the line is assigned to lane 1 and later worry about how to assign the lane index 1 to the appropriate lane. As we will soon see the lane with the shortest average assembly time should be assigned the index 1. Let $K_{\min}(t) = \min\{K_1(t), K_2(t)\}$ and $K_{\max}(t) = \max\{K_1(t), K_2(t)\}$. Then the indices of the kits stored in the output buffer, if any, are $K_{\min}(t) + 1, \ldots, K_{\max}(t) - 1$, and therefore the number of kits in the output buffer is $K_{\max}(t) - K_{\min}(t) - 1$. Because the buffer capacity is b, we have $K_{\max}(t) - K_{\min}(t) - 1 \le b$. $K_i(t)$ increases as t increases and $K_i(t) \to \infty$ as $t \to \infty$. Hence to construct a stationary Markov process for this system we look at $N_i(t) = K_i(t) - K_{\min}(t)$, $t \ge 0$, $i = 1, 2$. Then $\{(\mathbf{N}(t), \mathbf{X}(t)), t \ge 0\}$ is a Markov process on the state space $\mathcal{S} = \{(0, j, 1, 1), j = 1, \ldots, b + 1; (0, b + 1, 1, 0); (j, 0, 1, 1), j = 1, \ldots, b + 1; (b + 1, 0, 0, 1)\}$. Let $q(\mathbf{n}, \mathbf{x}) = \lim_{t \to \infty} P\{\mathbf{N}(t) = \mathbf{n}, \mathbf{X}(t) = \mathbf{x}\}$ be the stationary probability distribution of this process.

The flow balance equations for $\mathbf{q}$ are

$$(\mu_1 + \mu_2)q(0, 1, 1, 1) = \mu_1 q(0, b + 1, 1, 0) + \mu_2 \left\{ \sum_{j=1}^{b+1} q(j, 0, 1, 1) + q(b + 1, 0, 0, 1) \right\}$$

$$(\mu_1 + \mu_2)q(0, j, 1, 1) = \mu_2 q(0, j - 1, 1, 1), \qquad j = 2, \ldots, b + 1$$

$$\mu_1 q(0, b + 1, 1, 0) = \mu_2 q(0, b + 1, 1, 1)$$

$$(\mu_1 + \mu_2)q(1, 0, 1, 1) = \mu_1 \left\{ \sum_{j=1}^{b+1} q(0, j, 1, 1) \right\}$$

$$(\mu_1 + \mu_2)q(j, 0, 1, 1) = \mu_1 q(j - 1, 0, 1, 1), \qquad j = 2, \ldots, b + 1$$

$$\mu_2 q(b + 1, 0, 0, 1) = \mu_1 q(b + 1, 0, 1, 1) \qquad (9.53)$$

Solving these equations for $\mathbf{q}$ one obtains

$$q(0, j, 1, 1) = \left(\frac{\mu_1}{\mu_2} \right) \left(\frac{\mu_1 + \mu_2}{\mu_2} \right)^{b+1-j} q(0, b + 1, 1, 0), \qquad j = 1, \ldots, b + 1$$

$$q(0, b + 1, 1, 0) = \frac{\mu_2^{b+2}}{(\mu_1 + \mu_2)^{b+2} - \mu_1 \mu_2^{b+1}}$$

$$q(j, 0, 1, 1) = \left(\frac{\mu_2}{\mu_1}\right)\left(\frac{\mu_1 + \mu_2}{\mu_1}\right)^{b+1-j} q(b+1, 0, 0, 1), \qquad j = 1, \ldots, b+1$$

$$q(b+1, 0, 0, 1) = \frac{\mu_1^{b+2}\left(1 - \left(\frac{\mu_2}{\mu_1 + \mu_2}\right)^{b+1}\right)}{(\mu_1 + \mu_2)^{b+2} - \mu_1 \mu_2^{b+1}} \tag{9.54}$$

Because the throughput of this station is $\mu_1(1-q(b+1, 0, 0, 1)) + \mu_2(1-q(0, b+1, 1, 0))$ one obtains after some algebra

$$TH = (\mu_1 + \mu_2) - \frac{\mu_1^{b+3} + \mu_2^{b+3} - \mu_1^2\left(\frac{\mu_1 \mu_2}{\mu_1 + \mu_2}\right)^{b+1}}{(\mu_1 + \mu_2)^{b+2} - \mu_1 \mu_2^{b+1}} \tag{9.55}$$

Now suppose we assign the lane index 1 to the lane with assembly rate μ_2 (i.e., whenever both lanes are free at the same time, lane 2 is assigned the kit in the head of the line). Let $\widehat{TH}$ be the corresponding throughput. Then

$$\widehat{TH} = (\mu_1 + \mu_2) - \frac{\mu_1^{b+3} + \mu_2^{b+3} - \mu_2^2\left(\frac{\mu_1 \mu_2}{\mu_1 + \mu_2}\right)^{b+1}}{(\mu_1 + \mu_2)^{b+2} - \mu_1 \mu_2 \cdot \mu_1^{b}} \tag{9.56}$$

It is easily seen that if $\mu_1 > \mu_2$, then $TH > \widehat{TH}$. Therefore assigning the kit at the head of the line to the fastest server provides a higher throughput. If we fix $\mu_1 = \mu$ and change μ_2 observe that $TH = 0$ when $\mu_2 = 0$, $TH = (2 - (1/2)^{b+1})\mu$ when $\mu_2 = \mu$ and $TH \to (b+2)\mu$ as $\mu_2 \to \infty$. Conversely if we fix $\mu_2 = \mu$ and change μ_1 we have $TH = 0$ when $\mu_1 = 0$, $TH = (2 - (1/2)^{b+1})\mu$ when $\mu_1 = \mu$ and $TH \to (b+3)\mu$ as $\mu_1 \to \infty$ (see Figure 9.15). Because if we use only lane 1 the throughput is μ_1 we can easily observe that for some $\mu_2 < \mu_1$, the throughput with both servers is less than μ_1. In this case it is better to use lane 1 alone rather than adding lane 2 to that station. When the workers at both lanes are equally efficient (i.e., $\mu_1 = \mu_2 = \mu$) we

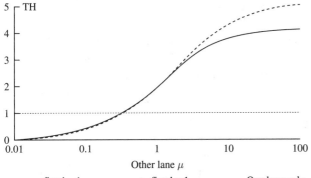

Figure 9.15   Effect of Lane Service Rates on Throughput. ($b = 2$)

have $TH = (2 - (1/2)^{b+1})\mu$. Hence we see that with as little as five buffer spaces almost the full throughput of 2μ can be achieved.

Now consider the preceding system with storage space dedicated to each lane (see Figure 9.16). Assume that the storage space assigned to each of the two lanes is k. Observe that if there are any kits stored in the output buffer of lane 1, the output buffer storage of lane 2 must be empty and vice versa. Therefore the dynamics of the kit flow through this system is the same as that of the common buffer case. Therefore the throughput of this system is the same as that obtained earlier for the common buffer case. When there are more than two lanes such an equivalence will not hold. Observe, however, that even in the two-lane system the actual buffer space needed in the dedicated storage case is $2k$ (twice that needed for the common buffer case).

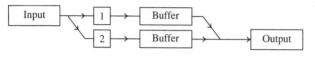

Figure 9.16 Assembly Station with Two Lanes and Dedicated

Though in principle we can formulate a Markov process model for more than two lanes, explicit results for the throughput cannot be easily obtained. An upper bound on this throughput can be obtained using the two-server model. We will illustrate this for the case in which all s lanes are equally efficient (i.e., $\mu_i = \mu$, $i = 1, \ldots, s$), and a common output buffer is used. Let $K_i(t)$ be the index of the kit being processed by lane i, $i = 1, \ldots, s$ and let $K_{[i]}(t)$ be the ith order statistic of $(K_j(t), j = 1, \ldots, s)$, $i = 1, \ldots, s$. Observe that the number of kits in the output buffer is $K_{[s]}(t) - K_{[1]}(t) - 1$, which needs to be less than or equal to b, the buffer capacity. Also let $X_i(t)$ be the status of the workers processing the $K_{[i]}(t)$th kit at time t, $i = 1, \ldots, s$. $\{(\mathbf{K}(t), \mathbf{X}(t)), t \geq 0\}$ is a (nonstationary) Markov process, and the throughput of the system is given by $TH = \lim_{t \to \infty} K_i(t)/t$, for any $i = 1, \ldots, s$. Therefore if we modify the preceding Markov process such that the values of $K_i(t)$, $i = 1, \ldots, s$, are increased faster than the original process then the throughput obtained by the modified process will be higher. Observe that in the original Markov process, if assembly to kit $K_{[i]}(t)$ is completed *and* $K_{[s]}(t) - K_{[1]}(t) - 1 < b$ (i.e., there is space in the buffer) then kit $K_{[i]}(t)$ will be transferred to the buffer and that server is assigned with kit $K_{[s]}(t) + 1$ for assembly. Hence the new state of the original Markov process at this time will be $K'_{[j]}(t) = K_{[j]}(t)$, $j = 1, \ldots, i - 1$; $K'_{[j]}(t) = K_{[j+1]}(t)$, $j = i, \ldots, s - 1$, and $K'_{[s]}(t) = K_{[s]}(t) + 1$. In our modified process we set $K''_{[j]}(t) = K_{[s]}(t) - s + j + 1$, $j = 2, \ldots, s$ and $K''_{[1]}(t) = K_{[1]}(t)$. Clearly $K'_{[j]}(t) \leq K''_{[j]}(t)$, $j = 1, \ldots, s$. It can be shown using sample path arguments that the modified process $\{(\hat{\mathbf{K}}(t), \hat{\mathbf{X}}(t)), t \geq 0\}$ is such that $\hat{\mathbf{K}}(t) \geq \mathbf{K}(t)$. Now defining $\hat{N}_{[i]}(t) = \hat{K}_{[i]}(t) - \hat{K}_{[1]}(t)$, $i = 1, \ldots, s$ one may construct the balance equations for the stationary distribution q of the Markov process $\{(\hat{\mathbf{N}}(t), \hat{\mathbf{X}}(t)), t \geq 0\}$ and compute the throughput. The throughput function is

$$\widehat{TH} = \left\{ s - \left(\frac{s-1}{2} \right) \left(\frac{s-1}{s} \right)^{b+1} \right\} \mu \qquad (9.57)$$

Note that when $s = 2$, this agrees with the exact results. This is no surprise, because if we have only two lanes the modification we performed does not affect the dynamics of the system. In general, however,

$$TH \leq \widehat{TH}$$

Earlier we indicated that because of the preassignment of instructions and components, it is required that the different types of assembly kits arrive at each of the m assembly stations in the same sequence as they were input to the system. This strict sequence requirement may be relaxed if the number of accessible instructions and components is not restricted to one, the first one in sequence. Specifically, if at assembly station j it is possible to access the instructions and components for the first $\sum_{i=1}^{j-1}(s_i - 1) + 1$ assembly kits in sequence, $(j = 2, \ldots, m)$, then there will be no blocking, and there is no need for any sequencing requirements.

In the preceding analysis we assumed that all assembly kits arriving at the assembly station are distinct. In reality, however, the number of different types of kits available for assembly is finite. So suppose the types of kits fed into the system have a cycle length of r. That is, if K_i is the index of the type of assembly kit in the ith position of the sequence, $K_i = K_{i+r}$, $i = 1, 2, \ldots$. It is then easily verified that if we provide an output buffer of size $b_j = (r - 1)s_j + 1$ for assembly station j, then one can guarantee that it is possible to observe the strict kit sequence requirement at assembly station $j + 1$ without having to block any lane at assembly station j, $j = 1, \ldots, m - 1$.

9.9 IMPLICATIONS OF MODELS

Why flexible assembly systems?   The models developed in this chapter have shown the advantage of a central buffer/dispatcher. The throughput of such a system is higher than that of an assembly system with a local buffer for each assembly station, although this statement may have to be qualified if the material movement and handling times are longer in the central buffer system. Of particular value is the fact that the throughput of a central buffer assembly system is unaffected by the form of the assembly-time distributions as long as the mean assembly times are fixed. The throughput of the central buffer system is unaffected by the processing time job diversity, whereas the throughput of the local buffer system degrades with increasing processing-time job diversity. It follows that the central buffer assembly systems are likely to be most attractive when there is a substantial variation in job processing times, either because of the variety of jobs handled or the unpredictable nature of processing times such as can occur in diagnosis, repair, or rework. They are also more appropriate when there are wide differences in worker skill levels, particularly if these differences are not easily compensated for by improved training. This is typical of tasks that involve high levels of skill, and substantial mental processing such as design or programming.

Material handling bottlenecks.   We have shown that flexible assembly systems can easily lead to material handling bottlenecks, and these bottlenecks need not be at work stations. They can be at crossing points or intersections of material handling paths. Thus careful attention to the details of layout is required. In AGV systems it is also necessary to consider the dynamic behavior of subloops in material handling paths and ensure that there is appropriate control logic to avoid lock-up in the subloop and hence lock-up of the whole system.

Central and local storage.   To achieve the benefits of central storage and in particular the insensitivity to processing-time job diversity, local storage should be kept as small as possible. Practically, however, it may be necessary to have more storage than just for the job in process at the station because otherwise transport and material handling delays in moving jobs to and from the station may lower throughput. Thus one input space and one output space are often provided at the station.

Number of material handling units.   If material handling is by unit load devices that are dedicated to the system (such as AGVs) then it is necessary to have the appropriate number of units in the system. Too few can result in stations being starved, whereas too many can result in stations being blocked. Desirably the number should be such that the central storage space is as fully used as possible without creating any blockage at the stations.

Incentives and equity.   A central dispatch system allows one to implement an equitable kit allocation to workers. We showed, however, that if worker capability or efficiency differs considerably the overall throughput can diminish with an increasing number of workers. It is sometimes better to reduce the number of workers in the system and maintain equity among the remaining workers. The two easy-to-implement kit allocation policies for equitable assignment are (1) cyclic and (2) random assignment. We have seen that the performance under cyclic assignment is better than that under random assignment.

From a broader perspective, if the requirement of equity is imposed in the cooperative systems described in the introduction, then this will impose increasing loss in throughput with increasing size, unless there is very careful selection and training of cooperative members, and also social and other incentives (or sanctions) to ensure continuing high-performance levels from each member. Having built-in rules to determine when a cooperative should split up thus would seem to be necessary.

Kit delivery sequence.   If it is required that strict kit sequence be maintained going into each work station it is also possible that throughput can go down with an increasing number of parallel lanes at each station. A certain amount of part storage space at the input of each work station can remove the need to maintain strict sequence of kits through the line. Alternatively, through providing some storage for kits at the output of the station, the loss in throughput resulting from the strict sequence requirement can be reduced.

Assignment of workers and tasks.   When one is not restricted to equitable kit allocation, the maximum throughput and minimum inventory are achieved by assigning kits to the fastest of the available workers. In a conveyor-serviced assembly system this implies that the workers should be arranged in decreasing order of efficiency in the direction of conveyor flow. That is, the fastest worker will be placed next (in the conveyor flow direction) to the loading point. When there are multiple products, the relative efficiencies of the workers may change with changing product mix. Hence having the sequence of delivery fixed by the conveyor may lead to a reduction in the throughput. With a central dispatcher, however, this problem can be overcome by the dispatcher. He may choose the fastest worker among those available with respect to the type of product to be dispatched next. With systems with strict sequencing requirements it is also true that each arriving kit should be assigned to the fastest available worker.

9.10 BIBLIOGRAPHICAL NOTE

The earliest attempt at stochastic models of conveyor systems was by Mayer [40]. Disney [18] [19] identified the key features of representing conveyors as ordered entry queues, a direction pursued by several other authors in the 1960s. Gregory and Litton [26] [27] developed results to describe the overflow process in conveyor systems with uniform spacing of kit locations. An excellent review of both stochastic and deterministic models of conveyors is contained in Muth and White [47]. Hausman, Schwarz, and Graves [29] originated the use of stochastic models of automatic warehousing systems such as AS/RS systems. Koenigsberg and Mamer [34] were the first to describe and model (using closed queues) flexible assembly systems with central dispatch and information feedback. Yao [57] developed models showing the impact of limited local buffers in flexible manufacturing systems and the insensitivity to processing-time distributions. The first description and a simulation model of assembly lines with strict sequencing requirements (the resequencing problem) is in Udomkesmalee and Daganzo [53], and the first stochastic model is in Buzacott [13]. The results in this chapter that are motivated by equity and other behavioral considerations owe much to the empirical observations of Ragotte [51].

PROBLEMS

9.1 Consider a system consisting of three single server stations arranged along a conveyor (as described in Section 9.4). There is no buffer space at any of these three stations. Processing times at the stations are exponential with mean $1/\mu_i$ at station i. Kits arrive at the system in accordance with a Poisson process with rate λ.

   (a) Using a Markov process model develop expressions for the throughput of each station and the total throughput of the system.

   (b) Determine the mean time between arrivals at each station and the overflow rate from the last station.

   (c) Suppose $\mu_1 \geq \mu_2 \geq \mu_3$ and that it is possible to rearrange the stations. Determine (1) the arrangement giving the maximum total throughput and (2) the arrangement that minimizes the maximum difference between the throughputs of any pair of stations. What is the difference between the total throughput in (1) and (2)?

9.2 A system consists of m identical stations arranged along a conveyor. The number of buffer spaces at each station is b, including the space at the station. If kits arrive at the system in accordance with a Poisson process with rate λ and the processing times at each station are exponentially distributed with mean $1/\mu$, determine the throughput of the system and the throughput of each station using a Markov process model.

9.3 Consider the system defined in problem 1.
  (a) Using equations 9.6 and 9.7 determine the Laplace transform of the interarrival times at each station and also of the interoverflow times.
  (b) From the Laplace transforms determine the first and second moments of the interarrival times at each station. Check the first moments with the results obtained in problem 1(b).
  (c) Determine the squared coefficient of variation of the arrivals at stations 2 and 3. How are the mean and SCV of the arrivals at station 3 affected by interchanging stations 1 and 2?

9.4 Consider the following first moment approximation for the performance of the systems described in problems 1 and 2. Arrivals at station i are assumed to be Poisson with mean interarrival time given by the mean interoverflow time from station $i-1$, $i = 2, \ldots, m$. Station i, $i = 1, \ldots, m$, is modeled by a $M/M/1/b$ queue.
  (a) For the system of problem 1 compare the throughput obtained using this approximation with the exact throughput. Determine for what combinations of the μ_i the approximation is likely to be particularly good or bad.
  (b) For the system of problem 2 compare the exact and approximate throughput, and indicate when the approximation is likely to be quite adequate.

9.5 Consider a second moment approximation for the performance of the system described in problem 1. Assume that the arrival process to station i, $i = 2, \ldots, m$, has a C_2 distribution with balanced means. (see Appendix A.6). Station i, $i = 2, \ldots, m$, is modeled by a $C_2/M/1/1$ queue. The mean and SCV of the overflow process at station i, $i = 1, \ldots, m$, are used to determine the parameters of the C_2 distribution, which represents the arrival process to station $i + 1$. Investigate the adequacy of this approximation.

9.6 Consider a conveyor with uniformly spaced locations for kits that travel at a constant speed v. Kit locations are spaced d apart. The conveyor delivers to uniformly spaced work stations located $l > d$ apart. Each work station has room for only one kit at a time, and processing times are exponential with parameter μ_i at station i. Initially, all locations on the conveyor when entering station 1 are occupied. Develop a model to determine (1) the overflow rate from station m and (2) the distribution of the spacing between kits on the conveyor after station m. Does the spacing of the work stations have any effect on the throughput of this system?

9.7 Consider a conveyor with locations for kits spaced d apart moving at a constant speed v. A worker has to put kits on the conveyor. They are collected from a very large stockpile, and the time required by the worker to collect a kit has an exponential distribution with parameter μ. Once the worker has collected a kit he waits until there is a free location and thus he does not start collecting the next kit until the kit has been placed in the free location as it passes him.
  (a) Suppose that all locations are empty when they arrive at the worker. Determine the distribution of the spacing between occupied locations leaving the worker.
  (b) Now suppose that not all locations are empty arriving at the worker, and the generating function of the interval between occupied spaces is $\tilde{G}(z) = \sum_{k=1}^{\infty} z^k g(k)$ where $g(k)$

is the probability of a spacing of k locations. Determine the generating function of the spacing between occupied locations leaving the worker.

9.8 Mayer [40] suggested the following model for a conveyor with uniformly spaced kit locations moving at a constant speed. Each worker has a large stockpile of kits so that if, when he becomes free, the next location is empty then he draws a part from his stockpile. All workers are identical so if p is the probability that the worker is free when a kit location reaches him and R_i is the rate at which filled kit locations arrive at worker i, the rate at which filled kit locations leave the worker is $R_{i+1} = R_i(1-p)$. The worker is assumed to become free at a rate pR_1 so the rate at which worker i draws kits from his stockpile is $pR_1 - p(1-p)^{i-1}R_1$.

(a) Develop expressions for the overflow rate and the total rate at which all worker stockpiles must be supplied, and determine the maximum value of p to ensure that overflow rate is greater than stockpile withdrawal rate.

(b) For a three-station system compare the throughput using this model with the throughput obtained using the approach described in problem 9.6. Comment on the value of Mayer's approximation.

9.9 Consider the recirculating conveyor described in Section 9.4.3. Suppose the conveyor has three identical stations with no buffer space at a station (apart from the space for the kit in process). Determine the throughput of each station as a function of b, the total space for storing kits at the buffer and the work stations. Plot the relationship between the throughput at each station and b and $\rho = \lambda/(3\mu)$.

9.10 Sometimes two-dimensional AS/RS use zoned storage of kits. That is, the AS/RS is divided up into r zones with the zones arranged so that zone 1 is the closest to the I/O point and zone r the most remote. Kits (or rather kit types) are divided into r classes with class i kit types stored in zone i. Within a zone all kits are located randomly, that is, if there is a demand for a kit type i it has a uniform distribution of possible locations in zone i. The relative demand rates of the r classes are $d_1 : d_2 : \cdots : d_r$, with $d_1 > d_2 > \cdots > d_r$. Consider the special case in which $d_i = (1/2)^{i-1}d_1$.

(a) Suppose all zones contain the same number of storage locations. Determine the throughput of the zoned storage system.

(b) Suppose the number of storage locations assigned to kit type i is proportional to $d_i^{1/2}$. Determine the throughput of the system.

9.11 Consider a carousel AS/RS that can rotate in only one direction (such as is found in many dry cleaners). If the carousel travels at a speed v and the total circumference is l, determine the average time to locate a specific kit if kits have a uniform distribution of possible locations. Would your answer be affected if the AS/RS were only half full? What would be the average time to locate a specific kit if the carousel can rotate in both directions?

9.12 Consider a large population of workers whose ratings or abilities are iid random variables with distribution function F. Suppose we choose n workers at random and let $R_1, \ldots, R_n$ be their ratings. Suppose we were to employ these workers in an assembly system in which equitable assignment must be retained among all workers. Find the throughput of the system as a function of n and F. If n is increased, what is its maximum value before the throughput of the system will start going down if F is a (1) normal distribution or (2) log-normal distribution? For this analysis you may use the following approximation for the mean r_l of the lth order statistics $R_{[l]}$ of $\{R_1, \ldots, R_n\}$:

$$\bar{F}(r_l) \approx \frac{n - l + 1/2}{n}$$

9.13 Consider two functionally identical assembly stations A and B to which kits are sent from a central dispatch point. Only the kit at the head of the line, however, is accessible at the central dispatch point. At each station there are $b_A = b_B = b$ storage locations (including the kit at the assembly station). Suppose the dispatcher has to allocate kits equitably, but the assembly stations have exponentially distributed processing times with parameters μ_A and μ_B, respectively. Assume that there are always kits available at the central dispatch point.

(a) Determine the throughput as a function of b for (1) cyclic and (2) random allocation. What value of b would you recommend if $\mu_A = \mu_B$?

(b) Consider the equitable allocation AABBAABB.... Determine the throughput.

(c) Compare the throughput obtained in (1) and (2) with the throughput achievable when the requirement for equitable allocation is removed.

9.14 Kits arrive at a rate λ at a central dispatch point in accordance with a Poisson process. They are then assigned and immediately dispatched to one of two assembly stations A and B. There is no limit on the number of kits that can wait for processing at an assembly station. Each assembly station processes kits in a time that has an exponential distribution with parameters μ_A and μ_B.

(a) Plot $E[N_A + N_B]$ where N_i is the number of kits at station i as a function of $\rho = \lambda/\mu$ if $\mu = \mu_A + \mu_B$ for different values of $r = \mu_A/\mu_B$ for (1) cyclic assignment and (2) random assignment.

(b) Suppose kits are allocated AABBAA.... Determine the average number of kits at each station.

(c) Consider the following equitable allocation policy. The dispatcher makes a preliminary selection of the station to which an arriving kit is allocated by tossing a coin. If this means, that the same station as the previous kit would be selected and it is less than τ units of time since that kit arrived then the kit is allocated to the other station. (Note that as $\tau \to \infty$ this is the same as cyclic allocation, whereas if $\tau = 0$ it is the same as random allocation.) Determine the performance of this allocation policy as a function of τ if $\mu_A = \mu_B$.

9.15 Compare the approximations for the throughput of a three-station finite local buffer assembly system with exponential processing times determined using equations 9.39 and 9.40 with the approximations suggested in Section 5.4. If the b_i are small, for what values of the μ_i do the approximations in this chapter give adequate results?

9.16 Formulate an optimization model to determine the optimal number of dispatchers needed for an assembly system with flexible routing described in Section 9.7. Using the upper bound for the throughput, given in equation 9.43, as an approximation for the throughput obtain the optimal number of dispatchers.

9.17 Consider a system consisting of a dispatcher and two single-server assembly stations A and B. There is no storage space at the stations apart from the job in process, but there is ample storage space at the dispatcher. Kits are routed to one of the two stations with probabilities r_A and r_B. On completion of processing at a station they leave the system and are immediately replaced by unprocessed kits that join the end of the queue at the dispatcher. Processing times at the stations and the dispatching time are exponentially distributed. If on completion of the dispatch the destination station is occupied then the dispatcher is blocked, and the kit is held until its destination is clear. The dispatcher always selects kits from the dispatch queue using the FCFS discipline.

(a) If kits are sent to their destination with no time lag, determine the throughput of the system as a function of n, the number of kits allowed in the system.

(b) Suppose once the destination becomes free dispatch activity is repeated on the same kit (with the same destination), determine the throughput as a function of n.

(c) Compare the throughputs obtained in parts a and b with the throughput obtained using equation 9.48.

9.18 Consider an assembly system consisting of a dispatcher and two assembly stations A and B with one server at each station. Processing times and dispatching times are exponentially distributed. The number of kits in the system is kept fixed at n. When a kit completes processing at a station it leaves the system and is replaced by another kit. The dispatcher knows which station each kit waiting in the dispatch area requires and the probability an arriving kit requires processing at a station is given by r_A and r_B respectively. Suppose there is no storage space at a station apart from the kit in process. Compare the throughputs for the following policies:

(a) The dispatcher waits until a station is free and he has a kit in the dispatch area for the free station before beginning dispatch activity. If both are free and there are kits in the dispatch area for both stations then he chooses (1) a kit intended for the faster station or (2) a kit intended for the slower station.

(b) The dispatcher starts dispatch activity whenever he is free. He chooses, if available, a kit for an idle station. If both stations are idle or both stations are occupied, and there are kits available for both stations then he chooses a kit intended for (1) the faster station or (2) the slower station.

(c) Compare the answers to parts a and b to the throughput given by equation 9.48.

9.19 In Chapter 2 it was shown that if there is a single repairman with exponentially distributed repair times and multiple machines then the performance of the system is independent of the time to failure distribution of the machines. What central dispatch, central store flexible assembly system is equivalent to this machine interference problem in the sense that it can be described by exactly the same model?

9.20 Consider a two-lane assembly station with strict kit sequence requirements. Suppose there is a common resequencing buffer of size b. If processing times are exponentially distributed derive an expression for throughput if the policy of allocating the next kit to the server who has been idle longest is used. Compare the throughput with this policy with the policy of allocating to the fastest server.

9.21 Consider the system in problem 20 but now suppose that $b = 0$, but the processing times are Erlang-2. By how much is the throughput improved if $\mu_1 = \mu_2$?

9.22 Consider a system with strict kit sequence requirements with no output buffer. Derive expressions for the throughput if $\mu_i = \mu$ for all i, $i = 1, \ldots, m$, for $m = 2, 3, 4$ by setting up the appropriate Markov process model. Compare your throughput results with the approximation given by equation 9.57.

BIBLIOGRAPHY

[1] A. AMIN. A model of the small firm in Italy. In E. Goodman, J. Bamford, and P. Saynor, editors, *Small Firms and Industrial Districts in Italy*, pages 111–122, Routledge, London, 1989.

[2] J. C. AMMONS and L. F. McGINNIS. Advanced material handling. *Appl. Mech. Rev.*, 39:1350–1355, 1986.

[3] ANONYMOUS. Automated guided vehicles move into the assembly line. *Modern Materials Handling*, 78–83, January 1985.

[4] ANONYMOUS. Conveyor systems for flexible assembly operations. *Modern Materials Handling*, 67–70, July 1985.

[5] ANONYMOUS. Flexible assembly—a boon for short production runs. *Modern Materials Handling*, 48–54, June 1984.

[6] ANONYMOUS. We achieved flexible flow to 250 assembly stations. *Modern Materials Handling*, 48–55, October 5 1984.

[7] J. BAMFORD. The development of small firms, the traditional family and agrarian patterns in Italy. In R. Goffee and R. Scase, editors, *Entrepreneurship in Europe: the Social Processes*, pages 12–25, Croom Helm, London, 1987.

[8] A. S. BASTANI. Closed-loop conveyor systems with breakdown and repair of loading stations. *IIE Trans.*, 22:351–360, 1990.

[9] C. S. BEIGHTLER and R. M. CRISP. A discrete-time analysis of conveyor-serviced production stations. *Operations Research*, 16:986–1001, 1968.

[10] Y. A. BOZER and J. A. WHITE. Travel time models for automated storage/retrieval systems. *IIE Trans.*, 16:329–338, 1984.

[11] S. BRUSCO. The Emilian model: productive decentralisation and social integration. *Cambridge Journal of Economics*, 6:167–184, 1982.

[12] S. BRUSCO. Small firms and industrial districts: the experience of Italy. In D. Keeble and E. Wever, editors, *New Firms and Regional Development in Europe*, pages 184–202, Croom Helm, London, 1986.

[13] J. A. BUZACOTT. Abandoning the moving assembly line: models of human operators and job sequencing. *Int. J. Prod. Res.*, 28:821–839, 1990.

[14] W.-M. CHOW. An analysis of automated storage and retrieval systems in manufacturing assembly lines. *IIE Trans.*, 18:204–214, 1986.

[15] W.-M. CHOW. Design for line flexibility. *IIE Trans.*, 18:95–103, 1986.

[16] E. G. COFFMAN, JR., E. GELENBE, and E. N. GILBERT. Analysis of a conveyor queue in a flexible manufacturing system. *Eur. J. Opnl. Res.*, 35:382–392, 1988.

[17] R. M. CRISP, JR., R. W. SKEITH, and J. W. BARNES. A simulated study of conveyor-serviced production stations. *Int. J. Prod. Res.*, 7:301–309, 1969.

[18] R. L. DISNEY. Some multichannel queueing problems with ordered entry. *J. Industrial Engineering*, 13:46–48, 1962.

[19] R. L. DISNEY. Some results of multichannel queueing problems with ordered entry—an application to conveyor theory. *J. Industrial Engineering*, 14:105–108, 1963.

[20] A. R. EL SAYED, C. L. PROCTOR, and H. A. ELAYAT. Analysis of closed-loop conveyor systems with multiple Poisson inputs and outputs. *Int. J. Prod. Res.*, 14:99–109, 1976.

[21] E. A. ELSAYED and B. W. LIN. Transient behaviour of ordered-entry multichannel queueing systems. *Int. J. Prod. Res.*, 18:491–501, 1980.

[22] E. A. ELSAYED and C. L. PROCTOR. Ordered entry and random choice conveyors with multiple Poisson inputs. *Int. J. Prod. Res.*, 15:439–451, 1977.

[23] K. R. FITZGERALD. Automated handling boosts flexible assembly safety. *Modern Materials Handling*, 60–63, February 1986.

[24] S. C. GILLESPIE. The use of transporters on the sewing floor. In M. Gaetan, editor, *Sewn Products Engineering and Reference Manual*, Bobbin Publications, Columbia, SC, 246–252, 1977.

[25] S. C. GRAVES, W. H. HAUSMAN, and L. B. SCHWARZ. Storage-retrieval interleaving in automatic warehousing systems. *Management Science*, 23:935–945, 1977.

[26] G. GREGORY and C. D. LITTON. A conveyor model with exponential service times. *Int. J. Prod. Res.*, 13:1–7, 1975.

[27] G. GREGORY and C. D. LITTON. A Markovian analysis of a single conveyor system. *Management Science*, 22:371–375, 1975.

[28] S. K. GUPTA. Analysis of a two-channel queueing problem with ordered entry. *J. Industrial Engineering*, 17:54–55, 1966.

[29] W. H. HAUSMAN, L. B. SCHWARZ, and S. C. GRAVES. Optimal storage assignment in automatic warehousing systems. *Management Science*, 22:629–638, 1976.

[30] R. H. HOLLIER, editor. *AGVs*. IFS, Bradford, UK, 1986.

[31] M. KAMATH and J. L. SANDERS. Analytical methods for performance evaluation of large asynchronous automatic assembly systems. *Large Scale Systems*, 12:143–154, 1987.

[32] M. KAMATH, R. SURI, and J. L. SANDERS. Analytical performance models for closed-loop flexible assembly systems. *Int. J. FMS*, 1:51–84, 1988.

[33] E. KOENIGSBERG. The analysis of AS/RS performance. In J. A. White and I. W. Pence, Jr., editors, *Progress in Materials Handling and Logistics,* volume 1, pages 133–150, Springer, New York, 1989.

[34] E. KONIGSBERG and J. MAMER. The analysis of production systems. *Int. J. Prod. Res.*, 20:1–16, 1982.

[35] I. P. KREPCHIN. Flexibility helps company cope with rapid growth. *Modern Materials Handling*, 54–58, August 1987.

[36] T. T. KWO. A theory of conveyors. *Management Science*, 6:51–71, 1959.

[37] J. R. MANNING. How not to schedule a stitching room. *J. Opl. Res. Soc.*, 36:697–703, 1985.

[38] M. MATSUI and J. FUKUTA. A queueing analysis of conveyor-serviced production station with general unit-arrival. *J. Opns. Res. Soc. Japan*, 18:211–227, 1975.

[39] W. L. MAXWELL and R. C. WILSON. Dynamic network flow modelling of fixed path material handling systems. *AIIE Trans.*, 13:12–21, 1981.

[40] H. E. MAYER. An introduction to conveyor theory. *Western Electric Engineer*, 4:42–47, 1960.

[41] P. B. MIRCHANDANI and S. H. XU. Optimal dispatching of multipriority jobs to two heterogeneous workstations. *Int. J. FMS*, 2:25–41, 1989.

[42] W. T. MORRIS. *Analysis for Materials Handling Management*. Irwin, Boston, MA, 1962.

[43] T. MUELLER. *Automated Guided Vehicles*. IFS, Bedford, UK, 1983.

[44] E. J. MUTH. Analysis of closed-loop conveyor systems. *AIIE Trans.*, 4:134–143, 1972.

[45] E. J. MUTH. A model of a closed-loop conveyor with random material flow. *AIIE Trans.*, 9:345–351, 1977.

[46] E. J. MUTH. Modelling and system analysis of multistation closed-loop conveyors. *Int. J. Prod. Res.*, 13:559–566, 1975.

[47] E. J. MUTH and J. A. WHITE. Conveyor theory: a survey. *AIIE Trans.*, 11:270–277, 1979.

[48] B. POURBABAI and D. SONDERMAN. A stochastic recirculation system with random access. *Eur. J. Opnl. Res.*, 21:367–378, 1985.

[49] A. A. B. PRITSKER. Application of multichannel queueing results to the analysis of conveyor systems. *J. Industrial Engineering*, 17:14–21, 1966.

[50] C. L. PROCTOR, E. A. EL SAYED, and H. A. ELAYAT. A conveyor system with homogeneous and heterogeneous servers with dual input. *Int. J. Prod. Res.*, 15:73–85, 1977.

[51] M. J. RAGOTTE. *The effect of human operator variability on the throughput of an AGV system—a case study: General Motors car assembly plant—door AGV system.* Master's thesis, University of Waterloo, Department of Management Sciences, 1990.

[52] I. L. REIS, J. J. BRENNAN, and R. M. CRISP, JR. A Markovian analysis for delay at conveyor-serviced production stations. *Int. J. Prod. Res.*, 5:201–211, 1967.

[53] N. UDOMKESMALEE and C. F. DAGANZO. Impact of parallel processing on job sequences in flexible assembly. *Int. J. Prod. Res.*, 26:73–89, 1989.

[54] A. WASHBURN. A multiserver queue with no passing. *Operations Research*, 22:428–434, 1974.

[55] G. S. WILKIE. Hands–off warehousing system. *Industrial Engineering*, 5:12–19, 1973.

[56] D. D. YAO. The arrangement of servers in an ordered-entry system. *Operations Research*, 35:759–763, 1987.

[57] D. D. W. YAO. *Queueing Models of Flexible Manufacturing Systems.* PhD thesis, University of Toronto, Department of Industrial Engineering, 1983.

10

Multiple-Cell Manufacturing Systems

10.1 INTRODUCTION

The idea that manufacturing within a plant should be organized around loosely linked cells seems to come from the different sources associated with the concepts of *group technology* and *sociotechnical systems*. More recently advocates of *just-in-time* approaches have also advocated cellular forms of organization.

10.1.1 Group Technology

Group technology (GT) developed in the former Soviet Union in the 1940s and 1950s as a means of reducing the difficulties that arise in managing large job shops when different components have to be machined and the job shop has the traditional grouping of machines by function in a process layout (Petrov [56]). The large job shop was replaced by specialized cells producing groups of job types that were identified as being similar in design and production sequence. This resulted in improvements in quality and on-time delivery. The specialized cells were less dependent on the performance of other cells (in contrast to the high dependency between different machine groups in traditional job shops owing to the large job flow between them). Cells resulted in higher labor productivity because each work place could be designed for a specific operation on a specific set of job types, resulting in lower set-up times and hence lower inventories. The disadvantages were mainly evident if there were changes in product mix or overall demand because this would result in fluctuations in the relative work load of different cells and, in particular, poor equipment utilization.

A further benefit of grouping job types into families was that it then became possible sometimes to produce the family using flow-line organization. Sometimes this would then enable automated equipment to be justified. Further work by Mitrofanov [51] in the late 1950s resulted in the development of methods of machining groups of similar job types on a common setup, although such group machining approaches were not only used within dedicated cells but could also be used on a machine that would be used for other (nongroup) tasks.

Thus, the essential idea of group technology was the division of a large job shop into a number of cells with reduced product variety in each cell. The goal was to minimize the movement of jobs between cells, although because some individual machines might be too lightly loaded if present in every cell where an operation using them had to be performed, some machines would be shared between cells even though this would make some movement between cells necessary. The cells themselves could be organized as job shops or flow lines, where because of the diversity of parts produced by a flow line not every part would necessarily visit every machine or work station in the line. The identification of 'similar' parts to be produced in a cell was aided by a variety of different coding and classification schemes which looked at similarities in shape, material, and processing sequence.

The idea of Group Technology spread to the UK in the 1970s and then elsewhere, although it was found that it could be applied most effectively if design and manufacture were closely linked organizationally. Often small changes in design greatly simplify formation of groups. As a consequence group technology has not lent itself to adoption by branch plants of multinational firms unless the branch plant has a mandate to design and make a product line.

10.1.2 Sociotechnical Systems

The concept of sociotechnical systems was originally developed by a group of researchers at the Tavistock Institute in the UK ([24]). After their initial experiments in the early 1950s in coal mines in the UK and weaving sheds in India, the concept attracted a great deal of interest in Scandinavian countries in the early 1970s as a solution to problems of high absenteeism and labor turnover. Subsequently, a number of manufacturing systems were designed from these principles and were widely publicized (e.g., Volvo Kalmar [29] [28]). The success of these systems led to them being copied by others although the underlying sociotechnical principles may not have been appreciated.

The sociotechnical systems approach views a manufacturing system as consisting of a technical system (the machines, material, processes) and a social system (the workers, support services such as maintenance or quality control and management). The social system is constrained by the nature of the technical system and thus changes in the technical system will influence such behavioral responses as quality, motivation, absenteeism and turnover. The sociotechnical systems perspective has challenged the Taylorist view that the role of the worker is restricted to performing the prescribed production task with any brain work to be done by management or staff support functions like personnel or quality control, thus implying that the responsibility for recognizing problems and dealing with them lies almost entirely with management. The sociotechnical systems approach begins by recognizing that workers have a variety of psychological needs that should be satisfied through the work situation, such as the need for social support and recognition, the need for some minimal amount of decision-making responsibility, the need for some learning on the job to occur, the need for the job content to be reasonably demanding, the need to be able to relate what the worker does and what is produced to the worker's social life, and the need to feel that the job leads to some sort of desirable future. Closely linked to the sociotechnical systems approach are a number of ideas from cybernetics concerning the nature of regulation and control. In particular, Ashby's Law of Requisite Variety ([4]) shows that in order to respond to disturbances there has to be sufficient "variety" or range of actions to neutralize the effect of possible disturbances. The major implications for system design are:

- Each work group has to be able to deal with problems that arise within its boundaries. For example, the group should be able to repair failed machines or have cross-trained workers so that absenteeism does not disrupt production.

- There should be buffers between groups to reduce the impact of disturbances arising elsewhere. The most effective buffers are physical and social isolation. This is only achievable if the groups do not exchange products and material, and do not share common support services. If there is material flow between the groups, however, then the buffers would be physical inventories.

- The group has to be provided with goals and performance measures that are consistent with the overall goals of the organization, yet are meaningful to the group.

The sociotechnical systems approach led to the ideas of organizing production by tasks in "flow groups," in which workers follow parts through the whole process and thus have long task times, and "product shops" in which are made a range of related products that can share common equipment but that are small enough to have simplified planning and control.

The essential ideas of the sociotechnical systems approach are the following:

- To have a production system that is "loosely joined," that is, the different cells are loosely connected.

- To introduce motivational approaches that will integrate all activities toward the system goals and also create increased job satisfaction.

- To internalize to cells the control and responsibility for dealing with disturbances arising within the cell, so that these disturbances will not spread to other cells and also so that dealing with the disturbance will not require outside managerial intervention.

- To ensure that there is "requisite variety" in coordinating and regulatory activities whenever disturbances occur.

It can be seen that both group technology and the sociotechnical systems approach lead to the idea of dividing up manufacturing into cells that are separated from cells that feed them by inventory banks. Cells specialize to make a variety of related job types, but they can be organized as job shops, flow lines, or could be automated as flexible machining or assembly systems. To achieve autonomous operation of the cells they should not be controlled from outside; any control has to be achieved by controlling the inputs and outputs of jobs, material, and information, and by ensuring that goals are set for the cell that are consistent with the overall system goals. There is a bias toward cells in parallel as this should result in less interaction between cells; however, cells can be in series or in more complex network configurations. Nevertheless, the network should not be strongly connected, otherwise disturbances could propagate through it by a variety of different paths, and hence make it difficult to control and stabilize the system.

The introduction of just-in-time ideas has resulted in a close examination of the use of inventories as buffers. To avoid inventory buffers, the emphasis is placed on developing other means of reducing the magnitude of disturbances, for example, by cross-training workers, by improving maintenance and quality, or by providing close manufacturing and industrial engineering support to the work group. Further, the production tasks would be allocated to work groups so as to minimize interactions and thus the need for inventories where groups interact.

It is also possible to view manufacturing systems that consist of components owned by different organizations as multiple cell systems. When different cells are owned by independent organizations, however, it is possible to use prices and other economic rewards and penalties as a means of coordinating the cells. This approach is beyond the scope of this book. Nevertheless, if prices are set by long-term contracts and specific shipments or deliveries can be determined independently, prices do not then function as coordinating mechanisms, and thus the coordination has to be achieved by the means discussed in this chapter.

10.2 ISSUES IN DESIGN, PLANNING, AND OPERATION

10.2.1 Design

Scale and scope of cells. The major issue in the design of a multiple-cell manufacturing system is the allocation of functions and products to cells. This will determine the overall structure of the system in terms of which cells will be connected by flows of parts and products. At the design stage it is not necessary to allocate specific job types; rather it is necessary to decide on the attributes of the job types and the degree of variation between job types within a cell. Allocation of functions means that it is also necessary to decide on the size of the cell in terms of number of machines and workers. Obviously, if cells are very small then the system will have to consist of many cells, some of which might interact considerably. Alternatively, if cells are large they are also likely to have a wide variety of job types produced in them, and the management of the cells might then be difficult. As mentioned earlier the preference is to seek a parallel structure so as to reduce interactions and remove the need for the inventory buffers between cells necessary in a series-structured system. Parallel structures, however, can mean that opportunities to exploit economies of scale are diminished, and they may require machines to be duplicated even though their utilization is low. Thus, the issue of system structure is closely linked to ideas of size and scope of individual cells.

Coordination.   The other key design issue is the determination of the way in which different cells are to be coordinated. There are two extremes, central control in which all information is processed centrally and decisions made at a central point, and decentralized approaches in which each cell communicates only with those to which it is linked by material flow. Control is then based on control of the flow along these links by the cells connected by the link. The latter tends to be simpler to implement. Nevertheless it is always necessary for clear and relevant goals to be set for each cell from the center.

Fixed/adaptable cells.   Occasionally, another design issue that arises if equipment is inherently flexible is whether to have permanent cells with a fixed grouping of stations into cells or whether the grouping is temporary and easily adaptable to new production requirements. This would require a flexible material handling system to permit easy change of cell makeup, and also an easily modified coordination and control system.

10.2.2 Planning

Job/task allocation.   Given the cells and the functions that they can perform, it is necessary to decide what categories of jobs will be allocated to different cells and also the tasks that each cell will be required to perform. Different cells may be able to perform similar required tasks so the issue is whether and how to partition the set of jobs and tasks among the cells. This requires not only considering the capability of the specific cells but also the impact on movement between cells and on the ease of coordination and control.

Coordination mechanism.   At the design stage the issue of central versus local control has been decided, whereas at the planning stage it is necessary to decide on the precise mechanism for coordination (i.e., what information will be transmitted from one cell to another and when will the information be sent). For example, each cell may produce to stock and send signals to supplying cells when its input inventory passes through certain critical levels. In other cases cells produce whenever sufficient raw materials are available at their input. More complex schemes involve looking at work-in-process levels upstream or downstream (i.e., over a number of cells) and producing up to a target, whereas other schemes involve looking at forecasts of future demand as well. The coordination mechanism will be implemented by some form of production authorization system (see Section 4.5). Thus, the planning decisions will require deciding between which pairs of cells there will be a flow of order tags and requisition tags.

Incentives or goals.   Depending on the external environment, it is necessary to assign to cells goals and performance measures by which they can measure the success of their operation. These goals will usually change with time and circumstance, for example, sometimes it could be that the number of jobs produced is critical, at other times it could be the dollar value of jobs that counts, or it could be the success in meeting promised due dates. Also, the goals for the cell have to be related to the performance measures and incentives of the workers in the cell.

10.2.3 Operation

While each cell will have to be operated in accordance with the coordination mechanism and toward meeting the goals, the system will also need to have operational parameters set.

Coordination system parameters.   The parameters of the coordination system will have to be determined. For example, it is necessary to decide on the number of order or requisition tags on each link between cells and how tags are to be transmitted (e.g., one at a time, at fixed intervals, or in fixed packet sizes). It would also be necessary to decide on any time lags between order and requisition tag movement.

Buffer sizes.   At input and output of cells there may be inventory buffers. The maximum (and minimum) sizes of these buffers will have to be determined, trying to weigh the costs of larger inventories versus their merits in achieving decoupling of cells. Smaller inventories impose more strains on the coordinating mechanism and require that boundary problems be resolved so that disturbances in one cell do not affect others.

Cell operation.   Each cell has to be operated to achieve the required performance goals, using the information provided by the coordination mechanism. This will involve consideration of the issues discussed in earlier chapters appropriate to the particular type of manufacturing system represented by the cell.

10.3 MODELING ISSUES

For modeling purposes each cell is identified by the common control of operations (internal information and material flows) exercised by an individual or a team managing that cell. A cell may consist of a single multipurpose machine or it may be a flow line, a transfer line, a job shop, a flexible machining cell, or a flexible assembly cell. It may even be a storage/retrieval system. The feature that distinguishes one cell from another is that the individual or a team that controls the operations of one cell may not control the operations of other cells except to control the material and information flow to and from his or their cell to the others.

Therefore, for the purpose of modeling, we view each cell as a work center and model the multiple-cell manufacturing system as one of the multiple work-center manufacturing systems studied in Chapters 5 to 9. The important feature one has to look for in this modeling is the nature of material and information flow that occurs in between cells.

It is clear that the modeling of a multiple-cell manufacturing system requires representation of (1) the aggregation of stations into cells and the interaction of cells to create the integrated system, and (2) the coordination mechanism.

Aggregation.   Models will have to provide insight into how the performance is affected by the scale and scope of individual cells, and also how the overall performance is modified as the variety of products and tasks assigned to cells changes. It is desirable to develop models that will guide in allocating products and tasks to cells so as to maximize overall performance. Given models to describe how cells would perform if isolated from the rest of the system, it is necessary to develop means of using these cell models to arrive at an aggregated model of the system, considering any physical inventory buffers and also how cells are coordinated.

Coordination.   Because there are a variety of different coordination mechanisms, it is desirable to have models that will enable the impact of different mechanisms to be found and the choice of parameters optimized. Although this may prove to be difficult for general networks of cells, it is useful to be able to do this for series and parallel arrangements of cells and also for such common situations as an assembly cell fed by component-producing cells. Another aspect of coordination is related to the determination of appropriate individual cell goals that are consistent with the overall system goals. Models that provide insight into this are needed.

We will focus our attention on four types of interaction between cells. They are the following:

1. Free flow.
2. Free flow limited by blocking.
3. Dispatch controlled flow.
4. Coordination using PA cards.

A multiple-cell system may have some or all of the preceding types of interaction between cells; however, we will consider each one separately.

10.4 MULTIPLE CELLS WITH FREE FLOW

Consider a manufacturing system consisting of m cells. Operations in each cell are controlled separately. Parts completing processing at cell i are automatically routed to cell j with probability p_{ij}, $i (\neq j)$, $j = 1, \ldots, m$. A part routed from cell i to cell j is automatically accepted by cell j (i.e., cell j has no control capability to reject or block the arrival of a part into it). Hence, we call this a free-flow system. Suppose cell i can be modeled by one of the models described in Chapters 5 to 9, and let $\mu_i(k)$ be the throughput rate of cell i when there are k parts in it. The multiple-cell system can then be viewed as a m-stage open or closed queueing network (depending on whether parts arrive at the system from outside or a fixed number of parts are circulating through the system) with state-dependent service rates $\mu_i(k)$ $i = 1, \ldots, m$. If required, one may also formulate a multiple-class queueing network model by modeling each cell by a multiple-class model and using state-dependent service rates that depend on the number of jobs of different classes.

Consider a multiple-cell dynamic job shop consisting of m cells. Cell k consist of $m^{(k)}$ work centers $\{i^{(k)}, i = 1, \ldots, m^{(k)}\}$, each of which in turn may be composed of single or several identical machines, $k = 1, \ldots, m$. Multiple classes of jobs flow through this job shop, and we wish to evaluate the performance of such a system and study the effects of design changes in one or more cells on the performance of this job shop. Because we may not be dealing with design alteration of the entire system, but only with one or few cells, it is advantageous to formulate an aggregate model that represents the cells as single-machine work centers with state-dependent service rate. Suppose we have carried out the aggregation procedure described in Section 7.3 for the job flow through this system and represented it by the flow of a single type of job. Specifically, suppose, based on this aggregation, we have represented the job flow through the system as follows: jobs arrive at the multiple-cell job shop according to a Poisson process with rate λ. An external job on its arrival joins work center $i^{(k)}$ with probability $\gamma_i^{(k)}$, $i = 1, \ldots, m^{(k)}$; $k = 1, \ldots, m$. A job completing service at work center $i^{(k)}$ is next routed to work center $j^{(l)}$ with probability $p_{i,j}^{(k)(l)}$, $i = 1, \ldots, m^{(k)}$; $j = 1, \ldots, m^{(l)}$; $k, l = 1, \ldots, m$. The processing requirement of a job at work center $i^{(k)}$ is exponentially distributed with mean $1/\mu_i^{(k)}$ and the service rate at that work center when $n_i^{(k)}$ jobs are there is $r_i^{(k)}(n_i^{(k)})$, $i = 1, \ldots, m^{(k)}$; $k = 1, \ldots, m$. Observe that the system in its entirety can be modeled by a single-class open Jackson queueing network. In this section we will develop an aggregate model that is easy to study when one is interested in understanding the effects that one or few cells have on the performance of the system. For this we will

1. Model each cell and obtain an aggregate single-machine work center that is equivalent to that cell.

2. Capture the material flow in between cells.

3. Develop an aggregate model of the multiple-cell job shop.

4. Analyze the aggregate model.

Modeling each cell.   Let $\lambda_i^{(k)}$ be the job arrival rate to work center $i^{(k)}$, $i = 1, \ldots, m^{(k)}$; $k = 1, \ldots, m$. Then the following set of equations uniquely determine these values (see Section 7.5):

$$\lambda_j^{(l)} = \lambda \gamma_j^{(l)} + \sum_{k=1}^{m} \sum_{i=1}^{m^{(k)}} \lambda_i^{(k)} p_{i,j}^{(k)(l)}, \quad j = 1, \ldots, m^{(l)}; l = 1, \ldots, m$$

Then if $v_i^{(k)}$ is the expected number of time an arbitrary job visits work center $i^{(k)}$, we have

$$v_i^{(k)} = \lambda_i^{(k)}/\lambda, \quad i = 1, \ldots, m^{(k)}; k = 1, \ldots, m$$

The rate at which jobs enter work center $j^{(l)}$ from outside of cell l is

$$\hat{\lambda}^{(l)}\hat{\gamma}_j^{(l)} = \lambda\gamma_j^{(l)} + \sum_{\substack{k=1 \\ k \neq l}}^{m} \sum_{i=1}^{m^{(k)}} \lambda_i^{(k)} p_{i,j}^{(k)(l)}, \quad j = 1, \ldots, m^{(l)}; l = 1, \ldots, m$$

where the total rate at which jobs enter cell l from outside of cell l is

$$\hat{\lambda}^{(l)} = \sum_{j=1}^{m^{(l)}} \{\lambda\gamma_j^{(l)} + \sum_{\substack{k=1 \\ k \neq l}}^{m} \sum_{i=1}^{m^{(k)}} \lambda_i^{(k)} \sum_{i=1}^{m^{(l)}} p_{i,j}^{(k)(l)}\}, \quad l = 1, \ldots, m$$

and $\hat{\gamma}_i^{(l)}$ is the probability that a job arriving at cell l from outside of cell l will join work center $i^{(l)}$ first. The internal job transfers within cell l of course are governed by the transfer probabilities $p_{i,j}^{(l)} = p_{i,j}^{(l)(l)}$, $i, j = 1, \ldots, m^{(l)}$. We will now model cell l in isolation as an open Jackson queueing network with external job arrival rate $\hat{\lambda}^{(l)}$, probability of joining work center $i^{(l)}$ equal to $\hat{\gamma}_i^{(l)}$ and internal job transfer probabilities $p_{i,j}^{(l)}$. The processing requirements and the service rates of the work centers are assumed to be the same as that in the original system.

Let $\overline{\lambda}_i^{(l)}$ be the job arrival rate to work center $i^{(l)}$ in this isolated network. Then

$$\overline{\lambda}_j^{(l)} = \hat{\lambda}^{(l)}\hat{\gamma}_j^{(l)} + \sum_{i=1}^{m^{(l)}} \overline{\lambda}_i^{(l)} p_{i,j}^{(l)}, \quad j = 1, \ldots, m^{(l)}$$

It is easily verified that

$$\overline{\lambda}_j^{(l)} = \lambda_j^{(l)}, \quad j = 1, \ldots, m^{(l)}$$

Then the probability distribution of the number of jobs in this cell can be represented by that in a single-server $M/M(n)/1$ queueing system with Poisson arrival process with rate $\lambda^{(l)}$, and state-dependent service rates $TH^{(l)}(n)$, $n = 1, 2, \ldots$ (see Section 8.4). Here $TH^{(l)}(n)$ is the throughput rate of this Jackson queueing network when there are n jobs in it. That is

$$TH^{(l)}(n) = \lambda^{(l)} \frac{\sum_{\mathbf{n}^{(l)} \in \mathcal{S}_{n-1}^{(l)}} \prod_{j=1}^{m^{(l)}} \left(\frac{\lambda_j^{(l)}}{\mu_j^{(l)}}\right)^{n_j^{(l)}} f_j^{(l)}(n_j^{(l)})}{\sum_{\mathbf{n}^{(l)} \in \mathcal{S}_n^{(l)}} \prod_{j=1}^{m^{(l)}} \left(\frac{\lambda_j^{(l)}}{\mu_j^{(l)}}\right)^{n_j^{(l)}} f_j^{(l)}(n_j^{(l)})}$$

where

$$\mathcal{S}_n^{(l)} = \{\mathbf{n} : \mathbf{n} \in \mathcal{N}_+^{m^{(l)}}, |\mathbf{n}| = n\}$$

and

$$f_j^{(l)}(0) = 1$$

$$f_j^{(l)}(n) = f_j^{(l)}(n-1)/r_j^{(l)}(n), \quad n = 1, 2, \ldots$$

We will therefore represent the aggregated cell l by a single machine with state-dependent service rate $TH^{(l)}(\cdot)$.

Representation of intercell job flow. We will now obtain a representation of job flow between cells. Observe that the rate of job flow from cell k to cell l is $\sum_{i=1}^{m^{(k)}} \sum_{j=1}^{m^{(l)}} \lambda_i^{(k)} p_{i,j}^{(k)(l)}$. Therefore because the rate at which jobs arrive at cell k is $\sum_{i=1}^{m^{(k)}} \lambda_i^{(k)}$ the probability that a job departing cell k will go next to cell l is

$$p^{(k)(l)} = \frac{\sum_{i=1}^{m^{(k)}} \sum_{j=1}^{m^{(l)}} \lambda_i^{(k)} p_{i,j}^{(k)(l)}}{\sum_{i=1}^{m^{(k)}} \lambda_i^{(k)}}, \quad k, l = 1, \ldots, m$$

The probability that a job from outside joins cell k for its first operation is trivially

$$\gamma^{(k)} = \sum_{i=1}^{m^{(k)}} \gamma_i^{(k)}, \quad k = 1, \ldots, m$$

Aggregate model. Now we are ready to specify the aggregate model of the multiple-cell job shop. The model consists of m single-machine work centers $\{1, \ldots, m\}$ with the single-machine work center l representing cell l. The service rate of machine l is $TH^{(l)}(n)$ when there are n jobs at that machine. The processing requirement of a job at each work center it visits has an exponential distribution with mean 1. Jobs arrive from outside according to a Poisson process with rate λ and join work center k with probability $\gamma^{(k)}, k = 1, \ldots, m$. Jobs are transferred from one work center to another according to the job transfer probabilities $p^{(k)(l)}, k, l = 1, \ldots, m$.

Analysis of aggregate model. Let $\lambda^{(k)}$ be the job flow rate into work center k in this aggregated model. Then

$$\lambda^{(l)} = \lambda \gamma^{(l)} + \sum_{k=1}^{m} \lambda^{(k)} p^{(k)(l)}, \quad l = 1, \ldots, m$$

It can be readily verified that

$$\lambda^{(l)} = \sum_{j=1}^{m^{(l)}} \lambda_j^{(l)}, \quad l = 1, \ldots, m$$

Now let $p(\mathbf{n})$, $\mathbf{n} \in \mathcal{N}_+^m$ be the stationary distribution of the number of jobs at these m aggregated work centers. Then from the results in Section 7.4 presented for the open

Jackson queueing network one has

$$p(\mathbf{n}) = K \prod_{k=1}^{m} (\lambda^{(k)})^{n^{(k)}} f^{(k)}(n^{(k)}), \quad \mathbf{n} \in \mathcal{N}_+^m$$

where K is a normalizing constant and

$$f^{(l)}(0) = 1$$
$$f^{(l)}(n) = f^{(l)}(n-1)/TH^{(l)}(n), \quad n = 1, 2, \ldots$$

Suppose now we want to obtain the performance measures of the multiple-cell job shops when one of the cells, say cell k, has been modified. To do this all we have to do is compute the new state-dependent service rates for the aggregated (new) cell k and use the previous results for the other cells along with this aggregate model. It is, however, important to see whether the performance measures computed through this aggregate model will match with those obtained by a model that represents the entire system.

Let $p(\mathbf{n}^{(k)}, k = 1, \ldots, m)$, $\mathbf{n}^{(k)} \in \mathcal{N}_+^{m^{(k)}}$ be the stationary distribution of the number of jobs at the work centers belonging to the cells in the original system. Then from the open Jackson queueing network results of Section 7.4 one has

$$p(\mathbf{n}^{(k)}, k = 1, \ldots, m) = \hat{K} \prod_{k=1}^{m} \prod_{i=1}^{m^{(k)}} \left(\frac{\lambda_i^{(k)}}{\mu_i^{(k)}} \right)^{n_i^{(k)}} f_i^{(k)}(n_i^{(k)}), \quad \mathbf{n}^{(k)} \in \mathcal{N}_+^{m^{(k)}}$$

where $\hat{K}$ is a suitable normalizing constant and $f_i^{(k)}(n)$, $n = 0, 1, \ldots$ is as defined earlier. Then the marginal distribution of the total number of jobs at each cell is

$$\hat{p}(n^{(k)}, k = 1, \ldots, m) = \sum_{\mathbf{n}^{(1)} \in \mathcal{S}_{n^{(1)}}^{(1)}} \cdots \sum_{\mathbf{n}^{(m)} \in \mathcal{S}_{n^{(m)}}^{(m)}} p(\mathbf{n}^{(k)}, k = 1, \ldots, m)$$

$$= \overline{K} \prod_{k=1}^{m} \sum_{\mathbf{n}^{(k)} \in \mathcal{S}_n^{(k)}} \prod_{i=1}^{m^{(k)}} \left(\frac{\lambda_i^{(k)}}{\mu_i^{(k)}} \right)^{n_i^{(k)}} f_i^{(k)}(n_i^{(k)}), \quad \mathbf{n} \in \mathcal{N}_+^m,$$

where $\overline{K}$ is a normalizing constant. Because

$$f^{(k)}(n^{(k)}) = \frac{1}{\prod_{j=1}^{n^{(k)}} TH^{(k)}(j)} = \frac{\sum_{\mathbf{n}^{(k)} \in \mathcal{S}_{n^{(k)}}^{(k)}} \prod_{i=1}^{m^{(k)}} \left(\frac{\lambda_i^{(k)}}{\mu_i^{(k)}} \right)^{n_i^{(k)}} f_i^{(k)}(n_i^{(k)})}{(\lambda^{(k)})^{n^{(k)}}}$$

one has

$$\hat{p}(n^{(k)}, k = 1, \ldots, m) = \overline{K} \prod_{k=1}^{m} (\lambda^{(k)})^{n^{(k)}} f^{(k)}(n^{(k)}), \quad \mathbf{n} \in \mathcal{N}_+^m$$

Now it can be easily seen that $\hat{p}(n^{(k)}, k = 1, \ldots, m) = p(n^{(k)}, k = 1, \ldots, m)$. That is, the aggregate model gives the same exact performance measures as the model for the entire system.

10.5 MULTIPLE CELLS WITH FREE FLOW LIMITED BY BLOCKING

Consider a multiple-cell system where each cell has only a limited amount of storage space available. A part processed by cell i is routed to cell j with probability p_{ij}. A part routed to cell j from cell i may be returned immediately to cell i if and only if all the storage space in cell j is occupied. Otherwise, cell j will accept the part and process it when resources there become available. As before such a system can be modeled by a multiple-cell system with blocking. We will next illustrate this by a case study of an automobile assembly system composed of several cells.

10.5.1 Case Study

This case study is based on part of a car plant that was totally reequipped in 1986–87 with extensive automation involving robots working together with AGVs and transfer lines. The plant is organized into numerous production areas separated by inventory banks. Figure 10.1 illustrates the simplified plant schematic. Work flows from the underbody assembly area to the framing line where left- and right-side frames are merged with the underbody subassembly. The side-build area is supplied by the wheel house and rear-quarter production areas. From the framing line jobs go to metal finishing before entering the paint department. After painting, bodies proceed to the trim area. There doors are taken off (and sent to the door assembly area), instrument panels installed, and doors are put back on again. Following trim, car bodies are sent to the chassis area where vehicles are completed. Chassis gets components from the underbody chassis line, which in turn gets engines from the engine dress area.

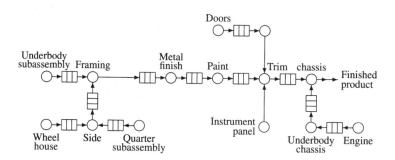

Figure 10.1 Simplified Schematic of Car Assembly Plant

The plant is designed to maintain a certain hourly net throughput rate. Equipment unreliability resulting from the high level of automation, however, implies a gross production rate higher than the design hourly net throughput. Depending on the level of automation found in each production area, various gross production rates are to be encountered throughout the plant. Areas with more automation are likely to incur more down time and hence should have a higher gross production rate to meet a given net rate. To achieve a well-balanced production system, appropriate gross production rates for the different plant areas must be determined.

Inventory banks separating different plant areas also play a significant role in obtaining a given net production rate. Because of the costs and space restrictions associated with inventory banks, the requirement is to determine the minimum buffer capacity necessary to maintain the desired throughput.

Macrosimulation models were developed for major segments of the plant, such as body in white (upstream of metal finish), paint, and trim and chassis (see [41]). Because of the complexity of the models, however, it is desirable to have analytical models to validate the simulations and also to explore design alternatives more readily. Our case study describes the analysis of the underbody subassembly area or supercell that feeds the AGV-based framing system.

Cellular structure of the underbody subassembly.   We will now choose the underbody subassembly supercell and illustrate how its cellular structure can be used to develop an aggregated procedure and analyze its performance using the results of Chapters 5 to 9. The underbody subassembly cell consists of five transfer line cells (1,2,3,4,5) and five storage/transport cells (1',2',3',4',5') (see Figure 10.2). The material (or part) flow from one cell to another is restricted by finite storage capacities in the storage/transport cells. The material flow from storage/transport cells 3' and 4' to cell 5 is restricted by the need to match one part from cell 3' with one part from cell 4' before dispatch to cell 5. In addition, material flow between cells can be hampered by the different cycle times implemented at the different cells. The basic approach taken to analyze this subassembly is (1) to aggregate each cell and represent it by an equivalent single-stage cell, consisting of a single machine or store with no transport or access delay, (2) to represent the coordination of these stages by incorporating the material flow control exercised between the cells, and finally (3) to analyze the aggregated model using the results presented in Chapters 5 to 9.

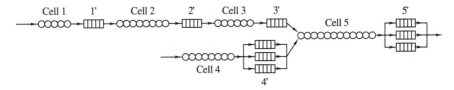

Figure 10.2   Underbody Subassembly System

Aggregation of cells

Transfer Line Cells.   Consider a generic cell consisting of a m-stage transfer line with no intermediate buffer storage space. Let the cycle time of this line be τ, and assume that the up times of machine i are geometrically distributed with mean $1/a_i$, and the repair times of machine i are geometrically distributed with mean $1/b_i$, $i = 1, \ldots, m$.

Now we wish to represent this m-stage transfer line by a single machine with mean up times $1/\hat{a}$ and mean down time $1/\hat{b}$. If $m = 2$ then from the maximum likelihood estimate of the parameters of an equivalent single machine given in Section 6.6.3, we see that $\hat{a}$ and $\hat{b}$ can be set equal to a_{12} and b_{12}, respectively, given by

$$a_{12} = a_1 + a_2 - a_1 a_2$$

$$b_{12} = \frac{a_{12}}{\frac{a_1}{b_1} + \frac{a_2}{b_2} - \frac{a_1 a_2}{b_1 + b_2 - b_1 b_2}}$$

For $m > 2$, applying this aggregation procedure sequentially to machines $1, 2, \ldots, m$, one gets

$$\hat{a} = a_{1m}; \qquad \hat{b} = b_{1m},$$

where

$$a_{1j} = a_{1j-1} + a_j - a_{1j-1} a_j, \qquad j = 2, \ldots, m$$

and

$$b_{1j} = \frac{a_{1j}}{\frac{a_{1j-1}}{b_{1j-1}} + \frac{a_j}{b_j} - \frac{a_{1j-1} a_j}{b_{1j-1} + b_j - b_{1j-1} b_j}}, \qquad j = 2, \ldots, m$$

Storage/Transport Cell.   Consider a generic storage/transport cell that has a storage capacity of z and a transport time s to move a part through the storage cell. Let τ be the minimum of the cycle times of the transfer line cells. Then we approximate the effectiveness of the buffer storage with capacity z and a transport time s by a buffer storage with capacity $\hat{z} = z - s/\tau$ and no transport/access time (see [16]).

Coordination.   We will next look at the coordination of the aggregated stages.

Different Cell Cycle Times.   Suppose the cycle times of the different transfer line cells are different. We wish to look at the material flow from one cell to another as if it occurs with respect to a common cycle time $\hat{\tau}$. Naturally, the common cycle time should be smaller than the cycle time of all the cells in the system being studied. Hence we will choose $\hat{\tau}$ to be the minimum of the cycle times of all the cells. Once we choose to coordinate the material flow through a common cycle time, the effective up and down times of the aggregated transfer line cells should be updated. So consider a single machine that has cycle time τ. Suppose the up times of this machine have a geometric distribution with mean $1/a$ and the down times have a geometric distribution with mean $1/b$. Then the first and second moments of the interdeparture times of parts from the machine if provided with an infinite supply of raw parts and infinite output buffer are (see Section 6.6):

$$E[D] = \left(1 + \frac{a}{b}\right)\tau$$

and

$$E[D^2] = \left(1 + \frac{a}{b} + \frac{2a}{b^2}\right)\tau^2$$

If $1/\hat{a}$ and $1/\hat{b}$ are the mean up and down times of an equivalent machine with cycle time $\hat{\tau}$, we would like the equivalent machine to have the same mean and second moment of interdeparture time, that is,

$$E[D] = \left(1 + \frac{\hat{a}}{\hat{b}}\right) \hat{\tau}$$

and

$$E[D^2] = \left(1 + \frac{\hat{a}}{\hat{b}} + \frac{2\hat{a}}{\hat{b}^2}\right) \hat{\tau}^2$$

Solving the preceding two equations for $\hat{a}$ and $\hat{b}$ we get

$$\hat{a} = \left(\frac{E[D]}{\hat{\tau}} - 1\right) \hat{b}$$

and

$$\hat{b} = \frac{2\left(\frac{E[D]}{\hat{\tau}} - 1\right)}{\frac{E[D^2]}{\hat{\tau}^2} - \frac{E[D]}{\hat{\tau}}}$$

Coordination of Merging.   Consider the assembly cell 5 where one part from cell 3' and one part from cell 4' are merged before it can begin production. Observe that each time a part is taken from cell 4' (and matched with a part from cell 3') a "hole" is created in cell 4'. When cell 4' is full of holes (i.e., empty of parts), cell 5 will be starved even if parts are available in cell 3'. We may view this event as cell 5 being blocked by cell 4' because it is full (of holes). The preceding description justifies us in viewing the coordination of parts merging at cell 5 as being equivalent to placing cell 4' after cell 5 in series.

Analysis of aggregated model.   The aggregated model is then a five-stage transfer line with four intermediate buffer storage spaces (see Figure 10.3). Note that because of the merging of parts at cell 5, the buffer cell 4' and the transfer line cell 4 come after cell 5 in the aggregated model. The throughput of this aggregated cell can be obtained using the approximate procedure described in Chapter 6. Some numerical results comparing the approximation to the simulation results of the underbody assembly cell are presented in Table 10.1.

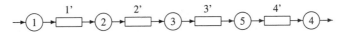

Figure 10.3   Aggregated Model of Underbody System

10.6 MULTIPLE CELLS WITH CONTROLLED DISPATCH

We now consider a multiple cell system where parts completing processing at cell i can be dispatched to one of c functionally identical cells. Cell i has dispatch control over

TABLE 10.1 COMPARISON OF APPROXIMATION AND
SIMULATION OF UNDERBODY SYSTEM

	Simulation		Analytical approx.
	Mean	95% conf. int.	
Av. contents 1'	11.14	(11.07,　11.21)	12.00
Av. contents 2'	12.80	(11.60,　13.98)	13.75
Av. contents 3'	16.81	(15.18,　18.43)	17.61
Av. contents 4'	62.77	(61.16,　64.38)	62.97
Throughput (jph)	76.93	(76.30,　77.55)	77.54

where to send the parts. In an equitable dispatch control one may either cyclically assign the parts from cell i to these c cells or randomly assign the parts from cell i to these cells with equal probabilities. Alternatively, if the parts dispatched from cell i to these c cells can be classified into different types, the dispatch control may be to send different part types to different cells. We will next address this issue of allocating different part types to different cells. For the purpose of illustrating the basic principle behind this allocation we will consider a simplified version of this problem.

10.6.1 Allocation of Part Types among Multiple–Single-Machine Cells

Consider parts arriving at a dispatch area according to a Poisson process with rate λ. The probability that the part arrived is of type j is q_j, $j = 1, \ldots, r$ ($\sum_{j=1}^{r} q_j = 1$). The parts are dispatched to one of the c functionally identical single machine cells as soon as these parts arrive at the dispatch cell i. This dispatch is to be carried out solely based on the part type and independent of any information that may be gathered from the c cells. The generic processing time of a type j part is S_j, $j = 1, \ldots, r$. We are therefore interested in the rate $\lambda_i^{(j)}$ of part type j dispatched to cell i, $j = 1, \ldots, r;\ i = 1, \ldots, c$ that will minimize the mean inventory of parts in all of these c cells. We require, however, that this assignment be equitable to all c cells. That is, the work load allocated to each cell should be the same. Formally, the problem is then

$$\min_{\lambda_i^{(j)}} \sum_{i=1}^{c} \left\{ \frac{\sum_{j=1}^{r} \lambda_i^{(j)} \sum_{k=1}^{r} \lambda_i^{(k)} E[S_k^2]}{2(1 - \sum_{j=1}^{r} \lambda_i^{(j)} E[S_j])} + \sum_{j=1}^{r} \lambda_i^{(j)} E[S_j] \right\} \tag{10.1}$$

subject to

$$\sum_{i=1}^{c} \lambda_i^{(j)} = \lambda^{(j)}, \qquad j = 1, \ldots, r$$

$$\sum_{j=1}^{r} \lambda_i^{(j)} E[S_j] = \rho, \qquad i = 1, \ldots, c$$

where $\rho = (\lambda/c)\sum_{j=1}^{r} q_j E[S_j]$ is the work load dispatched per cell per unit time, and $\lambda^{(j)} = \lambda q_j$ is the total dispatch rate of type j parts to the c cells; $j = 1, \ldots, r$. From the $M/G/1$ results (see Chapter 3) we see that $\sum_{j=1}^{r} \lambda_i^{(j)} \sum_{k=1}^{r} \lambda_i^{(k)} E[S_k^2]/[2(1 - \sum_{j=1}^{r} \lambda_i^{(j)} E[S_j])] + \sum_{j=1}^{r} \lambda_i^{(j)} E[S_j]$ is the mean number of parts in cell i for an allocation of dispatch rates $\lambda_i^{(j)}$, $j = 1, \ldots, r$, to cell i. We will first consider a simplified version of the problem whose solution will lead to a solution to the preceding problem. Suppose we have two cells (1 and 2) and r types of parts ($r = 2n$ for some $n \in \{1, 2, \ldots\}$) such that the work load $\lambda q_j E[S_j]$ dispatched per type j is the same for all r types of parts. We wish to group these part types into two sets A and $\overline{A}$ such that $|A| = n = |\overline{A}|$ and parts in A are assigned to cell 1 and parts in $\overline{A}$ are assigned to cell 2. The problem is then

$$\min_{A} \left\{ \left(\frac{\sum_{j \in A} \lambda^{(j)} \sum_{k \in A} \lambda^{(k)} E[S_k^2]}{2(1 - \rho)} + \rho \right) + \left(\frac{\sum_{j \in \overline{A}} \lambda^{(j)} \sum_{k \in \overline{A}} \lambda^{(k)} E[S_k^2]}{2(1 - \rho)} + \rho \right) \right\}$$

(10.2)

where $\rho = (\lambda/2)\sum_{j=1}^{r} q_j E[S_j]$. Without loss of generality assume that the part types are ordered so that $E[S_1] \le E[S_2] \le \cdots \le E[S_r]$. Then $\lambda^{(1)} \ge \lambda^{(2)} \ge \cdots \ge \lambda^{(r)}$. Now suppose the processing times of the different part types satisfy the following agreeability condition for the mean residual processing times $E[S_1^2]/(2E[S_1]) \le E[S_2^2]/(2E[S_2]) \le \cdots \le E[S_r^2]/(2E[S_r])$. Observe that here we require that if the mean processing time of a part type is smaller than that of another part type, then the same relationship holds with respect to the mean residual processing time. This is not a very restrictive condition because a large family of service time distributions will satisfy this condition. For example, if the processing times are of scaled values so that $S_j \stackrel{d}{=} E[S_j]X$, where X is any random variable (e.g., S_j is exponentially distributed with mean $E[S_j]$) then the preceding condition is trivially satisfied. When the preceding condition is satisfied one has

$$\lambda^{(1)} E[S_1^2] \le \lambda^{(2)} E[S_2^2] \le \cdots \le \lambda^{(r)} E[S_r^2]$$

(10.3)

Let $A^* = \{1, \ldots, n\}$ and $\overline{A}^* = \{n+1, \ldots, r\}$ (recall that $r = 2n$). Then for any partition $(A, \overline{A})$ of $\{1, \ldots, r\}$ such that $|A| = n = |\overline{A}|$ and $k, l \in \{1, \ldots, r\}$ one has

$$\sum_{j \in A} \lambda^{(j)} \cdot \lambda^{(k)} E[S_k^2] + \sum_{j \in \overline{A}} \lambda^{(j)} \cdot \lambda^{(l)} E[S_l^2] \ge$$

$$\sum_{j \in A^*} \lambda^{(j)} \cdot \min\{\lambda^{(k)} E[S_k^2], \lambda^{(l)} E[S_l^2]\} + \sum_{j \in \overline{A}^*} \lambda^{(j)} \cdot \max\{\lambda^{(k)} E[S_k^2], \lambda^{(l)} E[S_l^2]\} \quad (10.4)$$

Because the first n minimum values of $\lambda^{(j)} E[S_j^2]$ are for $j = 1, \ldots, n$ (see equation 10.3), it is immediate from equation 10.4 that

$$\sum_{j \in A} \lambda^{(j)} \cdot \sum_{k \in A} \lambda^{(k)} E[S_k^2] + \sum_{j \in \overline{A}} \lambda^{(j)} \cdot \sum_{k \in \overline{A}} \lambda^{(k)} E[S_k^2] \ge$$

$$\sum_{j \in A^*} \lambda^{(j)} \cdot \sum_{k \in A^*} \lambda^{(k)} E[S_k^2] + \sum_{j \in \overline{A}^*} \lambda^{(j)} \cdot \sum_{k \in \overline{A}^*} \lambda^{(k)} E[S_k^2] \quad (10.5)$$

Because ρ is fixed, from equations 10.2 and 10.5 one sees that the part types with the n smallest mean processing times should be grouped together and allocated to cell 1, and the part types with the n largest mean processing times should be grouped together and allocated to cell 2. This agrees with the traditional wisdom of group technology that similar part types should be grouped together.

Now consider our original problem of allocating the dispatch rates (not grouping part types) to two cells. Again, assume that the part types are ordered so that $E[S_1] \le E[S_2] \le \cdots \le E[S_r]$. Suppose $\rho^{(j)} = \lambda^{(j)} E[S_j]$, $j = 1, \ldots, r$ are rational. Then there exists a $\hat{\rho} > 0$ such that $n_j = \rho^{(j)}/\hat{\rho}$ is an integer, $j = 1, \ldots, r$. For some $m \ge 1$ divide part type j into $2m \cdot n_j$ artificial part types, each with a work load input rate $\hat{\rho}/2m$. Then in the new classification there are a total of $2m \cdot \sum_{j=1}^{r} n_j$ part types, each with a work load dispatch rate of $\hat{\rho}/2m$. The optimal grouping of the part types is $\{1, \ldots, m. \sum_{j=1}^{r} n_j\}$ and $\{m. \sum_{j=1}^{r} n_j + 1, \ldots, 2m \cdot \sum_{j=1}^{r} n_j\}$. That is, the original part types $\{1, \ldots, i\}$ with dispatch rates $\lambda^{(1)}, \ldots, \lambda^{(i-1)}, \lambda_1^{(i)}$ are allocated to cell 1, and original part types $\{i, \ldots, r\}$ with dispatch rates $\lambda_2^{(i)}, \lambda^{(i+1)}, \ldots, \lambda^{(r)}$ are allocated to cell 2. Here

$$i = \min \left\{ l : 2m \sum_{j=1}^{l} n_j \ge m \sum_{j=1}^{r} n_j; l = 1, \ldots, r \right\} \qquad (10.6)$$

$$\lambda_1^{(i)} = \left(m \sum_{j=1}^{r} n_j - 2m \sum_{j=1}^{i-1} n_j \right) \hat{\rho}/(2m E[S_i])$$

and

$$\lambda_2^{(i)} = \lambda^{(i)} - \lambda_1^{(i)}$$

That is, part type i is the $m \cdot \sum_{j=1}^{r} n_j$th artificial part type, and $\lambda_1^{(i)}$ is the total dispatch rate of the artificial part types assigned to cell 1 that are of part type i. Because $n_j = \rho^{(j)}/\hat{\rho}$, the identities in equation 10.6 may be rewritten as

$$i = \min \left\{ l : 2 \sum_{j=1}^{l} \rho^{(j)} \ge \rho; l = 1, \ldots, r \right\} \qquad (10.7)$$

$$\lambda_1^{(i)} = \left(\rho/2 - \sum_{j=1}^{i-1} \rho^{(j)} \right) / E[S_i]$$

and

$$\lambda_2^{(i)} = \lambda^{(i)} - \lambda_1^{(i)}$$

Observe that this optimal grouping is independent of m. The work load brought in by each class is $\hat{\rho}/2m$. Therefore as $m \to \infty$, $\hat{\rho}/2m \to 0$, and the part type grouping problem becomes the same as the dispatch rate allocation problem. Hence the independence of the solution of the grouping problem to the value of m implies that this solution is optimal to the dispatch rate allocation problem as well. Now using pairwise application of this result to c cells, one concludes that the optimal dispatch rates allocated to c cells is the following:

Allocate part types $(i_l, \ldots, i_{l+1})$ with dispatch rates $(\lambda_l^{(i_l)}, \lambda^{(i_l+1)}, \ldots, \lambda^{(i_{l+1}-1)}, \lambda_l^{(i_{l+1})})$ to cell l, where $i_1 = 1$,

$$i_{l+1} = \min \left\{ k : \sum_{j=1}^{k} \rho^{(j)} \geq l\rho, k = 1, \ldots, r \right\} \qquad (10.8)$$

$$\lambda_l^{(i_{l+1})} = \left(l\rho - \sum_{j=1}^{i_{l+1}} \rho^{(j)} \right) / E[S_{i_{l+1}}]$$

and

$$\lambda_{l+1}^{(i_{l+1})} = \lambda^{(i_{l+1})} - \lambda_l^{(i_{l+1})}, \qquad l = 1, \ldots, c$$

As in grouping part types we see that in an optimal dispatch rate allocation scheme parts requiring similar processing times are dispatched to the same cell. Hence we see that our model supports the traditional wisdom of forming cells to process similar items and to reduce wide fluctuations in processing requirements.

Once such a dispatch control policy is determined the resulting system can be viewed as a multiple-cell system with free flow of materials in between cells. Consequently the modeling techniques described earlier can be applied here after the dispatch control policies are determined.

10.6.2 Dedicated versus Multipurpose Cells

In the previous section we saw that, in an equitable part allocation, similar parts should be allocated to the same cell. Suppose it has been possible to carry out an idealized allocation where each of the m cells is allocated with one type of part. That is, cell j is allocated only part type j, $j = 1, \ldots, m$. It is therefore of interest to see whether the performance of this multiple-cell system can be improved by forming a single multipurpose cell that can produce all these m different types of parts. To make the comparison fair we will assume that the multipurpose cell is m times faster than each of the dedicated cells.

First, consider the system with m dedicated cells. Each cell consists of just one machine and suppose that parts arrive at cell j according to a Poisson process with rate λ_j and the processing times are iid random variables with mean $E[S_j]$ and second moment $E[S_j^2]$, $j = 1, \ldots, m$. Because these assignments are equitable, we assume that $\rho_j = \lambda_j E[S_j] = \rho$, $j = 1, \ldots, m$. Then from the results we have for the $M/G/1$ queueing system we see that the average number of parts in cell j is (see Section 3.3.2)

$$E[N_j] = \frac{\lambda_j^2 E[S_j^2]}{2(1-\rho)} + \rho, \quad j = 1, \ldots, m$$

and the average total number of parts in the system is

$$E[N] = \sum_{j=1}^{m} \frac{\lambda_j^2 E[S_j^2]}{2(1-\rho)} + m\rho$$

Now, consider the consolidated multipurpose cell. Assume that this cell consists of a single NC machine that is m times faster than the machines in the dedicated cell. That is, the processing time of a type j part at this machine has a mean $E[S_j]/m$ and a second moment $E[S_j^2]/m^2$, $j = 1, \ldots, m$. So the second moment of the service time of an arbitrary part type is $(\sum_{j=1}^m \lambda_j E[S_j^2]/m^2)/(\sum_{j=1}^m \lambda_j)$. Then the average number of parts in this system is

$$E[\hat{N}] = \frac{\sum_{k=1}^m \lambda_k \sum_{j=1}^m \lambda_j E[S_j^2]/m^2}{2(1-\rho)} + \rho$$

We will now compare these two performance measures for the case of deterministic processing times (that is, $E[S_j^2] = E[S_j]^2$, $j = 1, \ldots, m$). Then $E[N] = m\rho^2/[2(1-\rho)] + m\rho$ and $E[\hat{N}] = \sum_{k=1}^m \lambda_k \sum_{j=1}^m E[S_j][\rho/m^2]/[2(1-\rho)] + \rho$. It can be verified that $E[N] \le E[\hat{N}]$ iff

$$\sum_{k=1}^m \lambda_k \sum_{j=1}^m E[S_j] \ge m^3\rho + 2m^2(m-1)(1-\rho)$$

Because

$$\sup\left\{ \sum_{k=1}^m \lambda_k \sum_{j=1}^m E[S_j] \mid \lambda_j E[S_j] = \rho; \lambda_j \ge 0; E[S_j] \ge 0, j = 1, \ldots, m \right\} = +\infty$$

it is clear that there exists a mix of product types such that the performance of the entire system is better under a multiple-cell system with dedicated cells than with a consolidated single multipurpose cell. When the product types are uniform (that is, $E[S_j] = E[S]$ and $E[S_j^2] = E[S^2]$, $j = 1, \ldots, m$), however, a much-improved performance is obtained by consolidating the multiple cells into a single multipurpose cell.

10.7 MULTIPLE-CELL SYSTEMS COORDINATED BY PA CARDS (PAC SYSTEMS)

In a multiple-cell system each cell produces some set of parts, subassemblies, or finished products. When the cell completes manufacture of whatever it produces it will deliver its products to storage. Each distinct product, component, or subassembly that can be the output of a cell within the overall system has a unique storage location. Each such location can be replenished by some subset of the cells (because a product could be produced by more than one cell), and in turn each storage location can provide input raw material or parts for some other subset of the cells (because more than one cell could use the part). Thus, in terms of material flow, the manufacturing system consists of cells and storage locations with a cell feeding a storage location if the cell produces the product kept at that location and a storage location feeding a cell if the part kept at that location is required by the cell.

As in the single-cell system, coordination and control of material and part flow within the system is achieved by the use of order tags, requisition tags, and PA cards. We will call this the *PAC system*. *Requisition tags* control the movement of parts from stores to cells. That is, a cell issues a requisition tag to a store for an item of a particular part. On receipt of the requisition tag, the store delivers the requested item to the cell. If no items are in stock then the number of requisition tags at the store represents the backlog of unmet demand for that part.

At each store there is a set of *process tags*. Process tags will eventually become PA cards and so the maximum number of process tags is equal to the maximum number of PA cards. *Order tags* are sent from a cell to a store. When an order tag arrives at a store, it is matched with a process tag, and the match generates the PA *card*. The order tag would be retained at the store until the associated requisition tag arrives. The PA card is then in turn transmitted to the cell according to some release policy, authorizing the cell to manufacture an item of the product for each PA card. A natural release policy for store i is to transmit a packet of r_i PA cards as soon as r_i cards are accumulated. Because the cell may require component parts and raw materials, on receipt of the PA card the cell would then in turn generate order tags and requisition tags for the required raw materials and parts, and send them to the appropriate component part stores. Although order tags will be generated and sent to the appropriate part stores as soon as the cell receives the PA card, requisition tags need not be sent at the same time as the order tag. The cell may introduce some delay between the transmission of an order tag to a part store and the transmission of the associated requisition tag. Once the part store receives the requisition tag then the associated order tag can be destroyed. The requisition tag will accompany the part back to the cell as a delivery advice note. Once the cell receives the delivery advice notes for the requisite parts to make a unit of product then it can authorize manufacture to begin. On completion of manufacture by the cell the product is delivered to the appropriate store, the PA card that authorized its manufacture is converted to a process tag, and it returns to the product store location to wait with the other process tags.

From an overall system design and operation perspective there are thus three parameters associated with each store i: (1) its inventory z_i when there are no active PA cards generated by store i in the system, (2) the maximum number of process tags k_i, and (3) the packet size r_i for transmittal of PA cards. Note that $k_i \geq r_i > 0$. If we start the system at time zero with no PA cards in the system, z_i will be the initial inventory in store i. Therefore we will call z_i the initial inventory. z_i can be a positive or a negative number, although negative values will not give good service. Further, for each store there may also be a preferred delay τ_i between receipt of an order tag and the corresponding requisition tag. Note that it is possible for more than one cell to supply a given store and for a given store to supply more than one cell. The latter is determined by the product design, but the former introduces the need to specify to which cell the PA card would be sent once it is generated. One can envisage a variety of different policies for this, such as cyclic or random assignment, or, more generally, policies based on relative cost.

Within the general framework of a multiple-cell operation it is the responsibility of the cell to generate order and requisition tags for parts from PA cards for products.

Further, although we have described the previous system as if there is a unique store for each part, it might in some cases be considered appropriate to distinguish stores containing the same product on the basis of geographical location or perhaps ownership, so again the cell might have to choose between supplier stores.

Because of the range of complexity of possible system structures and operating rules, we will consider the special case of a series arrangement of cells and stores.

10.7.1 Cells in Series

We will now consider a PAC system consisting of multiple cells in series. Demand from customers is advised to store m by means of order tags, followed after a time lag of τ_m by a requisition tag authorizing shipment to the customer. Let z_i be the initial inventory of the output store i for cell i, and suppose that PA cards are transmitted from store i to cell i in batches of size r_i, $i = 1, \ldots, m$. The number of process tags assigned to store i is k_i, $i = 1, \ldots, m$. Cell i will generate and transmit an order tag to store $i - 1$ as soon as cell i receives a PA card. The corresponding requisition tag will be transmitted to store $i - 1$ after a delay of τ_{i-1} units of time.

We will now look at the dynamics of the PA cards and material flow through the system. For this, first, we will look at cell m and its output store very carefully and see how the demand process is converted into issuing PA cards to the last cell m (see Figure 10.4). Let A_k be the time at which the order tag for the kth item arrives at the output store from a customer. We do not restrict A_{k+1} to be strictly larger than A_k, $k = 1, 2, \ldots$ (that is, we allow more than one order tag to arrive at the same time). The requisition tag for the kth item then arrives at time $A_k + \tau_m$, where τ_m is the delay time between the order tag and requisition tag arrival times at the output store. Whenever a process tag becomes available a PA card will be generated for each order tag. Suppose at time zero there are k_m process tags available at the output store. Then the first k_m order tags, on their arrival, will immediately generate PA cards. Let $G_k^{(m)}$ be the time at which the kth PA card is generated at the output store. Then

$$G_k^{(m)} = A_k, \quad k = 1, \ldots, k_m \tag{10.9}$$

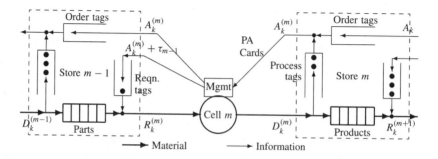

Figure 10.4 Information and Material Flow for Cell m

Now suppose the kth process tag movement from cell m to the output store occurs at time $D_k^{(m)}$, $k = 1, 2, \ldots$. Because the generation of the kth PA card will occur only after the arrival of the order tag for the kth item (at time A_k) and the $(k - k_m)$th movement of a process tag from cell m to the output store (at time $D_{k-k_m}^{(m)}$) one has

$$G_k^{(m)} = \max\{D_{k-k_m}^{(m)}, A_k\}, \quad k = 1, 2, \ldots \tag{10.10}$$

where we set $D_k^{(m)} = 0$, $k = 0, -1, -2, \ldots$. The release policy of PA cards to cell m requires that we accumulate r_m PA cards before releasing a packet of r_m PA cards. Suppose the kth PA card transmittal to cell m occurs at time $A_k^{(m)}$. Then

$$A_k^{(m)} = G_{\lceil \frac{k}{r_m} \rceil r_m}^{(m)}, \quad k = 1, 2, \ldots \tag{10.11}$$

because the kth PA card transmittal occurs as soon as $(\lceil k/r_m \rceil r_m)$th PA card is generated. This is the same as the transmission time of the $\lceil k/r_m \rceil$th packet of PA cards to cell m. Combining the preceding two equations we get

$$A_k^{(m)} = \max\{D_{\lceil \frac{k}{r_m} \rceil r_m - k_m}^{(m)}, A_{\lceil \frac{k}{r_m} \rceil r_m}\} \tag{10.12}$$

The PA cards transmitted to cell m in turn will generate order tags and requisition tags for store $m - 1$. Because an order tag is generated and transmitted to store $m - 1$ as soon as a PA card is received at cell m and the corresponding requisition tag will be generated and transmitted τ_{m-1} units of time later, we see that the kth order tag arrives at store $m - 1$ at time $A_k^{(m)}$ and the kth requisition tag arrives at store $m - 1$ at $A_k^{(m)} + \tau_{m-1}$. Suppose at time zero we have z_{m-1} items in store $m-1$, and the store fills the requisitions one by one. Then the first z_{m-1} requisition tags will be "filled" immediately on their receipt at store $m - 1$. Therefore if $R_k^{(m)}$ is the time at which the raw material needed to process the kth PA card order at cell m is received from store $m - 1$, one has

$$R_k^{(m)} = A_k^{(m)} + \tau_{m-1}, \quad k = 1, \ldots, z_{m-1} \tag{10.13}$$

To fill the requisition tags that came later (after the z_{m-1}th requisition tag) we need raw materials to be moved from cell $m - 1$ to store $m - 1$. So if $D_k^{(m-1)}$ is the time at which the kth raw material is moved from cell $m - 1$ to store $m - 1$, one has

$$R_k^{(m)} = \max\{D_{k-z_{m-1}}^{(m-1)}, A_k^{(m)} + \tau_{m-1}\}, \quad k = 1, 2, \ldots, \tag{10.14}$$

where we set $D_k^{(m-1)} = 0$, $k = 0, -1, -2, \ldots$.

Depending on the service process and the number of servers at cell m, we suppose that the input process $\{R_k^{(m)}, k = 1, 2, \ldots\}$ is converted into the output process by the relationship $D_k^{(m)} = \phi_k^{(m)}(\mathbf{R}^{(m)})$, $k = 1, 2, \ldots$. For example, if cell m consist of just one machine (i.e., $c_m = 1$) and the service policy is FCFS, then

$$D_k^{(m)} = \max\{D_{k-1}^{(m)}, R_k^{(m)}\} + S_k^{(m)}, \quad k = 1, 2, \ldots, \tag{10.15}$$

where $S_k^{(m)}$ is the processing time of the product for the kth PA card.

Extending these relationships (equations 10.12 and 10.14) to the other cells one gets

$$A_k^{(i)} = \max\{D_{[\frac{k}{r_i}]r_i-k_i}^{(i)}, A_{[\frac{k}{r_i}]r_i}^{(i+1)}\}, \quad k = 1, 2, \ldots$$

$$R_k^{(i)} = \max\{D_{k-z_{i-1}}^{(i-1)}, A_k^{(i)} + \tau_{i-1}\}, \quad k = 1, 2, \ldots$$

$$D_k^{(i)} = \phi_k^{(i)}(\mathbf{R}^{(i)}), \quad k = 1, 2, \ldots \tag{10.16}$$

for $i = 1, \ldots, m$, where $r_{m+1} = 1$, $\mathbf{A}^{(m+1)} = \mathbf{A}$, and $D_k^{(0)} = 0$, $k = 1, 2, \ldots$. $\phi_k^{(i)}(\mathbf{R}^{(i)})$ is the kth departure time from cell i for an input process $\mathbf{R}^{(i)}$. For example, if cell i has c_i parallel servers we have $\phi_k^{(i)}(\mathbf{R}^{(i)})$ equal to the kth element in the increasing rearrangement of $\{\tilde{D}_k^{(i)}, k = 1, 2, \ldots\}$ where

$$\tilde{D}_k^{(i)} = \{R_k^{(i)} \vee D_{k-c_i}^{(i)}\} + S_k^{(i)} \tag{10.17}$$

and $S_k^{(i)}$ is the processing time of a part for the kth cell i PA card to arrive at cell i. Finally, if $R_k^{(m+1)}$ is the time at which the kth requisition tag from a customer is filled, we have

$$R_k^{(m+1)} = \max\{D_{k-z_m}^{(m)}, A_k + \tau_m\} \quad k = 1, 2, \ldots \tag{10.18}$$

Impacts of choice of parameters. As we will soon see, by appropriate choices of the parameters (z_i, k_i, r_i, τ_i), $i = 1, \ldots, m$, we may obtain the standard material-flow control procedures such as material requirements planning (MRP), Japanese card (KANBAN), optimized production technology (OPT), and constant work-in-process control (CONWIP). First we will look at the effects of the parameters (z_i, k_i, r_i, τ_i) on the performance of the system.

The following properties of the departure process (and the shipment to customers process) can be easily verified from equations 10.16 and 10.18 (e.g., see Section 5.4 for similar results):

Property 1. Departure times are decreasing in z_i, $i = 1, \ldots, m$.

Property 2. Departure times are decreasing in k_i, $i = 1, \ldots, m$.

Property 3. Departure times are increasing in r_i, $i = 1, \ldots, m$.

Property 4. Departure times are increasing and convex in τ_i when $c_i = 1$, $i = 1, \ldots, m$.

Property 5. Departure times are increasing and convex in $S^{(i)}$ when $c_i = 1$, $i = 1, \ldots, m$.

Let $L_k = R_k^{(m+1)} - (A_k + \tau_m)$, $k = 1, 2, \ldots$, be the shipment delay, that is, the delay in filling a requisition from a customer. Then it follows that L_k will also have properties 1, 2, 3, and 5, but property 4 will become (because $D_{k-z_m}^{(m)}$ is independent of τ_m) the following:

Property 4'.   Shipment delays are increasing and convex in the τ_i, $i = 1, \ldots, m - 1$ but decreasing and convex in τ_m when $c_i = 1$, $i = 1, \ldots, m$.

It follows that the optimal assignment of delays between order tags and requisition tags will have $\tau_i = 0$, $i = 1, \ldots, m - 1$ and τ_m as large as possible.

Suppose we have c_1 parallel servers at cell 1. Then it is easily seen that the departure process from cell 1 is invariant with respect to k_1, as long as $k_1 \geq c_1$. To see this observe that $\tilde{D}_k^{(1)} = \max\{A_k^{(2)}, D_{k-k_1}^{(1)}, D_{k-c_1}^{(1)}\} + S_k^{(1)}$. For $k_1 \geq c_1$ we have $D_{k-k_1}^{(1)} \leq D_{k-c_1}^{(1)}$ and hence $\tilde{D}_k^{(1)} = \max\{A_k^{(2)}, D_{k-c_1}^{(1)}\} + S_k^{(1)}$. That is $\tilde{D}_k^{(1)}$ is independent of k_1 and thus $\{D_k^{(1)}, k = 1, 2, \ldots\}$ is not affected by the actual value of k_1 as long as it is larger than or equal to c_1.

We will next look at a natural constraint on k_i, before analyzing the special cases of the parameter choice.

The $(k + z_i)$th item will be released to cell $i + 1$ (at time $R_{k+z_i}^{(i+1)}$) only after the kth item is received in store i (at time $D_k^{(i)}$). Furthermore the $(k + z_i)$th item can be released from cell $i + 1$ to store $i + 1$ (at time $D_{k+z_i}^{(i+1)}$) only after it is received from store i and processed. Hence

$$D_k^{(i)} \leq R_{k+z_i}^{(i+1)} < D_{k+z_i}^{(i+1)} \tag{10.19}$$

The time at which the $(k + z_i)$th process tag is transmitted to store $i + 1$ is $D_{k+z_i}^{(i+1)}$. Hence the generation of the $(k + z_i + k_{i+1})$th PA card (at time $G_{k+z_i+k_{i+1}}^{(i+1)}$) and its transmission to cell $i + 1$ (at time $A_{k+z_i+k_{i+1}}^{(i+1)}$) can only occur after time $D_{k+z_i}^{(i+1)}$. That is,

$$D_{k+z_i}^{(i+1)} \leq G_{k+z_i+k_{i+1}}^{(i+1)} \leq A_{k+z_i+k_{i+1}}^{(i+1)} \leq A_{k+z_i+k_{i+1}}^{(i)} \tag{10.20}$$

Therefore $D_k^{(i)} < A_{k+z_i+k_{i+1}}^{(i)}$ and thus the maximum number of PA cards present at cell i at any time is less than or equal to $(k + z_i + k_{i+1}) - k = z_i + k_{i+1}$. Therefore without any effect on the performance of the system we may restrict $k_i \leq z_i + k_{i+1}$. It is important, however, to remember that the performance of the system will be affected by the choice of k_i within the preceding restriction. The other restriction we place for a cell consisting of c_i parallel machines is $k_i \geq c_i$ to ensure that all c_i servers will be used. Otherwise, if $k_i < c_i$ only k_i servers will ever be used in the processing of parts.

We will now consider the optimal placement of initial inventories at various product stores. Consider a system that has assigned an initial inventory of size z_i for product store i (recall that we assume that the system starts with no PA cards in the system), $i = 1, \ldots, m$. Rewriting the equations governing the flow sequences given by equation 10.16 one obtains

$$A_{k-\sum_{j=i}^{m} z_j}^{(i)} = \max\{D_{\lceil \frac{k-\sum_{j=i}^{m} z_j}{r_i} \rceil r_i - k_i}^{(i)}, A_{\lceil \frac{k-\sum_{j=i}^{m} z_j}{r_i} \rceil r_i}^{(i+1)}\}$$

$$R_{k-\sum_{j=i}^{m} z_j}^{(i)} = \max\{D_{k-\sum_{j=i-1}^{m} z_j}^{(i-1)}, A_{k-\sum_{j=i}^{m} z_j}^{(i)} + \tau_{i-1}\}$$

$$D_k^{(i)} = \phi_k^{(i)}(\mathbf{R}^{(i)}), \quad k = 1, 2, \ldots; \quad i = 1, \ldots, m \tag{10.21}$$

Consider another system that has assigned an initial inventory of size $\hat{z}_i$ for store i, $i = 1, \ldots, m$, such that

$$\sum_{j=i}^{m} \hat{z}_j \geq \sum_{j=i}^{m} z_j, \quad i = 1, \ldots, m \tag{10.22}$$

Let $\hat{A}_k^{(i)}$, $\hat{R}_k^{(i)}$ and $\hat{D}_k^{(i)}$ be the corresponding flow sequences. That is (see equation 10.21)

$$\hat{A}_{k-\sum_{j=i}^{m}\hat{z}_j}^{(i)} = \max\{\hat{D}^{(i)}_{\lceil \frac{k-\sum_{j=i}^{m}\hat{z}_j}{r_i} \rceil r_i - k_i}, \hat{A}^{(i+1)}_{\lceil \frac{k-\sum_{j=i}^{m}\hat{z}_j}{r_i} \rceil r_i}\}$$

$$\hat{R}_{k-\sum_{j=i}^{m}\hat{z}_j}^{(i)} = \max\{\hat{D}^{(i-1)}_{k-\sum_{j=i-1}^{m}\hat{z}_j}, \hat{A}^{(i)}_{k-\sum_{j=i}^{m}\hat{z}_j} + \tau_{i-1}\}$$

$$\hat{D}_k^{(i)} = \phi_k^{(i)}(\hat{\mathbf{R}}^{(i)}), \quad k = 1, 2, \ldots; \quad i = 1, \ldots, m \tag{10.23}$$

Then using an induction hypothesis it can be shown by substitution into equations 10.21 and 10.23 that for $k = 1, 2, \ldots; i = 1, \ldots, m$,

$$\hat{A}_{k-\sum_{j=i}^{m}\hat{z}_j}^{(i)} \leq A_{k-\sum_{j=i}^{m}z_j}^{(i)}$$

$$\hat{R}_{k-\sum_{j=i}^{m}\hat{z}_j}^{(i)} \leq R_{k-\sum_{j=i}^{m}z_j}^{(i)}$$

$$\hat{D}_{k-\sum_{j=i}^{m}\hat{z}_j}^{(i)} \leq D_{k-\sum_{j=i}^{m}z_j}^{(i)} \tag{10.24}$$

Because the time at which the kth requisition at the output store is filled is

$$R_k^{(m+1)} = \max\{D_{k-z_m}^{(m)}, A_k + \tau_m\} \tag{10.25}$$

for the first system and

$$\hat{R}_k^{(m+1)} = \max\{\hat{D}_{k-\hat{z}_m}^{(m)}, A_k + \tau_m\} \tag{10.26}$$

for the latter system, it is clear from equation 10.24 that

$$\hat{R}_k^{(m+1)} \leq R_k^{(m+1)}, \quad k = 1, 2, \ldots \tag{10.27}$$

Thus placing more initial inventory at the latter stages of the system provides a better service to customer demands than placing more initial inventory at the early stages. If the value added of the product at various stages is negligible, then it is optimal to provide all the initial inventories at the final stage. That is, the optimal initial inventories are

$$z_m^* = \sum_{j=1}^{m} z_j; \quad z_i^* = 0, \quad i = 1, \ldots, m - 1 \tag{10.28}$$

We will next use the preceding result and the properties described earlier to obtain bounds for the performance of PAC systems.

10.7.2 Bounding of Backlog Performance of PAC Systems

In this section we will develop an approach to bound the time needed to meet a demand in a PAC system. We will do this by constructing a modified PAC system that has a better service performance than the original PAC system. Of course this improved service performance is achieved at the expense of higher inventories. To be precise first fix a $j \in \{1, \ldots, m\}$, and consider a series cell system with cells $\{j, \ldots, m\}$. Order tags arrive at this system from outside according to the arrival process $\{A_k, k = 1, 2, \ldots\}$ same as that in the original system. These arrivals generate PA cards similar to that in the original system but are moved downstream through the system. This is achieved by providing an unlimited number of process tags at all stores, that is, we set $k_i = +\infty$, $i = 1, \ldots, m$. The following dynamics of the PA cards in the modified system will make this clearer: Let $\hat{A}_k^{(i)}$ be the time at which the kth PA card arrives at cell i, $k = 1, \ldots$; $i = 1, \ldots, m$. Then

$$\hat{A}_k^{(i)} = \hat{A}_{[\frac{k}{r_i}]r_i}^{(i+1)}, \qquad k = 1, 2, \ldots; i = 1, \ldots, m-1 \tag{10.29}$$

Clearly $A_k^{(i)} \geq \hat{A}_k^{(i)}$, $k = 1, 2, \ldots$; $i = 1, \ldots, m$. Now let $\hat{R}_k^{(i)}$ be the time at which a raw part is moved to cell i for the kth time from storage $i - 1$, $i = 1, \ldots, m$. Suppose the processing times at cells 1 to $j - 1$ are treated as zero in the modified system. That is,

$$\hat{D}_k^{(i)} = \hat{R}_k^{(i)}, \quad i = 1, 2, \ldots, j-1 \tag{10.30}$$

where $\hat{D}_k^{(i)}$ is the kth time epoch at which a part is processed by cell i (and is ready to be shipped to cell $i + 1$), $i = 1, \ldots, m$ (cell $m + 1$ is to be interpreted as the output buffer). Then from equation 10.16 one sees that for this system

$$\hat{R}_k^{(j)} = \max_{1 \leq i \leq j} \{A_k^{(i)} + \tau_{i-1}\}, \quad k = 1, 2, \ldots \tag{10.31}$$

where we assume that $\hat{z}_i = z_i$, $i = 1, \ldots, j-1$; $\hat{z}_i = 0$, $i = j, \ldots, m-1$ and $\hat{z}_m = \sum_{i=j}^m z_i$. In addition, if we assume that $\hat{\tau}_i = 0$, $i = j, \ldots, m-1$, then for $i \geq j$ one has from equation 10.16,

$$
\begin{aligned}
\hat{R}_k^{(j)} &= A_k^{(j)}, & k &= 1, 2, \ldots \\
\hat{R}_k^{(i)} &= \hat{D}_k^{(i)}, & k &= 1, 2, \ldots; i = j+1, \ldots, m \\
\hat{D}_k^{(i)} &= \phi_k^{(i)}(\hat{\mathbf{R}}^{(i)}), & k &= 1, 2, \ldots; i = j, \ldots, m
\end{aligned} \tag{10.32}
$$

The preceding dynamics of material flow is clearly equivalent to a free-flow series cell system with cells $\{j, \ldots, m\}$ and parts arriving at cell j according to the arrival process $\{\hat{R}_k^{(i)}, k = 1, 2, \ldots\}$. Parts arriving at each cell are processed and made ready to move to subsequent cells (at times $\phi_k^{(i)}(\hat{\mathbf{R}}^{(i)})$ by cell i). The output from cell m are moved to the output buffer from which the demand is met. For the modified system we have assumed that $\hat{k}_i = +\infty$, $i = 1, \ldots, m$; $\hat{\tau}_i = 0$, $i = j, \ldots, m - 1$, $\hat{z}_i = z_i$, $i = 1, \ldots, j - 1$; $\hat{z}_i = 0$, $i = j, \ldots, m - 1$; $\hat{z}_m = \sum_{i=j}^{m} z_i$ and that the service requirements are zero at cells $1, \ldots, j - 1$. If $\phi_k^{(i)}(\mathbf{x})$ is nondecreasing in $\mathbf{x}$ and the service times (as is the case in a single– or multiple–parallel-server cell or even a flow line or a job shop) for each $i, i = j, \ldots, m$, then it can be easily seen from the properties discussed earlier that

$$\hat{D}^{(i)}_{k - \sum_{i=j}^{m-1} z_i} \leq D_k^{(i)}, \qquad k = 1, 2, \ldots; i = j, \ldots, m \tag{10.33}$$

Therefore the time needed to meet the requisition tag arriving from outside at time $A_k + \tau_m$ (corresponding to the order tag arrived at time A_k) is

$$\left[D_{k - z_m} - A_k - \tau_m \right]^+ \geq \left[\hat{D}_{k - \hat{z}_m} - A_k - \tau_m \right]^+ \tag{10.34}$$

Observe that the preceding bound can be computed for each value of j, $j = 1, \ldots, m$. This leads to m bounds. One would therefore choose the maximum of these bounds as the final bound. We will later illustrate this for a special case of the PAC system.

10.7.3 Parameter Choice for Different Coordination Schemes

In this section we will demonstrate how through the appropriate choice of parameters the PAC system can be specialized into a wide variety of classical coordination approaches. Some of the distinctions in the literature are based on whether the coordination scheme uses forecast information or not, that is, *pull* versus *push*. Because our approach is essentially to ignore the difference by treating forecasts as equivalent to orders for future delivery, we will sometimes have one coordination scheme that appears equivalent to more than one conventional approach.

Produce-to-order system.   In produce-to-order systems no work is begun on any component of the product until receipt of the order. There are a variety of different produce-to-order systems, however. The simplest is that where there is no attempt at controlling the level of work in process at any cell. Then the choice of parameters that will achieve produce-to-order operation is as follows:

1. $z_i = 0$, $i = 1, \ldots, m$, that is, initially the manufacturing system is empty of initial inventory and of PA cards.
2. $k_i = \infty$, $i = 1, \ldots, m$. PA cards are generated at all product stores as soon as the order tag from the customer arrives.
3. $r_i = 1$, $i = 1, \ldots, m$. PA cards are moved in packets of size 1.
4. $\tau_i = 0$, $i = 1, \ldots, m$, that is, no delay between order tags and requisition tags.

In such a system the arrival of an order will immediately trigger a PA card at all cells, and this will immediately result in each cell releasing a requisition tag. The requisition tag will wait at the input part store until the parts are available. Manufacture of parts will thus begin as soon as the required parts are available irrespective of what is happening at other cells.

Variant with $\tau_i \geq 0$, $i = 1, \ldots, m$.   Although produce-to-order manufacture in the past was coordinated in this way, it would be commoner now to represent manufacture by a critical path network in which each activity corresponds to the manufacture of a product at a cell and with activity duration an estimate of the flow time through the cell. Events in the critical path network are the delivery of product by a cell to a product store. Extra events and (dummy) activities are introduced to denote the input of all required parts to a cell. From the critical path network the latest start time of each activity can be found. Then the parameters z_i and k_i would remain at 0 and ∞, respectively, but the delay τ_i in issuing a requisition tag at a cell would be set at the latest start time of the event in the critical path network that corresponds to the beginning of manufacture at the cell.

Base stock system.   In the base stock system the following parameter choices are made:

1. $z_i > 0$, $i = 1, \ldots, m$, that is, there is an initial inventory in each store.
2. $k_i = \infty$, $i = 1, \ldots, m$, that is, no delay in generating PA cards on arrival of an order tag at a store.
3. $r_i \geq 1$, $i = 1, \ldots, m$.
4. $\tau_i = 0$, $i = 1, \ldots, m$, that is, no delay between order tags and requisition tags.

When $r_i = 1$, $i = 1, \ldots, m$ (point 2) implies that as soon as there is a customer demand all cells will simultaneously receive a PA card and issue requisition tags for the required parts. Provided inventory is available they will all begin production to replenish their output product store. The size of the initial inventory in each store is determined by weighing the penalties of a product store being empty and hence not being able to supply another cell against the inventory costs associated with high levels of inventories in the product stores.

It is clear with base stock control that a bottleneck cell will have a large number of PA cards and hence a large amount of work in process. Increasing the upstream z_i will not benefit the system performance in any significant way as it would just increase the congestion of waiting material at the bottleneck cell.

Material requirements planning.   In an MRP-controlled system the parameter choices are as follows:

1. $z_i \geq 0$, $i = 1, \ldots, m$. z_i can be identified with the safety stock or the planned minimum on-hand inventory in conventional MRP calculations.
2. $k_i = \infty$, $i = 1, \ldots, m$, that is, no delay in generating PA cards on arrival of an order tag at a store.

3. $r_i \geq 1$, $i = 1, \ldots, m$.

4. $\tau_i \geq 0$, $i = 1, \ldots, m$. A delay between order tags and requisition tags. The amount of delay is determined by the lead times used in the MRP calculations.

In conventional MRP calculations work release to a stage is offset from the time when the work is required at the downstream store by the stage "lead time." This lead time is supposed to be related to the flow time of a typical job through the stage. Because flow times are inherently variable, however, it is better to think of the stage lead time as a management set parameter. Let l_i be the lead time for product i on cell i. Next, construct the critical path network corresponding to all the production activities required to produce the end products of the production system and in this critical path network set the duration of the activity corresponding to the production tasks in cell i on product i equal to l_i. Then analyze the network, and determine the latest start time of each activity. Then the amount of delay in issuing a requisition tag at a cell is set equal to the latest start time of the corresponding activity in the critical path network. This delay also applies to the final production activity, shipment to the customer, that is, the requisition tag authorizing shipment to the customer is delayed from the customer's order tag by the total duration of the corresponding project on the critical path network.

In the single–end-product series system consisting of cells $1, 2, \ldots, m$ with lead time l_i associated with cell i, the delay between order tag and requisition tag at cell i is given by $\tau_i = \sum_{j=1}^{i-1} l_j$ with $\tau_0 = 0$. At the output product store this implies that the requisition tag authorizing shipment to the customer must be delayed by an amount $\tau_m = \sum_{j=1}^{m} l_j$.

MRP systems are almost invariably driven by a master schedule, which for our purposes can be considered to be a forecast of future receipts of requisition tags. That is, the assumption is that customers will not be sending in order tags, only requisition tags, so therefore system management has to generate the order tags. If τ_m is the overall time to complete the equivalent project on the critical path network that uses the l_i as the activity durations, then each order tag has to be generated at a time τ_m before the forecast receipt of the requisition tag or customer demand, that is, τ_m is the forecast time horizon. We assume that these forecasts are perfect; otherwise the PAC system should be modified to ensure that changes in forecasts are transmitted through the system as correction or cancellation notices as appropriate.

For a given forecast time horizon τ^*, it is not difficult to show that the best delivery service will be obtained by setting all $l_i = 0$ except for those cells that deliver to final product stores (for which their lead time should be set equal to τ^*).

KANBAN system.   In a conventional KANBAN system parameter choices are as follows:

1. $z_i > 0$, $i = 1, \ldots, m$.

2. $k_i = z_i$, $i = 1, \ldots, m$, if only one cell can supply one product store.

3. $r_i \geq 1$, $i = 1, \ldots, m$.

4. $\tau_i = 0$, $i = 1, \ldots, m$.

Note that the limit on k_i influences the system in two ways. First, it means that the work in process at any cell is limited to k_i. Next, the limit on k_i means that information about the demands for final products does not pass back to earlier stages of manufacture immediately. Order tags can be waiting at product stores for their matching PA card to be generated and thus the information does not pass back immediately. If a cell becomes a bottleneck owing to machine failure or labor problems, then upstream cells have no knowledge of the existence of demands. Thus when the machine failure ends, these upstream cells can be swamped with work and in turn become bottlenecks.

Local control.   By local control we mean a cell management policy in which a cell produces product whenever the following conditions are met: (1) parts are available; (2) machine and labor capacity are available; and (3) the product store(s) to which the cell delivers product are not full, where full denotes some preset maximum permitted level. Condition (2) is quantified by assuming that cell i can be regarded as equivalent to c_i parallel servers.

The parameter choices that will result in the system operating as if it had local control are as follows:

1. $z_i > c_i$, $i = 1, \dots, m$.
2. $k_i = c_i$, $i = 1, \dots, m$.
3. $r_i = 1$, $i = 1, \dots, m$.
4. $\tau_i = 0$, $i = 1, \dots, m$.

To show that this choice of parameters will lead to local control, observe that from equation 10.16 when $\tau_{i-1} = 0$ and $r_i = 1$

$$R_k^{(i)} = \max\{D_{k-z_{i-1}}^{(i-1)}, D_{k-k_i}^{(i)}, A_k^{(i+1)}\}$$

If $k_i \leq z_i$ then $D_{k-k_i} \geq D_{k-z_i}$ and

$$\max\{D_{k-k_i}^{(i)}, A_k^{(i+1)}\} = \max\{D_{k-k_i}^{(i)}, D_{k-z_i}^{(i)}, A_k^{(i+1)}\}$$
$$= \max\{D_{k-k_i}^{(i)}, R_k^{(i+1)}\}$$

and hence
$$R_k^{(i)} = \max\{D_{k-z_{i-1}}^{(i-1)}, D_{k-k_i}^{(i)}, R_k^{(i+1)}\} \tag{10.35}$$

The three terms of equation 10.35 imply, respectively, that a job is not released into cell i unless (1) there is a job available in store $i - 1$, (2) there are fewer than k_i jobs in process in cell i, *and* (3) there are fewer than z_i jobs in cell i and store i combined. That is, because no information from downstream cells is used apart from the availability of space in the cell and its output store, this is equivalent to local control.

Multiple-stage finite buffer system.   If $k_i \leq c_i$, then the condition (2) implies that no released job is ever waiting for its processing to begin, and the combination of conditions (2) and (3) imply that store i need not have space for more than $z_i - c_i$ jobs because when this store is full completed jobs can be held in the cell on a machine without delaying release or processing of other jobs. By setting $k_i = c_i$ for $i = 1, \ldots, m$ such a system is equivalent to a multiple-stage finite buffer system with c_i machines per stage and an interstage buffer capacity of $z_i - c_i$. The remaining storage space required for product store i is achieved by letting jobs be blocked and held on a machine when they have completed processing. The system is identical to the produce to stock flow-line system considered in Chapter 5.

Series system with shared buffer space. If $z_i \geq k_i > c_i$, then the series system with local control becomes equivalent to a system with a buffer (buffer $(i, i+1)$) of size $z_i - k_i$ between cells i and $i + 1$ which can only store jobs that have completed processing at cell i but have not yet begun processing at cell $i+1$. At cell i there is space for k_i jobs to be stored (including space on the machines), and this space can be used either by jobs waiting processing by the machines at cell i, or by jobs that have been processed by cell i and are waiting for space to become available in the buffer $(i, i + 1)$ between cells i and $i + 1$. That is, at cell i there are the c_i spaces on the machines and $k_i - c_i$ spaces in the common buffer (i) associated with cell i. When a machine in cell i completes processing of a job, then the job would enter buffer $(i, i + 1)$ if space in it is available. Otherwise, if buffer $(i, i + 1)$ is full, then a machine at cell i that has just completed processing a job would swap this completed job with any as yet unprocessed job in the common buffer (i). Once there are no unprocessed jobs left in the common buffer (i) then no further swaps can occur, however, and machines at cell i would be blocked. Jobs are moved from buffer $(i, i + 1)$ to cell $i + 1$ as soon as space becomes available in buffer $(i + 1)$. Jobs not yet processed by cell $i + 1$ move from buffer $(i + 1)$ to a machine in cell $i + 1$ whenever a machine becomes unoccupied or as a result of a swap with a machine that has completed a job. Jobs processed by cell $i + 1$ that are waiting in the common buffer $(i + 1)$ move to the buffer $(i + 1, i + 2)$ as soon as there is space. This means that jobs that have been processed by machines in cell i, but have not yet been processed by machines in cell $i + 1$ can be found in either buffers (i), $(i, i + 1)$, or $(i + 1)$. At cell m buffer $(m + 1)$ is assumed to be the potential customers and hence its size is unlimited. Jobs do not move from buffer $(m, m + 1)$ until there is a customer requisition.

Integral control. By integral control we mean release rules that take into account the total inventory over a range of cells through which work flows. That is, instead of, as in local control basing the decision purely on the inventory position in the product store associated with the cell, the release decision is based on the total inventory in a number of cells and their respective product stores.

Integral control for the series of m cells as a whole can be implemented using the following choice of parameters in the system:

1. $z_i \geq 0$, $i = 1, \ldots, m$.
2. $k_i = z_i + k_{i+1}$ for $i = 1, \ldots, m - 1$, $k_m = z_m$.

3. $r_i = 1, i = 1, \ldots, m.$
4. $\tau_i = 0, i = 1, \ldots, m.$

We will now demonstrate that this choice of parameters will indeed result in integral control of the m cells as a whole. That is, we will show that the total number of items in the system will be always equal to $\sum_{i=1}^{m} z_i$. First we show that as soon as an order tag arriving from outside is converted into a PA card (by matching it with a process tag), order tags from store to store will be immediately propagated into the raw material (i.e., input) store in front of cell 1. That is, we have

$$A_k^{(m)} = A_k^{(m-1)} = \cdots = A_k^{(1)}, \quad k = 1, 2, \ldots \tag{10.36}$$

To see this note that from equation 10.16 one has

$$A_k^{(i)} \geq D_{k-k_i}^{(i)}$$

and

$$R_k^{(i)} \geq D_{k-z_{i-1}}^{(i-1)}$$

Hence

$$D_{k-k_i}^{(i)} \geq R_{k-k_i}^{(i)} \geq D_{k-k_i-z_{i-1}}^{(i-1)}$$

Because $k_{i-1} = k_i + z_{i-1}$ one sees that

$$A_k^{(i)} \geq D_{k-k_i}^{(i)} \geq D_{k-k_{i-1}}^{(i-1)}$$

So from equation 10.16 it is easily seen that

$$A_k^{(i-1)} = \max\{D_{k-k_{i-1}}^{(i-1)}, A_k^{(i)}\} = A_k^{(i)}$$

Because $k_m = z_m$, a process tag will be available at the output store m, if and only if a product is available at the output store m. Hence each time a PA card is generated at store m, a final product will leave the system (meeting the requisition tag corresponding to the order tag that generated the PA card). At the same time orders will be propagated to the raw material store 0 and a raw material will be injected immediately into the system. Therefore the total number of items in the system will always remain the same at $\sum_{i=1}^{m} z_i$.

Using an analysis similar to the preceding one it can be seen that if we set, for a fixed $i < j$,

$$k_j \geq z_j; \quad k_l \geq k_{l+1} + z_l, \quad l = i, \ldots, j - 1,$$

then the total inventory of items in cell and store l, $l = i, \ldots, j$, will be less than or equal to $\sum_{l=i}^{j} z_l$. Thus we can exercise integral control among the subsets of cells and stores in the system.

OPT.   In the context of a situation in which batch sizes are one so that batch splitting is irrelevant, the key idea of OPT is to control the level of work in progress upstream of any bottleneck facility. This can be achieved by setting $k_i \geq z_i + k_{i+1}$ for all cells i upstream of the bottleneck facility j^* and having $\sum_{i=1}^{j^*-1} z_i + k_{j^*}$ equal to the work in progress limit over cells $1, 2, \ldots, j^*$. If the bottleneck facility can be considered as equivalent to c^* servers in parallel it would be desirable to have $k_{j^*} \geq c^*$ and to avoid the parameters of further downstream stages, limiting ability to achieve the total inventory limit upstream of the bottleneck $k_{j^*} < z_{j^*} + k_{j^*+1}$ (and desirably $k_{j^*} \leq z_{j^*}$).

CONWIP.   When we applied the integral control to the entire system, we saw that we have a constant work-in-process equal to $\sum_{i=1}^{m} z_i$. Now if we want to maximize the customer service subject to this CONWIP, from earlier results we know that we should set

$$z_i^* = 0, \quad i = 1, \ldots, m-1; \quad z_m^* = \sum_{i=1}^{m} z_i$$

This optimized system is known as CONWIP.

10.7.4 Bounds for Performance of Base Stock System

In this section we will develop bounds and approximations for the performance of the base stock system described earlier.

For a special case of the base stock system described previously we will explicitly compute the bound developed earlier for the time needed to meet a demand. Suppose cell i has one server, the processing times at cell i form an iid sequence of exponential random variables with mean $1/\mu_i$, $i = 1, \ldots, m$ and customer demands arrive according to a Poisson process with rate λ. Let $\hat{D}_k^{(i)}$, $k = 1, 2, \ldots$; $i = j, \ldots, m$ be as defined earlier for the modified series cell system with cells $\{j, \ldots, m\}$. Then $\hat{D}_k^{(m)} - A_k$ is the time spent in the system by the kth customer arrived at the modified series cell system (which in this special case is a series of $m - j + 1$ exponential servers). Therefore if $\hat{T}_\infty^{(j)} \overset{d}{=} \lim_{k \to \infty} \{\hat{D}_k^{(m)} - A_k\}$, one has (from the results in Section 5.4)

$$E[e^{-s\hat{T}_\infty^{(j)}}] = \prod_{i=j}^{m} \left(\frac{\mu_i - \lambda}{\mu_i - \lambda + s}\right)$$

Now consider the time needed to meet the kth demand

$$\left[\hat{D}_{k-\sum_{i=j}^{m} z_i}^{(m)} - A_k\right]^+ = \left[\left(\hat{D}_{k-\sum_{i=j}^{m} z_i}^{(m)} - A_{k-\sum_{i=j}^{m} z_i}\right) - \left(A_k - A_{k-\sum_{i=j}^{m} z_i}\right)\right]^+$$

Clearly the quantities in the two brackets are independent and as $k \to \infty$, the first term is equal in distribution to $\hat{T}_\infty^{(j)}$ and the second term is an Erlang random variable with $(Z_j = \sum_{i=j}^{m} z_i)$ phases with mean $1/\lambda$. Therefore if $\hat{L}_\infty^{(j)} \overset{d}{=} \lim_{k \to \infty} [\hat{D}_{k-\sum_{i=j}^{m} z_i} - A_k]^+$, one has

$$E[e^{-s\hat{L}_\infty^{(j)}}] = \sum_{l=j}^{m} a_l \prod_{i=l}^{m} \left(\frac{\mu_i - \lambda}{\mu_i - \lambda + s}\right) + a_{m+1}$$

where

$$a_j = \left(\frac{\lambda}{\lambda + \mu_j}\right)^{Z_j}$$

$$a_{j+1} = \sum_{l_{j+1}=1}^{Z_j} \left(\frac{\lambda}{\lambda + \mu_j}\right)^{Z_j - l_{j+1}} \left(\frac{\lambda}{\lambda + \mu_{j+1}}\right)^{l_{j+1}} \left(\frac{\mu_j}{\lambda + \mu_j}\right)$$

$$a_l = \sum_{l_i=1}^{Z_j} \cdots \sum_{l_{j+1}=0}^{Z_j - l_i - \cdots - l_{j+2}} \left(\frac{\lambda}{\lambda + \mu_j}\right)^{Z_j - l_i - \cdots - l_{j+1}}$$

$$\prod_{i=j}^{l-1} \left(\frac{\lambda}{\lambda + \mu_{i+1}}\right)^{l_{i+1}} \left(\frac{\mu_i}{\lambda + \mu_i}\right), \qquad l = j+2, \ldots, m$$

$$a_{m+1} = \sum_{l_{m+1}=1}^{Z_j} \sum_{l_m=0}^{Z_j - l_{m+1}} \cdots \sum_{l_j=0}^{Z_j - l_{m+1} - \cdots - l_{j+1}} \left(\frac{\lambda}{\lambda + \mu_j}\right)^{Z_j - l_{m+1} - \cdots - l_j}$$

$$\prod_{i=j}^{m} \left(\frac{\lambda}{\lambda + \mu_i}\right)^{l_i} \left(\frac{\mu_i}{\lambda + \mu_i}\right) \tag{10.37}$$

and $Z_j = \sum_{i=j}^{m} z_i$, $j = 1, \ldots, m$. The mean time to meet the demand $E[L_\infty^{(j)}]$ is bounded from below by

$$E[\hat{L}_\infty^{(j)}] = \sum_{i=j}^{m} \sum_{l=j}^{i} a_l \cdot \left(\frac{1}{\mu_i - \lambda}\right)$$

Note that a_{m+1} bounds from below the probability that a demand will be met immediately by a part available in the output buffer. These results can therefore be used to find the *minimum* base stock target levels $z_i, i = 1, \ldots, m$ needed to achieve desired demand service levels such as the fraction of customer demand met immediately by parts from the output buffer and the average delay in meeting a demand.

10.7.5 Performance Evaluation of PAC Systems

We will now develop an approximation procedure to evaluate the average inventory, delivery performance and the production capacity for a special case of a PAC system with no time lags between order tags and requisition tags. For this special case we assume that each cell consists of a single server and the service times at cell i form an iid sequence of exponential random variables with mean $1/\mu_i$, $i = 1, \ldots, m$. There is an infinite supply of raw material in front of the first cell. The parts in buffer m that are output from cell m are removed by a demand process that we assume occurs according to a Poisson process with rate λ.

   Suppose the system is observed at time t and the following quantities are measured for $i = 1, \ldots, m$. $N_i(t)$: the number of parts waiting for processing at cell i, including the part, if any, being processed by the cell; $N_i^r(t)$: the number of requisition tags waiting to be filled at store i; $N_i^p(t)$: the number of process tags at store i; $N_i^m(t)$: the number of material tags at store i (i.e., the number of parts in the store); $N_i^o(t)$: the number of order tags waiting at store i to generate PA cards. It follows that $k_i - N_i^p(t)$ is the number of process tags of store i that have been converted into PA cards. Alternatively, since the number of PA cards active at cell i is $N_i(t) + N_{i-1}^r(t)$,

$$N_i^p(t) + N_{i-1}^r(t) + N_i(t) = k_i, \quad i = 1, \dots, m$$

Define $Z_i(t) = N_i^m(t) - N_i^r(t)$, $i = 1, \dots, m$, as the inventory position at store i. Next, let $D^{(i)}(t) = \max\{n | D_n^{(i)} \le t\}$, $R^{(i)}(t) = \max\{n | R_n^{(i)} \le t\}$ and $A^{(i)}(t) = \max\{n | A_n^{(i)} \le t\}$, $i = 1, \dots, m$. Then $N_i^r(t) = A^{(i+1)}(t) - R^{(i+1)}(t)$, $N_i^p(t) = k_i + D^{(i)}(t) - A^{(i)}(t)$, $N_i^m(t) = z_i + D^{(i)}(t) - R^{(i+1)}(t)$ and $N_i^o(t) = A^{(i+1)}(t) - A^{(i)}(t)$. Hence the following identity applies

$$z_i - N_i^m(t) + N_i^r(t) = k_i - N_i^p(t) + N_i^o(t), \quad i = 1, \dots, m \tag{10.38}$$

Alternatively, using equation 10.38 and defining $N_i^e(t) = z_i - Z_i(t)$ (so $N_i^e(t) \ge 0$)

$$N_i^p(t) - N_i^o(t) = k_i - N_i^e(t), \quad i = 1, \dots, m \tag{10.39}$$

The basis of the approximation is the recognition that in the multiple-cell system there are two types of process dynamics occurring:

- *Circulation through cell* i. A PA card goes from store i to cell i where a requisition tag is immediately generated and transmitted to store $i - 1$. Once it is filled it then returns to cell i, accompanying a part. It then generates a process tag which accompanies the part through processing at cell i until the part and the process tag are delivered to store i. The arrival of the process tag at store i will eventually enable another PA card to be generated. We will show that this circulation can be represented by a cyclic queue with k_i customers and three service centers: store i where a process tag waits until it generates a PA card, store $i - 1$ where a requisition tag waits to be filled, and cell i where actual manufacturing operations are carried out on a part.

- *Inventory position fluctuations at store* i. The inventory position increases as parts are delivered from cell i to the store and decreases as requisition tags are met and parts are delivered to cell $i + 1$.

Obviously store i serves to link the circulation through cell i and the circulation through cell $i + 1$ so the key to the approximation is the development of the relationship between the two types of processes. First we model the circulation through a cell, then we model the inventory position fluctuations at a store and lastly we describe how the two types of processes are interrelated.

Circulation through cell i. If $N_i^p(t-) = 0$ (i.e., store i has no process tags) and a service completion occurs at cell i at time t, then $N_i^p(t)$ will become 1 if there are no order tags queued up at store i (i.e., $N_i^o(t-) = 0$) and will remain zero otherwise. Let $q_{i:d}(t)$ be this probability (that is $q_{i:d}(t) = P\{N_i^o(t-) = 0 | N_i^P(t-) = 0$ and a service completion occurred at time $t\}$). Conversely, if $N_i^P(t-) > 0$ then a service completion at cell i will increase $N_i^P(t)$ by one (i.e., $N_i^P(t) = N_i^P(t-)+1$). Observe that whenever $N_i^P(t-) > 0$, then there will be no waiting order tags from cell $i + 1$ at store i. Hence $N_i^o(t-) = 0$. Therefore whenever $N_i^P(t-) > 0$ and an order tag (and also a requisition tag) arrives at store i at time t, then $N_i^P(t)$ will decrease by one (i.e., $N_i^P(t) = N_i^P(t-) - 1$). Let $v_{i:d}(n_i^P)$ be the rate at which requisition tags arrive at store i when there are n_i^P process tags at store i.

Now let us look at the number of requisition tags, $N_{i-1}^r(t)$ at store $i - 1$. Suppose $N_{i-1}^r(t-) = 0$ and a requisition tag for store $i - 1$ is generated at time t. If there are no parts available at store $i - 1$ at time t, then the number of requisition tags at store $i - 1$ at time t will become one (i.e., $N_i^r(t) = 1$). Let $q_{i-1:u}(t) = P\{N_{i-1}^m(t-) = 0 | N_{i-1}^r(t-) = 0$ and a requisition tag for store $i - 1$ was generated at time $t\}$) be the probability that this occurs. Then with probability $1 - q_{i-1:u}(t)$, the number of requisition tags at store $i - 1$ at time t will remain zero. Conversely if $N_{i-1}^r(t-) > 0$ and a requisition tag for cell $i - 1$ is generated at time t, then the number of requisition tags at store $i - 1$ will increase by one at time t (i.e., $N_{i-1}^r(t) = N_{i-1}^r(t-) + 1$). Let $1/v_{i-1:u}(n_{i-1}^r)$ be the average time for cell $i - 1$ to fill a requisition tag when n_{i-1}^r tags are waiting at store $i - 1$.

Now assuming that $q_{i:d}(t) = q_{i:d}$ and $q_{i-1:u}(t) = q_{i-1:u}$ are time invariant and the times needed to generate requisition tags for store i when there are n_i^P process tags waiting are iid exponential random variables with mean $1/v_{i:d}(n_i^P)$ and that the times needed to fill requisition tags at store $i - 1$ by cells upstream when there are n_{i-1}^r requisition tags waiting are iid exponential random variables with mean $1/v_{i-1:u}(n_{i-1}^r)$, we can model $\{(N_i(t), N_i^P(t), N_{i-1}^r(t)), t \geq 0\}$ as a Markov process with state space $S = \{(n_1, n_2, n_3) : 0 \leq n_1 \leq k_i; 0 \leq n_2 \leq k_i; 0 \leq n_3 \leq k_i, n_1 + n_2 + n_3 = k_i\}$. The transition diagram of this Markov process is shown in Figure 10.5. It can be easily verified that the stationary distribution $p^{(i)}(n_i, n_i^P, n_{i-1}^r) = \lim_{t\to\infty} P\{N_i(t) = n_i, N_i^P(t) = n_i^P, N_{i-1}^r(t) = n_{i-1}^r\}$, $(n_i, n_i^P, n_{i-1}^r) \in S$ is given by

$$p^{(i)}(n_i, n_i^P, n_{i-1}^r) = K \cdot \left(\frac{1}{\mu_i}\right)^{n_i} \prod_{v=0}^{n_i^P} \left(\frac{1}{v_{i:d}(v)}\right) \prod_{w=0}^{n_{i-1}^r} \left(\frac{1}{v_{i-1:u}(w)}\right) f_p(n_i^P) f_r(n_{i-1}^r),$$

$$(n_i, n_i^P, n_{i-1}^r) \in S \qquad (10.40)$$

where K is a normalizing constant given by

$$1/K = \sum_{(n_i, n_i^P, n_{i-1}^r) \in S} \left(\frac{1}{\mu_i}\right)^{n_i} \prod_{v=0}^{n_i^P} \left(\frac{1}{v_{i:d}(v)}\right) \prod_{w=0}^{n_{i-1}^r} \left(\frac{1}{v_{i-1:u}(w)}\right) f_p(n_i^P) f_r(n_{i-1}^r)$$

$$v_{i:d}(0) = 1, \qquad v_{i-1:u}(0) = 1$$

$$f_p(l) = \begin{cases} 1, & l = 0 \\ q_{i:d}, & l = 1, \ldots, k_i \end{cases}$$

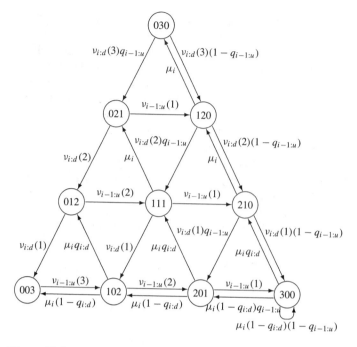

Figure 10.5 Transition Diagram for Circulation through Cell i

and

$$f_r(l) = \begin{cases} 1, & l = 0 \\ q_{i-1:u}, & l = 1, \ldots, k_i \end{cases}$$

Dynamics of store i. At store i the inventory position increases from $Z_i(t-)$ to $Z_i(t-) + 1$ if cell i completes a part at time t while it decreases from $Z_i(t-)$ to $Z_i(t-) - 1$ if a requisition tag (and an order tag) arrives from cell $i + 1$ at time t. Let $\mu_{i:u}(n_i^e)$ be the rate at which inventory position increases if $Z_i(t-) = z_i - n_i^e$ and $\mu_{i:d}(n_i^e)$ be the rate at which the inventory position decreases if $Z_i(t-) = z_i - n_i^e$. It follows that $Z_i(t)$ is a random walk on the integers $z_i, z_i - 1, \ldots, 1, 0, -1, -2, \ldots, -k_{i+1}$, and hence the stationary distribution $p^{(i)}(n_i^e) = \lim_{t\to\infty} P\{Z_i(t) = z_i - n_i^e\}$, $n_i^e \geq 0$, is given by

$$p^{(i)}(n_i^e) = \frac{\prod_{v=0}^{n_i^x-1} \mu_{i:d}(v)}{\prod_{w=1}^{n_i^e} \mu_{i:u}(w)} p^{(i)}(0), \quad n_i^e \geq 0$$

with $p^{(i)}(0) = \lim_{t\to\infty} P\{Z_i(t) = z_i\}$ determined by $\sum_{n_i^e=0}^{\infty} p^{(i)}(n_i^e) = 1$.

Relationship between stores and circulation through cells. The parameters needed to characterize the service processes as tags circulate through cell i, and the parameters needed to describe the dynamics of stores i and $i - 1$ are interrelated.

Cell parameters.   From equation 10.39 it can be verified that

$$\{N_i^o(t-) = 0\} \text{ and } \{N_i^p(t-) = 0\} \equiv \{N_i^e(t-) = k_i\}$$
$$\{N_i^p(t-) = 0\} \equiv \{N_i^e(t-) \geq k_i\}$$

Therefore one has

$$q_{i:d}(t) = P\{N_i^o(t-) = 0 | N_i^p(t-) = 0 \text{ and a service completion occurred at time } t\}$$
$$= P\{N_i^e(t-) = k_i | N_i^e(t-) \geq k_i \text{ and a part is delivered to store } i \text{ at time } t\}$$

That is,

$$q_{i:d} = \frac{p^{(i)-}(n_i^e = k_i)}{p^{(i)-}(n_i^e \geq k_i)}, \quad i = 1, \ldots, m \tag{10.41}$$

where $p^{(i)-}(n_i^e = v)$ is the probability that an arriving part sees an inventory position of $z_i - v$ in store i. Similarly, it can be shown that

$$q_{i:u} = \frac{p^{(i)+}(n_i^e = z_i)}{p^{(i)+}(0 \leq n_i^e \leq z_i)}, \quad i = 1, \ldots, m \tag{10.42}$$

where $p^{(i)+}(n_i^e = v)$ is the probability an arriving requisition tag at store i sees an inventory position of $z_i - v$.

Next, if n_i^p process tags, $0 < n_i^p \leq k_i$, are waiting at store i then it follows that the inventory position in the store is $z_i - k_i + n_i^p$ and $n_i^e = k_i - n_i^p$. Hence it follows that

$$v_{i:d}(n_i^p) = \mu_{i:d}(k_i - n_i^p), \quad 1 \leq n_i^p \leq k_i, i = 1, \ldots, m \tag{10.43}$$

Similarly if n_i^r requisition tags, $0 < n_i^r \leq k_{i+1}$, are waiting at store i the inventory position in the store is $-n_i^r$ and $n_i^e = z_i + n_i^r$. Hence

$$v_{i:u}(n_i^r) = \mu_{i:u}(z_i + n_i^r), \quad 1 \leq n_i^r \leq k_{i+1}, i = 1, \ldots, m \tag{10.44}$$

At cell 1, the assumption of adequate raw material implies that $1/v_{0:u}(n_0^r) = 0$ for $n_0^r > 0$ and $v_{0:u}(0) = 1$.

Store parameters.   Given n_i^e at store i it follows that $n_i^r = \max(0, n_i^e - z_i)$. Now, considering the circulation process through cell $i+1$, suppose there are n_{i+1} parts in process at cell $i + 1$, n_{i+1}^p process tags at store $i + 1$ and $n_i^r = \max(0, n_i^e - z_i)$ requisition tags at store i. It follows that $n_{i+1} = k_{i+1} - \max(0, n_i^e - z_i) - n_{i+1}^p = k_{i+1} + \min(0, z_i - n_i^e) - n_{i+1}^p$. Then the rate at which requisition tags will be arriving from cell $i + 1$ will be $v_{i+1:d}(n_{i+1}^p)$ provided $n_{i+1}^p > 0$. If $n_{i+1}^p = 0$ then the rate at which requisition tags will be arriving at store i will be $\mu_{i+1}(1 - q_{i+1:d})$. Hence, provided $n_i^e < z_i + k_{i+1}$

$$\mu_{i:d}(n_i^e)$$

$$= \frac{\sum_{v=1}^{k_{i+1}+\min(0, z_i - n_i^e)} v_{i+1:d}(v) p^{(i+1)}(k_{i+1} + \min(0, z_i - n_i^e) - v, v, \max(0, n_i^e - z_i))}{\sum_{v=0}^{k_{i+1}+\min(0, z_i - n_i^e)} p^{(i+1)}(k_{i+1} + \min(0, z_i - n_i^e) - v, v, \max(0, n_i^e - z_i))}$$

$$= \frac{\sum_{v=0}^{k_{i+1}+\min(0, z_i - n_i^e)-1} \left(\frac{1}{\mu_{i+1}}\right)^{k_{i+1}+\min(0, z_i - n_i^e)-1-v} \prod_{w=0}^{v} \left(\frac{1}{v_{i+1:d}(w)}\right) f_p(v)}{\sum_{v=0}^{k_{i+1}+\min(0, z_i - n_i^e)} \left(\frac{1}{\mu_{i+1}}\right)^{k_{i+1}+\min(0, z_i - n_i^e)-v} \prod_{w=0}^{v} \left(\frac{1}{v_{i+1:d}(w)}\right) f_p(v)}$$

$$= \frac{G_{i+1:d}(k_{i+1} + \min(0, z_i - n_i^e))}{G_{i+1:d}(k_{i+1} + \min(0, z_i - n_i^e) - 1)} \qquad (10.45)$$

where

$$G_{i+1:d}(n) = \frac{1}{\sum_{v=0}^{n} \left(\frac{1}{\mu_{i+1}}\right)^{n-v} \prod_{w=0}^{v} \frac{1}{v_{i+1:d}(w)} f_p(v)}$$

and $G_{i+1:d}(0) = 1$. That is, $G_{i+1:d}(n)$ is the normalizing constant for the closed queue consisting of the two service centers corresponding to the cell $i + 1$ and store $i + 1$ and with n customers circulating in this two-service center system. Note that if $n_i^e \geq z_i + k_{i+1}$ then $\mu_{i:d}(n_i^e) = 0$ (and hence the maximum reachable n_i^e is $z_i + k_{i+1}$). Also note that $\mu_{i:d}(n_i^e) = \mu_{i:d}(z_i) = G_{i+1:d}(k_{i+1})/G_{i+1:d}(k_{i+1} - 1)$ for $0 \leq n_i^e \leq z_i$. At store m it is obvious that $\mu_{m:d}(n_m^e) = \lambda$, $n_m^e \geq 0$.

Next, to determine $\mu_{i:u}(n_i^e)$, observe that $n_i^p = \max(0, k_i - n_i^e)$. The rate at which process tags (and products) will be arriving at store i from store $i - 1$ will be μ_i if $n_i > 0$ and 0 otherwise. When $n_i = v$ then $n_{i-1}^r = \min(k_i, n_i^e) - v$ for given n_i^e. Hence

$$\mu_{i:u}(n_i^e) = \mu_i \frac{\sum_{v=1}^{\min(k_i, n_i^e)} p(v, \max(0, k_i - n_i^e), \min(k_i, n_i^e) - v)}{\sum_{v=0}^{\min(k_i, n_i^e)} p(v, \max(0, k_i - n_i^e), \min(k_i, n_i^e) - v)} \qquad (10.46)$$

$$= \frac{G_{i:u}(\min(k_i, n_i^e))}{G_{i:u}(\min(k_i, n_i^e) - 1)}, \qquad n_i^e > 0 \qquad (10.47)$$

with

$$G_{i:u}(n) = \frac{1}{\sum_{v=0}^{n} \left(\frac{1}{\mu_i}\right)^{n-v} \prod_{w=0}^{v} \frac{1}{v_{i-1:u}(w)} f_r(v)}$$

and $G_{i:u}(0) = 1$. $G_{i:u}(n)$ is the normalizing constant in the closed queue consisting of two service centers, corresponding to cell i and store $i - 1$ and n customers circulating in the two service-center system. Note that if $n_i^e = 0$ $\mu_{i:u}(0) = 0$. Also note that $\mu_{i:u}(n_i^e) = G_{i:u}(k_i)/G_{i:u}(k_i - 1)$ if $n_i^e \geq k_i$. At store 1, $\mu_{1:u}(n_1^e) = \mu_1$ for $n_1^e > 0$.

It follows that to analyze the system it is necessary to solve the $\sum_{i=2}^{m} k_i$ equations for the $\mu_{i:u}(n_i^e)$, $i = 2, \ldots, m$, $n_i^e = 1, \ldots, k_i$; the $\sum_{i=1}^{m-1} k_{i+1}$ equations for the $\mu_{i:d}(n_i^e)$, $i = 1, \ldots, m - 1$, $n_i^e = z_i, z_i + 1, \ldots, z_i + k_{i+1} - 1$; the $\sum_{i=1}^{m-1} k_i$ equations for $v_{i:d}(w)$,

$i = 1, \ldots, m-1$, $w = 1, \ldots, k_i$; the $\sum_{i=1}^{m-1} k_{i+1}$ equations for $v_{i:u}(v)$, $i = 1, \ldots, m-1$, $w = 1, \ldots, k_{i+1}$; the m equations for $q_{i:d}$, $i = 1, \ldots, m$ and the m equations for $q_{i:u}$, $i = 1, \ldots, m$. Further, the boundary conditions $v_{m:d}(w) = \lambda$, $1 \leq w \leq k_m$; $v_{1:u}(v) = \mu_1$, $1 \leq v \leq k_2$; $\mu_{m:d}(n_m^e) = \lambda$, $n_m^e \geq 0$ and $\mu_{1:u}(n_1^e) = \mu_1$, $0 < n_1^e \leq z_1 + k_2$ are used.

Performance measures

Service Level.   The service level is measured by p_{ND} the probability an arriving demand is met immediately and $\overline{D}$ the average delay in meeting a demand. We have

$$p_{ND} = p^{(m)+}(n_m^e < z_m) = p^{(m)}(n_m^e < z_m) \tag{10.48}$$

because Poisson arrivals see time averages. Also

$$\overline{D} = \sum_{n_m^e = z_m + 1}^{\infty} (n_m^e - z_m) p^{(m)}(n_m^e)/\lambda \tag{10.49}$$

Inventory Levels.   There are two inventory-level measures of interest, $\overline{N}_i^m$, the average inventory of parts in store i, and $\overline{N}_i$, the average work in process inventory in cell i. We have

$$\overline{N}_i^m = \sum_{n_i^m = 1}^{z_i} n_i^m p^{(i)}(z_i - n_i^m) \quad i = 1, \ldots, m, \tag{10.50}$$

$$\overline{N}_i = \sum_{n_i = 1}^{k_i} n_i \sum_{n_i^p = 0}^{k_i - n_i} p^{(i)}(n_i, n_i^p, k_i - n_i - n_i^p), \quad i = 1, \ldots, m \tag{10.51}$$

Throughput.   To find the throughput it is assumed that there are always requisition and order tags waiting at store m. Hence $1/v_{m:d}(n_m^p) = 0$, $n_m^p > 0$. It follows that $\mu_{m-1:d}(n_{m-1}^e) = \mu_m$, $0 \leq n_{m-1}^e < z_{m-1} + k_m$, and $v_{m-1:d}(n_{m-1}^p) = \mu_m$, $1 \leq n_{m-1}^p \leq k_{m-1}$. To find TH, the throughput, observe that

$$TH = \mu_m \sum_{n_m = 1}^{k_m} p^{(m)}(n_m, 0, n_m^p = k_m - n_m)$$

After some manipulation it can be shown that

$$TH = \mu_m (1 - p^{(m-1)}(z_{m-1} + k_m))$$

Alternatively, it can be shown that

$$TH = \mu_1 (1 - p^{(1)}(0))$$

Examples

Flow Line. This example is included to show that the approximation is the same as that given in Chapter 5 (equations 5.69 to 5.74). In a series flow line we have that $k_i = 1$ for $i = 1, \ldots, m$. Hence $N_i^e(t)$ can only take on the values $0, 1, \ldots, z_i, z_i + 1$ for $i = 1, \ldots, m - 1$. Also because $k_i = 1$, it follows that $\mu_{i:u}(n_i^e) = G_{i:u}(1) = 1/(1/\mu_i + q_{i-1:u}/v_{i-1:u}(1))$ for $0 < n_i^e \leq z_i + 1$, $i = 1, \ldots, m - 1$. Similarly $\mu_{i:d}(n_i^e) = G_{i+1:d}(1) = 1/(1/\mu_{i+1} + q_{i+1:d}/v_{i+1:d}(1))$ for $0 \leq n_i^e \leq z_i$, $i = 1, \ldots, m - 1$. Defining $\overline{\rho}_i = \mu_{i:d}(1)/\mu_{i:u}(1)$ it follows that $p^{(i)}(n_i^e) = \overline{\rho}_i^{n_i^e} p^{(i)}(0) = \overline{\rho}_i^{n_i^e}(1 - \overline{\rho}_i)/(1 - \overline{\rho}_i^{z_i+2})$. Because in this case $p^{(i)-}(n_i^e) = \overline{\rho}_i p^{(i)-}(n_i^e - 1)$ and $p^{(i)+}(n_i^e) = \overline{\rho}_i p^{(i)+}(n_i^e - 1)$ it follows that $q_{i:d} = p^{(i)}(n_i^e = 1)/p^{(i)}(1 \leq n_i^e \leq z_i + 1) = (1 - \overline{\rho}_i)/(1 - \overline{\rho}_i^{z_i+1})$, $i = 1, \ldots, m - 1$ and $q_{i:u} = \overline{\rho}_i^{z_i}(1 - \overline{\rho}_i)/(1 - \overline{\rho}_i^{z_i+1})$, $i = 1, \ldots, m - 1$. At store m we have $p^{(m)}(n_m^e) = \overline{\rho}_m^{n_m^e} p^{(m)}(0) = \overline{\rho}_m^{n_m^e}(1 - \overline{\rho}_m)$ and $q_{m:d} = 1 - \overline{\rho}_m$. Noting that $b_i = z_i$ it can be seen that we have arrived at the same set of equations to be solved as equations 5.69 to 5.74 except for $i = m$.

Base Stock Systems. In a base stock system in which $k_i = \infty$, $i = 1, \ldots, m$, arriving demands are immediately signaled to all stores, hence it follows that $\mu_{i:d}(n_i^e) = \lambda$, $i = 1, \ldots, m$, $n_i^e \geq 0$, and $p^{(i)+}(n_i^e) = p^{(i)}(n_i^e)$ because Poisson arrivals see time averages. Also $v_{i:d}(n_i^p) = \lambda$, $i = 1, \ldots, m$, $n_i^p > 0$. Define $\rho_i = \lambda/\mu_i$ and $\rho_{i:u}(n_i^e) = \lambda/\mu_{i:u}(n_i^e)$. Then we have that $p^{(1)}(n_i^e) = \rho_1^{n_i^e}(1 - \rho_1)$ and $q_{1:u} = \rho_1^{z_1}(1 - \rho_1)/(1 - \rho_1^{z_1+1})$. Next, because $\mu_{i:u}(n_i^e) = G_{i:u}(n_i^e)/G_{i:u}(n_i^e - 1)$ it follows that $p^{(i)}(n_i^e) = (\lambda^{n_i^e}/G_{i:u}(n_i^e))p^{(i)}(0)$. Now

$$\frac{\lambda^n}{G_{i:u}(n)} = \sum_{v=0}^{n} \rho_i^{n-v} \frac{\lambda^v}{\prod_{w=0}^{v} v_{i-1:u}(w)} f_r(v)$$

$$= \rho_i^n + \sum_{v=1}^{n} \rho_i^{n-v} q_{i-1:u} \prod_{w=1}^{v} \rho_{i-1:u}(w)$$

$$= \rho_i^n + \sum_{v=1}^{n} \rho_i^{n-v} \frac{p^{(i-1)}(z_{i-1} + v)}{p^{(i-1)}(0 \leq w \leq z_{i-1})}$$

That is,

$$\frac{p^{(i)}(n_i^e)}{p^{(i)}(0)} = \rho_i \frac{p^{(i)}(n_i^e - 1)}{p^{(i)}(0)} + \frac{p^{(i-1)}(z_{i-1} + n_i^e)}{p^{(i-1)}(0 \leq w \leq z_{i-1})}$$

Note that $p^{(i)}(1)/p^{(i)}(0) = \rho_i + p^{(i-1)}(z_{i-1} + 1)/p^{(i-1)}(0 \leq w \leq z_{i-1})$. It is possible to develop a recursive calculation procedure to determine the performance measures of interest (see Buzacott, Price, and Shanthikumar [11]).

Numerical example. Figure 10.6 shows the relationship between P_{ND} and z_1 for a two-stage base stock system with $\lambda = 1$, $\rho_1 = 0.8$ and $\rho_2 = 0.6$ in which $z_1 + z_2 = Z_1$ is held constant. The figure shows the upper bound, the approximation, and simulation results. It is of interest to note that P_{ND} is relatively insensitive to z_1 until $z_1 > z_2$.

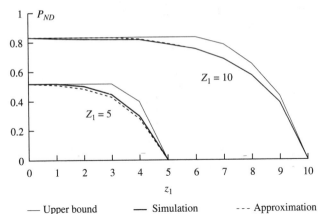

Figure 10.6 P_{ND} in Two-Stage Base Stock System ($\rho_1 = 0.8$, $\rho_2 = 0.6$)

— Upper bound   — Simulation   --- Approximation

KANBAN.   In a KANBAN system $z_i = k_i$, $i = 1, \dots, m$. Now at store m, $\mu_{m:u}(n_m^e) = G_{m:u}(z_m)/G_{m:u}(z_m - 1)$ for $n_m^e \geq z_m$. Hence it follows that $q_{m:d} = 1 - \lambda G_{m:u}(z_m - 1)/G_{m:u}(z_m)$. Consider now a two-stage system. Then define $\overline{\rho}_1 = G_{2:d}(k_2)/\mu_1 G_{2:d}(k_2 - 1)$ and $\overline{\rho}_2 = \lambda G_{2:u}(k_2 - 1)/G_{2:u}(k_2)$. Then $q_{2:d} = 1 - \overline{\rho}_2$ and $q_{1:u} = \overline{\rho}_1^{k_1}(1 - \overline{\rho}_1)/(1 - \overline{\rho}_1^{k_1+1})$. Further

$$\overline{\rho}_2 = \frac{\sum_{v=1}^{k_2} \rho_2^{k_2-v} \rho_1^v q_{1:u} + \rho_2^{k_2}}{\sum_{v=1}^{k_2-1} \rho_2^{k_2-1-v} \rho_1^v q_{1:u} + \rho_2^{k_2-1}}$$

and

$$\overline{\rho}_1 = \rho_1 \frac{\sum_{v=1}^{k_2-1} \rho_2^{k_2-1-v} q_{2:d} + \rho_2^{k_2-1}}{\sum_{v=1}^{k_2} \rho_2^{k_2-v} q_{2:d} + \rho_2^{k_2}}$$

$$= \rho_1 \frac{1 - \overline{\rho}_2 + \rho_2^{k_2-1}(\overline{\rho}_2 - \rho_2)}{1 - \overline{\rho}_2 + \rho_2^{k_2}(\overline{\rho}_2 - \rho_2)}$$

By an iterative approach $\overline{\rho}_1$ and $\overline{\rho}_2$ can be found and then the performance measures determined. If $k_1 = k_2 = 2$ then we have the following equations to solve:

$$\overline{\rho}_1 = \rho_1 \frac{1 - \overline{\rho}_2(1 - \rho_2) - \rho_2^2}{1 - \overline{\rho}_2(1 - \rho_2^2) - \rho_2^3}$$

$$\overline{\rho}_2 = \rho_2 + \rho_1 \frac{\rho_1 q_{1:u}}{\rho_1 q_{1:u} + \rho_2}$$

and

$$q_{1:u} = \frac{\overline{\rho}_1^2(1 - \overline{\rho}_1)}{1 - \overline{\rho}_1^3}$$

Because $p^{(2)}(1) = (\rho_2 + \rho_1 q_{1:u})p^{(2)}(0)$ and $p^{(2)}(n) = \overline{\rho}_2^{n-1} p^{(2)}(1)$ it follows that $p_{ND} = (1 - \overline{\rho}_2)(1 + \rho_2 + \rho_1 q_{1:u})/(1 - \overline{\rho}_2 + \rho_2 + \rho_1 q_{1:u})$ and $\overline{D} = (\rho_2 + \rho_1 q_{1:u})\overline{\rho}_2^2/((1 - \overline{\rho}_2)(1 - \overline{\rho}_2 + \rho_2 + \rho_1 q_{1:u}))$. Table 10.2 compares the approximation with simulation results. Note

TABLE 10.2 COMPARISON OF APPROXIMATION AND SIMULATION FOR KANBAN SYSTEM WITH $\lambda = 1$ AND $k_1 = k_2 = 2$. (***: UNSTABLE)

P_{ND}

ρ_2		ρ_1 = 0.1	0.3	0.5	0.7	0.9
0.1	Approx.	0.990	0.982	0.937	0.796	0.447
	Sim.	0.990	0.981	0.927	0.750	0.350
0.3	Approx.	0.910	0.898	0.846	0.707	0.378
	Sim.	0.911	0.902	0.847	0.677	0.301
0.5	Approx.	0.750	0.737	0.678	0.530	0.201
	Sim.	0.752	0.744	0.692	0.532	0.182
0.7	Approx.	0.510	0.496	0.435	0.282	***
	Sim.	0.515	0.507	0.454	0.300	***
0.9	Approx.	0.190	0.177	0.117	***	***
	Sim.	0.196	0.187	0.131	***	***

$\overline{D}$

ρ_2		ρ_1 = 0.1	0.3	0.5	0.7	0.9
0.1	Approx.	0.001	0.003	0.029	0.242	2.20
	Sim.	0.001	0.006	0.067	0.549	5.418
0.3	Approx.	0.039	0.047	0.102	0.376	2.52
	Sim.	0.039	0.046	0.127	0.689	6.064
0.5	Approx.	0.251	0.276	0.419	1.02	6.73
	Sim.	0.251	0.266	0.402	1.23	10.5
0.7	Approx.	1.14	1.22	1.68	3.90	***
	Sim.	1.14	1.19	1.55	3.70	***
0.9	Approx.	7.30	7.97	13.49	***	***
	Sim.	6.94	7.45	11.4	***	***

that when ρ_1 is large and ρ_2 is small the approximation tends to overestimate P_{ND} and underestimate $\overline{D}$. This is probably because the time between arrivals of requisition tags at store 1 is then more variable than the exponential distribution assumed in calculating q_{1u} for store 1. It is preferable to set P_{ND} in a KANBAN system equal to the minimum of the approximation and the base stock upper bound.

In the above three examples we saw how the general approximation for the PAC system can be specialized for specific cases. The general approximation itself was implemented and the results obtained are compared with the simulation results below.

Probability of no delay (P_{ND}) and the average delay to meet a demand ($\overline{D}$) in a two-stage PAC system with $\rho_1 = \rho_2 = 0.7$ and $\lambda = 1.0$ are given in Table 10.3.

TABLE 10.3 COMPARISON OF APPROXIMATION AND
SIMULATION RESULTS FOR A TWO-STAGE PAC SYSTEM WITH
$\rho_1 = \rho_2 = 0.7$ AND $\lambda = 1.0$. (*** UNSTABLE).

P_{ND}			$z_1 = z_2$				
$(k_1 = k_2)$			0	1	2	3	4
1	Approx.		0***	0***	0.183	0.448	0.641
	Sim.		0***	0***	0.166	0.413	0.602
2	Approx.		0***	0.070	0.282	0.496	0.661
	Sim.		0***	0.076	0.291	0.496	0.656
3	Approx.		0***	0.114	0.322	0.520	0.673
	Sim.		0	0.125	0.336	0.530	0.673
4	Approx.		0	0.133	0.341	0.533	0.679
	Sim.		0	0.143	0.359	0.549	0.691
25	Approx.		0	0.153	0.366	0.551	0.691
	Sim.		0	0.165	0.386	0.567	0.704

In Table 10.3, note that $k_1 = k_2 = 1$ corresponds to local control, $k_1 = k_2 = \infty$ (25 is large enough) corresponds to base stock system, $k_1 = z_1 = k_2 = z_2$ corresponds to the KANBAN system and $z_1 = z_2 = 0$ corresponds to the produce-to-order system. As expected (from the properties of PAC systems given in Section 10.7.1), the customer service level improves as z_1, z_2, k_1 and k_2 increases.

Table 10.4 contains the probability of no delay (P_{ND}) and the average delay in meeting a demand ($\overline{D}$) for a two-stage MRP system with $\rho_1 = 0.8, \rho_2 = 0.6$ and $\lambda = 1$ (for details of the approximation, see [11]). The customer service level improves as l_1, l_2, τ_1, τ_2 increases.

10.8 MULTIPLE-CELL SYSTEMS WITH ASSEMBLY CELLS

In this section we consider multiple-cell systems in which some of the cells are assembly cells. In these assembly cells parts arriving from other cells are "assembled," and the assemblies are then moved out of these cells. First we will consider a system with free flow of parts between cells.

TABLE 10.3 (*CONTINUED*): COMPARISON OF
APPROXIMATION AND SIMULATION RESULTS FOR A
TWO-STAGE PAC SYSTEM WITH $\rho_1 = \rho_2 = 0.7$ AND $\lambda = 1.0$.
(*** UNSTABLE).

$\overline{D}$			$z_1 = z_2$				
$(k_1 = k_2)$			0	1	2	3	4
1	Approx.		∞***	∞***	7.70	2.52	1.23
	Sim.		∞***	∞***	9.68	3.40	1.79
2	Approx.		∞***	11.12	3.90	1.95	1.09
	Sim.		∞***	10.50	3.95	2.08	1.28
3	Approx.		∞***	5.67	2.98	1.70	1.01
	Sim.		12.30	5.15	2.91	1.70	1.07
4	Approx.		7.69	4.37	2.59	1.56	0.96
	Sim.		7.01	4.03	2.40	1.47	0.97
25	Approx.		4.67	3.12	2.04	1.32	0.86
	Sim.		4.69	3.06	1.99	1.29	0.81

TABLE 10.4 COMPARISON OF UPPER BOUND, APPROXIMATION AND
SIMULATION RESULTS FOR TWO-STAGE MRP SYSTEM WITH $\rho_1 = 0.8$, $\rho_2 = 0.6$
AND $\lambda = 1.0$.

P_{ND}	(τ_1, τ_2)	$(0, 0)$	$(0, 1)$	$(1, 1)$	$(1, 2)$
(z_1, z_2)	(l_1, l_2)	$(0, 0)$	$(0, 1)$	$(1, 0)$	$(1, 1)$
$(0, 0)$	UB	0	0.062	0	0.188
	Approx.	0	0.062	0	0.156
	Sim.	0	$0.064 \pm .033$	0	$0.168 \pm .005$
$(0, 1)$	UB	0.080	0.188	0.188	0.319
	Approx.	0.080	0.188	0.151	0.299
	Sim.	$0.082 \pm .004$	$0.192 \pm .006$	$0.160 \pm .006$	$0.312 \pm .008$
$(1, 0)$	UB	0	0.188	0	0.319
	Approx.	0	0.147	0	0.222
	Sim.	0	$0.167 \pm .006$	0	$0.246 \pm .008$
$(1, 1)$	UB	0.192	0.313	0.313	0.436
	Approx.	0.144	0.289	0.201	0.378
	Sim.	$0.159 \pm .005$	$0.307 \pm .010$	$0.219 \pm .009$	$0.398 \pm .011$

TABLE 10.4 (*CONTINUED*) COMPARISON OF LOWER BOUND,
APPROXIMATION AND SIMULATION RESULTS FOR TWO-STAGE
MRP SYSTEM WITH $\rho_1 = 0.8$, $\rho_2 = 0.6$ AND $\lambda = 1.0$.

$\overline{D}$		(τ_1, τ_2)	$(0, 0)$	$(0, 1)$	$(1, 1)$	$(1, 2)$
(z_1, z_2)		(l_1, l_2)	$(0, 0)$	$(0, 1)$	$(1, 0)$	$(1, 1)$
$(0, 0)$	LB		5.50	4.52	4.52	3.64
	Approx.		5.50	4.52	4.61	3.69
	Sim.		$5.46 \pm .17$	$4.48 \pm .17$	$4.56 \pm .17$	$3.64 \pm .16$
$(0, 1)$	LB		4.58	3.71	3.71	2.96
	Approx.		4.58	3.71	3.77	2.99
	Sim.		$4.54 \pm .17$	$3.67 \pm .17$	$3.71 \pm .17$	$2.95 \pm .16$
$(1, 0)$	LB		4.58	3.71	3.72	2.96
	Approx.		4.70	3.77	3.99	3.11
	Sim.		$4.63 \pm .18$	$3.71 \pm .18$	$3.91 \pm .18$	$3.05 \pm .17$
$(1, 1)$	LB		3.77	3.02	3.02	2.40
	Approx.		3.84	3.06	3.19	2.46
	Sim.		$3.78 \pm .18$	$3.02 \pm .17$	$3.13 \pm .17$	$2.44 \pm .17$

10.8.1 Three-Cell System with Assembly Cell and Free Flow of Parts

Consider a three-cell system in which cells 1 and 2, process part types 1 and 2, respectively. Cell 3 assembles part types 1 and 2 (one of each) to obtain the final product. Raw material for parts 1 and 2 flows freely from outside to cells 1 and 2, respectively. Parts processed at cells 1 and 2 are immediately transferred to cell 3. We assume that each cell has a single server. Let $A_k^{(j)}$ be the arrival time of the kth raw part to cell j, $k = 1, 2, \ldots$; $j = 1, 2$, and let $S_k^{(j)}$ be the processing time of the kth type j part by cell j, $k = 1, 2, \ldots$; $j = 1, 2$. Then if $D_k^{(j)}$ is the time at which the kth type j part is moved from cell j to cell 3, one has

$$D_k^{(j)} = \max\{D_{k-1}^{(j)}, A_k^{(j)}\} + S_k^{(j)}, \quad k = 1, 2, \ldots; j = 1, 2 \qquad (10.52)$$

Therefore the time at which the kth pair of (one type 1 and one type 2) parts are available for assembly is given by

$$A_k^{(3)} = \max\{D_k^{(1)}, D_k^{(2)}\}, \quad k = 1, 2, \ldots \qquad (10.53)$$

If $S_k^{(3)}$ is the time needed to assemble the kth pair of parts, one sees that the kth completed assembly will be available at time

$$D_k^{(3)} = \max\{D_{k-1}^{(3)}, A_k^{(3)}\} + S_k^{(3)}, \quad k = 1, 2, \ldots \qquad (10.54)$$

From equations 10.52, 10.53, and 10.54 it is easily seen that the completion times at all three cells are improved (in the increasing convex ordering) if the service or arrival times are improved (in the increasing convex ordering).

Now suppose the injection of raw parts is triggered by customer orders, say arriving at times A_k, $k = 1, 2, \ldots$. Then we have $A_k^{(j)} = A_k$, $j = 1, 2$. In this case $D_k^{(1)}$ and $D_k^{(2)}$ will be dependent. If $\{A_k = k/\lambda, k = 1, 2, \ldots\}$ is a deterministic arrival process, however, then $D_k^{(1)}$ and $D_k^{(2)}$ will be independent and the distribution of $\max\{D_k^{(1)}, D_k^{(2)}\}$ can be computed. In particular

$$P\{A_k^{(2)} > x\} = P\{D_k^{(1)} > x\}P\{D_k^{(2)} > x\}, \quad k = 1, 2, \ldots$$

Therefore computing the departure times of each cell j ($j = 1, 2$) will be sufficient to compute the availability time of the kth pair of parts to cell 3. This way all the three cells can be analyzed in isolation as three single-stage queueing systems using the results developed in Chapter 3.

10.8.2 Three-cell System with Assembly Cell with Part Flow and Blocking

Consider the three-cell system described earlier, and assume that there is an infinite supply of raw parts available for both cells 1 and 2. The output store j for cell j, however, has a limited storage capacity $b_j - 1$, $j = 1, 2$. Processing of parts of type j is halted as soon as a processed type J part cannot be moved to store j (i.e., if store j is full), $j = 1, 2$. The part completion times are then (assuming that the stores are all empty at time 0)

$$D_k^{(j)} = \max\{D_{k-1}^{(j)}, D_{k-b_j}^{(3)}\} + S_k^{(j)}, \quad k = 1, 2, \ldots; j = 1, 2 \tag{10.55}$$

and

$$D_k^{(3)} = \max\{D_{k-1}^{(3)}, D_k^{(2)}, D_k^{(1)}\} + S_k^{(3)}, \quad k = 1, 2, \ldots \tag{10.56}$$

where $D_k^{(3)} = 0$, $k = 0, -1, -2, \ldots$. Now consider a three-stage flow line with two intermediate buffer capacities of $b_1 - 1$ and $b_2 - 1$, respectively (see Figure 10.7). Let $\hat{S}_k^{(j)}$ be the service time of the kth job at stage j and $\hat{D}_k^{(j)}$ be the service completion time of the kth job at stage j, $k = 1, 2, \ldots; j = 1, 2, 3$. Then assuming that there is an unlimited supply of jobs for stage 1, stage 2 and its input buffer are empty, and stage 3 and its input buffer is full, it can be seen that

$$\hat{D}_k^{(1)} = \max\{\hat{D}_{k-1}^{(1)}, \hat{D}_{k-b_1}^{(2)}\} + \hat{S}_k^{(1)}, \quad k = 1, 2, \ldots$$

$$\hat{D}_k^{(2)} = \max\{\hat{D}_{k-1}^{(2)}, \hat{D}_k^{(3)}, \hat{D}_k^{(1)}\} + \hat{S}_k^{(2)}, \quad k = 1, 2, \ldots$$

$$\hat{D}_k^{(3)} = \max\{D_{k-1}^{(3)}, \hat{D}_{k-b_2}^{(2)}\} + \hat{S}_k^{(2)}, \quad k = 1, 2, \ldots$$

Now by induction it can be verified that if

$$\hat{S}_k^{(1)} = S_k^{(1)}, \quad \hat{S}_k^{(2)} = S_k^{(2)} \text{ and } \hat{S}_k^{(3)} = S_k^{(3)}, \quad k = 1, 2, \ldots$$

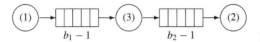

Figure 10.7   Equivalent Flow Line

then
$$\hat{D}_k^{(1)} = D_k^{(1)}, \quad \hat{D}_k^{(2)} = D_k^{(2)} \text{ and } \hat{D}_k^{(3)} = D_k^{(3)}, \quad k = 1, 2, \ldots. \tag{10.57}$$

Hence it is clear that the throughput of the three-cell system with an assembly cell and blocking is the same as that of the flow line defined by Figure 10.7.

10.9 INSIGHTS

The models in this chapter provide insight into the two fundamental aspects of the design and coordination of multiple-cell manufacturing systems: aggregation and coordination.

Aggregation.   We have shown that if a manufacturing system makes a variety of products it will often be appropriate to divide it up into cells that specialize to some subset of the parts produced by the system. Note that, in contrast to the usual behavioral arguments for division into cells, our justification is purely based on the stochastic nature of demands and processing, and a desire to minimize overall work in process. We have also shown that if the products are diversified in processing times the formation of a supercell even with proportionally increased processing capabilities may degrade the performance of a multiple-cell system with uniform part type produced in each cell. This agrees with the basis of group technology. What we have not shown because it is outside the scope of the book is how uncertainty about future demand for products modifies this conclusion. Even so, when facilities are provided with the *capability* to produce a variety of different products because of uncertainty about product life cycles and future demands, the models suggest that it may be appropriate not to use the full capability and specialize when appropriate.

Coordination.   A wide variety of material flow control mechanisms, ranging from base stock systems, MRP control, to KANBAN systems, can be accommodated by the PAC production authorization system. The PAC system has the flexibility to be adapted to the environment of the system at hand through the appropriate choice of the system parameters. We have also shown how changing the parameters modifies performance and demonstrated the placement of initial inventories that optimizes service levels.

In assembly operations the coordination of the order and requisition tag release to stores containing the components and the eventual coordination of PAC cards at those cells where the components are produced perform better than the basic PAC system or an uncoordinated system. That is, when the PAC mechanism is not enough to ensure optimal coordination, there also should be direct communications between cells that produce parts that eventually are to be used in the same product. In other words, the commonly used coordination systems such as MRP and base stock do not address this aspect of coordination, and thus are likely to result in problems when used in practice.

10.10 BIBLIOGRAPHICAL NOTE

The first attempts at using models to compare base stock and KANBAN systems were by Tabe, Muramatsu, and Tanaka [67], and Kimura and Terada [40]. The latter has the first comprehensive account of the KANBAN system. A variety of authors has since developed Markov or queueing models of multiple-stage KANBAN systems, for example, [57], [22] and [49]. There have also been several attempts to develop a general framework for describing and modeling the different approaches to cell coordination, in particular de Koster [20] and Buzacott [10]. On some of the other approaches to cell coordination, Lambrecht, Muckstadt, and Luyten [42] showed the relationship between MRP systems and base stock systems, and Spearman, Woodruff, and Hopp [64] introduced the term CONWIP and demonstrated its potential benefits.

The literature on modeling work flow in multiple-cell systems is rather sketchy. Stidham [66] showed the superiority of one fast server over several slower servers in a single-class queueing system, and Harrison [30] developed results that indicated the necessity of coordination of the supply of parts to an assembly cell. Our discussion of multiple-class systems in Section 10.6.1, with the attendant opportunity to have cells that usually specialize on particular job classes, appears to be new.

PROBLEMS

10.1 Consider the system shown subsequently consisting of three machine groups. Machine group 1 has two identical machines, each with service rate equal to 1 job per hour, machine group 2 has two identical machines each with service rate equal to 0.8 jobs per hour, and machine group 3 has one machine with service rate 1.2 jobs per hour. One-third of arriving jobs go to machine group 2, and the remainder go to machine group 1. When jobs are completed at machine group 1 one-half goes to machine group 2, and one-half goes to machine group 3. When jobs are completed at machine group 2, one-third go to machine group 3, and the rest leave the system. All jobs completed at machine group 3 leave the system. Now suppose machine groups 1 and 3 are combined into a single-cell A, whereas machine group 2 forms a second cell B. There is free flow of jobs between cells and no limit on storage space at each cell.

(a) Determine the state dependent service rates of the single machine equivalent to cell A.

(b) If the arrival rate of jobs from outside is 1.5 jobs per hour, what is the average number of jobs in each cell?

(c) Suppose machine group 2 is increased from two machines to three machines. Verify that your answer to part a is unchanged. What is now the average number of jobs in each cell?

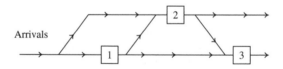

10.2 A product is assembled out of two parts 1 and 2. Part 1 is made in cell 1, part 2 in cell 2, and assembly occurs in cell A. There is no parts storage permissible except in buffer 1' for part type 1 and buffer 2' for part type 2. Each of these buffers has capacity for three parts. Suppose the manufacturing time of each part has an exponential distribution with parameter μ_i, $i = 1, 2$, and the assembly time has an exponential distribution with parameter μ_A.

(a) Develop a Markov chain model to determine the throughput of the system.

(b) Show that the system is equivalent to a system consisting of three stations in series separated by buffers of capacity three, and develop an approximation for the throughput.

10.3 Now consider a situation similar to the previous problem but instead the product is assembled out of *three* parts, 1, 2, and 3. The buffer capacity limit for each part remains at three.

(a) Is this system equivalent to four stations in series separated by buffers of size three?

(b) Develop an approximation for the throughput of the system. If possible check your approximation using a simulation model.

10.4 Consider the system of problem 2, but suppose that after the assembly cell A there is a finishing cell F. There is a buffer of capacity three between cells A and F. Finishing time has an exponential distribution with parameter μ_F.

(a) Is there an equivalent system consisting of cells in series separated by buffers of capacity three?

(b) Develop an approximation procedure by which the throughput of the system can be obtained.

10.5 A manufacturing system is required to produce two part types A and B. The usual processing time of part type A is 12 min, whereas for part type B it is 2 min. On average the ratio of demands of part type B to part type A is 6 to 1. Demands arrive in accordance with a Poisson process. Up to now both part types have been produced on a single machine. Current utilization of the machine is 0.8. It is anticipated, however, that demand will double. Three alternatives for meeting the increased demand are being considered: (1) Replace the machine by a machine that will process the parts at double speed, that is, the time to produce a part type A will become 6 min whereas the time to produce a part type B will become 1 min; (2) acquire another machine identical to the present machine, and continue to produce both part types on both machines; and (3) acquire a machine identical to the present machine, but have one machine dedicated to part type A and the other machine dedicated to part type B. Determine the average level of work in process of each part type for each of the following alternatives:

(a) At current demand levels when arriving demands are processed on a FCFS basis.

(b) At future demand levels if alternative (1) is adopted.

(c) At future demand levels if alternative (2) is adopted and arriving demands are allocated to a machine by tossing a fair coin.

(d) At future demand levels if alternative (2) is adopted and arriving demands are allocated cyclically to a machine (i.e., the first arrival goes to machine 1, the second to machine 2, the third to machine 1, and so on).

(e) At future demand levels if alternative (3) is adopted.

10.6 Consider problem 5 but suppose the processing times have exponential distributions instead of being constant (the mean processing times are unchanged). Repeat parts a to e.

10.7 In problems 5 and 6 rank your solutions to parts a to d, noting any ties (the ranking should be the same for both problems). Now consider a generalization of the problem where $\rho_A = \rho_B$ but otherwise arrival rates and service times for A and B can be chosen at will with "future" arrival rates being twice "current" arrival rates.

(a) Show that the ranking of your solutions (including ties) to parts a to d will be unchanged.

(b) An improved allocation rule with alternative (2) is to defer allocating parts to machines until a machine is idle and then the waiting part that arrived first is allocated to the idle machine. Show that alternative (1) will still be superior.

10.8 In problems 5 and 6 consider the choice between alternatives (1) and (2). Let r be the ratio of the cost per unit time of holding one unit of A as in-process inventory to the cost per unit time of holding one unit of B as in-process inventory. Also, let $\rho_A = \rho_B = \rho$ be the utilization of a machine (the same in both alternatives (1) and (3)).

(a) For deterministic processing times determine over what range of r and ρ alternative (1) will be preferred to alternative (3) because of lower inventory holding costs.

(b) Repeat part a assuming exponentially distributed processing times. Comment on the difference between parts a and b results.

(c) One obvious reason why the firm may choose to acquire a machine identical to the present machine rather than replacing the present machine by a machine of double speed, despite the reduction of inventory holding costs, is that the capital costs will be less. What other reasons might make it desirable to have two machines and ensure that either machine can produce both part types?

10.9 Within the framework of problems 5 and 6 it becomes more attractive to choose alternative (3) over alternative (1) if the utilization is high. Although with alternative (3) there is little opportunity to improve its performance, alternative (1) creates the opportunity to use other approaches besides FCFS for sequencing the different part types. Suppose the rule of always giving preference to any waiting part type B over part type A when the machine becomes free is used. How does this change the choice between alternatives (1) and (3), and how does the holding cost ratio affect the choice?

10.10 A manufacturing system is required to process three part types A, B and C, and there are three different processes required, processes 1, 2, and 3. Part type A only requires process 1, and the processing time has a mean of 1. Part type B requires process 1 followed by process 2. The mean processing time at 1 is 1, whereas the mean processing time at 2 is 2. Part type C requires process 1, then process 2, and then process 3, and the mean processing times are 1, 1, and 3, respectively. The arrival rates of demands for the three part types are equal and arrivals are Poisson. Assume all processing times are exponentially distributed. Two alternative system designs are being considered. Alternative 1 has three identical cells with each cell consisting of a machine to perform process 1, a machine to perform process 2, and a machine to perform process 3. Arriving demands for the part types would be allocated at random to the cells. Alternative 2 would also have three cells, but each cell is dedicated to a specific part type. The cell for part type A would have one machine to perform process 1 only. The cell for part type B would have one machine to perform process 1 and two machines to perform process 2. Lastly, the cell for part type C would have one machine for process 1, one machine for process 2, and three machines for process 3. Obviously both alternatives 1 and 2 have the same total number of machines for each process.

(a) Determine the total work in process of each alternative as a function of the total arrival rate of jobs, and determine which alternative has the lower total number of jobs in process. (This may require using the approximations from Chapters 5 or 7.)

(b) Compare the total work in process for the alternatives 1 and 2 with the total work in process of a system with three cells in series with each cell consisting of three machines dedicated to the same process. All jobs waiting processing by machines in a cell wait in a common waiting area, and they are served in FCFS sequence.

(c) What do you conclude about the relative merits of different cell formation strategies?

10.11 Consider the problem of allocating part types among multiple–single-machine cells considered in Section 10.6.1. Suppose now we wish to decide on the optimal allocation of part types among an optimal number of single-machine cells. Specifically we wish to solve the following problem:

$$\min_{\lambda^{(j)},c} \left\{ h \sum_{i=1}^{c} \left[\frac{\sum_{j=1}^{r} \lambda_i^{(j)} \sum_{k=1}^{r} \lambda_i^{(k)} E[S_k^2]}{2(1 - \sum_{j=1}^{r} \lambda_i^{(j)} E[S_j])} + \sum_{j=1}^{r} \lambda_i^{(j)} E[S_j] \right] - cr \right\}$$

where h is the holding cost per part per unit time and r is the per unit time cost of a single-machine cell. Find the optimal allocation of part types and the optimal number of cells.

10.12 Suppose a PAC-controlled–series cell system introduces the additional rule that cell i does not send a requisition tag to store $i - 1$ unless one of the c_i machines in the cell is free, $i = 1, \ldots, m$. Assuming $r_i = 1$ and $\tau_i = 0$, $i = 1, \ldots, m$,

(a) Write down the modified equations corresponding to equation 10.16.

(b) What can be said about the departure times as a function of z_i and k_i, $i = 1, \ldots, m$? What can be said if $k_i > c_i$?

(c) What can be said about the relationship between the departure times from a system controlled using the rules of this problem and the departure times of a system controlled using the rules of Section 10.7.1, given both systems have the same parameters z_i and k_i, $i = 1, \ldots, m$?

10.13 In the variant of the PAC-controlled system described in the previous problem, requisition tags and order tags are not always sent from cell i to store $i - 1$ simultaneously. Now suppose that the system of the previous problem is modified by requiring that an order tag and a requisition tag are always sent simultaneously from cell i to store $i - 1$, that is, order tags can also be delayed if all c_i machines in the cell are occupied.

(a) Write down the modified equations corresponding to equation 10.16.

(b) What can be said about the departure times as a function of z_i and k_i? What happens if $k_i > c_i$? Is there a z_i^* for any (or all) i, $i = 1, \ldots, m$, such that the departure times do not change if $z_i > z_i^*$?

(c) How does the performance of the PAC-controlled system of this problem compare with the PAC-controlled system of the previous problem for the same choice of z_i and k_i, $i = 1, \ldots, m$.

(d) If $c_i = 1$, $i = 1, \ldots, m$, can any conclusions be drawn about the comparative performance of this system and a series flow line? What if some restrictions are placed on the choice of z_i or k_i?

10.14 Consider a series cell system controlled using the PAC system variant described in the previous problem. Suppose $c_i = 1$ and $z_i = k_i$, $i = 1, \ldots, m$. Assume that the service times at cell i are exponentially distributed with mean $1/\mu_i$, $i = 1, \ldots, m$, and demands are Poisson with rate λ.

(a) Extend the approximation procedure of Section 10.7.5 to this system.

(b) For a two-cell system develop a Markov model of the system for $z_i = k_i = 2$.

(c) Compare the approximation with the Markov model. Note that it will be necessary to truncate the backlog in the Markov model to restrict the number of equations that have to be solved.

10.15 Consider a base stock–controlled system with one machine in cell i with exponentially distributed service times with mean $1/\mu_i$, $i = 1, \ldots, m$. Suppose the μ_i are such that $\mu_j \neq \mu_k$, $j \neq k$, $j, k = 1, \ldots, m$.

(a) Using the approach of Section 10.7.4, write down a formula by which the bounds on the service-level measures P_{ND}, the probability of no delay, and $\overline{D}$, the average delay, of a system with $m = 3$ can be determined.

(b) For what values of z_1, z_2 and z_3 is it possible to determine an exact formula for P_{ND} and $\overline{D}$?

(c) Next, using the approach of Section 10.7.5, determine an approximation for the service-level measures P_{ND} and $\overline{D}$ when $m = 3$.

(d) Now specializing your results from parts a to c to a two-cell system, suppose $z_1 + z_2 = Z_1$ where Z_1 is a constant. Determine for what values of z_1 and z_2 the difference between the bounds and the approximations are greatest.

(e) Section 10.7.4 gives an upper bound on the service-level (i.e., an upper bound on P_{ND} and a lower bound on $\overline{D}$). Suggest an approach to find a lower bound on the service level, and compare your bound with the approximation, giving particular attention to values of the z_i where either the difference between approximation and upper bound is greatest, or where there is no easily computable exact solution.

10.16 Consider a system of m-series cells controlled by the base stock system. If any demands arrive when the store m is empty, however, then these demands are lost.

(a) Write down the equations corresponding to equation 10.16.

(b) In a base stock–controlled system with back-orders the number of jobs in process at any cell can be arbitrarily large. Is this also true with lost sales? If not, determine the maximum number of jobs in process at each cell $i = 1, \ldots, m$.

(c) Suppose that $m = 2$ and demands are Poisson with parameter λ. Assuming that each cell consists of one machine and service times are exponential with mean $1/\mu_i$, develop a Markov model by which the service-level measures of the system can be found when $z_1 = z_2 = 3$.

(d) Modify the approximation procedure in Section 10.7.5 for the lost sales case, and compare it with the results obtained in part c.

10.17 Consider a base stock–controlled manufacturing system consisting of two cells in series. It is desired to gain some insight into the release and departure processes at the cells when demands arrive according to a Poisson process with rate λ, and each cell has a single exponential server with parameter μ_i for cell i, $i = 1, 2$. (Demands are back-ordered). At cell 1 the release process is Poisson with parameter λ if there is always sufficient raw material available, and the departure process from the cell will also be Poisson.

(a) Determine the distribution of time between releases into cell 2, and show that it has a squared coefficient of variation c_R^2 given by

$$c_R^2 = 1 - 2\rho_1^{z_1+1}\frac{(1 - \rho_1)}{1 + \rho_1}$$

and determine, as a function of z_1, the cell 1 utilization ρ_1^* which gives the minimum c_R^2. (*Hint*: Compute the probability distribution of the inventory position of store 1 just after a release from store 1 to cell 2.)

(b) Cell 2 will thus behave like a $G/M/1$ queue. Determine the (approximate) queue length distribution and squared coefficient of variation of the departure process from cell 2.

What would be the squared coefficient of variation of the departure process conditional on there being at least one job in the cell?

(c) Store 2 can then be considered as a $M/G/1$ queue with arrivals at the queue corresponding to demands and service corresponding to the departure process from cell 2, conditional on there being at least one job in the cell. Using the approximation for the $M/G/1$ queue length $p(n) = \rho\sigma^{n-1}$ for $n \geq 1$ and $\sigma = (E[N] - \rho)/E[N]$, determine the service level measures p_{ND} and $\overline{D}$.

(d) Extend the approach used in part a to find the squared coefficient of variation of departures from store 2. Note that this requires finding the distribution of the maximum of the exponential interarrival time of demands and the (approximate) C_2 distribution of time between departures from cell 2, conditional on there being at least one job in the cell.

(e) Compare the approximations for the service level in a two-stage base stock–controlled system determined in problem 15 with the approximations given in part c. If possible compare these results with either simulation results or results obtained using a Markov model when the z_i are small.

10.18 Consider a m-cell series system coordinated using the KANBAN system with $z_i = k_i$, $i = 1, \ldots, m$.

(a) Show that the delay in filling a requisition tag at store i is identical to the delay between cell $i + 1$ and cell i receiving information about occurrence of a demand at store m, that is, $L_k^{(i+1)} = R_k^{(i+1)} - A_k^{(i+1)} = A_k^{(i)} - A_k^{(i+1)}$.

(b) Show that the following recursive relationship connecting the delays at store i and store $i - 1$ can be shown:

$$L_k^{(i+1)} = \max\{0, L_{k-k_i}^{(i)} + F_{k-k_i}^{(i)} - \tau_{k-k_i,k}^{(i+1)} + L_{k-k_i}^{(i+1)}\}, \quad i = 1, \ldots, m$$

where $F_{k-k_i}^{(i)}$ is the time spent by job $k - k_i$ in cell i and $\tau_{k-k_i,k}^{(i+1)}$ is the time between the arrival of requisition $k - k_i$ and requisition k at cell i ($L_k^{(1)} \equiv 0$).

(c) Show that $L_k^{(i+1)}$ is equivalent to the waiting time to enter a queueing system, and identify the arrival and service processes of the system.

10.19 In the series cell system described in problem 14 where parts required by cell i are not drawn from store $i - 1$ until the cell i has a free machine, the service time of a job in cell i can be regarded as equivalent to the sum of a delay (if necessary) waiting for parts after the machine becomes free and the actual service time at the cell. Consider a two-cell system with one machine per cell where the cell service times are exponential with mean $1/\mu_i$, $i = 1, 2$ (thus the equivalent service time at cell 2 will have a C_2 distribution). Use the insights gained from the previous problem and a $M/C_2/1/k_2$ model to determine an approximation for the performance of this modified KANBAN system.

BIBLIOGRAPHY

[1] R. H. AHMADI and H. MATSUO. *A mini-line approach for pull production.* Working Paper 90-10-04, ICC Institute, University of Texas, Austin, Texas, 1990.

[2] R. AKELLA, Y. CHOONG, and S. B. GERSHWIN. Performance of hierarchical production scheduling policy. *IEEE Trans. on Components, Hybrids and Manufacturing Technology*, CHMT-7:225–240, 1984.

[3] T. ALTIOK and H. G. PERROS. Open network of queues with blocking: split and merge configurations. *IIE Trans.*, 18:251–261, 1986.

[4] W. R. ASHBY. *An introduction to cybernetics.* Chapman and Hall, London, 1956.

[5] B. J. BERKLEY. Tandem queues and Kanban-controlled lines. *Int. J. Prod. Res.*, 29:2057–2081, 1991.

[6] G. R. BITRAN and L. CHANG. A mathematical programming approach to a deterministic Kanban system. *Management Science*, 33:427–441, 1987.

[7] J. T. BLACK. An overview of cellular manufacturing systems. *Industrial Engineering*, 15:36–48, November 1983.

[8] J. A. BUZACOTT. *Generalized Kanban/MRP systems.* Technical Report, Department of Management Sciences, University of Waterloo, April 1989.

[9] J. A. BUZACOTT. Kanban and MRP controlled production systems. In *Proceedings Fifth International Working Seminar on Production Economics,* Igls, Austria, pages 269–306, February 1988.

[10] J. A. BUZACOTT. Queueing models of Kanban and MRP controlled production systems. *Engineering Costs and Production Economics*, 17:3–20, 1989.

[11] J. A. BUZACOTT, S. M. PRICE and J. G. SHANTHIKUMAR. Service level in multistage MRP and base stock controlled production systems. *New Directions for Operations Research in Manufacturing*, (ed. G. Fandel, T. Gulledge and A. Jones), Springer, 445–463, 1992.

[12] J. A. BUZACOTT and J. G. SHANTHIKUMAR. A general approach for coordinating production in multiple-cell manufacturing systems. *Production and Operations Management*, 1:34–52, 1992.

[13] D. W. CHENG. *Second order properties of system throughput and system size of a tandem queue with general blocking.* Working Paper, School of Business, New York University, New York, 1991.

[14] D. W. CHENG. *Tandem queues with general blocking: modeling, analysis and gradient estimation.* Working Paper, School of Business, New York University, New York, 1991.

[15] D. W. CHENG and D. W. YAO. *Tandem queues with general blocking: a unified model and comparison results.* Working Paper, School of Business, New York University, New York, 91.

[16] C. COMMAULT and A. SEMERY. Taking into account delays in buffers for analytical performance evaluation of transfer lines. *IIE Trans.*, 22:133–142, 1990.

[17] Y. DALLERY, Z. LIU, and D. TOWSLEY. *Equivalence, reversibility and symmetry properties of fork/join queueing networks with blocking.* Research Report 1267, INRIA, Rocquencourt, France, 1990.

[18] M. B. M. DE KOSTER. Approximate analysis of production systems. *European J. Operational Research*, 37:214–226, 1988.

[19] M. B. M. DE KOSTER. Approximation of flow lines with integrally controlled buffers. *IIE Trans.*, 20:374–381, 1988.

[20] M. B. M. DE KOSTER. *Capacity Oriented Analysis and Design of Production Systems. Lecture Notes in Economics and Mathematical Systems*, no. 323. Springer, New York, 1989.

[21] R. DE KOSTER and J. WIJNGAARD. Local and integral control of workload. *Int. J. Prod. Res.*, 26:43–52, 1989.

[22] J.-L. DELEERSNYDER, T. J. HODGSON, H. MULLER, and P. J. O'GRADY. Kanban controlled pull systems: an analytic approach. *Management Science*, 35:1079–1091, 1989.

[23] E. V. DENARDO and C. S. TANG. Linear control of a Markov prduction system. *Operations Research*, 40:259–278, 1992.

[24] F. E. EMERY and E. L. TRIST. Socio-technical systems. In C. W. Churchman and M. Verhulst, editors, *Management Science: Models and Techniques,* vol. 2, pages 83–97, Pergamon, New York, 1960.

[25] C. C. GALLAGHER and W. KNIGHT. *Group Technology.* Butterworth, London, 1974.

[26] S. B. GERSHWIN, R. AKELLA, and Y. F. CHOONG. Short-term production scheduling of a manufacturing facility. *IBM J. Res. Dev.*, 29:392–400, 1985.

[27] S. C. GRAVES, H. C. MEAL, S. DASU, and Y. QUI. Two-stage production planning in a dynamic environment. In S. Axsater, Ch. Schneeweiss, and E. Silver, editors, *Multi-stage Production Planning and Inventory Control*, pages 9–43, Springer-Verlag, New York, 1986.

[28] P. G. GYLLENHAMMAR. How Volvo adapts work to people. *Harvard Business Review*, 102–113, July-August 1977.

[29] P. G. GYLLENHAMMAR. *People at Work.* Addison-Wesley, Reading, MA, 1977.

[30] J. M. HARRISON. Assembly-like queues. *J. Applied Probability*, 10:354–367, 1973.

[31] W. J. HOPP and J. T. SIMON. Bounds and heuristics for assembly-like queues. *Queueing System: Theory and Applications*, 4:137–156, 1989.

[32] W. J. HOPP, M. L. SPEARMAN, and I. DUENYAS. Economic production quotas for pull manufacturing systems. *IIE Trans.* in press.

[33] N. L. HYER and U. WEMMERLOV. Group technology in the U.S. manufacturing industry: a survey of current practices. *Int. J. Prod. Res.*, 27:1287–1304, 1989.

[34] T. IYAMA and T. ODAWARA. A study of an automatic transfer line system with a sub-line. *IIE Trans.*, 22:204–214, 1990.

[35] U. KARMARKAR. Getting control of just-in-time. *Harvard Business Review*, 122–131, September-October 1989.

[36] U. S. KARMARKAR. *Integrating MRP with Kanban pull systems.* Working Paper QM86-15, Graduate School of Management, University of Rochester, 1986.

[37] U.S. KARMARKAR and S. KEKRE. Batching policy in Kanban systems. *J. Manufacturing Systems*, 8:317–328, 1989.

[38] S. KEKRE. Performance of a manufacturing cell with increased product mix. *IIE Trans.*, 19:329–339, 1987.

[39] T. M. KIM. J.I.T. manufacturing systems: a periodic pull system. *Int. J. Prod. Res.*, 23:553–562, 1985.

[40] O. KIMURA and H. TERADA. Design and analysis of pull system: a method of multi-stage production control. *Int. J. Prod. Res.*, 19:241–253, 1981.

[41] D. KOSTELSKI, J. A. BUZACOTT, K. MCKAY, and X.-G. LIU. Development and validation of a system macro model using isolated micro models. In *Proceedings Winter Simulation Conference,* Atlanta, 669–676, 1987.

[42] M. R. LAMBRECHT, J. A. MUCKSTADT, and R. LUYTEN. Protective stocks in multi-stage production systems. *Int. J. Prod. Res.*, 22:1001–1025, 1984.

[43] R. C. LEACHMAN, J. F. GONCALVES, and Z. K. XIONG. *Rate-based Materials Requirement Planning with Noninteger Lead Times*. Technical Report RAMP 89-3/ESRC 89-4, ESRC, University of California, February 1989.

[44] R. C. LEACHMAN, M. SOLORZANO, and C. R. GLASSEY. *A queue management policy for the release of factory workorders*. Technical Report ESRC 88-19, ESRC, University of California, Berkeley, CA., December 1988.

[45] R. LEONARD and K. RATHMILL. Group technology—a restricted manufacturing philosophy. *Chartered Mechanical Engineer*, 24:42–46, 1977.

[46] E. H. LIPPER and B. SENGUPTA. Assembly-like queues with finite capacity: bounds, asymptotics and approximations. *Queueing Systems: Theory and Applications*, 1:67–83, 1986.

[47] X.-G. LIU and J. A. BUZACOTT. Approximate models of assembly systems with inventory banks. *European J. Operational Research*, 45:143–154, 1990.

[48] X.-G. LIU and J. A. BUZACOTT. A zero-buffer equivalence technique for decomposing queueing networks with blocking. In *Queueing Networks with Blocking* (editors, H. Perros and T. Altiok), pages 87–104, North Holland, Amsterdam, 1989.

[49] D. MITRA and I. MITRANI. Analysis of a Kanban discipline for cell coordination in production lines, I. *Management Science*, 36:1548–1566, 1990.

[50] D. MITRA and I. MITRANI. Analysis of a Kanban discipline for cell coordination in production lines, II. *Operations Research*, 39:807–823, 1991.

[51] S. P. MITROFANOV. *Scientific Principles of Group Technology* [English Translation]. National Lending Library, Boston Spa, Yorkshire, 1966.

[52] Y. MONDEN. *Toyota Production System*. Industrial Engineering and Management Press, Atlanta, GA, 1983.

[53] R. MURAMATSU, K. ISHII, and K. TAKAHASHI. Some ways to increase flexibility in manufacturing systems. *Int. J. Prod. Res.*, 23:691–703, 1985.

[54] H. OPITZ and H.-P. WIENDAHL. Group technology and manufacturing systems for small and medium quantity production. *Int. J. Prod. Res.*, 9:181–203, 1971.

[55] J. A. ORLICKY. *Material Requirements Planning*. McGraw-Hill, New York, 1975.

[56] V. A. PETROV. *Flowline Group Production Planning*. Business Publications, London, 1968.

[57] K. M. REGE. Approximate analysis of serial manufacturing lines with buffer control. *Information and Decision Technologies*, 14:31–43, 1988.

[58] R. J. SCHONBERGER. Frugal manufacturing. *Harvard Business Review*, 95–100, September-October 1987.

[59] R. J. SCHONBERGER. Plant layout becomes product-oriented with cellular just-in-time production concepts. *Industrial Engineering*, 15:66–71, November 1983.

[60] A. SEIDMAN. Regenerative pull (Kanban) production control policies. *European J. Operational Research*, 35:401–413, 1988.

[61] R. K. SINHA and R. H. HOLLIER. A review of production control problems in cellular manufacture. *Int. J. Prod. Res.*, 22:773–789, 1984.

[62] W. SKINNER. The focussed factory. *Harvard Business Review*, 113–121, May-June 1974.

[63] K. C. SO and S. C. PINAULT. Allocating buffer storage in a pull system. *Int. J. Prod. Res.*, 26:1959–1980, 1988.

[64] M. L. SPEARMAN, D. L. WOODRUFF, and W. J. HOPP. CONWIP: a pull alternative to Kanban. *Int. J. Prod. Res.*, 28:879–894, 1990.

[65] M. L. SPEARMAN and M. A. ZAZANIS. Push and pull production systems: issues and comparisons. *Operations Research*, in press.

[66] S. STIDHAM. On the optimality of single server queuing systems. *Operations Research*, 18:708–732, 1970.

[67] T. TABE, R. MURUMATSU, and Y. TANAKA. Analysis of production ordering quantities and inventory variations in a multi-stage production system. *Int. J. Prod. Res.*, 18:245–257, 1980.

[68] K. TAKAHASHI, R. MURAMATSU, and K. ISHII. Feedback method of production ordering system in multi-stage production and inventory system. *Int. J. Prod. Res.*, 25:925–941, 1987.

[69] S. R. TAYUR. *Controlling serial production lines with yield losses using Kanbans*. Technical Report 947, School of Oper. Res. & Ind. Eng., Cornell University, 1990.

[70] S. R. TAYUR. *Structural properties of a Kanban controlled serial manufacturing system*. Technical Report 934, School of Oper. Res. & Ind. Eng., Cornell University, 1990.

[71] U. WEMMERLOV and N. L. HYER. Cellular manufacturing in the U.S. industry: a survey of users. *Int. J. Prod. Res.*, 27:1511–1530, 1989.

[72] U. WEMMERLOV and N. L. HYER. Research issues in cellular manufacturing. *Int. J. Prod. Res.*, 25:413–431, 1987.

[73] W. WHITEHEAD and K. RATHMILL. Group technology: the good and the bad. *Machinery and Production Engineering*, 135(3491), 12 Dec. 1979.

[74] R. WILD. Group technology, cells and flowlines. *Omega*, 2:269–274, 1974.

[75] W. E. WILHELM and S. AHMADI-MARANDI. A methodology to describe operating characteristics of assembly systems. *IIE Trans.*, 14:204–213, 1982.

[76] D. T. N. WILLIAMSON. The anachronistic factory. *Proc. R. Soc.* [London, A], 331:139–160, 1972.

[77] H. YAMASHITA, K. ITOH, and S. SUZUKI. Simulation study evaluating automatic production lines with different operating principles and configuration parameters. In *Recent Advances in Simulation of Complex Systems*, pages 470–476, JSST, July 1986.

[78] P. ZIPKIN. *A Kanban-like production control system*. Research Working Paper No. 89-1, Graduate School of Business, Columbia University, New York, 1989.

[79] B. I. ZISK. Flexibility is key to automated material transport system for manufacturing cells. *Industrial Engineering*, 15:58–64, November 1983.

11

Unresolved Issues: Directions for Future Research

11.1 INTRODUCTION

In this chapter we will describe the design, control, and coordination issues that need to be studied further. We will also provide some suggestions for developing models to address some of these issues. First, let us focus on the system-level design and coordination issues. Later, we will look at the cell-level design and control issues that need further study in flow lines, automatic transfer lines, job shops, FMSs, and flexible assembly systems.

11.2 SYSTEM-LEVEL ISSUES

We view the manufacturing system as consisting of several cells interconnected to one another by material handling systems, and the material (or job) flow controlled by a "coordination" module. There is a set of part types that needs to be manufactured by the system. There may also be a minimum (and a maximum) production rate of each part type that this system should be capable of achieving. Unlike in the cell-level design we do not have a choice of selecting a subset of part types to be manufactured. We do, however, have the choice of selecting the type or the combination of cells (such as flow lines or FMS) for producing and coordinating the manufacture of different part

530

types. Therefore at the system level one has to design the *cell structure,* the *number of replicated cells,* and the *types of each cell.* The cell structure allows one to partition tasks into groups and assign each of these groups to a cell. This way one provides a local control within each cell, thus reducing the propagation of disruptions between cells. Given the allocation of tasks to a cell one then has to select the type of cell to be formed. For example, it may be that either a flexible machining cell or a flow line is suitable for the functions of this cell. It is then possible that this cell is replicated such that a certain fraction of jobs, or certain types of jobs are routed to the FMS cell and others routed to a flow-line cell. Note that any two replicated cells need not be functionally identical. Alternatively, one may decide not to replicate, and choose between a FMS or a flow-line cell. A general methodology needs to be developed to address these design issues. Several parts of the results needed for this are already presented in the earlier chapters, and therefore the future developments can be carried out based on those results. To facilitate these developments we will outline some of the qualitative properties of different types of cells one can derive from the results presented in the earlier chapters.

To illustrate the relative performance of different types of manufacturing systems suppose that manufacture of a product requires m operations where each operation has to be performed on a unique machine that can only perform that operation. Label the machines and their associated operations by $i = 1, \ldots, m$. Further, we will assume that the machines have balanced capabilities so that the time to perform the operation i at machine i has mean $1/\mu$ and variance σ^2, the same for all i. Next, suppose that the operations can be performed in any sequence. Then this means we can consider two ways of organizing manufacture of the product: *flow line* or *job shop.* If the flow line is chosen the layout of the line would create a sequential arrangement of the machines and hence a fixed sequence of operations that would be the same for every unit of the product. Because the operations are balanced we can choose the sequence of machines arbitrarily as any sequence would result in the same performance. If a job shop is chosen we would determine an operation sequence for each item of product when the item becomes available for processing. In a job shop we can choose any one of $m!$ sequences for the operations. Once the operation sequence is chosen in conventional job shops the sequence will be adhered to, and it cannot be changed while the item is in the shop. In FMSs or flexible assembly systems, however, there is the potential to adjust the sequence dynamically in accordance with machine loading and other factors.

To compare the two forms of manufacturing organization we will consider produce-to-order situations in which demands for the product arrive in accordance with a Poisson process with parameter λ. As our measure of comparison we will use $\overline{N}$, the mean number of jobs in process. This will give exactly the same ranking of the different systems as the mean delay in meeting an order. We will compare performance using the approximations developed in Chapters 3, 5, and 7.

Flow line.   Assume the machines are arranged in sequence $1, 2, \ldots, m$. Also, it will be assumed that there is no limitation on storage space between machines. Then using the results from Section 5.4.2 we have

$$\overline{N} = \sum_{i=1}^{m} \left\{ \frac{\rho_i^2(1 + C_{S_i}^2)(C_{a_i}^2 + \rho_i^2 C_{S_i}^2)}{2(1 - \rho_i)(1 + \rho_i^2 C_{S_i}^2)} + \rho_i \right\}$$

where

$$C_{a_1}^2 = 1$$

and

$$C_{a_{i+1}}^2 = (1 - \rho_i^2)\frac{(C_{a_i}^2 + \rho_i^2 C_{S_i}^2)}{1 + \rho_i^2 C_{S_i}^2} + \rho_i^2 C_{S_i}^2, \quad i = 1, \ldots, m - 1$$

The equation for $C_{a_{i+1}}^2$ can be rewritten as

$$C_{a_{i+1}}^2 - 1 = (1 - \rho_i^2)\frac{(C_{a_i}^2 - 1)}{1 + \rho_i^2 C_{S_i}^2} + \rho_i^2(C_{S_i}^2 - 1)$$

Hence, using the fact that $\rho_i = \rho$ and $C_{S_i}^2 = C_S^2$ for all i and defining v by

$$v = \frac{1 - \rho^2}{1 + \rho^2 C_S^2}$$

where $0 < v < 1$ for $C_S^2 \geq 0$ and $0 < \rho < 1$, we get

$$C_{a_{i+1}}^2 - 1 = \rho^2(C_S^2 - 1)\frac{1 - v^i}{1 - v} \quad i = 0, \ldots, m - 1$$

$$\overline{N} = m\rho + m\frac{\rho^2(1 + C_S^2)}{2(1 - \rho)} + \frac{\rho^2(1 + C_S^2)}{2(1 - \rho)(1 + \rho^2 C_S^2)} \sum_{i=1}^{m}(C_{a_i}^2 - 1)$$

$$= m\rho + m\frac{\rho^2(1 + C_S^2)}{2(1 - \rho)} + \frac{\rho^2(1 + C_S^2)}{2(1 - \rho)(1 + \rho^2 C_S^2)}\rho^2(C_S^2 - 1)\left\{ \frac{m}{1 - v} - \frac{1 - v^m}{(1 - v)^2} \right\}$$

Job shop.   Suppose that arriving jobs are allocated a routing from the $m!$ possible routings at random. Then $p_{ij} = 1/m$, $i, j = 1, \ldots, m$, $j \neq i$ and $\gamma_i = 1/m$, $i = 1, \ldots, m$. Also, by symmetry, $C_{a_i}^2 = \hat{C}_a^2$ and $C_{d_i}^2 = C_d^2$, $i = 1, \ldots, m$. From equation 7.71 it follows that

$$\hat{C}_a^2 = 1 + \frac{(m - 1)}{m^2}(C_d^2 - 1)$$

or equivalently

$$\hat{C}_a^2 - 1 = \frac{m - 1}{m^2}(C_d^2 - 1)$$

and also

$$C_d^2 - 1 = \rho^2(C_S^2 - 1) + \nu(\hat{C}_a^2 - 1)$$

Hence

$$\hat{C}_a^2 - 1 = \left(\frac{m-1}{m^2}\right)\left\{\frac{\rho^2(C_S^2 - 1)}{1 - (m-1)\nu/m^2}\right\}$$

Thus

$$\overline{N} = m\rho + m\frac{\rho^2(1 + C_S^2)}{2(1 - \rho)} + \frac{\rho^2(1 + C_S^2)}{2(1 - \rho)(1 + \rho^2 C_S^2)}\rho^2(C_S^2 - 1)\frac{(m-1)/m}{1 - (m-1)\nu/m^2}$$

It can be shown that

$$\frac{(m-1)/m}{1 - (m-1)\nu/m^2} < \frac{m}{1-\nu} - \frac{1 - \nu^m}{(1-\nu)^2}$$

and hence $\overline{N}$ for a job shop is less than $\overline{N}$ for a flowline if $C_S^2 > 1$, whereas if $C_S^2 < 1$ a flow line has lower $\overline{N}$.

It is easier to get some improvement on the job shop. For example, suppose that job routings are allocated to arriving jobs so that the first operation is cyclically allocated, whereas subsequent operations are allocated at random. Then from equation 7.71 it can be verified that

$$\hat{C}_a^2 - 1 = \left(\frac{m-1}{m^2}\right)(C_d^2 - 2)$$

$$= \left(\frac{m-1}{m^2}\right)\left\{\frac{\rho^2(C_S^2 - 1) - 1}{1 - (m-1)\nu/m^2}\right\}$$

The condition for a job shop to be better than a flow line is then

$$\frac{m-1}{m}\left\{\frac{\rho^2(C_S^2 - 1) - 1}{1 - (m-1)\nu/m^2}\right\} < \rho^2(C_S^2 - 1)\left\{\frac{m}{1-\nu} - \frac{1-\nu^m}{(1-\nu)^2}\right\}$$

It can be shown that this condition will always be satisfied if $C_S^2 \geq 1$ for any m and if $m = 2$ for any $C_S^2 \geq 0$. Otherwise it is most likely to be satisfied if ρ is small and C_S^2 is relatively large. For example, with $m = 9$, the above condition is satisfied when $C_S^2 = 0$ and $\rho < 0.174$ or when $C_S^2 > 0.875$ and ρ is near 1. The critical values of C_S^2 for which the job shop is preferable to a flow line for $m = 5$ and $m = 9$ for various values of ρ are given in Figure 11.1.

Obviously if adaptive routing of jobs is used such as is feasible in a flexible assembly system with central dispatch the circumstances in which the flow line is preferable become even more limited.

It is also possible to use these results to determine when a synchronous flow line will be preferable to an asynchronous flow line. Assuming that a synchronous flow line

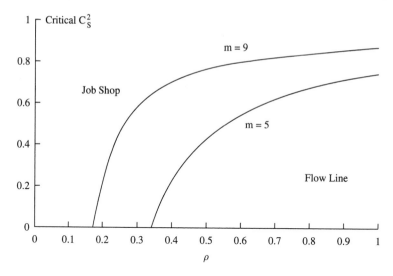

Figure 11.1   Condition for job shop to be preferable to a flow line

always has m jobs in process, one at each machine, it is easy to show that if $C_S^2 = 0$ that

$$\overline{N} = m + \frac{\rho^2}{2(1 - \rho)}$$

whereas in an asynchronous line

$$\overline{N} = m\rho + \frac{1 - (1 - \rho^2)^m}{2(1 - \rho)}$$

The synchronous line will be preferable for ρ near 1, for example, when $m = 5$ if $\rho > 0.82$ while if $m = 9$ if $\rho > 0.89$. Once $C_S^2 > 0$ and hence the synchronous line's cycle time is increased above $1/\mu$, however, the performance of the synchronous line relative to the asynchronous line deteriorates rapidly.

The next step would be to set up a "coordination" module that will coordinate the manufacture of parts from different cells. Though there are several different approaches, such as KANBAN, MRP, and OPT, we recommend the use of the PAC system described in Chapter 10. The PAC system involves setting up the right levels of inventory, look-ahead window length, number of PACs for each cell, along with the rules for card transmittal between cells. With the appropriate choice of the parameters for the PAC system one can capture the essential features of KANBAN, MRP, and OPT. Therefore the design of the coordination system is reduced to the optimal choice of the parameters and a rule for card transmittal between cells of the PAC system. It should, however, be noted that cell PAC system should be set up to complement the PAC system for the entire manufacturing system. This will involve the choice of protocols for service initiation at each service center of each cell.

11.3 CELL-LEVEL ISSUES

In this section we will discuss the design and control issues in flow line, transfer lines, job shops, FMS, and FAS that need to be addressed. Though we will give some specific systems as examples, by no means do we claim to cover all the different types of systems. First, we will discuss the design issues.

11.3.1 Design

Flow line.   During the last 25 years several design issues of flow lines have been addressed in the literature. The most commonly studied issues are (1) the allocation of work load to work stations, (2) the allocation of buffer storage in between work stations, and (3) the assignment of workers. Most of these studies have looked at the qualitative properties of the optimal allocations or assignments by simulating the flow lines. The analytical studies of these issues have provided some useful, but only partial, characterization of the optimal solutions. In Chapter 5 we have illustrated how approximations can be used to characterize fully the "optimal" allocation of work load or buffer capacity for flow lines with exponential processing times. It is, however, essential that we develop efficient algorithms (either using the exact results or approximations) to obtain the optimal design of one or more of the following system parameters to maximize the throughput rate or some other objective. The system parameters are (1) number of parallel servers per stage and the number of stages, (2) number of inspection stations and their location, (3) allocation of tasks to work stations, (4) allocation of buffer storage space, and (5) division of zone length (in a paced assembly line).

Automatic transfer line.   Most of the studies on automatic transfer lines have focused on developing approaches to evaluate the performance such as the production rate of the transfer line. There is a limited literature on the optimal allocation of a given total bank capacity among the set of possible locations between work stations. A related issue is the location of a set of banks with prespecified capacities among the possible sites in between work stations. Efficient algorithms need to be developed to obtain optimal locations of such buffers in general automatic transfer lines. The speed of a transfer line or a segment of such a line is often dictated by the required production rate and the wear-out of tools as a function of the line speed. It is therefore important to relate the throughput of a transfer line to the speed through the tool life function (as a function of the speed). Such a relationship can then be used to design the optimal speed of each segment of the line.

Job shop.   Most of the studies on job shops have centered around the choice of scheduling or service protocol policies. The design issues, particularly pertaining to the selection of the number of machines of each type and the assignment of workers to work centers, have been neglected. Optimization models using the queueing network models of job shops presented in Chapter 7, should be developed. Efficient algorithms need to be devised to solve these optimization problems.

FMS. During the past few years, more focus has been directed to the design of flexible machining cells. These studies have concentrated mainly on choosing either one or two of the system parameters such as (1) the set of part types, (2) number of machines, (3) number of pallets, (4) storage space, (5) number of AGVs and (6) work-load allocation. A general design methodology incorporating all these variables together should be developed.

Flexible assembly system. We have seen in Chapter 9 that the production rate of a central store assembly system with equitable part allocations is controlled by the slowest server. Therefore it is possible to partition the part types and workers into groups so that two or more distinct central store assembly cells are formed such that the overall system throughput is maximized. An optimization model can be developed for this purpose. In the context of assembly lines with strict sequence requirements we have modeled only a single-stage system. When there is more than one stage in such an assembly line, to avoid blocking of servers one needs to provide an output buffer space for each station where departing customers can be resequenced before dispatching to the downstream work station. It is necessary to develop a model for an assembly line with strict sequencing requirements and several stages. Such a model should then be used to design the optimal resequencing buffer spaces between the work stations.

11.3.2 Control

We will next consider some of the control issues that need to be addressed in these systems.

Flow line. It is common that paced assembly lines are used to assemble a variety of products. Because the line is paced, dispatching those products that require a large processing time one after the other may cause the worker to be unable to complete processing these products within the (paced) cycle time. Therefore to reduce the number of products that will not be completely processed on the line, one has to choose a proper dispatch control. All studies on this problem have assumed deterministic arrival and processing times. It is useful to obtain an optimal dispatch control policy when the arrival and processing times are random.

It is usually assumed in these assembly lines that when a worker reaches the end of his work space he abandons the item he is working on and will move on to the item next to it. It is possible that the performance of the flow line may be improved by allowing the worker in such a situation to jump back two or more positions and continue working. This question may be addressed through formulating a control problem.

Transfer line. The major control issue in transfer lines relates to the assignment of the repair crew to one (or more) of the failed stations. In most of the modeling of transfer lines it is assumed that there are ample repairmen available to repair all the stations even when all of them are failed at the same time. When this is not true one needs to devise an optimal control policy by which the limited number of people in the repair crew are allocated the right stations to repair so that the throughput or some other performance measure of the transfer line is maximized.

Job shop.   The control problems one faces in job shops are enormous. Most of the earlier studies have used simulation as a tool to investigate the merits of different control policies in job shops. Despite the many studies carried out in the past, any basic approach for control has not evolved. We recommend the use of the queueing network models developed in Chapter 7 to obtain control policies for (1) acceptance of a job to the system, (2) release to the shop, (3) job routing within a work center when the machines are of different ages or capable of different speeds or accuracies, (4) scheduling control at each work station.

FMS.   Because of the flexibility inherent in the machines, parts can have several alternative routings that they can follow. Consequently in a FMS one has to choose the path of a part dynamically, based on the information about the work load on all the machines. There is very little that has been done on devising optimal routing strategies. Some guidelines to the routing of parts can be obtained using the closed queueing network model developed in Chapter 8 for FMS. One needs, however, to formulate more elaborate control models to obtain an optimal routing strategy. A related problem that needs to be studied is the release of parts (the types and when) to the system.

Flexible assembly system.   In the central store assembly system with equitable part allocation we saw that the throughput is limited by the slowest worker. If the equitable assignment is "count" based (i.e., the number of items allocated to each worker should be the same in the long run) and there are different types of parts available for assembly, one may choose the part to be allocated to different workers. Indeed an optimal control policy needs to be devised so that the throughput of the system is maximized.

In the context of assembly lines with strict sequencing requirements, one may use a similar strategy of choosing the part types to be allocated to each of the workers. Therefore one may choose to allocate the part that requires the least amount of work to the slowest worker. A preassignment of part types to workers can cause problems, however. For example, there may be a situation in which the slowest worker is idling while the part type allocated to him may not be available yet other part types are available for processing. Therefore the decision to allocate part types to workers should be carried out on a dynamic basis. So an optimal control problem should be formulated and solved so as to minimize the average inventory or maximize the throughput.

11.3.3 Coordination

The coordination of the processing of items at different work stations is controlled by the general production authorization system. To obtain the best possible coordination system so as to maximize the benefits (considering the inventory carrying cost) one needs to obtain the optimal parameter values described in Chapter 10. This requires two steps: (1) development of a performance evaluation model and (2) development of an optimization model. We have fully illustrated this for single-stage systems (see Chapter 4) and for some special cases of multiple cells in series (see Chapter 10). It is important that general performance models incorporating the PAC mechanism be developed. It is of equal importance that we develop optimization models and efficient solution procedures to obtain the optimal parameter values.

A

Standard Probability Distributions

In this section we present some probability distributions of non-negative random variables commonly used in the modeling of manufacturing systems.

A.1 EXPONENTIAL (*E*)

$$F_Y(x) = 1 - e^{-\mu x}, \quad x \geq 0$$

$$E[Y] = 1/\mu$$

$$C_Y^2 = 1$$

A.2 ERLANG-*K* (*E*$_K$)

$$F_Y(x) = 1 - \sum_{l=0}^{k-1} e^{-k\mu x} \frac{(k\mu x)^l}{l!}, \quad x \geq 0$$

$$E[Y] = 1/\mu$$

$$C_Y^2 = 1/k$$

Note that for $k = 1$, Erlang-*k* is an exponential random variable, and as $k \to \infty$, Erlang-*k* becomes (degenerate) deterministic.

539

A.3 HYPEREXPONENTIAL WITH m PHASES (H_m)

$$F_Y(x) = 1 - \sum_{i=1}^{m} p_i e^{-\mu_i x}, \quad x \geq 0$$

$$E[Y] = \sum_{i=1}^{m} p_i/\mu_i$$

$$C_Y^2 = \frac{\sum_{i=1}^{m} 2p_i/\mu_i^2}{(\sum_{i=1}^{m} p_i/\mu_i)^2} - 1, \quad (\geq 1)$$

where $(p_1, \ldots, p_m)$ is a probability vector.

A.3.1 Two-Phase Hyperexponential (H_2)

Note that the *two-phase hyperexponential* distribution is

$$F_Y(x) = 1 - p_1 e^{-\mu_1 x} - p_2 e^{-\mu_2 x}, \quad x \geq 0$$

A.3.2 Two-Phase Balanced Hyperexponential ($H_{2:b}$)

When $p_1 = p_2 = 1/2$ we get a *two-phase balanced hyperexponential* distribution

$$F_Y(x) = 1 - \frac{1}{2}e^{-\mu_1 x} - \frac{1}{2}e^{-\mu_2 x}, \quad x \geq 0$$

A.3.3 Generalized Exponential (GE)

When $1/\mu_2 = 0$, we get a *generalized exponential* distribution,

$$F_Y(x) = 1 - pe^{-\mu x}, \quad x > 0$$

where

$$E[Y] = p/\mu$$

and

$$C_Y^2 = \frac{2-p}{p}$$

Obviously the case $p = 1$ corresponds to the exponential distribution.

A.4 PHASE TYPE WITH m PHASES (PH)

Let $\{X(t), t \geq 0\}$ be a continuous time Markov process on $\{1, \ldots, m+1\}$ with an infinitesimal generator matrix $\mathbf{Q}$, where

$$\mathbf{Q} = \begin{bmatrix} \mathbf{T} & -\mathbf{Te} \\ \mathbf{0} & 0 \end{bmatrix}$$

T is an $m \times m$ matrix, $\mathbf{e} = (1, \ldots, 1)'$ is a column m-vector of ones, and $\mathbf{0} = (0, \ldots, 0)$ is a row m-vector of zeros. Note that state $m + 1$ is an absorbing state. Define

$$Y = \inf\{t : X(t) = m + 1\}$$

Suppose the probability distribution of $X(0)$ is $\alpha = (\alpha_1, \ldots, \alpha_m)$ concentrated on $\{1, \ldots, m\}$. Then

$$F_Y(x) = 1 - \alpha e^{\mathbf{T}x}\mathbf{e}, \quad x \geq 0$$

$$E[Y] = -\alpha\mathbf{T}^{-1}\mathbf{e}$$

Note that the representation $(\alpha, \mathbf{T})$ fully characterizes this phase-type distribution (see Neuts [5]).

All the preceding distibutions (A.1–A.3) are special cases of phase-type distributions. Particularly for

1. Exponential, $\mathbf{T} = (-\mu)$ and $\alpha = 1$.
2. Erlang-k.

$$\mathbf{T} = \begin{bmatrix} -k\mu & k\mu & & & \\ & -k\mu & k\mu & & \\ & & & \cdot & \cdot \\ & & & & -k\mu \end{bmatrix}$$

and $\alpha = (1, \ldots, 0)$, and for

3. Hyperexponential with m phases

$$\mathbf{T} = \begin{bmatrix} -\mu_1 & & & \\ & -\mu_2 & & \\ & & \cdot & \\ & & & -\mu_m \end{bmatrix}$$

and $\alpha = (p_1, \ldots, p_m)$.

The next special case of *PH* distribution is widely used (e.g., see Altiok [1] and Yao and Buzacott [9]).

A.5 TWO-PHASE COXIAN (C_2)

Suppose

$$\mathbf{T} = \begin{bmatrix} -\mu_1 & \mu_1 p_2 \\ & -\mu_2 \end{bmatrix}$$

and $\alpha = (1, 0)$. Then the LST of F_Y is

$$\tilde{F}_Y(s) = p_1\frac{\mu_1}{\mu_1 + s} + p_2\left(\frac{\mu_1}{\mu_1 + s}\right)\left(\frac{\mu_2}{\mu_2 + s}\right)$$

$$E[Y] = \frac{1}{\mu_1} + p_2 \frac{1}{\mu_2}$$

$$C_Y^2 = \frac{\mu_2^2 + p_2(1 + p_1)\mu_1^2}{(\mu_2 + p_2\mu_1)^2} \qquad \left(\geq \frac{1}{2}\right)$$

where $p_1 = 1 - p_2$.

A.6 BALANCED MEAN TWO-PHASE COXIAN DISTRIBUTION ($C_{2:b}$)

Suppose in the two-phase Coxian distribution we set $1/\mu_1 = p_2/\mu_2 = 1/\mu$. Then

$$\tilde{F}_Y(s) = p_1 \left(\frac{\mu}{\mu + s}\right) + p_2^2 \left(\frac{\mu}{\mu + s}\right)\left(\frac{\mu}{\mu p_2 + s}\right)$$

$$E[Y] = 2/\mu$$

$$C_Y^2 = \frac{1}{2p_2}$$

Given the mean and squared coefficient of variation of a random variable, the preceding two relationships can be used to find a balanced mean two-phase Coxian distribution (provided $C_Y^2 \geq 1/2$). It should be noted that the squared coefficient of variation of a PH random variable cannot be smaller than $1/m$. For fitting distributions of PH-type see Johnson and Taaffe [3].

A.7 GENERALIZED PHASE-TYPE (GPH)

All the preceding distributions can be put as a mixture of Erlang random variables (Shanthikumar [7]). Let $(p_n, n = 1, 2, \ldots)$ be a discrete probability distribution. Then the GPH distribution is

$$F_Y(x) = \sum_{n=1}^{\infty} \frac{e^{-\lambda x}(\lambda x)^n}{n!} P_n$$

where $P_n = \sum_{k=1}^{n} p_n$, $n = 1, 2, \ldots$, is the discrete cumulative distribution.

$$E[Y] = \frac{1}{\lambda} \sum_{n=1}^{\infty} n p_n$$

B

Some Notions of Stochastic Ordering

In this section we present some notions of stochastic ordering between two non-negative random variables X and Y. For more details consult Marshall and Olkin [4], Ross [6] and Stoyan [8]. Note that increasing/decreasing are not used in the strict sense.

B.1 LIKELIHOOD RATIO ORDERING ($\geq_{lr}$)

X is larger than Y in the *likelihood ratio* ordering (denoted $X \geq_{lr} Y$) if $f_X(x)/f_Y(x)$ is increasing in x.

B.2 HAZARD RATE ORDERING ($\geq_h$)

X is larger than Y in the *hazard rate* ordering (denoted $X \geq_h Y$) if $\bar{F}_X(x)/\bar{F}_Y(x)$ is increasing in x.

B.3 REVERSED HAZARD RATE ORDERING ($\geq_{rh}$)

X is larger than Y in the *reversed hazard rate* ordering (denoted $X \geq_{rh} Y$) if $F_X(x)/F_Y(x)$ is decreasing in x.

B.4 USUAL STOCHASTIC ORDERING ($\geq_{st}$)

X is larger than Y in the *usual stochastic ordering* (denoted $X \geq_{st} Y$) if $\bar{F}_X(x) \geq \bar{F}_Y(x)$ for all $x \geq 0$. Note that

$$X \geq_{st} Y \quad \Leftrightarrow \quad Ef(X) \geq Ef(Y)$$

for all *increasing functions* f.

B.5 INCREASING CONVEX ORDERING ($\geq_{icx}$)

X is larger than Y in the *increasing convex ordering* (denoted $X \geq_{icx} Y$) if $\int_x^\infty \bar{F}_X(y)dy \geq \int_x^\infty \bar{F}_Y(y)dy$ for all $x \geq 0$. Note that

$$X \geq_{icx} Y \quad \Leftrightarrow \quad Ef(X) \geq Ef(Y)$$

for all *increasing convex functions* f.

B.6 CONVEX ORDERING ($\geq_{cx}$)

X is larger than Y in the *convex ordering* (denoted $X \geq_{cx} Y$) if $X \geq_{icx} Y$ and $E[X] = E[Y]$. Note that

$$X \geq_{cx} Y \quad \Leftrightarrow \quad Ef(X) \leq Ef(Y)$$

for all *convex functions* f.

The following implications are strict:

$$(\geq_{lr}) \quad \Rightarrow \quad (\geq_h) \quad \Rightarrow \quad (\geq_{st}) \quad \Rightarrow \quad (\geq_{icx})$$

C

Nonparametric Families of Distributions

We present nonparametric families of distributions that exhibit some properties that are useful.

C.1 LOG-CONCAVITY (LOG-CONVEXITY)

A random variable X (or its density) is said to be log-concave [log-convex] if $\log f_X(x)$ is concave [convex] in x.

$$X \text{ is log-concave [log-convex]} \Leftrightarrow$$

$$\{X - s | X > s\} \geq_{lr} [\leq_{lr}]\{X - t | X > t\}, \quad 0 \leq s \leq t$$

Exponential and Erlang-k density functions are log-concave, whereas hyperexponential density functions are log-convex.

C.2 INCREASING FAILURE RATE (IFR) (DECREASING FAILURE RATE [DFR])

A random variable X is said to have an increasing failure rate [decreasing failure rate] if $\log \bar{F}_X(x)$ is concave [convex] in x. Equivalently the hazard rate $h(x) = f_X(x)/\bar{F}_X(x)$

is increasing [decreasing] in x. Note that

$$X \in IFR \quad [DFR] \quad \Leftrightarrow$$

$$\{X - s | X > s\} \geq_{st} \quad [\leq_{st}] \quad \{X - t | X > t\}, \quad 0 \leq s \leq t.$$

C.3 DECREASING MEAN RESIDUAL LIFE (DMRL) [INCREASING MEAN RESIDUAL LIFE (IMRL)]

A random variable X is said to have decreasing [increasing] mean residual life if $E[X - x | X > x]$ is decreasing [increasing] in x.

C.4 NEW BETTER THAN USED IN EXPECTATION (NBUE) [NEW WORSE THAN USED IN EXPECTATION (NWUE)]

A random variable X is said to be new better [worse] than used in expectation if $E[X] \geq [\leq] E[X - x | X > x]$ for all $x \geq 0$.

The following implications are strict (e.g., see Barlow and Proschan [2]):

$$\text{Log-concavity} \quad \Rightarrow \quad IFR \quad \Rightarrow \quad DMRL \quad \Rightarrow \quad NBUE$$

$$\text{Log-convexity} \quad \Rightarrow \quad DFR \quad \Rightarrow \quad IMRL \quad \Rightarrow \quad NWUE$$

BIBLIOGRAPHY

[1] T. ALTIOK. On the phase-type approximations of general distributions. *IIE Transactions,* 17:110–116, 1985.

[2] R. E. BARLOW and F. PROSCHAN. *Statistical Theory of Reliability and Life Testing: Probability Models.* Holt, Rinehart and Winston, New York, 1975.

[3] M. A. JOHNSON and M. R. TAAFFE. An investigation of phase-distribution moment-matching algorithms for use in queueing models. *Queueing Systems: Theory and Applications,* 8:129–148, 1991.

[4] A. W. MARSHALL. *Inequalities: Theory of Majorization and Its Applications.* Academic Press, New York, 1979.

[5] M. F. NEUTS. *Matrix-Geometric Solutions in Stochastic Models—An Algorithmic Approach.* Johns Hopkins University Press, Baltimore, 1981.

[6] S. M. ROSS. *Stochastic Processes.* John Wiley and Sons, New York, 1983.

[7] J. G. SHANTHIKUMAR. Bilateral phase type distributions. *Naval Research Logistics Quarterly,* 32:119–136, 1985.

[8] D. STOYAN. *Comparison Methods for Queues and Other Stochastic Models.* John Wiley and Sons, New York, 1983.

[9] D. D. YAO and J. A. BUZACOTT. Queueing models for a flexible machining station: Part II. The method of Coxian phases. *Eur. J. Opnl. Res.* 19:241–252, 1985.

Index